THE MATTER OF ÇATALHÖYÜK

REPORTS FROM THE 2009–2017 SEASONS

Edited by
Ian Hodder

BRITISH INSTITUTE AT ANKARA
Monograph 54
Çatalhöyük Research Project Series Volume 14
2021

Published by
British Institute at Ankara
10 Carlton House Terrace, London SW1Y 5AH
www.biaa.ac.uk

This book is available from
Oxbow Books
10 Hythe Bridge Street, Oxford, OX1 2EW
www.oxbowbooks.com

ISBN 978 1 912090 50 1

The publication of this volume has been generously supported by

TEB
BNP PARIBAS JOINT VENTURE

Typeset by Abby Robinson
Printed by Short Run Press Ltd, Exeter

Contents

Online supplementary material

Supplementary material available online (https://doi.org/10.18866/BIAA/e-14) comprises additional elements for chapters 7, 8 and 13. Colour versions of selected figures are available for chapters 4, 5, 6, 9, 10, 12 and 16.

7. Large clay balls at Çatalhöyük East: 16 figures (S7.1–S7.16) and eight tables (S7.1–S7.8).
8. The small geometric clay objects at Neolithic Çatalhöyük: 18 figures (S8.1–S8.18).
13. The ground stone technologies at Neolithic Çatalhöyük: six figures (S13.1–S13.6) and one table (S13.1).

Contributors

Lise Bender Jørgensen	Department of Historical and Classical Studies, Norwegian University of Science and Technology
Lucy E. Bennison-Chapman	Netherlands Institute for the Near East (NINO), Leiden University
Gesualdo Busacca	Department of Anthropology, Stanford University
Lindsay Der	Department of Community, Culture and Global Studies, University of British Columbia Okanagan
Davide D'Errico	Department of Archaeological Sciences, Leiden University; LTFAPA Laboratory, Department of Classics, Sapienza University of Rome
Sean Doyle	Independent researcher
Virginia García-Díaz	Faculty of Archaeology, Leiden University
Ian Hodder	Department of Anthropology, Stanford University
Ceren Kabukcu	Department of Archaeology, Classics and Egyptology, University of Liverpool
Cristina Lemorini	LTFAPA Laboratory, Department of Classics, Sapienza University of Rome
Dominik Lukas	Department of Anthropology, University of Chicago
Arkadiusz Marciniak	Faculty of Archaeology, Adam Mickiewicz University
Camilla Mazzucato	Department of Anthropology, Stanford University
Carrie Nakamura	Independent researcher
Serap Özdöl Kutlu	Department of Tourism, Ege University, Izmir
Antoinette Rast-Eicher	ArcheoTex and University of Bern
Matilda Siebrecht	Arctic Center, University of Groningen
Duygu Tarkan	Research Center for Anatolian Civilizations, Koç University
Christina Tsoraki	School of Archaeology and Ancient History, University of Leicester
Milena Vasić	Independent researcher
Rena Veropoulidou	The M.H. Wiener Laboratory for Archaeological Science, American School of Classical Studies at Athens; Hellenic Ministry of Culture and Sports
Willeke Wendrich	Cotsen Institute of Archaeology, UCLA

List of figures

Chapter 14

List of tables

Acknowledgements

This publication of four volumes (Volumes 12, 13, 14 and 15 in the Çatalhöyük Research Project Series) would not have been possible without the help of a large number of individuals, institutions and sponsors. We have attempted to list everyone who assisted the Çatalhöyük Research Project during the period of excavation and post-excavation work (2009–2018) covered in these volumes and apologies are extended to anyone who has been inadvertently overlooked. Special thanks to Türk Ekonomi Bankası (TEB) for their support of this volume.

Funding for the field research was provided by a wide variety of corporate and academic bodies. The main corporate sponsors were Boeing, Yapı Kredi Bankası, Shell and Koçtaş. I am particularly grateful to them for providing support over most of the 25-year period of the project. They sustained the project through crises and celebrations and provided incalculable scientific, cultural, social and economic benefits to many. The John Templeton Foundation also provided key support for the period covered by these volumes, and I am particularly grateful for the advice and help provided by Paul Wason. During this period of research funding was also provided by the following: British Institute at Ankara, Global Heritage Fund, Foundation for Polish Science, Free University Berlin, Hedef Alliance, Humboldt Foundation, Imitatio (Thiel Foundation), Kaplan Foundation, Konya Çimento, Konya Şeker, National Geographic Society, National Science Foundation, Polish Heritage Council, Polish Ministry of Science and Higher Education, Polish National Science Center, Stanford Archaeology Center, Stanford University, SUNY Buffalo, TAV, Turkish Cultural Foundation, University College London, University of Gdansk, University of Poznan, US Embassy in Ankara.

The project worked in Turkey with a permit from the Ministry of Culture and Tourism, General-Directorate of Cultural Heritage and Museums. Over the period covered by these publications much support and advice was given by the department and Director Generals, through their representatives on site (bakanlık temsilcileri). The project worked under the auspices of the British Institute at Ankara and I would like to thank in particular Lutgarde Vandeput, Gülgün Girdivan, Tamar Hodos, Stephen Mitchell and Shahina Farid. Additionally, I am grateful for the support of the BIAA committee members and for the assistance of the Ambassadors and staff at the Turkish Embassy and consulates in London, Washington and Los Angeles and the British Ambassadors in Ankara. The patrons of the project are Professor Lord Renfrew of Kaimsthorn and Sir David Attenborough.

In the region, help and support were provided at many levels, in particular by the Konya Vali, the Konya Koruma Kurulu, the Cultural Director, the Konya Museums Director Yusuf Benli and his staff, officers at the Emniyet, our bank managers at Konya Yapı Kredi, the Çumra Kaymakam and Belediye Başkan. We would also like to extend our thanks to the Jandarma commitants, and the managers of the Dedeman Hotel in Konya, and Asim Kaplan from Karavan. Since 2009, the project managers have been Shahina Farid, Banu Aydınoğlugil, Yıldız Dirmit and Bilge Küçükdoğan. The latter in particular played a major role in extracting us from the site in 2016 and in overseeing the winding up of the project at Stanford and in Turkey. I am also forever grateful to the Assistant Director Serap Özdöl for her guidance and contributions to many aspects of the project over many years.

We owe an enormous debt to those who managed the project on site, particularly the camp manager Levent Özer, whose wise advice steered the project through many trials and successes. Our guards at the site were our close companions over the years: Ibrahim Eken and Mustafa and Hasan Tokyağsun. These volumes are dedicated to the memory of Ibrahim, whose long-term devotion to the project and gentle manner warmed the hearts of all who interacted with him.

From the local village of Küçükköy we would like to thank the people and their mukhtar. Those who worked at the site and contributed directly to the project in various ways are included in the list of team members that follows.

Doğu Furkan ACARER, Donovan ADAMS, Sabrina AGARWAL, Rifat AHSAN, Sam AINSWORTH, Atiye AKBULUT, Bünyamin AKBULUT, Hanafi AKCAN, Hasan AKÇAY, Enver AKGÜN, Kiraz AKOĞLU, Mustafa AKYURT, Moussab ALBESSO, Sophie ALCOCK, Richard ALLEN, Thomas ALLEN, Mehmet ALTINAY, Emma ANDERSON, Jennie ANDERSON, Veysel APAYDIN, Renata ARAUJO, Theodore ARNOLD-FORSTER, Mehmet ARSLAN, Numan ARSLAN, Monique ARNTZ, Althea ASARO, Gemma ASHBURY, Eleni ASOUTI, Mert ATALAR, Sonya ATALAY, Soner ATEŞOĞULLARI, Christopher ATKINSON, Quentin ATKINSON, Deanna AUBERT, Jeffrey AVISS, Gianna AYALA, Fahri AYÇİN, Banu AYDINOĞLUGİL, İnan AYDOĞAN, Melike AYHAN, Sema BAĞCI, Mustafa BAHÇECİ, Jack BAIGENT, Roseleen BAINS, Daniella BAR-YOSEF MAYER, Marek Zbigniew BARAŃSKI,

Judit BARASTEGUI, Alexandra BARMETTLER, Daniela A. BARRANTES, Marta BARTKOWIAK, Isabel BARTLEY, Célia BASSET, Rachel BASSINGER, Purnur Ece BAŞ, Emmeline BATCHELOR, Erin BAXTER, Alexandra BAYLISS, Umut BAYRAM, Tolga BAYRAM, Carlos BAZUA, Joel BEATH, Menna BELL, Cristina BELMONTE SANTISTEBAN, Brenda BENAVIDES, Lise BENDER JØRGENSEN, Lucy BENNISON CHAPMAN, Åsa BERGGREN, Johanna M. BERGKVIST, Mary BERMAN, Julia BEST, Barbara BETZ, Elisa BIANCI-FIORI, Peter BIEHL, Patrycja BIELSKA, Rachel BINGHAM, Tom BIRCH, İsa BİLGİÇ, Serdar BİLİŞ, Stephanie BLACK, Emmy BOCAEGE, Amy BOGAARD, Nikita BOGDANOV, Patrick BOLL, Jennie BORGSTROM, Sezgin BOŞLAMAZ, Hannah J. BOWDEN, Garrett BOYD, Mathew BOYD, Başak BOZ, Ahmet BOZGEYIK, Malwina Ewa BRACH-MANSKA, Henry BRADFORD, Jacob BRADY, Maxime BRAMI, Matthew BRITTEN, Kelly BROWN, Nicholas BROWN, Hallvard BRUVOLL, Laura BUCCIERI, Eniko BUDAK, Mikolaj BUDNER, Bayram BULUT, Narcis BURGUES, Oliver BURTON, Gesualdo BUSACCA, Emine BÜLÜÇ, İsmail BÜLÜÇ, Numan BÜLÜÇ, Jennifer BYRNS, Agnieszka BYSTRON, Katarzyna BZDUCH, Tiffany CAIN, Kelly CALDWELL, Stefano CAMPANA, Erica CAMURRI, Gözde CAN, Frank CARPENTIER, Christopher CARLTON, Robert Bergman CARTER, Tristan CARTER, Julie CASSIDY, Gianluca CATANZARITI, Rebecca CESSFORD, Merve CEYLAN, Benjamin CHAN, Ian CHANNELL, Michael CHARLES, Jessica CHATBORN, Claire CHRISTENSEN, Kimberly CHRISTENSON, Angeliki CHRYSANTHI, Maciej CHYLENSKI, Piotr CIESIELSKI, Marguerite CLARKE, Christopher CLEERE, Julia CLINE, Andrew COCHRANE, Jon COGDALE, Alana COLBERT, Tara COPPLESTONE, Grant COX, Michelle CREPEAU, Kyle CROSSET, Caitlin L. CURTIS, Cassy CUTULLE, Lech CZERNIAK, Agata CZESZEWSKA, Duygu ÇAMURCUOĞLU, Hatice ÇELİK, Lokman ÇELIK, Mahmut ÇELİK, Mustafa ÇEŞŞUR, Elif S. ÇIPLAK, Leyla E. ÇIPLAK, Mehmet ÇIRAK, Davide D'ERRICO, Nihan Dilşad DAĞTAŞ, Nevio DANELON, Julie DAUJAT, Anna DAVENPORT, Antonia DAVIDOVIC WALTHER, Neil DAVIES, Tudur DAVIES, Danielle DE CARLE, Alysha DE SOUZA, Funda DEĞER, Nicolo DELL'UNTO, Mateusz DEMBOWIAK, Marvin DEMICOLI, Burcu DEMİR, Arzu DEMİRERGİ, Işıl DEMİRTAŞ, Meghan DENNIS, Lindsay DER, Emma DEVEREUX, Paola DI GIUSEPPANTONIO, Charlotte DIFFEY, Bela DIMOVA, Emilie DINGLER, Sermin DİNC, Ayşe Ş. DİNÇER, Filiz DİRİ, Yıldız DİRMİT, Triantafyllia Eirini DOGIAMA, Chris DOHERTY, Irene DORI, Sean DOYLE, Kelly DU RAND, Güneş DURU, Graeme EARL, David EBNER, Daniel EDDISFORD, Leslie EDMONDS, Selma EFELER, Erol EKEN, Fadimana EKEN, Fatma EKEN, Selda EKEN, Saliha EKEN, Turgut EKEN, Ümmügülsüm EKEN, Sophie EKSTRAND, İzettin ELALMIŞ, Hermione ELDERTON, Nada ELIAS, Mustafa Özgür ELMACIOGLU, Erica EMOND, Claudia ENGEL, Ahmet ERDOĞAN, Burçin ERDOĞU, Kerim E. ERGEN, Duygu ERGENÇ, Müge ERGÜN, Gunhild ERIKSDOTTER, Rebecka ERNTELL, Tuğçe ERTABAK, Duygu ERTEMİN, Osman ERTÜRK, Cumhur ERTÜZÜN, Üğür EYİLİK, Catherine FAIRLESS, Chris FARIA, Shahina FARID, Sayeh FATTAHI, Michelle FEIDER, Haşim FERAHKAYA, Lauren FIELD-FIDLER, Rose FIGURA, Clara FILET, Dragana FILIPOVIC, Patrycja FILIPOWICZ, Ashley FISHER, MAX FORREST, Maurizio FORTE, Hayley FOSTER, Katrina FOXTON, Jenna FOWLER, Sheelagh FRAM, Tom FRANKLAND, Ingmar FRANZ, Dorian FULLER, Fabrizio GALEAZZI, Michelle GAMBLE, Eleonora GANDOLFI, Mary GANIS, Virginia GARCÍA-DÍAZ, Aroa GARCIA-SUAREZ, Katrina GARGETT, Evan GAROFALO, Gary GIBBONS, Bonnie GLENCROSS, Andrew GOLDMAN, Sarah GONZAGA, Lara GONZÁLEZ CARRETERO, Donna Rae GOULD, Nuriye GÖKÇE, Juan Jose GARCIA GRANERO FOS, Sarah GRANT, Laura GREEN, Rachel GREENBERG, Haskell GREENFIELD, Janet GRIFFITHS, Daniel GRISWOLD, Lisa M. GUERRE, Hilal GÜLTEKIN, Burçin GÜMÜŞ, Ramazan GÜNDÜZ, Gülgün GÜRCAN, Erkan GÜRÇAL, Sevgi GÜRDAL, Simge GÜREŞ, Nergis GURSES, Küpra GÜVEN, Anna HABERLAND, Piraye HACIGÜZELLER, Remi HADAD, Scott D. HADDOW, Christoffer HAGBERG, Lori HAGER, Cordelia HALL, Julie HAMILTON, Anette HANSEN, Katarzyna Weronika HARABASZ, Karen HARDY, Menekşe HAREMKAHYA, Beth HARLEY, Karl HARRISON, Laura HARRISON, Caroline HEBRON, Juliette HEMELAAR, Andrew HENDERSON, Liz HENTON, Xose HERMOSO-BUXAN, Lucia HERRERO, Simon HILLSON, Rachel HODARA, Claire HODSON, Kerrie HOFFMAN, Milicent HOLMAN, John HOLSTON, Phillip HOLT, Braxton M. HOOD, Jedrez HORDECKI, Rosemary HOSHINO, Michael HOUSE, Helen HUMAN, Susan HYDEN, Trevor ILIFF, Rachel IRESON, Justine ISSAVI, Graham ISTED, Resul İBİŞ, Tunç İLEDA, Bianca JACKSON, Mark JACKSON, Antonia JAMES, Rosemary JEFFREYS, Emma JENKINS, Erik

JOHANSSON, Emily JOHNSON, Karolina JOKA, Jennifer JONES, Kimberly JONES, Sarah JONES, Sian JONES, Kristina JONSSON, Emma JORDAN, Rosemary JOYCE, Friederike JÜRCKE, Ceren KABUKCU, Tuukka KAIKKONEN, George KAMBOUROGLOU, Till S. KAPPUS, Aydan KARADEMIR, Akrivi KATIFORI, Ali KAVAS, Kevin KAY, Ramazan KAYA, Vahap KAYA, Nurcan KAYACAN, Nuray KAYGAZ, Courtney KEMNITZ, Sheena KETCHUM, Kübra KILIÇ, Katy KILLACKEY, Hyunyoung KIM, Laurie KING, Ian KIRKPATRICK, Galip KİRAZ, Arkadiusz KLIMOWICZ, Adam KLUPS, Christopher KNÜSEL, Georgia KOROMILA, Vasileios KOURTIS, Vasiliki KOUTRAFOURI, Tomasz KOZLOWSI, Marcin KRZEWICKI, Milena KUBIACZYK, Gülbin KULBAY, Cansu KURT, Nejla KURT, Sevim KURTULDU, Sıla KURTULUŞ, Aldona KURZAWSKA, Melek KUŞ, Orhan KUŞÇUOĞLU, Bilge KÜÇÜKDOĞAN, Ahmet KÜRKMEN, Ditte Kannegaard KVIST, Jacquelyn KYLE, Florence LAINO, Carla LANCELOTTI, Clark LARSEN, Mikael LARSSON, Jinok LEE, Christina LEMORINI, Amanda LEON, Nicola LERCARI, Andrzej LESZCZEWICZ, Xuelei LI, Amanda LINDSEY, Arzu LINGA, Ashley Morgan LINGLE, Mikolaj LISOWSKI, Yan LIU, Alexandra LIVARDA, Rafael LIZERRALDE, Catherine LONGFORD, Serena LOVE, Jackie LOW, Leilani LUCAS, Dominik LUKAS, Julius LUNDIN, Stella MACHERIDIS, Helen MACKAY, David MACKIE, Marco MADELLA, Richard D. W. MADGWICH, Wiebke MAINUSH, Anna MARCHLEWSKA, Arkadiusz MARCINIAK, Elizabeth MARGOLIN, Darko MARICEVIC, Gemma MARTIN, Louise MARTIN, Jack MARTINEZ, Michele MASSA, Wendy MATTHEWS, Richard MAY, Camilla MAZZUCATO, Graeme MCARTHUR, Romy MCINTOSH, Claudia MCKENZIE, Katherine MCKUSTER, Sanaz MEHRAN, Teddy MENDOZA, Mehmet MERTEK, Lynn MESKELL, Gamze MEŞE, Alison MICKELL, Danica MIHAILOVIC, Eva Maria MIHAN, Marco MILELLA, Marina MILIĆ, Slobodan MITROVIĆ, Olja MLADJENOVIĆ, Lauren MONKS, Lucie MONO, Sophie MOORE, Gianfranco MORELLI, Colleen MORGAN, Jacob MORIS, Stephanie MOSER, Chiara MOTTOLESE, Elmas MOTUK, Mehmet Ali MOTUK, Jacqui MULVILLE, Charlene MURPHY, Daniel MURPHY, Inbal NACHMAN, Carolyn NAKAMURA ALDRICH, Goce NAUMOV, Adam NAZAROFF, Kate NELSON, Alexandra NEUMANN, Kelly NGUYEN, Bjorn NILSSON, Dorthe NISTAD, Antoni NOWAK, Selin E. NUGENT, Katie O'CONNELL, Aslı OFLAZ, Jessica OGDEN, Sinan OMACAN, Llonel ONSUREZ, David ORTON, Sonia OSTAPTCHOUK, Sara OUENES, Lütfi ÖNEL, Yasemin ÖZARSLAN, Mihriban ÖZBAŞARAN, Onur ÖZBEK, Özlem ÖZBEN, Ozan ÖZBUDAK, Serap ÖZDÖL KUTLU, Füsun ÖZER, Hakan ÖZER, Levent ÖZER, Özlem ÖZHABES, Özge ÖZKAN, Hembo PAGI, Francesca PAJNO, Philip PARKES, Kamilla PAWLOWSKA, Aruna PAWSON, Halle PAYNE, Chloe PEARCE, Jessica PEARSON, Daniela PEDROZA, Marta PERLINSKA, Sara PERRY, Anda PETROVIC, Paul PETTERSSON, Matteo PILATI, Camile PILLIOUGINE, Marin PILLOUD, Sharmini PITTER, Charles PIVER, Marek POLCYN, Marta PORTILLO RAMIREZ, Adrienne POWELL, Tera PRUITT, Laia PUJOL-TOST, Liz PYE, Lyla PYNCH-BROCK, Joanna PYZEL, Ling QIN, Jason QUINLAN, Antoinette RAST-EICHER, Flavia RAVAIOLI, Roddy REGAN, Katarzyna REGULSKA, Emily RICHARDSON, Megan RIDSDALE, Jana ROGASCH, Nolwen M. ROL, Kate ROSE, Elizabeth ROSEN, Eva ROSENSTOCK, Jamie ROWE, Abel RUIZ-GIRALT, Nerissa RUSSELL, Eugen RUZI, Philippa RYAN, Anna RYBARCZYK, Freya SADARANGANI, Josh SADVARI, Hannah SAINSBURY, İsmail SALMANCI, Nicole SAM, Carlos G. SANTIAGO MARRERO, Judre SAPRANAUSKAITE, Marta SAJ, Jill SAUNDERS, Billy SAWOYO SANKEI, Melania SAVINO, Heeli C. SCHECHTER, Sophie SCHMIDT, Eline SCHOTSMANS, Mesa SCHUMACHER, Jessica SCORRER, Mitchell SCOTT, Jerrod SEIFERT, Uğurcan O. SELÇUK, Recep Yunus SERİN, Gülay SERT, Kent SEVERSON, Harish SHARMA, Daniel SHAW, Russell SHEPTAK, Hannah SHILLING, Lisa-Marie SHILLITO, Anna SHOEMAKER, Ruth SIDDALL, Matilda SIEBRECHT, Maroles SIJSTERMANS, Ahmet SİVAZ, Ebru SİVAZ, Fadimana SİVAZ, Havva SİVAZ, Keziban SİVAZ, Mevlüt SİVAZ, Saliha SİVAZ, Zekeriya SİVAZ, Arne SJÖSTROM, Cassie SKIPPER, Dean SMITH, Kierstyn SMITH, Mehmet SOMEL, Tiffany SOULE, Muhammet SÖKEN, Abdurrahman SÖNMEZ, Charlotte SPIERING, Mira STEVANOVIC, Shannon STEWART, Ivana STOJANOVIĆ, Helen STOKES, Weronika STOSIK, Marketa STOVÍCKOVÁ, Elizabeth A. STROUD, Kristian STRUTT, Amy STYRING, Thomas SUTCLIFFE, Lauren SWEET, Martyna SZYMCZAK, Melike ŞAHİN, Neriman ŞAHİN GÜÇHAN, Elmas ŞENER, Esra ŞENER, Nevriye ŞENER, Muhsin ŞENOL, Ayşegül TABAKOĞLU, Wang TAO, Duygu TARKAN, Dena TASSE-WINTER, James S. TAYLOR, Beliz TERCELI, Kilian TEUWSEN, Gregory THOMA, Johanna THUNBERG, Belinda TIBBETTS, Jenna TINNING, Hatice TOKYAĞSUN, Mavili

TOKYAĞSUN, Yusuf TOKYAĞSUN, Margaret TOMASZCZUK, Angela TORNEY, Jovana TRIPKOVIC, Hoang Anh N. (Elizabeth) TRINH, Christina TSORAKI, Gemma TULLY, Mevriziye TUNCAY, Burcu TUNG, Didem TURAN, Özge TUTAR, Mustafa TUTUMLULAR, Talu TÜNTAŞ, Ali TÜRKCAN, Ülcan TÜRKKAN, Katheryn TWISS, Hakkı UNCU, İdris USLU, Özgür Can USLU, Bilgehan USTA, Oktay UZUN, Ekin ÜNAL, Petra VAIGLOVA, Anne VAN GIJN, Mirjam VAN SAANE, Milena VASIĆ, Maria VAYANOU, Renee VD LOCHT, Rena VEROPOULIDOU, Owen VINCE, Alice VINET, Sophie VULLINGS, John WAINWRIGHT, Sam WAKEFORD, Gillian WALKER, Jiajing WANG, Johnathan WANG, Marcin WAS, Amanda WATTS, Sadie WEBER, Willemina WENDRICH, Elizabeth WESSELLS, Joanne WESTBROOK, Lucy WHEELER, Harvey WHITEHOUSE, Jade WHITLAM, Patrick WILLETT, Alice WILLIAMS, Chelsea WISEMAN, Jessie WOLFHAGEN, Karen WRIGHT, Nurcan YALMAN, Thaer YARTAH, Mustafa YAŞ, Hasan YAŞLI, Hatice YAŞLI, Hulusi YAŞLI, Hüseyin YAŞLI, İsmail YAŞLI, Lokman YAŞLI, Metin YAŞLI, Mustafa YAŞLI, Osman YAŞLI, Rabia YAŞLI, Senay YAŞLI, Tulin YAŞLI, Lisa YEOMANS, Gökhan YEŞIL, Gülay YILANKAYA- ERDOĞU, Nejat YÜCEL, Onur YÜKSEL, Eren YÜNCÜ, Mustafa ZEYTIN and Bright ZHOU.

Volumes 12–15 in this series would not have been produced without the work of Scott Haddow, Jason Quinlan, Kathryn Killackey, Dominik Lukas and Camilla Mazzucato. I am very grateful to them for their commitment.

1. Moving matter at Çatalhöyük

Ian Hodder and Serap Özdöl Kutlu

Çatalhöyük is a 9,000-year-old tell site in central Turkey. First excavated by James Mellaart (e.g., Mellaart 1967) in the 1960s, a new project began in 1993 (Hodder 1996; 2000). The site is of international significance because it is large (13.5ha) at an early date, had dense population (probably over 1,000 people), has rich symbolism and sub-floor burial, and was occupied for 1,500 years; the Neolithic Çatalhöyük East Mound dates from 7100 BCE to 5900 BCE, with the Chalcolithic Çatalhöyük West Mound overlapping in time in the last quarter of the seventh millennium BC and continuing on until 5600 BCE (Orton et al. 2018). The well-preserved buildings and rich art in the Neolithic mound give a unique insight into early village life. The site allows study of many of the main questions dealing with the early formation of settled villages/towns and the early intensification of agriculture. The site was inscribed on the UNESCO World Heritage list in 2012.

This volume is one of four which discuss interpretation of the material obtained during excavation of the site from 2009 to 2017. Early work on the surface of the mound in 1993–1995 is published in Volume 1 (Hodder 1996), and the methodology is described in full in its own volume in Volume 2 (Hodder 2000; for updates see Hodder, Marciniak 2015 and Berggren et al. 2015). The excavation and the interpretation of contexts and buildings excavated in 1995–1999 are described in Volumes 3–6 (Hodder 2005; 2006a; 2006b; 2007), and Volumes 7–10 describe the excavation and interpretation of material obtained in the 2000–2008 seasons (Hodder 2013a; 2013b; 2014a; 2014b). Volume 11 is an account of the excavation of Building 3 by a team from the University of California at Berkeley (Tringham, Stevanović 2012).

The new series of four volumes (Volumes 12 to 15) presents reports on the results of excavations in 2009–2017, although many contributors also take the chance to summarise and compare results over the 25-year span of the Çatalhöyük Research Project (ÇRP). Volume 12 describes the excavation results regarding architecture and the deposits uncovered in the 2009–2017 seasons. Volume 13 reports on the ways in which humans engaged in their material and biotic environments, using a wide range of archaeological evidence. The current Volume 14 turns from 'ecofacts' and biology to material artefacts recovered from the site in these seasons, including a range of clay-based objects (ceramics, clay balls, tokens, figurines) as well as those made of stone, shell and textile. Volume 15 is also based on the 2009–2017 excavations, but its aim is synthetic, drawing on material from Volumes 12 to 14 to deal with broad themes. Data from architecture and excavation contexts are linked into broader discussion of topics such as social networks, seasonality, curation, colour, community engagement and outreach.

Future volumes will describe other excavations and analyses undertaken over recent years under the permit provided to the Çatalhöyük Research Project. Volume 16 will describe the excavations on the West Mound (Anvari, Biehl forthcoming), and future volumes will describe the work in the TP and TPC Areas (Marciniak et al. forthcoming) and on the historic materials found on and off the mound (Jackson, Moore forthcoming).

Research questions

The long (25-year) aim of the Çatalhöyük Research Project has been to situate the elaborate symbolic production at the site within its full environmental, economic and social context. Why did this rich outpouring of symbolism and ritual occur at this time and this place? Can we understand why the art and symbolism were produced and can they help us understand why people formed large agglomerations and intensified resource procurement?

In the earliest phase of the current project (1993–1995), minimal excavation took place. The work concentrated on regional survey (Baird 2002; 2005), on planning and studying the surface of the mounds, conducting surface pickup, drawing eroded profiles of the earlier excavation trenches and using geophysical prospection. The project also undertook a re-evaluation of the material in museums that had been excavated by Mellaart. All this work was published in 1996 (Hodder 1996).

During the second phase of excavation (1996–1999), analysis and publication (2000–2002), the research aim focused on individual buildings, asking questions about site formation processes and trying to understand whether the mudbrick buildings on the site were sometimes shrines (as Mellaart had suggested) or whether they were all houses (even if at times with much symbolic and ritual content). For the results of this work, see Hodder (2005; 2006a; 2006b; 2007).

In the third phase of excavation (2000–2008), followed by analysis and publication (2009–2012), the research aims turned from individual houses to the social geography of the settlement as a whole and larger community structure. The focus on social geography had a number of more specific components. One component concerned social and economic differentiation. Mellaart argued that he had discovered an area of the site inhabited by priests and their shrines and houses. But right from the first phase of the current project we had found uniformity across the site. Surface survey, geophysical work, surface scraping and excavation all showed that wherever one dug on Çatalhöyük, one tended to find the same thing: houses containing evidence of domestic activity varying only slightly in size and elaboration, and areas of refuse or midden. There seemed little on which to base any argument for social and economic differentiation.

So the fourth phase of excavation (2009–2017), followed by analysis and publication (2018–2019), focused on trying to understand how social and economic organisation varied through time. The excavation in 2000–2008 had concentrated on earlier and later levels of occupation on the East Mound, but it became increasingly clear that there was evidence for major shifts in many aspects of the evidence, especially around 6500 BCE, and also in the latest levels as occupation became more dispersed, including spread to the West Mound. It became critical to examine the Middle period occupation levels, especially South M–O (Mellaart's Levels VII and VI), and to open more contiguous buildings of the same time period in the North Area (North G). At the same time, Arek Marciniak undertook excavations on the latest levels of occupation on the East Mound (Marciniak et al. 2015b), followed by related work by Marek Barański (Barański et al. 2015; Barański 2016) in the adjacent GDN Area.

The overall trajectory of the project has thus been from the study of individual houses to communities and larger-scale social geography. But we have so far given the impression of a limited set of research questions that animated the project as a whole. In fact, the size of the project meant that many different sources of funding had to be sought and a range of different research questions asked. For example, funding for detailed dating was obtained by Hodder and Bayliss from NSF and AHRC, while NSF funding was also obtained by Twiss and Bogaard in order to study the integration of faunal and botanical data. A large number of graduate students have focused on particular research questions. While all such projects worked within the general overarching aims just described, the project has benefited from the diversity of perspectives brought by these individuals with their different funding sources (see Hodder, Marciniak 2015).

This is particularly true of a series of grants obtained from the John Templeton Foundation that all dealt with various aspects of the rich symbolism from the site in asking questions about religion and cognition. These grants were also important in adding to the multivocality and interdisciplinarity which have always been at the heart of the project. The Templeton projects involved bringing specialists in religion, philosophy and anthropology to spend a week each year engaging with the archaeologists at the site and ultimately producing a series of edited volumes.

In 2006, research funding for a three-year period from the John Templeton Foundation was directed towards the following four questions: (1) How can archaeologists recognise the spiritual, religious and transcendent in early time periods? (2) Are changes in spiritual life and religious ritual a necessary prelude to the social and economic changes that lead to civilisation? (3) Do human forms take on a central role in the spirit world in the early Holocene, and does this centrality lead to new conceptions of human agency that themselves provide the possibility for the domestication of plants and animals? (4) Do violence and death act as the foci of transcendent religious experience during the transitions of the early Holocene in the Near East, and are such themes central to the creation of social life in the first large agglomerations of people?

Answering this first set of Templeton questions proved very productive and resulted in a publication entitled *Religion in the Emergence of Civilization: The Example of Çatalhöyük* (Hodder 2010). One of the results of this work was the identification of different general house types at Çatalhöyük based on burial and architectural and symbolic elaboration. The most notable were 'history houses' (Hodder, Pels 2010), which are architecturally elaborate buildings with multiple burials that endure for generations and undergo numerous rebuilding phases. Houses can also have multiple burials whether we have evidence they were part of a long sequence or not. There are also elaborate houses without many burials and re-buildings, and there are non-elaborate houses. These four different house types and their associated burials at Çatalhöyük allowed us to explore social differentiation in relation to the symbolic and religious features in the houses.

In a second Templeton project (2009–2012), entitled 'Religion as the basis for power and property in the first civilizations. The analysis and publication of Çatalhöyük' another set of four questions explored the relationship between symbolism and differentiation in relation to the already excavated data from 2000 to 2008. The four research questions were as follows, the first two relating to a focus on social geography and the second two relating to change through time: (1) At Çatalhöyük what is the relationship between religious and other symbolism and the control of production in the different house types? For example, are there differences in the productive activities associated with history houses, elaborate houses and other houses? Is there evidence for differences in health and in skeletal markers of mobility and workload between history houses, elaborate houses and other houses? (2) Is there secure evidence that important symbolic and religious objects were handed down in houses, and preferentially in 'history houses', at Çatalhöyük? For example, what are the sequential processes by which some buildings became transformed into history houses while others declined? Is there evidence that objects were handed down from building to building through time creating histories, and does such transfer occur preferentially in the history houses? (3) Through time how does the 'history house' system change, and is there a concomitant decline in the role of religious symbolism? This question relates directly to the question identified above about changes in the social geography of Çatalhöyük through time. For example, what are the economic changes evident at the site in the upper levels? Were cattle domesticated, for example? What are the social changes associated with the upper levels? Do households become larger, more integrated and with less focus on continuity with the past? (4) Is religion related to power and property elsewhere in the emergence of civilisation in the Neolithic of the Middle East? For example, how does the changing role of ritual and differentiation at Çatalhöyük play a part in larger and longer-term changes across the Middle East and Anatolia during the Neolithic? This second Templeton grant allowed us to pursue the overarching aim of opening extensive areas in the middle phases of occupation at the site. The results were published as *Religion at Work in a Neolithic Society: Vital Matters* (Hodder 2014c).

A third Templeton project (2012–2015) asked a further four questions that followed on from the results of the second project and explored the hypothesis that religion had a primary role in the origin of settled life because it allowed the production of the two main struts of that life – historical depth and attachment to place. The first question considered whether Çatalhöyük could be considered a 'history town'. If groups of houses were tied together by history houses, was it possible that the settlement as a whole was constructed so that the inhabitants of could 'read' the history of their relationships with each other in the layout of their settlement? A second question extended the same point to the landscape as a whole. Was that too, including the organisation of fields, linked into the arrangement of houses in the settlement? Third, how did the 'history house', 'history town' and 'history landscape' at Çatalhöyük emerge? Fourth, what is the relationship between religion and history making at other sites in the Neolithic of the Middle East? The results of this project were published as *Religion, History and Place in the Origin of Settled Life* (Hodder 2018).

A fourth Templeton project (2015–2017) shifted focus from religion to aspects of cognition, in particular, consciousness, creativity and notions of self. Again four questions were asked: (1) Have levels of consciousness increased over time, leading to greater abstraction and objectification of thought? (2) Did rates of innovation and creativity increase during the Neolithic as seen through the sequence at Çatalhöyük? (3) Were there forms of cognitive change in the Neolithic that might have been related to a greater awareness of an integrated self? (4) Did any observed cognitive changes occur suddenly or were they stretched out over time? The answers to these questions were published in *Consciousness, Creativity and Self at the Dawn of Settled Life* (Hodder 2020).

Overlapping with these projects on religion and cognition at Çatalhöyük, between 2013 and 2016 funding was obtained from Imitatio, a group of scholars following the mimetic theory of René Girard. The general hypothesis explored was that there is a link between violence and the sacred. Specifically at Çatalhöyük it was argued that the process of mimesis whereby humans desire the same things as each other leads to violence that can easily break out into violence of all against all. This type of rampant violence can be resolved by the participants turning against one – the scapegoat. The latter may then take on a sanctified position as the one that brought peace to society, and through time the sacrifice of the scapegoat is repeated and re-enacted in ritual in order to sustain peaceful order. This hypothesis was seen to be relevant to the ritual treatment of the bull at Çatalhöyük. This work was published as *Violence and the Sacred in the Ancient Near East: Girardian Conversations at Çatalhöyük* (Hodder 2019).

Overall, then, the Templeton and Imitatio projects, as well as funding from research foundations and commercial sponsors, allowed excavation and post-excavation work to pursue the main questions of the third and fourth phases of excavation at Çatalhöyük. As the main research concerns shifted from individual buildings to overall social geography and change through time, the funding allowed large areas of excavation to take place in the North Area, and it allowed deep excavations to take place in the South Area. And it did so by involving the issue of variation in symbolism and ritual. In this way the larger aim of the project as a whole since its inception could be effectively pursued: 'to situate the elaborate symbolic production at the site within its full environmental, economic and social context' (see above).

Regional trends

The above research questions can be situated within wider understanding of the Neolithic in Turkey and the Middle East. A full account of the regional trends within which Çatalhöyük is situated is provided in Chapter 2; only a brief summary is provided here.

In Central Anatolia, by the second half of the ninth millennium there are both small settlements as at Boncuklu and Pınarbaşı (Baird 2007b) and the large highly agglomerated village at Aşıklı Höyük (8400–7400 BC) (Esin, Harmankaya 1999; Özbaşaran 2011). At the latter site there is very little evidence of storage facilities in domestic houses. As a result of large-scale area excavation, a clear social geography has been identified, with public ritual buildings (frequently renewed) separated from other dwellings by a monumental street. The dwellings are organised into sectors by narrow spaces or narrow streets, and there are collective common middens. The area with public ritual buildings and perhaps storage is distinctive in having larger numbers of pressure-retouched projectiles and higher percentages of cattle bones. There is evidence of butchering and sharing of meat. Overall, the evidence is interpreted in terms of the collective and communal rather than in terms of centralised elites.

In cultural and social terms, Çatalhöyük emerges from a strong local tradition on the Konya Plain, as seen at Boncuklu (Baird 2007a). The parallels with this site of the late ninth and early eighth millennia include below-floor burial, painting and wild animal installations, chipped stone point typologies and, most tellingly, a separation in houses between 'clean' floors (with burial and symbolism) and 'dirty' floors (with hearths and activity traces). There are of course many differences between the sites, in terms of layout and economy (for example, the shift to a heavy dependence on sheep at Çatalhöyük), but the socio-cultural continuities are strong, especially when contrasted with Cappadocian sites such as Aşıklı Höyük. It is fascinating to note that the north/south, dirty/clean division within houses at Boncuklu is inverted through most of the sequence at Çatalhöyük, where the north part of the main rooms sees clean floors and the south part 'dirty' build-up of occupation debris near ovens and hearths (Hodder, Cessford 2004). But the earliest levels see ovens and hearths in the northern part of buildings (in Buildings 161 and 17).

It is possible to conceive of the early population at Çatalhöyük being largely derived from small sites such as Boncuklu and Sancak and others identified by Baird in the Konya Plain survey (Baird 2005). Can Hasan III is another possible source (French et al. 1972), and there are undoubted similarities in settlement organisation and use of a microblade lithic tradition. Although there are also similarities in lithic tradition with Cappadocia, the Konya Plain sites differ markedly from Aşıklı Höyük, where there is less symbolism in houses. Beyşehir-Suğla basin could be another possible source (Duru 2012; Özdöl 2012; Özdöl-Kutlu et al. 2015). Süberde, in the basin, is an aceramic site dated to the eighth millennium BC (Bordaz 1973). Both Süberde and Erbaba, in the same basin, have similar settlement organisation (Düring 2006), pottery technology (Özdöl 2012) and lithic technology to Çatalhöyük. There are also similarities to Gökhöyük, also in the same basin, in terms of architecture and pottery.

However, new genomic evidence (summarised in Volume 13, Chapter 18) indicates closer links between Çatalhöyük and Aşıklı Höyük than between Çatalhöyük and Boncuklu. On the other hand, the aDNA data indicate that, beyond a general derivation from the east, the Cappadocian-Konya Plain grouping is relatively distinct from the Levant, indicating a somewhat separate trajectory. Epipalaeolithic and aceramic Neolithic central Anatolians (as seen at Pınarbaşı and Boncuklu) shared traits with groups in the Caucasus, Zagros and Europe. But the genomic profiles in central Anatolia in the early Neolithic are closely related to Epipalaeolithic populations in the same region (Lazaridis et al. 2016; Feldman et al. 2019; Chyleński et al. 2019), suggesting the spread of technology, ritual and domestic species rather than predominantly people.

From the aceramic period to the ceramic Neolithic period, human movement between Anatolia and neighbouring regions increases. Populations of the PPNB Levant (for example, 'Ain Ghazal) show higher genetic affinity to Neolithic Anatolians than did the Natufians who lived in the same region 2,000 years earlier (Kılınç et al. 2017). Similarly, the population of Tepecik-Çiftlik had more links to Levantine and Zagros populations than did people at aceramic Boncuklu, people who lived in the same broad region 2,000 years earlier. These changing genetic affinities can be explained by some gene flow from the Levant into Anatolia, alongside the cultural influences (Cauvin 2000).

There is also evidence of genetic affinities between central Anatolia and the early Neolithic in northwest Anatolia and Europe. Indeed, it is tempting to argue that Çatalhöyük was one source of the spread of early farming into these areas. As we will see below, there is evidence of population increase and higher densities of occupation, as well as stress and disease on human bodies, in the Middle phases of Çatalhöyük from about 6600 to 6500 BCE. After 6500 BCE there is evidence at the site for greater mobility and dispersal. Regional survey has found a scarcity of sites contemporary with the main ceramic layers at Çatalhöyük (Baird 2005). It is almost as if the site sucked in population and then spat it out again in the later levels of occupation, culminating in an increase of Chalcolithic sites on the plain. Seen at a larger scale, it is possible that such forces stimulated more distant dispersal. In Chapter 2, Marciniak makes the claim that the spread of farming into western and northwestern Anatolia, as well as new developments in Cappadocia, can be seen as the result of migration from Çatalhöyük. The examples he gives include Gökhöyük, the houses of which are indeed remarkably similar to Çatalhöyük, with the same internal division of clean and dirty space. Other examples are Erbaba, Bademağacı, Höyücek, Hacılar IX–VI and Kuruçay in the Lake District and adjacent areas. In northwestern Anatolia, cultural influence is seen at Uğurlu, Pendik, Fikirtepe, Aktopraklık, Menteşe and Barcın (Erdoğu 2013; Özdoğan 2010; Karul, Avcı 2011; Roodenberg et al. 2003; Gerritsen et al. 2013), as suggested by the dark burnished ware of the Fikirtepe culture, which is parallel to the dark burnished wares of Central Anatolia (Özdöl 2012; Özdöl-Kutlu et al. 2015). In western Cappadocia, cultural influence is seen at Tepecik-Çiftlik and Köşk Höyük (Bıçakçı et al. 2012; Öztan 2012).

Central Anatolia is located at a cusp between east and west in terms of the spread of farming (Özdoğan 2010). In many aspects the cultural assemblage from Çatalhöyük has similarities with the Marmara and Balkan region (bone spoons, bone belt hooks) rather than with the southeast of Turkey. This difference also includes the social organisation of settlements and ritual. As Özdoğan has argued, to the east, special buildings are more commonly found and a general focus on ancestors and history making. But to the west of Çatalhöyük, special buildings do not dominate, and there are extramural cemeteries (for example, in the Fikirtepe culture sites with rectangular architecture in the Marmara region) (Özdoğan 2011), although many of these regional differences are cross-cut by change through time.

Overall, then, Çatalhöyük is part of a distinct local tradition on the Konya Plain, certainly linked culturally and by the exchange of obsidian with Cappadocia but with its own distinctive focus on symbolic elaboration. Through time there is increasing human dispersal to and from the Konya Plain. There is also much evidence of a wider cultural koine. There are similarities in myth and symbolism (Hodder, Meskell 2011) over enormous areas and time periods, and overall general trends from northern Mesopotamia and the Levant into Anatolia. Throughout the region, houses often increase in size from the Epipalaeolithic onwards into the large settled villages of the MPPNB. Çatalhöyük is one of a number of 'megasites' (Verhoeven 2011) that, as at Çatalhöyük itself, often break up into smaller more dispersed units in the ceramic Neolithic. There are intriguing similarities in this process over extensive areas. In Volume 13, Chapter 15, Haddow et al. describe a wider process of increased secondary burial and separation of burial from houses as seen at both Çatalhöyük and Tell Sabi Abyad (Akkermans 1987), and there are indeed marked similarities in the multi-mound structure of the latter and of later Çatalhöyük. The emergence of elaborately painted pottery in northern Syria is reminiscent of the painted pottery at Çatalhöyük West. Çatalhöyük can thus provide an insight into important regional trends in social and economic processes, as well as a potential impetus for the spread of farming and a source for population dispersal.

Background and method

Çatalhöyük was first excavated by James Mellaart in 1961–1965 (Mellaart 1962; 1963; 1964; 1965; 1966; 1967). His work focused on the southwest area of the East Mound where over 160 buildings were excavated in 14 levels numbered from 0, I, II to XII. Level VI was subdivided into VIA and VIB. These excavations established the importance of the site as a large and densely packed Neolithic 'town'. No further work was conducted at the site until the present project began in 1993. As noted above, the first years of the project were spent in surface survey, and excavation began in earnest in 1995. This volume is concerned with presenting the results of specialist analysis of material from two areas of excavation at Çatalhöyük (fig. 1.1) known as South and North.

The excavations in the South Area (fig. 1.2) between 1995 and 1999 were initially focused upon a 20m x 20m area in the southwest part of the mound, which was investigated by James Mellaart in the 1960s. The main aim of our work in this area was to investigate the earlier phases at the site, and the lowermost deposits above the lake marl were excavated and designated Level Pre XII.A-E. This work was reported on in Volume 3. In 2000–2008, work continued in the South Area under a shelter. The foundation trenches for this shelter were first excavated (in 2002), and then excavation concentrated on later levels within the shelter in the upstanding deposits around the edges of the original 20m x 20m area. The aim was to reduce these surrounding deposits so that we could safely return to excavation of the lowest levels in the 20m x 20m area.

In the 1995–1999 period, our excavations in the North Area (fig. 1.3) aimed to contribute to the research aim of understanding the formation processes in individual buildings. In the northern area, buildings were excavated in great detail (Buildings 1 and 5 discussed in Volume 3). In 1997, a team from the University of California, Berkeley began excavating

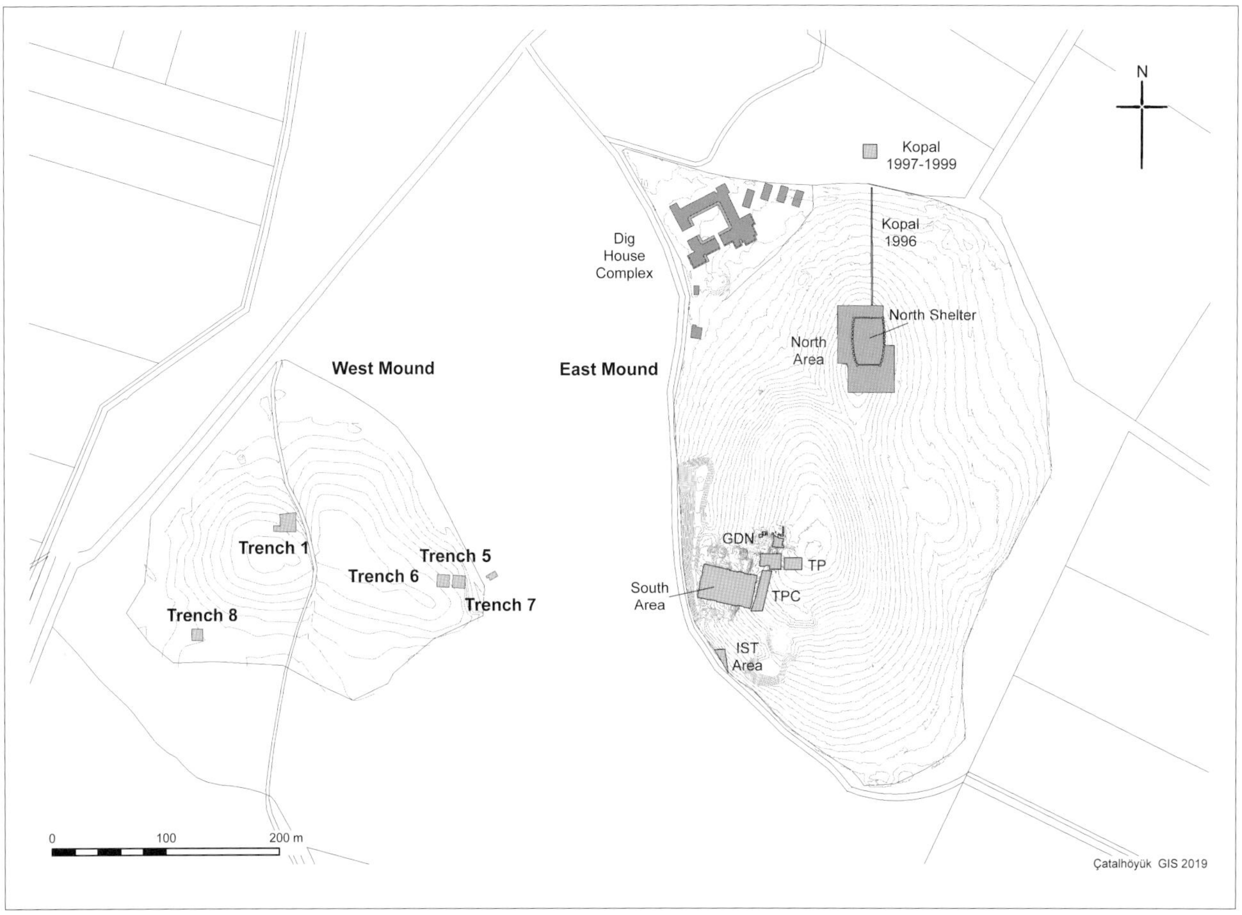

Figure 1.1. Map of excavation areas on the East and West Mounds at Çatalhöyük (map: Camilla Mazzucato).

Building 3 in the BACH Area and continued until 2002 (published in Tringham, Stevanović 2012). The KOPAL excavations consisted of a long trench across the northern flank of the mound to investigate site formation processes (1996–1997) and an off-site area to the north of this to determine what natural and cultural deposits were present (Boyer et al. 2006). In the 2000–2008 period of work, the smaller excavations in the northern part of the mound were extended in order to understand the overall social geography of settlement in this area. A 40m x 40m trench was identified as being relatively undisturbed by later historic activity on the basis of surface research conducted in 1993–1995 (Hodder 1996). In 2004–2008, the 4040 Area was investigated in order to understand the social geography and change through time in this part of the site. In 2007, the foundations for a shelter covering the 4040 Area, Buildings 3 and 5 were excavated. The shelter was then built, allowing excavations in 2008 to take place in a more controlled setting.

Excavations by other teams, especially the TP (Team Poznan) team led by Arek Marciniak of Poznan University and Lech Czerniak of the University of Gdansk in Poland and the IST team led by Mihriban Özbaşaran of Istanbul University, allowed further exploration of the upper levels of the East Mound. On the Chalcolithic West Mound, excavation by three teams (University of Thrace at Edirne led by Burçin Erdoğu, Selçuk University at Konya led by Ahmet Tırpan and Asuman Baldıran, and Berlin University and SUNY Buffalo led by Peter Biehl and Eva Rosenstock) allowed an increased understanding of the developments in the sixth millennium BC. Apart from the inclusion of the IST results in Volumes 7–10, all the excavations by other teams (TP, West Mound) are being published in separate volumes.

In the period 2009–2017, excavation has continued under the direction of Arek Marciniak on the late levels found between TP and the South shelter. The area is designated TPC (Team Poznan Connect), and targeted excavation in Mellaart's old trenches (GDN) from levels 0 to III has been undertaken by a team led by Marek Barański (the latter reported on in Volume 12). The main focus of excavation has been in the South and North Areas. In the South Area, excavations concentrated on middle level buildings in order to fill gaps in the sequence.

Figure 1.2. Map of buildings in the South Area (map: Camilla Mazzucato).

Figure 1.3. Map of buildings in the North Area (map: Camilla Mazzucato).

It also proved possible to return to the earliest levels in this part of the site. In 1999, the deep sounding had discovered open areas and animal penning deposits above the Pleistocene marl and we had not found the earliest houses at the site. In the period since 1999, it was hoped to return to these early deposits beneath Buildings 17 and 43 in the hope of finding houses. We did manage to achieve this and dig below these two buildings in 2017, but again open areas were uncovered. In the North Area the focus was on understanding the overall plan of buildings and exploring the variation between buildings and their inter-relations.

A modified form of single context ('unit' in the Çatalhöyük Research Project terminology) excavation and recording was employed, combined with an intensive and extensive sampling regime, including both wet and dry sieving. The unit forms the basic element of a nested hierarchical system that includes features (groups of related units), spaces (spatially bounded entities generally defined by the walls of buildings), buildings (groups of spaces forming a structural entity), areas (spatially discrete locations where excavation has occurred) and mounds. Chronological grouping is provided by phases/subphases (temporal divisions within a particular space or building, usually reflecting identified changes in the spatial organisation or arrangement of a space of building) and levels (broadly contemporary groups of spaces and buildings – see below).

Units are divided into nine primary data categories of fill, floor, construction, midden, activity, arbitrary, cut, cluster and skeleton. For the numbers of units excavated in these categories see table 1.1. A cut is any recognisable event that led to the removal of another deposit, and an arbitrary layer is a subdivision of a layer or a grouping of layers carried out for practical reasons during excavation. Clusters are any deposits defined primarily by their material inclusions such as obsidian, bone, botanical material etc. Skeletons are effectively a particular type of cluster, one composed of human remains. The other data categories represent types of layer. Fills are layers that have been deposited in any form of negative space, be that a cut, the limits of an upstanding feature, or a whole space or building. Construction/make-up/packing layers are deposits related to the construction of buildings and/or features, while a floor is any deposit inside or outside a structure that forms a surface upon which activities of any sustained duration occur. Middens are deposits primarily related to the deliberate disposal of cultural material, again either outside or inside a structure. All these deposits seem to have been unroofed and more recent research has shown they were used for a variety of activities including disposal of waste. The research conducted by Justine Issavi summarised in Volume 15, Chapter 16 indicates that the term 'midden' should be replaced with 'open area' (although the 'midden' term is found in the database at www.catalhoyuk.com and is retained in table 1.1 in order to retain consistency). Activity deposits relate to a heterogeneous range of activities that produce distinctive remains, including fire spots. These primary data categories are then subdivided according to whether they are in situ, their location, their basic description, the material they are composed of, the nature of their deposition and whether they are a basal deposit.

During excavation, selected units were prioritised for study based on discussions between the excavators and laboratory specialists. Every two to three days during the excavations members of the laboratory teams (faunal, botanical, chipped stone, phytolith, ceramic etc.) went on 'priority tours' of the excavation areas in order to (a) identify units and features for quick processing in the laboratories and (b) discuss with the excavators the information obtained from the quick processing of earlier material. These discussions form the basis for the detailed discussion of individual units, features and buildings in this volume. The aim was to identify a list of units that would be studied by all labs during post-excavation so that data could be compared. A list of 456 priority units was selected for analysis, including 71 units from the TPC Area. During the first stage of the selection process each laboratory team provided a 'wish-list' of preferred units for study. Depending on how many labs chose the same unit, units were given different weights. The final list, which contained both priority units (that is, units identified as priority during excavation) and other units prioritised in the post-excavation process included: (a) units selected by two or more labs, (b) units from certain buildings and spaces that were prioritised (based on Level attribution and type of context), such as B. 77, 79, 80, 160, 131, 132 and Sp. 489, 490, and (c) for the final publication cycle it was deemed necessary to provide a more contextualised and integrated study of burials and therefore a list of units from burial contexts was selected by the Human Remains team and these were analysed by all labs. To ensure comparability of results among different materials, all labs were requested to provide the same consistent information for all units selected as part of this process. This includes basic taphonomic information, count, weight and basic identification of species/artefact type (or size class for faunal material) for each priority unit. (For further discussion of the selection of priority units, see Chapter 16.)

Bayesian analysis of all reliable radiocarbon determinations from Çatalhöyük using BCal indicates that the main occupational sequence at Çatalhöyük East revealed so far probably lasted from 7100 to 6000 cal BC with

	2000–2008		2009–2017		Total (1993–2017)	
North Area	*Percentage*	*No. of units*	*Percentage*	*No. of units*	*Percentage*	*No. of units*
Activity	1%	31	1%	40	1%	80
Arbitrary	5%	141	2%	66	5%	391
Cluster	4%	93	1%	31	2%	162
Construction/make-up/packing	28%	709	39%	1,088	32%	2,525
Cuts	10%	250	13%	374	11%	910
Fills	24%	624	21%	589	25%	1,979
Floors	15%	371	14%	400	15%	1,189
Middens	6%	142	2%	55	3%	236
Skeletons	7%	177	6%	178	7%	530
Total	**100%**	**2,538**	**100%**	**2,821**	**100%**	**8,002**
South Area						
Activity	4%	86	2%	43	2%	171
Arbitrary	4%	79	1%	26	4%	295
Cluster	3%	74	3%	68	3%	231
Construction/make-up/packing	37%	795	37%	934	36%	2,752
Cuts	10%	211	11%	281	10%	725
Fills	17%	364	23%	577	20%	1,480
Floors	15%	321	17%	441	17%	1,269
Middens	7%	149	3%	89	6%	451
Skeletons	3%	68	4%	90	3%	204
Total	**100%**	**2,147**	**100%**	**2,549**	**100%**	**7,578**
TOTAL		**4,685**		**5,370**		**15,580**

Table 1.1. Excavated units by data category.

considerable overlap with the West Mound that continued into the early sixth millennium. In Volume 7, Chapter 3 Bayliss et al. describe the long history of radiometric dating at Çatalhöyük East culminating in their renewed large-scale project integrating Bayesian statistical analysis. This renewed radiometric programme is ongoing and until it is complete we do not have accurate links between the northern and southern areas of excavation. A summary of the current state of the programme is found in Volume 12, Chapters 3 and 19 (see also Bayliss et al. 2015; Marciniak et al. 2015a), but for the moment we do not have precise correlation between the deposits in the North and South Areas and to some degree we remain reliant on comparisons of the cultural material between levels in the North and South (see below).

However, we have since the 2000–2008 period come to realise that the grouping of buildings and deposits into levels numbered from 0 to XII by Mellaart no longer captures the complexity of the stratigraphy of Çatalhöyük (as discussed by Farid in Volume 7, Chapter 2). The main problems identified are as follows. Mellaart himself vacillated as he tried to fit buildings and open areas into a clear sequence of overarching levels. In particular, he had to divide Level VI into two, VIA and VIB, and at various times he re-allocated buildings between Levels VII and VI. The problem that he was struggling against is the same that we have faced – that in fact there are few large-scale horizons of activity that stretch across large areas of the site. In fact, the stratigraphy at Çatalhöyük is mainly house-based. Each house is rebuilt at its own pace and on its own footings. There is very little inter-cutting from house to adjacent house. Although there are some entrance holes between houses, the columns of individual houses rarely inter-connect; each follows its own sequence. Houses are built onto open areas at times, but there are few large-scale events that link adjacent houses together.

In the earlier phases of our work in 1993–1999 we had managed to fit newly excavated buildings into the Mellaart system of levels in the South Area. This is partly because we were mainly working in the lower levels, where stratigraphical relationships seem more regular. But as we worked in the South Area in 2000–2008, particularly in the upper levels, the system of Mellaart's levels began to break down. For example, counting upwards from buildings excavated by Mellaart, B.65 in the 65-56-44-10 sequence should have been Level V. This would have placed Buildings 56–10 in Levels IV to II. But still higher levels have been excavated in TP where the Polish team had stratigraphical links to Mellaart's Levels 0–III. So, clearly something was wrong. In response to these difficulties, in 2008 we introduced a new system of levels based on the stratigraphical relations in our own overall Harris Matrix. Our matrices for the separate South, IST, TP and North Areas were not linked stratigraphically, so we developed separate sequences of levels for each area. In each separate sequence, levels are given letters from lower (nearer A) to higher (nearer Z). Thus South G is lower than South H. It is important to note that the letters given to levels in each sequence are not linked to levels in other sequences. Thus it is not the case that Level South G is equivalent in time to Level North G. The current understanding of the relationships between levels in the different sequences, based on ceramic and chipped stone data, is shown in table 1.2.

Temporal groupings of levels	*South*	*North*	*Cal BC*
Final	TP.O-R and TPC Trenches 1 and 2 (B109 and 115)		6300–5950 BC
Late	GDN	North.H,I,J and IST	6500–6300 BC
	South.T. TP.N. TPC B110 and B150		
	South.S. TP.M. TPC B150 and B122		
	South.R		
	South.Q		
	South.P		
Middle	South.O	North.F, G	6700–6500 BC
	South.N		
	South.M		
Early	South.L		7100–6700 BC
	South.K		
	South.J		
	South.I		
	South.H		
	South.G		

Table 1.2. Current understanding of the relationships between levels in the South and North Areas.

Table 1.3 gives the numbers of units excavated in each level in the North and South Areas. This shows a much more even spread of units than was achieved in the earlier campaigns (the differences between the unit totals in tables 1.1 and 1.3 result from the ways in which Levels are attributed via Space). While 100 building numbers have been assigned by the project, it is important to note that not all buildings were fully excavated. Many buildings could not be excavated for structural reasons, or could only be partially excavated as they had to be left on display. Table 1.4 provides the amount excavated per building for 81 at least partially excavated buildings. For the more fully excavated buildings, table 1.4 also shows the numbers of burials and an elaboration index. The latter measure of architectural complexity was devised by Hodder and Pels (2010) and it simply sums, for any one phase of a building, the numbers of floor segments, basins, benches, installations (protuberances on walls, including bucrania and other animal fixtures), pillars and paintings in the main room of the building. It could be objected that this measure mixes mundane architectural features such as floor segments and wall pillars, on the one hand, and symbolic features such as paintings and bucrania. However, it is difficult to discern which features are more or less symbolic – for example, floor segments seem to have played an important symbolic role in differentiating areas of activity. The simple sum of features does seem to provide a measure that correlates with other variables such as numbers of burials (see Hodder 2016a).

The project worked through changing and complex political changes in Turkey that impacted methods in various ways. For example, the attempted coup in July 2016 resulted in the season's work being curtailed and shifted to following years. Towards the end of the project it became increasingly difficult to export material from Turkey. For example, we were not able to export phytolith and starch samples until too late for full analysis that could be included in these volumes (see Volume 13, Chapter 7). At times, as in the case of macrobotanical remains, samples could be sent within Turkey and studied, for example, at the British Institute in Ankara. In other cases, all analysis had to be carried out in the labs at the site.

It should also be noted that most of the laboratory teams changed through the lifetime of the project. Many of the chapters in this volume mention such changes. Some of the changes were the result of individual decisions by lab members and heads, but we also made the decision to change lab heads from time to time in order to promote critical reflection and diversity of debate. We are extremely grateful to all the lab heads who through time contributed to constructive and productive research. It is the sum total of this work that is reported on here.

2000–2008		2009–2017			Total (1993–2017)		
Levels - North Area	*Number*	*Units*	*Spaces*	*Buildings*	*Units*	*Spaces*	*Buildings*
North ?F		17	3	1	50	7	2
North F	3	554	21	8	814	23	9
SCRAPE ?G	355						
North ?G	86	290	16	6	1,038	37	7
North G	936	1,744	44	12	4,584	63	14
SCRAPE ?H	2						
North ?H	11	13	1	0	73	2	0
North H	471	208	23	11	854	37	13
North I	224	15	6	2	223	12	3
North J	25				29	1	1
Total	**2,113**	**2,841**	**114**	**40**	**7,665**	**182**	**49**
Levels - South Area	*Number*						
South G	1				102	1	0
South ?H		18	1	0	18	1	0
South H		58	1	1	75	2	1
South ?I		41	1	0	41	1	0
South I					11	1	0
South ?J		14	2	1	15	3	2
South J		3	1	1	342	5	2
South ?K	1	59	3	2	510	4	2
South K		332	6	2	685	8	4
South ?L	25	2	2	1	432	12	2
South L	29	260	11	4	611	12	5
South ?M	160	21	5	3	785	28	10
South M		90	7	2	90	7	2
South ?N		30	1	1	48	2	1
South N		360	6	3	360	6	3
South O	1	1,428	10	7	1,440	10	7
South ?P	10				10	1	0
South P	334	221	9	2	570	12	2
South Q	429	9	4	2	470	11	3
South R	377				390	7	3
South S	493	1		1	572	5	1
South ?T	33				39	1	0
South T	27	9	1	0	73	4	1
Total	**1,920**	**2,956**	**71**	**33**	**7,689**	**144**	**51**
TOTAL	**4,033**	**5,797**	**185**	**73**	**15,354**	**326**	**100**

Table 1.3. The numbers of units excavated in each level.

Synopsis and integration: moving matter

As noted above, our main research aims in the 2009–2017 seasons were to understand how social and economic organisation varied through time, especially in the Middle period. There was also a continued focus on the social geography of the settlement and the role of the 'history houses' that were rebuilt more times, were more elaborate (see elaboration index above) and had more burials beneath the floors. During this period there was also an emphasis on cognitive questions and the organisation and social efficacy of ritual and religion. These questions led us to examine the entanglements between humans and their material worlds at various scales, from the overall use of the landscape around the site, to the arrangement of buildings on the site, and to the social lives of the inhabitants of the mounds.

The most obvious way that matter was moved at Çatalhöyük was in the bringing of clays and other materials into the site. The whole mound was constructed by obtaining clays, marls, sand and gravels from the surrounding landscape to build the 20m high mound. Botanical materials from wooden timbers and branches to reeds and seeds were brought in to build and nourish, as were shells, fish, animals, dung, ceramics, ground stone, obsidian, chert and a wide array of other materials. These were all moved to make the site and its inhabitants possible.

But these materials also moved each other. The weight of the roofs destabilised the walls; the soft soil in pits caused walls built above them to slump. Ovens, hearths and clay balls transformed raw into cooked food. Baskets, skins and pots were needed to carry clay, water, grains. Nets were used to catch fish, and obsidian projectile points were used to down and cut up animals so they could be brought in. Ground stone axes were used to cut trees and branches, to hone timbers and to grind cereals. Dung was brought in to make fuel for fires to cook food. One of us has talked elsewhere about all the entanglements that resulted from these thing-thing and human-thing interactions (Hodder 2012; 2016b).

We wish to add a further dimension to the idea of moving matter, which is that matter itself is unstable and always changing. This is certainly true of all the particles at the subatomic level, but we would argue that it is also true of matter at the level of human interaction. Perhaps the best example of this at Çatalhöyük is the house, and the way houses are built on top of each other. In excavating the site, we have often found it difficult to decide where one house finishes and another begins. This is because the layout of a new house often seems to be present in the way that a previous house is filled in. Does the new house start when the previous house stops being occupied (although it may be used to deposit refuse for a while – is this still a 'house' or has it become a midden?), or when its ovens and installations are abandoned, or when it is filled in (is the fill the ending of the old house or is it the start/foundation of the new house?), or when new walls are built and make-up layers deposited? In fact, the house is just a flow or process of actions held together by the intent to hold it together and to use it as a house. As archaeologists we infer a sequence or flow of events and this is what allows us to 'lump' a set of actions into a 'house' even though it is always changing.

Much the same can be said of all the material artefacts at Çatalhöyük. For example, in Chapter 15 the study of use-wear on obsidian tools indicates that they were frequently used for different purposes through time. Is the obsidian blade the same thing through these different activities, or is each use a different thing (hide scraper versus meat knife, for example)? We can always split or lump typologies in different ways. Or consider clay balls, discussed in Chapter 7. We actually find very few complete clay balls; they are nearly always fragmented and heat shattered. But we call them all 'clay balls'. Even when broken, they still seem to have been used to retain heat or to line the base of ovens and so on. What holds these items all together as 'clay balls' is our inference that they were used in sequence. The intents and purposes that flow through the balls vary, but there is also the flow of the matter itself as it changes and transforms. In the end, of course some clay balls become broken down into such small particles that they become part of the matrix of the mound – they can no longer be identified as clay balls, unless the particles are again brought together to make a new ball.

In this chapter, we want to explore this vision of matter as made up of a series of flows. We want to consider these different material sequences as lines of energy that interact, animate or constrain each other, leading to change. It is as if the lines are electromagnetic forces that attract or repulse each other. Or, another metaphor, material things are like stones thrown into a pond, causing ripples and waves that intersect with other ripples and waves to produce complex interference patterns. Perhaps the most obvious example at Çatalhöyük is the incorporation of the skulls and horns of wild animals into the structure of the house. Bloch (2010) argued that such installations animated or vitalised the house (see Keane 2010; Anspach 2019; and other chapters in Hodder 2010 and Hodder 2014c) and had impact on memories and social and economic relations. The humans buried beneath the floors of the houses seem to have 'attracted' painting on nearby walls; the incorporation of unused obsidian bifaces into the southern part of houses seems to have been part of a 'repulsion' of burial and ritual practices into the northern part of houses. Shells obtained from a distance were added to necklaces of beads in order to animate local processes of exchange (see below). In all these ways we can talk of 'flows' of matter that intersect and transform, moving each other, entangled with human aims and strategies.

Building	*Area*	*Level*	*% excavated*	*Complete in plan*	*Internal msq*	*Side room msq*	*Plat-form msq*	*Burnt Y/N*	*Elabration index (Eɪ)*	*Total no. of individuals*	*No. of known rebuilds*
1	North	North G	100	Y	37.96	9.83	11.91	Y (partially)	33	61	2 (B1-B.5)
2	South	South K	75	Y	22.13	5.14	6.27	N	16	0	no known rebuilds
3	North/ BACH	North G	100	Y	27.54	6.26	10.11	N	17	8	no known rebuilds
4	South	South L	50	N	NA	NA	NA	N	NA	0	no known rebuilds
5	North	North F	50	Y	39.04	10.37	2.59	N	10	4	2 (B1-B.5)
6	South	South L	25	Y	38.89	10.06	6.85	N	9	10	4 (S.VI.10-B.24-B.6-B.17)
7	South	South L	0	Y	NA	NA	NA	N	NA	0	5 (S.VI.8-B.20-B.7-B.16-B.18)
8	South	South M	0	Y	NA	NA	NA	N	NA	0	5 (S.VI.1-B.8-B.21- B.22- B.23)
10	South	South T	75	N	NA	NA	NA	N	NA	0	4 (B.10-B.44-B.56-B.65)
12	North	North ?H	15	N	NA	NA	NA	N	NA	1	2(B.12-B.77)
13	North	North H	15	N	NA	NA	NA	N	NA	0	no known rebuilds
16	South	South K	25	Y	23.73	NA	NA	N	12	0	5 (S.VI.8-B.20-B.7-B.16-B.18)
17	South	South K	75	Y	35.68	9.05	6.11	N	22	22	4 (S.VI.10-B.24-B.6-B.17)
18	South	South J	25	Y	24.15	4.16	NA	N	3	1	5 (S.VI.8-B.20-B.7-B.16-B.18)
20	South	South M	0	Y	NA	NA	NA	N	NA	0	5 (S.VI.8-B.20-B.7-B.16-B.18)
21	South	South L	0	Y	NA	NA	NA	N	NA	0	5 (S.VI.1-B.8-B.21-B.22-B.23)
22	South	South K	0	Y	26.32	NA	NA	N	6	0	5 (S.VI.1-B.8-B.21-B.22-B.23)
23	South	South J	50	Y	21.88	4.38	NA	N	6	2	5 (S.VI.1-B.8-B.21-B.22-B.23)
24	South	South M	25	Y	NA	NA	NA	N	12	0	4 (S.VI.10-B.24-B.6-B.17)
40	South	South M	<25	N	NA	NA	NA	N	NA	0	no known rebuilds
42	South	South R	50	N	18.75	NA	NA	N	NA	8	no known rebuilds
43	South	South L	75	Y	24.13	4.92	5.7	N	NA (very eroded + excavated by Mellaart)	9	4 (B.43-B.160-B.161-B.162)
44	South	South S	100	Y	25.72	NA	13.51	N	10	16	4 (B.10-B.44-B.56-B.65)

Table 1.4. Summary information about buildings excavated.

Infilling material	*Infill/ closure*	*Part occupation sequence only*	*Complete occupation sequence*	*Construction phase excavated*	*Walls excavated*	*Notes*
clean infill	Y	N	Y	Y	Y	
abandoned	Y	Y	N	partially	N	
clean infill	Y	N	Y	Y	Y	
clean infill	Y	N	Y	Y	Y	
clean infill	Y	Y	N	N	N	
unknown	NA	Y	N	Y	N	Mostly excavated in 1960s. Finished by Hodder Team in 1999
unknown	NA	N	NA	NA	N	Outline only. Excavated by Mellaart Team as Shrine E. VIII.8. Excavated in 1960s leaving extant walls only
unknown	NA	N	NA	NA	N	Outline only. Mellaart Shrine 1.VII. Complete in plan. Excavated in 1960s leaving extant walls only. Walls recorded by Hodder team.
NA	N	N	Y	Y	Y	Excavated by team from Thessaloniki (96 – 97)
N	N	N	N	N	Y	
N	N	N	N	N	N	
clean infill	N	Y	N	N	Y	Mellaart shrine E.IX.8
clean infill	Y	Y	N	Y	partially	Sp.182 has not been fully excavated
unknown	Y	Y	N	N	N	Mellaart shrine E.X.8
unknown	NA	NA	NA	NA	partially	Mellaart E.VII.1. Outline only
N	N	N	N	N	N ?	Mellaart shrine VIII.1
N	N	N	N	N	recorded	Mellaart shrine 1. E.IX.I Hodder Team recorded walls
uknown	N	Y	N	Y	N	Mellaart shrine E.X.1
clean infill	Y	Y	N	N	N	Mellaart S.10.VII
unknown	N	partially	N	partially	partially	Mostly excavated in 1960s. Mellaart house E.VII.2 and E.Vii.12
clean infill	Y	N	Y	Y	Y	Incomplete plan. Heavily eroded and collapsed area on S edge
unknown	N	N	Y	Y	Y	Mellaart E.VIII.27. Complete in plan. Fully defined and excavated by Mellaart as Shrine. E.VIII.27 down to occupation sequence. Re-opened by Hodder Team in 2004 with partial excavation of some features only. Reopened in 2012
clean infill	Y	N	Y	Y	Y	

Building	Area	Level	% excavated	Complete in plan	Internal msq	Side room msq	Plat-form msq	Burnt Y/N	Elabration index (Eι)	Total no. of individuals	No. of known rebuilds
45	North	North H	25	Y	46.84	11.57	18.49	Y	10	3	no known rebuilds
46	North	North H	0	N	NA	NA	NA	N	NA	0	2 (B.46-B.48)
47	North	North J	25	N	16.87	NA	3.83	N	NA	0	no known rebuilds
48	North	North G	25	Y	12.31	3.37	5.51	N	7	1	2 (B.46-B.48)
49	North	North G	100	Y	15.08	2.25	5.55	N	12	15	no known rebuilds
50	South	South M	75	Y	15.95	3.06	6.59	N	NA	15	no known rebuilds
51	North	North G	100	Y	11.06	NA	3.86	N	8	0	no known rebuilds
52	North	North G	85	Y	44.66	8.45	9.32	Y	42	18	no known rebuilds
53	South	South Q	50	N	NA	NA	NA	N	NA	8	no known rebuilds
54	North	North I	25	Y	28.69	7.53	NA	N	NA	12	no known rebuilds
55	North	North H	25	Y	42.79	NA	7.77	N	NA	0	no known rebuilds
56	South	South R	100	Y	30.98	4.67	13.01	N	7	4	4 (B.10-B.44-B.56-B.65)
57	North	North H	25	Y	26.83	5.8	7.21	N	7	3	no known rebuilds
58		North H	25	Y	26.2	6.79	8.39	N	10	9	no known rebuilds
59	North	North G	100	Y	50.05	29.54	14.53	N	13	1	3 (B.83-B.59-B.60)
60	North	North H	25	N	20.16	NA	NA	N	NA	10	3 (B.83-B.59-B.60)
64	North	North H	25	Y	31.68	NA	NA	N	2	0	no known rebuilds
65	South	South Q	100	Y	24.01	5.27	8.04	N	10	16	4 (B.10-B.44-B.56-B.65)
66	North	North G	25	Y	19.37	NA	NA	N	NA	0	no known rebuilds
67	North	North H	25	N	24.86	7.95	1.49	N	NA	0	no known rebuilds
70	North	North I	0	N	NA	NA	NA	N	NA	0	no known rebuilds
71	North	North I	25	N	NA	NA	NA	N	NA	0	no known rebuilds
75	South	South P	25	N	NA	NA	NA	N	NA	3	no known rebuilds
76	South	South O	50	N	27.49	NA	10.34	Y	NA	5	no known rebuilds
77	North	North G	100	Y	28.19	7.93	8.44	Y	20	37	2 (B.12-B.77)
79	South	South O	25	N	NA	NA	NA	Y	NA	0	no known rebuilds
80	South	South O	75	N	18.7	NA	9.37	Y	25	17	no known rebuilds
82	North	North G	50	Y	22.07	NA	NA	N	NA	0	no known rebuilds
88	North	North H	25	Y	13.77	NA	NA	N	NA	7	no known rebuilds
89	South	South N	75	N	35.11	NA	7.98	N	11	14	no known rebuilds
91	South	South M	25	N	NA	NA	NA	N	NA	4	no known rebuilds
92	South	South M	25	N	NA	NA	NA	N	NA	5	no known rebuilds
96	South	South O	25	N	NA	2.58	NA	N	NA	9	no known rebuilds

Table 1.4 (continued). Summary information about buildings excavated.

Infilling material	*Infill/ closure*	*Part occupation sequence only*	*Complete occupation sequence*	*Construction phase excavated*	*Walls excavated*	*Notes*
clean infill	Y	Y	N	N	N	
unknown	N	N	N	N	Y	Almost completely eroded
unknown	Y	Y	N	Y	Y	
clean infill	Y	Y	N	N	N	
clean infill	Y	N	Y	Y	N	
NA	N	N	N	Y	Y	Excavated by Mellaart as E.VII.9
clean infill	Y	N	Y	Y	Y	
rubble	Y	N	Y	N	N	
clean infill	Y	Y	N	Y	Y	Highly eroded building
unknown	N	Y	N	N	Y	
clean infill	Y	N	N	N	N	Excavated to last occupation sequence only. Truncated ın the NW corner
clean infill	Y	N	Y	Y	Y	
clean infill	Y	N	N	N	N	
clean infill	Y	Y	N	N	N	
clean infill	Y	N	Y	Y	Y	
unknown	N	Y	N	Y	Y	Very eroded, only the eastern part of the building survived (25% of building survived)
clean infill	Y	N	N	N	N	
clean infill	Y	N	Y	Y	Y	
abandoned	Y	N	N	N	N	
unknown	N	Y	N	N	N	
NA	N	N	N	Y	Y	Heavily eroded
clean infill	Y	Y	N	Y	Y	Truncated
unknown	Y (Mellaart)	N	Y	Y	partially	Excavated to the floor level by Mellaart. Re-opened in 2008, fully excavated ın 2009. Very eroded and damaged by Neolithic fire. Truncated to the north
rubble	Y	N	Y	Y	Y	
rubble	Y	Y	N	N	N	
rubble	Y	Y	N	N	partially	Side room truncated
clean infill	Y	Y	N	N	N	
clean infill	Y	Y	N	N	N	Eroded
clean infill	Y	Y	N	N	N	Underneath 76
clean infill	partially	Y	N	N	N	FT South
	partially	Y	N	N	N	FT South. Correlated to Mellarts's house E.VII.18
abandoned	Y	Y	N	N	N	

Building	Area	Level	% excavated	Complete in plan	Internal msq	Side room msq	Plat-form msq	Burnt Y/N	Elabration index (Eı)	Total no. of individuals	No. of known rebuilds
97	South	South O	100	N	21.84	3.89	8.28	Y	9	14	no known rebuilds
100	South	South M	25	N	NA	NA	NA	N	NA	1	no known rebuilds
101	North	North G	<25	N	NA	NA	NA	N	NA	2	no known rebuilds
102	North	North G	50%	N	NA	NA	NA	N	NA	36	no known rebuilds
108	North	North H	25	N		NA	NA	N	NA	9	no known rebuilds
112	North	North G	50	N	28	NA	NA	N	NA	0	2(B.112-B.119)
113	North	North G	0	Y	31.86	10.05	NA	Y	NA	0	no known rebuilds
114	North	North G	75	N	18.75	NA	1.65	N	11	16	no known rebuilds
118	South	South H	25	N	NA	NA	NA	N	NA	0	no known rebuilds
119	North	North F	75	Y	27.52	5.97	11.33	N	12	0	2 (B.112-B.119)
122	TPC	South S – TP.M								12	
128	North	North G	50	Y	16.76	1.62	NA	N	NA	12	no known rebuilds
129	North	North H	25	N	45.8	NA	NA	N	NA	23	2 (B.129-B.131)
130	South	South L	25	Y	NA	NA	NA	N	NA	0	no known rebuilds
131	North	North G	100	Y	46.6	15.7	10.13	Y	18	42	3 (B.129-B.131 - B.139)
132	North	North F	75	N	72	NA	8.75	N	15	11	no known rebuilds
139	North	North F		N						0	3 (B.129-B.131-B.139)
150	TPC	South T – TP.N								26	
160	South	South K	100	Y	22.8	6.7	3.07	N	5	4	4 (B.43-B.160-B.161-B.162)
161	South	South J	50	N	17	NA	NA	N	NA	8	4 (B.43-B.160-B.161-B.162)
162	South	South J	50	N	16.84	NA	NA	N	NA	0	4 (B.43-B.160-B.161-B.162)
163	North	to be assigned	0	N	NA	NA	NA	N	NA	0	no known rebuilds
165	North	to be assigned	0	N	NA	NA	NA	N	NA	0	no known rebuilds
166	TPC	to be assigned								6	
167	North	to be assigned	<25	N	NA	NA	NA	N	NA	4	no known rebuilds

Table 1.4 (continued). Summary information about buildings excavated.

Infilling material	*Infill/ closure*	*Part occupation sequence only*	*Complete occupation sequence*	*Construction phase excavated*	*Walls excavated*	*Notes*
rubble	partially	Y	N	N	partially	Excavated by Mellaart as House E.VIB.28
unknown	partially	Y	N	N	N	FT South
unknown	N	Y	N	N	N	FT North
rubble	Y	Y	N	N	N	Part excavated in FT 1, FT 2 and north beam slot
unknown	Y	Y	N	Y	N	
unknown	Y	Y	N	Y	N	Highly eroded
burnt/ unknown	partially	Y	N	partially	partially	Heavily eroded. The building has been cross-sectioned N-S midway across Sp.96 and Sp.95
rubble	Y	N	Y	N	N	Sp.87 occupation not fully excavated
clean infill	Y (partially)	Y (partially)	N	Y (partially)	N	Mellaart House 25 (Level XII). Space 510. The structure has been affected considerably by the 1960s excavation. In 2013 the CRP excavated a small targeted trench to establish the presence, or otherwise, of earlier architecture beneath the structure
clean infill	Y	Y	N	Y	Y	
NA	N	NA	N	Y	Y	Heavily eroded and trampled
NA	NA	NA	NA	Y	NA	Highly eroded building constructed on B.131. Possibly very short temporal break between the closure of 131 and the construction of 129
very clean	Y	Y	N	N	N	The building is very small
clean infill	Y	N	Y	Y	Y	Infill with a high amount of human remains. Highly processed in small pieces
clean infill	Y	Y	N	Y	partially	Sp. 511 only partially excavated
	Y	N	Y	Y	Y	
	Y	Y	N	Y	partially	
	Y	Y	N	N	partially	
N	unknown	N	N	N	N	
N	unknown	N	N	N	N	
Y	clean infill	Y	N	N	N	

Exchange

One example of this process is provided by objects that were procured from outside and brought into Çatalhöyük. These items were brought into a complex local set of meanings and strategies. In many ways, the site and its inhabitants were local and isolated. In Volume 13, Chapter 11, Pearson et al. discuss the results of the analysis of human dental strontium isotopes. The picture gained is that the majority (n=72) of the individuals buried at Çatalhöyük spent their childhood on the Konya Plain; only eight individuals fell outside this range. There is no evidence for patrilocal or matrilocal movement on marriage and very few other settlements are known on the plain during the main occupation of the site (Baird 2005). It is possible the site was endogamous and somewhat isolated and inward-looking. The grazing of animals, while it expanded in the Late levels, remained largely on the plain (Volume 13, Chapter 11). There is strontium evidence that up to half the wild cattle procured in the Early and Middle periods grew up on the limestone terraces in the Taurus foothills (about 15+ km south of the site), with the rest growing up on the Konya Plain. But by the Late period at Çatalhöyük, almost all grew up on the Konya Plain (Volume 13, Chapters 8 and 9). Much plant exploitation was local, with fields intensively managed close to the site (Volume 13, Chapters 5 and 6).

So although in the Late period there is evidence from the measurement of human femur mid-shafts of greater mobility, a shift supported by a wide range of data (see Volume 8, Chapter 1 and Volume 9, Chapter 1), much activity, even in the later levels, was local. On the other hand, Çatalhöyük was keyed into wider networks of trade and exchange, especially in the Late period. In Volume 9, a diversity of examples is provided, from obsidian to ground stone, ceramics, beads, shells and speleothems and timbers obtained from the Taurus. It is also clear that the site was linked to wider cultural influences, as described by Marciniak (Chapter 2). Many of the symbolic themes found at Çatalhöyük occur over wide areas (Hodder, Meskell 2011) and, for example, the types of projectile point occurring at the site include those identified as widely found PPNB forms such as Byblos, Amuq, Herzeliya, Jericho and Nizzanim (Chapter 14). While the analysis of ancient genomes from the site suggests a largely local development on the Konya Plain from small, scattered sites such as Boncuklu (Volume 13, Chapter 18), there was nevertheless movement of people and ideas, and Çatalhöyük must be seen as closely linked to a wider koine.

An important example of widespread links is provided by study of obsidian. As Doyle (Chapter 14) argues, five different variants of PPNB point types are found at the site, and they continue throughout the occupation. Widespread connections are also indicated by obsidian sourcing. Through much of the Çatalhöyük sequence there is a heavy reliance on Cappadocian sources, first Göllü Dağ and later Nenezi Dağ, but in the later levels there is also use of distant sources in eastern Turkey (Volume 9, Chapter 17). More recent work has demonstrated a rather different picture for the use of chert – a small component of the Çatalhöyük chipped stone repertoire (Nazaroff et al. 2013; 2015). As argued by Marciniak (Chapter 2), there is a continuity of exploitation and use of regional chert resources. An important early example was Konya radiolarite, the source of which occurs on the southwestern fringe of the Konya Plain. Another important early source was Sınanlı, located 90km south of Ankara. Throughout the sequence there was use of the Akdere source about 250km northwest of the Konya Basin.

Nazaroff et al. (2013; 2015) argued that the source of the radiolarite chert was close to the sources of other materials, especially clay for pottery, ground stone, red ochre, pumice, carnelian and timber, and may have been tied up in their exploitation. In Chapter 13, Tsoraki suggests that the andesites used for grinding tools probably derive from the volcanic Erenler-Alacadağ area ca 60–70km to the southwest of Çatalhöyük (Doherty 2013; Doherty, Tarkan 2013; Doherty 2017; and Tarkan, this volume, Chapter 4). Stone resources from this upland volcanic region were used from early on. The transportation of large andesite boulders for quern production would have involved considerable effort.

The tension between local, regional and distant procurement is seen in a most interesting way with shell. In Chapter 10, Veropoulidou notes that local shells like *Unio* which are commonly found at the site are rarely used to make shell artefacts. Most shell artefacts are made from shells derived from a distance. For example, marine shells were obtained from the Mediterranean coast probably through networks of exchange and there is a steady increase in numbers into the later levels – reinforcing the picture of greater long-distance procurement through time. Fossil shells also increase through time, coming from the Taurus, the Hatay and İskenderun basins (more than 300–400km distant).

As regards stone beads, in Chapter 9, Vasić notes that different forms of limestone and marble could have been procured locally, within an area of 15 to 20km from the site, but that carnelian (see above), fluorapatite and turquoise were sourced from significantly further away. The shell and stone beads allow interesting insights into the nature of trade and exchange. The differences in wear of beads in the same composite ornament (such as a necklace) suggest a process of collecting from various sources. Strings in burials frequently combined beads made of locally available materials with others that were harder to procure (for example, limestone and marine shells or carnelian). Even if these strings of beads were assembled solely for the purpose of burial, they suggest complex networks of associations, near and far. It is possible to imagine that the distant sourcing created a prestige that was inserted into local networks of exchange, vitalising and revitalising them.

The complexity of these exchange links is also suggested by the caches of obsidian that occur in the Early and Middle periods. The artefacts in these sometimes very large hoards of bifaces do not seem to have been used (see Chapters 14 and 15). Placed beneath the floors, often in the southern part of the house, the bifaces in the caches seem to have been obtained as part of the social processes that founded and maintained a house. We know that the distant exchange or procurement strategies were very targeted, as indicated by the shift from the Göllü Dağ to Nenezi Dağ sources despite their being no apparent functional distinction between them. The evidence suggests shifts in exchange networks over time. The bifaces held some special symbolic importance in creating the house in the earlier levels; their otherness somehow contributed to the vitalisation of the house (in perhaps a similar way to the installations of wild bull horns, as argued by Bloch (2010)). In other words, the distant was used to create local meanings. In the same way, distant rather than local shells were preferred to give value and meaning to necklaces, and distant stones had the same effect.

The house and the community

We have already seen that complex exchange networks and flows of value along them played a part in establishing the house. The caching of unused obsidian bifaces below the floors of houses in the Early and Middle periods is parallel to many examples of depositional events associated with the founding and abandonment of houses.

In Chapter 13, Tsoraki notes examples of intentional incorporation of large and small ground stones in walls and other architectural features. She also describes the varied clusters of material, often including a large amount of ground stone, in clusters beneath platforms and in other such contexts. It had earlier been thought that ground stones were intentionally destroyed on these occasions. However, the variation of weathering on the broken ground stone fragments, for example, deposited in the abandonment of B.77, as well as the lack of refits, suggest that the ground stone fragments had different histories and were purposefully assembled at the time of the abandonment. It seems, then, that a wide group of people participated in the abandonment process and contributed 'gifts' to the abandoned house. It is possible that a similar process occurred with regard to the necklaces described above – with people contributing beads with different wear and different life histories to a burial. Abandonment and burial were thus opportunities to create links and liaisons.

Thus different flows of materials were brought together in clusters, reinforcing or adding to each other, and having impacts on the house and those living in it. An insight into the range of activities that took place in the house is provided in Chapter 15 by Lemorini and D'Errico's work on chipped stone use-wear. They describe all the evidence for meat processing in and around B.131 and the range of craft and other activities in B.132, such as the treatment of hide and the working of wood, the scraping of antler and the processing of herbaceous plants, and perhaps the making of baskets. They also identify differences between houses. For example, in B.77 they see a restricted range of activities presented. There is less evidence of use-wear – and what there is largely concerns the production and shaping of chipped stone tools, and projectile points among them, as well as hide processing.

It is possible that some of the ritual activities that so dominated B.77 'repulsed' some of the activities that might have been expected in such a building. This notion of different material flows and entanglements that repulse each other is clearly seen in the arrangement of activities inside the house. It has long been recognised that the southern parts of the main rooms in houses have what we have called 'dirty' floors, containing the sweepings or residues from a wide range of domestic tasks, while the northern and eastern parts of main rooms contain more evidence of adult burial, clean white floors, and paintings (Hodder, Cessford 2004). In Chapter 3, Vasić confirms the pattern that higher platforms in rooms are cleaner of heavy residue traces than the lower general-height floors. And in Chapter 13, Tsoraki finds that the ground stone data confirm the use of the southern dirty floors for the production and maintenance of andesite grinding tools and other forms of tool production.

While Tsoraki finds this pattern throughout the sequence, and there is much evidence for a sustained north-south difference in main rooms in the East Mound, recent work has shown that in the earlier levels the separation is often less marked. In Buildings 17 and 160, for example, ovens are found in the northeast of the main room, and in early buildings such as these and B.132, platforms and floor divisions are less clearly defined and demarcated. Indeed, it is in the Middle period that the classic north-south divide is most clearly present.

The caches of obsidian bifaces are almost always found in the central or southern part of the house, and they occur only in the Early and Middle periods (Chapter 14). Shifts through time in the internal organisation of the house are seen most clearly in the placing of paintings in the main rooms. In Chapter 12, Busacca finds a lack of spatial patterning in the painting locations in the Early period, then a concentration in the north and east of the main room in the Middle Period, and then in the Late period a wider dispersal again, but with a focus in the southwest. There is some evidence for a spatial association between painting and burial, with paintings often associated with burial platforms. It is thus of interest that the 'flows' of painting and burial reinforce each other through time. In Volume 13, Chapter 15, Haddow et al. show that burial locations in the main rooms change in tandem with the shifting paintings. In the Early period, burials occur throughout the main room, but in the Middle period, burial is focused along the north and east walls. In the Late period, burial concentrates in the central east area but also in the side room, often in the west or southwest.

These shifts not only give meaning to each other and change together over time, they are also entangled with other flows of energy and material running through the house. In Chapter 12, Busacca notes a specific shift from paintings linked to burial in the Middle period to paintings linked to ovens and hearths in the Late period. In the Early and Middle periods, ovens (and the pottery and clay balls associated with them) are never decorated. By the Middle period, the ovens occur in the south part of main rooms associated with the dirty floors. But in the Late period, some ovens are actually themselves decorated. This shift is part of an overall process of change that we have long noticed. In the Early and Middle periods, before 6500 BCE, the houses are working concerns, but there is a strong emphasis on burial, memory-making and ritual. But after 6500 BCE there is less focus on ritual and greater emphasis on domestic production. Bull horn and other installations still occur (for example, in the IST Area) but the frequency of installations declines (Russell, Meece 2005). In the upper levels there is greater investment in sheep and cattle and changes in farming techniques that all stimulate domestic activity in the house. Decorating an oven, and the general shift in the Late period towards elaborating the southern oven area in the main rooms, can be seen as responses to these wider changes, or they can be seen as actively manipulated in order to steer social and economic change in new directions. So again the different material flows reinforce or restrain each other.

Cooking

Central to these changing flows are the practices surrounding cooking. These help us to understand the changing arrangement of activities in the house and also the larger social and economic changes through time in which cooking played a central role.

Bennison-Chapman (Chapter 7) finds much evidence, especially secondary colour change and sooting on clay balls, plus their heat-shattered form, of clay balls being used in fires/cooking. The balls are also often found near ovens and hearths in the southern part of main rooms. They very much concentrate in the Early and Middle periods, with a clear decline afterwards. Bennison-Chapman suggests that the balls were mainly used on floors or in hearths in order to roast meat or plant foods. This contrasts with earlier suggestions (Atalay 2013) that the clay balls were used in wet cooking in baskets. However, in Chapter 13, Tsoraki argues that the fairly large numbers of fire-cracked natural stones found on the site were probably involved in wet cooking with heated stones, based on their size and depositional context near hearths etc. and because they decrease through time in parallel with the clay balls and in contrast to the rise in cooking pottery.

But during the Middle period a shift to a new form of cooking gradually emerged. The flow into the site of dark gritty pottery fabrics from volcanic areas to the southwest of Konya gradually had a major impact on many other flows at the site. In particular, these wares became the first pots used for cooking, given their grit-tempered thin fabric that was efficient for transferring heat without cracking. Up to this point most pot forms, used as containers, were small-sized bowls (Chapter 4). But with the appearance of the Dark Gritty Ware, the hole-mouth jar emerges. Residue analysis as well as the sizes of broken-up sheep bones suggests intensive processing of meat products in these pots. Cut bone dimensions and vessel sizes are consistent with the boiling of meat in pots (Yalman et al. 2013: 158). Evidence of this shift from roasting meat using clay balls to boiling meat with clay pots is also found in the study of cut marks. Demirergi et al. (2014) showed that in the Early and Middle periods, sheep- and cow-sized animal bones are dominated by consumption cutmarks. These marks occur when meat is cooked on the bone and are the result of cutting away the meat after cooking (Russell, Martin 2005: 85). Such marks are mainly found as a result of processing baked and roasted meats. Boiled and stewed meat separates from the bone very easily and rarely produces such cutmarks (Demirergi et al. 2014). The decrease in consumption cuts over time at Çatalhöyük suggests a decrease though time in the roasting and baking of meat. There is a corresponding increase in the frequency of filleting cut marks on sheep- and cow-sized animals through time (Demirergi et al. 2014). As noted by Bennison-Chapman in Chapter 7, filleting indicates the processing of meat into small pieces, for drying and storage or to create manageable pieces of meat to be cooked and eaten off the bone (Demirergi et al. 2014). The decrease in consumption cuts alongside the increase in filleting cuts implies a change in consumption practices, with meat more likely to be prepared in a liquid form, stewed or boiled in the Late period as opposed to the roasting of large joints in the Early and Middle periods.

Evidence from the analysis of plant food remains shows a similar trend. Lara González Carretero (Volume 15, Chapter 19) demonstrates a shift in cooking practices around 6500 BCE. The Early and Middle periods are dominated by wheat, with food traces of bread as the main source of grain at this time. With the introduction of cooking pots, a barley-based 'porridge' appears in the archaeological record. This liquid meal is prepared in pottery vessels using direct heat. The Late and Final periods of occupation see a dominance of barley, suggesting this meal formed an important element of the diet in the later half of occupation. Thus both the faunal and botanical evidence suggest a shift from dry baking and roasting in the Early and Middle periods to boiling and stewing in pots in the Late and Final periods. We see clearly, then, how the flow into the site of a particular form of clay and pottery causes ripples that gradually build up energy, causing fundamental change in the house and in ways of life at Çatalhöyük.

The cooking of liquids in pots may have had further reverberations. Previous work by Pitter et al. (Volume 9, Chapter 11) showed that while cooking pots were used for processing meat products, they also started to be used for milk products by the Middle phase. In the later West Mound, the extraction of ancient proteins from ceramics has confirmed the processing of cereals, pulses, meat and dairy products (Hendy et al. 2018). These findings have been made yet more intriguing by research on the proteomic analysis of human dental calculus from the site (Hagan et al. 2019). This work supports the evidence for the consumption of milk, but it also provides some evidence for milk proteins from the genus *Bos*. Since some of these samples are from levels in which cattle remain biometrically wild, it appears that the early management of cattle may have involved milking. It is even possible, therefore, that the introduction of cooking pots in which liquids could be processed had the impact of allowing and encouraging the domestication of cattle – fully biometrically domestic cattle appear in the upper Late and Final periods. So here we see very clearly how the flow of one material – clay for making cooking pots – can become allied with other flows, such as the cooking of liquids and the management of cattle, to create fundamental change. We can also see here how these burgeoning entanglements came to transform the house into a hive of productive tasks by the Late and Final periods.

The organisation of production – specialisation of production

So how was production organised, and is there evidence of change through time? Did all production take place within the house? Can we think of production less in terms of humans organising to produce things they need and more as flows of things coming together to provide opportunities for productive work? To some extent, do things proffer production rather than the other way round?

We have already seen this process of flows of things entering the site and offering the potential for new energies in the way that the introduction of grit-tempered hole-mouthed cooking jars gave a kick-start to the boiling of stews, the cooking of gruel and the processing of milk. In turn, through time as an on-effect, there is evidence in the Late and Final periods of houses that are larger and with more space for storage and external yards. Houses became more self-sufficient in terms of production.

As noted above, use-wear studies on chipped stone tools indicate many phases of use and re-use for multiple purposes (Chapter 15). The impression is gained of intensive processing and hard work. Some of these tools were used to process plants, but many aspects of plant processing are difficult to discern. We are fortunate at Çatalhöyük that the intentional and controlled burning of some houses led to remarkable preservation of organic materials, including wooden bowls (see Volume 13, Chapter 4).

In Chapter 11, Jørgensen et al. note that the textiles from Çatalhöyük are amongst the earliest woven textiles known. They provide a fascinating insight into the ways in which plant fibres were brought into the settlement to create beautiful textiles, though at the expense of a lot of work. The woven textiles seem very complex, with evidence of borders, hems and even decoration in a herringbone pattern. But this complexity depends on other things. Woven textiles such as these require a loom, either a backstrap system or a simple frame. The lack of evidence of loom weights or spindle whorls from the Neolithic layers at Çatalhöyük makes it difficult to talk about the organisation of production, and the vagaries of preservation of organic remains on the site make it difficult to assess the relative importance of hides and textiles for clothing. But all the evidence points to a weaving process that probably took place at a household level – and it adds to all the other evidence of pressure on the household. The processing of basts into thread, the spinning and weaving, and the high levels of skill involved all suggest a considerable amount of time was spent on this task. There are figurine depictions of textile clothing and we know that the dead were often buried in strands, ropes and textiles that encased the body. So many aspects of Çatalhöyük society were caught up in these threads.

Indeed much of the evidence for production comes at the house level. In Chapter 15, Lemorini and D'Errico identify possible traces of mirror production in B.131, the same building that had a mirror in a grave and with the building above (B.129) also having mirrors. Is this evidence of a specialisation in mirror production? As we shall see below, some buildings have concentrations of bead production, while B.150 in TPC has a concentration of stone figurines, and in buildings neighbouring B.150 there is concentrated evidence of ground stone production. In Chapter 5, Nakamura notes that B.150 not only had five stone figurines deposited within its walls, the building directly succeeding and on top of it also produced a stone human figurine of a similar form. But such evidence always occurs in the context of domestic houses with the full range of normal daily activities. There is often evidence of in situ production of a variety of different items (for example, beads and ground stone artefacts) in the same building. There is never evidence to suggest specialists that depend on their craft for economic survival – rather, the impression is of variation in skills that leads certain houses to concentrate on low-level productive activities of use to others in the community.

One interesting observation returns us to the use of clay balls. We have long noted that while ovens and hearths occur in all buildings, they may be absent in certain phases of use of a building. This suggests that for brief periods houses may have shared in food preparation. We have also seen from human isotopes (Volume 13, Chapter 11) that those buried in a house had also eaten together during their lives; since more people are buried in certain houses than could have lived in them, this suggests that groups of houses shared food. It is thus of interest that in Chapter 7, Bennison-Chapman finds some tantalising evidence for a possible association between large numbers of clay balls and less elaborate buildings. The sample is small, and the association does not hold true of all buildings, but there is some slight indication that in the earlier levels intensive cooking was associated with less elaborate buildings.

But for most productive activities, organisation was largely house-based. There is some evidence that in the earlier levels fields and herds were collectively managed, but the secondary and tertiary processing of meat and cereals was at the house level. There is much evidence that glume wheats were stored in the husk in houses and that the final stages of de-husking and grinding took place in side rooms and the southern parts of main rooms. We have already seen the evidence for intensive sheep (and to a lesser extent cattle) product processing in the house.

Another form of intense activity in the house was bead production (Chapter 9). Stone and shell assemblages indicate that a combination of percussive (flaking, pecking) and abrasive techniques (sawing, drilling, grinding, polishing) were used to transform the raw materials into varied bead forms. But despite the large number of beads found, there are relatively few blanks, preforms, roughouts, nodules and debitage. Production contexts remain scarce and small scale. The evidence of in situ bead production that occurs in a few buildings is usually limited to one or two types – there is no suggestion that a large range of bead types was produced in one building. Similarly, while there is evidence of obsidian production in many buildings, the evidence is scattered and does not indicate specialised workshops (Chapters 14 and 15). In Chapter 14, Doyle notes higher levels of knapping debris in Buildings 160 and 132, neither of which has a very high elaboration index.

It is perhaps with ground stone that we might have expected to see indications of specialist production. But again the evidence indicates dispersal and house-based production. Tsoraki in Chapter 13 finds widespread evidence of production debris in many houses and phases of houses. She examines the microdebitage from andesite grinding tools (since most grinding stones are made of andesite) and finds no concentration in special buildings. In the later levels in the TPC Area there are high densities of debitage in a group of buildings (B.142, B.140, B.81) and Mellaart interpreted B.140 as 'a stoneworker's shop'. But the lack of standardisation and other factors lead Tsoraki to conclude that there is not good evidence for 'specialised workshops' in a strict sense. Similar results occur for plastering activities. Wall plastering toolkits have been identified at Çatalhöyük, consisting of assemblages of stone materials along with bone scapulae, plaster burnishers and polishers of various smoothness, and with trace wear and traces of plaster in them. These are widely found, suggesting again a lack of specialisation (ibid.).

So there was a lot of work done in and around the house. This involved everything from de-husking and grinding grain, to cooking, making textiles, making bone tools and making beads. But it is of interest that while *Unio* were locally available, people preferred to make shell beads and other artefacts from marine shells, recent and fossil, obtained from the Mediterranean coast. The daily practices inside the house were punctuated with objects that were given value and that can be seen as revitalising and giving meaning to those practices. Not only was the fabric of the house embellished with caches of obsidian beneath the floors and bull horns on platforms and walls, but shells, stone beads, chert, pottery, ground stone, oak and juniper timbers were brought from afar and embedded within local practices. Speleothems were brought from caves to the south and east at a distance of about 100km, probably from İncikini (Erdoğu et al. 2013). These objects all provided energy in the form of economic and social work that they could do. Often in some sense 'costly', they paid back in terms of how they moved the house and its inhabitants in new directions.

Building inequality

We need to be careful, however, in talking about the house in this way as if the 'house' was coincident with the building. We know that one of the most significant ways in which objects brought into the house gave vitality to the house was through the burial of ancestors beneath the floors. Buildings that were more able to attract burials from the inhabitants of other buildings often are more elaborate in terms of architectural detail (Hodder 2016a). They are more likely to have bull horn installations and paintings. We have come to term such buildings 'history houses' (Hodder, Pels 2010). Since some buildings have many burials and others few or none, it seems likely that such buildings became burial places for the inhabitants of other buildings. We also know that the people who were buried in history houses also had similar diets (Volume 13, Chapter 11). So it seems that a 'house' at Çatalhöyük often consisted of a group of several houses that formed a unit – a 'history house' in a larger sense.

We have spent much effort trying to establish whether individual history houses were able to turn their special role in burial and ritual into the control of production. We have seen above the very slight indication from clay balls that the less elaborate buildings in a history house group may have been more involved in food preparation in the Early and Middle periods. But in earlier research we were unable to find good supportive evidence (Volumes 8 and 9) for such differences. In particular, we have been able to find no correlation between the size of a building and the amount of architectural and symbolic elaboration or number of burials. Given that the elaboration index and the number of burials are the most direct measures of whether a building is a history house, this suggests that individual history houses were not larger. We have also failed to find a link between elaboration or number of burials and side room storage space.

We find similar patterns in this volume. For example, in Chapter 6, Der studies the distribution of faunal horns and clay figurine horns across a range of different buildings. She finds no statistically significant correlation with size. But she does find some statistically significant correlation between figural and faunal horn densities and the number of burials in houses and the elaboration index. In Chapter 5, Nakamura finds a lack of overall correlation between figurines more generally and the elaboration index. But figurines occur somewhat more frequently in the burial fills of more elaborate buildings, although not necessarily in those buildings with the most burials. In Chapter 8, Bennison-Chapman finds some correlation between a building's size and degree of internal elaboration (its elaboration index) and the number of clay objects it contains.

Figurines and figurine horns and clay objects (though we have little sense of the function of the latter) may be providing evidence of the symbolic elaboration of buildings. But what about evidence for production? Is that concentrated in the individual history houses? And what about markers of status? For the earlier periods, we have already noted the slight evidence that less elaborate buildings had a greater concentration of clay balls used in cooking. In Chapter 3, Vasić notes some slight evidence of lower heavy residue values in the more elaborate buildings, which could mean either that these buildings were more carefully cleaned or that less productive work was done there. But there is little other evidence in a similar vein. In Chapter 13, Tsoraki makes a concerted effort to explore whether ground stone production was concentrated in history houses. As regards plastering tools, she finds no correlation with buildings with high elaboration indices. Similarly there is no correlation between the elaboration index and the embedding of grinding slabs in floors. Tsoraki also studies the size of grinding stone surfaces which are consistent with use for plant processing and finds no changes in scale over time and no correlation with the elaboration index. She also finds no correlation between history houses and the presence of maceheads (often seen as an indicator of status in the later levels).

Beads might also be thought to be a possible indicator of status, but neither the evidence of bead production nor the numbers of beads in burials and fills are concentrated in elaborate buildings. As Vasić notes in Chapter 9, the occurrence of beads does not seem to be related to either the size or the elaboration of buildings. This is evident in the lack of correlation between the count of beads in burials, clusters and floors or other deposits and building size and elaboration index. Similarly, there is no correlation between bead density in floor deposits and elaboration and size of buildings, with large and elaborate buildings such as B.77, B.131 and B.52 having low bead density in comparison to not-so-elaborate and smaller houses.

In Volume 13, other evidence is discussed regarding the position of the history houses. In Chapter 16 in that volume, correlations are sought between indicators of health and well-being and elaboration in history houses. The authors find that the extent of building elaboration and inclusion in a history house sequence is unrelated to the living conditions of the buried individuals. Equally in Chapter 11 in that volume, Pearson finds no difference in diet related to building elaboration or to numbers of burials in a building. And in Chapter 17, Vasić confirms earlier evidence that building size and elaboration do not correlate with the presence of different aspects of burial treatment. The percentages of buried individuals with pigment associations, adornment or objects do not depend on the size of a building nor its elaboration. Similarly, no exceptional treatment has been observed in the burial assemblage in buildings defined as 'history houses'.

It seems necessary to conclude, therefore, that the history houses acted as foci for a group of houses, the inhabitants of which worked in the landscape and ate together, buried their dead together and probably conducted rituals together (in the history house buildings), but within the history house collective there were not marked differences in the status of individuals or houses. This conclusion is supported by the evidence that the more elaborate houses did not have more evidence of storage capacity or productive capacity (see Volume 8, Chapter 1 and Hodder 2016a).

If there was little differentiation within the history house between the status, work and health of individuals, there is some evidence of other forms of differentiation – by sex and age. On the whole at Çatalhöyük men and women lived very similar lives. They had close to identical diets as indicated by isotopic evidence (Volume 13, Chapter 11) and they were buried in similar ways (Volume 13, Chapters 15 and 17). Both men and women had their heads removed and circulated after death. However, an important interdisciplinary study by Sadvari et al. (2015) managed to tease apart some differences in activity. Study of skeletal evidence, ground stone and projectile points indicated that women had a greater role in grinding activities and men played a greater part in hunting activities, and these differences in daily life can also be glimpsed in wall paintings and figurines. So although mortuary practices, diet and relative status did not greatly differ among women and men at Çatalhöyük, the study by Sadvari et al. shows that habitual activity patterns may have differed.

But the major social differences within the history houses seem to have focused on age. Although some contradictory evidence is described in this volume in Chapter 10, Vasić in Chapter 9 finds that beads are more commonly found with adults. The older people were when they died, the more likely they were to be buried with beads. This could simply be the result of amassing beads through one's life, or amassing exchange relationships through life that resulted in more people contributing beads at death. The evidence does not necessarily indicate a higher status. However, there is clear evidence from isotope studies of human bone that older individuals had a different diet, with more C_4 plants and higher nitrogen, than younger individuals (Pearson, Meskell 2015); the same authors interpret the figurines from the site as often indicating maturity and age.

Change through time

So far we have seen that the centrality of the history houses within the history house group was established and animated by moving in bull horns and other installations as well as by burying the dead. Other artefacts were also moved in, such as obsidian caches, beads, paintings and figurines. But in these relatively egalitarian groups, all houses were involved in a wide range of productive tasks, and much production was organised at the house scale. The moving in of things also generated productive tasks that became numerous and demanding by the Middle period (as seen by skeletal markers of stress – see Volume 13, Chapter 16).

One of us has already indicated that the moving of new things into the house created new potentials and energies that moved the society in new directions. For example, the introduction of cooking pottery in the Middle period responded to the build-up of stress and allowed easier preparation of food. Cooking with clay balls is very efficient in terms of heat transfer, but the balls have continually to be re-heated and replaced. By shifting to cooking pots, the cook was freed to carry out more tasks while cooking was taking place. In addition, the use of pots allowed a shift to boiling gruel (less demanding than baking bread) and boiling stews (with the meat easier to remove from the bone and more intensive processing of fats and marrows possible). The new cooking technology in turn allowed a marked increase in the reliance on sheep and perhaps also the closer management of cattle and cattle milk. In general terms, in the Late and Final period houses are less continuous, there is more space around them, including yard space, and they often have more side rooms and are larger. There is more productive activity seen in ground stone; in this volume, Chapter 13, Tsoraki notes a marked increase in densities of edge tools (axes etc.) in the Late period as well as an increase in the number of palettes. There is more pottery and a wider range of wares (Chapter 4).

Through time there seem to be changes in the ways in which the flows and energies of things are brought into relation with each other. Throughout, the main conduits for energy flows are the house, wild animals and the human body. But the organisation of the flows seems to shift from a vertical to a horizontal emphasis.

In the Early and Middle periods, and in contrast to earlier sites such as Boncuklu, Can Hasan III and Aşıklı Höyük, more and more is brought into the house. This is when the evidence for the history house system seems most marked. There is a strong focus on continuity of the house and burial within the house. Houses seem to vie with each other to attract burials and bull heads and other installations. As Busacca shows in Chapter 12, by the Middle period the house walls are multiply plastered and often painted. And we have already seen that it is in this phase that the internal divisions in the house are most clearly marked, both in terms of the location of burial and painting and in terms of the architectural division of space into platforms, dirty and clean areas. The numbers of burials and installations in houses increase towards the end of the period, in Levels South N–O and North G. Houses are more tightly packed and population density and fertility are at their height (see Volume 13, Chapter 16), but so too is stress and disease (ibid.).

Levels South N–O and North G are 'classic Çatalhöyük': this is the time of Building 77 or, from Mellaart's excavations, the elaboration of Shrine 10 in E.VIA and E.VIB. This is the time of the greatest number of burials and the most elaborate history houses. It can be argued that the social and economic system at this point is very much dependent on the history house formulation, whereby individual rights of access to goods and services are related to membership of the history house as well as other neighbourhood and cross-cutting sodalities (see Volume 13, Chapter 1 and Mills 2014). It was the bringing in of bull horns and human burials that generated the vitality of the house and attracted members. The ritual system based in the house was used to create historical ties through time that connected people in the present. It was the bringing of things together in the house that created a social focus and a productive force. The house linked the present to the past and so established communities in the present.

As the houses expanded and competed with each other, new technologies such as cooking pottery were brought in to ease some of the tensions. But the new technologies themselves unleashed new potentials that took things in new directions. The more efficient cooking in the house and the shift to gruel and stews released some of the constraints on all the other activities that took place in the house, such as textile production, bead manufacture, bone tool manufacture and so on. There was greater investment in sheep herding and cattle management and people found themselves under less stress, but moving farther over the Konya Plain and more widely, such that skeletal markers of mobility increased (Volume 13, Chapter 16 and this volume, Chapter 10 regarding the wider sourcing of beads). These economic changes allowed houses to be more independent and to break away to some extent from the history house system. On the site there seems to have been some contraction and dispersal of settlement. It is important not to exaggerate the breakaway from the dominance of the history house system, since there are buildings that continued to be built on each other in the same place. The B.65-56-44-10 sequence is a good example, but even here there are changes. In the upper levels there is more space around individual buildings, and in the case of the B.65-56-44-10 sequence, at times there are doors through walls into surrounding yards. There is less plastering of walls and less painting, and less burial. And the burial that does occur gradually changes towards more secondary burial; that is, skeletons were more frequently moved around after initial burial.

It is as if rather than focusing on house continuity, in the upper levels the focus shifts to making links between houses. Skeletons are more frequently moved around, and this change to greater mobility and interaction between houses is also seen in the emergence of stamp seals – these often use the same motifs that had earlier been used in house wall paintings but are now placed on mobile objects – perhaps used to decorate human bodies or skins. Pottery too increasingly diversifies and includes examples of motifs such as bull heads that had earlier been confined to house walls. So one way of characterising this shift from the earlier to the later levels is from constructing horizontality out of verticality to constructing horizontality directly.

But we wish to follow Nakamura in Chapter 5 and argue that the change can also be seen as a shift from the house to the human, or perhaps from the animal to the human. After all, while in the earlier phases the continuity of the house was established through the burial of humans and the installation and re-installation of animal bones, in the later phases the human body was used to establish connections between houses, and there is also an increase in the number and diversity of zoomorphic figurines (Chapter 5).

In the upper levels there are increasing numbers of anthropomorphic figurines that replace the earlier abbreviated forms. In Chapter 5, Nakamura is surely right to argue that these new figurines show less emphasis on the individual and more on the human. There are several types that emerge. There are stone representations of bearded figures riding animals that may be associated with a continued emphasis on the social importance of hunting, and there are corpulent naked figures with rounded body forms sometimes seated or reclining and at rest. Nakamura discusses ethnographic examples in which excess fat of certain individuals could have been used as an outward sign of a group's success. So rather than representing the success of the house in terms of wild bulls and other animals, the success is now shown by the representation of fattening. The fact that these clay and stone figurines of corpulent and mature bodies are found in houses, in some cases intentionally buried beneath platform floors, consolidates the link between the human form and the house, but their mobile hand-held character emphasises the possibility of interaction and exchange.

It is possible to see this shift from the house to the human and from the vertical to the horizontal as a direct result of the economic changes brought about by technological shifts such as the more efficient cooking procedures. They can be seen, then, simply as ideologies responding to economic change. But in this chapter we have taken a different view and have shown that different material flows entering Çatalhöyük energised each other or contradicted each other in order to stimulate change. Thus the greater investment in domestic production in the upper levels allowed houses to be more independent and to break out of the history house system (thus from the vertical to the horizontal), and in such a context, fatness could be used as an expression of the domestic success of the productive group. In all the complex interactions that have been traced in this chapter, it becomes impossible to identify initial impulses and primary causes. For example, we have implied several times that the shift from clay balls to pottery cooking set in train numerous other changes, but it is also the case that socio-ritual competition between history houses was one of the lines of action that led to stress in the Middle period and so to the introduction of cooking pottery. Rather, we have tried to demonstrate the multiple strands that came together and interacted, leading to substantive change.

Conclusion

We have in this account presented a rather narrow view of changes through time at Çatalhöyük by focusing on the movement of materials into the site over a 1,100 year period. We have characterised the change at around 6500 BCE from the Early and Middle to the Late and Final periods as involving shifts from vertical to horizontal social relations, the house to the human, the ritual to the domestic. In Volume 13, on the other hand, wider considerations are brought into play. In particular, the shift from the earlier to the later levels is seen as involving climatic and environmental changes from wetter to drier, and the gradual expansion of the alluvial fan on which the site sits. In Volume 13, Chapter 5, archaeobotanical evidence is discussed that demonstrates linked changes in the way the landscape was farmed, also influenced by the dispersal of smaller settlement nodes on the site in the later periods. The shift from verticality to horizontality can be seen as a response to these changes or as contributing to their instigation.

Equally, earlier volumes provided other examples of comparable material changes through time. For example, in Volume 9, Chapter 21 Carter and Milić discuss the shift in the Late and Final periods from an obsidian flake industry to a blade industry dominated by the pressure-flaking technique (see also this volume, Chapter 14). This new technique involved higher levels of skill, but it also brought new energies. The reduction of obsidian cores into blades resulted in more workable edge being produced from a given amount of obsidian. We have seen the evidence for the increase in the frequency of filleting cut marks on sheep- and cow-sized animals through time. Filleting indicates the processing of meat into small pieces, for drying and storage or to create manageable pieces of meat to be cooked and eaten off the bone. It is also linked to the cooking of meat in small pots as stews. So here multiple lines converge – the shift to blade production responds to the need for intensive processing of smaller cuts of meat which is linked to the introduction of cooking pots. Obsidian blades, grit-tempered pots, meat stews reinforce each other as the inhabitants of houses invest in domestic labour and become more self-sufficient and independent, beginning to break out of the all-encompassing symbolic codes and ritual affiliations of the history houses.

As the material lines (obsidian blades, cooking pots, meat processing) intersect, they energise each other and stimulate change; but they also create friction in the form of conflicts and contradictions. As houses vie with each other in terms of their productive success, it becomes attractive to put energy into managing cattle more fully, ultimately leading to their domestication, but also into the demands of tending and penning cattle. So humans get drawn into yet more work. The houses also vie with each other in terms of presenting food in shows of food presentation and feasting. At the end of the occupation of the site and onto the West Mound the hearth often moves into the centre of the main room and pottery becomes highly elaborate and painted. So again, humans get involved in the extra work of producing different pot types, for cooking, storage and social display. And in the upper levels of the East Mound there is also a major increase in the prevalence of the phytoliths of *Phragmites*, an aggressive reed that thrives in disturbed environments (see Volume 8, Chapter 9). One interpretation of this change in the upper levels is that the increased focus on sheep and cattle herding involved keeping animals for part of the year near the site, as well as in pens on the site. Habitats near the site would have been disturbed, resulting in the proliferation of *Phragmites*: humans would have been drawn into cutting back the reeds and managing them.

As a final example of these complex intersections and energies, the charcoal evidence described in Volume 13, Chapter 4 shows a shift from oak to juniper in the upper levels. It is not easy to make sense of this shift, but one possibility is that juniper offers harder wood that may have been of greater value as houses increased in size in the upper levels. So again the move of a particular material onto the site may have empowered changes taking place there.

It seems possible to argue, then, that materials moved at Çatalhöyük, or rather that they moved Çatalhöyük. The materials afforded potentials that may only have been realised gradually. For example, pottery was introduced early in the sequence, but it was initially organic tempered and used as a container. Through time the fabric was changed and the pot was used for cooking, and by the time of the West Mound pots were also used for storage and social display. The affordances are only gradually realised but once realised they generate energy in terms of economic production as much as social and symbolic impact.

And finally we see that it is not so much the things themselves that generate change, but the lines of productive potential. For example, it is all the processes involved in the intention of cooking with a pot so that more activities can take place in the house. It is all the biotic, abiotic and human energies involved in using bast to make textiles to clothe the dead. It is all the heterogeneous processes involved in making a figurine to demonstrate the productive success of a house. And the lines or flows regarding, for example, cooking pottery intersect and bring together other lines such as the cooking of meat and the making of obsidian cutting implements, and all the other lines in the house, such as the weaving of textiles, the making of bone tools, the plastering of walls and so on. Rather than the interaction of things, we can talk instead of the flow of different electro-magnetic currents, different material eddies that intersect and generate change.

2. Neolithic Çatalhöyük and the outer world: the origin, rise and demise of the settlement in its regional context

Arkadiusz Marciniak

Introduction

Work carried out over the past 25 years has brought about in-depth understanding of different dimensions of the life of the inhabitants of the Neolithic settlement at Çatalhöyük (e.g., Hodder 2011). It covered major thresholds in its development as well as revealing the tempo of changes in subsequent centuries. This was also the period of new studies of the Neolithic across different parts of Anatolia (e.g., Bıçakçı et al. 2012; Gerritsen et al. 2013; Karul 2017). This work significantly contributed to a better understanding of the processes of the spread of Neolithic lifeways into previously uninhabited areas. While the research undertakings at both Çatalhöyük and contemporaneous sites have brought about numerous discoveries, very little has been done to date to examine mutual relationships among them in the seventh millennium BC in a comparative perspective. They mostly involve similarities and differences in material culture, such as ceramics and lithics, as well as changes in occupation of the region (see Baird 2012; Baird et al. 2012; Marciniak 2018; Özdöl Kutlu et al. 2015).

This chapter aims to examine the changing nature of connectedness between the inhabitants of Neolithic Çatalhöyük and contemporaneous communities in Central Anatolia and neighbouring regions throughout different stages of the settlement's occupation, stretching back from the last century of the eighth millennium BC to the very beginning of the sixth millennium BC. In particular, it will focus upon examining relations with adjacent communities and the social networks required for the acquisition of different types of resources. This will aim at understanding the development of the Çatalhöyük community through time, in which the relationships with non-local foreign groups were certainly integral to its existence. It will throw some light on the cultural continuity with the site's predecessors who practised hunting and gathering from the Konya Plain, as well as revealing parallel developments with groups inhabiting contemporaneous settlements. The chapter will focus on a number of fundamentally significant issues in relation to the very existence of the Çatalhöyük settlement, namely its origin, rise and dominance, restructuring and ultimate demise. It aims at recognising both causes and consequences of these changing relations of the settlement itself as well as of a wide range of communities in nearby and more distant regions.

The objectives of the chapter will be achieved by examining available evidence pertaining to these multifaceted processes, such as raw material procurement, production and technology, as well as everyday activities, lifestyles and after-life practices. This evidence will be scrutinised at two major scales of interaction, namely regional and pan-regional, and will cover the assortment of networks that transcended macro-scale interactions to include practices operating at meso- and micro-scales. However, as it is also not my intention to provide a comprehensive overview of synchronous developments in different parts of the regions, the chapter is intended to address convergent processes and activities. Considering the chapter's size and character, it will not provide in-depth theorising of the examined networks and entanglements. This task is further complicated due to differences between the excavation and publication standards adopted by teams investigating different sites examined in this chapter at different times during the past 50 years.

The developments at Çatalhöyük will be discussed in relation to four major occupation periods of the settlement: the beginning, rise, re-structuring and ultimate demise. It will be based upon the stratigraphic sequence of the South Area, which covers the entire span of occupation. It will adopt both absolute radiocarbon chronology and relative chronology in two interlinked stratigraphic sequences: Levels South G–T, covering the first, second and the third of these periods, and Levels TP N–R, covering the third and the fourth periods (see Hodder 2014d; Marciniak 2015a).

The Neolithic settlement at Çatalhöyük through time

The beginnings of the Çatalhöyük settlement

The Bayesian model indicates that the East Mound was settled in 7165–7085 cal BC (95% probability), probably in 7120–7085 cal BC (68% probability) (Bayliss et al. 2015). As revealed by the Çatalhöyük Research Project

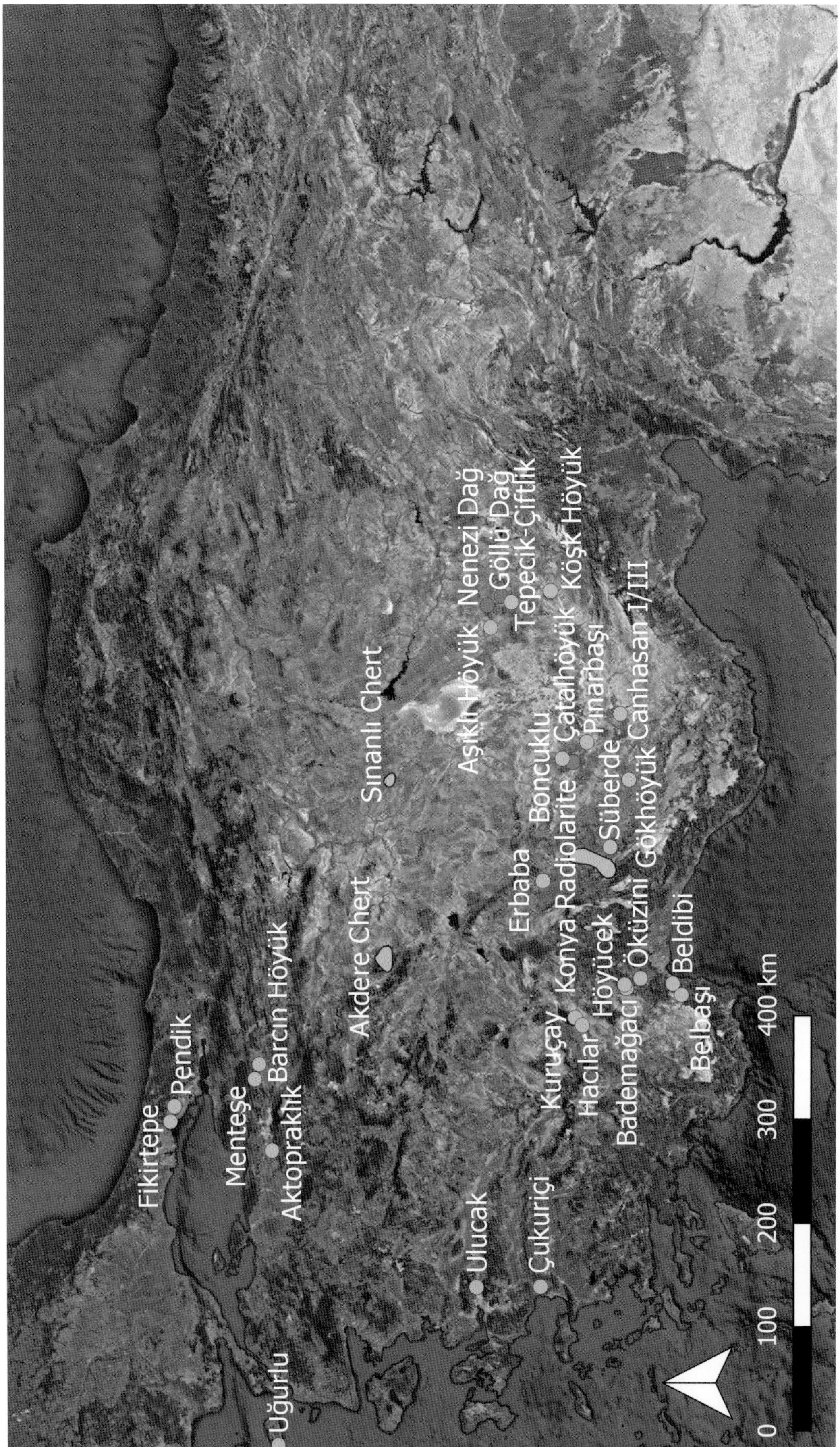

Figure 2.1. The sites and material sources connected with the occupation of Neolithic Çatalhöyük (map: Jędrzej Hordecki; map data: Google).

excavations of the bottommost part of the settlement, the start of Çatalhöyük dates close to 7100 BC, at least 300 years later than previously estimated (Cessford 2001). Most likely, it developed from one of the small settlements, such as Pınarbaşı (Baird et al. 2018), Canhasan III (French et al. 1972) or Sanak (Baird 2010b), scattered across the Çarşamba alluvial fan that extended out into the flat marl of the Konya Plain in the ninth and eighth millennia BC (Baird 2005).

The deposits from the lowest level, South G, were unearthed only on the edge of the settlement. They contain dumps of processed waste from daily consumption. The marl beneath these waste deposits was reached in a small trench, which can hardly be said to be representative of the whole site (Cessford 2007). This makes the recognition of the very beginnings of the settlement occupation inconclusive. It cannot be ruled out that earlier occupation exists in other parts of the mound.

The rise of the Çatalhöyük settlement in the Middle period

In subsequent centuries, the settlement of Çatalhöyük was growing and the population gradually rose until Levels South M, N and O (e.g., Hodder 2014). Houses continuously increased in size and came to be subdivided into rooms. The houses were clustered together and accessed through the roof. More space for storage and productive activity was introduced. Cooking took place in hearths and ovens placed in houses and was achieved with the aid of clay balls. The use of these early 'boilers' gradually decreased as the use of cooking pottery increased (Hodder 2014d). Vegetable-tempered pottery was introduced in Level South I or H (Last 2005a and see Chapter 4). This differs from the mineral-tempered clays reported for the initial Pottery Neolithic in Upper Mesopotamian (Nieuwenhuyse et al. 2010).

The development of the settlement was accompanied by the intensification of elaborate symbolic systems, as indicated by a steady rise in architectural reliefs and the diversity and number of objects interred in burials (Hodder 2011; 2014d; and see Volume 13, Chapter 17). They culminated in Levels South M to O. This peak of symbolic elaboration is also the time when most burials occurred beneath the floors of houses (Cessford 2005a).

The Çatalhöyük imagery in this period was concentrated around images of wild animals. The bull seems to have had a special position, as evidenced by installations and feasting deposits as well as wall paintings and figurines. Other important, frequently depicted animals were bear, leopard and vulture (Czeszewska 2014; Filipowicz 2019). An emphasis was also put on particular body parts such as navels, cranial parts, teeth, beaks, claws and paws. However, as indicated by installations and reliefs, the focus on animal heads remained of prime importance. Another connected, underlying theme was a desire for re-fleshing (Nakamura, Meskell 2009), expressed by modelling and re-plastering skulls, reliefs and the obese shapes of figurines.

The imagery of Çatalhöyük through this period was manifested through media on which those images appear, namely wall paintings, reliefs, installations and figurines (Filipowicz 2019). Reliefs comprise representations of bodies of animals or heads of animals moulded in clay on the house walls. The term installations refers to deliberately placed architectural elements made of animal parts. The most prominent example of installations are bucrania, that is, installations of cattle horns. They took different forms such as clay heads with cattle horns, horns set into the walls, benches with multiple horns and horns on platforms (Russell, Meece 2005; Twiss et al. 2009).

The Çatalhöyük clustered community in this period was predominantly collective and egalitarian (Hodder 2014d). It was engaged in developing and enhancing networks for the pooling of resources. The subsistence basis was provided by domestic caprines and plants. The ceremonial domain of its existence was focused upon hunting, killing and feasting on wild animals (Russell et al. 2013a; 2013b).

The restructuring of the Çatalhöyük community in the Late period

The Çatalhöyük population reached its highest point in Levels South N/O around 6600–6500 BC (e.g., Hodder 2014). The subsequent developments in Levels South P through South T and TP M–N marks the settlement's major re-structuring. It is manifested in a break in the continuity of houses and the creation of more open spaces. Some buildings became larger and had more rooms and spaces. They were an arena for increased activity and production. As buildings grew bigger and multi-roomed, they also increasingly made use of adjacent open areas. This involved increased encroachment onto adjacent midden areas and the creation of yards with ovens and hearths.

The houses still had numerous built-in structures, including platforms, benches, bins and fire installations. They were intensively used and rebuilt many times. Walls in Level TP-M were plastered and decorated with black and white geometric designs, which had the form of vertical and transverse sets of parallel lines. Such paintings finally disappeared at the end of the period, as did both reliefs and clay moulded features. These changes were accompanied by the emergence of stamp seals (Türkcan 2005). As in the preceding phases, the northern part of the house was of ceremonial character, while the southern part served more domestic purposes. Burial pits were placed in a wider range of contexts in houses (see Volume 13, Chapter 15).

The last houses of this character were built around 6300 BC. As previously, their abandonment was accompanied by filling with sand and refuse. However, instead of constructing a new house on top of these abandoned buildings, people started to re-occupy them by constructing different, irregular rooms and putting in fire installations of different character. Around 6100 BC, they made an attempt to reconstruct new houses, which differed significantly from the previous dwelling structures. These were solid, multi-roomed constructions with neither distinct floor divisions (platforms) nor burials underneath them. They were slightly bigger than the previous houses (ca 50–75m^2) and were composed of two to four distinct rooms. They appear to be unfinished. They were used for around 15–20 years, encompassing no more than one generation (Marciniak, Czerniak 2012; Marciniak 2015b).

The restructuring of the Çatalhöyük community involved independent production and the build-up of surpluses by some of the houses at the expense of the previously expected investments in ritual ties within and between clustered neighbourhoods. The increasing demands on the house were relieved by some degree of specialisation and increasing social differentiation. Specialisation and differentiation increased but they made their own demands on the productive system (Hodder 2014d; Marciniak 2016).

The increased focus on domestic production led to heavier investment in sheep herding and the increased management of cattle and, ultimately, to the adoption of domestic cattle (see Volume 13, Chapters 8 and 9). The combined impact of cattle domestication and increased exploitation of sheep and goats would have transformed the lives of the inhabitants and required shifts in organisation. The demand for food at the house and community level led to greater inputs of labour. This involved greater mobility and use of the landscape for a wider range of resources. Caprine herds may have been smaller and tended by family (or smaller group) shepherds, with less separation by age and sex (Pearson 2013; Russell et al. 2013a; 2013b).

The demise of the local community and the abandonment of the settlement in the Final period of the East Mound

While Çatalhöyük inhabitants went through a time of unease, a dramatic process of global significance took place thousands of kilometres away from Anatolia. Around 8,200 years ago, a massive release of glacial freshwater in the North Atlantic led to an abrupt climatic event that had impacts all over the world. This is known as the '8.2-kyBP event' and it has been recorded in the Greenland ice cores, stalactites/stalagmites (speleothems), lake pollen or tree rings in Europe and several lakes in Anatolia (e.g., Roffet-Salque et al. 2018).

A direct climate proxy for this climate event has been reported from the site (Roffet-Salque et al. 2018); its scale and the response of the settlement's inhabitants to it have not yet been satisfactorily determined (for an alternative view, see Chapter 1; Orton et al. 2018; Wainwright, Ayala 2019). In any instance, the hitherto-practised mode of life was significantly modified in the Final period. Level TP O marks a major discontinuity in the occupational sequence. Solid architectural structures were replaced by light shelters and open spaces. Unfinished solid buildings were re-used. Some of their parts were turned into hut-type constructions with light roofs, while others served as open areas. The latter were intensively used, as indicated by numerous fire spots. They were later turned into a continuously used kitchen midden, as demonstrated by numerous fire installations and rich occupational debris.

Around 6100 BC those inhabitants of Çatalhöyük who still lived at the settlement started to develop a new strategy of existence, which differed significantly from earlier periods. Levels TP Q and TP R are characterised by distinct, large multi-roomed dwelling structures, built directly on top of the open area (middens and infill layers) from the preceding level. The new houses were composed of a series of small, cell-like spaces surrounding a larger central 'living room' and lacking symbolic elaboration. The main room had a central hearth and hardly any other built-in structures. The houses had neither platforms nor intramural burials. A new form of funerary practice involved the construction of burial chambers inserted into the house interior following its abandonment (Marciniak, Czerniak 2007).

Çatalhöyük in the world – localness, connectedness and dependencies

The outer world and the beginnings of the Çatalhöyük settlement

The character of the Çatalhöyük occupation derives from its beginnings and origins in the region. The settlement is located in the southern part of the Konya Plain, in a region that had been occupied by Epipaleolithic and Aceramic communities for significantly longer than 1,000 years. This long period is represented by three sites, extensively excavated over the past three decades: Pınarbaşı, Boncuklu and Canhasan III. The founders of Çatalhöyük are believed to have been the cultural descendants of these local groups, belonging to a distinctive regional form of the Neolithic (Baird 2012).

However, the groups inhabiting Pınarbaşı and Boncuklu belonged to two distinct communities (Baird et al. 2018). They occupied largely similar environments and most likely interacted with each other but were characterised by different economies, cultural characteristics and material culture. The site of Pınarbaşı is located in the Karaman Province, ca 30km southeast of Çatalhöyük. It was occupied between ca 9800/9400 cal BC and ca 7800 cal BC. The community was based on the hunting of wild mammals and the exploitation of wetlands, and had a significant focus on nut exploitation. There was neither crop cultivation nor animal husbandry. Pınarbaşı houses had wattle and daub superstructures and were not internally divided into any distinct areas. Burials were most likely placed outside buildings, possibly in small cemetery areas. There is no evidence that ritual and symbolic practices took place in the buildings (Baird et al. 2018).

The settlement at Boncuklu is located 10km northeast of Çatalhöyük. It was occupied between 8300 and 7800 BC (Baird et al. 2012). However, the presence of chipped stone points in the latest levels – similar to Musular, Canhasan III and early Çatalhöyük, where they are dated to the second half of the eighth millennium BC – suggests a significantly longer occupation (Baird et al. 2018: 3078). The economy of the Boncuklu communities was characterised by low-level crop cultivation and developed animal management. Rudimentary herding was adopted by local foragers from approximately 8300 cal BC. This is believed to have been a local development rather than a result of introduction by farming groups from the Fertile Crescent. Cultivation appeared through adoption by indigenous foragers in the mid-ninth millennium cal BC, and it was accompanied by experimentation with sheep/goat herding. The buildings in Boncuklu have distinctive internal arrangements, with clean southeast floor areas with burials and dirty northwest kitchen areas around the main hearth. The walls were constructed of mudbrick. The clean areas were often decorated with paint and incorporated animal bones, mainly in the form of wild aurochs horns and skulls attached to the walls and floors (Baird 2010a; Baird et al. 2018).

Both communities co-existed in the period of approximately 8300–7800 cal BC. Despite their participation in shared exchange practices, most likely related to procurement of obsidian and exporting marine shells, they maintained their economic and cultural distinctiveness throughout this period. Both sites were at least partially contemporary with Levels 4 and 3 at Aşıklı Höyük in Cappadocia, and Pınarbaşı is probably earlier than and contemporary with Aşıklı Level 5 (Baird et al. 2018: 3079). The settlement at Aşıklı Höyük, 150km to the east in western Cappadocia, appears to be significantly different, with its more substantial mixed farming economy, including a wide range of crops and significant investment in herding (van Zeist 2003; Stiner et al. 2014).

The distinct ritual and symbolic practices at Boncuklu, manifested in particular in structured and repetitive use of domestic space, is believed to prefigure practices at Çatalhöyük, with its north/south division between clean and dirty areas in houses. This may imply that the community at Boncuklu was a direct antecedent to that at Neolithic Çatalhöyük (Baird 2012). The supposed embeddedness of the early Çatalhöyük milieu in the local Central Anatolia tradition can only be proved indirectly, considering the lack of temporal overlap between the occupation of both settlements. However, one cannot rule out the possibility that Çatalhöyük was in fact inhabited earlier than the recognised and dated sequence (see Bayliss et al. 2015) and/or Boncuklu was occupied longer than the preserved sequence.

The demise of the Boncuklu settlement was certainly not accompanied by an abandonment of the Konya Plain. A number of Late Aceramic sites from the second half of the eighth millennium BC are reported from across the region (Baird 2005: 65). Unfortunately, it is impossible to establish their precise chronological position due to a dearth of radiocarbon dates. One such site was Sancak, known from its characteristic lithic material (Baird 2010b).

The only known and excavated settlement from the second half of the eight millennium BC is Canhasan III. It is located in Karaman Province, close to the foothills of the Taurus and ca 30km away from Çatalhöyük. The settlement was ca 1ha in size and composed of rectangular buildings with one or two rooms. The walls were made of mudbrick. Subsequent houses were built one on top of the other, creating a stack of buildings ca 4m high (French et al. 1972). The settlement's chronology is highly disputed. It was originally dated back to the period between ca 7600 and 6500 BC. Hence, the onset of activity at Canhasan III coincided with the transformations of material culture visible in the mid to late eighth millennium cal BC at other sites on the Konya Plain (Baird 2010a). However, its persistence into the seventh millennium BC is not corroborated by available evidence, in particular the presence of pottery. Moreover, Byblos points and Canhasan/Musular points found at Canhasan III are only known from the very first levels at Çatalhöyük East and disappeared after 7000 cal BC (Carter et al. 2005a). The Canhasan III houses were much smaller than those at Çatalhöyük in the first half of the seventh millennium BC. Since the duration of the occupation of Canhasan III has not been satisfactorily recognised, it remains possible that the settlement was in use during the foundation of Çatalhöyük. Nonetheless, Canhasan III, with its structured occupation and distinctive architecture, was certainly embedded in the Central Anatolia tradition developed earlier at Boncuklu, and it anticipated later developments at Çatalhöyük.

The embeddedness of early Çatalhöyük in the Central Anatolia Aceramic Neolithic tradition is further corroborated by the character of its lithics assemblage. The assemblage from the lowest level, South G, is characterised by the presence of a highly distinctive prismatic microblade tradition with obliquely truncated and backed asymmetric triangular microliths (scalenes), which are reported from the Cappadocian Aceramic Neolithic site at Acıyer as well as from Canhasan III (Carter et al. 2005a). They belong to a long regional tradition, as they have also been found at the Epipalaeolithic site of Pınarbaşı (Baird 2012). Comparable items have come from the Epipalaeolithic cave sites of Öküzini, Belbaşı and Beldibi on the Mediterranean coast near Antalya (Carter et al. 2005a). The embeddedness of early inhabitants of the Çatalhöyük settlement in the Konya Plain network, probably stretching back to the Epipaleolithic, is further corroborated by the continued exploitation and use of chert resources, mainly from the local Konya radiolarite and Sınanlı sources (see Nazaroff et al. 2013). The archaeological evidence has recently been confirmed by new genome-wide data from Pınarbaşı and Boncuklu indicating that these populations derived a large fraction of their ancestry from the Epipaleolithic Anatolian population, implying that farming was adopted locally by the hunter-gatherers of Central Anatolia. This in turn is indicative of the formation of the distinct Anatolian early farming population (Kılınç et al. 2016; Feldman et al. 2019).

The origin of Çatalhöyük within the Central Anatolian Neolithic does not exclude its belonging to the Near Eastern Neolithic koiné. This is clearly manifested in the presence of domestic sheep from the lowest levels, which stands in stark contrast to Boncuklu. Domestic sheep and goats were present in Central Anatolia from the mid eighth millennium cal BC and spread into Cappadocia and the Konya basin (Düring 2010; Arbuckle et al. 2014). The introduction of sheep appears to have been sudden, and they may have originated from unspecified areas in the Fertile Crescent (Baird 2012: 440). It has to be mentioned that sheep were managed at Aşıklı Höyük but were not fully domesticated (Buitenhuis 1997). It cannot be ruled out, however, that domesticated sheep were present at Canhasan III (French et al. 1972).

The rise of the Çatalhöyük settlement and its relations with the outer world

The first half of the seventh millennium BC brought about an unprecedented development and growth of the Çatalhöyük settlement. It developed from a small site into a large mega-site, arguably occupied by thousands of inhabitants (see Cessford 2001). These arrangements, in what is often labelled as its classic phase, are often presented as representative for the character of this world-renowned site.

The direct relations of Çatalhöyük in the period of its dynamic growth with contemporaneous communities in the region are not known. Except for Canhasan III, the settlement appeared to be the only site in the Konya Plain (Baird 2005: 66). Its attractiveness seemed to draw neighbouring groups from the area into it, resulting in significant regional depopulation. The relationships of the settlement's occupants with the world, however, can be recognised by examining the modes of acquisition of innovative technologies and the procurement of strategic resources.

One of the major developments in this period was the emergence of pottery. The take-up in its use was a long process, as indicated by a slow increase in the numbers of sherds (Last 2005a; Nieuwenhuyse et al. 2010; Yalman et al. 2013; Tarkan, this volume, Chapter 4). However, the idea may have spread very quickly shortly after the settlement was founded and the adoption took place within a context of widespread dissemination of myths and ideas about symbolism (Hodder, Meskell 2011), as well as the circulation of obsidian and the spread of caprines. The earliest pottery at Çatalhöyük made use of local clays (Doherty, Tarkan 2013).

Radiolarian cherts remained major components of the Neolithic non-obsidian chipped stone assemblage at Çatalhöyük (Carter et al. 2005a; Bezić 2007). They were procured from multiple sources across Central Anatolia throughout the entire span of the settlement's existence. In the first half of the seventh millennium BC, in Levels South G–P, the assemblage was dominated by the Akdere chert (Nazaroff et al. 2013). Its source is located 250km northwest of the Konya Basin. This was accompanied by the Konya radiolarite, the source of which occurs on the southwest fringe of the Konya Plain. Its frequency was significantly higher at the beginning of the settlement occupation, and it steadily declined from its highest concentration in Level South G to an insignificant percentage by Level South P (Bezić 2007). The third source was the Sınanlı chert, located 90km south of Ankara. The majority of this material comes from the earliest Neolithic occupation – Level South G, while only a minimal amount is present in later levels. It was used to produce ordinary objects used in the daily lives of the inhabitants (Nazaroff et al. 2015).

The Konya radiolarite is located close to the sources of several other materials procured by the inhabitants of Çatalhöyük, such as clay, ground stone and timber (Baysal 1998; Asouti 2005; Doherty 2008). A limited range of locally available raw materials was also used for stone beads in Levels South G to M (Bains et al. 2013). The steady decline of the exploitation of the Konya radiolarite chert may have been linked with changes in the procurement regimens. The rise in the exploitation and use of Akdere chert in later levels was arguably triggered by the increase of interregional contacts, in particular emerging contacts with northwest Anatolia.

The dominant domestic species were sheep and goats, present at Çatalhöyük from the lowest level, South G (Russell et al. 2013a). The cattle in this period were morphologically wild and were hunted throughout the occupation of the site with increasing degrees of management (see Volume 13, Chapter 8) until the appearance of fully domesticated cattle in the Final levels. Most likely, the emergence of domestic cattle at Çatalhöyük was part of a larger process of the abrupt appearance of cattle in Central Anatolia after the middle of the seventh millennium BC (Arbuckle, Makarewicz 2009). Similar to caprines some thousand years earlier, they originated from areas outside the region.

The belonging of Çatalhöyük to the broader Near Eastern Neolithic is clearly manifested by the character of the imagery deployed at Çatalhöyük. Recently, Hodder and Meskell (2011) pointed out that the major themes are violence and masculinity, as the emphasis is often put on the dangerous parts of animals such as teeth, beaks or claws, which is best evidenced at Göbekli Tepe. The abundance of imagery at Çatalhöyük is in stark contrast with earlier sites in the region (Filipowicz 2019). At Aşıklı Höyük, figurines and other symbolic objects were rare and restricted to animal figurines, most likely boar or ox. Other finds included shaft straighteners with incised motifs (Özbaşaran 2012: fig. 18). At Boncuklu, imagery of Near Eastern origin is hardly present and was restricted to red-painted floors, a wall painting depicting probably a part of a wild animal, and a considerable number of stones with incised geometric decorations (Baird et al. 2012). Interestingly, such items were also found at the site of Pinarbaşı, which is otherwise very different to Boncuklu (Baird 2012). However, there is direct evidence for detachment and circulation of human crania at Boncuklu. In addition, two cattle skulls placed into the wall have been found.

The connectedness of Çatalhöyük inhabitants with the outer world has recently been explored by the analysis of ancient DNA. The newly generated mitochondrial genomes of a group of individuals from Level South O revealed that while falling within the genomic diversity of Near Eastern Neolithic populations, the Çatalhöyük population is set apart from the groups from the Fertile Crescent (Chyleński et al. 2019). A close affinity of the studied individuals with genome data from Boncuklu seems to corroborate the Central Anatolian origin of Çatalhöyük inhabitants. The role of Central Anatolia in this process was also supported by whole genome data from several individuals from Boncuklu as well as from Late Neolithic Tepecik-Çiftlik (Kılınç et al. 2016).

Relations with the outer world at the beginning of the settlement's disintegration

Two centuries after the peak of the Çatalhöyük settlement there were significant changes. The restructuring of the Çatalhöyük community is manifested in the material from levels South Q–T and TP M–N, that is, towards the end of the Late period. This period is marked by the habitation of previously unoccupied zones in the central, western, southwestern and northwestern parts of Anatolia, as seen in the emergence of numerous settlements. The Çatalhöyük inhabitants must have started building up relations with these emerging and rapidly growing communities. The increased productive capacity of individual houses was intimately linked to the extension of domestic production into the landscape (Hodder 2014d).

Around 6600/6500 cal BC, the site of Gökhöyük in Seydeşehir (Gündüz 2016) was founded. At roughly the same time, a few sites emerged in the Beyşehir-Suğla basin, directly west of Çatalhöyük, including Süberde and Erbaba (Bordaz, 1973; Bordaz, Bordaz 1976; 1982; Özdöl 2012). The Neolithic was also well established shortly before 6500 cal BC in the Lake District in southwest Anatolia (Duru 2012), where it is represented by four excavated settlements: Bademağacı EN I (7–5)–II, Höyücek ESP–ShP, Hacılar IX–VI, and Kuruçay Levels 13–11 (see E. Özdoğan 2015: fig. 6). In western Anatolia, fully developed farming groups arrived around 6700 cal BC from upper Mesopotamia and/or the southern Levant, as indicated by the materials from Çukuriçi XIII and Ulucak VI (Horejs et al. 2015). A similar situation has been revealed in northwest Anatolia, where the beginnings of the Neolithic are dated back to 6600 cal BC (Gerritsen et al. 2013), as seen at Uğurlu V, Pendik, Archaic Fikirtepe, Aktopraklık C, Menteşe (Basal, Middle) and Barcın Höyük (VId–c) (Roodenberg et al. 2003; Özdoğan 2012; Erdoğu 2013; Gerritsen et al. 2013; Karul, Avcı 2013). Western Cappadocia is represented by two Late Neolithic settlements: Tepecik-Çiftlik Levels 9–4 and Köşk Höyük Levels V–III (Bıçakçı et al. 2012; Öztan 2012).

The Gökhöyük settlement is located ca 80km southwest of Çatalhöyük in a strategic location at the gateway to the Beyşehir-Suğla basin. Four Neolithic levels were unearthed during the excavation campaign in the 2000s. Houses were built of mudbrick and some of them were constructed on stone foundations. The internal layouts were very similar to Çatalhöyük, with clean and dirty areas. The houses were placed in small clusters (Gündüz 2016).

The other nearby contemporaries of Çatalhöyük are Süberde and Erbaba from the Suğla and Beyşehir basins. They reveal striking similarities to the Çatalhöyük architecture in its classic phase. Süberde is located 11km southeast of Seydeşehir in the Taurus Mountains, in proximity to Lake Suğla (Bordaz 1965: 31). The second and third layers, identified by the excavator as Neolithic, contained remains of mudbrick houses with plastered floors and clay bins (Bordaz 1966: 32). Erbaba is located 10km northwest of the town of Beyşehir and 1.5km east of Lake Beyşehir on a natural hill. The site contained three levels, all belonging to the Neolithic period. They contained a sequence of clustered houses, which were square to rectangular in plan, with thick foundations and entrance provided through the roofs (Bordaz 1982: 87).

The relations with the Lake District are complex. The area is located further away from Çatalhöyük, in a region that is difficult to reach due to the high ridge of the Taurus Mountains. The Neolithic Levels IX–VI at Hacılar are dated to the second half of the seventh millennium BC (Thissen 2010: 273). Due to the small number of radiocarbon dates, it is impossible to reliably recognise changes through time. The first well-preserved phase is the burnt settlement designated Hacılar VI (Mellaart 1970: 10–22). It is composed of a number of L-shaped structures consisting of two houses each. No sub-floor burials existed, which recalls the practice known from the latest 300 years of the Çatalhöyük occupation. Three isolated burials were found in Hacılar Level VI (Mellaart 1961: 40).

The site of Kuruçay is located northeast of Hacılar. Levels 13–12 are attributed to the Early Neolithic, Level 11 to the Late Neolithic and Level 10 to the Early Chalcolithic (Duru 1987). However, Thiessen (2010: 273) argues that Levels 12–8 are related to one another and can be treated as a single building unit. Due to a dearth of radiocarbon dates, the absolute chronology of the settlement occupation remains largely unknown. The best recognised remains come from Level 11. They are represented by the 26m-long and 1.1m-thick stone wall situated on the east–west axis (Duru 2008). Altogether, seven intramural burials were found at the site. Four of them were related to Level 11 and two of those were children (Duru 1994: 101).

Three major Neolithic phases were distinguished in Höyücek. The Early Settlement Phase was assigned to Early Neolithic I, the Shrine Phase to Early Neolithic II and the Sanctuaries Phase to the Late Neolithic. While the Early Settlement Phase does not contain any architectural remains, the following Shrine Phase is represented by five mudbrick structures attached to each other. The interior dimensions of these buildings are 7.20 x 4.50m (Duru, Umurtak 2005). The structures are known to have contained a wide range of different items, such as deer antlers, mandibles of large animals, astragali, a marble bowl containing a terracotta bowl and stone chisel, and a boot-shaped vessel. One newborn was interred outside the wall of one of the buildings from this phase. Of significance are special ceramic forms such as kidney-shaped, boot-shaped and bird-shaped vessels, and small jars with flat ledge-rims on the interior (Duru, Umurtak 2005: pls 33–43; Duru 2012).

The site of Bademağacı is represented by nine Neolithic horizons. According to Duru (2012: 14), Levels 9–5 are dated back to Early Neolithic I and the following Levels 4–3 to Early Neolithic II, while Levels 2–1 belong to the Late Neolithic. As only eight radiocarbon dates originated from the deposits, the absolute chronology of the Neolithic sequence is only tentative. The Early Neolithic, according to Duru (2007: 349), took place in the second half of the seventh millennium BC, corresponding mostly to the Late and Final levels at Çatalhöyük. No dates from the Late Neolithic strata are available. Early Neolithic Levels 9–5 have revealed no trace of architectural elements, except for a floor made of solid limestone (terrazzo) with traces of red and black paint encountered in Level 8. Mudbrick houses appeared in Level 4 and were in use until the end of the Neolithic. These were rectangular, single-roomed houses constructed of regular or plano-convex bricks, without stone foundations. The entrance was at the floor level in the centre of the longer wall. The ovens were usually placed opposite the door (Duru 2008: fig. 42). Buildings from Level 3 were free-standing constructions. The layout of the settlement, however, remains unknown (Duru 2012: 5). The houses were accompanied by individual storage silos constructed outside or between them (Duru 2012). Altogether, 27 burials were found in Early Neolithic strata. These were mostly infant burials, while adult burials were decidedly rare. The skeletons were placed in a pit in crouched position, lying on the side. One headless skeleton has been found (Duru 2008: 151).

For the Lake District, aside from single and doubtful radiocarbon dates from Hacılar and Bademağacı, we do not have any reliable evidence for the Neolithic before 6500 BC. Hence, we can reliably assume that the beginning of the Neolithic occupation in this area started around that time. The roots of the Lake District Neolithic have been sought either in the Beyşehir-Suğla basin or in the Antalya region, not in the Konya Plain. However, there are many indications, especially in the light of recent research at Çatalhöyük, that the Neolithic of the Lake District might have been somehow related to that of Central Anatolia.

The links and dependencies have recently been revisited through in-depth comparative analysis of the imagery (Filipowicz 2017; 2019). Buildings from the Lake District had no wall paintings, reliefs or in-house installations. Instead, a large number of diverse mobile decorated objects were present. In the following Early

Chalcolithic period, this was enhanced by the sudden introduction and subsequent rapid increase of ceramics painted with geometric motifs. The images employed often resemble those known from Çatalhöyük wall paintings. They include zoomorphic, anthropomorphic and geometric motifs present across different media, either on painted or relief pottery. This is indicative of an inheritance of the Çatalhöyük tradition by the Lake District communities. This could either have been caused by some kind of unspecified influx of settlers originating from Çatalhöyük into the Lake District or, most likely, by intense contacts and relations between these communities.

More direct proof of contacts between these two regions stems from the lithic production. The reported similarities are attested primarily through the common reliance on pressure blade industries. It is most likely that this mode of tool production was introduced to Lake District populations from Central Anatolia, potentially via Çatalhöyük itself (Balkan-Atlı 2005: 136). This was a part of the western expansion of pressure blade technologies. Molluscs *Borlumastus yildirimi* found in the TP Area are endemic to the Lake region around Isparta of Turkey, and they live in limestone rocks and debris, on calcareous soil (Kurzawska et al. in press).

The first farmers in northwest Anatolia inhabited two different environments, the coasts and the plains (Karul 2019). In the first part of the second half of the seventh millennium cal BC, a wide range of settlements were present. These comprise Uğurlu V, Pendik, Archaic Fikirtepe, Aktopraklık C, Menteşe (Basal, Middle) and Barcın Höyük (VId–c) (see detailed discussion of this evidence in Marciniak 2018). Different elements of the classic Çatalhöyük tradition appeared in this period at northwest Anatolian sites. These comprise in particular (a) ceremonial consumption of cattle and deer, revealed at Aktopraklık C, (b) burials underneath the floor at Aktopraklık C (burials with grave goods) and Basal Menteşe (female skeleton under the house floor) and (c) a slightly modified tradition of burying the dead in courtyards at Basal Menteşe and Barcın VIb from the final phase of the Late Neolithic (Karul 2019; Özbal, Gerritsen 2019).

The relations between central and northwest Anatolia have been most often discussed in terms of connection between the core and periphery, where the latter region was arguably colonised by groups from the former via terrestrial routes. These relations got shaped in the context of a complex web of interactions between the Neolithic groups across different areas of the Near East. Different explanations of the arguably complicated mechanisms of this process have been offered, including leapfrogging movement, acculturation, infiltration etc. (e.g., Özdoğan 2010; Thiessen 2010; Düring 2013).

It has been argued that a major impulse involved the movement of groups from central to northwest Anatolia around the middle of the seventh millennium BC. This was indicated by dark-coloured and burnished ware from Fikirtepe culture, which 'strongly resembles the dark burnished wares of Central Anatolia, suggesting a transfer of technology from this region' (Brami, Heyd 2011: 178). The relations between both regions were further corroborated by a shared absence of domestic pigs (Arbuckle et al. 2014: 8). These similarities implied the spread of farming groups with a distinctive cultural tradition and the movement of an animal economy characterised by domestic caprines and cattle.

The dominant form of architecture in both southwest and northwest Anatolia reveals striking similarities. It comprised rectangular free-standing buildings with floor-level entrances, which are made of wattle and daub. These multi-roomed houses were composed of a series of small, cell-like spaces surrounding a larger central 'living room' with street-level exterior entrances and lacking symbolic elaboration. Some houses had horned benches and installations. They were most often associated with different forms of empty spaces such as courtyards, storage areas or alleyways. They were linked by external yards, ovens and hearths where activities took place. The houses in the majority of instances lacked intramural burials (E. Özdoğan 2015: 43). They were revealed at Aktopraklık C, Ilipinar X, Menteşe – Basal Menteşe and Barcın VId–b (e.g., Roodenberg et al. 2003; Erdoğu 2013; Gerritsen et al. 2013; Karul, Avcı 2013).

The relationships between central and northwest Anatolia have recently been examined by the new whole genome data as well as mitochondrial genomes. They seem to favour the idea that the Neolithic in the latter region emerged as a result of expansion, potentially originating in Çatalhöyük (Kılınç et al. 2016; Chyleński et al. 2019).

The Neolithic site at Tepecik-Çiftlik in western Cappadocia was recognised in Levels 9–4. The final Neolithic level, Level 4, is dated back to the period between 6400/6300 and 6000 cal. BC (Bıçakçı et al. 2012: 90). As the lowest levels were only recognised in a small sondage, details of the character of the settlement and its reliable chronological position are unavailable. A large rectangular structure comes from Level 5, which contained as many as 60 burials. Adjacent to this structure was an open space. The preserved structures from Level 4 comprise a multi-roomed building complex of ca 100m^2. Its walls were made of large and flat stones. The occupation area had numerous ovens and kilns. The dominant form of burial practice involved inhumation, both inside buildings and in open courtyards, as well as pot-graves (ibid. 93–95).

The neighbouring settlement at Köşk Höyük has three Neolithic levels, V–III. Its beginnings are set to around 6300 BC, while the end of the following Early Chalcolithic Level II is dated back to 5600 BC (Öztan 2012: 42). The houses were made of limestone, or less commonly of mudbrick. They were composed of two to four rectangular or square rooms. The built-in structures comprised benches and stone platforms as well as hearths, ovens and box-like installations of clay. The houses in later periods were significantly reconstructed by putting in partition walls, adding annexes, altering the door location and placing hearths or ovens outside. The walls in one of the houses from Level III were painted, depicting 20 figures dancing or hunting around a deer. This recalls paintings from Çatalhöyük. Burials were only recognised in Levels III and II. The funerary practice was dominated by inhumation and pot-graves. Some of the interred individuals had plastered skulls (ibid. 32–37).

The origin of western Cappadocian sites at Köşk Höyük and Tepecik-Çiftlik appears to be different from that of Canhasan I and Pınarbaşı. This is implied by a significantly different lithic technology. While at Çatalhöyük the pressure-blade traditions dominated, the Cappadocian sites were characterised by flake and percussion blade industries (Bıçakçı et al. 2012: 98–101; Öztan 2012: 42–44). This is further characterised by the production of large spearheads, most likely indicative of hunting in the region, in stark contrast to what we see in the Late levels at Çatalhöyük.

The external contacts of the Çatalhöyük community in the period included the procurement of raw materials, where direct contacts with local communities did not necessarily take place. The period between Levels M and P marked the gradual shift from East Göllü Dağ to Nenezi Dağ obsidian (Carter, Milić 2013a). The material from these two sources was delivered to the site from the distance of ca 130km either by direct access or through exchange links. From Level South P onwards, the range of raw materials used for stone beads increased and some of them originated from greater distances. Shell beads may show a related pattern, with a greater diversity of shell materials being brought into the settlement from a wider range of locations (Bains et al. 2013). Greater human mobility, most likely both local and pan-regional, from South P onwards is also indicated by a femoral midshaft index (Larsen et al. 2013). As compared with the preceding period, the pattern of chert procurement changed as well. The Akdere source maintained a steady rate of consumption throughout the Neolithic (cf. Bezić 2007), which is indicative of a growing significance of networks and connectivity with communities northwest of Çatalhöyük. At the same time, the significance of the Konya radiolarite and Sınanlı chert, which were acquired directly from their primary deposits (Nazaroff et al. 2013: 357), declined. Pottery from South M marked a major switch from local to predominantly non-local clays, probably obtained from volcanic regions to the west (Özdöl-Kutlu et al. 2015).

The demise of the Neolithic community and the East Mound settlement abandonment

The final 300 years of occupation (6250–5950 BC) of Neolithic Çatalhöyük (the Final levels), were accompanied by dynamically accelerating developments in other regions, near and far. This is manifested by numerous settlements from this time. Southwest Anatolia is represented by Bademağacı EN II–LN?, Hacılar V–III, Höyücek SP and Kuruçay 9 (see E. Özdoğan 2015: fig. 6). A number of settlements from the previous period in northwest Anatolia show uninterrupted occupation, including Classic Fikirtepe, Pendik, Yenikapı, Aktopraklık B, Ilıpınar X–IX, Menteşe (Upper) and Barcın Höyük (VIb–a) (Roodenberg et al. 2003; Erdoğu 2013; Gerritsen et al. 2013; Karul, Avcı 2013; Özdoğan 2013). The Cappadocian settlements at Tepecik-Çiftlik and Köşk Höyük were also uninterruptedly occupied.

The end of occupation of the Neolithic settlement on the Çatalhöyük East Mound is accompanied by the emergence of the West Mound settlement on the other side of the Çarşamba River (e.g., Biehl et al. 2012). As its lowermost strata have not been excavated, the early development of this settlement and its relations with the existing Çatalhöyük East Mound settlement remain largely unknown. New dating for the excavated deposits on the West Mound (Orton et al. 2018) clearly shows that there was contemporary occupation on the two mounds. As the latest excavated house on the East Mound is very probably later than the earliest excavated house on the West Mound, the presence of ca 5m of anthropogenic deposits below this West Mound building strongly suggests that the two mounds overlapped for a significantly long period (see also Bayliss et al. in press), probably around 200 years.

The closest known contemporary of Çatalhöyük at this time is Canhasan I. It is located in Karaman Province, in proximity to Aceramic Canhasan III (see above). There was certainly no continuation between these two sites, and Canhasan I appears to be a resettlement. Altogether, four levels, 7–4, were dated to the Late Neolithic. Level 3 is dated to the Early Chalcolithic. The rectangular and square mudbrick houses from Levels 7–4 were accessed from the roof. Their internal surfaces were plastered over and often painted (French 1998: 20). The dwelling structures from Levels 5–4 were occupied significantly longer, and their internal arrangements got more complex. One of the buildings was composed of three rooms, and it was enlarged by adding rooms during its life. These free-standing houses were arranged in structured complexes (ibid. 24–25).

Level 5 dates provide convincing absolute evidence that it was contemporary with the early occupation at Çatalhöyük West between ca 6000 and 5900 BC (Orton et al. 2018: 629–30). The onset of activity at Level 5 also precedes the terminal phases of occupation at Çatalhöyük East, which are modelled to 5975–5865 cal BC (95.4% probability; Marciniak et al. 2015a). The strata beneath Level 5 remain undated. The pressure blades common in Canhasan I Levels 7–4 (Ataman 1988: 242) only become common at Çatalhöyük East after ca 6400 cal BC (Carter and Milić 2013), and the carinated and S-profiled jars and bowls common in Canhasan Levels 7–4 (French 2005: 118–29) comprise 18–25% of the assemblage in Çatalhöyük East's final occupation levels (Özdöl-Kutlu et al. 2015). Hence, it is possible that Canhasan I was founded somewhere in the 63rd century BC. The close cultural and architectural affinities between both settlements point to strong ties between them, with the most likely possibility being that Canhasan's resettlement was by people from Çatalhöyük.

At the same time, a seasonal satellite camp at Pınarbaşı, close to the foot of Karadağ, was settled (Baird et al. 2011). It is located 30km from Çatalhöyük, in proximity to Epipaleolithic Pınarbaşı (see above). It was probably frequented by task groups, possibly sheep herders, from Çatalhöyük during the last 200 years of its occupation (Baird 2007b). The chipped stone assemblage, in particular tool types and reduction strategies, is over 90% obsidian, like contemporary Çatalhöyük (Conolly 1999: 18). These close ties are further corroborated by the fauna implying it was a seasonal hunter-herder campsite. The emergence of this site and contacts with the Çatalhöyük community indicate increasingly intensive and extensive use of the landscape.

The well-established relations with the Lake District in the preceding period continued in the last centuries of the seventh millennium BC. Due to a dearth of reliable radiocarbon dates, the corresponding developments at these sites cannot be reliably specified (hence see the section on the Lake District above). However, according to Alex Bayliss (personal communication July 2017), most C14 determinations from these four sites overlap with the later parts of the Çatalhöyük sequence or postdate it (Filipowicz 2019).

The relations between Çatalhöyük and settlements from northwest Anatolia remained intense, albeit less pronounced and more idiosyncratic. The reported similarities are in ceramic production and styles. These comprise, in particular, dark-slipped pottery, mainly hole-mouth vessels and S-shaped jars and bowls (Brami, Heyd 2011; Özdöl 2012: 92). As at Final period Çatalhöyük, there were no burials underneath the floors, and funerary practices transitioned from house interment to extramural cemeteries and burials in courtyards. At the same time, the number of differences increased. The ceramics from northwest Anatolia were dominated by neckless jars with straight or inturned rim and a deep globular body with vertically perforated lug, while at Çatalhöyük the assemblage was dominated by a relatively small percentage of S-profiled vessels, very little carination, but lots of rounded profiles, an absence of tubular and vertical strap handles and consistent eschewal of decoration (see Pyzel in press and Tarkan, this volume, Chapter 4). Changes in spatial arrangements led to the emergence of well-organised villages constructed in a circular plan and with public areas. New forms of dwelling structures, unknown in Central Anatolia, include round wattle and daub huts with semi-subterranean floors (see Marciniak 2018). The presence of obsidian of Cappadocian origin provides firm evidence of contacts with the Konya Plain (Milić 2014). Hence, the most conceivable scenario is a migration from Çatalhöyük, as discussed above. Barcın Höyük, however, is a site with possible impact from the Lake Region (Özbal, Gerritsen 2019).

Unfolding Çatalhöyük in the changing Neolithic world

The available evidence indicates that local groups founded the Çatalhöyük settlement at the very end of the eighth millennium BC. Most likely, the founding was achieved by one of the communities that inhabited the Konya Plain in the second half of the eighth millennium BC. A solid economic footing and access to fundamental raw materials provided necessary sustainability and prosperity. The groups were well familiarised with Central Anatolia, as indicated by effective exploitation and procurement of strategic resources. From the early years of the settlement's existence, its inhabitants, similarly to their predecessors, continuously exploited and procured raw materials, including obsidian from the Cappadocian sources, the Konya radiolarite and the Sınanlı chert, and the Akdere chert from the northwestern part of the region (see Carter et al. 2005a; Nazaroff et al. 2013; 2015). In particular, the exploitation of the Akdere chert, which is located between the Konya Plain and northwest Anatolia, from the very beginning of the seventh millennium BC is indicative of a relatively good recognition of the zone west and northwest of the settlement, much earlier than the establishment of the early farming occupation zone in northwest Anatolia around 6600 BC (Marciniak 2018; 2019).

The settlement from its very beginning certainly remained in close contact with the inhabitants of the Canhasan III settlement in neighbouring Karaman Province. One cannot rule out the possibility that it may have served as an ancestral place for the new Çatalhöyük settlers. A deep embeddedness in local traditions, as indicated by convincingly proven parallels with much earlier Boncuklu, was further strengthened by sheep husbandry, which is proven from the very beginning of Çatalhöyük (Russell et al. 2013a). Sheep provided a solid economic footing for the emerging community, while imagery made it possible to embed the site's inhabitants within the Neolithic cultural world. The economic and socio-cultural footing was further consolidated by pottery, which was introduced around 100 years after the foundation of the settlement (Last 2005a).

An impulse appears to have been given by ongoing contacts with the Fertile Crescent. The Çatalhöyük settlers maintained these relations, which stretched back to the previous millennium. However, links appear to have been intensified following the demise of the large centres of the PPNB and the ultimate reconfiguration of the Neolithic mode of life at the turn of the eighth millennium BC. The most important of them was arguably the arrival of an imagery (Hodder, Meskell 2011), implying that local communities were exposed to powerful modes of ritual and ceremony from the Fertile Crescent. However, the nature of these relations remains unclear. One needs to exclude an influx of immigrants from the Fertile Crescent, as recently corroborated by genetics (e.g., Özdoğan 2010; Chyleński et al. 2019). The shift involved the spread of ideas rather than specific techniques and functions (Bayliss et al. 2015).

The systematic provision of important raw materials, a thorough recognition and effective exploitation of the vastness of Central Anatolia, the sheep-based economy, the embeddedness in secure and sustainable local traditions, further enhanced by impulses from the Fertile Crescent, in addition to favourable conditions on the Çarşamba fan, all contributed to the success of the newly founded settlement. Parallel developments, albeit not without important differences, took place to the east and southeast in the upper Euphrates, as marked by the emergence of powerful pottery cultures that eventually led to the emergence of a large Halaf centre (e.g., Campbell 2007).

The exploitation of locally available resources known by the settlers from the very beginning, in particular the Konya radiolarite and Sınanlı chert, defined the character of lifeways of this community and its mode of interaction with the surrounding landscape in the first centuries of its existence. It provided access to a range of new resources, resulting in unprecedented growth of the settlement and the number of its inhabitants. As argued by Bezić (2007), the use of chert, unlike obsidian and other chipped stone materials, was intimately entwined in a range of other practices. The acquisition of Konya radiolarite was interlinked with the procurement of other nearby woodland resources (Nazaroff et al. 2013), which witness a shift in exploitation through time, from local catchments to zones further away in the region (Asouti 2005).

This locally based procurement model proved to provide satisfactorily sustainable foundations for the growth of the site and its inhabitants. But both chert and obsidian were also procured from areas outside the immediate zone exploited by the Çatalhöyük community. Interestingly, maintaining permanent contacts with these areas did not result at this time in the emergence of local farming communities in Cappadocia, the Beyşehir area or the Marmara region.

Around 6600–6500 BC, Çatalhöyük was established as a demographically large and culturally prolific community of unprecedented scale and regional position. It was certainly a major trigger of regional developments. The relations with the outer world were significantly reshaped. The procurement of raw materials, mainly clay and chert, saw a significant shift towards the external areas west of the site at the expense of the hitherto dominant contacts with different parts of the Konya Plain. This involved exporting clay from the Beyşehir area at the expense of previously dominant local resources (Özdöl Kutlu et al. 2015). The Akdere chert was now decidedly dominant, while the significance of locally procured Konya radiolarite and Sınanlı chert ceased. Obsidian was continuously supplied from the Cappadocian sources but involved a shift from the previously dominant Gölü Dağ to Nenezi Dağ sources. The procurement zone was further expanded by targeting lithic sources from the Lake Van area, 650–800km to the east of the site (Carter et al. 2008).

A number of sites nearby to Çatalhöyük were founded, including Gökhöyük (see above). The regions of Cappadocia, Beyşehir and Marmara, well known from at least 400 years of exploitation of obsidian and chert, were eventually chosen for settlement, perhaps by a portion of the Çatalhöyük community. Numerous settlements emerged in the area south of the Marmara Sea, in particular around Iznik Lake. The process also involved contacts with indigenous foragers, both in this area and in the Eskişehir region. An outcome of the intertwined processes is the emergence of a very powerful version of the Neolithic in northwest Anatolia, which means that previous long-lasting relations with foragers in relation to chert and obsidian exploitation paid off. This process can justifiably be described as the spread of the Çatalhöyük farmers out of the Konya Plain towards Beyşehir-Suğla and then northwest Anatolia (e.g., Marciniak 2018). Spreading out around Levels South M/N/O was possible thanks not only to the considerable demographic potential of the Çatalhöyük community but mainly to the newly emerged social arrangements based upon households facilitating independent production and the build-up of their own surpluses (Marciniak 2015b).

The Lake District communities emerged around this time and established contacts with the Çatalhöyük community. The Beyşehir-Suğla Basin could be considered an intermediate region between these two settlement zones (Duru 2012; Özdöl 2012). Contact may have been further enhanced by the exploitation of the Akdere chert. In particular, the Lake District communities introduced imagery (Filipowicz 2019) that not only shaped their original character but made them the part of the Çatalhöyük cultural realm. Direct contacts are also documented by the exports of some shells into Çatalhöyük (Kurzawska et al. in press).

Around the same time, two distinct settlements at Tepecik-Çiftlik and Köşk Höyük were established on the western fringe of Cappadocia. They most likely resulted from a local trajectory of development in relation to the procurement and distribution of obsidian resources, but they were significantly shaped by the Çatalhöyük world, as indicated by similar modes of ceremonial life and distinct imagery directly resembling that of Çatalhöyük. The settlement remained a continuous point of reference for the migrating groups. Many symbolic elements originating from it appeared in the Late Neolithic and Early Chalcolithic ceramics of these Niğde-Aksaray settlements (Bıçakçı et al. 2012; Öztan 2012). This is manifested in the placing of motifs such as bulls, upraised splayed figures, spiral motifs etc. on movable objects such as pots. These can be viewed as a range of signifiers mobilised out of the Çatalhöyük repertoire and believed to be good markers of supra-individual identities (Meskell 2007: 25).

Hence, it is justifiable to argue that both the Lake District and western Cappadocia were significantly shaped by and remained under the influence of the Çatalhöyük cultural realm. They epitomise the powerful character of the cultural dominion developed at Çatalhöyük. These complex relations between inhabitants of Çatalhöyük and neighbouring regions from the beginnings of the second half of the seventh millennium BC onwards are part of a wider phenomenon of dispersal of Neolithic groups from the hitherto exploited centres into new ecological zones, which ultimately led to building up a complicated network of relations between them (Özdoğan 2010).

The relations of the Çatalhöyük community shifted again in the 63rd century BC. The central role of the settlement towards the end of the Late Neolithic was retained, but it became less pronounced and significant. Individual groups were increasingly differentiated, and different local traditions were more visible. The relations with the Beyşehir-Suğla Basin and northwest Anatolia were maintained. But these local groups were becoming increasingly more independent and detached from the arrangements inherited from ancestral settlers. The same process took place in the Lake District and Cappadocia. While these groups remained enmeshed in the cultural realm of Çatalhöyük, they developed into strong and dynamic local entities.

The turmoil triggered by deteriorating climatic conditions may have forced some portions of the Çatalhöyük population to exploit non-farming resources, as seen at nearby site of Pınarbaşı, and return to the arguably ancestral place at Canhasan I. If David French (1998) is right in arguing that Canahasan III may have been continuously visited until around 6500 BC, it is likely that its significance was not forgotten and may have been reestablished around 6300–6200 BC. In the period that brought about an acceleration of developments in different parts of the region, the Çatalhöyük groups inhabiting the East Mound became stuck between the need to catch up with the dynamically changing outer world and the need to abide by the traditions of bygone ancestors. While communities in western, southwestern and northwestern Anatolia were booming and their dynamism accelerated, the inhabitants of Çatalhöyük East remained embedded in their own traditions. They avoided stone architecture, painted pottery and new forms of vessels, and they continued to exploit wild cattle in ceremonial settings. When coupled with decreasing demographic potential, there was no way to move forward. The part of the local community that remained eventually moved to the neighbouring West Mound and the East Mound settlement collapsed. However, despite these difficulties, until its final years, the Neolithic settlement at Çatalhöyük remained above all a product of the developments taking placed in its localised historical, social, environmental and economic setting. And it continued to respond to emerging challenges posed by the dynamically changing outer world in its own specific way.

3. Heavy residue

Milena Vasić
(with contributions by Jovana Tripković)

Although initially just a by-product of flotation, the primary goal of which is retrieval of plant remains, heavy residue (henceforth HR) processing is now recognised as a valuable tool in archaeology. Based on the hypothesis that microartefacts are likely to reflect in situ activities (Schiffer 1987), the Çatalhöyük Research Project (henceforth ÇRP) has invested a significant amount of time and labour into designing the HR strategies that would systematically collect as much data as possible from every single excavated deposit.

The main aim of HR processing is to obtain further information about the use of the internal spaces that were kept meticulously clean and to provide an insight into the dynamics of the use of buildings at Çatalhöyük. However, due to complex processes related to site formation, this has proved to be a complex and not so straightforward approach (Cessford 2003). Given that building floors at Çatalhöyük generally contain small quantities of material, differentiation between in situ activities and 'background noise' (for example, material used for floor make-up and packing) is possible in only some cases. For example, a concentration of chert microblades and microdrills in a floor deposit (16565) in B.75 was recovered from a HR sample, whilst it was not noted during the excavation of this unit, and these artefacts typically do not occur on site (Carter et al. 2008b). B.75 also has evidence of bead production (Bains et al. 2013), and Siebrecht's recent analysis has confirmed that the bead roughouts found in the same floor deposit were perforated by chert drill bits (Siebrecht in preparation; this volume, Chapter 10).

Despite not always reflecting traces of in situ activities, HR processing represents an extremely expedient instrument in archaeology. It enables a higher recovery rate of material than hand collection or dry sieving provide, with the possible retrieval of fragments of biofacts and artefacts that are as small as 1.1mm. HR sorting is particularly useful for recovery of various microartefacts, but also microfauna remains, fish bones and manufacture debris. For example, the majority of beads are smaller than 3mm when complete (Chapter 10), and HR sorting facilitates the retrieval of even tiny bead fragments. In addition, the entire microfaunal assemblage at Çatalhöyük derives from HR (Jenkins, Yeomans 2013).

Due to the sheer amount of data that 25 years of excavations at Çatalhöyük have produced and the time constraints, recording of every single artefact that was recovered was practically impossible. Whilst smaller datasets (for example, figurines or clay objects) were relatively easy to handle, other material that exists in very large quantities (for example, faunal remains or ground stone) had to be sampled in order to be studied in detail. As a result, quantification of material had to be done based on the selected sub-sample (for the purpose of this volume, as well as previous rounds of publications, the selection was based on the list of priority units as described in Chapter 1). On the other hand, HR has been fully processed from each excavated unit, thus providing more accurate means of exploring the distribution of different material categories. It could thus be argued that the analysis of material from flotation samples represents a more precise method for the exploration of contextual and spatio-temporal distributions of artefacts than the analysis of the bulk material recovered during the dry-sieving on site could offer.

Although it represents a very small portion of the entire excavated material assemblage, processing of the HR samples has yielded over 1,500 kg of material (excluding building materials such as bricks and mortars). Over 25 years of the ÇRP excavations, 13,565 flotation samples were taken, and 240,029 litres of soil were floated from 11,003 deposits, thus producing the largest HR assemblage in the world thus far.

This chapter builds upon the previous publication (Mitrović, Vasić 2013) by incorporating the datasets from two segments of the excavations (2000–2008 and 2009–2017) of the Neolithic East Mound. As the sampling and sorting procedures in the first segment of excavations (1995–1999) underwent significant modifications, these samples were not included, in order to avoid the potential bias caused by the different methods (for example, the use of a riffle box was introduced in 1999, see Çatalhöyük Diary Database, entries by Fairbairn and Kennedy on 01.05.99, 26.05.99, 31.05.99, 01.06.99 and 02.06.99 for further discussion).

Newly added samples enable a better understanding of thc temporal variations in the distribution of material, especially given that the samples from the early levels in the South Area were considerably under-represented in the previous publication (see Mitrović, Vasić 2013: Table 2.5). We now have a much higher number of samples from the Early period in the South Area, as well as the Middle period in both the North and the South Areas (tables 3.1 and 3.2). In addition, two new areas were excavated (TPC and GDN), and although they were not explored to the same extent as the North and South Areas, they significantly contribute to our knowledge of the occupation of the site in the Late period (table 3.3).

The results presented in this chapter are based on 6,506 selected samples from 5,831 Neolithic contexts in the North, South, TPC and GDN Areas, excavated from 2000 onwards. Samples from fill, construction/make-up/packing (henceforth construction) and floor deposits form the largest part of the dataset, reflecting the total number of excavated deposits (table 3.3). For TPC, only units allocated to the Level TPC M have been included, whilst the samples from the Final period will be published elsewhere (Joka in preparation). The average volume of samples from the studied dataset is 40.5l, though volumes vary, and samples from burial fills and midden deposits tend to be larger than samples from the other types of deposits (table 3.4). In total, over 610kg of material, mainly animal bone (408kg), were recovered from 117,555.7l of soil processed (table 3.5).

This chapter explores the distribution of the HR material across different types of deposits, areas of the settlement and levels of the occupation. A special focus is given to the floor and midden deposits in order to establish variations in the use of external spaces and individual buildings. Burial fills also represent an interesting type of deposit, and this chapter compares the burial fills, midden and fill deposits so as to determine their similarities and differences, with the intention of establishing whether the material from burial fills represents burial goods or if the retrieved artefacts originate from the soil used to fill the burial feature. For that reason, burial fills were separated from other fills (for example, room, bin or pit fills) and assigned a 'data category' status (see Farid, Hodder 2014).

	Early	*Middle*	*Late*	*Unknown*	*Total*
North		2,638	423	12	3,073
South	440	1,004	925	48	2,417
GDN			25		25
TPC			315	1	316
Total	440	3,642	1,688	61	5,831

Table 3.1. Number of sampled units in each excavation area and period.

Methods

Samples were collected and processed following the standard flotation and HR strategies at Çatalhöyük (Mitrović, Vasić 2013; Cessford 2005b; Hastorf 2005), with some minor changes in the HR recording procedure (see below). The samples were sieved using the stack of 4mm, 2mm and 1mm sieves; when necessary (that is, in

	Early	*Middle*	*Late*	*Unknown*	*Total*
Activity	9	58	103		170
Floors (use)	76	957	276	2	1,311
Construction/make-up/packing	188	1,472	610	30	2,300
Fill	113	843	385	15	1,356
Midden	19	99	219	12	349
Burial fill	35	213	95	2	345
Total	440	3,642	1,688	61	5,831

Table 3.2. Sampled units belonging to different data categories and periods.

	Activity	*Floors (use)*	*Construction/ make-up/packing*	*Fill*	*Midden*	*Burial fill*	*Total*
North	57	752	1,161	729	176	198	3,073
South	100	529	987	514	164	123	2,417
TPC	13	26	145	103	5	24	316
GDN		4	7	10	4		25
Total	170	1,311	2,300	1,356	349	345	5,831

Table 3.3. Number of units of different data categories in each excavation area.

the case of larger samples), smaller fractions were split using a riffle box; organic and inorganic artefacts were separated from the remaining soil and then weighed using a digital scale with 0.01g accuracy and 200g capacity. The artefacts were recorded under 16 different material categories. Plant, Bone, Mollusc, Eggshell, Obsidian, Flint and Beads were retrieved from each of the three fractions; Bone Diagnostic was collected only from the 2mm and 1mm fractions; whereas Stone, Pottery, Figurine, Shaped Clay (non-diagnostic pieces of objects and natural material), Clay Object, Clay Ball, Worked Bone and Worked Stone were recovered only from the 4mm fraction.

Information stored in the HR database is composed mainly of weights of each material category; nonetheless, as previously argued, counts of artefacts rather than their weights are more meaningful for some material categories in the HR assemblage (Mitrović, Vasić 2013; Cross 2005; Cessford 2005b). For that reason, both counts and weights were recorded for Beads, Clay Object, Shaped Clay, Clay Ball, Clay Figurine, Worked Bone and Worked Stone. However, as this change was implemented halfway through the last segment of excavations, the obtained data were not used in this analysis. Similarly, weight of Stone started being recorded in 2013, whereas in prior years, only its presence was noted. As a result, Stone weights were recorded for only just over a third of the studied dataset, and consequently, similarly to the previous papers, this chapter considers only the presence of Stone across the deposits.

Deposits	*Average vol.*	*Median vol.*	*Total vol.*
Activity	10.9	5	1,845.1
Construction	15.9	14	36,505.2
Fill	20.8	22	28,146.1
Floor	15.2	12	19,995.5
Midden	35.3	27	12,310.2
Burial fill	54.4	28	18,753.6
All deposits	40.5	17	117,555.7

Table 3.4. Volumes of samples from each data category.

Material	*Weight (kg)*
Worked bone	0.32
Beads	0.99
Flint	2.16
Clay figurines	2.41
Eggshell	4.37
Shaped clay	7.12
Bone diagnostic	12.08
Worked stone	>13.28
Mollusc	17.69
Obsidian	26.32
Clay ball	30.60
Plant	38.55
Clay object	60.63
Pottery	126.74
Stone	>205.84
Bone	812.16

Table 3.5. Overview of total weight of each material category in HR samples.

Another change in the protocol over the last few years was a more thorough checking of the sorted material, a process to which all specialists who studied the material, as well as the finds officers, actively contributed. Whereas previously the central Finds Database was regularly being updated after any misidentified material was returned by the specialists to the Finds Lab, this information was not being fed into the HR database. As a consequence, the HR Database did not contain updated information. During the last years of excavations, we continually modified the records and the weights as entered according to the information from the specialists and the finds officers.

There are several issues with the implemented HR procedure that were recognised from early on (see Çatalhöyük Diary Database, entries by Fairbairn and Kennedy on 01.05.99, 26.05.99, 31.05.99, 01.06.99 and 02.06.99). For example, HR records offer only partial information regarding the distribution of Plant, given that plant remains are recovered from both light and heavy fractions of flotation samples. Ideally, despite the different sorting strategies of two fractions, as well as the level of specialist information obtained from each of them, the data should be combined for the overall analysis of the distribution of plant remains.

Another difficulty with HR protocol concerns the recovery of microfaunal remains. The sorting team was supposed to collect only diagnostic fragments of animal bone from 1mm and 2mm fractions and not the small non-identifiable fragments of large mammals. Despite their meticulous sorting and their ability to recognise and pull out even the smallest fragments of artefacts, given that the sorting team was composed of local workers and not trained zooarchaeologists, this was a rather difficult task. Over the years, there were several attempts to address this issue, including collection of all bone fragments, which were then further sorted by the microfaunal team. However, this did not yield good results, as it was extremely time consuming for both the sorting team and microfauna specialists. Testing of different strategies with the aim of improving the procedure and providing more accurate data resulted in an inconsistent dataset. For that reason, for the purpose of this analysis, category Bone Diagnostic was not included.

Analyses of the contextual, spatial and temporal distribution of materials were conducted using the following statistical tools: presence/absence and density. Presence is reported as a percentage of excavated contexts containing the material in question, regardless of the fraction of sample in which it was found.

'Standardised' weights were utilised in order to calculate densities. Due to the large volume of the sampled soil that had to be processed and time limitations, larger samples were split before sorting. The largest fraction (4mm) was always completely sorted, whereas smaller fractions (2mm and 1mm) were sometimes split using a riffle box, and only part (1/2, 1/4 or 1/8) of the flotation sample was further processed, whereas the remaining soil was stored in the depots. As a result, in these cases, there is no information on the actual weight of different material categories. For that reason, the weights were standardised under the assumption that if, for example, half of the sorted sample contained a certain weight of material, the complete sample would contain twice as much. Although it could be argued that this method does not provide completely accurate weights, it does enable the rough estimation of the quantity of material. Density was calculated as gram per litre; that is, the standardised weight of each material category was divided by the volume of the flotation sample. Whereas Bone and Stone exist in higher quantities, densities of material are typically in hundredths of grams (table 3.6). Densities were calculated per sample, rather than per fraction, using the method described above. As a consequence, Obsidian densities reflect to a great extent their quantities in the 4mm fraction, whereas their densities in 2mm and 1mm fractions are comparatively small (table 3.7). In contrast, due to their high fragmentation rate, larger quantities of Mollusc and Eggshell exist in smaller fractions, as once broken into small fragments, Eggshell survives relatively well (Sidell 1993) and so does Mollusc. As previously demonstrated, although representing an arbitrary division, smaller fractions (1mm and 2mm) have different depositional patterning than the large fraction (4mm) (Cessford 2005b; Mitrović, Vasić 2013) although it is not certain how meaningful these differences are. Nonetheless, smaller fractions do not necessarily reflect in situ activities and relate as much to intentional deposition as they do to accidental inclusion or 'background noise'.

All material categories exhibit a highly skewed distribution, with an exceedingly high frequency of samples with low densities on one hand, and a smaller number of samples with exceptionally high densities on the other (see also Chapter 14). Given that the densities are not normally distributed, values were transformed using the Log10 transformation. Nevertheless, as the Log10 transformation does not allow 0 values, box plots generated using Log10 values do not include samples that did not contain certain material. For that reason, material densities were explored using both the median of the real values and Log10 values. Once the logged densities were calculated, and a normal distribution obtained, a two-sample T-test with the assumption of unequal variance between the samples was applied in order to determine if the differences in the observed data were statistically significant.

	All units		*Units containing the selected material*	
	Average density	*Median density*	*Average density*	*Median density*
Bone	2.99	0.83	3.03	0.86
Clay ball	0.23	0	2.58	0.42
Pottery	0.15	0	0.89	0.23
Plant	0.12	0.02	0.13	0.03
Clay object	0.10	0	1.39	0.15
Obsidian	0.10	0.03	0.12	0.04
Worked stone	0.07	0	4.46	1.37
Mollusc	0.06	0.03	0.07	0.03
Shaped clay	0.04	0	0.50	0.08
Eggshell	0.02	0.00	0.03	0.01
Flint	0.01	0	0.12	0.01
Figurine	0.01	0	0.21	0.04
Beads	0.00	0	0.02	0.00
Worked bone	0.00	0	0.07	0.01
Stone	N/A	N/A	N/A	N/A

Table 3.6. Densities of material categories (standardised weight/litre of soil).

Both presence and density of ubiquitous categories were analysed in order to explore the similarities and differences in their spatio-temporal distribution across different types of deposits and areas of the settlement, whereas, given their rarity, only presence was considered for other material categories. Correlations between the presence or densities of different material categories across the excavated deposits were explored using the Pearson correlation coefficient (Pearson's r).

Presence

As previously demonstrated (Fig 2.1. in Mitrović, Vasić 2013), Animal Bone (including the category Bone Diagnostic), Plant, Mollusc, Obsidian, Stone and Eggshell are the ubiquitous material categories on site and occur in the large majority of samples, whilst the other categories (Clay Ball, Clay Figurine, Clay Object, Shaped Clay, Flint, Pottery, Beads, Worked Bone and Worked Stone) are less frequent (fig. 3.1). As in previous years, Obsidian is much more common than Flint, but in contrast to the previous publication, which showed an almost equal presence of Mollusc and Eggshell, with the newly added samples, it is now clear that Mollusc is significantly more common on site than Eggshell is.

The studied dataset does not reveal strong correlations between the occurrences of different material categories in samples; nonetheless, some moderate positive correlations (r=0.3–0.4) have been observed,

	Fill	*Burial fill*	*Construction*	*Floor*	*External floor*	*Midden*	*Activity*	*All*
Bone 4mm	1.308	2.092	0.430	0.316	6.091	9.042	4.442	0.738
Bone diagnostic 2mm & 1mm	0.043	0.059	0.022	0.020	0.080	0.156	0.085	0.033
Bone all fractions	1.448	2.383	0.507	0.400	6.176	9.638	4.530	0.830
Mollusc 4mm	0.008	0.011	0.003	0.002	0.007	0.008	0.000	0.005
Mollusc 2mm & 1mm	0.020	0.018	0.020	0.020	0.027	0.015	0.015	0.019
Mollusc all fractions	0.033	0.035	0.030	0.030	0.042	0.028	0.025	0.030
Eggshell 4mm	0.000	0.000	0.000	0.000	0.002	0.004	0.000	0.000
Eggshell 2mm & 1mm	0.004	0.004	0.002	0.002	0.013	0.018	0.001	0.003
Eggshell all fractions	0.005	0.006	0.002	0.003	0.019	0.025	0.001	0.004
Obsidian 4mm	0.022	0.030	0.004	0.004	0.089	0.198	0.031	0.013
Obsidian 2mm and 1 mm	0.010	0.012	0.004	0.006	0.026	0.047	0.018	0.008
Obsidian all fractions	0.037	0.046	0.014	0.018	0.121	0.259	0.067	0.025

Table 3.7. Median density of material in different data categories (standardised weight/litre of soil).

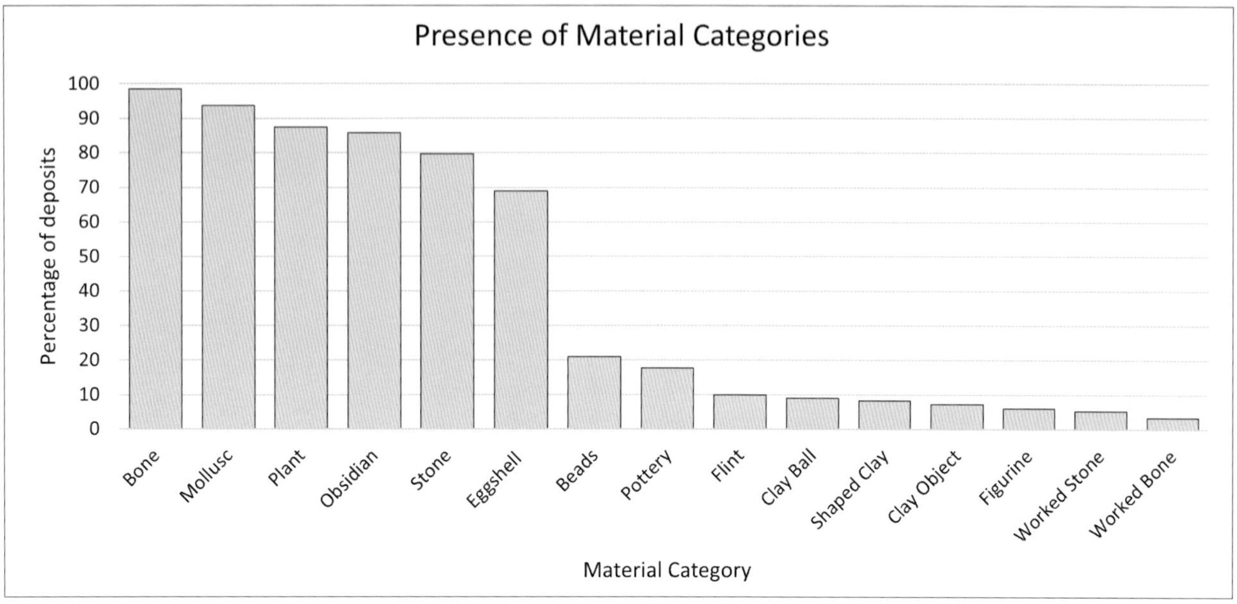

Figure 3.1. Presence of material categories in HR samples.

perhaps the most notable being a relatively common co-occurrence of Figurine and Clay Object and Figurine and Shaped Clay categories (135 out of 356 units that contained Figurine also had Clay Object category, whereas all three categories were found in 77 units – for definition of clay object categories, see Chapters 7 and 8). These units do not represent evidence of simultaneous production of these artefacts; nonetheless, they could point to them potentially being used and/or discarded together.

Midden deposits tend to contain almost all non-ubiquitous materials more commonly than other types of deposits, but also Eggshell, Obsidian and Plant (fig. 3.2). Animal Bone (henceforth Bone), as the most ubiquitous material, occurs equally across the deposits and so does Mollusc, with the exception of activity contexts, which contain this material less frequently than the other deposits. Obsidian, Stone and Eggshell are less frequently found in the activity, construction and floor deposits than they are in fills, burial fills and middens. Some categories, such as Beads, Flint, Clay Ball and Stone, have a very similar presence in middens and burial fills.

Density

Densities of material in different types of contexts correspond to the observed patterns in the presence of material categories, with middens, but in some cases activity deposits as well, containing the highest quantity of Obsidian, Plant, Eggshell and Bone (as well as Pottery and Flint) (fig. 3.3). Mollusc, on the other hand has almost identical densities across the deposits, with its density being slightly higher in the burial fills. Equal quantities of Mollusc in different kinds of deposits have also been confirmed by the T-Test, which shows that the difference in Mollusc densities is only statistically significant in the case of construction and floor deposits, and construction and fill deposits. In contrast, other ubiquitous material categories generally demonstrate statistical significance in their different distribution across the deposit types (table 3.8).

Notably, data categories reveal different correlations between the densities of materials. A strong positive correlation (r=0.7–0.8) between the densities of Mollusc, Obsidian and Eggshell has been noted in the floor deposits, which demonstrates that the floor deposits which contain a higher density of one of these materials tend to have a higher density of the other materials as well (table 3.9). Further investigation and spatial analysis need to be conducted in order to confirm if this tendency can be related to the observed differences between clean and dirty floors (Matthews 2005b; see also Hodder, Cessford 2004a).

Middens, on the other hand, show only moderate positive correlations, the highest being in the densities of Flint and Clay Ball (r=0.4) (table 3.10). Burial fills exhibit a moderate correlation between Obsidian density and the densities of the other ubiquitous materials, whereas the Plant density is also moderately correlated to Bone and Eggshell densities (r=0.4–05) (table 3.11). Non-burial fills exhibit a strong positive correlation between the density of Shaped Clay and the densities of Pottery and Bone (r=0.8–0.9). In addition, densities of Clay Object and Clay Ball also seem to be moderately correlated in fills (r=0.6) (table 3.12).

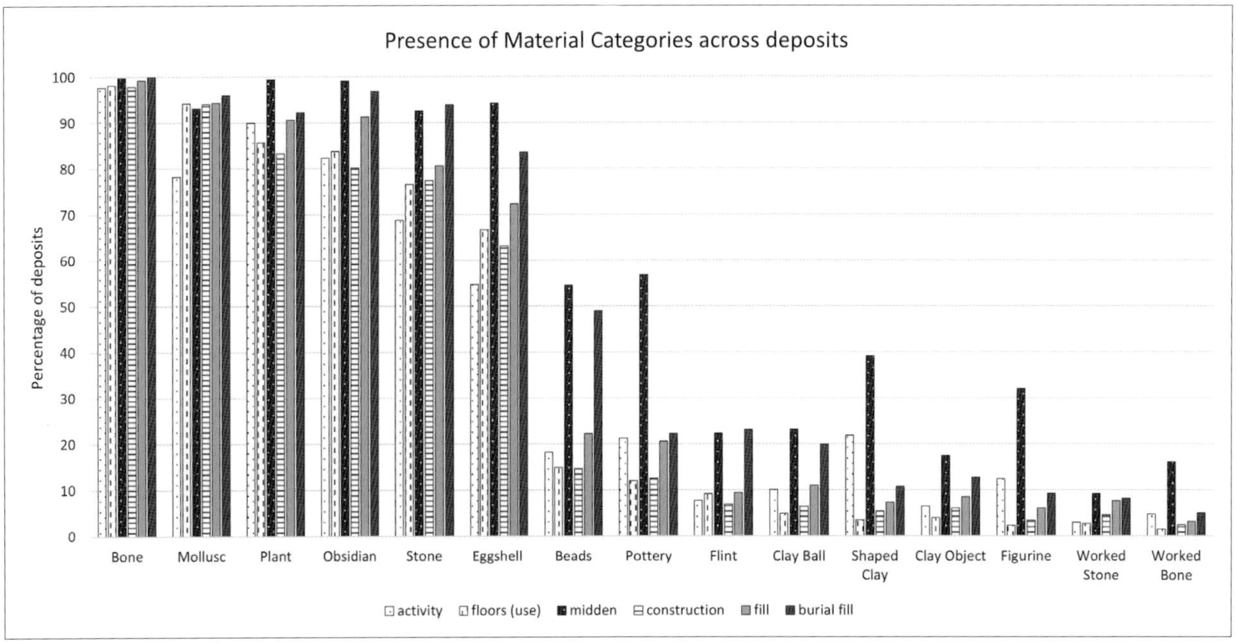

Figure 3.2. Presence of material categories across deposits.

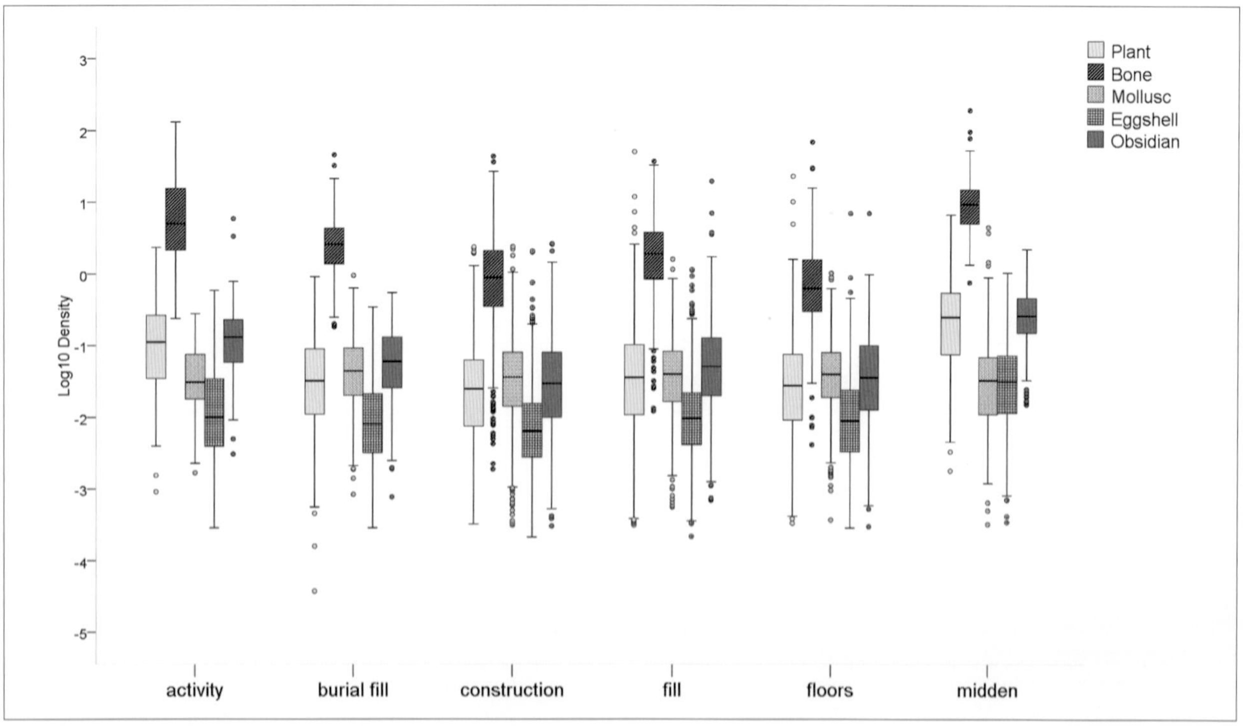

Figure 3.3. Log10 density across different contexts.

p-values	*Plant*	*Bone*	*Mollusc*	*Eggshell*	*Obsidian*	*Flint*	*Pottery*
Floor & midden deposits	0.00	0.00	0.24	0.00	0.00	0.00	0.00
Construction & floor deposits	0.00	0.10	0.01	0.00	0.00	0.74	0.57
Construction & fill deposits	0.00	0.00	0.00	0.00	0.00	0.27	0.03
Fill & midden deposits	0.00	0.00	0.12	0.00	0.00	0.02	0.00
Burial & midden deposits	0.00	0.00	0.06	0.00	0.00	0.01	0.00
Burial & room fill deposits	0.41	0.00	0.06	0.84	0.08	0.27	0.00
Burial & non-burial fill deposits	0.11	0.00	0.42	0.11	0.69	0.70	0.00

Table 3.8. Comparison of densities between different deposits (T-Test).

Construction and floor deposits have similar densities of materials, which tend to be the lowest on site. This corresponds to the presence of material categories (see fig. 3.2), the majority of which occur least commonly in these contexts. Amongst the internal floor types, the highest densities of material are usually found in bins and/or general building floors, though variations do exist (fig. 3.4). Previous HR analysis has confirmed that the material is accumulated differently in floors and platforms, with the latter containing a lower density of artefacts, whereas amongst the general height floors, white clay floors contain a significantly lower quantity of material than dirty floors do (Cessford 2005b). The new dataset further confirms the tendency of general height floors to contain a higher density of materials than platforms, although the density of Mollusc is quite similar. Similarly, there is a noted difference between ovens and hearths, with ovens having a higher density of all materials except Bone. This could be a consequence of superstructure that makes ovens harder to clean than hearths, but it could also be related to the different ways in which hearths and ovens were used (Ketchum in preparation). Another interesting result is a relatively high density of Eggshell in basins, although it should be noted that the median density of Eggshell from the 4mm fraction is 0 across all types of internal floor deposits.

	Clay ball	Figurine	Clay object	Shaped clay	Pottery	Stone	Worked bone	Worked stone	Beads	Plant	Bone	Mollusc	Eggshell	Obsidian	Flint
Clay ball	1.0														
Figurine	0.0	1.0													
Clay object	0.0	0.0	1.0												
Shaped clay	0.0	0.0	0.1	1.0											
Pottery	0.0	0.0	0.0	0.0	1.0										
Stone	0.0	0.0	0.0	0.0	0.1	1.0									
Worked bone	0.0	0.0	0.0	0.0	0.0	0.0	1.0								
Worked stone	0.0	0.0	0.0	0.0	0.0	0.0	0.0	1.0							
Beads	0.0	0.0	0.0	0.0	0.0	0.0	0.0	0.0	1.0						
Plant	0.0	0.0	0.0	0.0	0.0	0.0	0.0	0.0	0.0	1.0					
Bone	0.0	0.0	0.0	0.0	0.1	0.1	0.0	0.0	0.1	0.0	1.0				
Mollusc	0.0	0.0	0.0	0.0	0.0	0.3	0.0	0.0	0.0	0.3	0.0	1.0			
Eggshell	0.0	0.0	0.0	0.0	0.0	0.1	0.0	0.0	0.0	0.4	0.0	0.8	1.0		
Obsidian	0.0	0.0	0.0	0.0	0.1	0.0	0.1	0.0	0.0	0.3	0.1	0.7	0.8	1.0	
Flint	0.0	0.0	0.0	0.0	0.0	0.0	0.0	0.0	0.1	0.0	0.0	0.0	0.0	0.1	1.0

Table 3.9. Correlation of presence of material in floor deposits.

	Clay ball	Figurine	Clay object	Shaped clay	Pottery	Stone	Worked bone	Worked stone	Beads	Plant	Bone	Mollusc	Eggshell	Obsidian	Flint
Clay ball	1.0														
Figurine	0.0	1.0													
Clay object	0.0	0.1	1.0												
Shaped clay	0.0	0.1	0.0	1.0											
Pottery	-0.1	0.0	-0.1	0.1	1.0										
Stone	0.4	0.0	0.1	-0.1	-0.1	1.0									
Worked bone	0.0	0.0	0.0	0.0	0.0	0.0	1.0								
Worked stone	0.0	0.0	0.0	0.0	0.0	0.0	0.0	1.0							
Beads	0.0	0.0	0.1	0.0	0.0	0.0	0.0	0.0	1.0						
Plant	0.0	0.1	0.1	0.2	0.1	0.0	0.0	0.0	0.0	1.0					
Bone	0.1	0.1	0.0	0.0	0.3	0.1	0.0	0.0	0.0	0.2	1.0				
Mollusc	0.1	0.0	0.0	0.0	-0.1	0.0	0.0	0.0	0.0	0.0	0.0	1.0			
Eggshell	0.0	0.0	0.0	0.0	0.0	0.1	0.0	0.0	0.0	0.1	0.3	0.0	1.0		
Obsidian	0.3	0.0	0.1	0.1	0.2	0.2	0.0	0.0	0.1	0.3	0.3	0.1	0.1	1.0	
Flint	0.4	0.0	0.0	0.0	0.0	0.3	0.0	0.0	0.0	0.0	0.2	0.0	0.0	0.0	1.0

Table 3.10. Correlation of presence of material in midden deposits.

	Clay ball	Figurine	Clay object	Shaped clay	Pottery	Stone	Worked bone	Worked stone	Beads	Plant	Bone	Mollusc	Eggshell	Obsidian	Flint
Clay ball	1.0														
Figurine	0.1	1.0													
Clay object	0.0	0.1	1.0												
Shaped clay	0.0	0.0	0.0	1.0											
Pottery	0.0	0.1	0.1	0.1	1.0										
Stone	0.3	0.0	0.2	0.0	0.1	1.0									
Worked bone	0.0	0.0	0.0	0.0	0.0	0.0	1.0								
Worked stone	0.0	0.0	0.0	0.0	0.0	0.0	0.0	1.0							
Beads	0.0	0.0	0.0	0.1	0.0	0.0	0.1	0.0	1.0						
Plant	0.0	0.1	0.3	0.0	0.0	0.1	0.0	0.0	0.0	1.0					
Bone	0.1	0.3	0.2	0.1	0.4	0.1	0.0	0.0	0.1	0.4	1.0				
Mollusc	0.3	0.0	0.0	0.0	-0.1	0.2	0.0	0.1	0.0	0.2	0.2	1.0			
Eggshell	0.1	0.1	0.3	0.0	0.0	0.1	0.0	0.1	0.0	0.4	0.2	0.3	1.0		
Obsidian	0.1	0.2	0.2	0.1	0.3	0.2	0.0	0.1	0.1	0.5	0.4	0.4	0.5	1.0	
Flint	0.0	0.0	0.0	0.0	0.0	0.0	0.0	0.0	0.0	0.0	0.0	0.0	0.0	0.0	1.0

Table 3.11. Correlation of presence of material in burial fills.

	Clay ball	Figurine	Clay object	Shaped clay	Pottery	Stone	Worked bone	Worked stone	Beads	Plant	Bone	Mollusc	Eggshell	Obsidian	Flint
Clay ball	1.0														
Figurine	0.0	1.0													
Clay object	0.6	0.0	1.0												
Shaped clay	0.0	0.3	0.0	1.0											
Pottery	0.0	0.3	0.0	0.8	1.0										
Stone	0.3	0.0	0.2	0.0	0.0	1.0									
Worked bone	0.0	0.0	0.0	0.0	0.0	0.0	1.0								
Worked stone	0.0	0.0	0.0	0.0	0.0	0.0	0.0	1.0							
Beads	0.0	0.0	0.0	0.0	0.0	0.0	0.0	0.0	1.0						
Plant	0.0	0.0	0.0	0.1	0.1	0.0	0.0	0.0	0.0	1.0					
Bone	0.0	0.3	0.0	0.8	0.9	0.0	0.0	0.0	0.0	0.1	1.0				
Mollusc	0.0	0.0	0.0	0.0	0.0	0.1	0.2	0.0	0.0	0.0	0.0	1.0			
Eggshell	0.0	0.0	0.0	0.0	0.0	0.0	0.0	0.0	0.0	0.0	0.0	0.2	1.0		
Obsidian	0.0	0.1	0.0	0.1	0.1	0.0	0.0	0.0	0.0	0.0	0.2	0.1	0.1	1.0	
Flint	0.0	0.0	0.0	0.0	0.0	0.0	0.1	0.0	0.0	0.0	0.0	0.1	0.1	0.0	1.0

Table 3.12. Correlation of presence of material in non-burial fills.

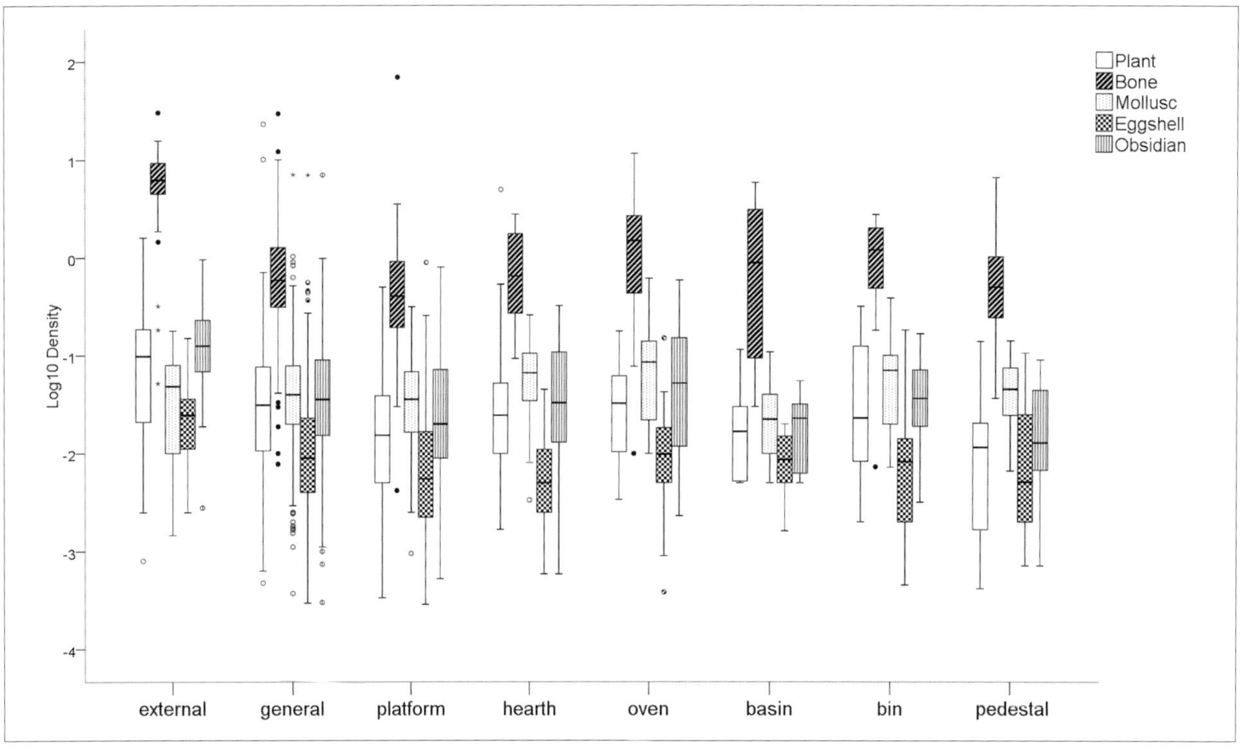

Figure 3.4. Log10 density of ubiquitous material in different floor types.

When building and external floor deposits are compared, external floors contain a significantly higher density of all material categories (fig. 3.4). The higher density of material in the external areas, witnessed both in the midden and floor deposits, reflects both the use of internal and external spaces and the discard strategies on site (Martin, Russell 2000; Hodder, Cessford 2004a; Bogaard et al. 2014; Matthews 2005b; 2005a). The low presence and density of artefacts in the floor deposits is a site-wide pattern and a result of regular sweeping of building floors. As a consequence, material that gets trampled into the floor surface or is occasionally left on the floor exists in a significantly smaller quantity than is witnessed in the external floor deposits. Continual sweeping and keeping the living spaces tidy is clearly the main reason behind the small quantities of material in the HR samples from floor deposits, but the latter could also partly be a result of the methodology employed. When artefacts (especially complete, elaborate or fragile objects) are found in situ, they tend to be hand picked and not included in flotation samples, thus significantly reducing the density of material from that unit. Secondly, over the years, with so many excavators working at the site, the data category 'floors (use)' has been used in slightly different ways. Some excavators assigned this category to the thin and fine plaster surface on the floors, some defined bottom layers of, essentially, building fill (that is, a few centimetres above the plaster surface) as floor deposits, while others lumped the floor surface in with the packing/make-up deposits below. Consequently, some of the fine differences in the HR densities could have been missed due to the different interpretations of the deposits during excavations. Another problem is represented by the category 'Cluster'. Clusters are defined as concentrations of objects or organic material grouped together that clearly stand out from the surrounding matrix, and many of which reflect deliberate deposition (Farid, Hodder 2014; Nakamura 2010). However, if a group of objects is found on a floor deposit and defined as a cluster, analysis of the distribution of material in floor deposits would not include these groups of objects, thus again reducing the overall density of material from that unit. For that reason, bringing the datasets together and a comparison of the HR material with the artefacts obtained through dry-sieving or hand collection is essential for a better understanding of the material distribution in the floor deposits, although it should be noted that the finding of larger artefacts on floors was relatively rare.

Amongst the intramural contexts, all material categories occur most frequently in burial fills. Whilst previously it has been suggested that the majority of material retrieved from the burials represents 'background noise' (that is, material that comes together with the soil used to fill the burial features), it is now recognised that this is not always the case. Complexity of funerary practices and the high level of the Neolithic disturbances of burials have resulted in a lot of material that might originally have been placed together with the interred individuals ending up in the fills of burial features (see Volume 13, Chapter 17). Whilst in some cases, a burial cuts into a midden, and one might assume that this is the reason why its fill is rich with material (for example, burials in B.17), a lot of burials are not located stratigraphically above midden deposits. Furthermore, there is a clear difference between the overall densities of material in burial fills and construction deposits, as well as between burial fills and the floor deposits, and thus the 'richness' of burial fill units cannot always be simply explained by accidental inclusions that come from the surrounding matrix.

All material categories tend to occur more frequently in burial fills than they do in other types of fill deposits (fig. 3.2); this is especially visible in the presence of Beads and Flint. Nonetheless, they do not display much variability in terms of material density, and burial and non-burial fills contain similar densities of Obsidian, Mollusc, Plant, Eggshell and Bone. The largest difference is seen in different densities of Pottery, which is at its lowest in burial fills, but higher in fills of other features. This corresponds to the overall pattern noted on site that pottery was almost never placed with the dead (Volume 13, Chapter 17). Whilst figure 3.2 shows the presence of Pottery in over 20 % of the excavated burial fills, it is plausible that pottery sherds indeed came together with the soil used to fill the burial feature.

As seen in table 3.8, the difference in density of material between the non-burial and burial fills has a statistical significance only when it comes to Bone and Pottery. However, if these deposits are compared in individual areas, and when only contemporary deposits are sampled, it becomes clear that it is not that straightforward. If deposits from the Middle period are tested, burial and non-burial fills display a statistically significant difference in the densities of Bone and Mollusc; in addition, densities of Plant and Eggshell are different in these deposits in the South Area, whereas in the North Area, the difference in the Obsidian density between samples from burials and fills of other features proved to be significant. When the content of burial fills in individual buildings is compared to their surrounding matrix, there is also a noted variation in the densities, as seen, for example, in burial fills in B.76 compared to platforms and room fills above them and the room fill of the building below, B.89 (fig. 3.5). Due to the coarse nature of the HR data, it is not possible to make firm conclusions based on this evidence. A more detailed investigation needs to be conducted and specific burial fill deposits examined and compared with the surrounding matrix, but it is plausible that the soil used to fill up burial features is not the same soil that was dug out from the platforms in which interments took place. Although it is possible that the material indeed came with the soil used to fill up the burial features, this soil could have come from somewhere else.

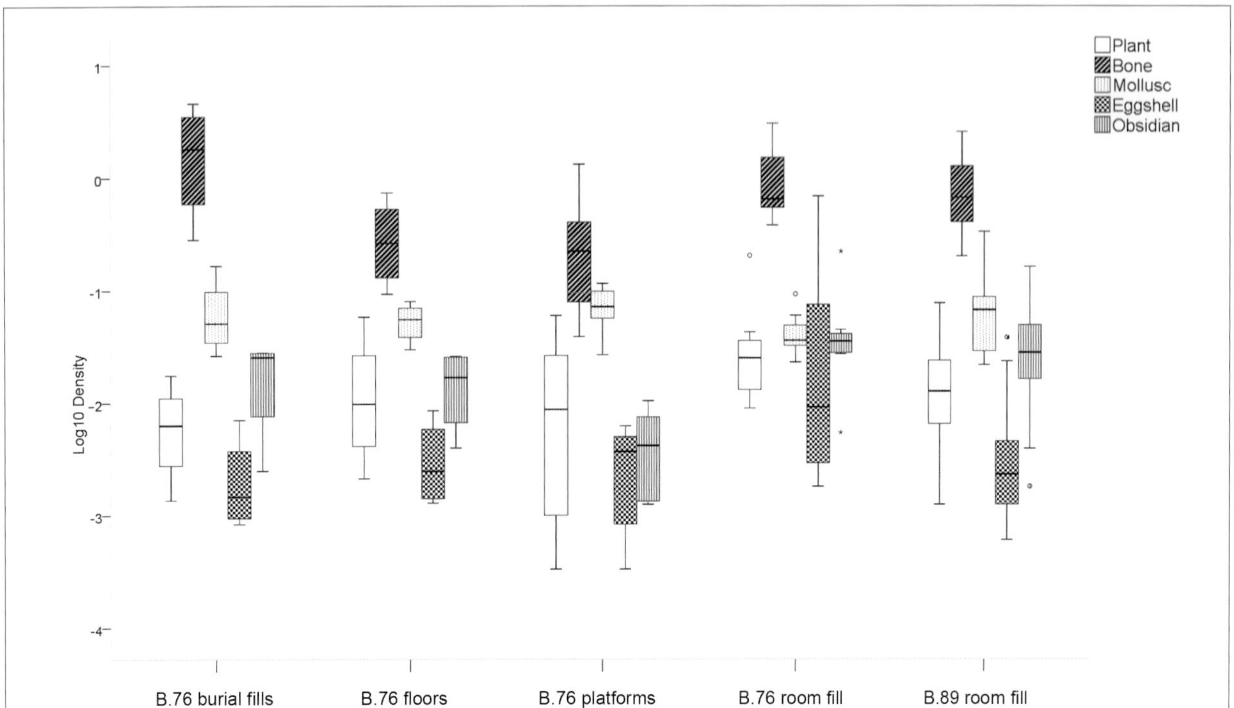

Figure 3.5. Comparison of densities (Log10) in burial and non-burial contexts in subsequent B.89 and B.76.

Similarly, it remains unclear where the soil used for infilling buildings in the process of abandonment came from. Although it has previously been noted that abandonment processes were by no means uniform (Russell et al. 2014), with quite a sterile soil being occasionally used for the room fill, the density of material categories is clearly higher in the midden deposits (including both those from external areas and those formed in abandoned buildings) than it is in the units interpreted as room fill deposits (fig. 3.6).

Contextual distribution in the North and South Areas

The overall distribution of Bone, Obsidian and Plant is very similar in the North and South Areas, with the exception of Plant, which occurs slightly less commonly in the floor deposits in the South Area (fig. 3.7). The distribution of Mollusc is also very consistent across both areas and different types of deposits, with the only difference seen in its relatively low presence in the activity deposits in comparison to other deposits in the South Area, as well as in the North Area. On the other hand, Eggshell displays the largest variability in spatial distribution. Whilst its presence is about the same in the activity deposits in the two areas, burial fills, construction, fill, floor and midden deposits in the North Area contain Eggshell more frequently.

Burial fills and middens are the types of deposits that contain almost all material categories more commonly than other contexts do. This has been noted in both the North and South Areas, with some differences in the proportions of deposits containing material. The most notable variability in the occurrence of non-ubiquitous categories in the two areas is witnessed in the midden deposits in the North Area, which have a higher presence of material. In contrast, other types of deposit sometimes contain material more commonly in the South Area than they do in the North Area (as seen, for example, in the occurrence of Beads and Pottery in the floor deposits or Clay Ball in burial fills). Nevertheless, most of categories tend to occur relatively frequently across the non-midden deposits equally in both areas.

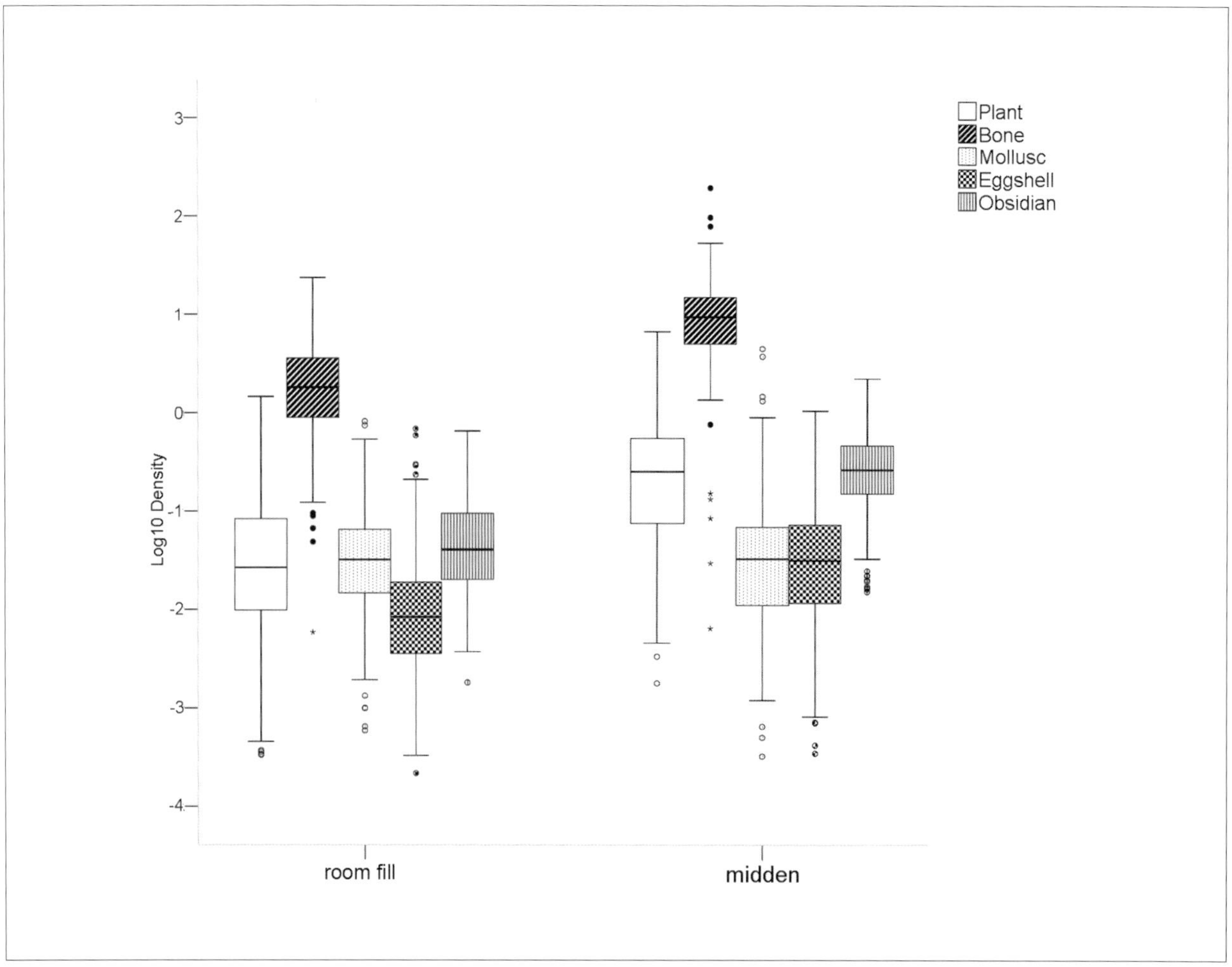

Figure 3.6. Log10 densities of ubiquitous material in midden and room fill deposits.

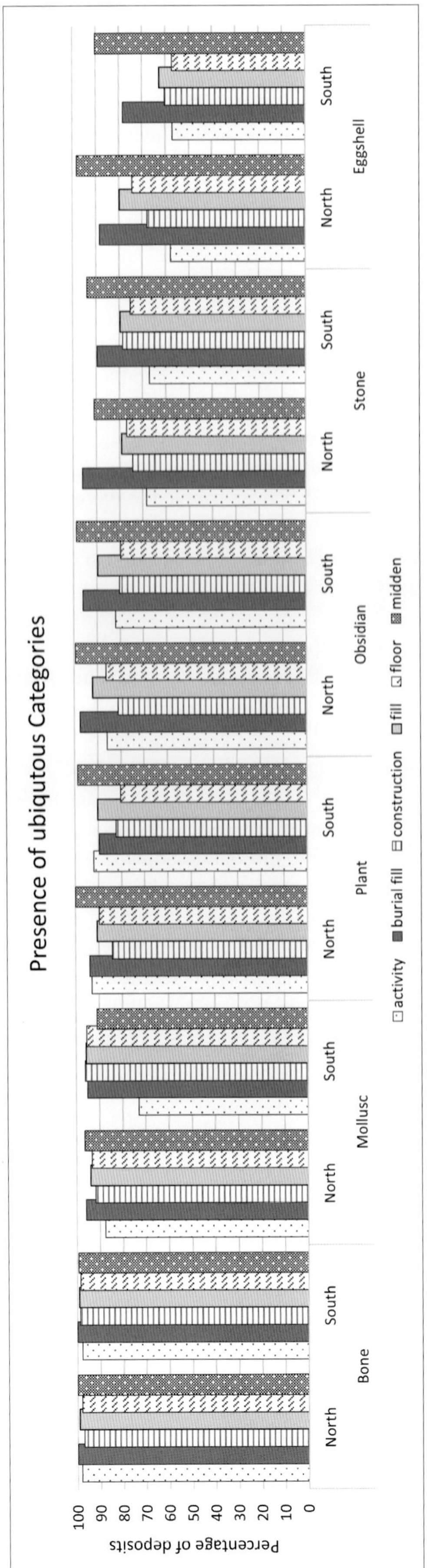

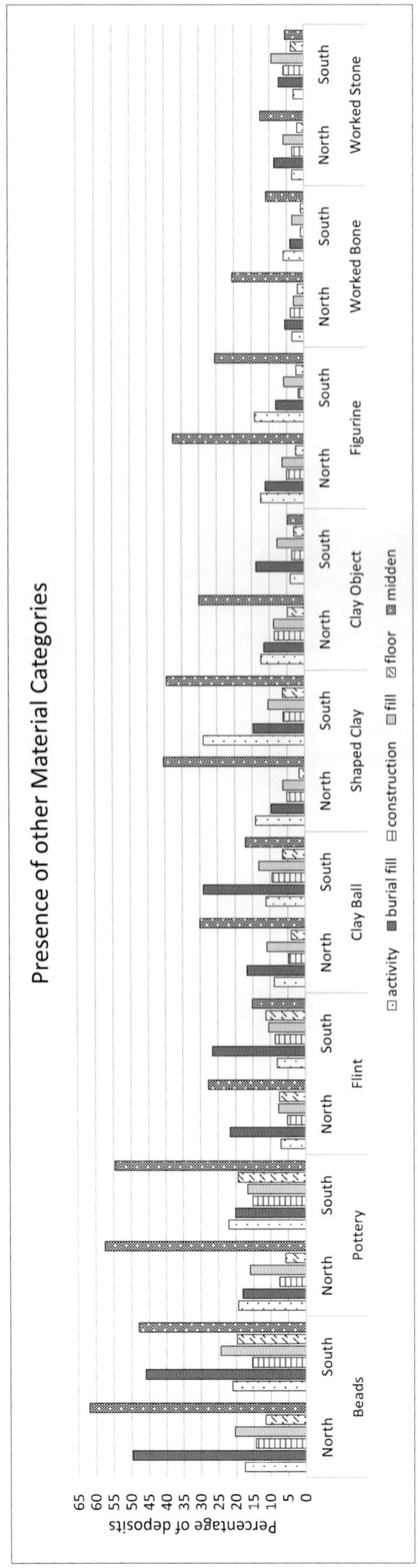

Figure 3.7. Presence of ubiquitous (above) and other (below) material categories in different deposits in the North and South Areas.

In order to avoid bias caused by the potential temporal differences between excavated deposits in the North and South, the investigation of the spatial distribution of material categories and similarities and differences in their densities in the different areas has been done only on contemporary deposits. Early period deposits have been explored only in the South Area, and for that reason, comparison of material densities in different parts of the settlement is not possible. On the other hand, the Middle period is represented by excavated contexts in the North and South Area, whereas the Late period deposits have been excavated in all four Areas of the settlement, albeit not to the same extent (table 3.3).

The distribution of the ubiquitous categories in the contemporary deposits belonging to the Middle period demonstrates that there is a slight variability between the North and South Areas (fig. 3.8). The overall pattern that has been observed is that burial and non-burial fills, floors and construction deposits in the South Area tend to contain material less commonly, as seen, for example, in the distribution of Plant, Obsidian and most notably Eggshell (due to the small number of samples from the South Area, midden units were excluded).

Densities also show a very similar distribution. The North Area floor deposits have a higher density of Plant and Eggshell, and platforms contain a higher density of Obsidian as well (fig. 3.9). Internal deposits in the South Area also tend to be less diverse than the deposits in the North Area (table 3.13). For example, the average count of material categories in the samples taken from the different floor types (general, platform, bin, hearth etc.) in B.76, B.80, B.89 and B.96 in the South Area ranges between 3.7 and 4.7, whereas samples from the floors in the North Area buildings B.3, B.77, B.131, B.132 and B.114 have an average count between 4.7 and 6.1 of different material categories.

A tentative difference between the Areas has also been observed in burial contexts. Burial fills in the North Area have a higher occurrence of Beads, Flint, and Figurine than these deposits do in the South Area. Similarly, they contain a higher density of Plant, Mollusc, Obsidian and especially Eggshell (also confirmed by the T-Test, p=0) (fig. 3.10).

Presence of materials in the Late period deposits also displays similarities in the two areas, with occasional variability observed in the occurrence of some categories (fig. 3.11). Strikingly, whereas northern deposits contained material more frequently in the Middle period, there is a notable shift in the Late period, and apart from middens, there is a tendency toward material being found in a higher proportion of all types of deposits in the South Area. The most outstanding difference observed in the floor deposits is seen in Pottery distribution. Pottery fragments were found in 44.57% of the floor deposits in the South Area, whereas they exist in only 19.35% of the northern floors. Flint, Stone and Shaped Clay also occur more frequently in the southern floor deposits.

Similarly, floor deposits in the South Area tend to contain higher densities of material, especially evident in Bone and Obsidian categories (fig. 3.12). In terms of the distribution in platforms, the only difference that has proved to be statistically significant is the higher density of Bone in the South Area, whereas Obsidian is very close to the threshold (p=0.06); however, this is likely to be due to a very small number of samples in the studied dataset. A higher density of material is also confirmed by the midden deposits (fig. 3.13) and the burial features (fig. 3.14). In addition, burial fills in the South Area contain Plant, Eggshell, Pottery and Shaped Clay more commonly, whereas a higher occurrence rate of Mollusc, Stone, Beads and Clay Ball exists in the northern burials.

Area	*Building*	*Floors*	*Construction*	*Fills*	*Burial fills*
North	B.3	5.95	6.65	6.71	7.86
	B.114	5.11	6.3	6.78	6.91
	B.132	6.08	5.9	6.75	8.73
	B.77	4.86	4.4	5.56	6.94
	B.131	4.72	4.31	5.51	8.82
South	B.76	4.74	4.87	5.06	7
	B.80	4.68	4.64	5.89	5.71
	B.89	4.62	4.2	5	6.67
	B.96	3.71	3.52	5.41	4.92

Table 3.13. Diversity (average count of material categories) of samples in contemporary buildings in the North and South Areas.

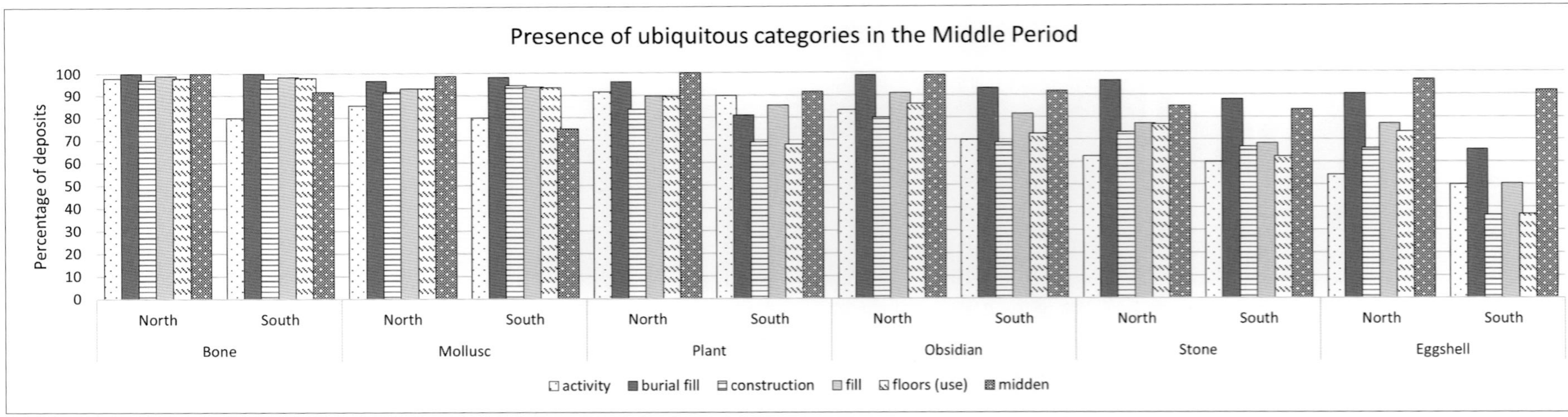

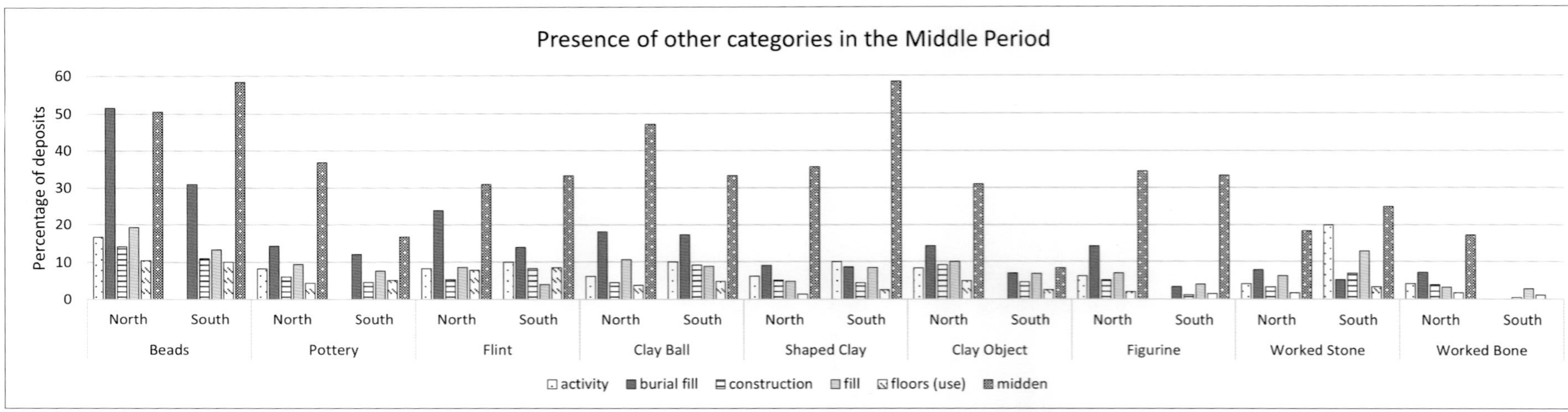

Figure 3.8. Presence of ubiquitous (above) and other (below) material categories in the Middle period deposits in the North and South Areas.

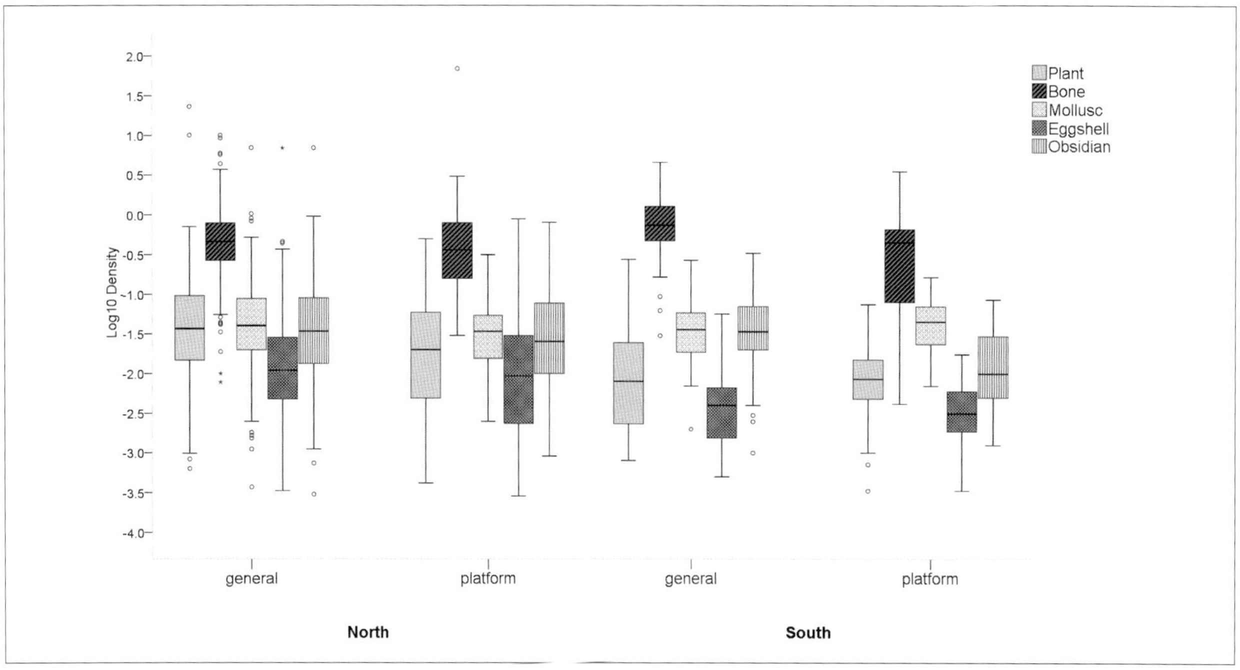

Figure 3.9. Log10 densities of ubiquitous material in general building floors and platforms from the Middle period in the North and South Areas.

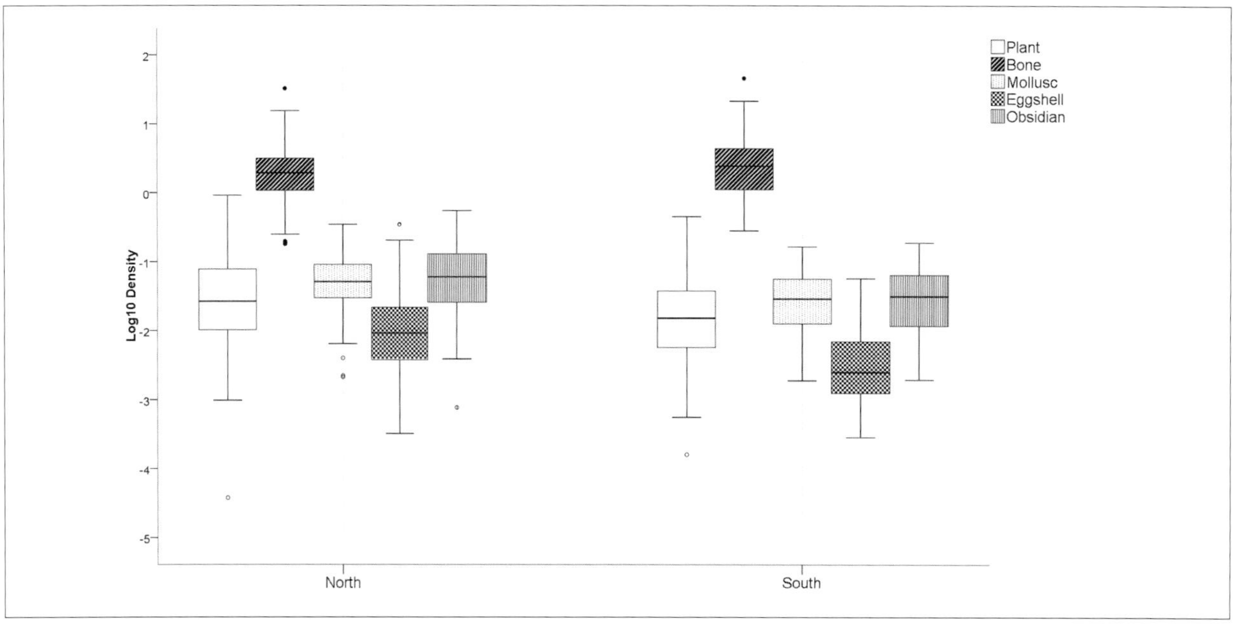

Figure 3.10. Log10 densities of ubiquitous material in Middle period burial fills in the North and South Areas.

The spatial and contextual distribution of different material categories does not exhibit major differences between the two areas of the settlement, which is suggestive of similar activities and use of material in both Areas of the settlement. However, it is important to note a tendency of the North Area deposits to contain more material in the Middle period, whereas in the Late period, higher densities have been observed across different deposit types in the South Area.

Comparison of individual buildings

The above-presented patterns do not take into account variability between individual buildings; rather, they show how the deposits were accumulated in the two areas. In order to get closer to the more nuanced differences among buildings, densities of material in the floor deposits of individual buildings were compared. As buildings were not excavated equally, only buildings from which at least 100 litres of soil were sampled were taken into consideration.

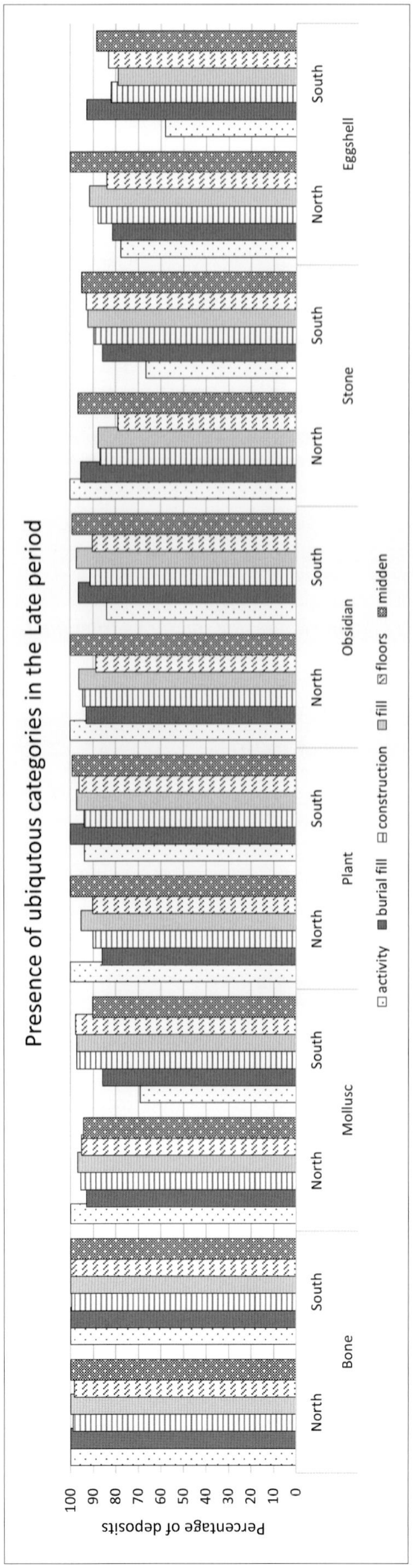

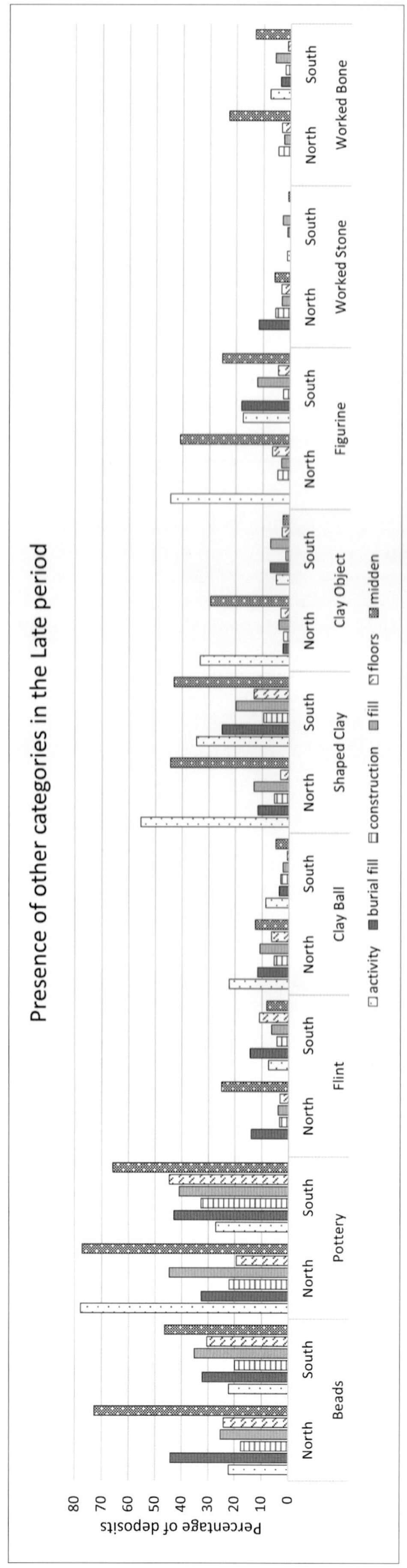

Figure 3.11. Presence of ubiquitous (above) and other (below) material categories in the Late period deposits in the North and South Areas.

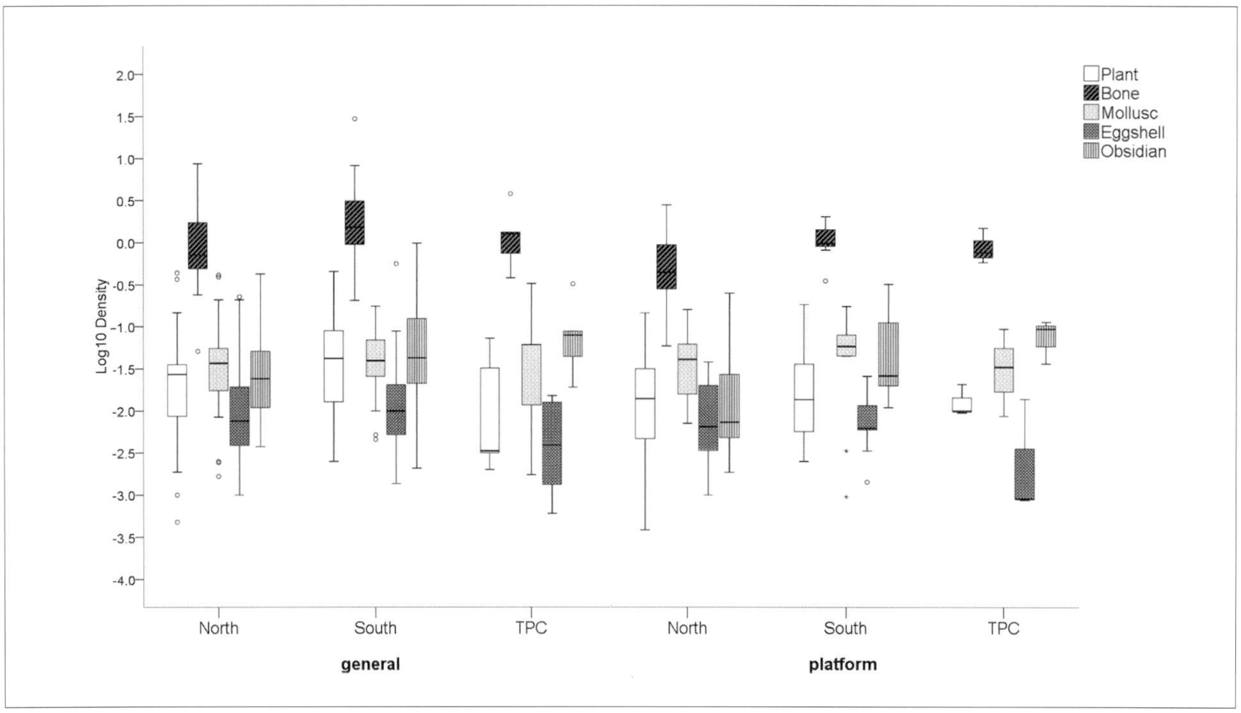

Figure 3.12. Log10 densities of ubiquitous material in general building floors and platforms from the Late period in the North and South Areas.

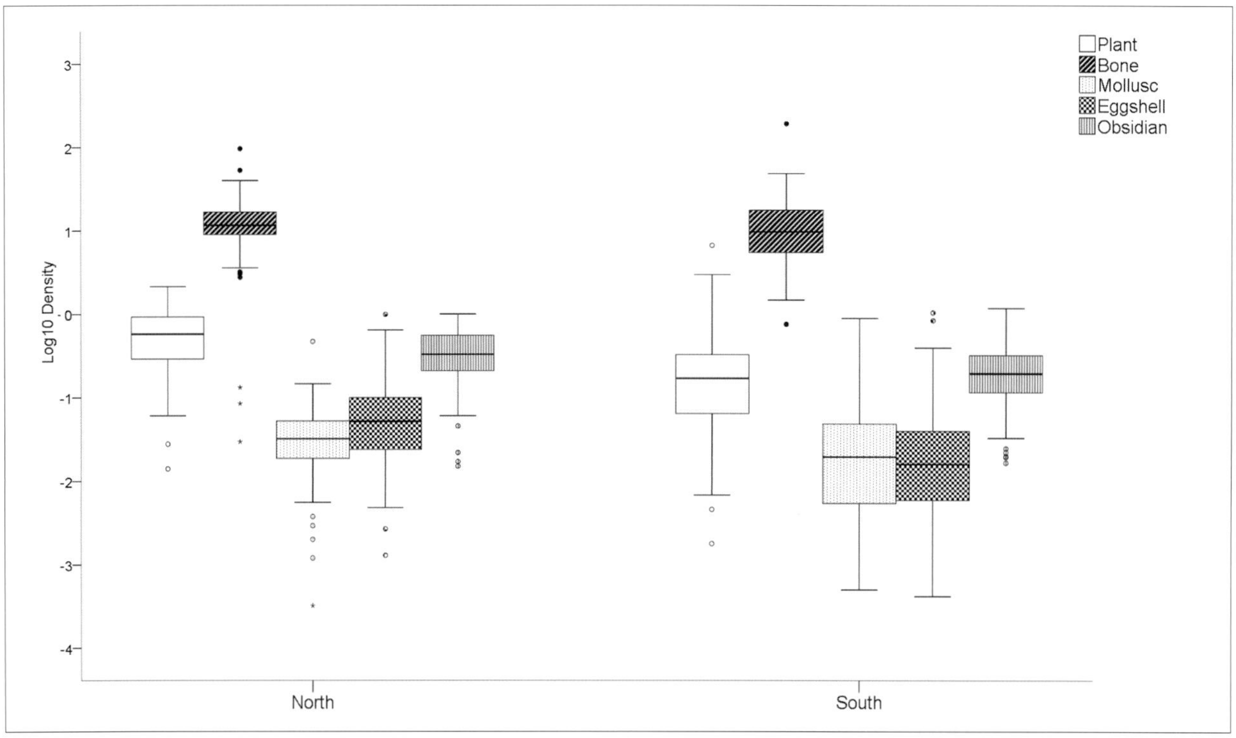

Figure 3.13. Log10 densities of ubiquitous material in Late period midden deposits in the North and South Areas.

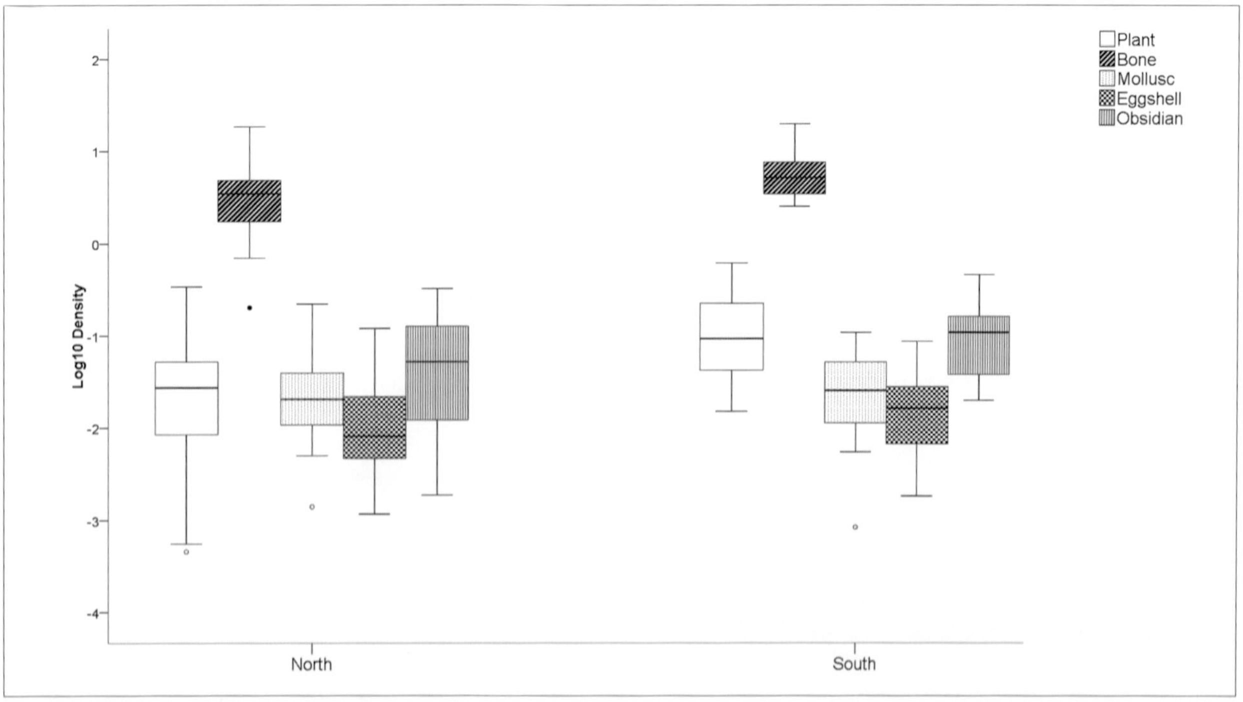

Figure 3.14. Log10 densities of ubiquitous material in Late period burial fills in the North and South Areas.

One of the aims of the previous segment of excavations was to establish whether there is any variability in different types of buildings; that is, to compare building inequality by comparing buildings defined as 'history houses' (Hodder, Pels 2010), buildings with multiple burials or elaborate buildings in order to see if any of them stand out in any way (Hodder 2013b; 2013c; 2015; 2014b). Whereas tentative patterns did emerge, various aspects of material culture did not produce any substantiate evidence that would argue for clear differentiation between the types of buildings (Hodder 2013a). Similarly, distribution of HR material did not produce any straightforward patterning, though there were signs of slight variability in different house types; more specifically, more elaborate buildings contained lower diversity and density of material in floors than other houses did (Mitrović & Vasić 2013). The addition of new samples in the recent excavations confirmed this; the median densities of ubiquitous materials in the floor deposits do not exhibit strong correlations with either the size or the elaboration of buildings. As a matter of fact, material densities display a negative correlation, albeit a weak one (r value is between -0.1 and -0.3), between the median density in the floor deposits and both size and elaboration of buildings (table 3.14). A moderate negative correlation (r=-0.5) has been observed between the median density of Obsidian and the Elaboration index; that is, floors in more elaborate buildings tend to have a smaller median density of Obsidian.

	Elaboration index	*Size*
Plant	-0.1	-0.0
Bone	-0.3	-0.2
Mollusc	-0.1	-0.1
Eggshell	-0.2	-0.1
Obsidian	-0.5	-0.2

Table 3.14. Correlation of material densities, Elaboration index and building size.

Correspondingly, the distribution of different material categories in the floor and platform deposits varies and does not seem to be strictly correlated to the area of the settlement or the period during which buildings were occupied (figs 3.15 and 3.16). However, neighbouring buildings belonging to the Middle period in the South Area tend to have smaller densities of the ubiquitous material, witnessed particularly in B.96, with the median densities of Plant, Eggshell and Obsidian being 0, whereas the lowest median density of Bone on site has also been noted in this building. This does not mean, however, that there these material categories are completely absent. Rather, floors of this building have produced an unusually low quantity of material.

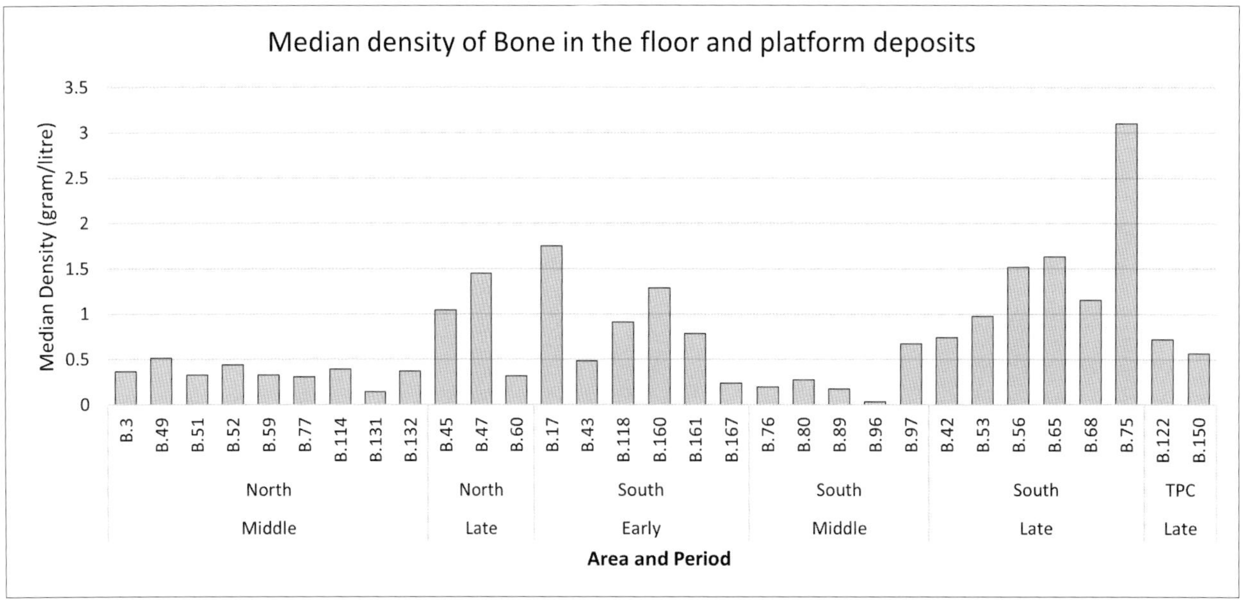

Figure 3.15. Bone density on floors and platforms in individual buildings.

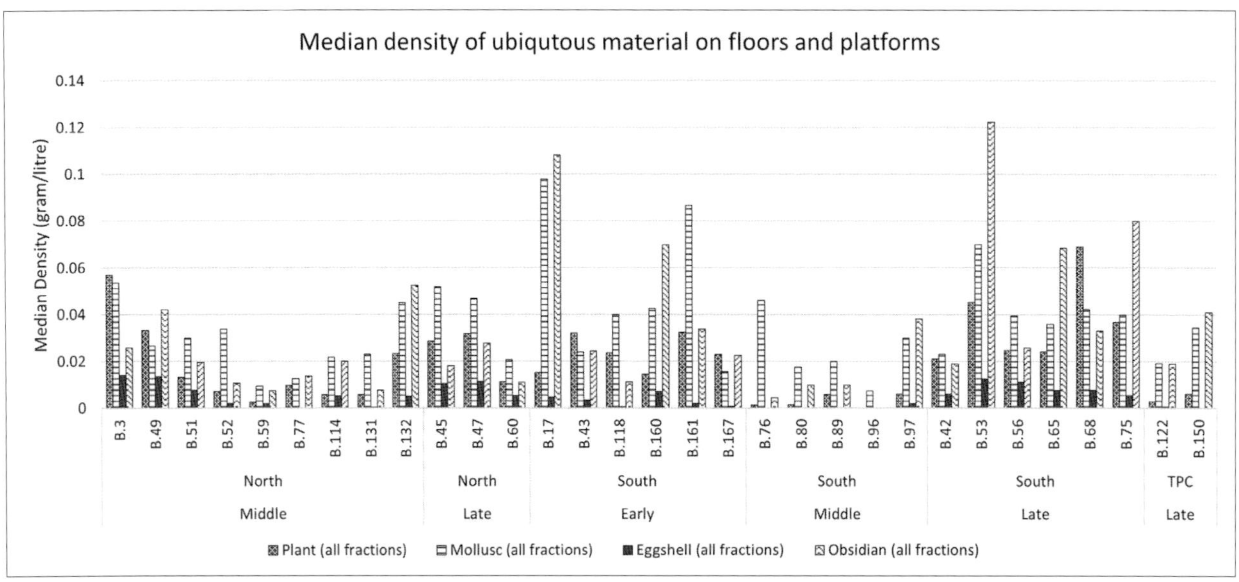

Figure 3.16. Plant, mollusc, eggshell and obsidian densities on floors and platforms in individual buildings.

Amongst the less common categories, median density of Clay Ball is above 0 only in B.161, belonging to the Early period. The median density of Pottery is above 0 in B.122 in the TPC Area, B.65 and B.68 in the South Area and B.45 in the North Area, all belonging to the Late period. This corresponds to the previously observed pattern that clearly demonstrates that the use of pottery intensifies in the Late period, whereas clay balls were used more frequently in the Early period. Noteworthy also is that the median density of Flint is above 0 in the Early period buildings B.17 and B.160. Surprisingly, the median density of Beads is the highest (and above 0) in B.17 and B.75, whilst although B.132 provides indirect evidence of bead manufacture (this volume, Chapter 10), the density of beads in their floors is the same as elsewhere (that is, median density=0).

However, these numbers only represent an overall 'inventory' of each building throughout its entire occupation and do not take into account nuanced differences that must have existed in the use of buildings from one occupation phase to another. Cessford observed that rare occasions when artefacts are left lying on the floors are usually from the period of the use immediately prior to the abandonment of the house (Cessford 2003; 2005). Heavy residue, on the other hand, demonstrates a slightly different picture, a consequence of a smaller size of material that had likely become trampled into floor deposits. There is much variability, but in some cases, it is actually the earlier phases of buildings that contain more material, as seen in the example of B.132. The earliest phases (B.132.1 and B.132.2) of this building contain significantly higher median densities of all ubiquitous materials but Mollusc than the last occupation phases do (fig. 3.17). Similarly, samples taken from the floors and platforms belonging to the early stages of the occupation of the roughly contemporary B.80 also have higher densities of Bone, Obsidian and Mollusc. B.49 with its complex occupation sequence also provides a good example of the observed variability throughout the occupation phases. General floors from the early phases contain a slightly higher Bone density, but on the other hand, its density increases on platforms through time, with phases 49.5 and 49.6 containing a higher density of Bone on platforms than exists in the floor deposits, which is unusual. Eggshell density is also higher in the later phases (49.5–49.7) in both platforms and floors. Whilst these observations are based on a very small number of samples, they illustrate the variation in the densities and point to the dynamic use of buildings at Çatalhöyük.

Variability in the presence and densities of material is not limited to buildings, and there are also variations across different external areas, as well as evidence of different activities taking place throughout their use-life (see Volume 12, Chapter 33 for details).

Diachronic distribution

Temporal distribution exhibits a certain level of consistency in the occurrence of the ubiquitous material categories through time with some minor variability, perhaps best seen in the increased presence of Eggshell in fill and floor deposits of the Late period (fig. 3.18). On the other hand, non-ubiquitous categories have a higher variability in their distribution over time, especially notable in the occurrence of material in burial fills and midden deposits. The Early period burial fills seem to contain artefacts such as Beads, Flint, Clay Ball, Clay Object, Shaped Clay, Worked Bone and Worked Stone more commonly, and the presence of these categories is significantly lower in the burials from the later periods (fig. 3.19). Not only is Flint more common in burial fills and midden deposits, but also in building floors.

The negative correlation between the use of Clay Ball and Pottery categories is perhaps the strongest pattern in the diachronic distribution on site (Atalay 2005), and this is also evident in heavy residue. Clay Ball is significantly more frequent in all deposits in the Early period, after which its presence drastically drops. In contrast, the presence of Pottery intensifies in the Late period, with, for example, over 70% of midden deposits containing Pottery fragments. The use of other clay artefacts, however, intensifies in the later levels, witnessed in the augmentation of the occurrence of clay beads, figurines and other clay artefacts (see Chapters 7, 8, 9 and 15), and this is also evident in the increased presence of Shaped Clay in the Late period non-burial deposits.

Although the presence of ubiquitous categories is fairly similar through time, the material densities vary across the deposits. The Early period floors contain higher densities of Obsidian and Mollusc than the Middle and Late period floors (fig. 3.20). On the other hand, density of Eggshell increases in the Middle period deposits (though it should be noted that the highest median density of Eggshell is 0.01 in each building). Although the median density of Plant is fairly similar, the Middle and Late periods seem to have a higher variation in their floor deposits, likely to be a consequence of a few individual buildings containing relatively higher or lower densities (figs 3.15 and 3.16). Bone displays a slightly different distribution, with its density being similar in the Early and

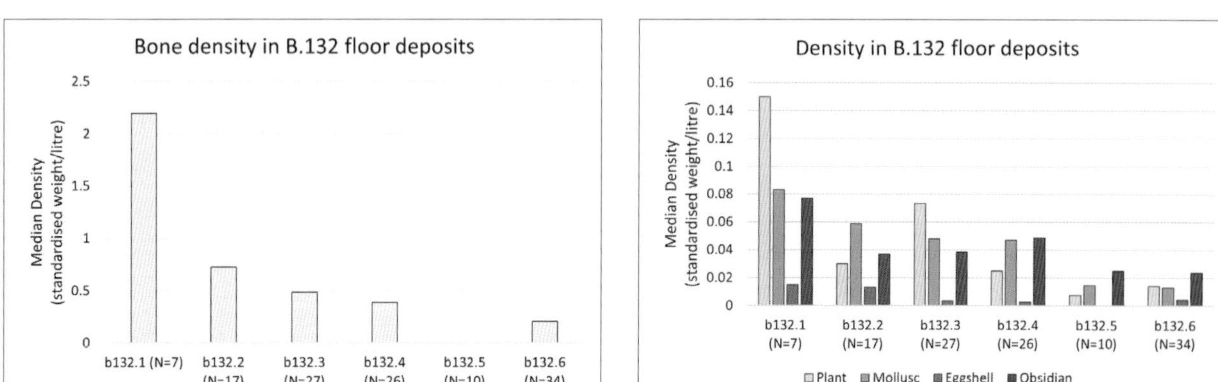

Figure 3.17. Density of bone (left) and plant, mollusc, eggshell and obsidian (right) in floor deposits of B.132 across its different occupation periods.

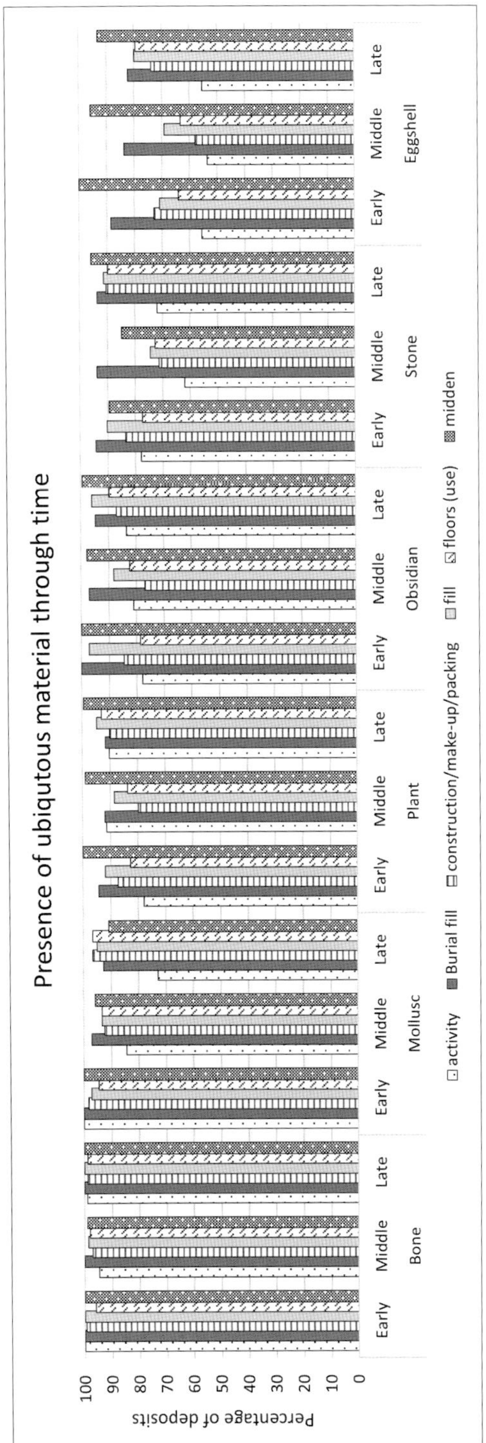

Figure 3.18. Presence of ubiquitous material through time in each deposit type.

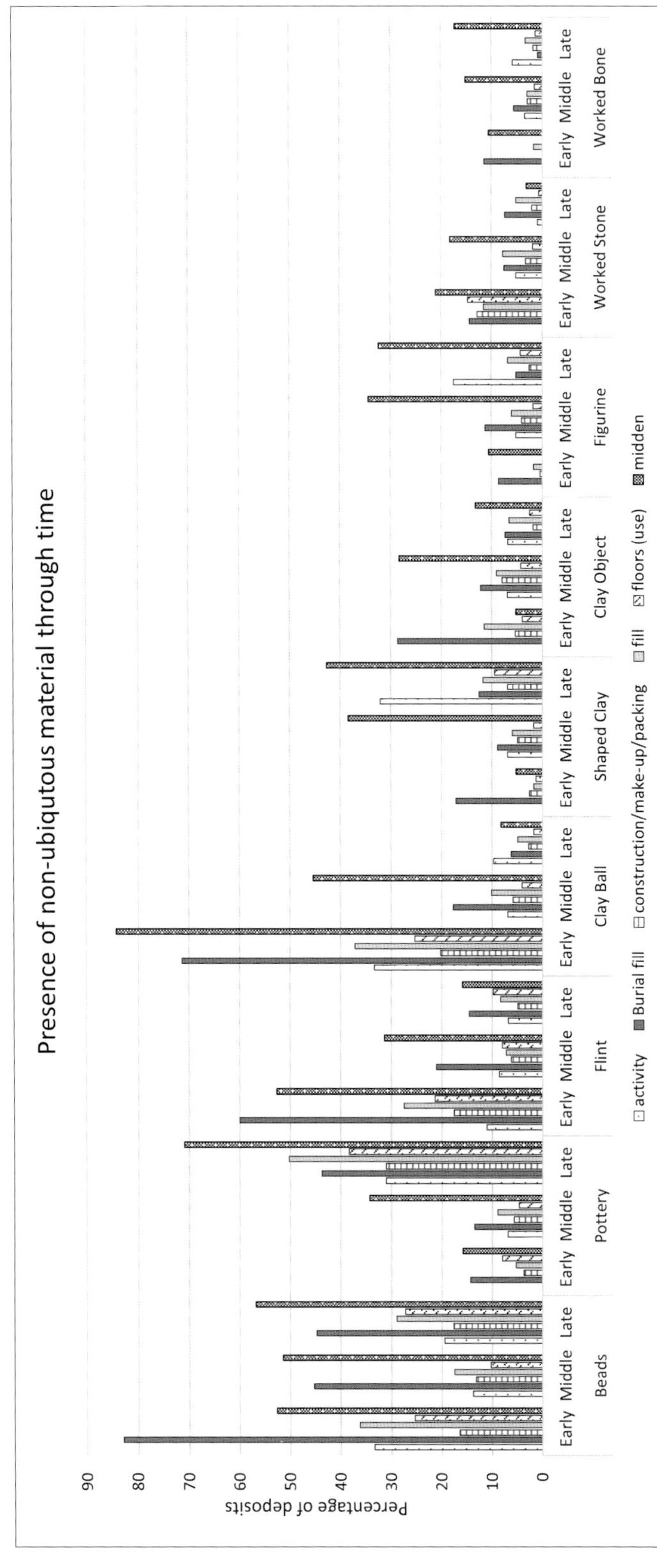

Figure 3.19. Presence of non-ubiquitous material through time in each deposit type.

Late periods, and considerably higher than was found in the Middle period floor deposits. As shown in figure 3.15, median density in buildings from the Middle period never exceeds 0.5g/l, whereas in the Late period, median density of Bone is, with the exception of B.60, always higher.

Different distributions of material were also noted in burial fills over time, with the Early period burials containing higher densities of Eggshell and Obsidian, and there is also a notable decrease in the Mollusc density through time (fig. 3.21). Whereas a decrease of the Mollusc density has also been observed in the fills of non-burial features, densities of other materials increase through time and are at their highest in the Late period fills. Similarly, there is a statistically significant increase ($p<0.04$) in Plant and Bone densities in the Late period midden deposits. In contrast, Obsidian density in the midden deposits is almost identical in each period, whilst samples from the Early period tend to contain a smaller density of Eggshell.

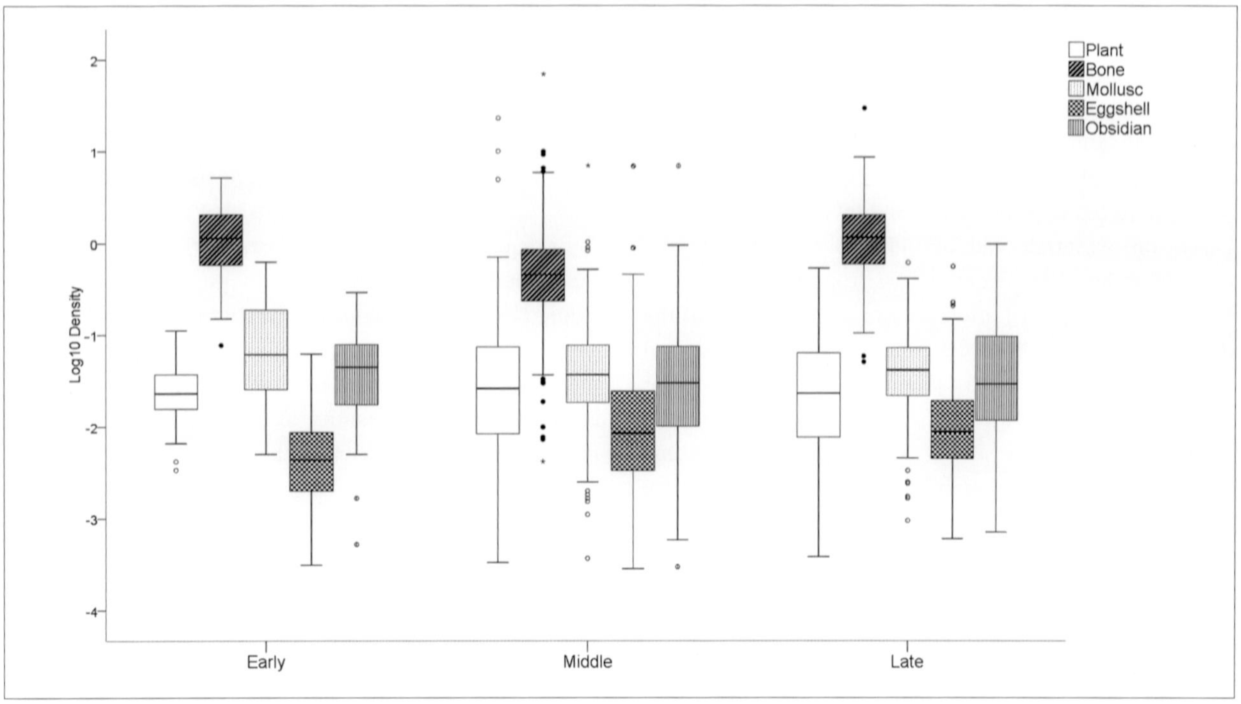

Figure 3.20. Log10 densities of material in the floor deposits through time.

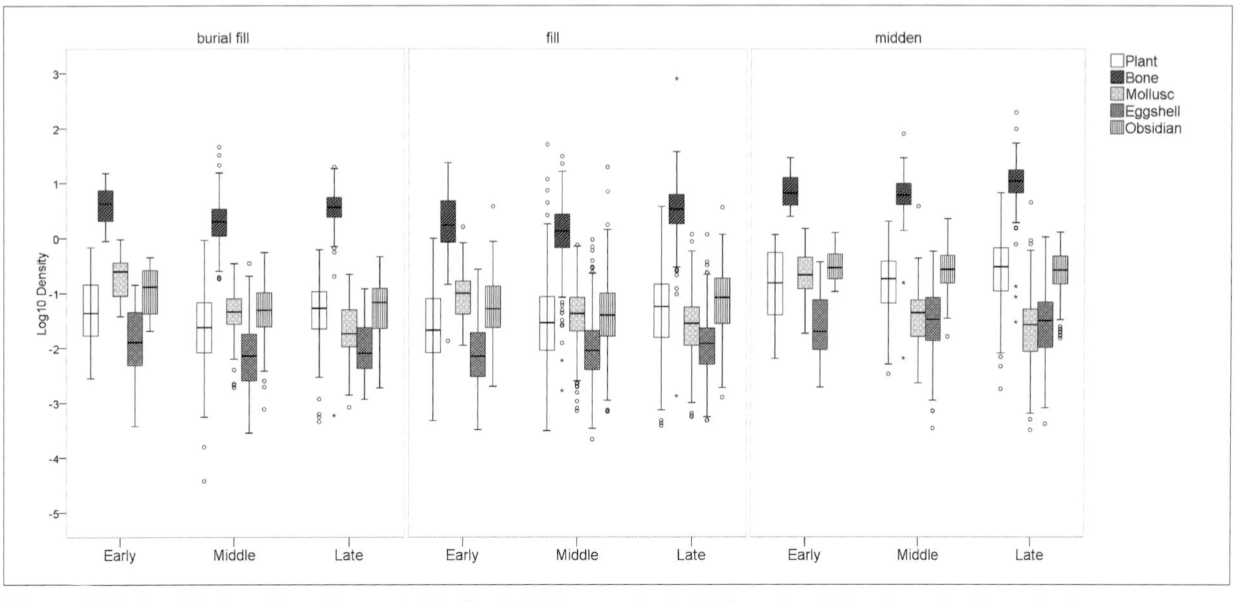

Figure 3.21. Log10 densities of material in burial fills, non-burial fills and midden deposits through time.

The diachronic distribution of the common material categories at Çatalhöyük demonstrates a relative continuity, but the slight variability observed in different data categories is noteworthy. Although a decline in the Mollusc density, most likely reflecting the decrease in the use of the most common species, *Unio*, is evident in each type of deposit, there is a variation in the distribution of other material categories across the deposits through time. These are the overall patterns through time; however, as discussed above, there is a noted variability between the areas of the settlement, as well as across individual buildings and throughout their occupation phases.

Discussion

Processing and analysis of HR Material provides an expedient instrument for comparing the densities of material across deposits, areas and levels; however, it is based on variation and similarities in quantities of material rather than on qualitative properties such as the composition, types of artefacts etc. Interpretation of the distribution of HR material is not straightforward, and different factors must be taken into consideration. For example, as Cessford notes, comparison of densities of material in floor deposits is over-simplistic, given that some floors lasted longer than others (Cessford 2005b). Consequently, higher densities of material could be a result of material accumulating over a longer period of time, rather than showing in situ activities. If this is true, then HR could also be a useful tool in determining the duration of use of building floors. This, of course, needs to be investigated further by correlating the HR densities with the information obtained from other datasets. For example, it has previously been suggested that buildings from the early levels had a longer duration of use, whereas certain Late period buildings were especially short-lived, such as the sequence B.65-B.56-B.44 in the South Area, in relation to which it is believed that all three buildings together lasted for only about half a century (Hodder 2006c). As figures 3.18 and 3.19 show, whilst some buildings from the Early period do have a relatively higher median density of some HR materials, the aforementioned B.65 and B.56 also contain larger quantities of material than expected. Therefore, although Cessford was right to bring up the duration of the use-life of buildings as an important aspect of the rate of accumulation of material, it is evident that the direct correlation cannot be simply assumed, as there are other factors that need to be considered.

Despite the aforementioned unresolved issues, flotation and HR are useful methods at Çatalhöyük, and processing of these samples enables a detailed analysis of the distribution of material across all excavated deposits. As discussed above, there are clear patterns in the contextual distribution, such as external midden and floor deposits having significantly more material than any type of internal floors and intramural contexts in general, with perhaps the exception of burial fills in some cases. Amongst the intramural deposits, burial fills have the highest densities of material, followed by fills of other features, whereas construction and floor deposits have both the lowest presence and the lowest density of each of the material categories. Surprisingly, construction and floor deposits overall do not exhibit major distinctions, which could point to the fact that the material recovered from the floor deposits does not necessarily reflect primary deposition. Rather, this material (especially in cases in which it occurs in lower quantities) could have formed part of the floor make-up. This needs to be explored further, as the construction data category also includes units that are not related to floor deposits (for example, plaster surfaces on walls).

Floors and platforms do not greatly differ in the studied assemblage, which is likely to be a consequence of the existence of different types of floor deposits, with the floors in the northern part of buildings being usually clean and a higher quantity of material having accumulated due to various activities taking place in the southern area of the building.

Furthermore, buildings show a variation in the density of materials that is not strictly related to the area of the settlement, or the period in which the buildings were occupied. In addition, there is a clear variation in the densities of material throughout the use of individual buildings; this is definitely worth exploring further, as it could potentially be linked to the intensity of activities taking place throughout the use-life of each building.

Another clear pattern is the continuity observed in the accumulation of material over time. Bone is by far the most ubiquitous material on site, evident both in its presence in almost every excavated deposit and in the quantity in which it exists. This is consistent throughout all levels of the occupation of the settlement. On the other hand, there is a noted decrease in Mollusc and an increase in Eggshell through time.

The lack of stronger differentiation across some deposit types or across individual buildings and between the areas of the settlement is quite an interesting pattern. Some variability does exist between the two excavation areas: the North Area deposits are richer in materials in the Middle period, whereas this changes in the Late period with the tendency of the South Area deposits to have higher densities of ubiquitous materials. Nonetheless, the observed differences are not large enough to reflect a clear-cut differentiation between the people occupying these areas. If, for example, residents from one area had more access to resources or were engaged in more intense production, we would expect a much higher difference in the densities, in either indoor or outdoor deposits, which is not the case. This corresponds well with the other lines of evidence that do not support any major distinction between the material culture of the North and South Areas (see Volume 13, Chapter 1).

Although the usefulness of HR processing at Çatalhöyük is manifold, HR material only rarely directly represents traces of in situ activities that can be interpreted as such. HR analysis works very well as a provider of contextual information about the distribution of material on site, as well as a check on the overall patterns observed in the different aspects of the distribution of material. However, on its own, HR analysis cannot offer more in-depth and meaningful information. Due to the coarse nature of the data, the results presented in this chapter can only produce tentative interpretations and the data need to be combined with qualitative information from other databases in order to obtain a better understanding of both the site formation processes and the past activities that were taking place in the settlement.

Acknowledgements

I would like to thank Ian Hodder and all the members of the Çatalhöyük Research Project, especially the heavy residue team. Heavy residue analysis would not have been possible without the diligence of the sorting team. Without Hatice Tokyağsun, Fatma Eken, Fadimana Sivaz and especially Ebru Sivaz, who also helped out with sieving, labelling and weighing, heavy residue processing would not have been completed by the end of the 2017 season. Thanks are due to Şenay Yaşlı and Saliha Sivaz who over the years trained the new sorting team members. I am also grateful to the Excavation, Finds and Archaeobotany teams for their support, especially Dragana Filipović, Elizabeth Stroud and Ceren Kabukcu for keeping everything in order, as well as Mevlüt Sivaz and Hüseyin Yaşlı for their excellent flotation. Finally, I would like to thank Jovana Tripković, Karolina Joka and Talu Emre Tüntaş for their help with heavy residue processing and organisation.

4. Pottery production technologies and quantified analysis

Duygu Tarkan

Introduction

Ceramics contain sensitive indicators of human decision-making processes. For this reason, exploring the nature of ceramic production is important in terms of understanding the knowledge/skills, identity and traditions of the potter. This chapter presents the results of recent studies on the ceramic production technology of Çatalhöyük, from the selection of raw materials to finished vessel. A comprehensive picture of the East Mound ceramics will be drawn up, with a review of the various studies over the last 25 years. In this framework, following detailed characterisation of pottery groups, a comparison of different wares and forms in different types of deposits from different levels will be presented to highlight temporal trends. This study documents materials retrieved from the North, South, and TPC Areas as well as the TP and IST Areas to provide material integrity. Material from the TPC and GDN Areas will be published in other volumes, although they are briefly included in this chapter for comparative purposes.

Background

James Mellaart, who led the first excavations in the 1960s, followed a traditional method in pottery studies: he placed the material, chronology and regional comparisons in the foreground (Mellaart 1966). Mellaart stated in his first reports that pottery was not seen between Levels VIb and VII, and that Level X and lower were completely pre-pottery; however, in the last season of excavation, he revised this interpretation to indicate that all levels of Çatalhöyük contained pottery (Mellaart 1966: 170). Mellaart's collection of only diagnostic pieces, such as rims and bases, demonstrates that the studied and published materials did not represent all ceramics recovered from the 1960s excavations. Methodologically, Mellaart evaluated the pottery via general definitions relying on differences in surface colour, fineness or coarseness, but he did not distinguish ware groups.

Jonathan Last examined the first pottery finds from the second period of excavations, started by Ian Hodder. Last studied 1,216 ceramic sherds recovered between 1995 and 1999 using the sherd-by-sherd recording method, wherein each piece was recorded according to ware and form specifications. Although he worked on a limited number of materials, his study added new features to the variety of recognised forms in the pre-Level VI layers of the site (Last 2005a).

Within the scope of her 2003 PhD thesis, Serap Özdöl studied all the ceramic materials, including complete vessels, deriving from Mellaart's excavations that are located in the Konya Archaeology Museum. A total of 750 ceramic sherds were classified as body sherds according to colour and property criteria, while diagnostic sherds were classified according to form properties (Özdöl 2006: 127).

Mellaart, Last and Özdöl provide an outline of Çatalhöyük Neolithic pottery and all obtained similar results. In none of the studies, however, were ceramics recorded in a database.

Nurcan Yalman, Hilal Gültekin and the author of this chapter carried out studies on Çatalhöyük ceramics between 2003 and 2011. Yalman et. al. (2013: 144) examined the ceramics obtained from the 2002–2008 excavations, as well as the earlier material from the 1960s and 1995–1999 excavations, previously studied and published by Last. The first goal was to establish a detailed database. Instead of seeing ceramic sherds as individual 'small finds', the authors sought to understand the logic behind the production; form and ware groups were delineated on this basis. Many questions arose during this process and new analytical methods were applied, among them experimental studies on organic residues (Pitters et al. 2013), petrographic analyses (Doherty, Tarkan 2013) and cooking functions (Yalman et al. 2013).

The database was designed to evaluate all ceramic sherds in different categories, no matter how small their size. In addition to sherd counts, information such as weight, condition, abrasion/weathering, secondary function and residues were recorded in detail.

Since 2011, Yalman and her colleagues' data have been maintained and developed. In addition, an MA thesis titled *The Investigation of Clay Use and Resources of the Neolithic Pottery at Çatalhöyük East Mound* (Tarkan 2015) helped to determine the details of pottery production. The aim of that study was to investigate the selection and use of various clay sources within the context of Çatalhöyük East Mound Neolithic pottery. Furthermore, the study looked at the environmental changes surrounding Çatalhöyük during the 1,100 years of its Neolithic occupation and how society responded to these changes through pottery. On the basis of this, 642 pottery samples were analysed by petrographic methods to understand their micro-textural features, whilst the raw clay materials used were identified in detail. By examining the mineralogical composition of the raw clay materials, it was possible to suggest their provenances. Results were interpreted within an archaeological context in which relationships between pottery, people and environment were explored in greater depth (Tarkan 2015).

Methodology

The data presented in this chapter aim to compile the results of the above-mentioned Mellaart, Last, Özdöl, and Yalman-Tarkan-Gültekin studies, as well as Tarkan's MA thesis and all ceramic analyses carried out since 2008. For this purpose, (1) all the body and diagnostic sherd data are used, (2) terms such as ware, form, surface treatment etc. are standardised, (3) general groupings are formed, which include all pottery production stages and features instead of having separate groups for fabric, wares, types and functions, and (4) all these data are characterised according to both temporal and spatial distributions.

All pottery obtained from the recent excavations was recorded with full detail in the database. The TPC material, however, was only counted and weighed according to Neolithic, Chalcolithic and Post-Chalcolithic categories, because the upper levels of the area were truncated for many reasons. Later, only the Neolithic TPC units selected for publication were examined in detail. The numbers of materials examined so far (including TPC Area materials) can be seen in table 4.1.

All the ceramic data recorded in the database were used in the section below on diachronic change. For this purpose, all pre-2009 database records were merged with the new database based on ware groups. To assess context-related variation, randomly selected sub-samples chosen by the specialist were used. Ceramic pieces smaller than 2cm were not included, because ware group attributions were not made for such pieces.

Ceramic Ware Groups

Ceramic production starts with decisions concerning intended function and, accordingly, the selection of raw materials. Pottery found in a settlement should have regionally and geologically characteristic fingerprints according to the geological structure of the region in which it is produced. These traces are chemical and mineralogical features typical for that region. We can say that the pottery produced in a certain settlement should most probably originate from one or more clay sources specific to that area (Yalçın, Seifert 1994: 17–18). In order to find the source(s) of raw materials, it is necessary to determine the parent minerals and rocks from which the clay is formed. Rocks are composed of the combination of one or more minerals. For this reason, mineral identification is vital. Following identification of minerals and affiliated clay(s), it is possible to write the story of production steps such as preparation, shaping and firing.

In work on ceramic production in 2011 and onwards, differences in raw materials were identified, and these differences were found to correlate with changes in ceramic function and food consumption habits. Ceramics were grouped into seven different wares, considering all similarities and differences (fig. 4.1 and table 4.2).

Light Silty Plant Tempered Ware (SILTY-P)

Ceramics collected under this group are the earliest examples in the site. They comprise roughly shaped shallow pots with straight and thick walls, which are characterised by chopped plants (mostly chaff) deliberately added to alluvial clay. In the sherd sections, 80% of the examined area is black/dark brown clay body; the remainder is horizontally shaped, differently sized pores ranging from 0.01 mm to 9 mm (fig. 4.2). These horizontal

	Total Weight		*Total Sherd Count*	
Neolithic	g	%	no.	%
Body sherds	223,497.4	30.0	19,824	50.2
Diagnostic sherds	148,402.0	19.9	4,912	12.4
Unidentified sherds	12,149.4	1.6	3,370	8.5
Non-Neolithic	360,847.5	48.4	11,391	28.8
Total	744,896.3		39,497	

Table 4.1. Total studied ceramics.

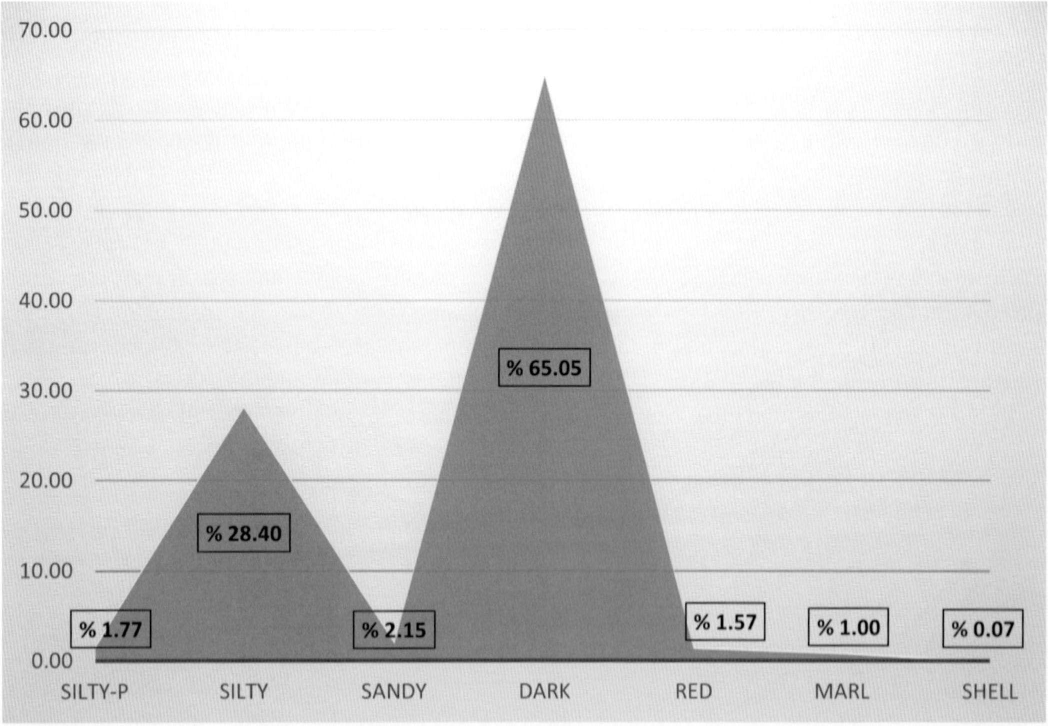

Figure 4.1. Ware group frequency in all ceramics.

Ware Group	*Abbreviation*	*Fabric Description*	*Clay Provenance*	*Form*	*Stratigraphy*
Light Silty Plant Tempered	SILTY-P	Calcerous silty clay with chopped plant (possibly straw)	In vicinity of the site	Roughly made, thick and straight-walled small jar (bowl) types	Only early levels
Light Silty	SILTY	Calcerous silty clay with chopped plant (possibly straw)	Catchment area of the Çarşamba and May Rivers	Bowl types	From later phases of early levels through final levels
Sandy	SANDY	Sedimentary calcerous clay, coarser grains than silty	Catchment area of the Çarşamba and May Rivers	Bowl types	Early to late levels
Dark Gritty	DARK	Non-calcerous volcanic clay	May River catch-ment area, Erenler-Alacadağ mountain complex region	Hole-mouth deep jars (cooking pots)	Midlle to late levels
Red Paste	RED	Metamorphic clay, phyllite-rich with vein quartz	Taurus Mountain foothills	Oval-shaped bowls with basket handles	Middle to late levels
Marly White	MARL	Soft calcerous marl including fossils	All around Konya basin	Small bowls and miniature vessels	Late to final levels
Shelly	SHELL	Clean clay filtered in the small ponds with included natural shells	Around the site	Very rare – no diagnostic sherds	Only final levels

Table 4.2. Ware groups of East Mound Neolithic ceramics.

pores, left behind by plant materials, are evident both on the surfaces and in the paste, and they can be clearly seen even with the naked eye. The burnt plant contribution was charred and preserved on the surface and in the paste, and the clay body is optically active. This group of ceramics wəs thus fired at temperatures lower than 500°C.

Surface colours tend to yellowish buff and cream. The colour of the body clay varies from blackish to light brown. This colour indicates that there are also natural organic materials in the contents. This type of clay is called smectite: it is rich in magnesium and iron and displays different pore thicknesses formed by the penetration of water and organic substances between clay layers (Akıncı 1968: 65). This causes shrinkage and deterioration in the pottery. For this reason, the creation of pottery without additives, such as straw or chopped plants, is very difficult in Neolithic technology. Also, because of the plasticity effected by this type of porous structure clay, it is not possible to make thin-walled pots with it (Doherty, Tarkan 2013: 180). There are no visible inclusions other than burnt plant remains. In other words, the paste textures of the sherds are quite homogeneous. SEM-EPMA and thin-section analyses showed that the minerals, which could not be identified by eye because of their small size, are quartz, chert and limestone (fig. 4.2). In addition, the chemical composition of the clay body (Si> Al> Ca> Mg> Fe> Na> K) also contains high levels of calcium.

During the KOPAL surveys conducted by Neil Roberts and his team, it was argued that the Çarşamba Delta had two main fillings: the upper and lower alluvia. According to their study, the lower alluvium occurred around 10,000–9500 BC, is brown/black in colour and is made of organic material-rich clay (Roberts et al. 1999: 625). In addition, between 2007 and 2009, Chris Doherty and Michael Charles studied cores taken from Çatalhöyük and its surroundings to investigate the location of different sediments in the immediate vicinity of the settlement. Doherty and Charles have proved the existence of upper and lower alluvia deposition, but argue that, contrary to Roberts' claim, the differences in stratigraphy and location of these soils may have occurred at different thicknesses due to topography rather than climate change (Doherty et al. 2008: 263). This explains the slight changes in the paste of the ceramics produced with local clay. Because of this, it is considered that minor differences may not have been made consciously as a property of the ware, and thus the ware has been evaluated as a single ware group which has not been subdivided. Lower alluvium or 'backswamp' clay was used in early level mudbrick at the site (Tung 2013: 75). For this reason, it can be said that the raw materials used in mudbricks may have been tried in pottery production as well.

These low-temperature-fired ceramics were very fragile, and all the pieces were found fairly worn. This makes it difficult to define the form types used in the group (fig. 4.3). However, diagnostic sherds suggest they comprise a single form named 'Straight Walled Bowls', consisting of flat-base, shallow, cylinder-shaped small bowls.

Light Silty Ware (SILTY)

This is the first group to use the natural state of the clay in the production of Çatalhöyük pottery, without the addition of temper. For this reason, the emergence of this ware group can be interpreted as a technological change (fig. 4.1).

The calcareous silty clay, the raw material that defines the group, derives from the alluvial soils formed by Çarşamba River sediments (Roberts 1999; Boyer et al. 2006; Doherty 2013; Charles et al. 2014). These soils are highly calcareous because they include high proportions of marl deriving from the Pleistocene Konya Lake (Doherty 2013: 55).

Macroscopic examinations reveal buff surface colours and grey-toned paste (fig. 4.4). Organic materials burned during firing have led to ash structures in the paste and colouration in lead tones. Inclusions observed in thin-section analysis have a distribution ratio of 25–60% throughout the paste. Common inclusions found in similar proportions in all samples include limestone, quartz, calcite, chert, phyllite, alkali feldspar, garnet,

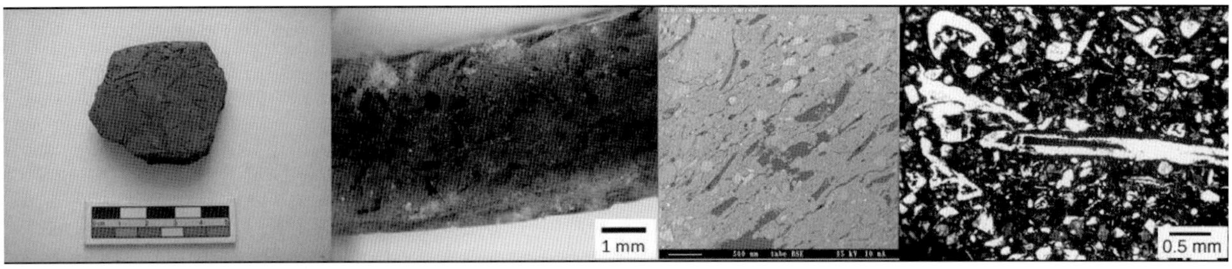

Figure 4.2. Light Silty Plant Tempered Ware (14604) (from left to right: sherd, under the stereo-microscope, SEM, polarised light microscope).

andesitic and dacitic rock fragments, plagioclase feldspar, calcium amphiboles, clinopyroxene, biotite mica, micro molluscs (Bar-Yosef 2013), marl pellets, clay pellets, sandstone and zircon. Minerals and rock fragments show no thermal degradation; hence, pots were fired in temperatures not exceeding 600°C (fig. 4.5).

Figure 4.3. Light Silty Plant Tempered Ware forms.

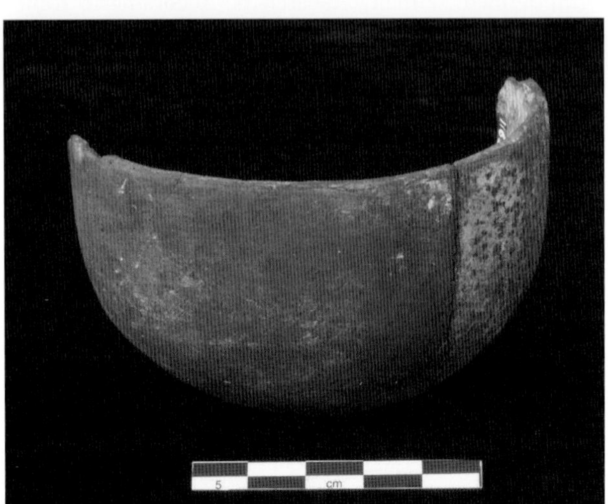

Figure 4.4. Light Silty Ware.

Two different types of limestone are encountered in the paste. One of these structures is 'dolomitic limestone', characterised by a 10–35% (by mass) magnesium carbonate ($McCO_3$) additive to the calcium carbonate ($CaCO_3$). Dolomitic limestones in this structure can also be seen in the vicinity of the settlement (van der Plas, Schoorl 1970: 213). The second limestone type includes quartz fragments in its texture. Such limestones are found in the Bozkır region, from where the Çarşamba River springs; that is, in the upper terraces of the river. Since the content sizes are very small, it is assumed that they were naturally transported over long distances. For this reason, it can be said that this group's clay accumulates sediments carried both from the place where the Çarşamba River springs and from a source very close to the settlement, which constitutes the lower terraces of the river. In SEM-EPMA analyses, clay pellets were found in several samples in different structures. Some of these clay pellets contain particles of human / animal bone. It is possible to say that the source of the raw materials from which the ceramics were produced is very close to the site, because of the inclusion of cultural materials likely occurring in the clay as a result of erosion. It is likely that the calcareous silty clay was used because the backswamp clay, which occurred very close to the base of the mound, had been depleted. Light Silty Ware is a dominant raw material for ceramics produced with local clay. It was also used throughout all levels.

Sandy Ware (SANDY)

Examples in this group are characterised by small amounts of clay and high sand content. They are easily distinguishable by their sandy texture and can even be recognised macroscopically. In thin-section analyses, all samples include limestone, quartz, chert, garnet, andesitic and dacitic rock fragments, biotite mica, mollusc shells (gastropod and ostracod), and marl and clay pellets (fig. 4.6). In terms of inclusion types, Sandy Ware overlaps with Light Silty Ware. Differences in grain sizes and proportions distinguish them, however. Rounded and semi-rounded inclusions are distributed through 80% of the paste. The remaining 20% body clay is not enough to hold the coarse particles together. This explains the tendency of some pieces to separate. Sand tempering is quite widespread in Neolithic pottery production. According to the results of this analysis, however, the regular size and distribution of sand grains in the paste shows that this clay is a natural formation.

This clay is likely to have come from lake sand dunes located around 10km away, within the boundaries of the local resource area. It is difficult to explain why it is preferred, given its unfavourable structure. It could have

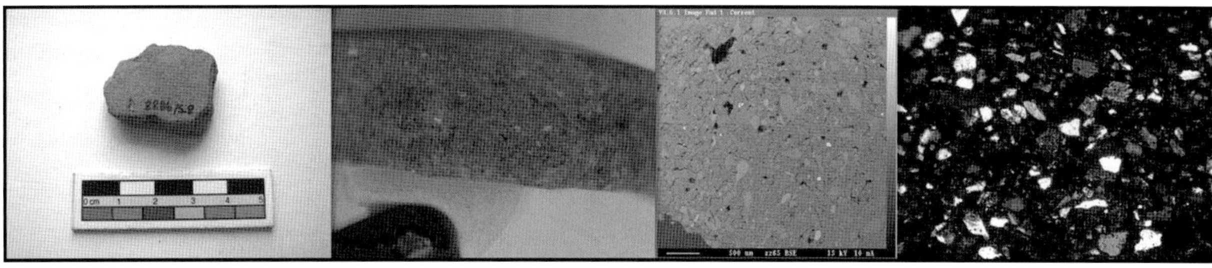

Figure 4.5. Light Silty Ware (8888) (from left to right: sherd, under the stereo-microscope, SEM, polarised light microscope).

Figure 4.6. Sandy Ware (10711) (from left to right: sherd, under the stereo-microscope, SEM, polarised light microscope).

been obtained when trying to reach calcareous clay; alternatively, it could have been the conscious preference of a social group. Again, it is difficult to say anything about form types because of the tendency of the pieces to separate, resulting in few full profiles (fig. 4.7).

Dark Gritty Ware (DARK)

This group is characterised by a highly coarse, granular paste texture made out of volcanic rocks and their minerals, dark brown tones of surface and paste colour and hole-mouth jar form. It is known that this gritty structure provides both technological and functional advantages in pottery (fig. 4.8). Technologically, producing pottery with coarse granular clay prevents cracking or distortion during the firing process (Rice 1987: 104–06). Functionally, it is usually preferred for cooking purposes, because the grains in the paste trap heat, gradually transporting it to food contents, which thus retain heat for a longer period of time after being removed from the cooking fire (Güner 1988). Lipid analysis results show that all Dark Gritty Wares were used as cooking pots (Pitters et. al. 2013). Without exception, all sherds are well burnished and thin walled (3–9mm).

The clay composition of the group is andesite and dacite volcanic rocks, and its minerals include zoned feldspar, dark brown and green calcium amphiboles, alkali feldspar, clinopyroxene and rhyolite. All inclusions range from 0.002mm to 1.5mm and form 65% of the clay body (fig. 4.9). Because the whole paste is composed of the same rock-related grains, the clay must be completely natural with no added temper. The chemical structure of the paste contains a high proportion of iron and magnesium (Si>Al>Fe>Mg>Na), and the lack of calcium in the clay differs significantly from calcareous sources and therefore the Light Silty Ware group.

The inclusions are varied in size and length and are semi-angular in form, indicating their derivation from natural colluvial sedimentation (fig. 4.9). It can be concluded that the colluvial materials were carried by a river with a low volume and were collected not far from their parent rock. The 'Erenler-Alacadağ' volcanics support this formation scenario and the mineralogical and chemical structure of the region. The Erenler-Alacadağ volcanic formations located in the western part of the Konya plain are of Miocene and Pliocene date and have been defined by many researchers. According to Keller (1974), they consist of andesite, dacite, rhyodacite, rhyolite-type lavas, tuffs and ignimbrites. Ota and Dinçel (1975) identified the rocks as andesite, pyroxene hornblende andesite, pyroxene hornblende biotite dacite, biotite dacite and hornblende biotite dacite through detailed petrographic studies. Peccerillo and Taylor (1976) indicate that they are located in the area of high-potassium andesite and high-dacitic dacite, and they are typical high-potassium series rocks (Ercan 1986: 128).

With all these data, the above-mentioned mineralogical character and chemical structure of the ceramic paste prove to be in conformity with the Erenler-Alacadağ volcanic formation. There are also other volcanic formations around Çatalhöyük. The mineralogical and chemical structure of these formations were compared with the ceramic samples, but they were found to be different. For instance, in the

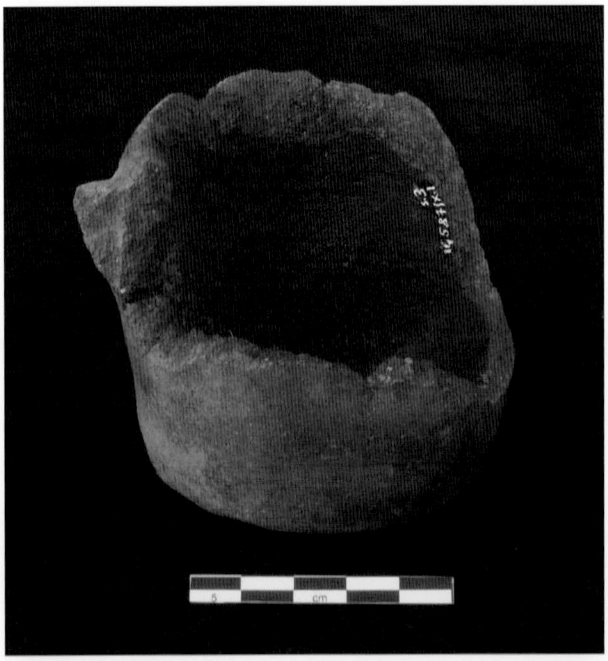

Figure 4.7. Sandy Ware bowl (14587.x1).

Figure 4.8. Dark Gritty Ware sherds (12268).

Figure 4.9. Dark Gritty Ware (11376) (from left to right: sherd, under the stereo-microscope, SEM, polarised light microscope.

Karaman volcanic formation, basaltic andesite and trachyandesites are observed (Bozyiğit, Güngör 2011: 176). In the Karacadağ volcanic formation augite-olivine basalt, augite andesite, olivine-augite andesite, augite-hornblende andesite, augite-hornblende-biotite dacite and hornblende biotite dacites are found (Ercan 1986: 128). In other words, all of these rocks and minerals are more acidic than the ones observed in the ceramics. For these reasons, the Erenler-Alacadağ volcanics were determined as the most likely source of materials for the Dark Gritty ware group. This region is about 70–80km southwest of Çatalhöyük. The May River, which begins in that region, merged with the Çarşamba River around the İçeri Çumra region after passing about 45km of volcanic rock-based sediments (de Ridder 1965). Because the May River merges with the Çarşamba River, a small percentage of the above-mentioned andesitic and dacitic volcanic rocks and associated minerals are also seen in the local clay in the vicinity of Çatalhöyük (fig. 4.10).

In the ceramic thin-section samples, however, there are no Çarşamba River sediments such as limestone or chert (see Light Silty Ware for the Çarşamba sediment types). In order to make sure that the Dark Gritty Ware group was not mixed with calcareous silty clay, the 'acid reaction' test was carried out on randomly selected sherds. In this test, hydrochloric acid (HCl) is added dropwise to sherd sections; the presence of calcite ($CaCO_3$) is marked by a strong foaming response. No such response occurred when HCl was applied to the Dark Gritty Ware samples. For this reason, it can be said its clay was obtained from the region upstream of the merging of the May and Çarşamba Rivers (fig. 4.10).

Concerning other characteristics, the Dark Gritty Ware group (fig. 4.1) appears to have been produced for the same function and in similar forms and dimensions as those previously found in the Middle levels (fig. 4.11).

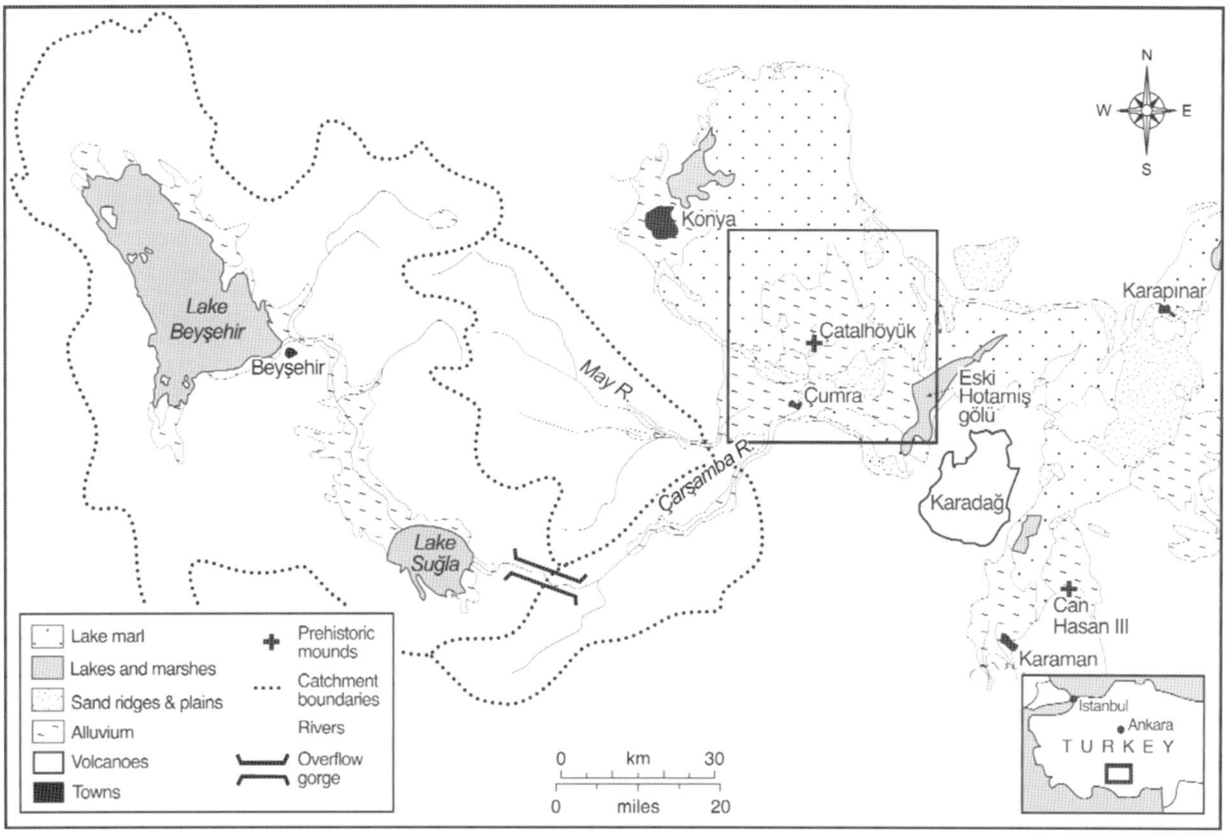

Figure 4.10. Çarşamba River fan (Boyer et al. 2006: 677).

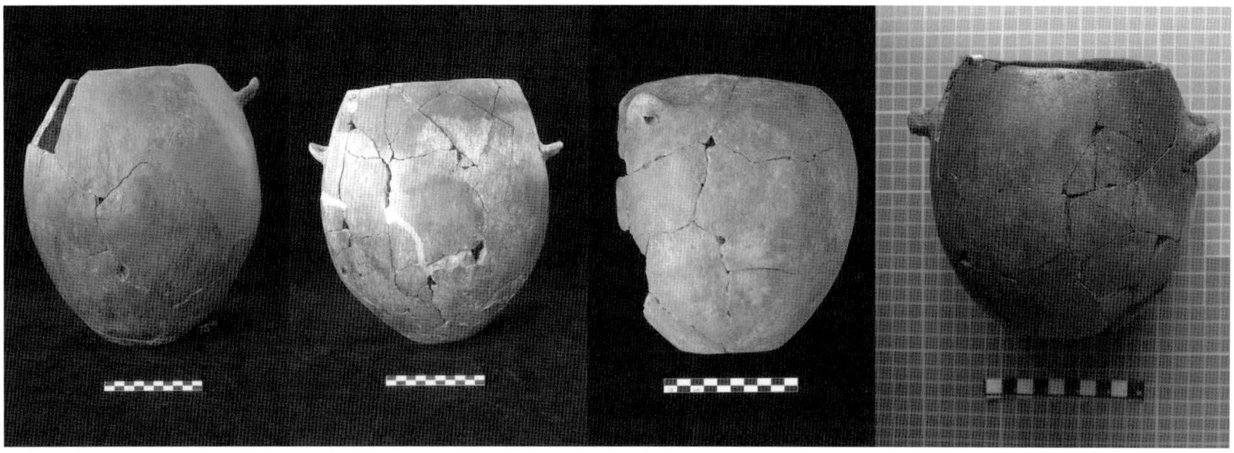

Figure 4.11. Dark Gritty Ware cooking vessels.

Red Paste Ware (RED)

This group can be easily distinguished because of its reddish surface and paste colour, as well as its coarse-grained texture with white and sometimes grey inclusions. Macroscopic observations suggest that these white grains are limestone. Thin-section analysis, however, shows that they are laminated metamorphics and likely the related minerals mica-schist, crystalline limestone and phyllite. In addition, mylonite, quartzite and titanite, which are fault-zone rock types, were observed in SEM-EPMA analyses (fig. 4.12). Such rocks with foliated structures are generally formed by the effects of temperature and pressure beneath the crust (Monroe, Wicander 2007: 184).

Neither the Çarşamba sediments nor the former Konya Lake marl traces were found in the paste. This indicates that the clay source is outside the boundaries of the plain. The metamorphic inclusions of schists, phyllites, marbles and quartzites mentioned above are found in different regions close to the Konya Basin (Bozyiğit, Güngör 2011:

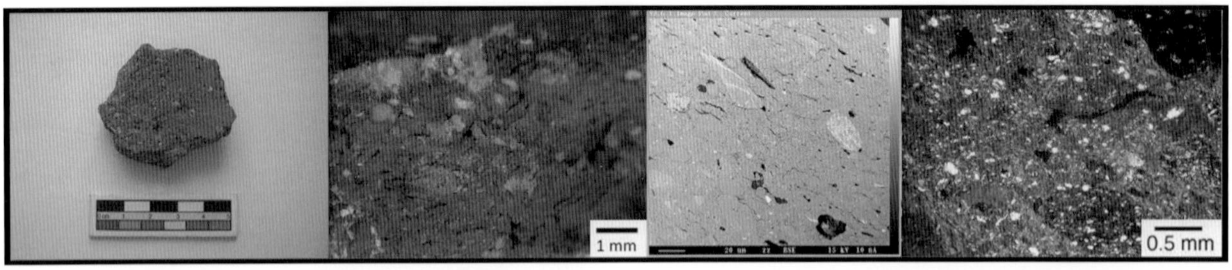

Figure 4.12. Red Ware (12508) (from left to right: sherd, under the stereo-microscope, SEM, polarised light microscope).

172). They are also encountered in the Bozkır region, where the Çarşamba River begins. However, in Çarşamba sediments, which can be found in the vicinity of the site, only some of these metamorphic rocks can be seen, and in very low proportions. For this reason, it can be postulated that they came from a region south of the Konya Basin, due to similarities established with the above-mentioned data. All this information points to the Southern Taurus as the clay source of the Red Paste Ware group.

The coarse grains in the paste provide advantages when such wares are used for cooking functions (see Dark Gritty Ware). None of the sherds of this group, however, has any indication of such use, such as traces of contact with fire or food remains (Yalman et al. 2013: 160). The rim and base parts of the vessels display an oval-shaped form. In some samples, small lugs, thought to be decorative, are found together with basket handles. In other words, it can be clearly said that this group is characterised by oval-shaped, basket-handled vessels (Yalman et al. 2013: 160). Before the appearance of Red Paste Wares, no oval-shaped vessels were found throughout the settlement. Considering the form and clay source, the ware can be interpreted as a technology brought to the site. Sherds of this ware are much more concentrated in the North area and can be interpreted as being the preference of a group of potters. Pottery produced from this raw material was also found in the Chalcolithic levels of the West Mound.

Marly White Ware (MARL)

As mentioned above, marl deposits formed in the former Konya Lake provide the calcareous clay structure seen in the ceramics. Marls are highly calcareous and fine-grained soils formed as sediments in still waters such as lakes. The original colour is white, due to the calcium carbonate content (30% $CaCO_3$). However, they can vary from grey to light brown because of the influence of other trace substances (Çamurcuoğlu 2015: 78). When we look at this group, it is possible to say that the entire clay body (85%) consists of marl and is named 'white' because of its surface and paste colours. The paste is composed of equally long, rounded Çarşamba sediment inclusions in a very dense calcareous clay. The inclusions are generally the same type of minerals and rocks, but they vary in proportion across thin-section samples (fig. 4.13). Because there are no signs of thermal degradation in the mollusc shells and carbonate grains, they must have been fired at very low (300ºC) temperatures and for a very short time.

This group is very rare (fig. 4.1). Although marl is a soil type frequently used for plastering, it is rarely used for making pots. The most important reason for this is the lack of iron content. Iron, an important transfer material for firing in low temperatures, adds strength to the paste (Rice 1987: 478). Pottery made from marly white clay and fired at low temperatures is rather unstable. This is another reason why it is rare. For all these reasons, it is possible that this group was made for its visual effect rather than functional utility (fig. 4.14). The well-known face-pot with relief human face and incised bucranium decoration is also made of this clay and hence supports this hypothesis (Yalman et. al. 2013: 160).

Shelly Ware (SHELL)

Even though its surface is very similar to Light Silty Ware, Shelly Ware can be distinguished when looking closely, because of the high amount of mollusc shell in its ceramic paste. Thin-section analysis reveals only shells set in a dense calcareous clay body. No other inclusions were observed (fig. 4.15). Although the paste only contains shells and gives the impression of tempering, it is evident that it derives from natural formations. The most important evidence leading to this conclusion is that the inclusions are crumbled in the field and integrated with the clay. Indeed, in SEM-EPMA analyses, the ratio of calcium in the chemical composition appears to be higher than aluminium (Si> Ca> Al> K> Mg). The delicate structure of the shells is not thermally degraded and therefore this ware is fired at low temperatures, between 300 and 500ºC.

According to archaeomalacologists, the shells derive from gastropods (land-snails), which make up the most populous class of molluscs living in fresh water and on land. The animal's body is covered with a helical (spiral-shaped) shell. At Çatalhöyük, such shells were used as beads, and the same type of shell was found in building materials (Gümüş, Bar-Yosef Mayer 2013: 81).

Figure 4.13. Marly White Ware (11648) (from left to right: sherd, under the stereo-microscope, SEM, polarised light microscope).

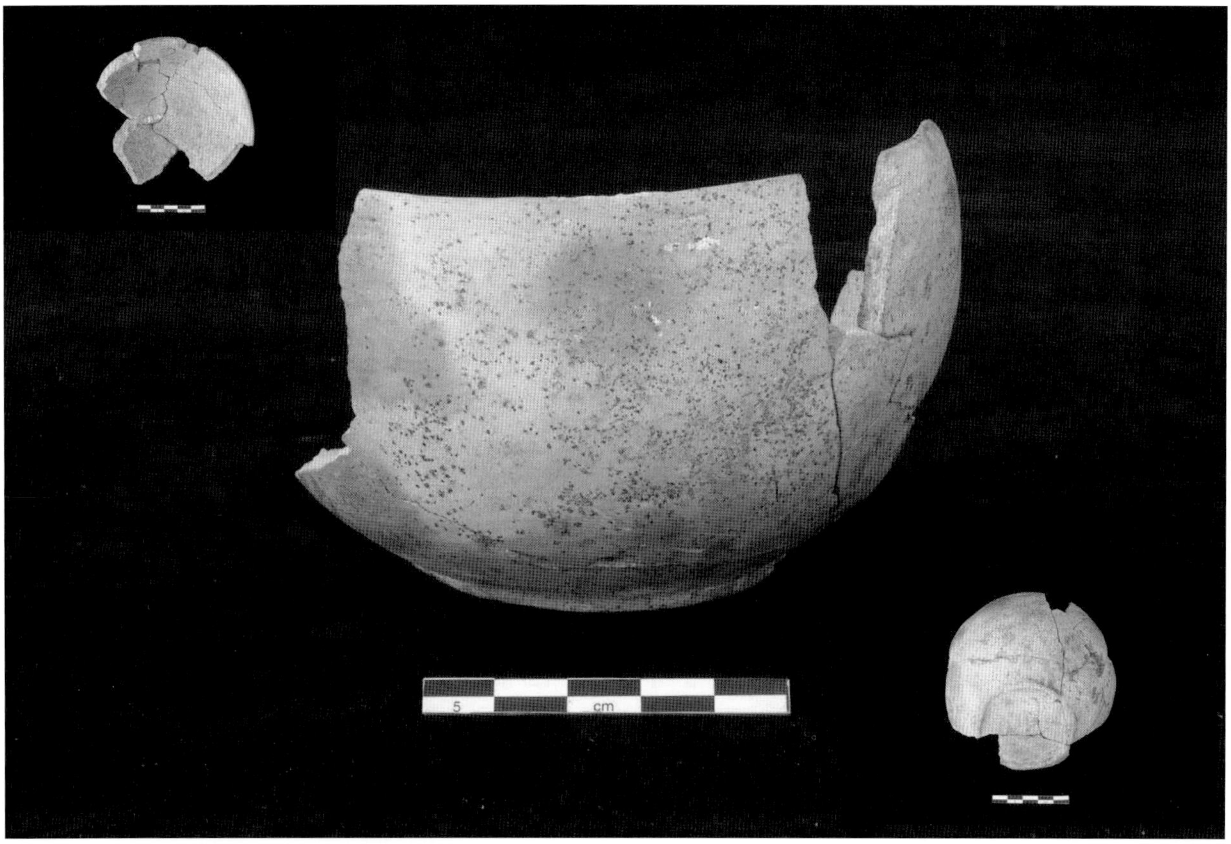

Figure 4.14. Marly White Ware (11985).

Figure 4.15. Shelly Ware (12980) (from left to right: sherd, under the stereo-microscope, SEM, polarised light microscope).

The possible scenarios of the formation of this clay source include the accumulation of micro-mollusc shells in small water ponds after seasonal river floods. They may also have accumulated by filling pits previously quarried for the collection of marl or clay. The clay source must be a region within the boundaries of the former Konya Lake, because the shells accumulated within the marl deposit (personal communication Doherty).

It is not possible to give any information about the function and forms preferred for this ware, because diagnostic sherds, such as rims, bases, or lugs/handles, have not been found. However, such wares must have fulfilled special functions, as they are made from a hard-to-reach raw material (Figure 4.1).

Form Types

Ceramic forms were evaluated under two main categories, as in the previous studies mentioned above. These are Jars, with deep and long bodies, and Bowls, with shallow and short bodies. Subgroups were simplified as much as possible with the method used for ware groups (fig. 4.16 and table 4.3).

Jars (figure 4.17)

Hole-mouth. This is a non-necked, deep, spherical or bag-shaped, hole-mouth cooking vessel. It is mostly frequently made in Dark Gritty Ware. The Late period examples are thicker and more uniform than those of the middle levels. The mouth rims have diverse shapes, such as thickened, pointed, or simple rounded etc. It was determined that these properties naturally occurred during freehand shaping, however, and were not technologically conscious decisions. For this reason, hole-mouth jars are not divided into subgroups.

Everted rim. We can also use the very light S-profile definition for these jars. The necks of the pots are slightly curled; the edges of the mouth are thinner; and the body starts to move slightly. It is seen quite rarely in the late levels.

Bowls (figure 4.18)

Straight walled. These are small bowls with upright, inner wall depths of about 5cm, flat bases and shallow, cylindrical shapes.

Inturned rim and Globular body. As can be understood from the name, these are globular or spherical bodied shallow pots with mildly inwardly closed rims.

Hemispherical: These resemble a globe shape divided into two halves. Late level samples have ring or footed bases as well.

Cone-shaped. These are inverted, triangular-shaped bowls extending from base to rim.

S-profiled. These are bowls with a slight 'S' curvature just below the rim, in the neck.

Oval. These are ellipsoid-shaped shallow bowls. Some samples in large pieces have been found with basket handles and lugs that appear to have had decorative purposes.

Diachronic change

According to Mellaart (1966), Çatalhöyük pottery changed according to his levels as follows: (1) Experimental phase: X–IX; (2) Intermediate: VII–VIB; (3) Full advanced stage: VIA and later.

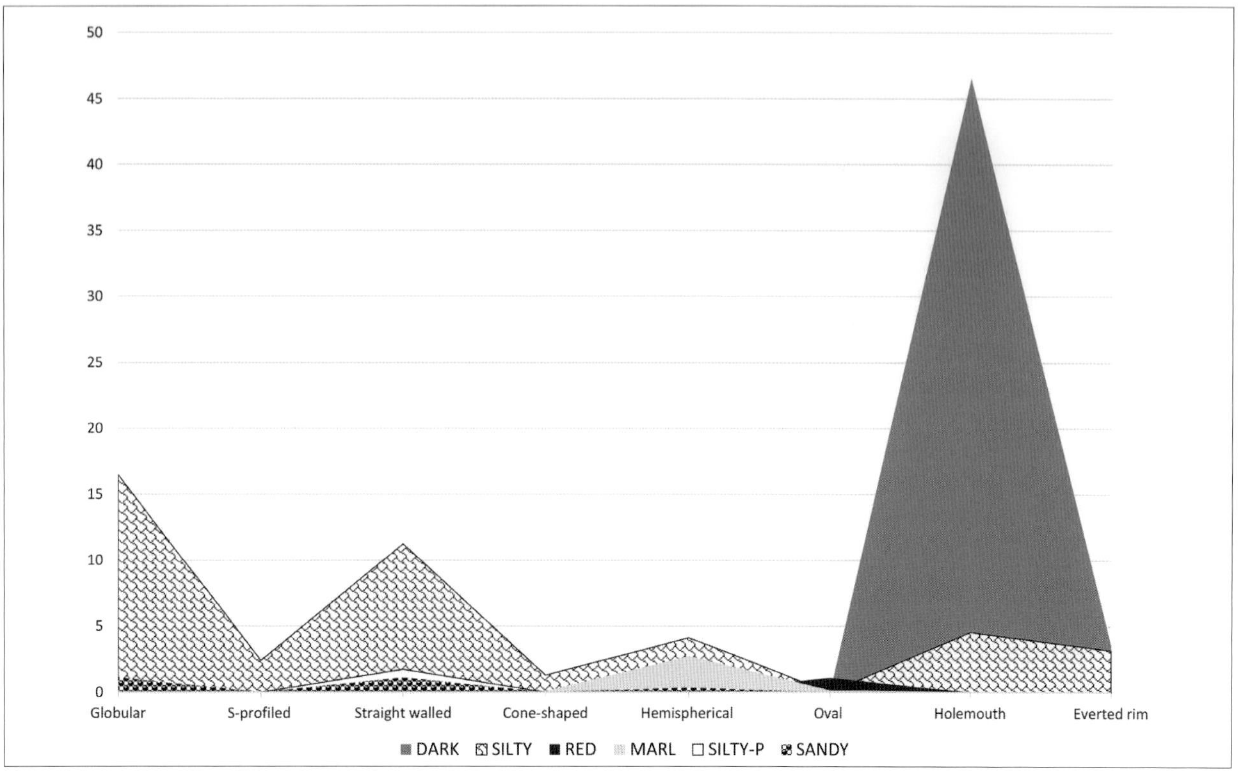

Figure 4.16. Total sherd frequency chart of ware groups by vessel form.

Ware & form	*Bowls*						*Jars*		*Total frequency*	*Total %*
	Globular	*S-profiled*	*Straight walled*	*Cone-shaped*	*Hemispherical*	*Oval*	*Hole-mouth*	*Everted rim*		
SILTY-P			n=20						20	
			1.72							1.72
SILTY	n=192	n=28	n=131	n=15	n=48	-	n=53	n=37	504	
	16.47	2.40	11.23	1.29	4.12		4.55	3.17		43.22
DARK							n=543	n=41	584	
							46.57	3.52		50.09
RED						n=13			13	
						1.11				1.11
MARL				n=1	n=32	n=2			35	
				0.09	2.74	0.17				3.00
SANDY	n=13		n=13		n=4				30	
	1.11		1.11		0.34					2.57
SHELL	No diagnostic sherd								0	0

Table 4.3. Total sherd frequencies of ware groups by vessel form.

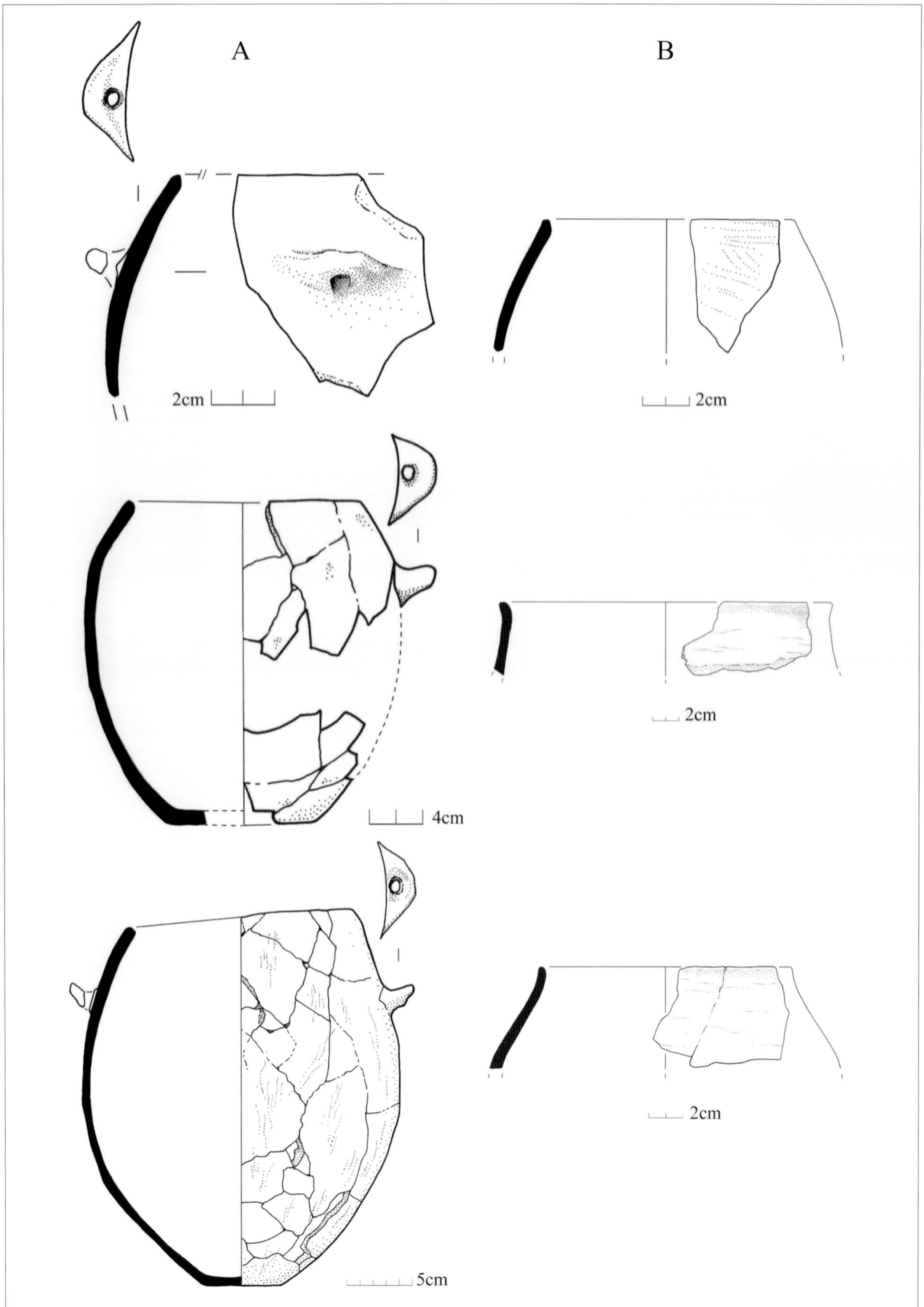

Figure 4.17. Jar forms: A) holemouth jar; B) jars with slightly everted rim.

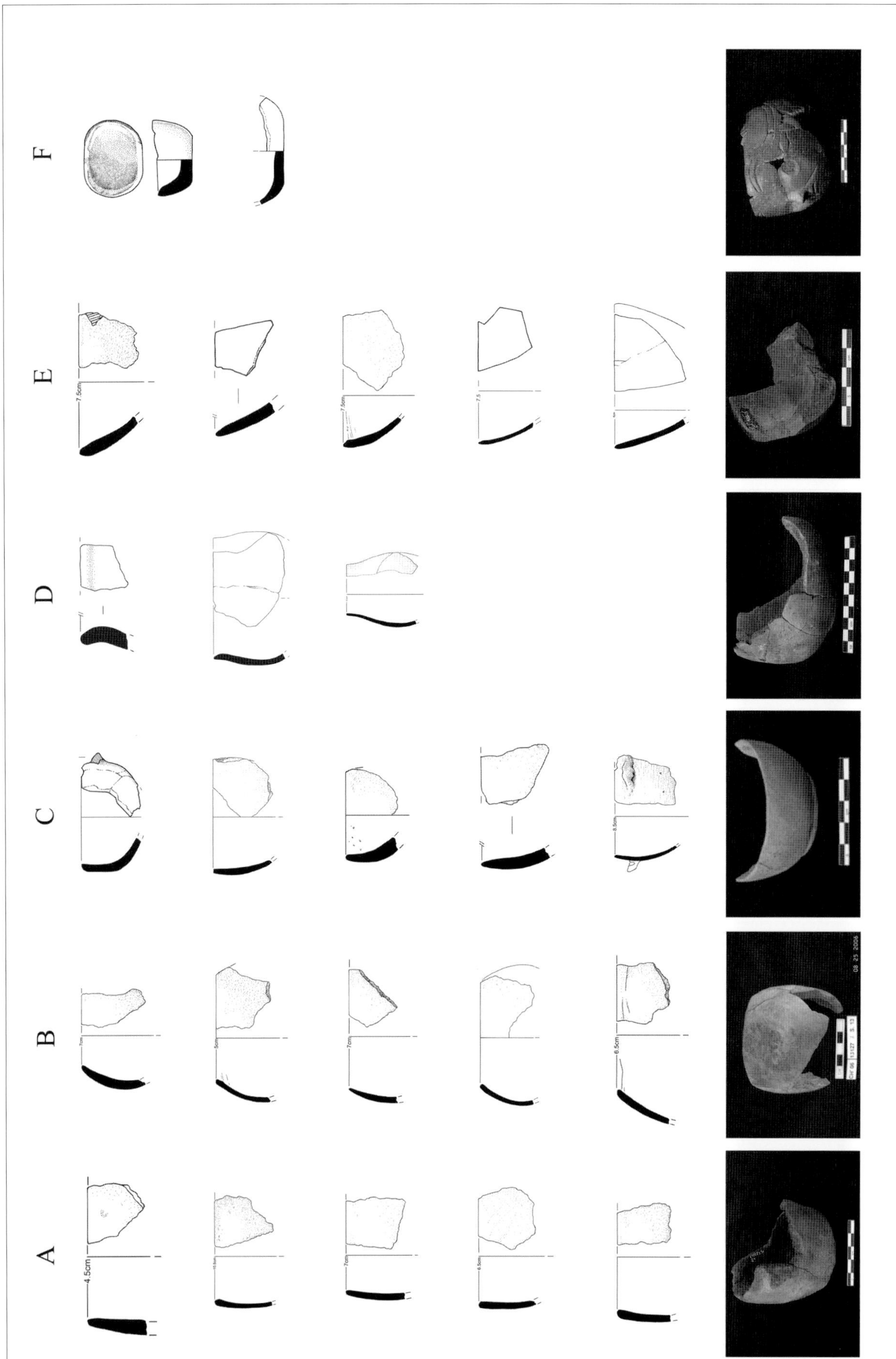

Figure 4.18. Bowl forms: A) straight-walled; B) globular; C) hemispherical; D) S-profiled; E) cone-shaped; F) oval.

Last (2005a) supported Mellaart's chronological sequence with a form typology. Özdöl (2006) suggested different phases of tradition based on the Mellaart excavations. Özdöl's table is as follows: (1) Early Neolithic Pottery Tradition: XII–VIII/VII; (2) Middle Neolithic Pottery Tradition: VII–IV/III; (3) Late Neolithic Pottery Tradition: IV/III–I/0.

Yalman et al. (2013) suggest that the major shift in pottery production and perception occurred during Mellaart Levels VIII–VI (South L–O), and that the overall pottery tradition changed gradually because of internal social factors.

Excavations from 2008 onwards focused mainly on early levels in both the South and North Areas. Here small changes in the early period pottery production were observed. In addition, excavations in the TP, TPC and GDN Areas in which late Neolithic levels were examined contributed to the creation of a more holistic picture (table 4.4).

Early period

During the first occupation of Çatalhöyük, the use of pottery seems to have been very rare (N=520). In the earliest ceramics, plant tempering (SILTY-P) is highly characteristic (fig. 4.19). These early samples are seen first in the level XII (South H) and remain the dominant group up to Level South M. In this context, the tradition of plant tempering appears dominant for Çatalhöyük Early period ceramics. The SILTY group, seen together with SILTY-P, is another early tradition produced with local clay. In the following levels (IX–South K), it can be seen that ceramics (SILTY) were produced without any deliberate addition of plants but with the same construction technology. Since these two groups do not differ in form, it is possible to interpret the continuation of the plant tempering as a desire to maintain tradition. As a matter of fact, SILTY-P gradually decreases from the Middle period, while SILTY continues up to the Late period in the TP, GDN and TPC Areas.

The typical form of the early levels is the thick and straight walled bowl (fig. 4.20). Lipid analyses of Light Silty Ware (SILTY) samples showed animal fat traces in the pottery surfaces (Pitters et al. 2013). It is thought, however, that this cooking process took place using additional tools. In fact, no trace of direct contact with heat was found on the outer surfaces of the sherds. In the clay inclusions, too, there is no thermal degradation due to heat. Based on this information, it is assumed that the cooking process was performed using 'clay balls' (Atalay 2013: 242) or heated stones.

The number of sherds in these levels is very small, however, and no diagnostic rim sherds were found in the South H-I-J levels. The only form seen in the early levels is the Straight Walled Bowl, which is the typical form of the Light Silty Plant Tempered ware (see SILTY-P).

SANDY Ware also starts to be seen in Early period levels (IX–South K). As mentioned above, this group was produced from a disadvantageous clay (see Sandy Ware). This explains why it is rare, but it remains difficult to explain what kind of motivation made society produce it. SANDY might be an experimental product. The Dark Gritty Ware (DARK), which also has a coarse-grained texture, started to be seen immediately afterwards and occurred in ceramic production as cooking vessels. Still, it does not explain why SANDY wares were still being produced in the Late period. Nevertheless, the appearance of fat-burning traces in a rare semi-profiled, cup-shaped bowl (fig. 4.7) may indicate that this group was produced for illumination functions such as an oil lamp.

Middle period

In the middle levels the Dark Gritty Ware first starts to appear as a new technology. This group is the dominant group of Çatalhöyük Neolithic period pottery, with 65% representation (N=12,017).

Dark Gritty Wares, as mentioned above, consist of volcanic rocks. These inclusions provide several advantages in both production and cooking. Because of the coarse grains, it is possible to produce thin-walled and large vessels with this raw material. These grains in the raw material facilitate evaporation and provide thermal shock resistance, which prevents cracking during drying and bursting during firing. In this respect, production loss must also be reduced. These thin-walled vessels are also advantageous in that they retain heat and thus reduce cooking time (Rice 1987: 104). The probable clay source is about 70–80km away from the site. Access to this clay requires more labour and time than the local clays. For this effort, there must be a motivation or driving force. Why did potters go to a distant source for volcanic clay if they were able to use the local silty clay?

Here it is possible to talk about a change in food consumption habits. All of the Light Silty Ware sherds found in the middle levels are in small-sized bowl forms. But with the appearance of Dark Gritty Ware, the hole-mouth jar appears among the form types. In addition, according to animal bone studies, meat was separated from the bones after having been cooked (Hodder 2013: 20). This process may only be carried out after boiling. In light of this interpretation, it was investigated how the meat could be put into all the vessels in experimental works performed jointly by ceramic specialists and Dr Arzu Demirergi. It was found that cut bone dimensions and vessel sizes were consistent (Yalman et al. 2013: 158). The Dark Gritty (DARK) group appears first in Level South M and becomes dominant in South N (fig. 4.22).

	SILTY-P		SILTY		SANDY		DARK	
	g	%	g	%	g	%	g	%
South.T			919	20.46	44.3	0.99	3,278.5	72.98
South.S			3475	16.39	365.8	1.73	17,078.05	80.54
South.R			1,703.1	17.07	221.9	2.22	7,807.6	78.27
South.Q			2,530.68	13.94	370	2.04	14,334.3	78.98
South.P	19.3	0.14	729.7	5.47	50.8	0.38	11,716.2	87.75
South.O	192.8	10.13	115.3	6.06	2.5	0.13	1,467	77.06
South.N	174.3	40.40	11.5	2.67			245.6	56.93
South.M	1,659.9	61.86	224.9	8.38	75.8	2.82	722.6	26.93
South.L			351.5	66.70	175.5	33.30		
South.K	2,070.7	48.78	1,405.3	33.10	769.2	18.12		
South.J	167.5	92.75	13.1	7.25				
South.I	47.7	70.46	20	29.54				
South.H	96.5	52.30	88	47.70				
North.J			1,072.5	38.33	31.6	1.13	1,667.2	59.58
North.I			14,761.9	18.12	1018	1.25	62,967	77.28
North.H	224	1.44	3,436.8	22.02	538	3.45	10,615.2	68.03
North.G	1,159.7	9.97	1,859.5	15.98	375.6	3.23	6,767.4	58.15
North.F	426.4	38.16	466	41.71	42.4	3.79	173.5	15.53
TP.R			6,685.2	69.92	213	2.23	2,585.2	27.04
TP.Q			3,766.5	58.64	239.5	3.73	2,564.6	39.93
TP.P			7,319	51.21	170	1.19	6,398	44.77
TP.O			10,511.5	53.01	290	1.46	9,153	46.16
TP.N			25,943	51.04	1,437	2.83	23,141.5	45.53
TP.M			3,415.5	18.96	476	2.64	13,948.5	77.45
IST.Unassigned			4,047.600	16.30	454.1	1.83	19,925.7	80.23
GDN (TP.N)			7,540	54.45			6,308.5	45.55
GDN (TP.M)			2,042.5	16.60			10,202	82.91

Table 4.4. Total weight of ware groups (g column) according to the area levels and percentages (% column) within each level.

The number of pottery sherds in the Middle period increases (N=1,308) compared to the Early period (N=520). Variation is seen in the ware groups. In addition to the Dark Gritty Ware, Red Ware and Marly White Ware appears during the Middle period. The earliest example of Red Ware is seen in the North Area first, unlike other ware groups, which start from the South Area. However, there are only a few examples (N=3, 2.06%), and it would be wrong to say something definite. Level North F, where the Red Paste Ware is seen, is similar to Level South L–M in terms of the distribution of ware groups. SILTY-P still shows the highest rate in these levels. The Red Ware Group first appears in the South Area in South O, and the distribution of ceramics looks similar to North G, with high ratios of DARK, less SILTY and decreasing numbers of SILTY-P (fig. 4.22). In Level North G, it is also observed that Marl pottery appears, which is the plaster material (see Marly White Ware). Considering that MARL Ware was used in the production of special

RED		*MARL*		*SHELL*		*Total sherd weight*	*Total sherd count*	*Total deposit vol.*
g	*%*	*g*	*%*	*g*	*%*	*g*	*#*	*lt*
34	0.76	22.8	0.51	193.9	4.32	4,492.5	361	8468.8
41.4	0.20	243.8	1.15			21,204.0	1421	63803.8
40.5	0.41	202.7	2.03			9,975.8	689	46728.2
618.6	3.41	294.7	1.62			18,148.3	1119	39024.1
816.9	6.12	18.8	0.14			13,351.7	876	99603.6
126	6.62					1,903.6	194	75249.3
						431.4	21	15015.0
						2,683.2	186	26541.9
						527.0	58	32640.7
						4,245.2	128	45441.1
						180.6	17	5393.3
						67.7	7	6106.4
						184.5	11	5938.3
					Total	77,395.5	5,088	469,954.5
27	0.96					2,798.3	124	2,461.61
1,459.9	1.79	1,267.1	1.56			81,473.9	4,846	93,813
721.3	4.62	69.2	0.44			15,604.5	681	141,618.5
1,234.7	10.61	240	2.06			11,636.9	818	762,211.1
9	0.81					1,117.3	89	75,969.6
					Total	112630.9	6558	1,076,073.81
		160	1.673395	10	0.104587	9,561.4	715	127199.9
23	0.358083	56	0.871853			6,423.1	711	836898.1
56	0.391855	318	2.225177	30	0.209922	14,291.0	1812	117716.7
		72	0.363114			19,828.5	1875	227017.6
82	0.16133	224	0.440706			50,827.5	2775	362281.9
33	0.1832263	137.5	0.763444			18,010.5	677	36504.4
					Total	118,942.0	8,565.0	1,707,618.6
272	1.10	136.9	0.55			24,836.3	1,419	490.8
					Total	24,836.3	1,419	490.8
			0			13,848.5	754	311,196.4
		60	0.49			12,304.5	499	6818.4
					Total	26,153	1,253	318,014.8

forms, it can be postulated that the ceramics in the upper levels of the Middle period played different roles.

A number of new form groups emerged parallel to the appearance of new ware groups in the Middle period. The Hole-mouth Jar produced with Dark Gritty Ware appears at varying ratios, ranging from 70% to 41% in the South and North levels. Straight Walled Bowls, produced in Light Silty Plant Tempered Ware in the early levels and in Light Silty Ware in the Middle period, become the second most common form type after Hole-mouth Jars. This is followed by Bowls with inturned rims and Globular bodies, again produced in Light Silty Ware, with 20% in South M and 60% in North F. With the addition of Red Ware and Marly White Ware to the ware group range, Hemispherical, S-profiled, and Oval bowl types produced from these ware groups begin to appear in North G (fig. 4.23). Although Marly White Ware was seen in the South Area first, it appears to exist only in the North Area in figure 4.22, because of the lack of a diagnostic sherd.

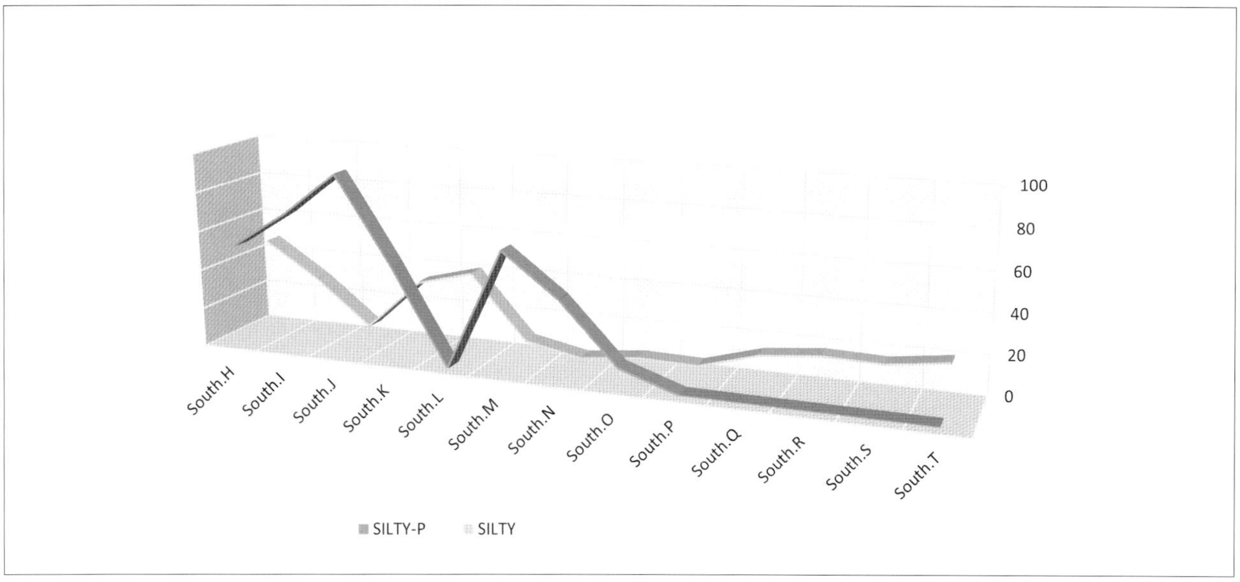

Figure 4.19. Frequency of SILTY and SILTY-P groups in South Area levels (n=520).

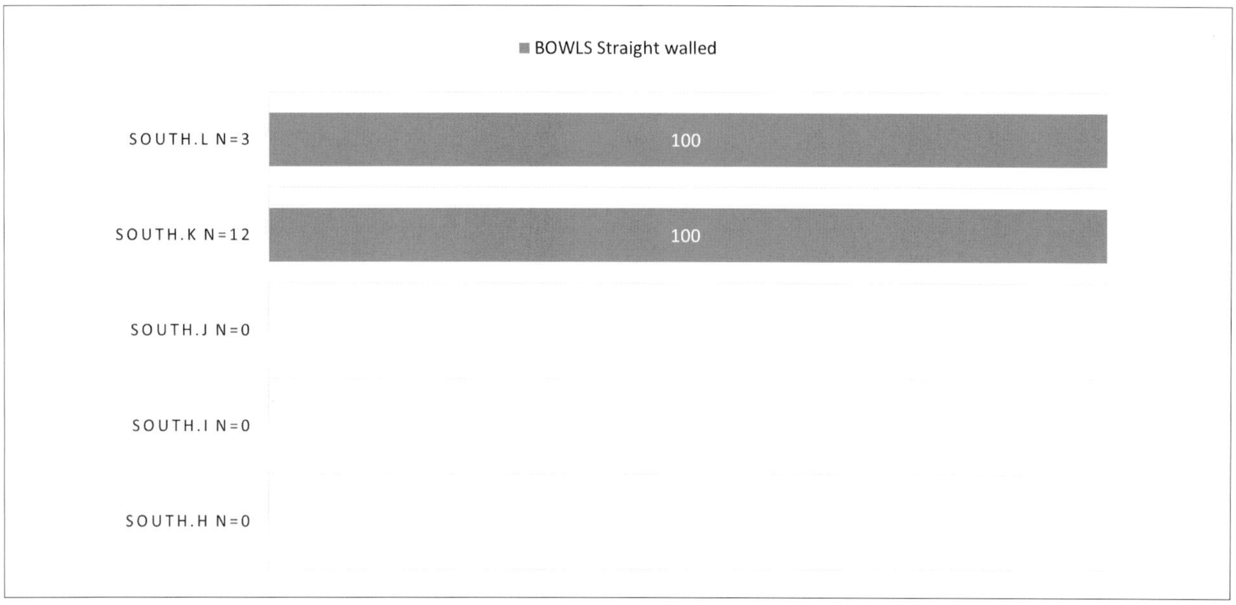

Figure 4.20. Forms in the Early period.

Late period
Several significant changes can be observed in ceramic production during the Late period occupation of the site. These include, first, an increase in diversification of ware groups and technology (figs 4.24 and 4.25) and, second, a change in vessel sizes (fig. 4.26).

With the addition of the MARL and SHELL group, it can be seen that all the ware groups are represented as a result of the wider use of the landscape. DARK wares show a decrease, even if it is small scale, towards the upper level. The rate of DARK is 73% in South T, which is the final level of the area, while it is 87% in South P. Only in the TP Area is the DARK ratio between levels very close. It is useful to remember here that the TPC ceramics evaluated in this chapter are randomly selected sub-sample units (for the entire ceramic assemblage please see the TPC Volume). In the North Area the picture is similar to the situation in the South: the DARK decreases from 68% in North H to 59% in North J. The most significant decrease is observed in the TP and GDN Areas. It is 82% in Level TP M, whereas in TP R it has decreased to 26%. However, the portion of local-origin ware groups seems to increase. This is most evident in the second most abundant ware group: SILTY ware. The

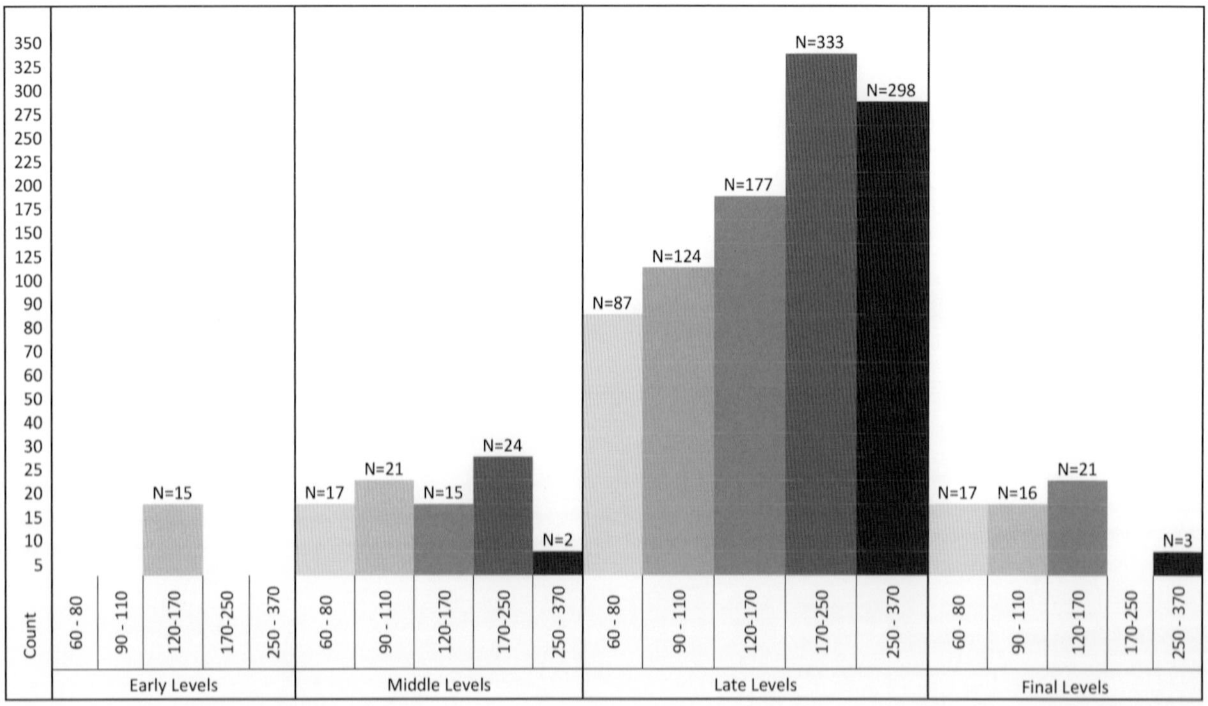

Figure 4.26. Vessel sizes through levels.

Temporal groupings of levels	*South*	*North*	*Cal BC*	*Buildings with complete vessels*
Final	TP.O-R and TPC Trenches 1 and 2 (B109 and 115)		6300–5950 BC	
Late	GDN	North.H,I,J and IST	6500–6300 BC	
	South.T. TP.N.			B.10 , B.45 , B.63
	South.S. TP.M.			B.44 , B.150, B.166
	South.R			B.42 , B.56
	South.Q			B.65
	South.P			
Middle	South.O	North.F, G	6700–6500 BC	
	South.N			
	South.M			
Early	South.L		7100–6700 BC	
	South.K			
	South.J			
	South.I			
	South.H			
	South.G			

Table 4.5. Buildings with complete vessels through time.

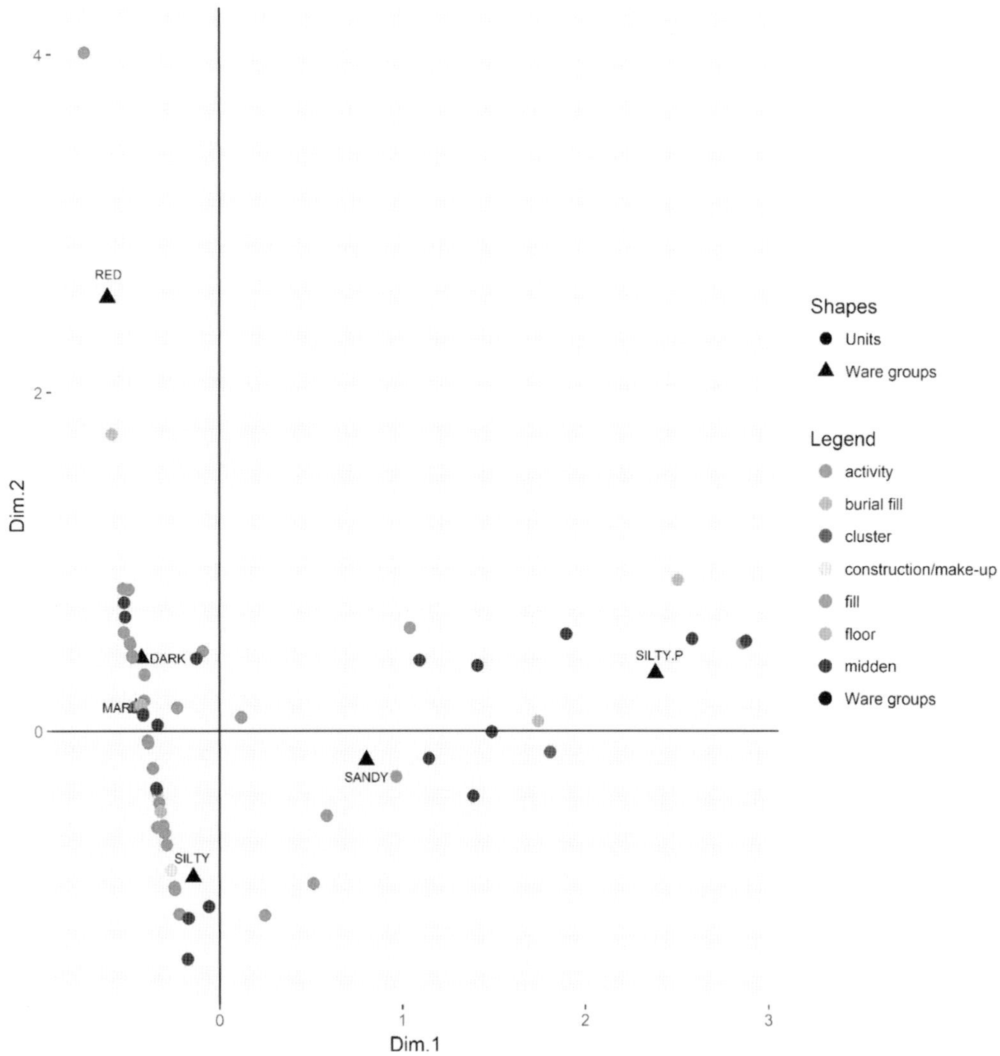

Figure 4.27. Multivariation by context type (for colour version, see online supplementary material).

Micromorphological analysis revealed that the pots may have been produced in the settlement. Matthews claims this through the analysis of early level fire spots (Matthews 1999). According to Shillito (2011), ash deposits indicate the craft activity of pottery production starting from the Middle period. The lithic assemblage also suggests a household production in these levels. The idea of outdoor activity areas, possibly for pottery firing, proposed by Shillito's research, also represents a change in technology, from low-fired clay material seen in the early levels, to large-scale open-air firing (Shillito 2011: 113–15). Although these studies make strong arguments, it is very difficult to make a definitive comment on this subject, since there is no other trace of pottery production in the mentioned contexts of the settlement. There is strong evidence that ceramics were produced inside the settlement in the post-Neolithic West Mound. From the earliest levels of the West Mound, dating to the Late Neolithic/Early Chalcolithic, unfired potsherds, half-made unfinished pottery pieces and production wastes are seen (Franz 2009: 45). Such traces were not found in the Neolithic East Mound.

Assuming that the pottery was produced in the settlement, it can be thought that the variety could indicate the pottery was produced by different households using different types of raw material or shaping technologies. Or this variation over time could be explained by new members coming into the settlement, bringing their own production techniques with them, which were then adopted by the current people.

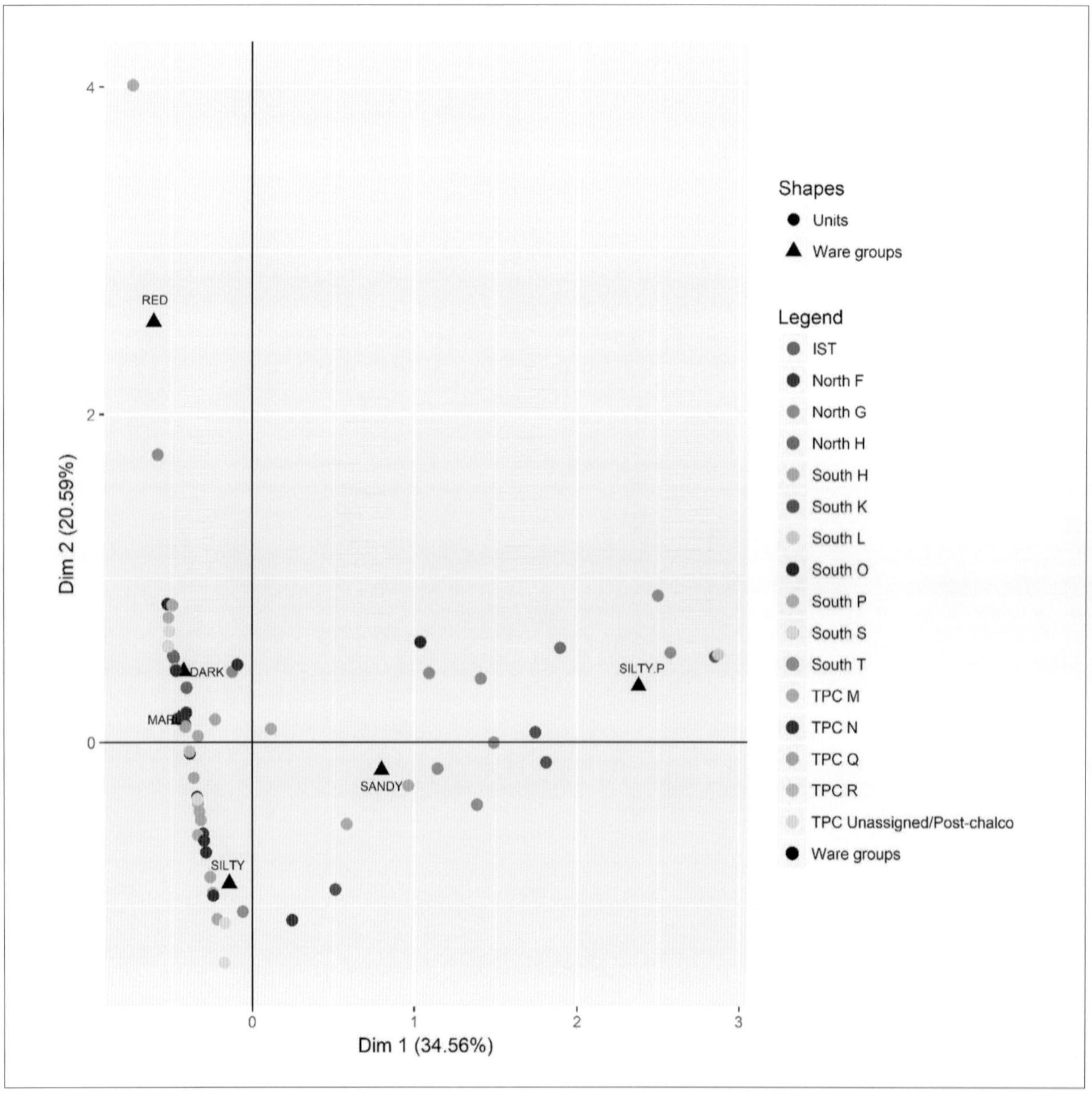

Figure 4.28. Multivariation by period (for colour version, see online supplementary material).

All the above-mentioned ware and form groups are different but consistent within themselves in terms of raw material types, forms and functions. It is thus hard to say whether one group displaced another. Over time, the use of one group began to decrease while that of another appeared and increased. Nevertheless, a previous group does not disappear and even continues to exist in some cases to the last building level. The changes in and longevity of ware production suggest that some groups in the community were innovative in pottery, while others were tightly bound to their traditions.

Supplementary material

For supplementary material related to this chapter, please visit https://doi.org/10.18866/BIAA/e-14. It comprises colour versions of figures 4.25, 4.27 and 4.28.

5. Figuring diversity: the Neolithic Çatalhöyük figurines

Carolyn Nakamura
(with contributions from Lynn Meskell)

After 13 seasons of documenting and studying Çatalhöyük materials, our initial impressions of the figurines have not significantly changed: we stand by our earlier observation that the diversity of the Çatalhöyük figurine assemblage resists any easy or simplistic categorisation of its practices (Meskell et al. 2008; Nakamura, Meskell 2013b). From the beginning, we have approached 'figurines' as processes rather than as end products, examining their production, circulation, practice and disposal, all of which enact a set of choices and processes embedded in a particular social world. Challenging the entrenched image of figurines that James Mellaart made famous as objects of distanced veneration, we have argued instead that these objects constituted a diverse and flexible suite of embodied actions and ideas. The significance of these objects unfolded through practice and action, not in distanced separation and contemplation. They were things to be used and even re-used.

Our approach to figurines has always treated figurines as any other archaeological material. In various Archive Reports, chapters and articles, we have studied figurine production, deposition and circulation, correlated artefact density with various features, buildings and contexts (Meskell et al. 2008); performed stone and clay experiments and analyses (with other specialists) in order to study manufacture techniques, heat exposure and material selection (Meskell, Nakamura 2009; Avis 2010; Meskell et al. 2016); examined bodily representations through quantifications of figurine features, qualities, measurements (Nakamura, Meskell 2009); and written about how the miniature, portable and multisensorial materiality of figurines affords particularly affective and tactile bodily relations with their makers and users and biographical inscription (Nakamura, Meskell 2004; Nakamura 2008; Meskell 2015). We have also worked across specialisations, examining potential correlations between figurine practices and the burial of human remains (Meskell, Nakamura 2009), as well as faunal (Martin, Meskell 2011) and isotopic data (Pearson, Meskell 2015). We have also carried out 3D scanning and replication studies and made videos of people handling figurines and the different perspectives (see https://web.stanford.edu/group/figurines/cgi-bin/omeka/collections/browse). We have also done comparative work across the Neolithic, both in Anatolia and more globally (Meskell, Nakamura 2009; Hodder, Meskell 2011; Nakamura, Meskell 2013b).

Figurine assemblages, like Çatalhöyük building plans, seem to conform to certain general guidelines, yet they also demonstrate remarkable flexibility and variation. As a social technology, the Çatalhöyük figurines suggest particular attention to certain themes or areas, namely animality, and bodily abbreviation, partability, excess, ambiguity and maturity. These themes, which often converge or overlap, underscore the diverse ways in which figurine practices mediated certain pressing concerns in Neolithic life: the negotiation of relationships with animals, forebears or the dead, and the success and survival of the community as mediated through ideas of body and flesh.

Since our last summary in 2008, the project has added 1,215 figurines excavated during the 2009–2017 seasons. With the exception of one remarkable building in the TPC Area that contained a number of impressive stone anthropomorphic figurines – including the only two figurines found in a definitively primary context during the Hodder excavations – these additional finds generally do not suggest any radically new figurine types, material or practices. Therefore, we will not substantively revisit various themes discussed in detail in other publications such as figurine recording, figurine as process, figurine fabrics and production techniques, headedness and headlessness, enfleshing and embedding, animality, or bodily maturity and sexuality (for summary with references, see Nakamura, Meskell 2013b). Instead, in this chapter, we will update the site-wide analyses of figurine distributions, describe a few notable examples, and look more closely at aspects of figurine practice that might mediate social power or world-making through various articulations of form, degree of elaboration/abbreviation, pose/posture and materiality.

Figurine distributions

The Çatalhöyük figurine database now contains 2,994 Neolithic figurines, including those from Mellaart's excavations and spoil heap (REC), and non-diagnostic and indeterminate forms; 2,027 from this corpus are figural forms excavated by the Hodder team between 1993 and 2017. The 3,145 figurines in the database further include West Mound (Chalcolithic) and post-Neolithic materials. The contextual analyses below will only consider the 2,027 Neolithic figural forms excavated by the Hodder team, while basic tallies will include all Neolithic figurines. Analyses involving levels and patterns over time will only consider the 1,830 figurines given level assignments. Analyses will focus on four main figural forms: anthropomorphic (humans), zoomorphic (quadrupeds and horns), abbreviated (generic forms with head, torso and base) (see Nakamura and Meskell 2008: 202 for detailed discussion of figural categories).

Preservation and fragmentation (table 5.1)

Fragments of figurines comprise the bulk of the corpus and are three times more common than complete or nearly complete figurines. Anthropomorphic (46% complete) and abbreviated (44% complete) forms are much more likely to be found mostly intact compared to zoomorphic forms, which largely occur as fragments (87% fragments). As before, fragmentation patterning generally suggests that figurines break at their weakest points, such as the neck, limbs, horns, ears etc. Very few (n=13) show some indication of intentional breakage or damage (ten quadruped, two abbreviated, one anthropomorphic).

Depositional practices (tables 5.2 and 5.3; figures 5.1 and 5.2)

The added materials from the last eight years of excavation have not significantly changed the distribution patterns of figurines across the site. Figurines are predominantly found in external contexts (78%) rather than in buildings (22%) (table 5.3) and are most common in midden contexts, followed by fill (fig. 5.1). As noted above, figurine practices across the site peak during the Middle occupation levels and then drop off dramatically in the Final period. One difference can be seen in numbers and densities across the Middle and Late periods: in external spaces, figurine practices are sustained through the Late period, whereas in buildings they drop by nearly fourfold (table 5.3).

General trends (figures 5.1–5.5). In terms of the distribution between internal and external areas, anthropomorphic forms occupy the largest percentage in buildings by a very slight amount, while quadrupeds comprise the largest percentage in external areas (table 5.2). Horn fragments are the dominant form in almost all contexts due to their ubiquity, small size and identifiable shape (see Chapter 6). They are particularly dominant in contexts of disposal (midden), infilling (building and feature infill, burial fill) or activity (fire installation, wall or platform construction).

Abbreviated forms, like horns, occur in every context in terms of overall use, but in fewer numbers (figs 5.1 and 5.2). In this way, horns are more comparable to abbreviated forms than quadrupeds, as quadrupeds and anthropomorphic figurines (as the more elaborated forms) are found in a more limited range of contexts. It might, therefore, be

	Anthropomorphic	*Zoomorphic*	*Abbreviated*	*Total*
Complete	61	75	124	260
Possibly complete	10	5	4	19
Nearly complete	37	122	165	324
Fragment > 50%	58	173	182	413
Fragment < 50%	53	832	144	1,029
Fragment	17	342	43	402

Table 5.1. Figurine preservation.

	Anthropomorphic		*Abbreviated*		*Quadruped*		*Horn*		*Total*	
Building	25	25%	68	22%	50	19%	128	23%	271	22%
External	77	75%	235	78%	216	81%	435	77%	963	78%
Total	102		303		266		563		1,234	

Table 5.2. Figurine forms: external spaces vs buildings.

Period	Buildings				External Spaces			
	Density	No. of figurines	Dry sieve volume (kL)	Total volume (kL)	Density	No. of figurines	Dry sieve volume (kL)	Total volume (kL)
Early	0.309	74	239.37	352.20	0.443	18	40.67	53.14
Middle	2.440	471	193.03	675.79	3.530	349	98.86	142.03
Late	0.575	81	140.80	216.29	3.667	333	90.80	132.40
Final	0.491	19	38.70	600.18	0.419	8	19.10	32.44

Table 5.3. Figurine densities and counts 2009–2017: external spaces vs buildings. These figures do not include the latest information from the final cleaning of the database, but should be close estimates. Volumes by level were calculated from the Çatalhöyük Project Database by Dominik Lukas on 25 October 2017.

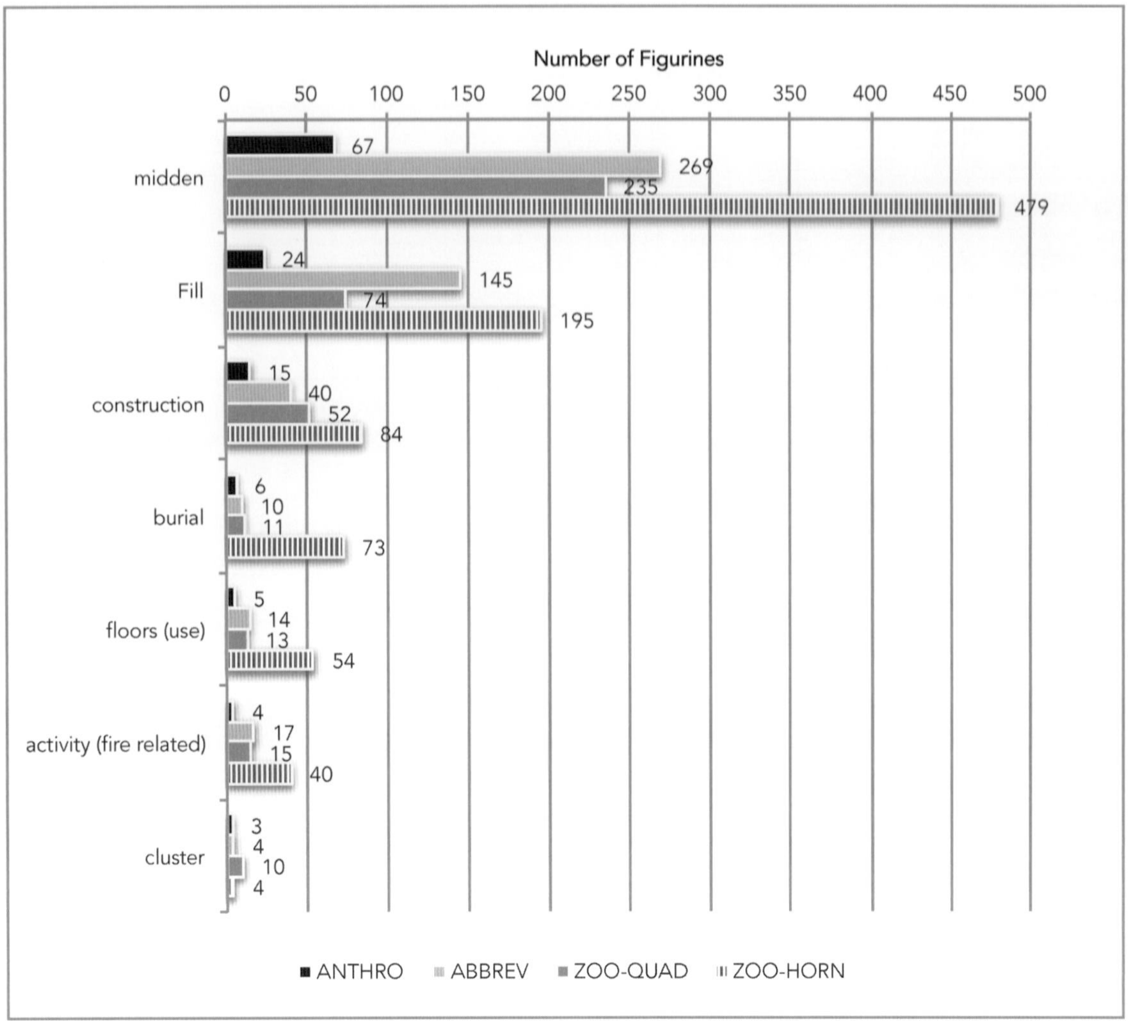

Figure 5.1. Figurine forms by excavation data category.

advisable to consider if figurine practices were organised by formal as well as representational considerations. Beyond the iconic and indexical relationship horns have to horned animals, they may also serve a metonymic function. Similar to how abbreviated forms, in many cases, served as expedient depictions of humans, some horns may have also been expedient representations of different animals. Since there is no gender-neutral body part that easily sustains a metonymic relationship to the human body like a horn to an animal, the generic form of two limbs (legs or feet), a torso and head does the job of evoking a bipedal human. Of course, there are several examples (although a minority) of horns and abbreviated figurines that do not fit this generalised scheme. However, it is worth considering the possible reasons for depicting more elaborated and abbreviated forms of both humans and animals.

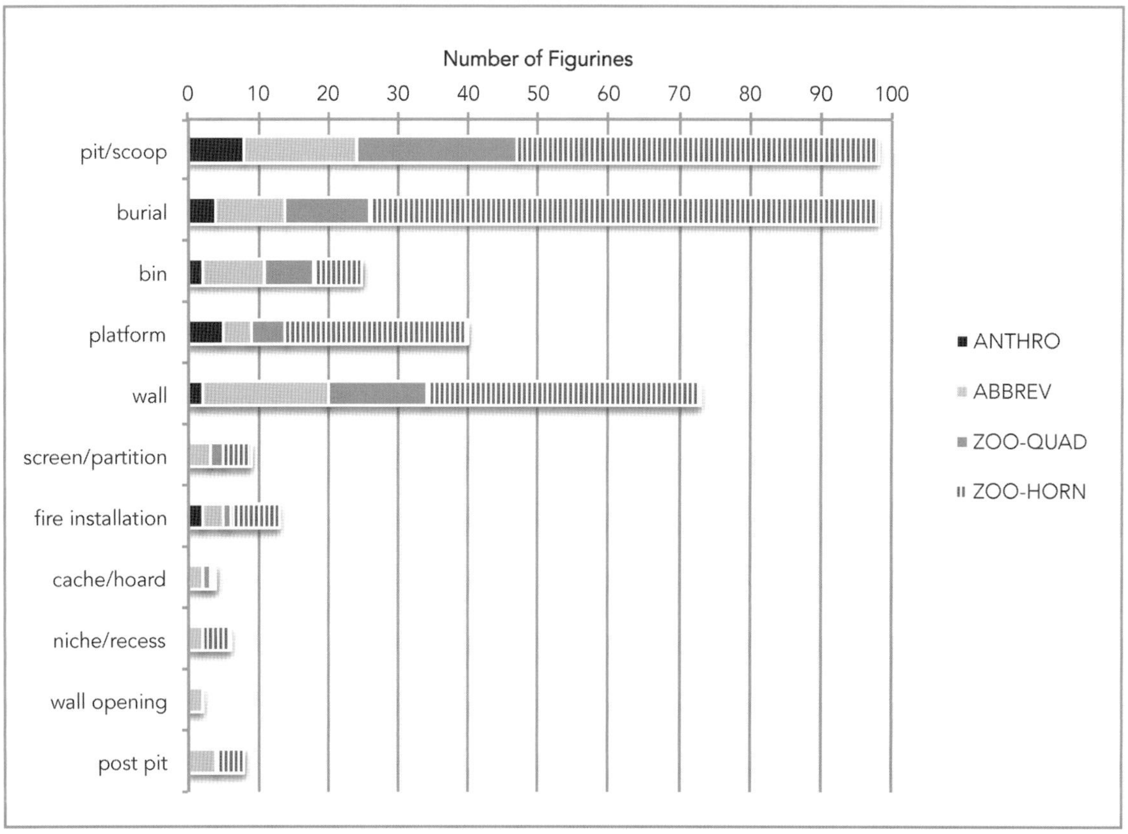

Figure 5.2. Figurine forms in features.

One possibility is that more generic forms were suitable for multiple and general purposes. The feature types associated with all figurine forms – pits/scoops, burial fill, bins, platforms, walls and fire installations (fig. 5.2) – comprise either construction or house-sustaining activities. In a way, these features provide a map of the activities and structures of social life: pits encompass activities of retrieving or emplacing, burials incorporate the dead within the house and mediate relations between the living and the dead, bins contain and store materials and thus articulate a future-oriented gesture, platforms mark and structure internal space, walls support and divide internal space, and fire installations sustain human life by providing heat to live, cook and work. Given that the house anchored the social world in Neolithic Çatalhöyük, it is not surprising that structures and activities that support and sustain the house would attract a variety of figural forms, both generic and elaborated. As we have noted before, there does not seem to be a correlation between house elaboration, number of burials and the presence of figurines (but see below). However, the features in which only some figurine types appear also provisionally suggest a more 'active' or meaningful function for certain filling practices. And while excavators have long observed a diverse range of disposal and infilling practices at the site, such practices remain under-theorised and observed and noted rather than seriously analysed (Nakamura, Meskell 2013b). It is possible that figurine types may be correlated with specific features, activities and materials, evoking a more intentional kind of action (see discussion below).

Figurines and funerary practices (fig. 5.1, table 5.4). While we have discussed the various contexts noted above in previous publications (Meskell et al. 2008; Nakamura, Meskell 2013b), we have not considered one context in particular: burial fill. We had previously observed that figurines were remarkably absent from burial contexts. While this pattern does hold true for the materials directly associated with skeletons, 100 figurines (mostly fragments) have been found in burial fill. All figurine types are represented, with horns and horn fragments being notably overrepresented at 73% (fig. 5.1). Compared to other fill and midden contexts, in which horn forms comprise around 45% of the corpus, this high figure in burials is quite striking; in fact, the overall figurine distribution in burial fill is more similar to those found on floor use and fire activity areas, where horns comprise more than 50% of the figurines. Burial features are, of course, activity areas that involve a sequence of events and

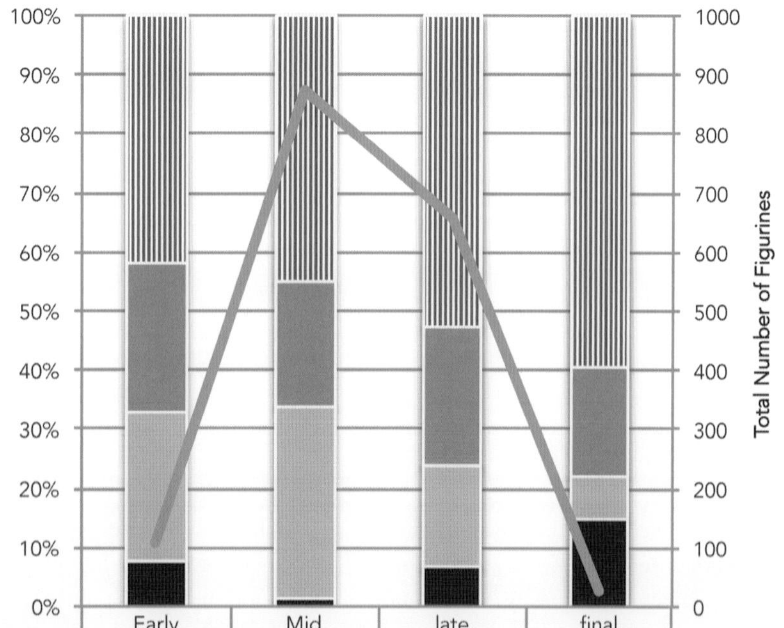

	Early	Mid	late	final
horn	45	393	345	16
quad	27	184	156	5
abbrev	27	286	112	2
anthro	8	10	44	4
fig total	107	873	657	27

Figure 5.7. Proportion of figurine types by period.

Figuring the human: a shift in concerns? Another suggestive pattern that has emerged is the elaboration of anthropomorphic forms over time. As we note above, there does appear to be a loose inverse correlation to the proportion of abbreviated and anthropomorphic figurines over time; this pattern could lend support to the suggestion that in a general sense, abbreviated forms could be placed on a spectrum of anthropomorphic forms (but see Nakamura and Meskell 2009 for more discussion) as more shorthand, expedient representations of humans. Abbreviated forms are comparatively well represented in the Early period and peak in the Middle period, while anthropomorphic forms maintain their modest number across the Early and Middle periods. But after the Middle period, abbreviated forms begin to contract in density and relative proportion, while anthropomorphic forms increase in both metrics and eventually overtake abbreviated forms in the Final period. If we place abbreviated and anthropomorphic forms on a spectrum, one possible interpretation of this trend is that it appears to show an increasing concern for elaborating the human form. However, we would still suggest that the two forms likely mobilised different activities and concerns, since abbreviated forms, while maintaining a somewhat standardised form, in the later levels also see some elaboration in detail. Many examples depict folded head elements from the Middle period onward, and there are a few instances of indicated facial features in the Late period.

Taking the figurine corpus as a whole over time, all forms, in fact, suggest an increasing attention to elaboration and variety in form. In the Middle levels (and intensifying in Late levels), we see the introduction of horses, deer, boars (although there is one example of a possible boar from early level South H, with a long trunk-like snout that turns up at the end and a ridged back – although unusual, it is likely that this depicts a boar and if so marks a singular, early example) and a long, large-tailed species (fox or reptile?) in addition to the standard cattle and sheep/goat quadrupeds we find in the Early levels (table 5.6). Although cattle and sheep/goat forms remain the majority of figurine quadrupeds throughout, this diversification may track with the changing status of wild animals, particularly cattle, in the Çatalhöyük lifeworld. Perhaps receding into communal memory, history or even myth, wild bulls in the transition to animal domestication are joined (or even replaced) by

Form	*Levels*	*No. of figurines*
Horse (equid)	South P/N,P, North G-I	9
Deer (cervid)	South N, P, North F, G, I	6
Boar (*sus*)	South P, North G, ?H, I	12
Long-tailed species (unidentified)	North ?G, North I	5

Table 5.6. Quadruped species introduced in Middle period.

new species (domesticated and wild) in certain areas of social life. Figurines here may have located a site of play, one that mediated a shifting lifeworld that the inhabitants of the site were embedded in and co-creating.

Diversification in the figurine quadrupeds may reflect not only a changing human relationship to animals more generally; it also may reflect a shift away from a more animal-focused world to a more human-focused world. The project team has documented a number of changes in the Late and Final levels that suggest an increase in household independence, commensalism and social exchanges, and a concomitant decrease in ritual cohesion and house-based history-making (for summary, see Hodder 2014d). Hodder further argues that by the later levels, more cohesive community bonds and concerns for continuity centred on the house gave way to a greater focus on movement, exchange of goods, and individuals; certain wild animals, especially wild bulls appeared to decline in ritual and social importance (ibid. 181).

Given Hodder's suggestion that these changes could relate to an increased focus on individuals in the Late levels, many researchers have been exploring their data for additional evidence for such a shift. Figurines in particular have often been mobilised in such arguments (Bailey 1994; Knapp, Van Dommelen 2008), despite the problematic application of liberal Western concepts such as 'individual' in ancient and prehistoric contexts (see Meskell 1999). At Çatalhöyük, we find no evidence from the figurine corpus that supports the idea of an increased concern with individuals over time, even with the most recent impressive stone additions. Instead, there is perhaps a greater concern for rendering the human body in particular ways. As we will argue below, such bodies (which were most likely idealised, see Nakamura, Meskell 2009) more likely served to articulate specific notions about the group or community rather than about certain individuals. Moreover, while the figurine data may also track with some of the site-wide changes in the Late levels, they do not necessarily support Hodder's conclusions that 'history-making through the use of houses became less possible or prevalent' (Hodder 2014e: 181). For instance, all of the stone figurines from the Middle to Late levels are found in houses, and the latter appear to be more deliberately deposited, which would seem to indicate an intense or particular focus on the house. The association of stone human figurines with certain houses, then, at least in one instance, could even constitute a repetitive (or 'history-making' in Hodder's terminology) practice: B.150 not only had five stone figurines deposited within its walls, the building directly succeeding and on top of it also produced a stone human figurine of a similar form. Even if these figurines circulated during their lives before deposition, in the final gesture they are associated with a house. Certainly, the flexibility of clay figurine practices – their relative expediency and ease of manufacture – easily enabled more personalised forms of expression, whether in gestures of memory, commemoration or more formal practices of history-making. It is worth noting, however, that portable media have been present throughout all levels, so the transfer of certain practices to more mobile forms was perhaps intensified, but not necessarily innovative in the Late and Final levels.

Human and animal isotope, chipped stone and material culture analyses have provided compelling evidence for increased contact with other communities, landscapes and materialities (see chapters in Hodder 2014b). While these data suggest an increase in diversifying and networking tendencies away from the house, they do not necessarily provide 'evidence for distinctive traits associated with individuals' (Hodder 2014e: 181). While it might be tempting to interpret the more elaborated renderings and forms of human figurines in the Late levels as supporting an argument for an increased focus on individuals, a more long-term, contextual view of anthropomorphic figurine practices and formal analysis of their representational conventions suggests a rather different story.

We would argue that the temporal patterning of figurine practice indicates an increased concern not with individuals, but with humans within the context of a historically animal-focused lifeworld. While a human-centred worldview is perhaps taken for granted by many contemporary researchers schooled in a Western liberal tradition, such a construction is far from given or inevitable. Many contemporary Indigenous peoples and some rural agricultural communities across the Americas, Oceania and Asia, for instance, maintain notions of the human that eschew human exceptionalism: not all human groups understand humans as radically apart from animals (or plants, soil, water etc.), nor have they understood personhood as something exclusive to humans, articulated through individual identity and status (cf. Viveiros de Castro 1998; Kimmerer 2013;

Watts 2013; Todd 2014; Lyons 2020). Moreover, scientific understandings and categories of animals often fail to capture the complex and nuanced relationships that humans share with animals in daily life (see Feldman, Norton 1995). We should therefore critically question rather than assume how Neolithic humans at Çatalhöyük understood their world and their place in it.

Turning to the figural forms, when considered alongside other data from the site, they do appear to track a shift in the focus on and depiction of the human form over time. Human figural forms increase in the Middle and Late levels while animal forms recede. While the densities of all figurine forms except abbreviated surge in the Late levels, anthropomorphic forms see the most dramatic increase (more than 600%, compared to the next highest increase, for horns, at less than 25%, see fig. 5.6). Suggestively, in the Final levels, when the figurines see a dramatic drop in number and density across the site, both anthropomorphic and horn forms constitute a larger percentage of the figurine corpus than they did in the Late levels, while quadrupeds and abbreviated figurines contract. Again, we should note that this result may in part be due to underrepresentation of certain figural forms due to excavation strategies employed in the last weeks of excavation. However, if less soil was sieved, it is odd that horn fragments, which are generally the smallest figural form, would be overrepresented. Over this time period, the density of the more elaborated animal forms – the quadrupeds – contracts the most, while the overall density of elaborated human forms contracts the least. If we consider some horns as possible 'abbreviated' forms of animals, one could read these overall trends as human representations becoming more elaborated (abbreviated to anthropomorphic), while animal forms are becoming more abstracted (quadruped to horns). In figurine terms, the human was beginning to demand more attention.

While this increased concern for human forms in figural practices seems hard to ignore (also given what we know of the Chalcolithic figural practices that follow), the premature impulse to focus on the emergence of individuals seems to have largely eclipsed this more fundamental shift of attention from animal to human bodies. Given that much of what we have found in the Early to Middle (and even much of the Late) levels at Çatalhöyük suggests a notion of the human as thoroughly embedded in if not beholden to the animal world (and particularly aligned with the power of certain wild animals), a shift towards a more human-centred and human-dominated social imaginary would likely have accompanied a significant social transformation and consequent tensions. Such upheaval, however, may not have immediately precipitated a more individual-focused consciousness; rather, as we will suggest in the next section, it more likely signalled a (contested) turn towards the human and away from the (wild) animal, and perhaps an incipient human-focused worldview, explored through specific conceptions of the human body. Hodder has pointed to the evidence for an increase in the size and diversity of beads and types of decoration in the Late levels as supporting a greater concern for differentiating the self and individual. However, these trends, while indicating a broadened network of human interaction that supported an increased mobility of various materials, images and ideas, might simply reflect a turn towards human-mediated or human-made social capital. Status and success could increasingly be pursued and attained through trade, control of managed resources and intensified contact with other human groups, not only through the economic and symbolic control of powerful wild animals. There is scant evidence to indicate that possession or display of unique, different or 'status' objects would be associated with a particular individual as opposed to a group.

Perhaps indirectly supporting this intensified interest in the human form, figurine trends over time also suggest a changing kind of attention to the animal world. Within the figurine corpus animals become increasingly abstracted in form over time, as practices seem to shift away from quadruped and towards horn forms. In terms of relative densities, quadrupeds progressively decrease as horns progressively increase from Early to Final levels (fig. 5.8), suggesting that not all horns were part of quadrupeds. Louise Martin and Lynn Meskell have noted at least two different cattle types represented in the zoomorphic quadrupeds (2012: 408–10). The most common type they interpret as wild aurochs, but examples from the later levels suggest a different type, possibly related to domesticates (see Twiss, Russell 2009). Some horns were made as stand-alone objects (not to be attached to a quadruped form), and the variety of horn forms we find demonstrate that the Neolithic makers were quite capable of accurately rendering the horns of different species. Therefore, further research on horn morphology over time could provide more insight into the relationship between wild and domestic animals and figurine practices. If clay animal horns came to evoke the power of wild aurochsen or other animals, this might suggest not a decrease in social importance of the wild, but rather its shift into the realm of history, memory and myth. By condensing the power of the wild and a 'mythological' past into a metonymic, miniature form (see Nakamura 2004; 2019; Nakamura, Pels 2014), figurines may have figured power that could be redirected in a multitude of ways. Alternatively if, in the later levels, clay horns were linked to actual animals within the domesticating process, they may have signalled a shift in certain ontological commitments; horn figurines could have performed a more expedient commemorative function as animals circulated within a more human-centred economy of networking and exchange. Perhaps most likely, given the non-standardisation of many practices, horn figurines brokered in power in a variety of social spheres.

Building correlations (tables 5.7–5.9, figures 5.10–5.12)
This diversity in figurine practice seems especially to bear out when we look more closely at specific building assemblages. Overall, there is a wide variance of figurine numbers and densities within buildings. But five buildings reveal relatively high figurine densities: B.3, B.49, B.65, B.131 and B.139 (buildings with >1.0 figurines/kL and more than 75% excavated, see table 5.7). All of these buildings except for B.65 are located in the North Area. While the figurine assemblages in each building are somewhat idiosyncratic (see table 5.8), a few general and suggestive patterns may be worth exploring. In buildings where figurines are found in a variety of contexts, such as B.3 and B.131, figurine distributions unsurprisingly follow the general site-wide pattern of horns being the most ubiquitous, followed by abbreviated, then quadrupeds and humans. In buildings with figurines primarily from midden and redeposited fill contexts, however, abbreviated forms slightly edge out horns (B.139 and 131/139 interface).

Although it requires further exploration and support, there is some suggestion – also by the overall patterning – that abbreviated figurines may be more particularly associated with certain filling practices than other forms. If we consider B.139 and its interface with B.131, middens within buildings may have had a different kind of status than those in external spaces; in a technical sense, they are part of discrete depositional events within building fill (even if after building 'closure'). More generally, it is also notable that the only excavation category for which abbreviated forms begin to approach the number of horns is fill (fig. 5.1). Abbreviated figurines also occur in the fill of bins, obsidian caches/hoards and wall openings (however, the sample size is quite small); horns also occur in the first two contexts. Buildings, bins and caches can all be considered discrete spaces of containment (different from burials, which are sometimes discrete deposits, but are frequently intentionally intercut and disturbed, and sometimes rearranged). Figurines in these spaces therefore draw attention to containment – more precisely, the state of being contained rather than the power to contain. Fill is different from features that are structural or activity based, such as walls, platforms, fire-installations, or pits dug to place or retrieve various items. Fill occupies more than it structures space (although it does often serve a structural purpose); it also provides a more passive form of support or action, as it inhabits, archives and preserves gestures made at transitional moments of or moments of concealment. Wall openings, while not containers, were likely filled during transitional moments of buildings' closure. That we find abbreviated forms in particular at these moments (along with horns), might suggest a more particular kind of apotropaic or commemorative function for such forms. At the same time, it may mark fill or filling practices in a specific way.

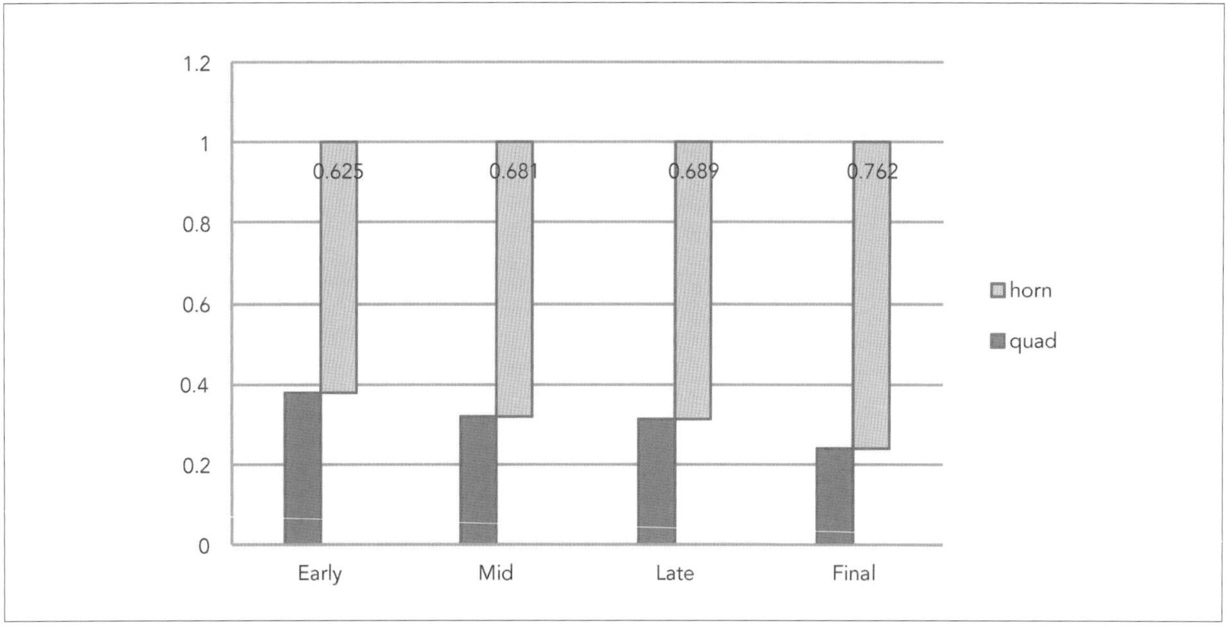

Figure 5.8. Relative overall densities of zoomorphic forms over time.

GID	*Material*	*Description*	*Level*	*Context (C) and associated materials (AM)*
20736.x1	Recrystallised limestone; 1041.63g, 10.96cm (h), 8.74cm (w), 17.20cm (th)		TPC M	C: platform make-up (F.3855) AM: obsidian blade, stone figurine 20736.x3
20736.x3	Pale yellow limestone; 54.40g, 6.93cm (h), 3.74cm (w), 2.36cm (th)		TPC M	C: platform make-up (F.3855) AM: galena, two blue beads, stone figurine 20736.x1
32806.x1	Soft limestone; 566g, 9.83cm (h), 8.46cm (w), 8.46cm (th)		TPC M	C: deposit in southwest corner AM: stone figurine 32806.x2, blue bead, bird wing bones, stone pounder polisher fragment, unworked clay, schist abrading tool, organic material, bright green material (malachite?) and burnt shell
32806.x2	Marble; 2725g, 24.70cm (h), 12.00cm (w), 6.46cm (th)		TPC M	C: deposit in southwest corner AM: stone 328076.x2, blue bead, bird wing bones, stone pounder polisher fragment, unworked clay, schist abrading tool, organic material, bright green material (malachite?) and burnt shell
23704.x7	Plastered or slipped clay; 9.19g, 2.50cm (h), 1.72cm (w), 2.58cm (th)	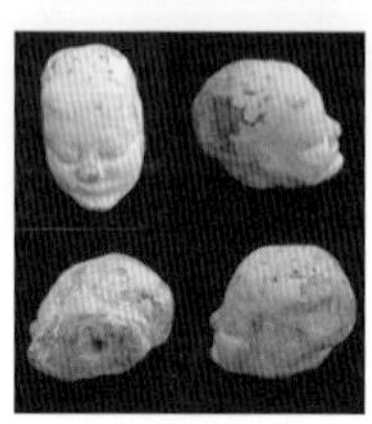	TPC M	C: northeast platform (F.3893), occupational layer, rich in materials AM (possible): astralagus, bone point, flint perforator/scraper, obsidan flake, worked greenstone, speleothem, polishing stone
31852.x3	Limestone	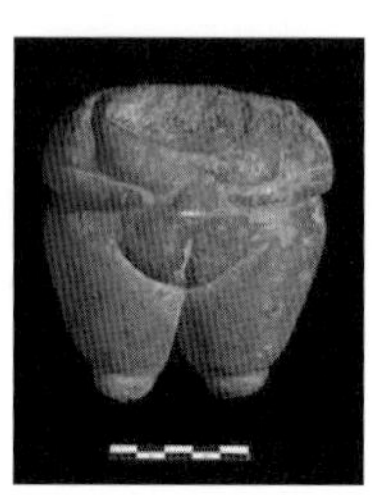	TPC N	C: in room rubble of first structure post-dating B.150, in association with fire spot AM (possible): cluster of worked bones and stones, flint and obsidian tools, a fragment of a horn core and two wings from a large bird

Table 5.9. Figurines from B.150 (and post-structure).

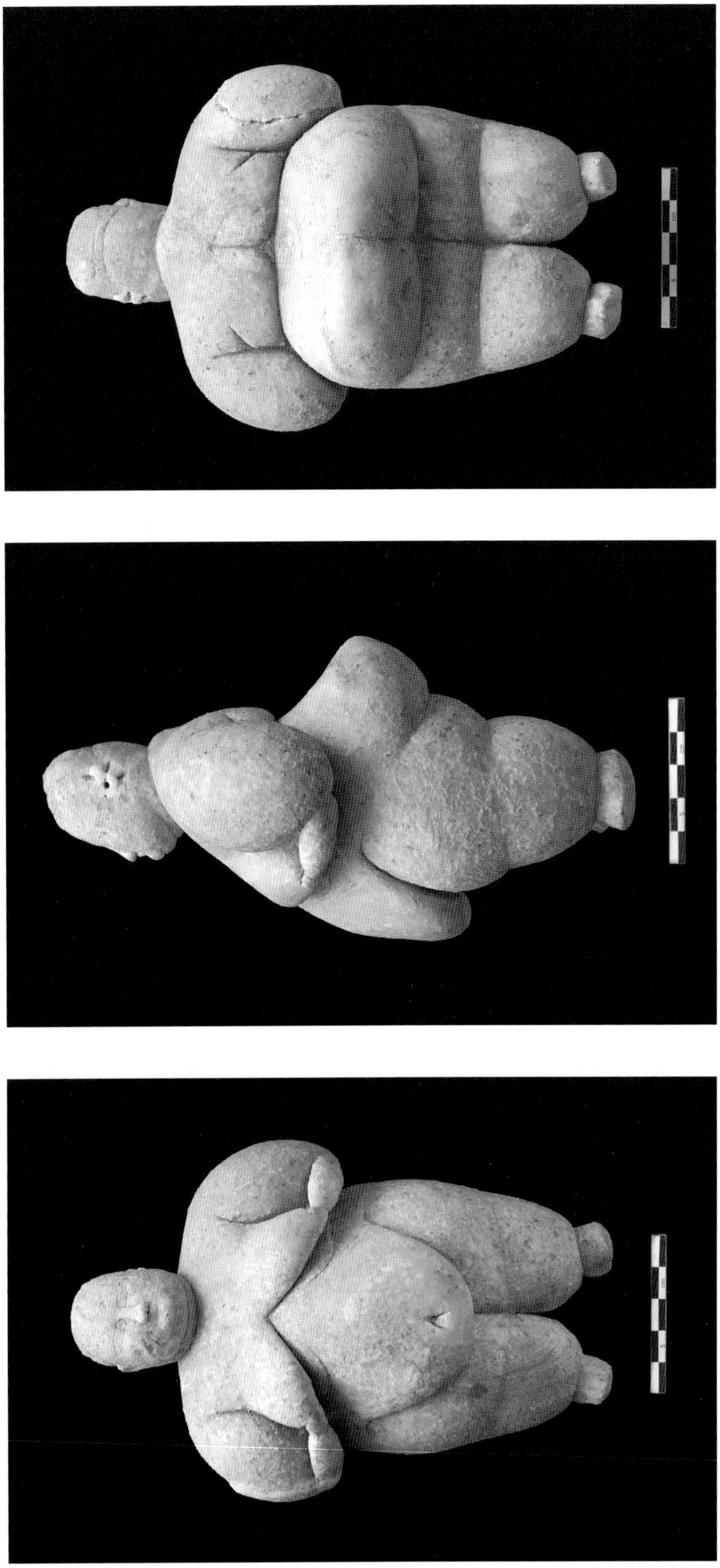

Figure 5.9. 20736.x1: (L) front view, (C) side view, (R) back view (photographs by Jason Quinlan).

Figure 5.10. 20736.x1 in situ (photograph by Jason Quinlan).

similarities between Mellaart's deposit and the one in (20736): the standing or extended postures, the differentials in size and technical skill or effort in rendering form, and the association with platforms. Mellaart's observation of finding speleothems with human figurines, while not corroborated by the (20736) deposit, is also interesting, given that speleothems have been found in B.150 (23765, 23993) and are also generally concentrated in the TP and TPC Areas (late to final occupation levels). Speleothems at Çatalhöyük appear in the form of stalactites, dog tooth spar, flowstones and calcitic concretions and only appear from North G levels onward. And it is perhaps significant that at least one anthropomorphic figurine excavated by the Hodder team is made from speleothem (10475.x2, see also a similar one found by Mellaart, CHC21). Erdoğu et al. (2013) have suggested that these materials likely derive from caves about 100km away (Incikini and Hatçenini caves). This association of human forms with speleothems, as well as other materials, may have had specific significance, a topic that we will return to below.

The northeast platform of B.150 also produced a remarkable clay figurine head (23704.x7, fig. 5.13). The head is very finely modelled, and its surface appears to have been plastered or slipped. The facial features resemble the head of stone figurine 20736.x3 with eyes (possibly closed or lidded), nose, mouth, ears and a pointed (delicate) chin. A hairline is also visible on the right side and back, extending across the forehead to above the ear and then behind the ear to the back of the neck. The head appears to have been attached to a body by a dowel as there is a dowel hole in the base. The function of this appears to be 'structural' or part of the production process, since the break with a clear line of where the slip ends suggests that it was permanently attached to a body rather than made as a piece that could circulate independently. This piece vaguely evokes the alabaster head found in VIA.44, which also sports a 'socket hole' at the base of the head. While this piece is made from stone and is quite large (5.4cm), it also suggests a tradition of making separate heads for human bodies. Notably, Mellaart noted that this alabaster head was found in the midst of a collection of pink, yellow and white stalactites and limestone concretions in the centre of the room.

Finally, (32806) in the southwest corner of B.150 consisted of an artefact-rich deposit that included two stone figurines (32806.x1 and x2, figs 5.15 left and 5.16 left). As these figurines were found in a cluster with many other materials (fig. 5.16), their find context is similar to

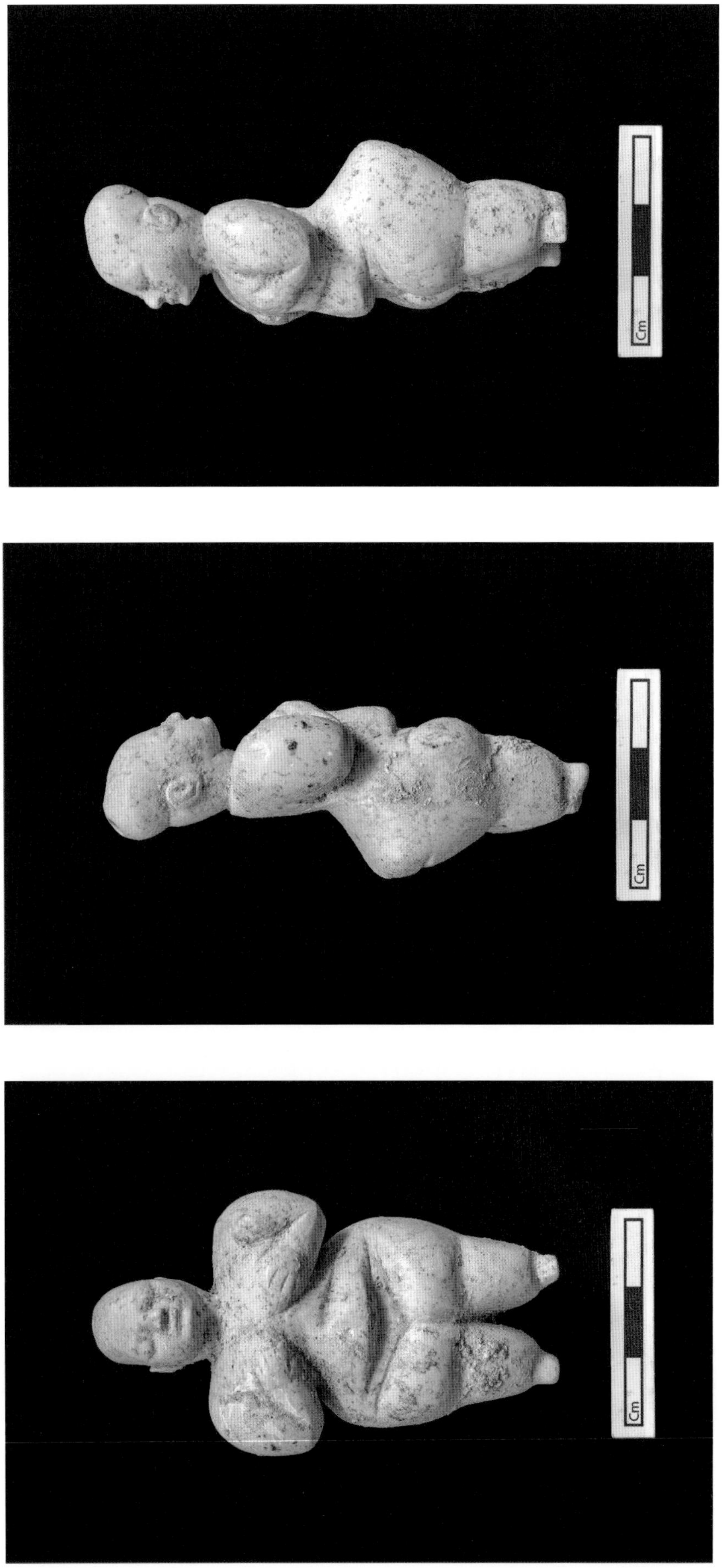

Figure 5.11. 20736.x3: (L) front view, (C) right side, (R) left side (photograph by Jason Quinlan).

Figure 5.12. 20736.x3 in situ (photograph by Jason Quinlan).

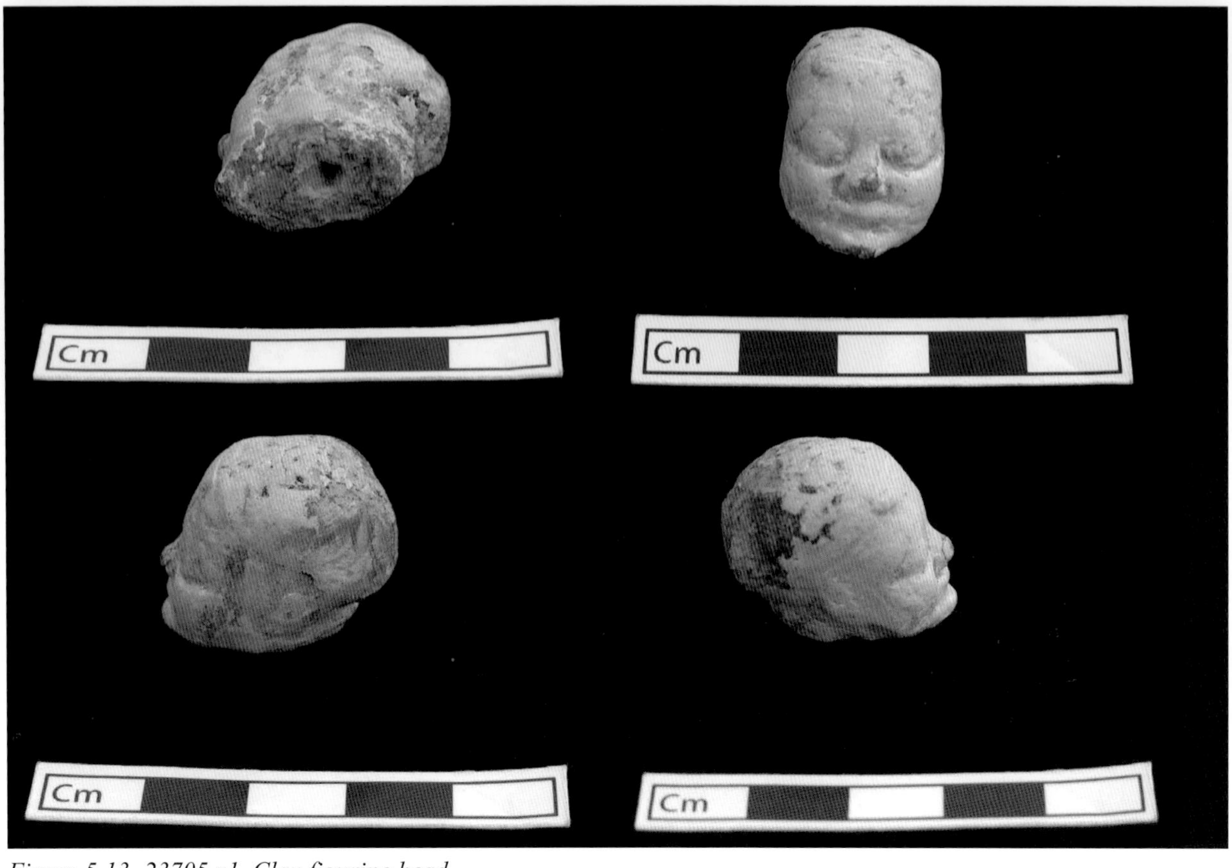

Figure 5.13. 23705.x1. Clay figurine head.

31852.x3, found in the rubble of a dwelling just above B.150. The larger 'statuette' (x2), placed on its back, was the focal point of the assemblage (fig. 5.14): a broken stone pounder/polisher (x5) was found broken side down just above the body within the lens of burnt material that sealed the entire assemblage; a palette/abrader (x7) was found northeast of the feet; a light blue-green bead (x3) was found underneath the buttock area; bones from a bird wing (x4, carpo-metatarsus) were found next to the left shoulder; an unidentified organic material, possibly hide or leather (32806.s4) was found under the body and appeared to wrap up around its right side; under this layer, a cluster of bright green malachite(?) was found under the torso and a burnt shell under the head area (not collected, but see fig. 5.17). 32806.x1 was found face down, about 10cm north of x2.

This arrangement of materials in this deposit suggests intentional placement, and the two figures vaguely resemble others found in Mellaart's 'Leopard Shrine'. 32806.x2 evokes CHC461/79-452-65 in Ankara (fig. 5.14, right). One notable difference between the Mellaart figure and 32806.x2 is that the latter is quite flattened on the back. While the overall body shape and features like the back of the ears and cap are skilfully depicted, back details are quite subtle or absent, and features such as the buttocks and head are very flattened compared to other figurines, as if it was meant to lie on its back or perhaps up against a wall. Also, like 20736.x3 above, 32806.x2 displays uneven manufacturing attention and/or skill. Again the head seems to have received more attention, as it is more finely rendered and smoothed. The head and face, depicting only the nose and ears carved explicitly, with only the suggestion of eyes and mouth, is very well executed and exceptionally smoothed with an off-white colour, showing minimal environmental wear. In contrast, the body and back of the head shows significant staining and erosion and unevenness in planning and care in execution. The lower body sports areas of damage and remnants of plaster and staining from the organic material upon which it was placed. 32806.x1 likewise vaguely evokes the form of CHC459/79-450-65 from VIA.44 (fig. 5.15 right). Both give the impression that the figure is seated, perhaps with the legs going underneath the body, although 32806.x1 is much more rounded with an elongated neck and upturned face.

The southwest corner of the building was exceptionally rich in artefacts, many of which seem to have been placed (intentionally?) in groups with other materials (see Marciniak et al. 2017). Excavators have suggested that deposits in this corner of the building may have been part of a ritual related to the closure or transition of a building phase or feature. There are indeed similarities to practices seen in the Building 65-56-44 sequence and the artefact-rich make-up deposits sealed under a platform in the southwest corner of B.49 (see above and Nakamura, Meskell 2013b: 212–14). But what is striking about the figurine deposit in this corner are the associated materials. In some sense they provide a kind of 'snapshot' of a world, not only referencing various activities such as processing and fabricating (pounder/polisher, palette/abrader, textile or leather), animals (bird wing), materials, colours and places (bead, malachite, shell), but perhaps associated/symbolic capacities and powers – the human power of forming, creating and decorating, and access to certain materials, groups and locations in the landscape. While there are no speleothems in this assemblage, such materials sourced from distant (and perhaps powerful) places may have also served a similar associative or presencing purpose in Mellaart's assemblage.

Emergent themes from the figurine corpus and future directions

The Çatalhöyük figurine corpus suggests a number of themes or questions that could be explored in future studies. Throughout our involvement in the project, most of our attention has been focused on the humble but ubiquitous finds of the figurine corpus: namely the clay materials, predominantly zoomorphic in form. As we have explored various themes and future directions of Neolithic clay figurine practice elsewhere (for summaries see Meskell et al. 2008; Nakamura, Meskell 2013b), now at the end of the project (which quite fortuitously produced several new examples), the stone figurines practices demand some attention. Excavations in the Late period levels doubled the number of stone figurines. One building in particular is responsible for this surge: B.150 (Level TPC M) and its afterlife produced five human figures rendered in stone. A cluster of other unusual objects, including a female figurine with its head intentionally removed (31852.x3), was found just above its southwest room in Sp.585 (Marciniak et al. 2015b). Only burned Building 79 (Level South O) contained more than one anthropomorphic stone figurine (18523.x1, 18545.x1), while B.58 (Level North H) contained one (10264.x1), and a burial under B.42 (Level South R) contained another (10475.x2). The others were found in Late or Final period non-building contexts (7814.x1/TP R, 12102.x1/ North I, and 15839.x10/TP O).

While the sample size remains small, a few incipient or suggestive patterns deserve some consideration. In the Late period, the number of human figurines rendered in stone increases and their deposition begins to appear more considered, as they appear not as single items, but in association with a mix of other materials. In particular, there may be some association between these figurines

Figure 5.14. Left: 32806.x2 (photograph by Jason Quinlan). Right: CHC461/79-452-65 (photograph by Belma Kulaçoğlu).

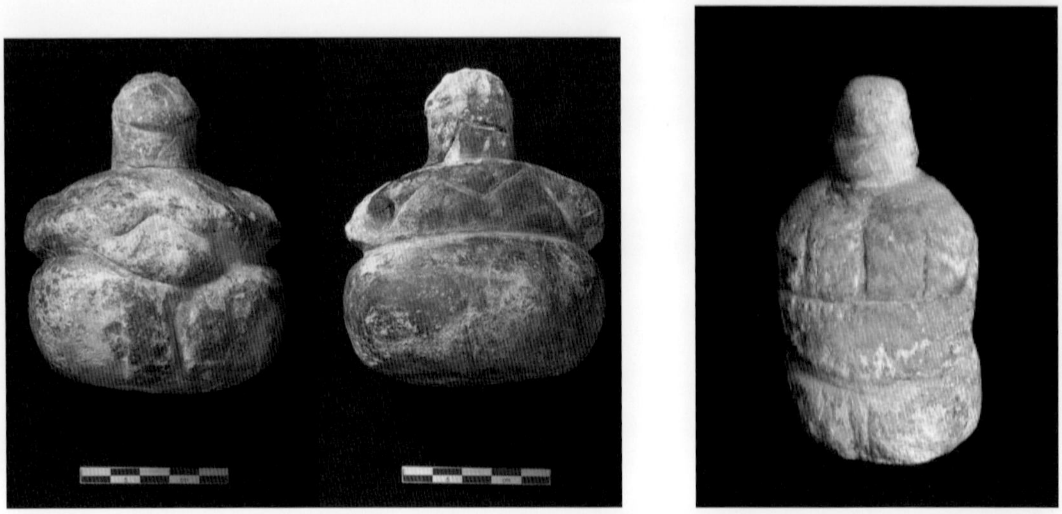

Figure 5.15. Left: 32806.x1 (front and back) (photograph by Jason Quinlan). Right: CHC459/79-450-65. 5.50cm (h) 3.00cm (w).

and materials that evoke broader networks and associations (speleothems, chipped stone, greenstone, bird wings etc.). The stone forms depict four main types (these become even more apparent when we consider Mellaart's finds) including backward-leaning figures, riding figures, abstracted (possibly seated) forms, and extended human bodies with a detailed abundance of fat and flesh (fig. 5.18). Strikingly, the Hodder excavations have only found the first two types in the Middle period (a and b), while the Late and Final periods give way to the latter two (c and d). In fact, by the Final period, only the last type (d) is found. In addition, there is a notable size differential – from miniature to statuette – amongst the stone figurines that may vaguely follow a temporal trend. The stone figurines begin palm- to hand-sized in the Middle period, with some becoming quite large in the Late period and then miniaturised by the Final period (fig. 5.19). Finally, when one considers all the stone human figural forms that have emerged from the Neolithic occupation levels of the site excavated by the Hodder team, there appears to be a decrease in the diversity of 'types' over time, settling around the corpulent, extended body form (fig. 5.18d).

While the diversity and occasionally idiosyncratic nature of the clay figurine corpus have prompted us in the past to eschew more iconographic accounts and focus more on the more fluid, ambiguous, metonymic and indexical aspects of figurine practice, the stone figurines draw attention to the whole body, particular forms and their expressive symbolic capacities. At the end of this project, it is perhaps time to speculate on what ideas and values some figurines might have expressed in the context

Figure 5.16. Group photo of (32604.x1–x7) (photograph by Jason Quinlan).

Figure 5.17. Green material and burnt shell (unrecorded) found under 32604.x2 (for colour version, see online supplementary material).

of Neolithic Çatalhöyük. The stone figurine assemblage, when considered more broadly within Çatalhöyük figurine practices, draws attention to three intersecting themes that we will discuss below: (1) elaboration/exaggeration and abbreviation as incipient formal conventions, (2) the human body as an expressive medium of power, and (3) stone figural practices as expressing a diversity of Neolithic values and affiliations.

Elaboration and abbreviation

First, large-scale patterns may suggest two different 'poles' of figural formal practice: abbreviated-abridged (horns and abbreviated bodies) on the one end and elaborated-exaggerated (animal and human bodies) on the other. While these are not strictly delimited conventions, there is some suggestion that, at least in terms of deposition, horns and abbreviated forms may have been mobilised in similar ways (see above). If this is the case, abbreviated forms may raise interesting questions about metonymic, specific and generic invocations of subjects and their relationship to power. The differential treatment of human and animal heads and bodies may also be related to such questions. Practices at Çatalhöyük demonstrate a particular attention to both animal and human heads: animal bucrania were plastered and installed on walls, benches and pilasters, while human crania were removed from interred bodies and possibly circulated. However, over time, the human head and face become a site of greater attention and elaboration, while the animal head (and possibly its associations) become visually condensed into horns. We have previously written about the possible significance of headedness and body partability for both animal and human bodies (see Meskell 2008; Nakamura, Meskell 2013b). In figurine practices, we begin to find separate human heads and bodies in Level South M/North G (n=3), which become more common in the Late period (n=16). And stone figurines in the Late period (B.150 mentioned above) demonstrate a particular attention given to the head and face. So in the figurine corpus, we find different kinds of attention given to various forms, some more detailed and elaborated, others more condensed or abbreviated. These different forms of attention do not necessarily translate into the retreat of animal power and the ascendance of human power in a Çatalhöyük worldview. Rather, the process was likely more complex. The increased visual interest in corpulent human forms and concomitant condensation of animals into horns could track with a changing social world in which interaction with other human groups and more distant places and materials was becoming more economically, politically and symbolically important. But animals were not disappearing, nor were they becoming less important (even if one argued for a decline in their symbolic importance, this might be balanced by an increase in their economic importance). Rather, human relationships to animals, and possibly ideas of the human itself, were undergoing a shift. Some figurine practices may have mediated changes in ideas and social orders; wild animals and the activities and powers associated with them were likely becoming recoded or renegotiated as their presence and status in daily and/or ritual life was challenged by the introduction of new technologies and social arrangements.

The basic difference in gestures of abbreviating and elaborating perhaps comes down to intensity and focus of concern. The ubiquitous presence of animals, their flesh, remains, imagery and installations on site suggest an in-depth knowledge and articulation of animals as symbolic and socioeconomic resources (see Meskell 2008; Twiss, Russell 2009; Meskell, Hodder 2011; Nakamura, Pels 2014). The importance and power of certain animals, for instance wild bulls, was well established by the Late period and their invocation in the expedient form of a horn (rather than as entire bodies) would still have been easily understood and recognisable. Given such familiarity with such animal power, more gestural invocations of these animals would be sufficient in many cases. While images of corpulent (or what many have interpreted as pregnant) female bodies are known since Palaeolithic times, the particular focus at Çatalhöyük on what we have interpreted as mature (even possibly ungendered) flesh rather than fertile female flesh, suggests a more specific concern for longevity and economic, social and reproductive success. This idea is also supported by our work on the burial data, where we found that some elderly women displayed some of the most elaborate, even personalised, burial assemblages (see Nakamura, Meskell 2013a). Unlike the figurines, which lack specific markers or details that would be associated with an individual, elaborate burial assemblages are quite specific, perhaps even biographical.

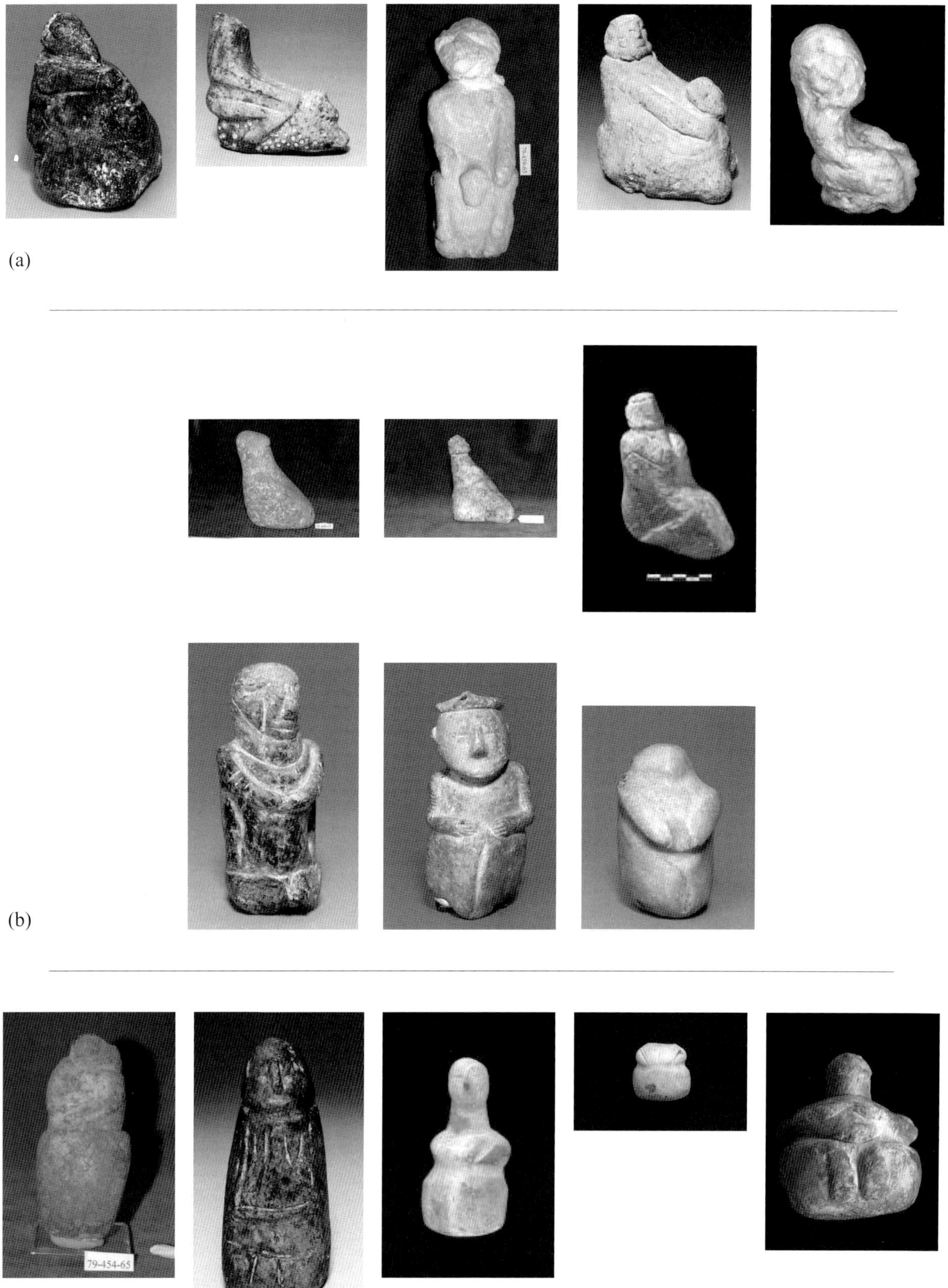

Figure 5.18a–c. Possible different anthropomorphic types including Mellaart examples: (a) riding figures: CHC160, CHC168, CHC468, CHC466, 18545.x1; (b) seated/reclined with knees up: top row (left to right): CHC476, CHC465a, 18523.x1; bottom row: CHC281, CHC169, CHC162; (c) abridged forms: CHC463, CHC167, 10264.x1, 12102.x1, 32806.x1.

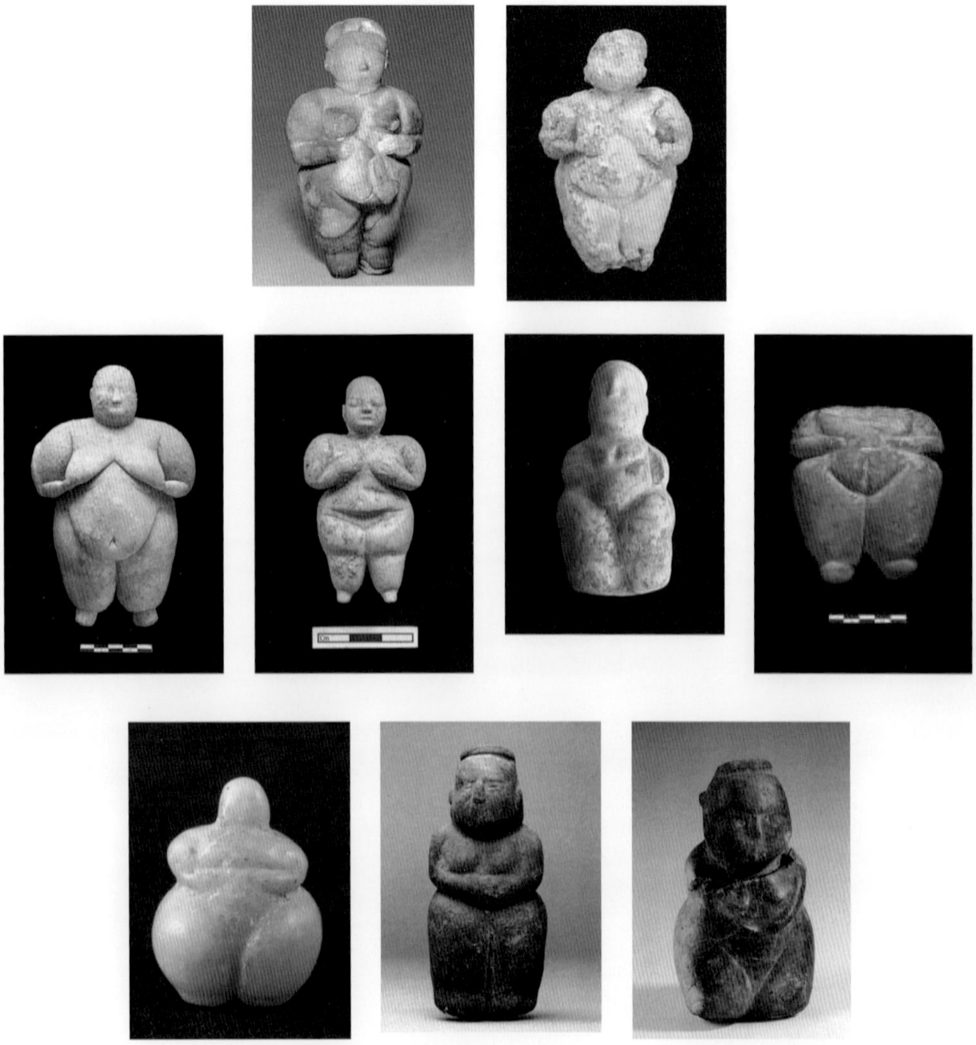

(d)

Figure 5.18d. Possible different anthropomorphic types including Mellaart examples: (d) fleshy bodies: top row (left to right): CHC21, 10475.x2; middle row: 20736.x1, 20736.x3, 32806.x2, 15839.x1; bottom row: 31852.x3, CHC461, CHC174.

While the human burial evidence may lend support to the differentiation of certain individuals, the fleshy figurine bodies express a more general but particularly human power and capacity. When rendered in stone, such figures might offer an impressive expression of the human in a form similar to the visible and physically imposing bucrania installations. Mobile rather than fixed to the house, entire body rather than body part, consuming rather than consumed, these human forms may have brokered or asserted another aspect, or even competing model, of social and symbolic power. Moreover, in an expanding landscape of social interaction and movement that afforded increased exchange and exposure to different materials and ideas, it is likely that questions of belonging and of power, of what it meant to be human or a person, would intensify.

Cultivating the human: flesh, nakedness and rest

Previously, we have elaborated on how many anthropomorphic figure bodies convey a strong impression of abundance (even excess) and maturity (Nakamura, Meskell 2009; Pearson, Meskell 2015); we now would like to push this idea further and examine how other aspects of these forms might articulate other emergent ideas of status and changing ideas of the human and power. We have argued that most renderings of breasts and bellies give the impression of bodies that are quite mature – possibly females who are not in but rather *beyond* their peak reproductive years. While generically evocative of the earlier female forms from the Palaeolithic ('Venus figurines'), these bodies display an overabundance of flesh more consistent with marks of maturity, which could emphasise a particular status

Figure 5.19. Stone figurines from Hodder excavations, earliest to latest (approximated relative size).

acquired after reproductive success and through longevity. The fat-bearing female body in particular, would have provided a potent figure of high status and success, as it could convey power and status on several levels: reproductive, economic and social success. Fatness as rendered in these figurines thus suggests a specific image of maturity, one that evokes endurance and accumulation (both nutritional and social) writ large in the flesh.

While the figurines most likely embody an idealised or specific body kind rather than actual individuals, they demonstrate some knowledge of how actual bodies accumulate fat. For instance, there is an emphasis on fleshy parts of the body that take on extra fat. Not only the breasts, belly and buttocks (3Bs), but also the upper legs and arms, as well as areas of the neck and back prone to weight gain, are emphasised, even exaggerated, compared to the hands and feet, which often appear disproportionately tiny in comparison. In this regard, the figurine bodies do make reference to real bodies – if not individuals who bear such an appearance, then bodies that accumulate fat in a physically accurate way. It is possible that the community bore or came into contact with individuals genetically predisposed towards obesity. Yet an obese bodily form is still something that requires support by certain social and economic conditions. In the Neolithic context, this support would presuppose a group or community agreement and effort. Pearson and Meskell (2013) have drawn attention to the importance of considering the physical or lived body and the representational body in tandem; specifically, they remind us that flesh is made and modified through food consumption, physical activity and health, which not only produces a specific physical appearance, but also reflects a range of lived conditions. For instance, an abundance of body fat comes not only from the liberal consumption of calorie-rich foods, but also from low levels of activity: energy ingested exceeds energy expended (ibid. 464). This metabolic process then lends itself to the construction of social status though such behaviour, especially in contexts in which food sources and procurement require substantial amounts of labour.

At Çatalhöyük a link between excess fat and status could likely arise from a confluence of factors. Both sharing and exchange economies could have, in theory, sustained perhaps a few individuals who were given good access to high-caloric food while maintaining low activity, and were thus able to cultivate fatness as a body project. Indeed, such individuals would likely have held a special status within the community and also provided an outwardly visible symbol of the group's success. It is important to emphasise, however, that in both sharing and exchange economies, the project of fat-cultivation would have been a communal or collective effort. Moreover, the individual labour of body fat accumulation can be in service to the community. In her research on Azawagh Arab communities, Rebecca Popenoe has noted how through practices of fattening, the female body becomes a site of 'anti-work'. While others produce and process, consumption itself is also a form of labour:

> Fattening is a sort of anti-work, but it is in a broader sense of the term women's main 'work' in life: their active contribution to the production of value in the society by their own act of consumption. By controlling and literally embodying the flow of milk … Azawagh Arab women express and safeguard their society's vision of itself, as well as its continuation and increase (Popenoe 2012: 123).

The project of fattening therefore can articulate a connection between the construction and care of the 'social body' through individual bodies. Similarly, in the Fijian context, Anne Becker has studied how core cultural values can become encoded in body shape and where a 'robust' shape is particularly associated with collective care: here the body is not articulated as a source of individual differentiation, but as a medium of community participation and identification:

> In Fiji, the cultivation of bodies is legitimated as a collective rather than a personal endeavour. Indeed, the community is embodied through its work on its members' bodies. ... Care, then, is concretized in anatomic space. The display of core cultural values through the medium of personal bodies is a primary expression of the ethos of care within the social body (Becker 1995: 84).

While the social world of Çatalhöyük was no doubt quite different from those of Azawagh Arab and Fijian communities, these intersecting valences of fatness, bodily cultivation and communal care might resonate with the Neolithic context more than liberal Western notions of the body as a vehicle for producing the self, personal identity, individual differentiation, status or achievement. Neither divine nor individual, robust human figurines may have instead mediated ideas of the collective care or cultivation of the social group: the individual body as a socially rather than a personally managed resource and symbol of power.

It is also worth drawing attention to the ways in which the stone (and many of the clay figurines Mellaart found) explicitly evoke images of rest and immobility, which might relate to an ideal or real cultivation of fatness. Almost all of the human figurine forms appear to be unclothed. A few clay examples from Mellaart's excavation bear some decorative markings, but all of the stone figurines depict an unclothed and unadorned body (except in some cases a possible hat or cap on the head). The choice to depict the human body naked, although common since Paleolithic times, was certainly not the only option. Wall paintings at Çatalhöyük show human figures in animal skins, and figurines elsewhere in the Anatolia Neolithic depict adornment and clothing (see examples in Hansen 2014). The unadorned body draws attention to the particular form, features and position of the body itself. In the case of Çatalhöyük, human figurines depict a form that is corpulent and robust, with certain exaggerated areas of fat accumulation and a bodily pose that evoke maturity, rest or even immobility.

In the Late period at Çatalhöyük, human figures are depicted with their legs extended (although unable to stand) or give the impression of being seated, even slightly reclining (fig. 5.19). Such postures all evoke a dignified but restful and inactive body. Here, immobility could be associated with both status and control. For instance, the more corpulent figures recall a social value in Tonga where

> a high-ranking girl symbolizes her status by her physical size, well-oiled skin, and associations with stasis. A stylized sitting posture is intended to convey a powerful image of immobility (Shore 1998: 158; see also Gifford 1929: 129–30).

The production of this female body is therefore the result of very specific social norms of beauty and collective care. Although in many ways incomparable to the egalitarian social world of Çatalhöyük, the Saharan and Polynesian examples draw attention to how tropes of fatness and stasis can contribute to a symbolic lexicon of communal status and identity that might be at work in the Neolithic community. The ability to support bodies that materialise excess – of flesh and rest – during a period of shifting alliances and economies would have provided a potent symbol of group success and status, one notably more focused on collective human effort than animal power.

Human, animals, materiality and social power

While clay figurine practices resist any singular interpretive narrative (and seem to be somewhat formally conservative rather than innovative over time), the stone examples register a discernible shift in attention to and development of the human form and its social significance. Across the site and over time, we find the same basic forms in clay: anthropomorphic forms, zoomorphic horns and quadrupeds, and abbreviated forms. While various examples and trends reveal some elaboration or diversification amongst these types (see discussion above), such 'innovations' maintain rather than transform or reinvent these four formal idioms. Furthermore, the differences in the distribution of clay figurine types across the site and in various contexts are subtle and only suggest, rather than clearly indicate, certain shifts or associations. Prior to 2008, when the project had only recovered four stone examples, we argued that aside from form (all stone examples depicted anthropomorphic bodies), there did not appear to be any clear differentiation between how stone and clay figurines were treated. Now with a total of at least 12 examples from the Hodder excavations, we would suggest that stone figurines may have participated in a more specific set of practices than their more varied clay counterparts. While the contexts for such figurines are limited, the choice of material, emergence and development of specific anthropomorphic types, and increasing focus on exaggerated corpulent forms may have articulated a significant ontological and/or symbolic shift in how members of the Neolithic community constituted social power. We would like to suggest that changes in how humans related to animals and the environment, under the turn towards animal domestication and agriculture and increased interactions and trade with other groups, coincided with a new or emergent worldview in which human power became more of a socio-symbolic focus.

From clay to stone. Perhaps the most obvious aspect of figurine practice from which such a turn could be argued is in the choice of rendering certain human forms in stone from the Middle levels onward. Clay figurines evoked rather expedient practices. Even finely rendered examples did not require a high investment of time for making such forms or developing the requisite skill set to do so. Moreover, given the dominance of clay in the Neolithic as the endlessly reusable material of daily life, it is notable that, at first, human figures were rendered in the same material that not only gave form to animal figures, mudbricks, pottery and cooking balls, but could easily be remixed, reformed, destroyed or repurposed. While the choice to render human figures in clay may have been for reasons of expedience, or lack of technical expertise, it is also possible that such a choice reflected an implicit or explicit notion of the human as part of, rather than apart from, this greater world of animals, substances and landscapes. However, it is intriguing that, just before we see evidence for a significant shift in circumstances and social organisation that likely coincided with an intensification of animal domestication and agriculture, we find the first examples of human figures rendered in limestone, marble and speleothem – materials that are not the stuff of local, remixable daily life, but enduring, hard substances that come from more distant sources (at least 20–100 km from site) and are comparatively rare on site (Bains et al. 2013; Erdoğu 2013).

The choice to render (or focus on) only certain figural forms in stone (while others remained in clay) was a gesture of setting apart, of drawing attention to some symbolic or even ontological difference. Such choices likely drew upon specific qualities of stone materials, including hardness, colour, capacity for smoothness and luminosity, but also their sources and biographies. Mellaart was inclined to interpret all anthropomorphic figurines as either gods or goddesses regardless of material. Several scholars have offered convincing critiques of Mellaart's ideas (Meskell 1995; Voigt 2000), so we will not rehearse them presently. We remain unconvinced that stone figurines, much less all anthropomorphic figurines, embody divinities, not least because most are found discarded or dumped, even if done with some consideration (for instance at transitional moments in the lives of buildings). However, we do suspect that the choice of rendering certain figural forms in stone was significant, and perhaps signalled a shift in the worldview (at least for some) in the Neolithic community of Çatalhöyük. It is almost assured that people during Neolithic times were well aware of the various properties and aspects (from sources to biographies) of different materials and mobilised these in their decisions to use and transform them. The choice to make or bring to site specific human forms not only in a new material, but in one that is rare, with distant origins, and requires considerable skill to transform, would appear to assert a different status for what the rendered form represents, namely the human.

Previous studies of various aspects of Çatalhöyük have pointed to the practical dominance of clay in Neolithic life (Stevanovic 1997; Tringham 2000; Doherty 2017), which likely inflected the symbolic realm as well. Moreover, the house and wild animals dominate social and symbolic registers for most of the Neolithic occupation; iconographically, humans share the stage with, or are even eclipsed by, wild animals until the Late period. It is therefore possible that the arrival of stone figurines at the end of the Middle period announces an emergent focus on the human as something radically different from the clay stuff of life. It is notable that figurines in stone display more bodily specificity even when it is harder to achieve, whereas figurines in clay tend not to exploit the full possibilities of this more malleable material – not only in adding details such as decoration or clothing, but even in the skilful rendering of the bodily form itself. Not only does stone offer a different set of material qualities and associations, it also requires a particular level of skill to effect its transformation. Stronger, more durable, with more varied qualities and properties than clay, and sourced from possibly mysterious or powerful places (for instance caves), stone materials offer a broadened range of qualities and associations and require a distinctly human hand to masterfully bring new form into being. As such, stone is an ideal material for humans to announce their aspirations for a new world and their status within it.

Competing allegiances or multiple ways of being human? But anthropomorphic forms still persisted in clay. So if not as materialisations of divine power, then how might we interpret these stone figures? Preliminary findings from the most recent excavations have prompted Hodder to suggest that the 'radically egalitarian' Neolithic society may have started to show signs of increasing specialisation and the emergence of individuals in later levels, possibly priming the way for social stratification (see discussion above). While it might be tempting to interpret the newly excavated stone figures as evidence of high-status individuals and therefore as providing support to such arguments, we resist this interpretation for a number of reasons. First, while such figurines do not show a high level of standardisation, they do conform to some general conventions of form and pose that appear to coalesce into four general stone types (see above, fig. 5.18); furthermore, visible idiosyncrasies of the different examples do not evoke distinctive characteristics of individuals (for example, ornamentation, details or features, or identifying markers), but instead likely indicate the maker's specific vision in negotiation with the raw form of the material. Second, attention to aspects of pose and bodily form over clearly sexed bodies suggests a more generic representation, one that is more appropriate for the representation of an ideal rather than an actual body. Third, the bodily forms may articulate notions of social power, but not along the gendered axis of divinity that Mellaart proposed; we suggest that they might instead reveal transitional or competing ideas and allegiances to different worldviews and ways of life.

Although the sample size is admittedly small, the stone figurine assemblage suggests an intriguing scenario. The earliest stone figurine examples (both from Building 79, Level South O) produced two types only seen in the Mellaart assemblage: the riding figure (18545.x1) and a seated, reclining bearded figure (18523.x1). Mellaart has interpreted both of these figures as male (see fig. 5.18a, b). However, like the corpulent 'female' forms, these two examples lack explicit sex features (see Nakamura, Meskell 2009), although when viewed from different perspectives they both evoke a phallic form. In the Late period, we see a distinct convergence around a kind of robust or corpulent bodily form, some with stunning detail and execution and others more abstract and gestural of fat and excess flesh (figs 5.15, 5.16). All of these figures are found in buildings, with B.150 alone producing five examples (see discussion above). By the Final period, these corpulent forms further become miniaturised and even stylised and are found in midden and burial fill, outside of buildings. This assemblage reveals two distinct idioms of social power: the riding figure articulates power through an image of the human control of a wild animal, while the reclined and corpulent human figures suggest power as a capacity for rest and excessive consumption. Given the small sample size (and also given that Mellaart recorded finding all stone figurine types together in the same building, see VIA.44) we are reluctant to interpret the temporal arrangement of figures as evidence for an evolution of social power in which the domestic comes to eclipse the wild, or the human eclipses the animal. But given this minimal suggestion of a shift, we do find it productive to consider how the specific forms of stone figurines might embody different relationships between humans and animals and ideas of the humans in the world more generally.

In the examples of seated or riding average-bodied individuals (fig. 5.18a, b), focus is drawn towards the body's position or activity. The riding figures depict a human controlling a presumably wild animal and the reclined/seated figures evoke inactivity or even immobility. Considering the former, even though the human is dominating the animal, the animal is present, which suggests a significant relationship between the two, and importantly the power not just of the human but also of the animal. Here the human draws its power from the control of a powerful being. The site has produced much evidence for hunting wild animals, particularly aurochs, as a focal point for social power. Furthermore, even as hunting (indicated by the presence of wild animals on site) became less common, hunting imagery and symbolism appear to intensify in the later levels (see Russell et al. 2014: 216). Hunting and wild bulls in particular appear to have maintained significant social prestige in the Late period, at least in the symbolic sphere. Human bodies evocative of power associated with the hunt would likely have been active, strong and lean, perhaps not unlike the two figurines found in Building 79.

In contrast, bodies evocative of economic power derived from herding, agriculture and accumulation of animal wealth and control of land would require a more abstracted figure of such power. Thinking speculatively in this direction, one could argue that the fleshy corpulent figural forms we find in the Late period onward provide a rather apt expression of such power. Hunting and herding capacities articulate different relationships between humans and animals, but both evoke a similar power in the ability to transform animal flesh into economic and social power. While the riding figure draws attention to the human appropriation of wild animal power, the corpulent figure at rest embodies the incorporation (though consumption) of animal flesh and other managed food resources into excessive human flesh and fat. In the latter, the focus is entirely on the human capacity to accumulate by consuming, controlling and managing animal and plant life.

While further studies may spar over whether in time the 'domestic' model of social power supersedes the 'hunting' model at Çatalhöyük, it is worth considering these instead as competing and even complementary modes of Neolithic life, perhaps revealing an added layer of complexity rather than assuming a problematic linear narrative of human 'progress'. It is possible, even likely, that households mobilised different pathways to success. For instance, future studies might explore possible relationships between corpulent human figurines, history houses, and control of land and animal wealth; they could also explore the idea that some non-history houses, such as B.49 or B.79, with strong associations with animals, may have been regarded as no less 'successful' or powerful, but instead have pursued economic and social power through different strategies. Such questions aside, it appears that stone figurines may have provided one medium in which different ideas of power, community and being human were expressed and negotiated. In this way, certain differences and tensions, which likely animated a social world that supported various lifeways (hunting, foraging, trade, agriculture and animal management) and even worldviews, may have been figured through the body. The introduction of stone human figures might suggest that with the gradual adoption of agriculture, animal management and increased trade and mobility, a more human-centred idea of the world had started to emerge or solidify. Regardless, the enduring importance of wild animals and their associated activities and powers would have cast a long shadow in the realms of human memory, myth, and for some, in daily life.

Supplementary material

For supplementary material related to this chapter, please visit https://doi.org/10.18866/BIAA/e-14. It comprises a colour version of figure 5.17.

6. Tracing human-animal relations across time and space through animal materialisations

Lindsay Der

Çatalhöyük is perhaps best known for its elaborated buildings, an array of iconography and a constellation of rich material culture, much of which is indicative of ritual and symbolic practice. For instance, geometric paintings on house walls were plastered and repainted over hundreds of layers. Inhabitants of some houses engaged in rituals around mortuary practices, such as the post-mortem manipulation of human body parts, including practices of defleshing, decapitation, cranial retrieval, curation and redeposition. Yet, in the ritual and symbolic sphere, it is actually the materialisations and visual expressions of animals, primarily wild, dangerous creatures, that are the most conspicuous.

This chapter examines the temporal and spatial patterning of animal materialisations during the Neolithic occupation of Çatalhöyük's East Mound. How do these materialisations speak to the active role that animals played in the lives of Çatalhöyük's residents? Was the relationship between humans and animals static or one marked by change through time? How do patterns within the corpus of animal materialisations compare to other evidence regarding social differentiation at the origins of agriculture? The data demonstrate that these materialisations were not purely aesthetic devices, but a means by which animals were embedded within the social bonds at the site.

Investigating animal materialisations at Çatalhöyük

Previous work

Numerous scholars have discussed the Çatalhöyük animal iconography situated within the context of Neolithic symbolism and/or religion (Mellaart 1967; Cauvin 2002; Relke 2007; Watkins 2007; Hodder, Meskell 2010; 2011; Stordeur 2010). Analytical studies on faunal materialisations outside of these contexts are relatively limited. The first study was conducted by Russell and Meece (2005), who published their extensive survey of the faunal remains in conjunction with zoomorphic iconography at the site. They noted a clear preference for *Bos Taurus* horn cores in faunal installations and special deposits over those of sheep and goat. Their observation that animal materialisations at Çatalhöyük overwhelmingly consist of wild, rather than domesticated beasts, in both the visual culture and the physical skeletal remains, is one that has since been repeatedly confirmed by other scholars examining the faunal and figurine assemblages. Seven years later, Martin and Meskell (2012) compared quadruped figurine morphologies to those found in the natural world, using zooarchaeological taxa identification methodologies. Their work found that although stylistically schematic, there is a consistent likeness to real animal bodies. While a range of species can be identified, the majority represent wild animals, based on physical attributes such as ear position and horn shape. With almost 50% of quadruped figurines identifiable as wild cattle, the pre-eminence of cattle types matches the centrality of cattle seen in other iconographic and symbolic practices at the site. The present study investigates the intersection of material culture, human-animal relations and social organisation through time, adding to the aforementioned body of work pioneered by Russell and Meece, and Martin and Meskell.

Materials and methods

The house was utilised as the base unit of analysis. All buildings at Çatalhöyük are considered to be residential, domestic places due to consistency of layout and architectural features, as well as evidence of food processing, consumption and tool production (Hodder, Cessford 2004; Hodder 2006c; 2010b; 2014f). Beyond simply a place of residence, the house is also a legitimate and tangible social unit (Joyce, Gillespie 2000; Hendon 2004; Düring, Marciniak 2005; Borić 2007). A total of 33 houses from the North Area and an additional 32 houses from the South Area were sampled. These two areas comprise the bulk of buildings excavated by the Çatalhöyük Research Project and allow for an intra-site comparison of spatially differentiated areas. Houses from the Final grouping of occupational levels and the TP/TPC Areas were not included in this analysis as they comprised an insufficient sample size. Houses were categorised into three chronological groups for each excavation area, based on where they fell within the broader temporal groupings of levels: Early, Middle, and Late (table 6.1). This analysis was conducted prior to the finalisation of the Çatalhöyük East Temporal Groupings by Levels. The differences between the groupings are minor and do not affect the overall trends seen at the site. Mainly, the approach taken here splits the earliest levels in the North into two time periods in order to capture variability across time for the North Area. Such a grouping is justified as there is a distinct break in economic and ritual practice that occurs in the middle levels, presenting a chronological point of reference. By grouping houses according to occupational levels, it is possible to detect patterns of change over time.

Qualitative and quantitative datasets were synthesised, drawing on the large quantity of data amassed by the Çatalhöyük Research Project's specialist labs, excavation data, archival material and museum collections. Zoomorphic figurines and taxa, frequency and density information for the faunal remains and architectural installations associated with houses were treated as the primary datasets. As well, secondary datasets derived and/or calculated from excavation records were used in order to incorporate data pertaining to individual houses, including the volume of soil removed in excavation, number of burials, the presence or absence of burning during house closure and building size (m^2). Qualitative data pertaining to the wall paintings, moulded wall reliefs and stamp seals was integrated in order to identify patterns through time and space.

In summary, the animal materialisations examined include:

a) Faunal installations (fig. 6.1a) inclusive of wall, bench and pillar features consisting of animal skeletal remains, such as horn cores and bucrania. These materialisations are differentially distributed among houses. Some faunal installations were abstracted into lumps through plastering, while others were more readily visible.

b) Structured faunal deposits, other than faunal installations, deemed special for both their treatment and placement and likely related to ritual practices, such as pit, platform, burial and foundation contexts (Russell, Martin 2005; Russell, Meece 2005).

c) Plastered wall reliefs (fig. 6.1b).

d) Wall paintings (fig. 6.1c).

e) Portable objects inclusive of figurines (fig. 6.1d), stamp seals (fig. 6.1e) and pottery.

A two-pronged approach comprising Geographic Information Systems (GIS) and multivariate statistical analyses was taken. Artefact densities were used whenever possible as frequencies are more susceptible to skew from excavation bias across occupational layers and geographic areas of the site. Densities calculated using total deposit volumes removed per unit (kl) prior to dry sieving are preferable, however, these quantities were only reliably recorded from the 2000 field season to present and some volumes have been lost or are unable to be recalculated (Mazzucato 2013: 31). Some areas of the site, including the former BACH Area, have little to no volumes recorded as they were run as partly independent excavations using variations of the overall recording methodologies. As such, whenever total deposit volumes removed were missing from the excavation data, a total estimated volume (m^3) excavated for each building was calculated instead (tables 6.2 and 6.3), drawing on an aggregation of data found primarily in the excavation records (wall height and length, area of spaces, levels) and publication descriptions for individual buildings (Hodder 2014a). Whenever possible, building size was also calculated.

Data collected was analysed for the following markers:

Levels	*Building*
North Area	
Early North (North F)	B.5, B.83, B.84, B.132
Middle North (North G, BACH G)	B.1, B.3, B.48, B.49, B.51, B.52, B.59, B.66, B.77, B.82, B.114, B.119, B.128
Late North (North H through J)	B.12, B.13, B.45, B.46, B.47, 54, B.55, B.57, B.58, B.60. B64, B.67, B.70, B.71, B.88, B.108
South Area	
Early South (South H through M)	B.2, B.4, B.6, B.8, B.16, B.17, B.18, B.20, B.23, B.24, B.40, B.43, B.50, B.104, B.118*
Middle South (South O)	B.76, B.79, B.80, B.86, B.87, B.96, B.97
Late South (South P through T)	B.10, B.42, B.44, B.53, B.56, B.65, B.68, B.75**

*Table 6.1. Buildings by time. *B.9 (South J) excluded as only one unit was recorded and measurement information is scarce; **B.380 (South P) excluded as unclear if this was actually an external space.*

Figure 6.1. Examples of animal materialisations: (a) faunal installation of horn cores in Building 77 (photograph by Jason Quinlan); (b) plastered wall relief of two leopards (photograph by James Mellaart); (c) wall painting of people baiting a wild bull (photograph by Ian Todd); (d) zoomorphic figurine (photograph by Lynn Meskell); (e) stamp seal with leopard spots (photograph by Jason Quinlan).

1) The exclusivity or non-exclusivity and the presence and absence of wild faunal species through time.

2) The exclusivity or non-exclusivity and the presence and absence of materialisation forms through time.

3) The correlation or lack of correlation between wealth accumulation and the consumption of wild and managed cattle as seen via contemporaneous changes in house layout, the form and make up of animal materialisations, and the amount of and taxa found in feasting deposits.

A closer look at figural and faunal horns

From the primary datasets of zoomorphic figurines and structured faunal deposits and installations, the body part representation of cranial appendages was selected for concentrated analysis. Cranial appendages, or horns, are found across all three assemblages. Body part representation has traditionally been used as a methodology in zooarchaeology for answering questions concerning subsistence strategies or economy, including butchering practices and carcass transport (Perkins, Daly 1968; Marean et al. 1992; Klein et al. 1999; Schoville, Otárola-Castillo 2014), as well as utility indices or meat procurement (Binford 1978; Metcalfe, Jones 1988; Morin, Ready 2013; Niven 2013). Still other scholars have recognised that body part representation patterning may also be a cultural outcome (Marciniak 1999; 2005; Russell 2012a; Sykes 2014) related to symbolic practice (Szynkiewicz 1990; Finlayson et al. 2012), ritual (Brown 2005; Davis 2008), social identity and status (Sykes 2014; Allentuck 2015), magic (Choyke 2010) and taboo (Politis, Saunders 2002).

It should be noted that this analysis makes use of a folk taxonomy classification of the cranial appendages. Folk taxonomies, unlike scientific classification systems, are phenomenological, rather than morphological, emphasising people's engagement with the world and relations with other things (Zedeño 2009: 407). Such a taxonomic system, with its fuzziness, may be more appropriate in considerations of the Çatalhöyük 'horns' than Linnaean classifications. Zooarchaeologists and figurines specialists alike have acknowledged that prehistoric ontologies were likely discordant from Western ones and that these ontologies would have informed interactions with faunal and zoomorphic objects (Marciniak 2005; Martin, Meskell 2012; Hill 2013; O'Connor 2013; Sykes 2014; Nanoglou 2009). Here, horns encompass pointed or tapering appendages that protrude from the head and were used by animals for defence and the assertion of dominance. This vernacular category includes antlers, tusks and bucrania, that is, cranial projections. The inhabitants of Çatalhöyük may have engaged with such an umbrella grouping of these body parts, rather than splitting each morphological type into discrete classes.

Building	Area (m^2)	Estimated Volume (m^3)
Early North		
5	50.50	37.96
83	46.20	14.55
84	14.00	2.10
132	3.42	0.27
	Total	54.89
Middle North		
1	48.60	46.17
3	31.40	36.90
48	11.76	2.48
49	14.00	15.51
51	11.61	2.81
52*	43.00	28.58
59	74.03	66.63
66	14.59	3.72
77	27.56	40.32
82	11.28	10.94
114	10.62	8.41
119	34.72	43.75
128	19.37	7.94
	Total	314.15
Late North		
12	30.24	26.19
13	3.20	71.04
45	54.02	3.40
46	14.15	9.24
47	34.79	16.65
54	37.41	58.03
55	65.20	15.56
57	24.50	20.28
58	26.00	18.74
60	33.47	25.98
64	31.68	5.91
67	22.292	4.02
70	10.30	1.17
71	4.66	5.37
88	13.77	13.61
108	29.93	71.04
	Total	296.28

*Table 6.2. Area and estimated volume for buildings in the North Area and across time. *B.52 volume has been adjusted to account for the construction of B.51 inside the footprint of B.52.*

Building	Area (m^2)	Estimated Volume (m^3)
Early South		
2	15.38	24.54
4	12.04	13.72
6	41.40	9.54
8	1.02	0.53
16	25.65	30.78
17	37.44	68.19
18	27.73	30.50
20		0.34
23	21.84	12.23
24	8.34	8.34
40	52.48	18.63
43		5.50
50	15.32	6.11
104	11.27	1.77
118	14.53	16.42
	Total	247.15
Middle South		
76	28.74	13.22
79	12.19	5.79
80	29.76	51.78
86	2.03	0.28
87	4.23	4.74
96		25.05
97		13.86
	Total	114.72
Late South		
10	28.71	25.98
42	21.50	18.49
44	29.68	53.82
53		5.38
56	35.77	37.20
65	27.97	25.17
68	13.31	7.45
75	12.09	6.23
	Total	179.74

Table 6.3. Area and estimated volume for buildings in the South Area and across time.

Such a grouping is not unusual to either folk or scientific taxonomies (Prothero, Schoch 2002; Bubenik, Bubenik 2012). This category seems especially appropriate to the situation at Çatalhöyük, given that in deer quadruped figurines, antlers are straight and devoid of tines, taking on a more horn-like form rather than a naturalistic one (Martin, Meskell 2012: 12).

Horns are pervasive at the site. The ubiquity of horns, as well as their nutritional deficiencies, suggests they constituted a distinct and culturally valued category at Çatalhöyük. They dominate architectural installations and special deposits (Russell, Martin 2005: 47–49) and are an overrepresented skeletal element in buildings (fig. 6.2). A disproportionate presence of horns is surprising considering that horns are heavy to transport, are susceptible to attritional processes and contain no meat (Russell, Martin 2005; Russell et al. 2014). The vernacular category of faunal horns at Çatalhöyük consists of horn cores of caprines and cattle, as well as the tusks of wild boar and the antlers of cervids, predominantly red deer. Cattle remains are predominantly found in feasting remains, faunal installations, or other structured deposits. Wild aurochsen appear to have been preferentially selected (Russell et al. 2013b). Likewise, the majority of caprine horn cores for which domestication can be ascertained have been identified as either morphologically wild or probably wild. Sheep and goat make up the minority of horn cores which have been found in faunal installations or structured deposits (Russell, Meece 2005: 219). For this analysis, 13,845 bone fragments for faunal horn cores, antlers, bucrania and tusks, quantified by number of identified specimens (NISP) were considered. Even though NISP counts are susceptible to bias from fragmentation, they can be applied at different scales and are most easily calculated. The faunal remains team has consistently used NISP as a recording method during the Hodder excavations, having ceased the practice of quantification by number of identified elements in 2007.

From the figurine assemblage, a total of 152 horn forms were analysed, with 86 derived from the North Area and another 66 excavated in the South Area. Nakamura (Chapter 5) has noted that horns account for more than 54% of all zoomorphic figurine forms, comprising the most common figural form type, or more than a third of the total figurine corpus. Figural horns have been recorded as one of eight form types: bucranium, curved horn, flat horn, horn/tail, horn(?), straight horn and tusk (fig. 6.3). Figural horn types can also be attributed to taxa, drawing upon Martin and Meskell's zooarchaeological taxa identification methodologies (2012). Curved horns, which have the highest frequency amongst figural horn types, are morphologically similar to the three wild cattle horn core types (fig. 6.4) identified by Russell, Martin and Twiss in the faunal assemblage (2009). Thus, curved horn types are indicative of cattle, rather than caprines. Tusk types differ from curved horns as they are slightly flattened and taper to a distinct point. Figural tusks are rare, with only four recorded figurines in the entire corpus. Straight horns have been interpreted as deer due to antlers for cervid quadruped figurines having this morphology. Flat horn types are caprine, mirroring the flatter cross-section that distinguishes bovids from caprines in faunal taxa identification. Quadruped figurines identified as sheep or goat display these flattened horns, although stand-alone flat horns are relatively rare, with a frequency of 26.

Although figurines, as a whole, tend to be found in external deposits, zoomorphic figurines are more equally distributed inside and outside houses (Meskell et al. 2008: 144). The secondary context of figurines suggests a high degree of circulation and use. These were not artefacts that were placed within houses for 'staging', but were situationally deposited, potentially reinforcing ties to specific houses, individuals or practices (Meskell et al. 2007; Nakamura, Meskell 2009). Figural horns were not 'special' objects, but more quotidian ones, engaged with the everyday lives of the people interacting with them. Thus, Meskell et al. argue that the importance of figurines was in their making, their coming into being, rather than the finished product holding some sort of inalienable power (Meskell et al. 2007; Meskell 2013; Nakamura, Meskell 2013b). On the other hand, the depositional context of faunal remains is highly variable, with special deposits appearing in middens, fills, house floors, architectural installations or foundation trenches (Russell, Meece 2005). Still, faunal remains used in architectural installations, like the figurines, underwent a life cycle of placement, dismantlement, burial and retrieval. Though they sometimes appear in house closure deposits, these were objects linked to the houses where they were used, even if their final depositional context does not reflect their active use (Russell, Wright et al. 2014).

Results

A total of 21 houses in the North Area were identified as having faunal and/or figural remains (table 6.4), while all houses in the South Area contained at least one of these object types (table 6.5). Figural horns are found across 13 buildings, while faunal horns come from 43 houses out of a possible 65 excavated or partially excavated structures in the North and South Areas.

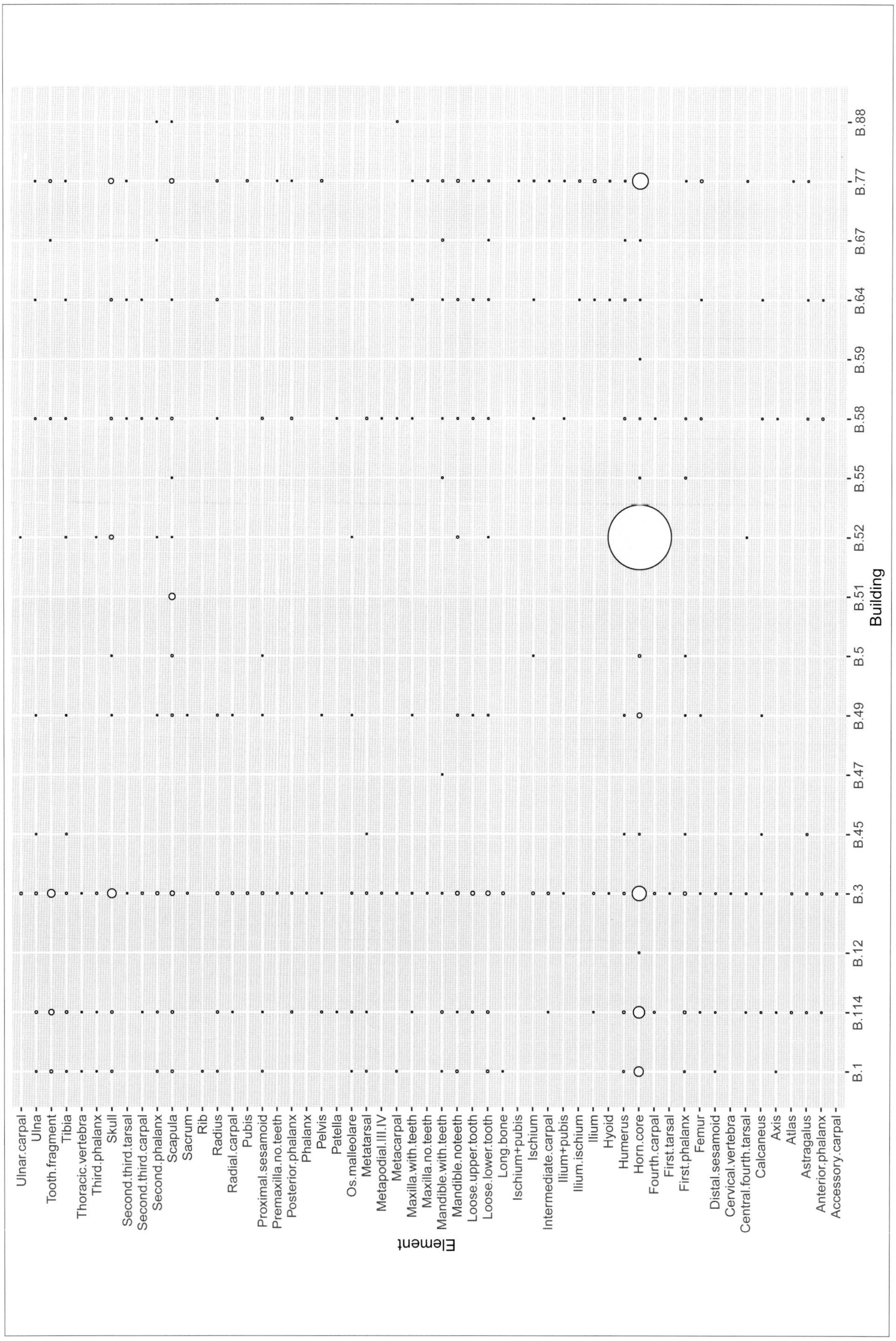

Figure 6.2. Relative proportion of Bos *body parts by NISP in the North Area, demonstrating the overrepresentation of horn cores in houses.*

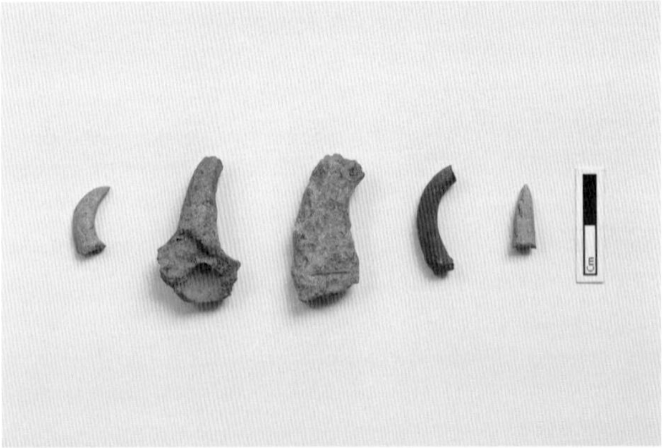

Figure 6.3. Figurine horn types, from left to right: tusk, bucrania, flat horn, curved horn, straight horn (photograph by Jason Quinlan).

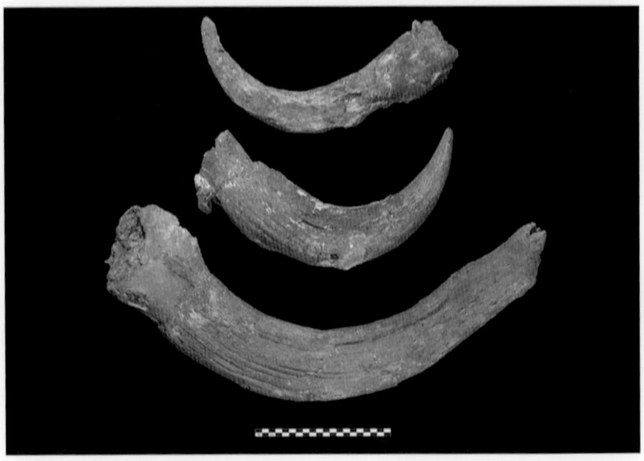

Figure 6.4. Russell and Twiss's (2009) three wild cattle horn types (photograph by Jason Quinlan).

Horn densities over time

Overall, densities of zoomorphic figurine forms increase from the Middle to Late periods. However, when looking specifically at house contexts, both densities of faunal (table 6.6) and figural horns (table 6.7) peak in the middle levels for both the North and South Areas. As well, the densities of figural and faunal horns are more widely distributed across houses in the later levels, with an increasing percentage of houses containing horns. Likewise, spatial mapping of the densities for the North Area (figs 6.5–7) demonstrates a positive relationship between the presence of faunal and figural horns in the middle levels with some spatial clustering of houses with high densities of both skeletal and figural horns. A Ward's Method dendrogram cluster analysis confirms the clustering of these houses by horn densities (fig. 6.8). This cluster consists of history houses, elaborate buildings and those structures within which multiple burials were interred (B.1, B.49, B.52, and B.77) (Hodder, Farid 2014).

Building	*Level*	*Fauna*	*Figurines*
1	North G	X	X
3	BACH G	X	X
5	North F	X	X
12	North H	X	
13	North H		
45	North H	X	
46	North H		
47	North J	X	X
48	North G		
49	North G	X	X
51	North G	X	
52	North G	X	X
54	North I	X	
55	North H	X	
57	North H	X	X
58	North H	X	X
59	North G	X	X
60	North H	X	X
64	North H	X	X
66	North G		
67	North H	X	
70	North I		
71	North I		
77	North G	X	X
82	North G	X	
83	North F		
84	North F		
88	North H	X	
108	North H		X
114	BACH G	X	X
119	North G	*	X
128	North G	**	
132	North F	***	

*Table 6.4. Absence or presence of faunal and figural remains in North Area houses. *B.119 (30127) faunal data excluded as appears to be rubble from a potentially collapsed wall; **B.128 (15406) faunal data excluded as unit is classified as an external dump and is part of Foundation Trench 3, (15411) faunal data excluded as part of Foundation Trench 3 and surface scrape (Scrape G); ***B.132 (20965) faunal data excluded as is a midden deposited post-infill and above collapsed roof/plaster remains.*

It is not unexpected to find correlation between the real animal body parts and the clay horns in these classes of buildings, considering that horn cores and bucrania were commonly used in house elaboration.

In contrast, the South Area displays less spatial patterning by horn densities. The only detectable pattern is houses with low densities of faunal horns also displaying low densities of figurine horns across all occupational levels (figs 6.9–11). In both areas of the site, buildings belonging to history house sequences in the later levels (B.60 in the North; B.10, B.44, B.56, B.65 in the South) do not seem to continue the correlation between high densities of faunal and figural horns that we see in the Middle period North Area history houses.

Bivariate correlation analysis using the non-parametric Kendall's tau returned a tau score of 0.6465 with a p-value of 0.005, indicating a strong, significant correlation between the faunal and figural horns for the Middle period North Area houses (table 6.8). This relationship is especially strong when only unburned houses are measured. Horns in these houses display a monotonic relationship, that is, a relationship that is nonlinear but follows a pattern of both variables increasing in measurement in tandem. There was no significant correlation for the Late period North Area houses, and the Early period North sample size was statistically insufficient for testing bivariate correlation. The South Area shows no statistically significant relationship between clay and skeletal horn densities. Of note, a Ward's Method cluster analysis on the South Area buildings created a positive grouping for houses burned during closure (Buildings 76, 79, 80 and 97).

Building	*Level*	*Fauna*	*Figurines*
2	South K	X	X
4	South L	X	
6	South L	X	X
8	South M	X	X
10	South T	X	
16	South K	X	X
17	South K	X	X
18	South J	X	X
20	South M	X	
23	South J	X	X
24	South M	X	
40	South M	X	X
42	South R	X	X
43	South L	X	X
44	South S	X	X
50	South M	X	X
53	South Q	X	X
56	South R	X	X
65	South Q	X	X
68	South Q	X	X
75	South P	X	X
76	South O	X	X
79	South O	X	X
80	South O	X	X
86	South O	X	
87	South O	X	X
96	South O	X	X
97	South O	X	X
104	South M	X	
118	South H	X	X

Table 6.5. Absence or presence of faunal and figural remains in South Area houses.

House group	*Frequency (NISP)*	*Density*	*% Houses with horns*
North Area			
Early North	16	0.42	25
Middle North	11,681	37.18	63.6
Late North	99	37.5	56.3
South Area			
Early South	503	2.04	75
Middle South	1,387	12.09	100
Late South	159	0.88	77.8

Table 6.6. Summary of faunal horn density in houses in the Early, Middle and Late periods.

House group	*Frequency (NISP)*	*Density*	*% Houses with horns*
North Area			
Early North	6	0.11	25
Middle North	75	0.24	61.5
Late North	5	0.02	25
South Area			
Early South	19	0.08	37.5
Middle South	27	0.24	85.7
Late South	20	0.11	44

Table 6.7. Summary of figural horn density in houses in the Early, Middle and Late periods.

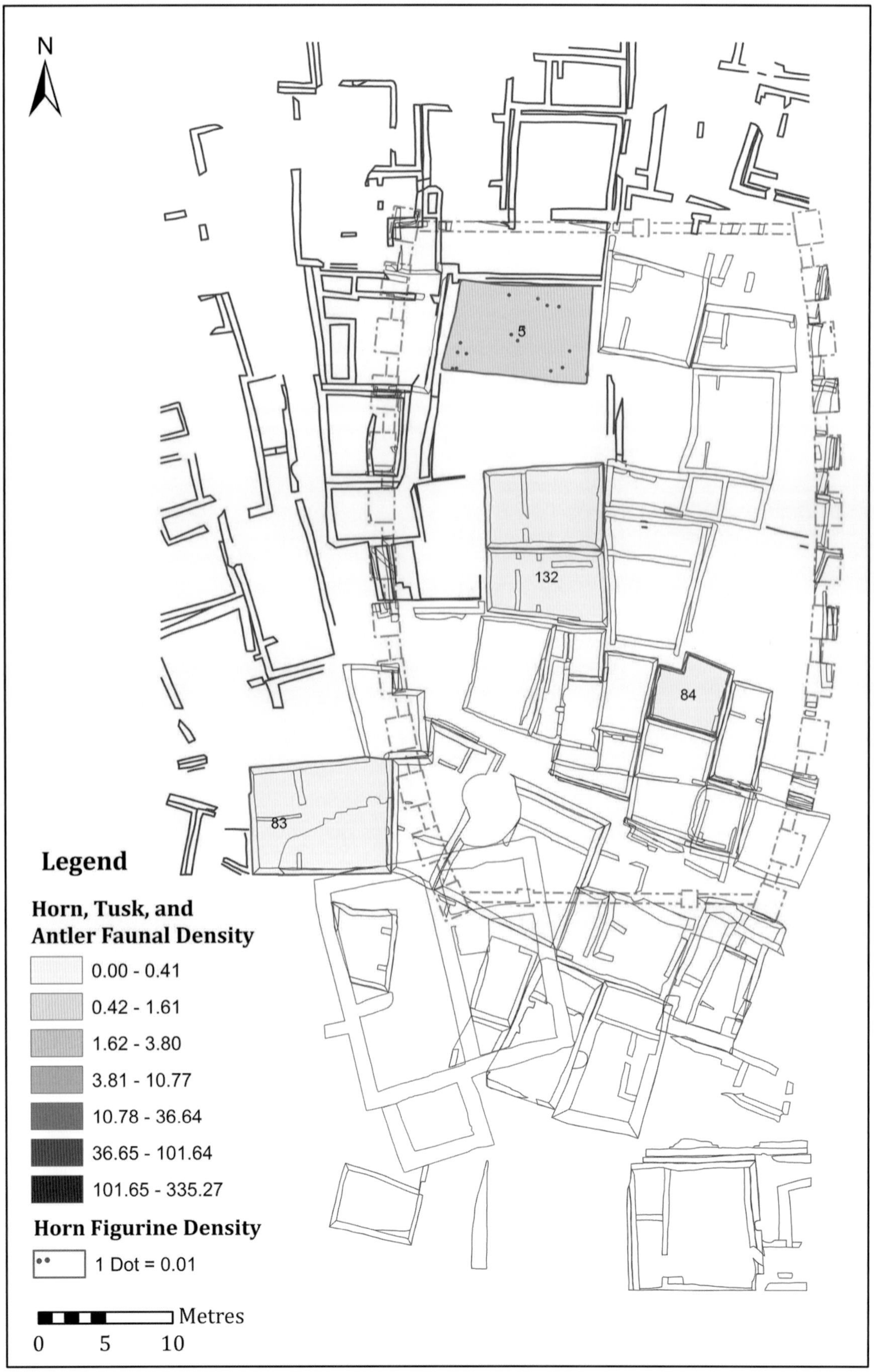

Figure 6.5. Early period North Area faunal horn versus figural horn, tusk and antler building densities (for colour version, see online supplementary material).

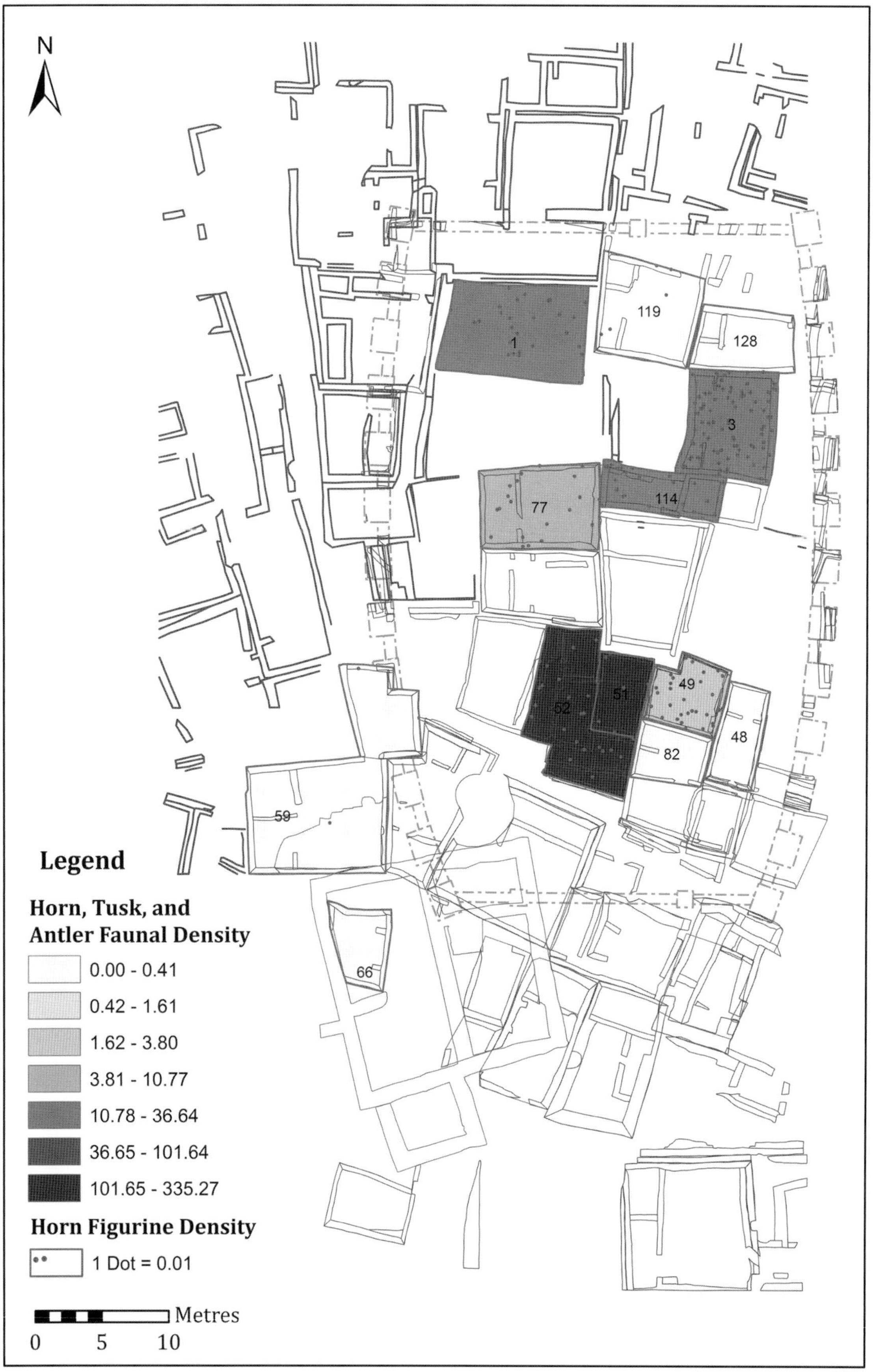

Figure 6.6. Middle period North Area faunal horn versus figural horn, tusk and antler building densities (for colour version, see online supplementary material).

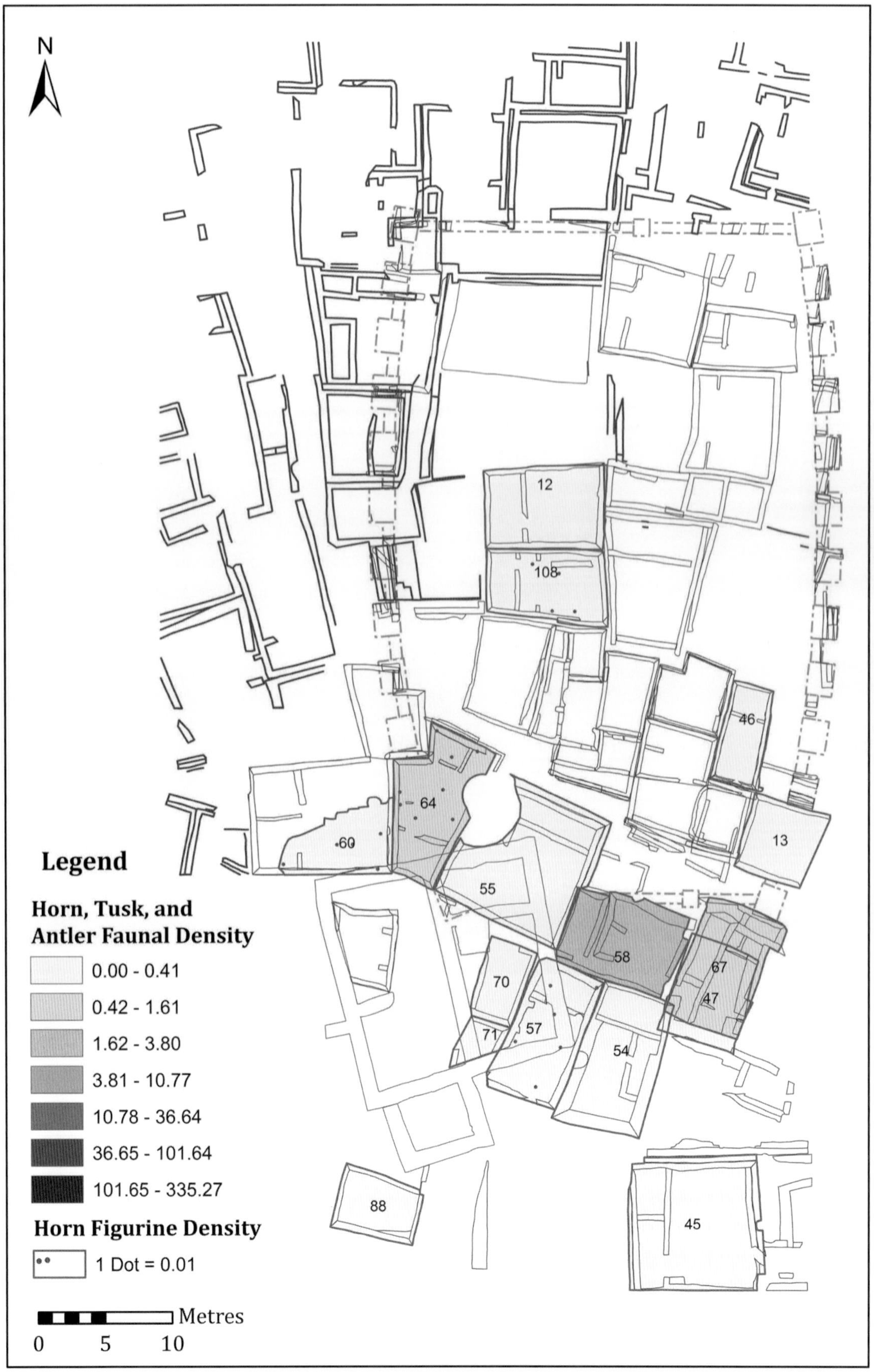

Figure 6.7. Late period North Area faunal horn versus figural horn, tusk and antler building densities (for colour version, see online supplementary material).

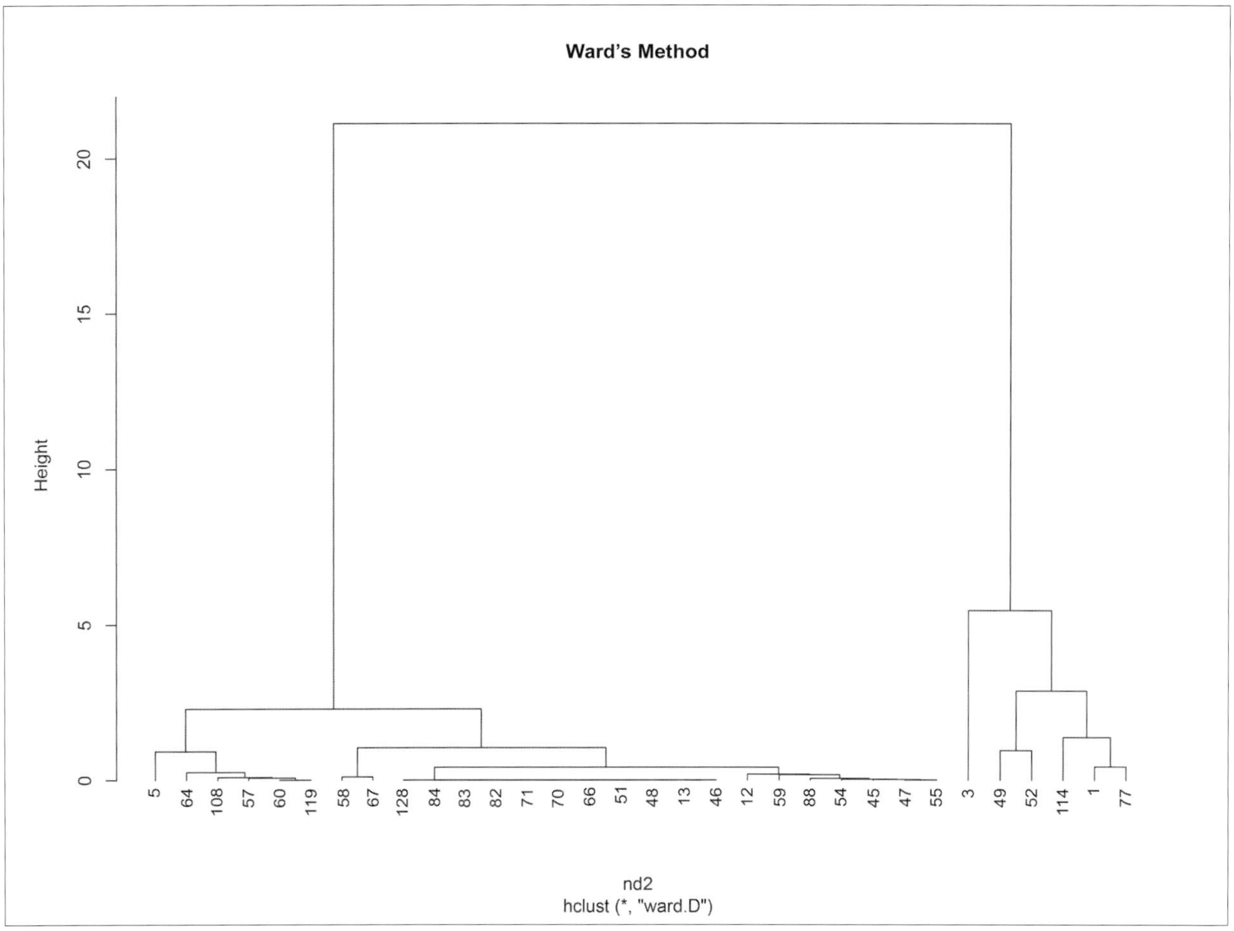

Figure 6.8. Ward's Method dendrogram cluster analysis of figural and faunal horn densities for all North Area levels.

House group	*Score*	*p-value*	*Significant*
North Area			
Early North	1	0.0832	
Middle North	0.6465	0.0051	X
Unburned buildings only	0.7401	0.0090	X
Late North	-2.3251	0.2890	
South Area			
Early South	0.2980	0.1582	
Middle South	-0.0476	1.0000	
Late South	0	1.0000	

Table 6.8. Kendall's-tau non-parametric correlation test of figural and faunal horn densities in the North and South Areas across time.

In sum, horns persist and even increase in density on a site-wide basis (see Chapter 5). Yet they appear to have been reduced to a minor role in house contexts in the later levels of the Neolithic occupation, suggesting that horns became less important to the daily activities that bonded household social groups. Instead, as Nakamura puts forward, horns may have functioned as an individualised mnemonic device at this time. The lack of a dependent relationship between the clay and faunal horns in the later levels in the North accords well with the overall trend at the site toward decreasing elaboration. The occurrence of both installations and special deposits of fauna peak in Level North G, with only one or two of each being found in each of the later levels (Russell, Meece 2005).

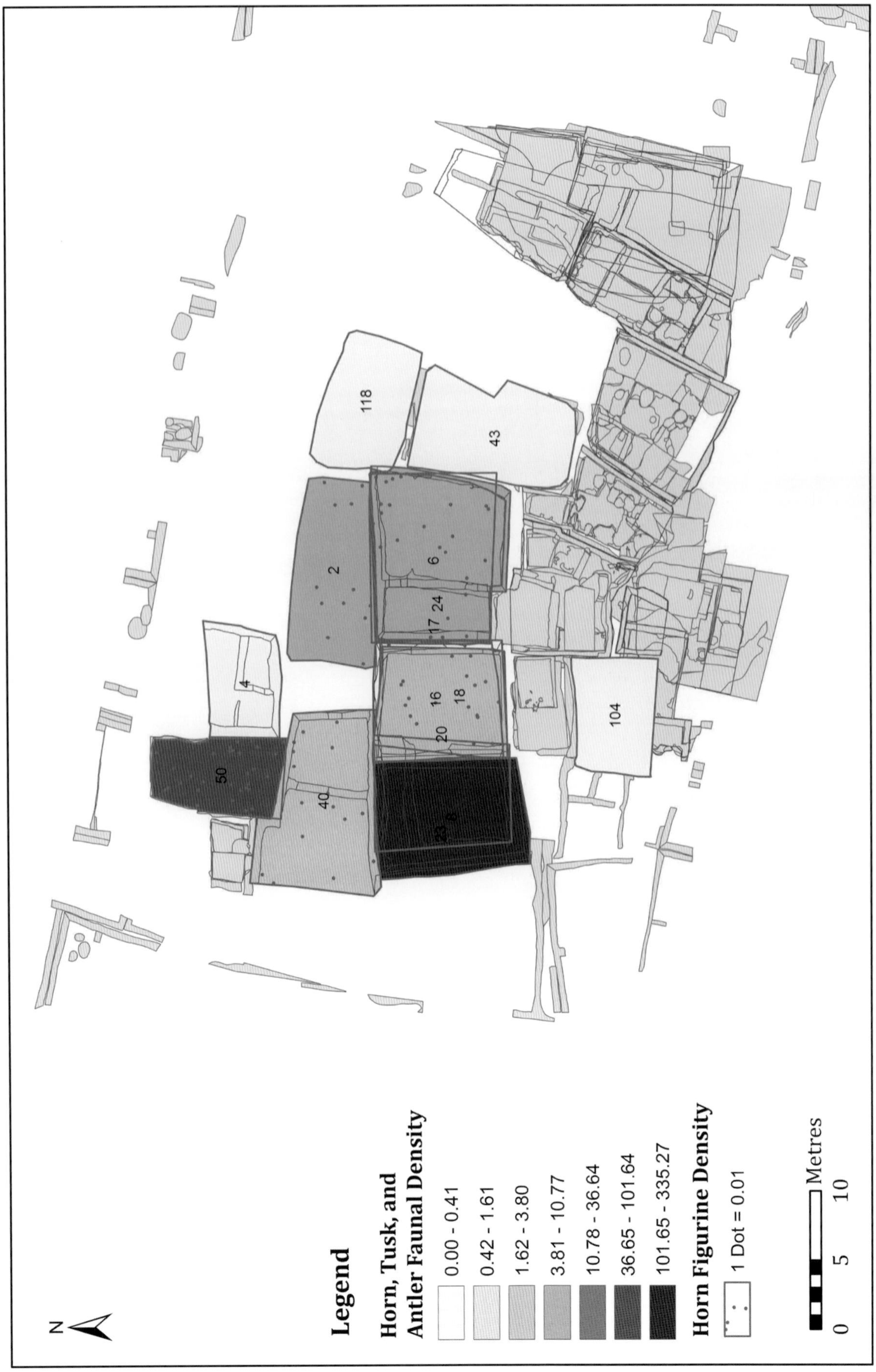

Figure 6.9. Early period South Area faunal horn versus figural horn, tusk and antler building densities (for colour version, see online supplementary material).

Figure 6.10. Middle period South Area faunal horn versus figural horn, tusk and antler building densities (for colour version, see online supplementary material).

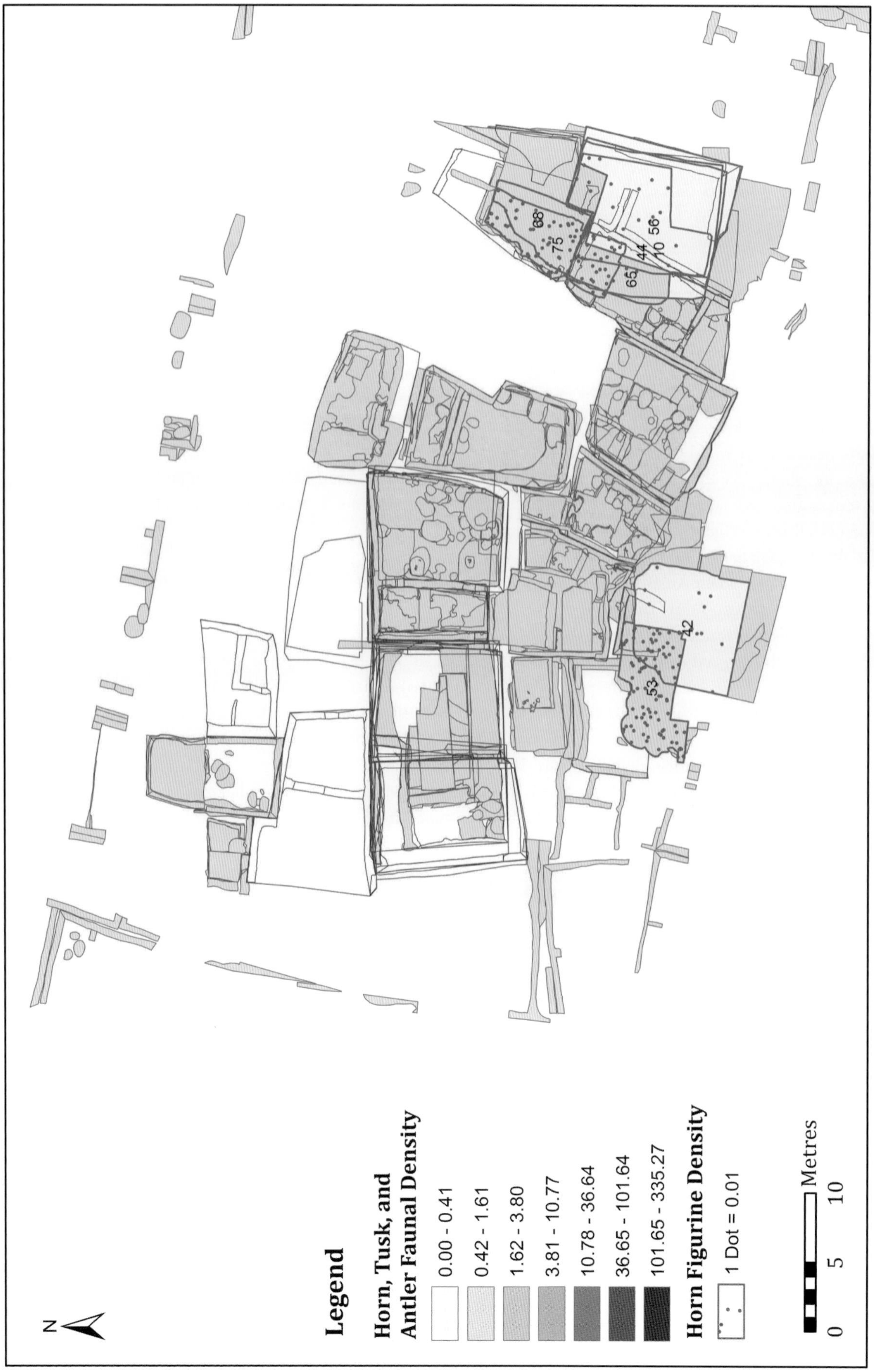

Figure 6.11. Late period South Area faunal horn versus figural horn, tusk and antler building densities (for colour version, see online supplementary material).

Representation by taxa

Proportional stacked bar charts for houses show that horns peak across all taxa in the Middle levels before declining in the Late period (figs 6.12, 6.13). *Bos* is the predominant taxa for the faunal horns in both the North and South Areas. Boar tusks are absent from buildings in the later levels, while sheep and goat horns and deer antlers are present in houses throughout the occupation. Caprine horns dominate the Late period North Area houses, though it is in the Early period that they are most prevalent in the South Area. Distribution of faunal horns is differential across houses, with only a single species found in some buildings. That only antlers were found in Building 54 is anomalous given that sheep, goat and cattle horns are much more commonly represented.

Figurine horns exhibit a similar patterning. The Middle period exhibits the greatest diversity in horn form types, as well as the peak density of figural horn deposition (table 6.9). Like the faunal assemblage, cattle figural horns (curved, bucranium and horn form types) dominate the corpus across all levels (figs 6.14, 6.15). Tusks are rare, with only one for each area (Building 3 in the North Area and Building 97 in the South Area) found in the Middle period. No caprine, or flat, horns are found in the early levels and their presence is relegated to Building 49 in the Middle period North Area and Buildings 80 and 75 in the Middle and Late period South Area, respectively. The appearance of antler figurine horns in buildings occurs throughout the South Area's occupation. These straight horns only appear from the Middle period onward in the North Area, though their absence in earlier levels may simply be an effect of the Early period North Area's small sample size of four houses.

Correlations with building size and number of burials

Multiple regression and Ward's Method cluster analyses on building size found no statistically significant relationship with the density of faunal or figural horns in houses. This finding accords with previous work that found little patterning between building size and other variables, such as figurine frequency across all forms (Nakamura, Meskell 2013b). Building size has little correlation with frequencies of burials and burial goods (Hodder, Pels 2010; Hodder, Farid 2014; Nakamura, Meskell 2013a). House elaboration is only weakly correlated with building size, but this relationship may be attributable to the practical requirements of space to accommodate architectural features (Hodder, Pels 2010). Kendall's-tau non-parametric correlation tests show statistically significant correlation between figural and faunal horn densities and the number of burials in houses in the Middle period North Area (tables 6.10, 6.11). In the South Area, this correlation is only statistically significant for figural horns in the Early period; there are no burials in Early period North Area houses. As no correlation was previously found between house burials and figurines, in general, it appears that horn forms had their own associations with other artefacts as compared to the remaining figurine forms (Nakamura, Meskell 2013b).

Discussion

Social organisation: previous models

Various models have been proposed for the social organisation of Neolithic Çatalhöyük. The notion of a priestly class that held religious authority via control of ritual activities dates back to Mellaart's original excavations in the 1960s (Mellaart 1967). Bender's classic model for social inequality as the prime mover for the shift to agricultural lifeways cited Çatalhöyük as a case study, using Mellaart's now defunct scheme of shrines differentiated from regular houses as evidence of a religious rank (1978). The possibility of a tiered social structure with religious leaders has been widely discussed in more recent years (Lewis-Williams 2004; Hodder, Meskell 2010; Hodder, Pels 2010; van Huyssteen 2010). Lewis-Williams (2004) directly linked the visual expressions of dangerous creatures to shamanistic status and power, arguing that each house, or family, had its own shaman.

	Bucranium		*Curved horn*		*Flat horn*		*Horn*		*Horn/Tail*		*Horn?*		*Straight horn*		*Tusk*	
	no.	*dens.*	*no.*	*dens.*	*no.*	*dens.*	*no.*	*dens.*	*no.*	*dens.*	*no.*	*dens.*	*no.*	*dens.*	*no.*	*dens.*
North Area																
Early North			5	0.09			1	0.02								
Middle North	3	0.01	30	0.10	1	0.00	30	0.10	1	0.00	1	0.00	8	0.03	1	0.00
Late North			3	0.01			1	0.00					1	0.00		
South Area																
Early South	1	0.00	12	0.05			5	0.02					1	0.00		
Middle South			12	0.10	1	0.01	11	0.10					2	0.02	1	0.01
Late South			5	0.03	2	0.01	10	0.06					3	0.02		

Table 6.9. Summary of figural horn form type frequencies and densities in houses in the North and South Areas across time.

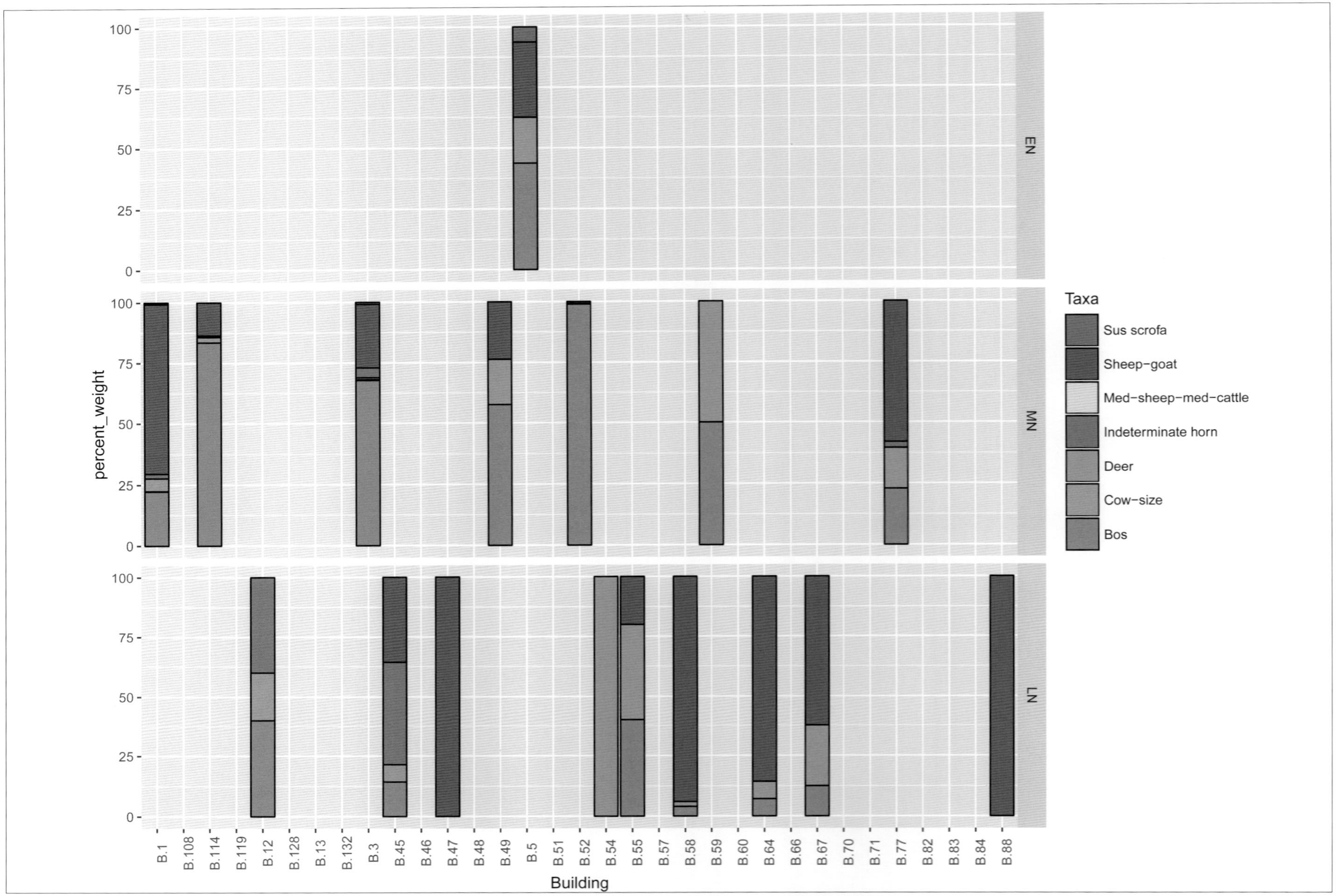

Figure 6.12. Proportional stacked bar charts for taxa (NISP) by building in the Early (EN), Middle (MN) and Late (LN) periods of the North Area (for colour version, see online supplementary material).

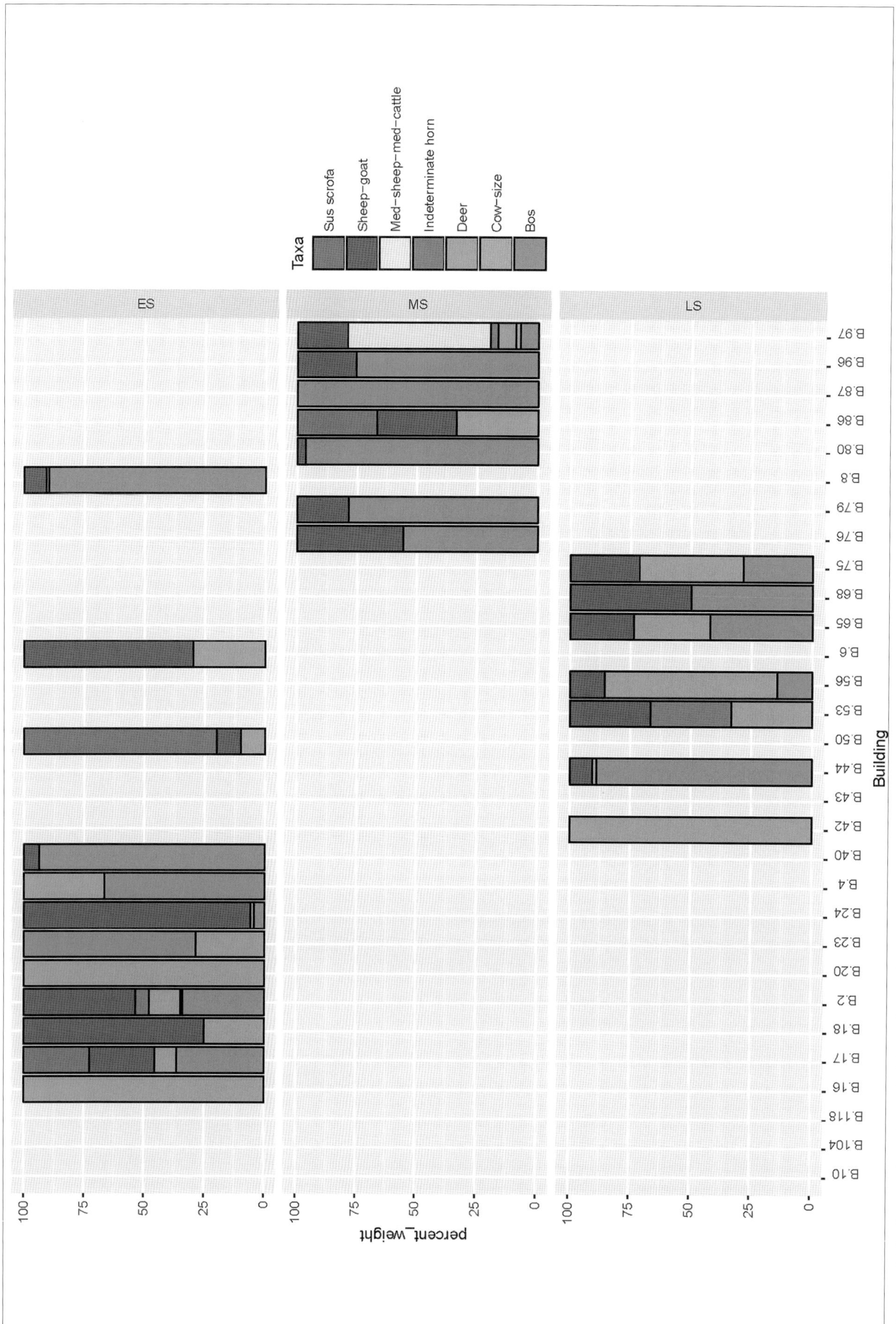

Figure 6.13. Proportional stacked bar charts for taxa (NISP) by building in the Early (ES), Middle (MS) and Late (LS) periods of the South Area (for colour version, see online supplementary material).

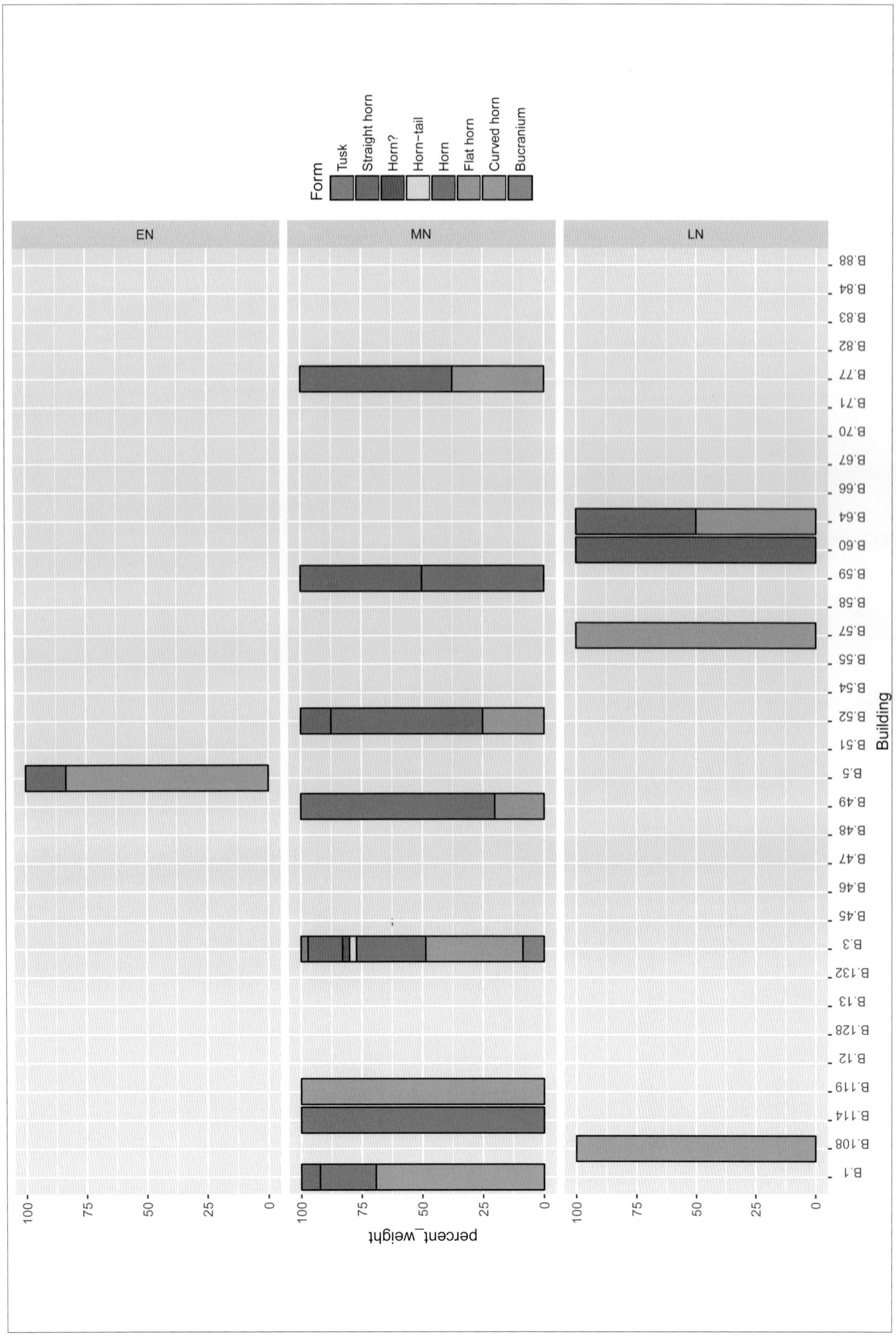

Figure 6.14. Proportional stacked bar charts for figural horn form types (count) by building in the Early (EN), Middle (MN) and Late (LN) periods of the North Area (for colour version, see online supplementary material).

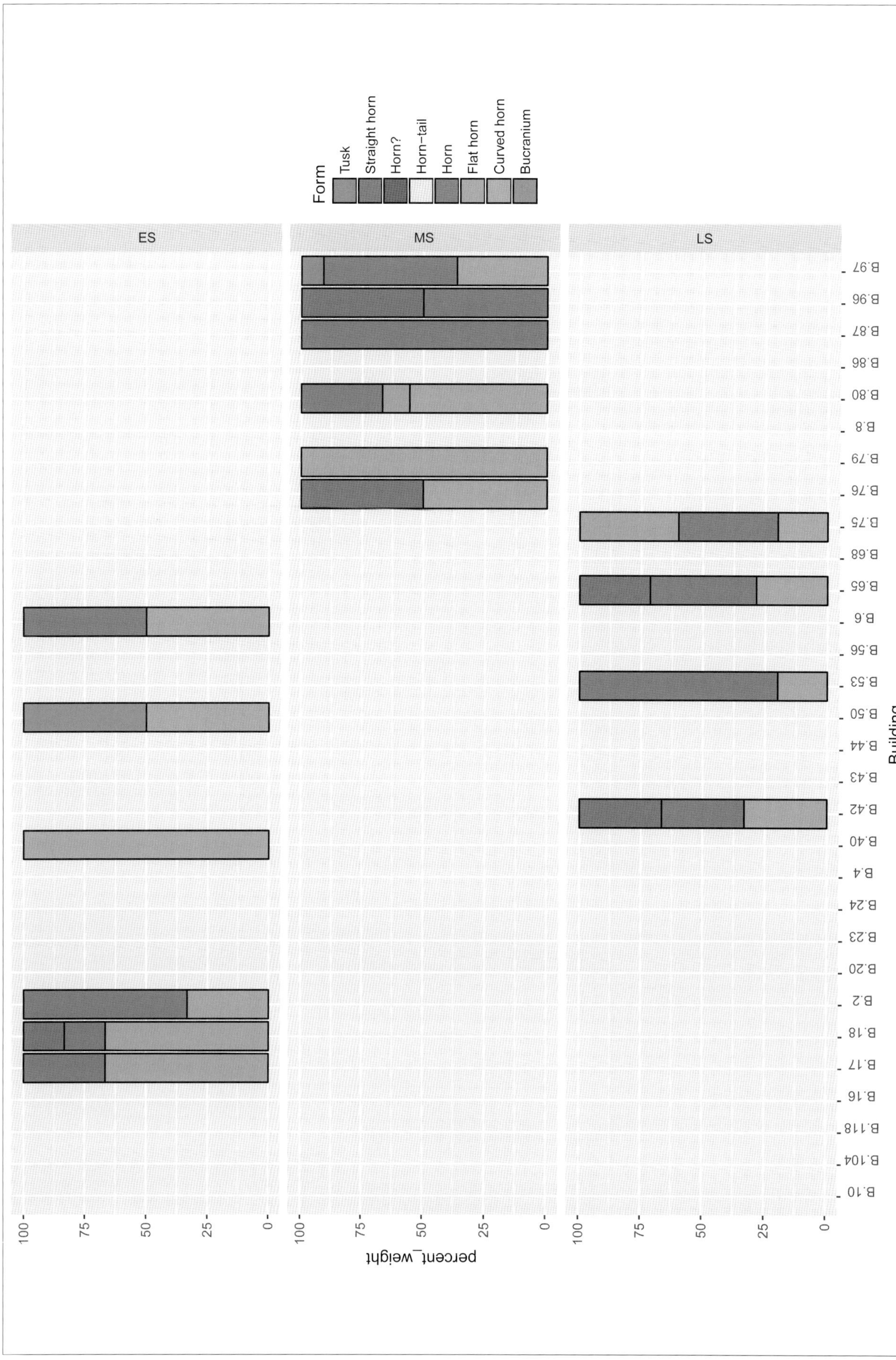

Figure 6.15. Proportional stacked bar charts for figural horn form types (count) by building in the Early (ES), Middle (MS) and Late (LS) periods of the South Area (for colour version, see online supplementary material).

House group	*Score*	*p-value*	*Significant*
North Area			
Early North	-	-	
Middle North	0.4890	0.0317	X
Late North	0.1712	0.4432	
South Area			
Early South	0.4928	0.0305	X
Middle South	0.2057	0.5305	
Late South	0.2927	0.3456	

Table 6.10. Kendall's-tau non-parametric correlation test for figural horns and number of burials.

House group	*Score*	*p-value*	*Significant*
North Area			
Early North	-	-	
Middle North	0.6875	0.0028	X
Late North	0.0632	0.7613	
South Area			
Early South	0.0596	0.7778	
Middle South	0.2057	0.5305	
Late South	0.14825	0.6152	

Table 6.11. Kendall's-tau non-parametric correlation test for faunal horns and number of burials.

Each shaman's ability to control spirit-animals translated to control over animals in the real world, precipitating social differentiation and a subsequent shift to domestication. The present excavation has found little evidence to support models of social hierarchy and agricultural lifeways triggered by religiosity, as there have been no individuals archaeologically identified as shamans or priests. Though the material culture at the site is indicative of ritual practice, ritual is not inherently associated with religion. Following Firth (1963), ritual is any repetitive, expressive practice that follows certain rules of behaviour (either social imposed or individual imposed). The ritual practices seen at Çatalhöyük are more quotidian in nature and not characteristic of an organised, administered and controlled religion or cult.

Alternative models have largely focused on systems of corporate groups appearing as clans (Russell 2003; Adams 2005; Mithen 2007; Bloch 2010; Hayden 2012; Russell 2012a; Wright 2014; Hodder 2016b), religious sodalities (Mills 2014) or lineages linked to houses (Hodder, Cessford 2004; Düring 2005; 2007a; 2014; Hodder, Pels 2010; Hodder 2016b). Numerous scholars have argued that houses formed clans that utilised feasting rituals as a means of aggrandisement (Adams 2005; Hayden 2012; Wright 2014). Other researchers (for example, Russell 2003) have postulated that there was a totemic system, though at this point in time this idea remains speculative. Heat maps for rare fauna species do exhibit differentiation by taxa across houses, but there does not appear to be any clustering by individual taxa as would be expected in a totemic social structure (figs 6.16, 6.17). It is possible that clans revolving around groups of multiple taxa or taxa in specific combinations may have existed, but further research and statistical testing is needed to verify such a scheme. Seeing parallels to Zuni pueblos, Mills (2014) instead contends that religious and medicinal networks structured house alliances, creating labour networks delineated by religious affiliations. She asserts that wealth was determined by human ability to perform labour and that elaborate houses were the centres around which lower-ranked farmers organised themselves. Moreover, horns in architectural installations are considered markers of meditative and/or ceremonial spaces associated with hunting rituals or healing. While Zuni pueblos and Çatalhöyük display similarities in their emphasis on wild, dangerous animals and the association of animal skeletal elements with benches, Çatalhöyük lacks the public and ceremonial spaces that this model for social organisation requires. In addition, there is little data to substantiate links between faunal installations and the performance of religious or medicinal practices.

Düring's 'lineage houses' model argued for a tiered system wherein groups of minor houses were dominated by an authoritative house with greater access to ancestors, animal species and rituals (2005; 2007a; 2007b). However, there is no evidence that houses with preferential access possessed higher status, rank and authority than their peers. Hodder and Pels (2010) took a different approach to the concept of house lineages. Following Lèvi-Strauss' premise of a house society (1982), their model positioned the house as the central force of social organisation, with each house embodying its own individual identity and the identity of associated individuals. This identity was produced and reproduced over multiple generations. Houses that engaged in the passing down of human body parts and the construction of histories involving events associated with wild animals, and that were subject to multiple constructions of at least four rebuilds utilising the same footprint and old wall stubs were pivot points for supra-household social groupings. This category of 'history houses' is problematic, however, with no correlation existing between these particular kinds of houses and the material objects associated with them (Hodder, Farid 2014).

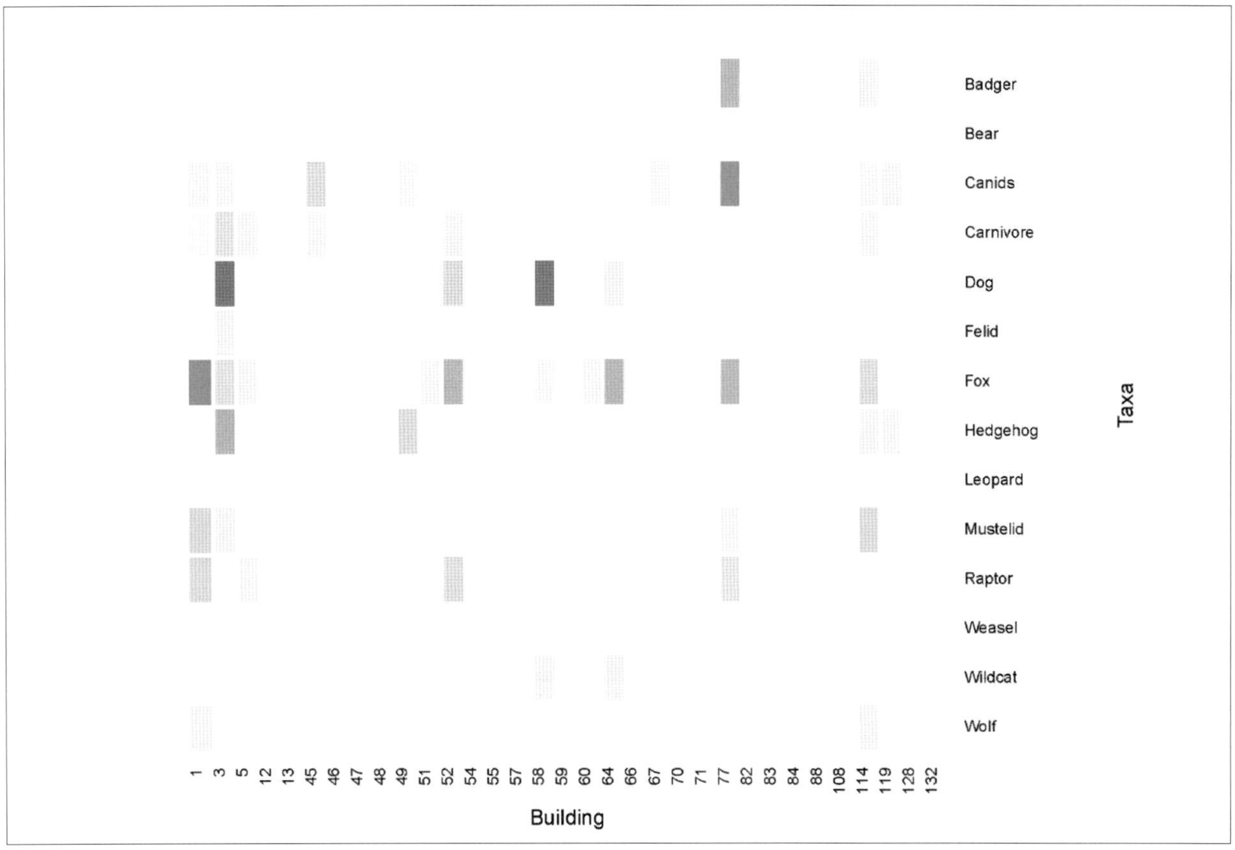

Figure 6.16. Heat map for frequencies of rare taxa in the North Area.

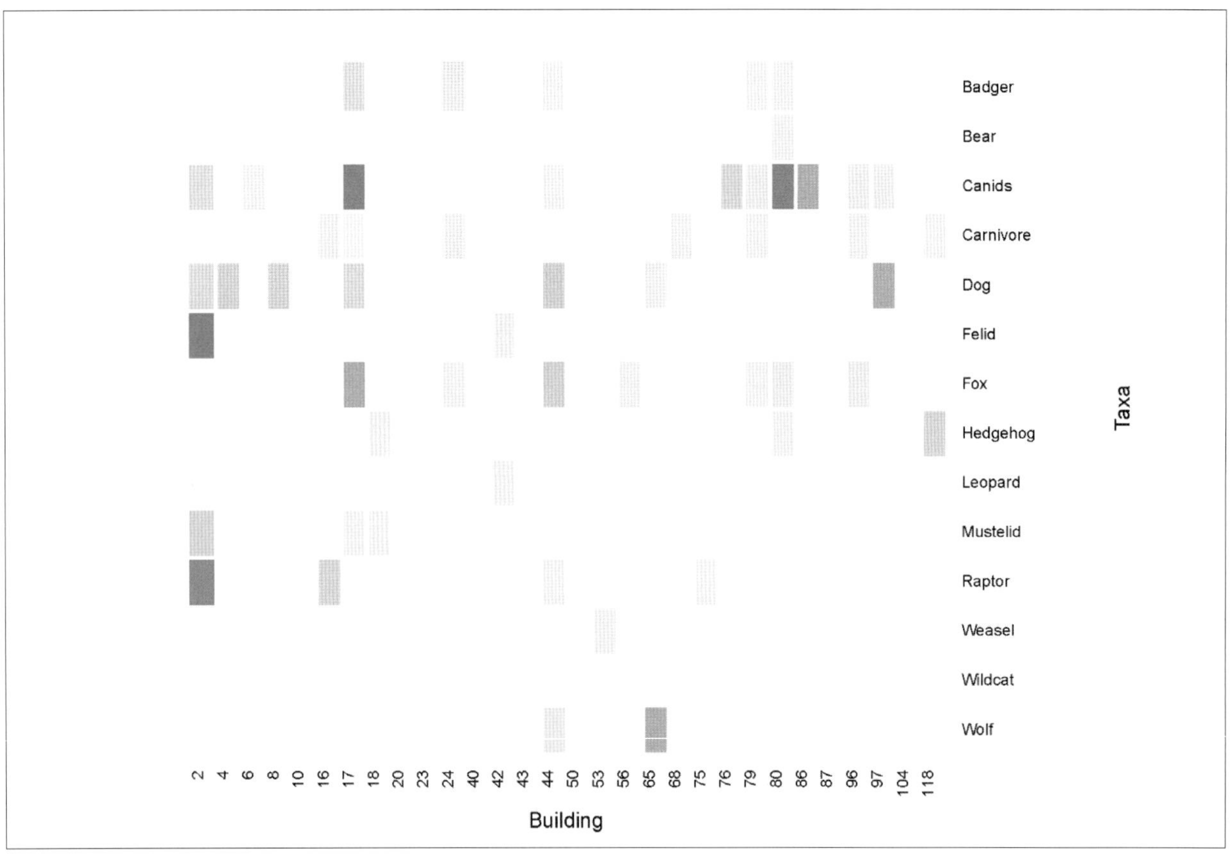

Figure 6.17. Heat map for frequencies of rare taxa in the South Area.

Heterarchy and nested social organisation

The results of this research have shown that the deposition of animal materialisations across the East Mound was not uniform, but rather subject to variability between both individual houses and different areas of the tell. This variability was not simply spatial, but also temporal. Taxa and the presence or absence of faunal and figural horns found in houses is variable. For instance, bar charts of the proportional representation of densities of taxa for horns, antlers and tusks show that both the South and North Areas subscribed to a pattern of *Bos* peaking in the Middle period and then dropping off in the later levels. The presence of boar tusks ceases after the Middle period in both areas. Additionally, there was differentiation between houses by taxa; for example, Buildings 16, 20 and 42 in the South Area only have deposits of cervid antlers and no horn cores.

Based on the patterning of animal materialisations, a heterarchical social structure is suggested. The concept of heterarchy is used cross-disciplinarily but was first applied archaeologically by Crumley (1979) to explain the regional distribution of sites via cross-cutting interactions as an alternative to top-down, core-periphery models. Heterarchy is a useful concept, as it involves the production of difference – how difference is produced and at what nodes, and the fluctuation thereof. At Çatalhöyük, these nodes took the form of animal materialisations, themselves intricately interwoven into the lived experience by way of their placement within the house as a mundane space. In many respects, heterarchy approaches push back against the assumption that hierarchy is the only social structure capable of creating order from chaos. In sum, social organisation in the first half of the Neolithic occupation can best be viewed under a heterarchical framework for the following reasons:

1) The site does not conform to social evolutionary models that claim population growth and density inevitably result in political centralisation. Despite its dense population and large settlement size, there is little evidence at Çatalhöyük for class or a central administration. Burial goods appear to have been distributed based on the age of an individual at death (Nakamura, Meskell 2013a) and the site has no material record of public projects like monumental buildings or a system for the accumulation and redistribution of resources.

2) Social organisation at Çatalhöyük was multivalent, with nested scales of social groupings consisting of households, corporate sodalities and the larger communities that made up the South and North excavation areas (fig. 6.18).

3) With its high carrying-capacity, population estimated in the thousands and sheer size, Çatalhöyük was a proto-urban mega-site (Der, Issavi 2017). As typical of Middle Eastern mega-sites, it would have required stabilisation to keep inequality in check and the community intact in the face of rising tensions from a mushrooming population and crowding (Byrd 1994; Bar-Yosef, Meadow 1995; Kuijt 2000; Simmons 2002; Düring, Marciniak 2005; Clare 2010; Goring-Morris, Belfer-Cohen 2010). Heterarchy's flexibility makes it an advantageous social system in unstable conditions (Brumfiel 1995; Ehrenreich 1995). As there is no evidence for distinct hierarchy, scalar stress would have instead been mitigated through ritual practices that fostered group decision-making, cooperation and integration; for example, the curation of human and animal body parts. Shared practices like hunting and the deposition of faunal and figural horns continued throughout the settlement's occupation, enduring temporal changes that other activities did not.

4) The flexibility of heterarchy to allow for contradiction accords well with the paradoxical quality of temporal changes occurring in the middle levels, for instance, the simultaneous expression of difference and forging of social bonds between houses.

Research on the biological affinity of burials associated with houses has found no evidence for social units delineated by biological kinship (Pilloud, Larsen 2011; Hillson et al. 2013; Hodder 2016a). Given that there is little evidence for social differentiation beyond variation in house elaboration and ritual, much of which centres on animal themes and associations, it is argued that, in the first half of the occupational sequence, the social positioning of wild animals was one factor which bound people together. Animal materialisations located within the milieu of the house would have allowed for the expression of difference on a horizontal rather than vertical plane, allowing the community to remain largely egalitarian. These animal expressions are only found in certain houses, so that these households where they are present are bonded by a commonality of practice. It seems that the residents of these houses had especially close ties to wild animals, so much so that they incorporated them into the lived environment. In this way, social difference between and amongst houses was communicated and reinforced. These were houses where wild animals loomed large in the everyday experience of life, where wildness was intentionally and visibly and invisibly interwoven into the domestic space. In sharp contrast, the social positioning of wild animals seems to have played a much more muted role, if any, for the households lacking animal materialisations.

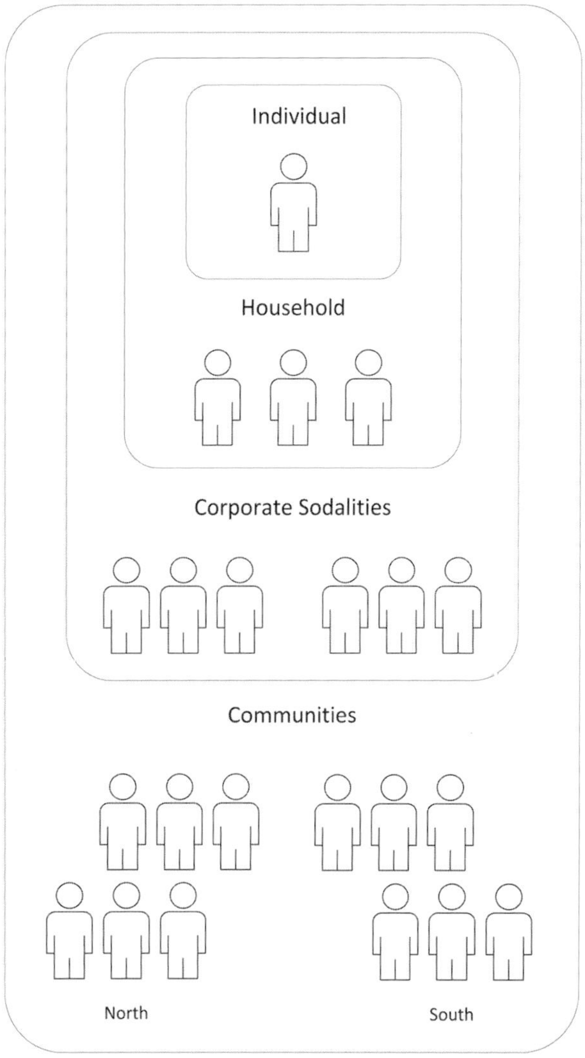

Figure 6.18. Proposed model of nested social organisation at Çatalhöyük.

Likewise, ephemeral and opportunistic shared practices and events served as a means for the formation and maintenance of intra-community social groups. People were brought together not only by burial events, but also by practices involving wild animals, such as feasting and hunting parties. Cross-cutting supra-household groups formed for the purposes of hunting would have acted as sites of coalescence and collaboration. Herding in the early levels, as indicated by isotopes and dental microwear studies, was a multi-household activity, with flocks of sheep being aggregated for pasturing (Henton 2014; Russell et al. 2014). Feasting deposits primarily consist of wild bulls, and feasts would have been provocative collective acts, along with the plastering of wild animals into houses. These are acts that would have brought people together through exchange and communal experience. Research conducted by the faunal remains team has shown that there was a nested food sharing system occurring between individuals at the household level, and then between multiple households at the sodalities level (Demirergi et al. 2014). Meat sharing was not done out of distributive necessity, as all households had relatively equal access to meat resources (Demirergi 2015). Thus, the portioning out of meat obtained as part of the kill was not just an economic act, but also a social one (Russell, Bogaard 2016). Feasting appears to have been a more communal activity, rather than an opportunity for aggrandisement, based on the insufficient amount of plant food surplus in storage (Bogaard et al. 2009). Participation in shared activities would have solidified social bonds, keeping sodalities together and undermining the emergent tensions associated with Middle Eastern mega-sites (Adams 2005; Atalay, Hastorf 2006; Hodder 2006c; Twiss 2008; Bogaard et al. 2009; Demirergi et al. 2014; Russell, Bogaard 2016).

Social groupings reached beyond the level of sodalities, with larger communities forming out of the spatially defined socialities on the East Mound – the North and the South Areas. The two areas display phenotypic variation and were likely made up of two different descent groups (Hillson et al. 2013). Although North and South Area communities partook in common practices, distinguishing idiosyncrasies do exist. For example, residents grazed their caprine herds in different locations (Pearson et al. 2007). In the South Area, burned houses cluster by figural and faunal horn densities, whereas in the North Area it is the unburned houses that show a high correlation with these two material types. The differentiation between burned and unburned buildings and their correlation or lack of correlation with the horns implies that the North and South Areas operated as separate communities with very different strategies. The relationship between burial frequencies in houses and horn densities is another example, significantly correlating with the figurines in the South and North Areas, while only in the North with the faunal horns. Nonetheless, these disparate strategies occurred under an umbrella of shared practices, as seen in the broader pattern of the rise and decline of horn deposition over the length of the site's occupation. Overarching social solidarity is also seen in the standardisation of animal depictions in zoomorphic quadruped figurines. Horn form types (for example, bucranium, straight, curved et cetera) are fairly uniform morphologically within their categories. Martin and Meskell's (2012) study of identifiable-to-taxon quadrupeds has noted how figurine makers were generally concerned with heads, horn morphologies and placement, and tails and that this focus remains steadfast throughout the entire occupation.

Similarly, a fixation with heads appears in the human remains and in anthropomorphic visual culture at the site. Thus, it can be argued that social and material organisation were not uniform across the settlement, but that there was variability between communities in the rhythm and complexion of their engagement with common practices, activities and traditions.

Temporal change and incipient social inequality
The Middle levels mark the beginning of a dramatic upheaval of the preceding way of life and the eventual disintegration of the site. This break accords well with general patterns of temporal change seen at other Neolithic Middle Eastern mega-sites, such as 'Ain Ghazal (Rollefson, Köhler-Rollefson 1993). As they reached their maximum carrying capacity, mega-sites would have had to deal with internal conflicts over competition for economic and social resources. Çatalhöyük would have faced such difficulties at this point in time, which was characterised by population and population density peaking, as indicated by regional survey, a highly packed settlement plan, mortuary data showing that fertility rates were at their maximum at this time and evidence of significantly high occurrences of osteoperiostitis (Baird 2005; Hillson et al. 2013; Hodder 2014f; Der, Issavi 2017). This period and the ability to maintain social bonds in the face of new and increasing tensions was critical to the trajectory of the settlement and its continuation or abandonment.

At Çatalhöyük, materialisations entangled with human-animal relations may have functioned as a levelling mechanism that stabilised and mitigated conflict, providing a means of expressing heterarchical difference and working to maintain an overall egalitarianism. Data has shown that there is a marked intensification of figural and faunal horn deposition occurring in the Middle period. This dramatic increase pertains to both the overall densities for horns and a surge in the percentage of houses with horn deposits found in the North and South excavation areas. In the Middle period, the percentage of houses with horn deposits increased by 30 per cent or more for fauna and by 50 per cent or more for figurines. This change in distribution may reflect a shift to the individual house as the dominant social unit. It is at this time, too, that taxa representation in the figural horns peaked, so that the diversity of horn forms is at its greatest. For example, in the North Area, all horn forms, that is, bucranium, flat horn, straight horn, curved horn, and tusk, appear in the Middle period. It is only during this time, when there is the peak of population stress, that we see the aforementioned practice of burning buildings, a practice that seems to be associated with installations or structured deposits of faunal horns.

From the Middle period onward, the settlement became increasingly reliant on an intensive farming economy. Biometric data and mortality profiles indicate that domesticated cattle were likely introduced in the North Area's later levels, though their use was not fully fledged and site-wide until the Final and later Chalcolithic occupations (Wolfhagen et al. Volume 13, Chapter 9). The adoption of domesticated cattle appears to have been intentionally delayed, as domesticates were known in the region as early as 7000 cal BC (Arbuckle, Makarewicz 2009; Arbuckle 2013; Arbuckle et al. 2014) and domesticated sheep and goat were consumed throughout Çatalhöyük's occupation. Hunting's prominence across economic, ritual and social dimensions at the site, and in particular the social capital it afforded, may have been the impetus for this delay (Arbuckle, Makarewicz 2009; Arbuckle 2015; Hodder 2016b). Indeed, Twiss et al. (Volume 13, Chapter 8) speculate that the gradual diminution in cattle size over time is indicative of overhunting and the need for increased management of cattle as a resource. In the Neolithic, the inhabitants of Çatalhöyük never ceased to hunt wild aurochsen. Contrastingly, when it came to caprines, the emphasis was on domesticates, with the exception of skeletal horns. Sheep and goat in feasting deposits and special consumption contexts significantly rise in the later levels, reinforcing an overall shift away from wild species in favour of herded animals (Russell, Twiss et al. 2014; Demirergi 2015).

Agriculture, and the generated associated surplus, enables social stratification and entrapment by requiring labour, time and investment (Hodder 2016b). At Çatalhöyük, we begin to see the roots of inequality in the latter half of the occupation, as residents turn away from sharing practices and households become more autonomous social units. There is burgeoning craft specialisation in stone bead and chipped stone production (Conolly 1999; Bains et al. 2014; Carter, Milić 2013a). House footprints change, with an expansion in size, increased compartmentalisation and the enclosure of middens and courtyards, converting them into household activity areas with external hearths and ovens (Düring, Marciniak 2005; Hodder 2014d). Building continuity and the tradition of history houses declines. In Levels X–VI, houses at Çatalhöyük were distributed in clustered neighbourhoods. After this time, these neighbourhoods broke apart, the settlement contracted and the Neolithic mound was eventually abandoned (Düring, Marciniak 2005; Düring 2006).

Change appears to have been difficult, with much evidence for rising social tensions. During the transitional period, breaks with the past were spaces of contestation. For instance, Middle period North Area B.52 intrudes upon a radial divider that formerly delineated neighbourhoods.

The subsequent construction of B.51 atop B.52 re-established this division of neighbourhoods, suggesting a resistance to spatial remodelling. Households became increasingly independent, controlling their own means of production by pasturing smaller flocks of sheep over different areas in the landscape (Pearson 2013; Russell et al. 2014). Management of herds was no longer a collective task but one undertaken at the household level, as indicated by the isolated pathologies common only to the sheep found in Building 65 (Russell et al. 2014). The scale of feasts was amplified in the later levels, and this increase is interpreted as a reaction against change, while also being a means of either the redistribution of resources to downplay difference or an assertion of difference and aggrandisement (Demirergi, 2015). Rising social tensions are further attested to by the appearance of elaborate andesite serving trays and maceheads, possibly weapons or objects used for the assertion of political authority, in the Middle and Late periods (Wright 2014). Similarly, knucklebones from worked sheep and goat astragali do not appear until Levels South S and North H, after which they are found frequently through the very end of the Neolithic occupational sequence and even in deposits from the Chalcolithic West Mound (Russell, Griffitts 2013). Thought to function as dice, due to the absence of the characteristic wear that marks burnishing tools, these objects were likely game pieces. Gaming is a competitive activity that pits individuals or teams against one another, potentially intensifying difference and fostering social tension.

The proto-domestication of cattle was consequential to the trajectory of the site and the complexion of social relations because such animals can be considered moveable, self-contained property (Burchell 1822; Comaroff, Comaroff 1990; 1991; Dawson 1998; Helmer et al. 2004). Following Renfrew's contention that the material symbol is constitutive of the concept (2001), the closer management of cattle in the later levels may have preceded the notion of economic inequality. Comaroff and Comaroff's work has shown that cattle can simultaneously be a form of symbolic, economic and social capital (1991). The value of wild animals at Çatalhöyük, especially *aurochs*, may have shifted away from symbolic concerns toward a novel emphasis on economic currency. This transformation in relations between humans and fauna can be seen in the animal materialisations. For example, faunal installations and special deposits, across all taxa, peak in Level North G before severely declining in frequency and body part representation in the later levels (Russell, Meece 2005). In sum, there was a revision of the category of the animal to predominantly a source of subsistence and even the emergent commodification thereof.

Further corroborating the relegation of wild animals to the economic sphere is a shift toward portable objects from the mid-point onward of the occupation (fig. 6.19). Frequencies of plastered reliefs and faunal architectural installations fall off after this point in time (Russell, Meece 2005). Türkcan's research on the stamp seals has shown that they do not appear until Level North G (Türkcan 2005; 2013). Animal motifs on stamp seals include rare wild species, like bear and leopard, mirroring the plastered reliefs that characterised earlier levels (fig. 6.19a, b). Stamp seals are commonly used to denote ownership (Mellaart 1962; Hodder 2006c; Skeates 2007; Türkcan 2013). Whether this assemblage was used for such a purpose is unknown at this time, though the presence of residue confirms that these were objects that were used. It is possible that the seals marked organic materials that have not survived in the archaeological record due to poor preservation value. Although this usage is speculative, some seals are associated with textile production artefacts (Türkcan 2013). Additionally, ethnographic examples of stamped bread are known from the city of Konya (Bogaard et al. 2014).

Similar to the stamp seals, a ceramic sherd decorated with a splayed animal is reminiscent of the three-dimensional effect of plastered wall reliefs of wild animals in early occupational levels (fig. 6.19c). Pot knobs that take the form of quadrupeds are another example of the way that animal materialisations transitioned from fixed to portable manifestations. Even though zoomorphic figurines are known over the duration of the occupation, the frequency of zoomorphic cattle quadruped figurines is highest in the later levels and especially prevalent in the North Area.

Other types of material culture also show signs of increasing concerns with signalling individual difference, such as the use of animal body parts for personal adornment. Beads and points fashioned from worked bone become more specialised, elaborate and varied in type (Russell, Griffitts 2013: 298; Vasić et al. 2014). The use of red deer beads in the North Area only occurs from the Middle levels onward. In the South Area, they are known from the beginning of the occupational sequence but intensify in frequency in the Middle and Later periods. Deposits of finished imitation red deer canine beads do not appear until Level South P. As adult male deer will only grow two of these upper teeth over their lifetime, the raw material to make genuine red deer canine beads is limited. These were objects that may have been used to signal identity, their rarity making them valuable (Choyke 2001; Russell, Griffitts 2013; Vasić et al. 2014). The amplification in the production of red deer canine beads, real or fake, indicates that supply outstripped demand, coinciding with an overall shift from heterarchy to increasing inequality at Çatalhöyük.

The desire for increased individualisation is also substantiated by the presence of mirrors in the second half of the occupation, with those excavated under Mellaart dating to Levels VIB through IV and the Hodder finds to Level South S (Vasić et al. 2014).

Animal materialisations and human-animal relations

It is doubtful that animal materialisations could have existed in isolation from the intricacies of contemporaneous human-animal engagements. Sykes succinctly equates these engagements to the '"mirrors and windows" to human ideology' (2014: 169). Given the preoccupation with animals, and especially wild animals, in the houses and iconography at the site, relations between people and animals cannot simply be reduced to economic models of production and consumption. While the density of horns may have fluctuated through time, it is a testament to their significance that they persist throughout the entire occupation of the Neolithic mound. Figural and faunal horns were multi-textured mediations of the intimacies enmeshed in encounters with animals. Animals actively shaped everyday human life to the extent that inhabitants intentionally embedded within domestic spaces their body parts, and materialisations of these parts, in a reification of this relationship. Material objects can be imbued with performativity, bolstered by their form and appearance as well as function (Pels 2010). The implications of performativity as reconstitutive are detailed in Butler's argument that performativity challenges taxonomies, reifies ontologies and has social repercussions (2010). These were objects which were presenced and which, in turn, would have tempered how people perceived and interacted with animals. The making and manipulation of these objects was in and of itself an interface with beastly beings and may have even been an attempt to reconfigure the complexion of these relations.

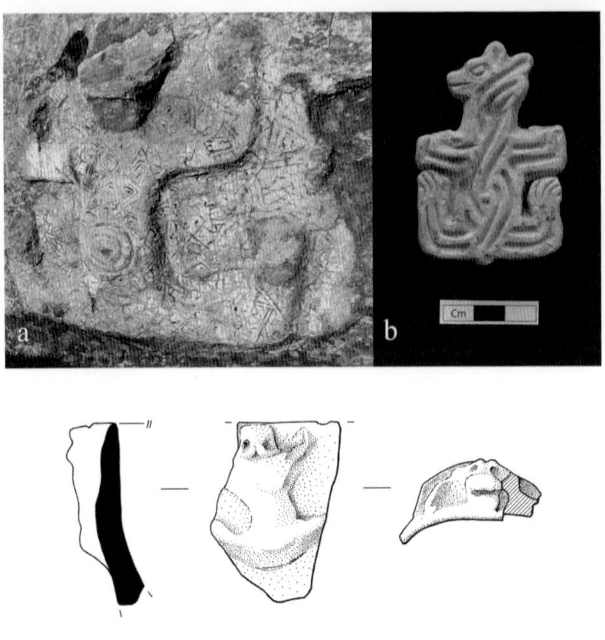

c

Figure 6.19. Wild animal motifs shift from fixed to portable media: (a) splayed quadruped plastered wall relief (photograph by Alan Mellaart); (b) bear stamp seal (photograph by Jason Quinlan); (c) splayed quadruped relief on pot sherd (illustration by Kathryn Killackey).

The transformation of the Neolithic East Mound from the Middle period onward may have been an unintended consequence of the destabilisation of the pre-existing classificatory system and the turn to an emphasis on the domesticated. As hunting became less frequent, the value of cattle shifted from symbolic to economic, in tandem with their increased management. As Cauvin asserts, the shift to fully agricultural lifeways was concurrent with a 'revolution of the mind', bringing forth a transformation of thinking and symbols (Cauvin, Watkins 2007). The affinity between hunters and their prey that previously placed human and non-human on equal footing came from a relationship of mutual respect. With the appearance of penning (Shillito et al. 2013; García-Suárez et al. Volume 13, Chapter 14) and incipient herding, this relationship underwent a drastic reconfiguration, becoming asymmetrical. Humans began to see themselves as being in a position of power and holding superiority over their faunal counterparts.

There is evidence for such a turn at Çatalhöyük. It is in the second half of the occupation, concurrent with the increased management of cattle, that numerous examples of human domination in the animal materialisations appear. The proportion of anthropomorphic forms in the figurine assemblage intensifies from Levels North J and South S through to the end of the Neolithic (Nakamura et al. 2014). Moreover, figurines of humans astride leopards or bulls date to the Middle period onward (Mellaart 1967) (fig. 6.20b). Similarly, the most famous figurine from Mellaart's excavations depicts a seated female figure with her hands resting atop the heads of the two animals flanking either side of her. Leopard skins are portrayed in the stamp seals and adorn human figurines and individuals appearing in wall paintings (fig. 6.20a). Wall paintings become more narrative and less iconistic in their subject matter, showing people teasing and baiting wild beasts. An especially prolific example is the painted hunting scene from Shrine A.III.I wherein a multitude of figures, with leopard skins around the waist, are depicted hunting deer with bows and arrows.

Figure 6.20. Humans dominating animals in faunal materialisations: (a) wall painting of hunter wearing leopard skin; (b) 'goddess' seated astride leopard (photographs by Arlette Meryem Mellaart).

Concluding remarks

At Çatalhöyük, faunal materialisations were both products and producers of relations between humans and animals. These materialisations were not simple aesthetic devices, but a means by which animals were embedded within the social bonds at the site. Human-animal relations were not static throughout the Neolithic, but subject to profound transformation. During the first half of the occupation, relations between people and the wild animals they hunted were largely symmetrical. Faunal skeletal remains that comprised structured deposits and architectural installations were subject to the same traditions of curation, circulation, defleshing and refleshing (by means of plaster) as human body parts. Wild animals were literally close at hand, consciously interwoven into daily life through their proliferate materialisation within the house. Hunted creatures were preferentially chosen over domesticates, as shown in the zoomorphic figurine assemblage (Martin, Meskell 2012; Hodder 2016b). Social organisation in these early occupational levels was heterarchical, marked by interdependencies, shared practices, traditions and activities between and amongst individuals, households, sodalities and the larger communities that comprised the South and North Areas. As such, social groupings were nested and multivalent. Wild animal materialisations were an outlet for the expression of horizontal social difference and a levelling mechanism for the maintenance of social bonds in the face of new problems sparked by increasing population and population density. These social bonds were forged and reinforced by acts of collective herding, feasting, hunting parties, and the curation and circulation of animal materialisations, such as the figural and faunal horns. Notably, figural horns endure throughout the Neolithic, even while quadrupeds and other forms dwindle (Chapter 5).

In contrast, the later levels mark a point in time when the emphasis on wild fauna gives way to a focus on domesticated and managed species. Relations between people and wild animals became asymmetrical with the introduction and uptake of proto-domesticated cattle, spurred by an ideological disjuncture with the past. Cattle were gradually reclassified to a category similar to domesticated sheep and goat, species that were never afforded much symbolic or intangible value, but were instead primarily relegated to daily consumption contexts. A preoccupation with being human is seen in the newfound objectification of wild animals. Nakamura (Chapter 5) argues that there appears to be a turn towards humanity, concurrent with the increased proportion of anthropomorphic figurines in the later levels. The appearance of stone figures marks a setting apart of people from their faunal counterparts, a reconfiguration of the worldview to one which emphasises the human, rather than animal. Animals, as Nakamura (Chapter 5) points out, even dominated, continued to play a vital role by affording humans power that could be transformed into economic and social capital. Such a major revision to human-animal relations would have impacted the social fabric of the settlement. The ability of animals to act as a kind of moveable property is reflected in the way that animal art affixed to the house drastically declined alongside the intensification of portable animal art forms. The ontological move toward animals as beings to be controlled is evidenced by narrative wall paintings showing animals surrounded and prodded by hunters, figural representations of people astride wild beasts and the way that animals were used and depicted as sources of raw materials for personal adornment. The communally driven social organisation, with its emphasis on shared practices, that characterised the earlier levels was replaced with an incipient hierarchical social system. There is a shift from general egalitarianism to incipient social inequality marked by the amplification of craft specialisation and household independence amid growing concerns with what it meant to be human, as well as expressions of individuality and identity.

Animal materialisations were conspicuous both in their physical presence and in the way that they were situated at the intersection of material culture, human-animal relations and social organisation. The social makeup of Çatalhöyük society was braided by relationships between people and animals and was thus correspondingly shaped by any reconfigurations to these relationships. The adoption of proto-domesticated cattle in the later levels and, eventually, domesticated cattle in the final occupational levels brought about a fundamental shift in human-animal relations and peoples' engagement with the world. Wild animals became a sort of nascent private property, their Othering ultimately giving rise to socialities of emergent inequality and a human-focused world.

Supplementary material

For supplementary material related to this chapter, please visit https://doi.org/10.18866/BIAA/e-14. It comprises colour versions of figures 6.5–6.7 and 6.9–6.15.

7. Large clay balls at Çatalhöyük East

Lucy E. Bennison-Chapman

Introduction

Since excavations began at Çatalhöyük in 1961, caches of mostly complete, well-formed, large clay balls were reported from the site (Mellaart 1962: 56; 1967: 63; Atalay 2005: 149). With many hundreds of examples of large clay balls recovered from inside buildings during the Hodder project's earlier years, clay balls have been subject to detailed study since the 1990s (Atalay 1998; 1999; 2000; 2001; 2003; 2005; 2008; 2009; 2012; 2013; Atalay, Hastorf 2005; 2006). Studies of food preparation form a fundamental part of archaeological research into human subsistence. Sonya Atalay's research into the function of Çatalhöyük's large clay balls places them in a central role within food preparation, as cooking 'stones'. In the earlier levels of occupation, a time when pottery suitable for cooking was not widespread, clay balls were heated and then transferred into baskets or skins full of liquid foodstuffs (Atalay 2005: 156–58; Atalay, Hastorf 2005: 118–19, 2006: 306–09).

Large clay balls (average 6.32cm diameter) remain a common find at Çatalhöyük. Thus, a new phase of clay ball study was initiated in 2015, which aimed to study all clay balls excavated from the site from 2009 until 2017. A total of 11,196 clay balls and clay ball fragments were excavated and studied during this period. Clay balls are abundant at the site, especially compared to quantities of other clay artefacts. During the same time frame, 1,879 small, geometric clay objects and 1,215 figurines (including stone examples) were recovered (Chapters 5 and 8). This chapter summarises the nature of the 2009–2017 large clay ball assemblage, highlighting unusual and unique examples within the collection. Details of form alongside contextual deposition are explored to seek support for the 'cooking' function. Other possible functional interpretations for their primary role, secondary use and re-use are explored.

Aims and research questions

Previous research

Reports of James Mellaart's initial Çatalhöyük excavations list 'sling-stones…of unbaked clay' amongst the clay artefacts recovered during his first season in 1961 (Mellaart 1962: 56). Mellaart later reported finding numerous examples of caches of 'sling ammunition' within the bins of various 'houses' and 'shrines' (Mellaart 1967: 63). These objects, which are not illustrated or described in further detail in subsequent reports (i.e., Mellaart 1963; 1964; 1966), were later identified as groups of largely complete examples of the artefacts which later became known as 'clay balls' (Atalay 2005: 149).

Under the Çatalhöyük Research Project led by Ian Hodder, the ubiquitous clay balls were studied for 15 years by Sonya Atalay (1996–2008), who published a number of reports and research articles on their form and distribution (Atalay 1999; 2000; 2001; 2005; 2008; 2009; 2012; 2013). Atalay's research argued, based on almost 3,000 examples, that Çatalhöyük's large clay balls were cooking tools, utilised primarily during the lower levels of settlement as heating stones to cook foodstuffs (e.g., Atalay 2005: 156–57; 2013; Atalay, Hastorf 2005: 118–19; 2006: 305–09, fig. 8, table 5). Atalay argues that in Çatalhöyük's lower levels, prior to the widespread introduction of pottery suitable for cooking, waterproofed baskets or animal skins were used to hold liquid foodstuffs (Atalay 2005; Atalay, Hastorf 2005; 2006). Clay balls were heated in ovens, then transferred into stews of 'carbohydrate-rich plants with meat and plant or animal grease', providing secondary heat (Atalay 2005; Atalay, Hastorf 2005: 118–19; 2006: 306, 308, table 5). Such cooking methods, she argues, are mirrored in the ethnographic record, where cooking stones are utilised to boil both food and water (Atalay, Hastorf 2006: 309). It has also been proposed that clay balls might have been heated and placed next to ovens, allowing for the grilling of meat (Atalay, Hastorf 2006: 309).

Aims

With clay balls continuing to be a common find, especially in the earlier half of Çatalhöyük's occupation, and one of the research project's overall aims being to reach virgin soil in the South Area, it was important that clay balls continued to be studied during the project's final research phase (2009–2017). The present author took over the study of Çatalhöyük's large clay balls in 2015. The overall aim of clay ball study during 2015–2017 was to record all incoming clay balls from the current excavation season, working backwards to 2009. Evidence was examined to support or refute the use of clay balls as tools to be heated and placed into liquid meals, acting as boilers. Taking Atalay's 'cooking ball' interpretation as the starting point, the goal was to gain an understanding of the bigger picture of clay ball presence and use at Çatalhöyük East. Rather than focusing on the study of specific units, buildings or areas of the site, a broad picture of clay ball distribution (by temporal period, stratigraphic level, site area and context type) along with changes evidenced in the *nature* of clay balls across these different elements was sought. Therefore, *all* known clay balls and fragments thereof excavated from 2009–2017 needed to be studied in as much detail as possible. This would enable broad contextual analysis of the count and the density, and changes in these two figures, of clay balls according to period, level, excavation area and context type.

Early on in the research, it was noted that a small yet significant proportion of clay balls differed from the 'typical' clay ball assemblage. Some had deliberate, seemingly meaningful incisions on their surfaces. Others were smaller than average or were oval in shape (Atalay 2005: fig. 6.1; Bennison-Chapman 2016). It might be thought that there would be no need to decorate boiling tools and nothing to be gained by deviating from the traditional spherical shape. Therefore, such examples were actively sought, especially amongst the fragments of clay balls. From this it was possible to accurately quantify the degree of artefact homogeneity within the clay ball assemblage and assess whether or not such non-typical examples were manufactured for a distinct function.

Specific research questions include:

• Are clay balls more common in/limited to the lower occupation levels?

- When do they begin to decline and fall out of use?

• What is the most common primary deposition context represented by clay balls?

- Are they really most commonly recovered in situ, in or around ovens inside buildings of the Early and Middle periods?

• What proportion of clay balls are recovered alone, as opposed to in clusters?

• When recovered in secondary contexts, does this represent their disposal or intentional re-use as rubble as part of structural elements?

• What proportion of clay balls are decorated (incised with crosses, dots and so on), where are these examples recovered, and does the decoration carry any functional or symbolic significance?

• Are the clay 'balls' of non-typical shape, and/or the smaller varieties, used in the same way?

• Is there any relationship between clay balls and other artefacts at the site?

• Do any of the above features change through time or across different occupation areas?

Particular attention was paid to recording detail of the clay objects and the context from which they were recovered. Many fragmentary clay balls were burnt on both their interior and exterior surfaces. This resulted in the questioning of their re-use and possible secondary function. Thus the research questions listed above sought to (1) identify the primary function of large clay balls, (2) identify secondary use/re-use and (3) to identify whether or not the non-typical clay balls held any specific symbolic value.

Methods

Commencing in 2015, all clay balls from the current excavation season were studied individually on site. Each clay ball or fragment thereof was assigned a unique find number, along with a corresponding unique database number (DB #). Each clay ball was studied in detail, with many varied aspects of form, appearance, manufacture, wear and indicators of post-depositional processes and activity studied. Clay balls were labelled *large clay ('cooking') ball* to differentiate them from the smaller spherical clay artefacts of the *clay object* category discussed in Chapter 8.

Definition

'Clay balls' are large, spherical-shaped artefacts made from clay. Their size (average 6.32cm diameter) and weight makes them distinct from the smaller spherical-shaped clay objects (previously defined as 'mini clay balls') (e.g., Atalay 2001; 2005; 2013; Bennison-Chapman 2015; 2016; 2017 and this volume, Chapter 8). Clay balls are far more homogenous than the smaller clay objects (Chapter 8; Bennison-Chapman 2015; 2016; 2017). Though largely present as fragments, clay balls are immediately recognisable due to a number of distinctive, shared characteristics. As their name suggests, the objects are spherical. This along with their large size make them easily recognisable, even when found in tiny fragments (fig. 7.1). Their outer surface is extremely smooth, burnished in most instances. This contrasts starkly with their coarse interiors. Clay balls are baked, possibly fired at low temperatures in order to achieve their characteristic appearance. The degree of completeness (either 100% complete, 99–75%, 74–50%, 49–25% or <25%) was recorded. By nature, clay balls are spherical, yet variations do occur. Shape definition was therefore recorded. Only when enough of the artefact was present to confirm that the *complete* artefact *definitely* represented a true sphere was 'sphere' recorded under 'shape definition'. Otherwise 'indicative fragment' was logged. A small proportion of artefacts classified as clay balls varied from the traditional shape. They were

Figure 7.1. Typical clay ball (fragments) from Çatalhöyük East (DB# 3420–3430/21642.m106 to m116).

classified along with clay balls due to the presence of most of the typical characteristics of the artefact category. Oblate spheroids (asymmetrical ovoid with flattened ends) and ovoids are included here.

Recording

A detailed, individual recording process was carried out for a large number of clay balls (a total of 2,339). However, as the quantity of clay balls recovered during the final few excavation seasons was vast, and the study of the 2009–2017 excavated clay balls began with a large backlog (study commenced in 2015), the recording strategy was amended to a two-tier system. After registration, clay balls were studied at *either* Level 1 or Level 2 detail. Level 1 logged clay ball fragments as a group per unit. The artefacts were not assigned individual find numbers but were counted and weighed collectively, by unit of excavation. This allowed for the distribution of all recovered clay balls to be analysed (according to area of site, broad temporal period, stratigraphic level and nature of context). If a particularly rare form of clay ball (decorated or complete, for example) was recovered during the Level 1 recording process, it would be put aside to be studied in detail. Specific units only (mostly the priority units) were then studied at Level 2. Once recording was complete, the data was analysed, looking at the detail of appearance, production and use of all Level 2 studied clay balls and the contextual deposition of *all* studied clay balls.

Artefact dimensions along with weight were recorded. Dimensions were taken from three angles (1. *length*, 2. *width* and 3. *height/thickness*), recording the dimensions of the complete sphere or remaining fragment. For complete examples, the length, width and height readings represent the radius from three angles. For fragmentary artefacts, the length and width represent the minimum and maximum dimensions of the remaining original face of the artefact and the thickness of the fragment, measured from the centre of the original face inwards. In addition, the circumference was recorded when a complete reading was possible (clay balls with 50% or more of the original artefact remaining).

Clay colour was logged, noting the dominant colour of the original exterior surface, the presence of a secondary colour on the exterior, and the interior colour where applicable. The presence of burning (blackened patches) was similarly recorded, including the location: *exterior*, *interior* or *both*. The presence of soot on an artefact (including whether this was on the interior or exterior surface) was also recorded. Evidence of the manufacturing process was sought, with the presence of fingerprints, deep finger impressions, basketry and/or matting impressions recorded. A small yet notable proportion of clay objects studied in the project's final phase of research was found to exhibit intentional elaboration on the exterior surface. The presence and detail of this was recorded when discovered. Lastly, the presence of adhering substances such as lichen, paint or other secondary substances was sought and noted.

Clay ball character

A total of 11,196 large clay balls (or fragments thereof) were studied during the 2015–2017 seasons, 100% of clay balls excavated during the final excavation phase of the Çatalhöyük Research Project (2009–2017). The vast majority of examples (n=11,190) were recovered from the East Mound and represent 79.9% of all clay balls recovered since Ian Hodder's excavations began in 1993 (table 7.1). N=8,852 were recorded by unit (Level 1) and n=2,338 were recorded individually, in detail (Level 2). Clay balls share many unique features in shape and manufacture, thus are easily recognisable, even when represented by a small fragment. Clay balls are overwhelmingly recovered as fragments, just n=41 (<2%) of clay balls recovered in this final phase of research were 100% intact. Clay balls are generally recovered highly fragmented, with larger fragments represented by distinctive 'wedge-shaped' pieces. More than three quarters were represented by fragments constituting <25% of the original, complete artefact (fig. 7.2). As the name suggests, clay balls are rounded in form. More than two thirds (n=1,585 of the Level 2 recorded, 2009–2017 excavated) are fragments *indicative* of being spheres. Definitive, *true* spheres and *indicative fragments* combined form 99.4% of the study assemblage (table 7.2). A very small proportion deviate from the norm. Two examples of oblate spheroids (one of which is laterally asymmetrical) and 13 ovoids were recovered (table 7.2, fig. 7.3, online fig. S7.1).

The fragmentary nature of the clay ball assemblage is reflected in the diameters data. The *average* diameters (*length*, *width* and *height*) of the total assemblage are all less than 5cm, ranging from just 0.3cm up to 9.5cm (table 7.3). If only the clay balls with *complete* measurements are considered (maximum diameter measurement of clay balls ≥50% intact), clay balls vary in size from 3.6cm to 9.5cm, with an average maximum diameter of 6.32cm (n=170 examples). Considering the complete clay balls only (n=41), the typical character of the clay ball as an artefact category is revealed, with the diameter ranging from 3.6–9.5cm and an average circumference of 21.8cm (table 7.3, fig. 7.4). The *true* spherical shape of the

Excavation season	*Level 1*	*Level 2*	*Total*
1993–2008 (database records)	264	88	352
1995–2003 (counts published, Atalay 2005; 2009)	n/a	n/a	2,811
2009–2017	8,857	2,339	11,196
2009–2017 East mound only	8,852	2,338	11,190
Total	9,121	2,427	14,006

Table 7.1. Counts of the clay balls recorded during previous research seasons (Atalay 1993–2008), the most recent phase of the Çatalhöyük Research Project (Bennison-Chapman 2009–2017) and in total.

3D shape definition	*Count*	*Percentage*
Sphere	738	31.57
Indicative fragment	1,585	67.79
Spheres & indicative fragments combined	2,323	99.36
Oblate spheroid	2	0.09
Ovoid	13	0.56
Total	2,338	100.00

Table 7.2. Detail of the shape distinctions found within the large clay ball assemblage from the current 2009–2017 phase of the Çatalhöyük Research Project.

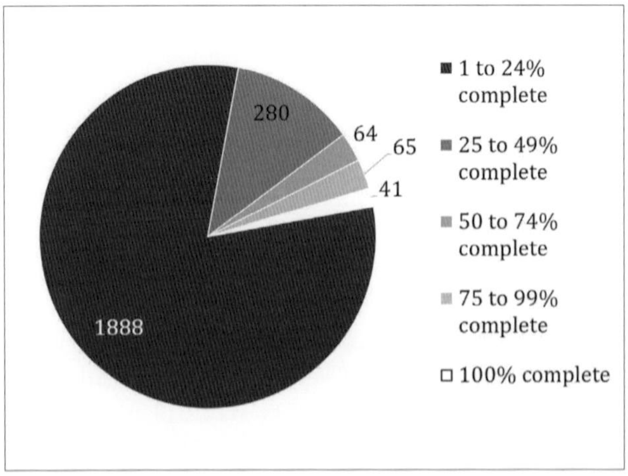

Figure 7.2. Degree of fragmentation evidenced within the 2009–2017 excavated clay ball assemblage.

majority of examples is highlighted when assessing dimensions of the complete clay balls, seen in the high degree of consistency across the three diameter measurements taken for each artefact. Notable amongst the 41 complete clay balls are the smaller examples. These remain larger than spherical forms of *clay objects* and display all of the typical characteristics of large clay balls aside from their reduced size, so were included in the clay ball category. The smallest example measures just 4.3 x 4.3 x 4.1cm (DB# 4959/22245.x2).

In weight, clay balls are dense, with a small fragment of clay ball weighing noticeably more than a small, geometric clay object or clay figurine of comparable size. The combined weight of all 2009–2017 excavated clay balls is 445,902.7g, with an average fragment/artefact weight of 40.0g (table 7.4). The lack of difference

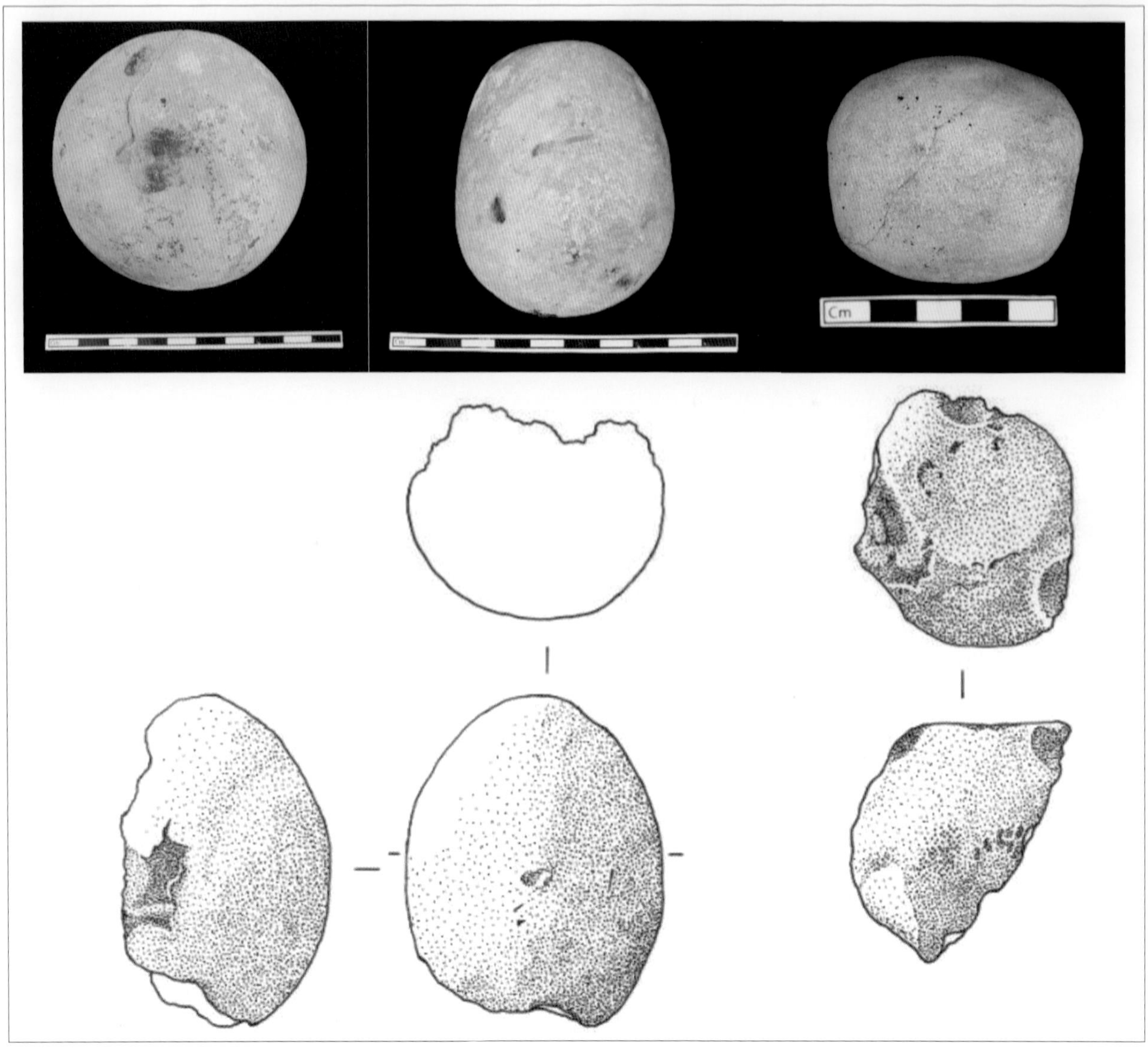

Figure 7.3. Examples of variation in clay ball shape. Top left: typical, true sphere (DB# 3839/22314.x1); centre: oblate spheroid (DB# 4304/22300.m101); right: ovoid (DB# 6936/21859.m1); drawing: asymmetrical ovoid (DB# 3685/21661.m173) (illustration by Jennie Anderson).

Dimensions	*Min. cm*	*Max. cm*	*Av. cm*
All clay balls (n=2,338)			
Diameter 1 (length)	0.6	28.1	4.6
Diameter 2 (width)	0.4	8.9	3.5
Diameter 3 (height/thickness)	0.3	8.9	2.7
Complete clay balls only (n=41)			
Diameter 1 (length)	4.3	9.5	6.7
Diameter 2 (width)	4.3	8.9	6.4
Diameter 3 (height/thickness)	3.6	8.9	6.4
Circumference	14.0	30.0	21.8

Table 7.3. Minimum, maximum and average measurements of the 2009–2017 excavated East Mound clay ball assemblage (Level 2 recorded, n=2,338). Top: all artefacts; bottom: complete examples only (n=41).

between the average fragment/artefact weight of the Level 1 and the Level 2 assemblage confirms that the latter is representative of the former. The relatively light average weight reflects the true nature of the clay ball assemblage, the vast majority of which are proportionally small fragments. Therefore, 91% of clay balls weigh less than 100g. This contrasts with the weights of the complete clay balls, which average 306.7g. The lightest complete clay ball is the sphere mentioned above (22245.x2), 79.8g. The heaviest, in contrast, weights 712.1g, demonstrating a large degree of variation in weight (fig. 7.5). Shape definition does not affect clay ball weight. The weight of the complete ovoids (n=3) and oblate spheroid (n=1) falls within the standard range, and the fragment weight of the other non-true spheres reflects that of similarly proportioned spherical fragments.

Manufacture. Çatalhöyük's clay balls are crafted from a fine, highly compacted clay, with few visible inclusions. The clay is dense in make-up. Clay balls typically have an extremely smooth, almost burnished exterior surface, often exhibiting many shallow fingertip depressions from manufacture. This along with the dense clay make-

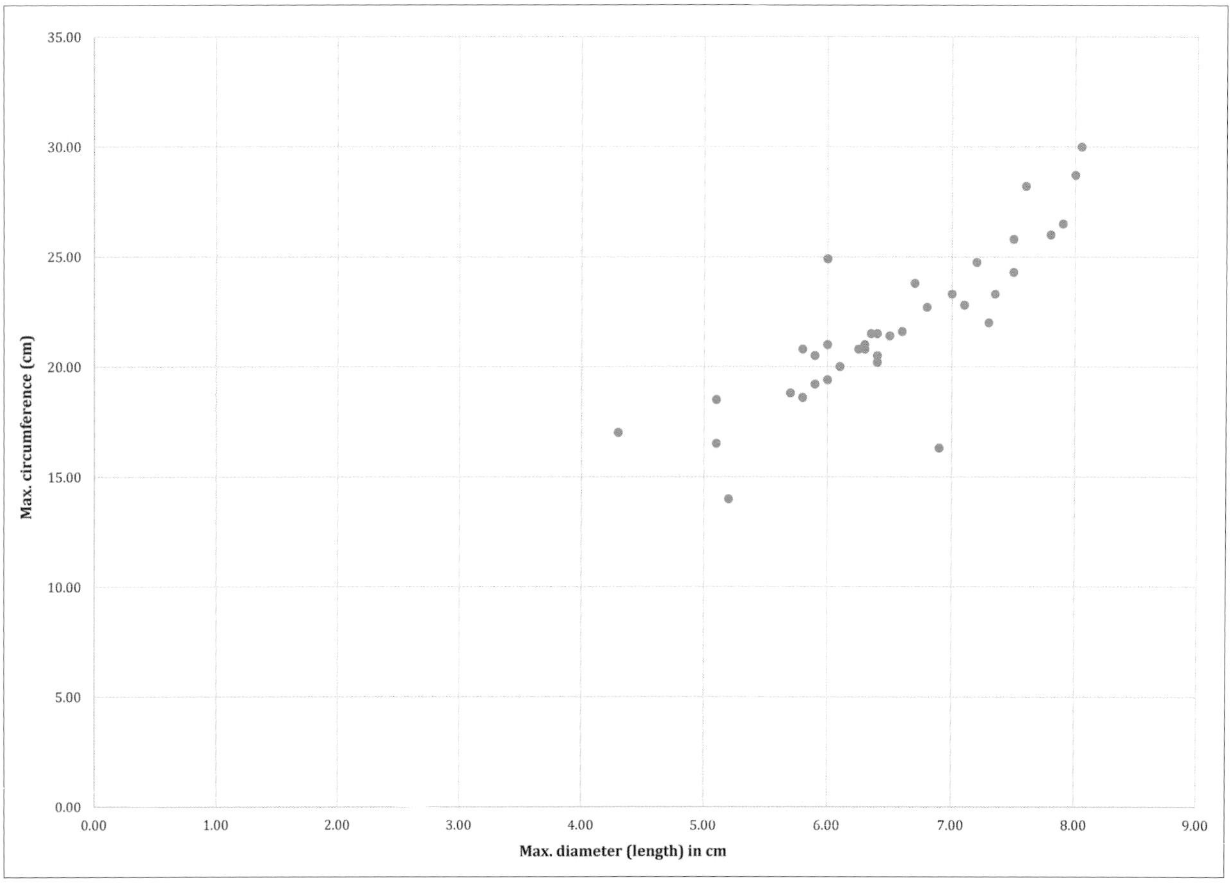

Figure 7.4. Comparison of the length/maximum diameter and circumference of the n=41 complete clay balls (excavated 2009–2017).

up suggests clay balls were manufactured by careful and lengthy manipulation within the hands (rather than rolling on a flat surface, for example), followed by exposure to heat.

Analysis of the clay colouring, texture and inclusions suggests clay balls are made from a limited range of widely locally available clays, namely backswamp, silty and sandy calcareous types. This reflects conclusions reached in previous research into the clay sourcing of figurines, clay balls and clay objects at Çatalhöyük (Doherty 2017: fig. 4.1, fig. 4.3, 44–47; also see Avis 2010; Bennison-Chapman 2013; Doherty 2013; Bennison-Chapman, this volume, Chapter 8). Clay balls are dominated by grey shades. The largest single colour category is 'mid-grey', 185 (n=425). Shades of mid-light grey combined are the primary colour of almost half of all clay balls recovered (40.7%, n=951). In contrast, very pale clays of beige, cream, light grey and white are extremely rare, representing just 12% of the assemblage (n=287) (table 7.5, online fig. S7.2). The range of clay colours reflects post-depositional practices and the secondary use of clay balls.

Record	*Combined weight (g)*	*Count*	*Weight (g) Av. per artefact/fragment*
Level 1	346,561.9	8,852	39.16
Level 2	99,340.8	2,338	43.21
All	445,902.7	11,190	39.99

Table 7.4. Weight per clay ball/fragment according to recording level of study (2009–2017 excavated East Mound artefacts only).

At least four clay balls are unusual for displaying extremely deep finger or fingertip depressions. Two have deep, pronounced fingertip depressions all over their surfaces (DB #3390/21660.m101, 3433/21660.m108). The third displays three long and flat finger impressions (DB #3839) and the final example has three deep, vertical finger depressions, thus resembling a bowling ball ((DB #3840/22314.m106), fig. 7.6). These examples likely represent mistakes in the production process, a lapse in care whilst moving clay balls from one place to another during manufacture (also see online fig. S7.3).

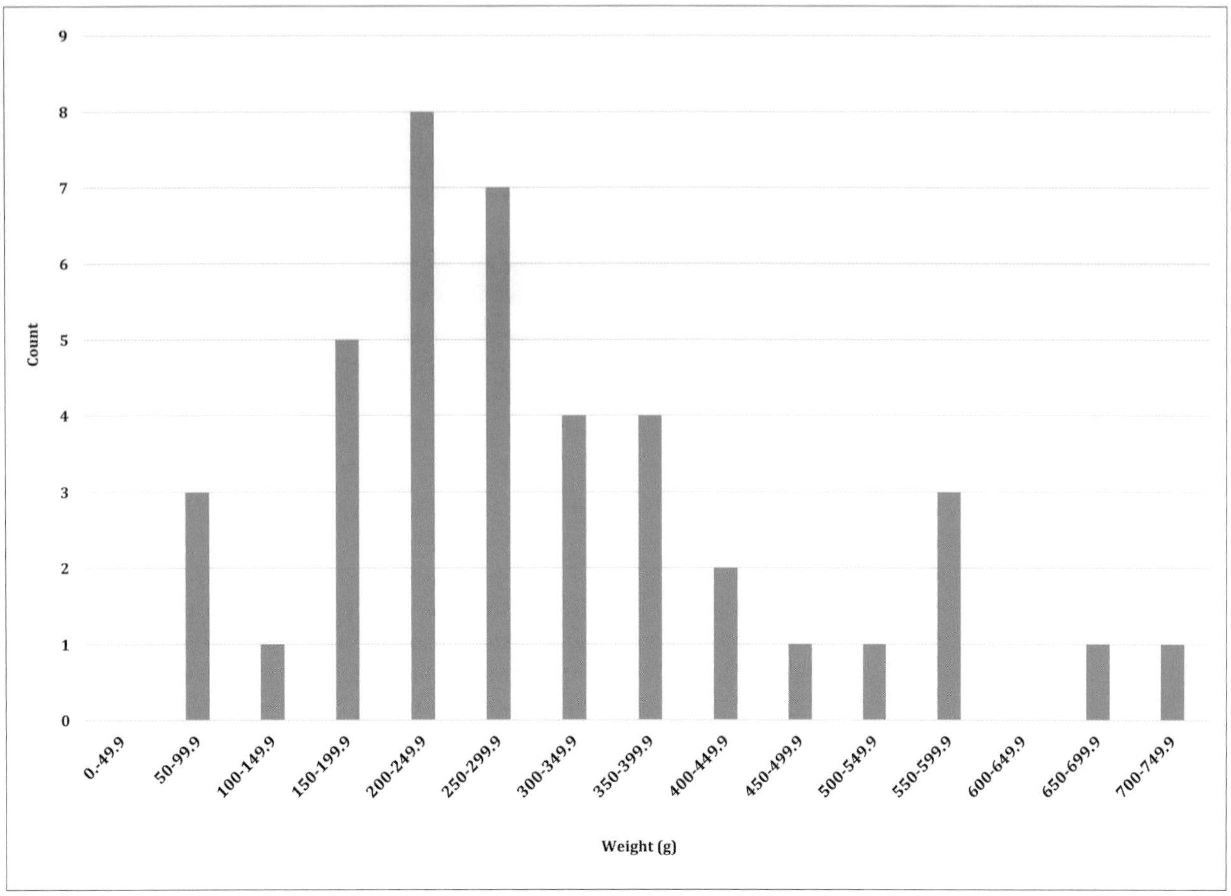

Figure 7.5. Clay ball weights (complete examples from 2009–2017 excavations only, n=41) distributed by 50g bins.

Primary clay colour (simplified)	*Count*	*Percentage*
Grey	951	40.68
Orange	427	18.26
Pale (white/cream/grey/beige)	287	12.28
Black	234	10.01
Brown	230	9.84
Other	209	8.94
Total	2,338	100.00

Primary clay colour	*Count*	*Percentage*
Mid-grey	425	18.18
Orange/cream	232	9.92
Orange/orange-brown	195	8.34
Dark grey	192	8.21
Light grey	177	7.57
Mid-brown	157	6.72
Mid-light grey	157	6.72
Black/v. dark grey	139	5.95
Light beige/cream	122	5.22
V. Light brown/beige	106	4.53
Black	73	3.12
Dark brown	47	2.01
V. Light grey-white	30	1.28
V. Light cream/white	29	1.24
Light brown	26	1.11
Black/dark brown	22	0.94
Red/red-brown	5	0.21
Red/pink	3	0.13

Table 7.5. Primary clay colour (main colour of the exterior surface) of clay balls, simplified and detailed (2009–2017, East Mound excavated, Level 2 recorded artefacts, n=2,338).

However, the depressions on DB# 3840 are far too deep to have been accidental and remain unnoticed. This example represents the production of a tool for a distinct use, perhaps enabling the better handling of the ball.

Also reflective of manufacture is the presence of basketry and matting impressions on the outer surface of clay balls. These are rare (26 definitive examples), yet when found are deep and clear, identifiable to three common forms: (1) 'twill plaited matting' (e.g., Wendrich 2005: fig. 15.3 right, fig. 15.5); (2) 'tabby weave' (Wendrich 2005: fig. 15.6; Jørgensen et al., this volume, Chapter 11); and (3) coiled basketry (Wendrich 2005: fig. 15.2–3) (table 7.6, fig. 7.7). The presence of these impressions suggests clay balls were stored in large groups inside baskets or piled up on matting on the floors of buildings. The rare examples with impressions likely

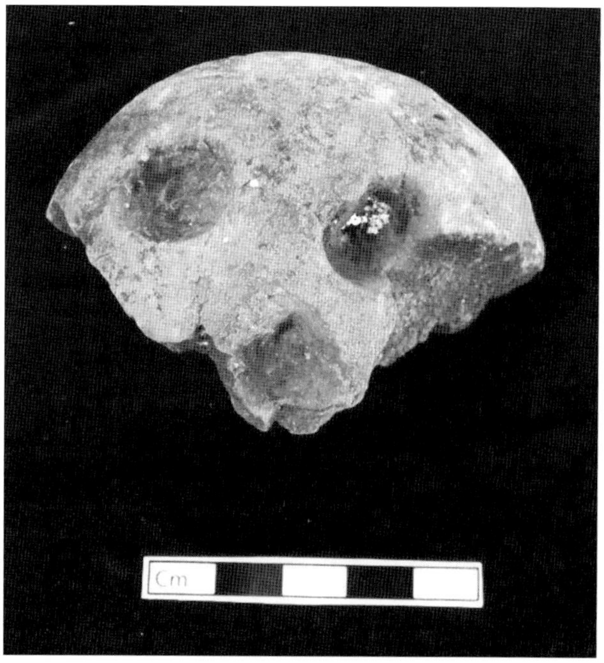

Figure 7.6. Clay ball fragment with three deep finger impressions (DB# 3840/22314.m106).

represent the premature stockpiling of clay balls before they had completely hardened. It is likely one batch of clay balls was prepared at a time, then stashed in a pile inside buildings before baking; thus the lowest received basketry or matting impressions, which occasionally were neglected when it came to smoothing them before they were placed into the oven.

Evidence of post-depositional processes. A small yet significant proportion of clay balls have a deep orange or orange/brown primary colour, 8% (n=195). When fragmented, these examples almost always have a grey interior, showing that the orange colour was not the original shade of the clay. The bright orange colour is identical to the colour of the clay floors and mudbrick walls of heavily burnt buildings at Çatalhöyük (online fig. S7.4). Therefore, it appears to be the result of exposure to high temperatures post-manufacture, likely during post-abandonment burning of a building.

More than half of clay balls (55%, n=1,158) have a *secondary* colour suggestive of heat exposure. Small patches of blackened surfaces, resulting from the direct exposure to flames, is common, seen on 93.4% of examples (n=1,949). A far smaller proportion are thoroughly and entirely burnt to a solid black colour, 10% (n=234, table 7.5, online table S7.1). Though blackening may occur during the initial firing or baking of clay balls in manufacture, direct exposure to heat continued as a result of the activities linked to their daily use. Indeed, daily/primary use involved the heating of clay balls to

Figure 7.7. Impressions on clay balls. Left: coiled basketry, DB# 3686/21661.m174 (4.9cm x 4.6cm x 2.9cm); centre: coiled basketry, DB# 6961/32673.m1; right: tabby weave, DB# 3893/22332.m101 (illustration: Jennie Anderson).

Impressions	*Count*	*Percentage*
Coiled basketry	8	30.77
Tabby weave matting	7	26.92
Twill plaited matting	5	19.23
Unidentified matting	6	23.08
Total	26	100.00

Table 7.6. Range of matting and basketry impressions identified on the surface of clay balls.

Adhering material	*Count*	*Percentage*
Brown lichen	3	8.33
Lenticular gypsum	14	38.89
Red paint	1	2.78
Yellow/orange lichen	17	47.22
Total	35	100.00

Table 7.7. Detail of the n=35 clay balls displaying adhering material (aside from soot).

high temperature, resulting in their initial fragmentation into large, wedge-shaped pieces. Clay balls are commonly blackened on both their exterior and interior surfaces, accounting for 73% of examples (n=1,705) (online fig. S7.5). Thus initial breakage did not lead to the disposal of clay balls or a change in use. Exposure to heat was integral to the function of clay balls, and even though a negligible proportion show no blackened surfaces, soot is found adhering to the surfaces of many of the clay ball fragments which otherwise show no clear sign of burning as evidence of heat exposure (n=51, 2.2%). Such examples were clearly placed alongside heavily burnt clay balls or other burnt items. Indicative of post-depositional processes, a small proportion of clay balls display material adhering to their original, outer surface. Thirty-four examples (1.5%) are covered in white or brown specks, demonstrating the growth of lichen and the adherence of lenticular gypsum from surrounding soils (Porta, Herrero 1990; Herrero, Porta 2000) (table 7.7).

Clay balls are almost exclusively plain in appearance. A single example (half of a clay ball, DB #5074/22301.m153) has a faint yet certain covering of red pigment. Twenty-four clay balls (1%) display markings on their exterior surfaces. Applied during initial manufacture, these marks appear to be decorative, possibly symbolic in function in some instances. The markings display a variety of application techniques, styles and motifs, yet just two examples are from complete clay balls. Some markings comprise a single element, others have a design consisting of up to 18 markings or more, with the design covering the entire remaining fragment (table 7.8, online table S7.2 and online fig. S7.6). Markings may have had a practical function, such as the large cross incised on one ball, which could have been grooves created specifically to hold two pieces of string in place in order to carry the ball or tie it to something (DB #5681/20965.m3. fig. 7.8 top). This could be particularly useful if the ball was to be transferred from location to location whilst hot. Similarly, the 18 notches covering one side of the surface of a complete oblate spheroid may be symbolic, retaining information and meaning (DB #4304/22300.m101. fig. 7.8 bottom). The use of small holes (made by inserting a thin implement into wet clay) is a common motif. The resultant hole can be found singly on small fragments, yet most likely

	DB no.	*Find no.*	*Marking(s) description*
1	3683	21661.m171	Total of five very clear markings: one set of three and one set of two on the opposing sides of this complete ball. The set of three form a triangle shape and are equally spaced 2.5cm apart. The two markings on the opposing side are the same.
2	3684	21661.m172	Two markings which cover almost the entire remaining surface. They are two deep, incised dots 0.5cm deep. One is in the fragmented edge.
3	4285	32106.m124	A total of seven markings. All are shallow, 0.5cm x 0.3cm, and run in two parallel rows. Located on a flattened part of this fragment. Possibly unintentional.
4	4287	32106.m126	Five clear, deep grooves running in parallel. They extend to the edge of the present fragment.
5	4300	31594.m103	Three deep holes closely spaced, forming a triangle shape. Partially in the fragmented section. Each is identical; they are 0.3cm wide and spaced 0.8cm apart from one another. Each is at least 1.0cm deep.
6	4304	22300.m101	Approximately 25% of this complete object has markings, a total of 18. Each is a deep, diagonal notch. They are generally placed in groups, closely spaced. In addition, there are three single, isolated long notches (ca 2.0cm long).
7	4307	22300.m104	Three faint notches can be seen on one side of the object. Similar to those seen on #4304 (22300.m101).
8	4905	22343.m106	One single deep, incised hole.
9	5129	32467.m127	Two rows of dots. Each row measures 1.2cm long. The two rows are spaced 1.0cm apart.
10	5568	32128.m.73	Total of 21 small incised dots, running in pairs, resembles the number '5'.
11	5616	19836.x.1	Two rows of small dots arranged in a zigzag design.
12	5657	30127.x.1	Three dots incised into the exterior surface. Each measures 0.3cm x 0.2cm wide. They are equally spaced, 1.2cm apart.
13	5671	19596.m.8	One single rounded hole, 0.3 cm deep and 0.3 x 0.3 cm wide. Located 1.5cm from the fragmented edge.
14	5681	20965.m.3	Two lines crossing each other.
15	5725	20965.m.15	One single deep fingertip depression. It is located on the fragmented edge, so possibly one of many. Measures 0.9cm x 1.0cm, 0.30cm deep.
16	5823	20965.m.113	Three deep fingertip depressions (first and second fingers and the thumb fit perfectly inside). Two are located on the fragmented edge – just half of each of these remains. The depressions measure 3.0cm x 1.9cm, 0.3cm deep. They are spaced approximately 1.5cm apart.
17	5833	20965.m.123	Three incised lines, two forming an 'L' shape, and one line inside running diagonally. The length of each line is 2.3cm (complete), 1.8cm (centre-fragmented) and 1.5cm (fragmented). Each is 0.2cm wide and 0.15cm thick.
18	5978	32692m.45	Three holes run in a line left to right, spaced 0.50cm apart. The third hole is in the fragmented edge. A further two holes are placed at a 90° angle to these, one of which is in the fragmented edge, thus barely visible. A total of five incised holes, three of which are complete.
19	5989	32692.m.56	Two incised dots located along the fragmented edge. Each measures 0.2–0.3cm (diameter). They are spaced 0.4cm apart.
20	6113	21841.m.22	Tiny fragment with two shallow, circular markings.
21	6211	19486.m.19	Two complete, clear and deep holes on the surface. Each measures 0.25cm x 0.25cm, 0.90cm deep. The holes are spaced 0.60cm apart. Two further holes are located along the fragmented edge (larger at 0.60cm).
22	7130	23143.m.29	There are eight incised holes running along the broken edges.
23	7175	23143.m.74	Two parallel incised lines.
24	7541	21634.m.1	Two crossing lines covering the entire remaining face.

Table 7.8. Detail of the 'markings' found on n=24 clay balls.

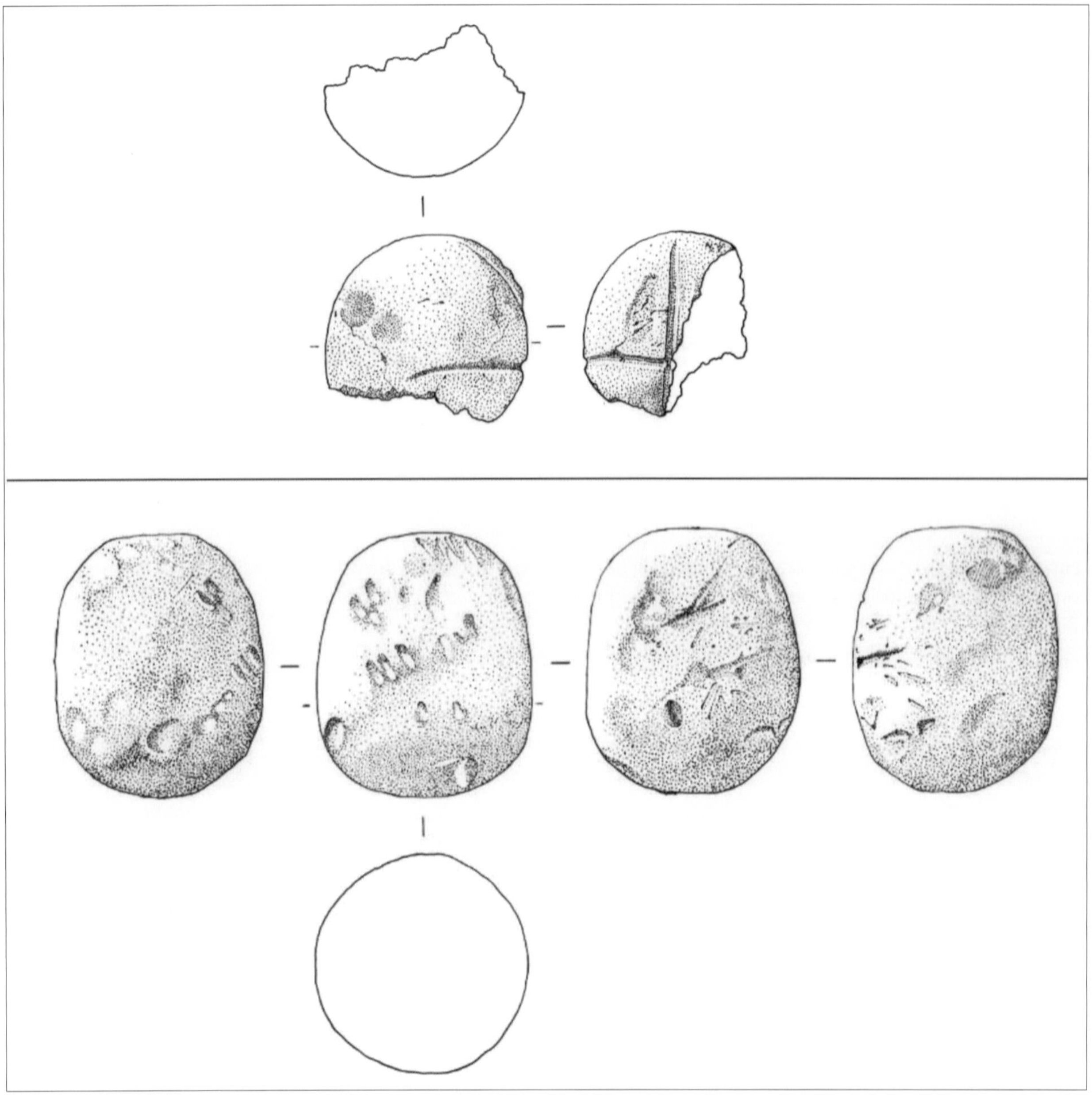

Figure 7.8. Clay balls with markings. Top: notched cross, DB# 5681/20965.m3 (5.5cm x 3.8cm x 5.3cm); bottom: oblate spheroid with 18 notches, DB# 4304/22300.m101 (6.0cm x 5.6cm x 7.2cm) (illustration by Jennie Anderson).

they were used in decorative combinations (online fig. S7.7). One complete ball has two triangles on opposing sides, each made from three small holes (DB #3683/21661.m171). A triangle formed with three dots is seen on other examples such as DB# 5705/20988.m11 and DB# 6955/23143.m3. Many other clay balls with markings display intricate designs made with the same technique, creating swirls, curves and zigzags (fig. 7.9, table 7.7).

Clay balls are highly uniform in manufacture and appearance, present in a limited range of clay colours reflecting the utilisation of a limited range of local clay types. Clay balls are easily recognisable, even as fragments, due to their dense clay make up, the shallow curve of their surface and their extremely smooth finish. Markings, matting and basketry impressions are distinctive yet exceptionally rare. The spherical shape of clay balls is highly regular, and though variants to shape do occur, they are infrequent. The overwhelming majority of clay balls are fragments. Over 80% (n=1,888) of the assemblage is fragments of less than 25% of the original artefact. Just 41 complete clay balls were recovered in the 2009–2017 seasons (1.8%). These examples averaged a 21.8cm circumference, 6.7cm diameter and 306.36g weight; thus they are extremely heavy and comparable to a grapefruit or baseball in size.

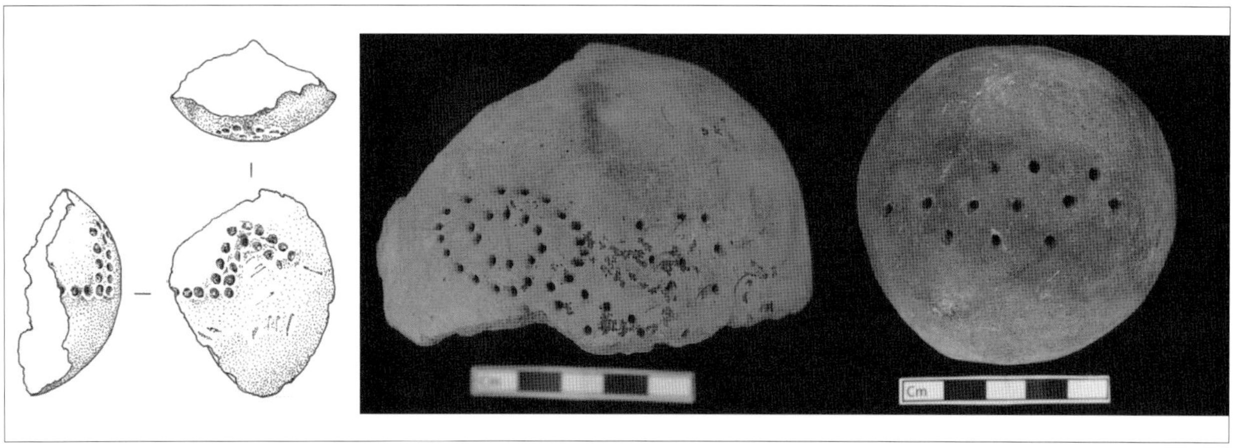

Figure 7.9. Clay balls with decorative markings. Left: zigzag design, DB# 5616/19386.x1 (6.5cm x 5.3cm x 3.0cm); centre: intricate design, DB# 69413/2616.m1; right: linear pattern, DB# 6963/32685.m1 (illustration by Jennie Anderson).

Though baked or fired during manufacture, the overwhelming majority of clay balls were, post-production, heated and re-heated time and time again in a practice related to their function.

Re-use

Detailed analysis of the small geometric clay objects (Chapter 8) and large clay balls, along with a general survey of all other clay artefacts and materials recovered at Çatalhöyük, has revealed the existence of a small number of artefacts which do not fit into any previously designated typological category. Twenty-six disc-shaped objects have been identified, manufactured via the re-use of fragments of large clay balls. As such, the discs are dense in clay make-up and their upper surface has a shallow convex curve. The upper surface is extremely smooth, as it is formed from the original exterior surface of a clay ball. The base surface in contrast is almost flat and is coarse in texture, being formed from the interior clay of a large clay ball (online fig. S7.8). The *re-used discs* are small in size, slightly larger than an average small geometric *clay object* at 4.73cm x 4.21cm x 2.23cm average dimensions. Yet due to their raw material, *re-used discs* are significantly heavier than *clay objects* (average weight of 4.65g), with an average weight of 41.93g. *Re-used discs* all share the characteristics of disc type-3, lenticular in shape with lightly convex upper and base surfaces, joined by sharp or rounded corners (Chapter 8). Viewed from above (plan view), they are a mixture of round and oval (table 7.9). It is unclear whether the *re-used discs* were deliberately crafted into their present shape for use in a specific activity, or whether their shape is simply the result of the re-use of clay ball fragments in a rubbing, polishing or grinding activity.

Contextual deposition

Area and temporal period

Clay balls are distributed almost exclusively across the two main excavation areas, North (n=4,981, 45%) and South (6,096, 54%), with a slightly increased count and density in the South Area (0.0200 and 0.0139 objects per litre respectively) (fig. 7.10, online table S7.3). Less than 1% of clay balls were recovered from TPC and GDN Areas combined (n=83/0.7%). This is the result of the broad temporal distribution of clay balls, which are found almost exclusively in the first half of settlement at Çatalhöyük, the Early (7100–6700 BC, n=5,222) and Middle (6700–6500 BC, n=5,274) periods, prior to the occupation of TPC and GDN (fig. 7.11 top).

Ninety-four percent of clay balls are recovered from stratigraphic levels pre-ca 6500 BC (the Early and Middle periods). The Late period (ca 6500–6300 BC) sees a sharp decline in clay ball counts to just n=360, representing only 3.3% of clay balls. This is despite continued settlement in both the North and South Areas. Clay ball counts continue to decline through time, with just four fragments of clay balls belonging to the Final period, ca 6300–5950 BC (fig. 7.11 top, online fig. S7.9). The temporal patterning of clay balls clearly demonstrates they were most important during Çatalhöyük's Early period of occupation, declining in use through time, with an abrupt end in their widespread use ca 6500 BC. This is supported by density analysis (fig. 7.11 bottom).

Period and stratigraphic level

The South Area is the longest-occupied zone at Çatalhöyük, spanning from the start of the Early period into the end of the Late period, ca 7100–6300 BC. Temporally,

	DB no.	*Find no.*	*3D shape*	*Registration*	*Weight (g)*	*L (cm)*	*W (cm)*	*H (cm)*	*Burnt?*
1	4249	32405.co.2	Disc 3	‘Re-used disc’ (*clay object* database)	32.8	4.6	4	2	No
2	4950	32495.co.1	Disc 3	‘Re-used disc’ (*clay object* database)	24	4	3.6	1.9	Yes
3	4951	32495.co.2	Disc 3	‘Re-used disc’ (*clay object* database)	30.2	4	4.4	2	Yes
4	4952	32495.co.3	Disc 3	‘Re-used disc’ (*clay object* database)	37.4	4	4.7	2.3	Yes
5	4953	32495.co.4	Disc 3	‘Re-used disc’ (*clay object* database)	50.7	5	4.6	2.2	Yes
6	4954	32495.co.5	Disc 3	‘Re-used disc’ (*clay object* database)	56.9	5.5	5	2.3	No
7	4955	32495.co.6	Disc 3	‘Re-used disc’ (*clay object* database)	67.1	6	5.8	2.4	No
8	5062	22301.co.1	Disc 3	‘Re-used disc’ (*clay object* database)	29	4	3.9	2.2	Yes
9	5063	22301.co.2	Disc 3	‘Re-used disc’ (*clay object* database)	48.7	4.1	4	3	No
10	5144	32496.co.1	Disc 3	‘Re-used disc’ (*clay object* database)	45.8	5.2	4	2.1	No
11	5292	31583.co.1	Disc 3	‘Re-used disc’ (*clay object* database)	37.2	4.3	4	2.4	Yes
12	5293	22670.co.1	Disc 3	‘Re-used disc’ (*clay object* database)	27.7	4	4	2.2	No
13	6366	21873.m.1	Disc 3	‘Re-used disc’ (*clay object* database)	78.9	5.4	5	2.9	Yes
14	6924	32632.co.1	Disc 3	‘Re-used disc’ (*clay object* database)	30.3	4.4	4	1.7	Yes
15	6925	31542.co.1	Disc 3	‘Re-used disc’ (*clay object* database)	45.8	4.9	4	2.35	No
16	6926	23226.co.1	Disc 3	‘Re-used disc’ (*clay object* database)	62	5.2	5	2.5	Yes
17	6927	23226.co.2	Disc 3	‘Re-used disc’ (*clay object* database)	30.3	4.1	4	2.2	No
18	6928	21863.co.1	Disc 3	‘Re-used disc’ (*clay object* database)	48.7	5	4.7	2.3	No
19	6942	23831.x.1	Disc 3	‘Re-used disc’ (*clay object* database)	42.4	4.4	4	2.5	Yes
20	6943	32745.co.1	Disc 3	‘Re-used disc’ (*clay object* database)	31.7	4.2	4	1.85	Yes
21	6969	31906.co.1	Disc 3	‘Re-used disc’ (*clay object* database)	72.2	5.6	5	2.4	Yes
22	5034	22301.m.113	Disc 3	’Re-used disc ‘(*clay ball* database)	68.2	7.42	4	2.7	Both
23	5565	32128.m.70	Disc 3	’Re-used disc ‘(*clay ball* database)	5.7	3.1	2	1.6	Yes
24	5690	20489.m.4	Disc 3	’Re-used disc ‘(*clay ball* database)	28.4	5	4	2.1	Both
25	6588	21816.co.1	Disc 3	‘Re-used disc’ (*clay object* database)	39.9	4.5	4	2.4	No
26	6592	32433.co.2	Disc 3	‘Re-used disc’ (*clay object* database)	18.3	5	3.7	1.6	Yes

Table 7.9. Detail of the n=26 disc-shaped objects made from the re-use of large clay ball fragments.

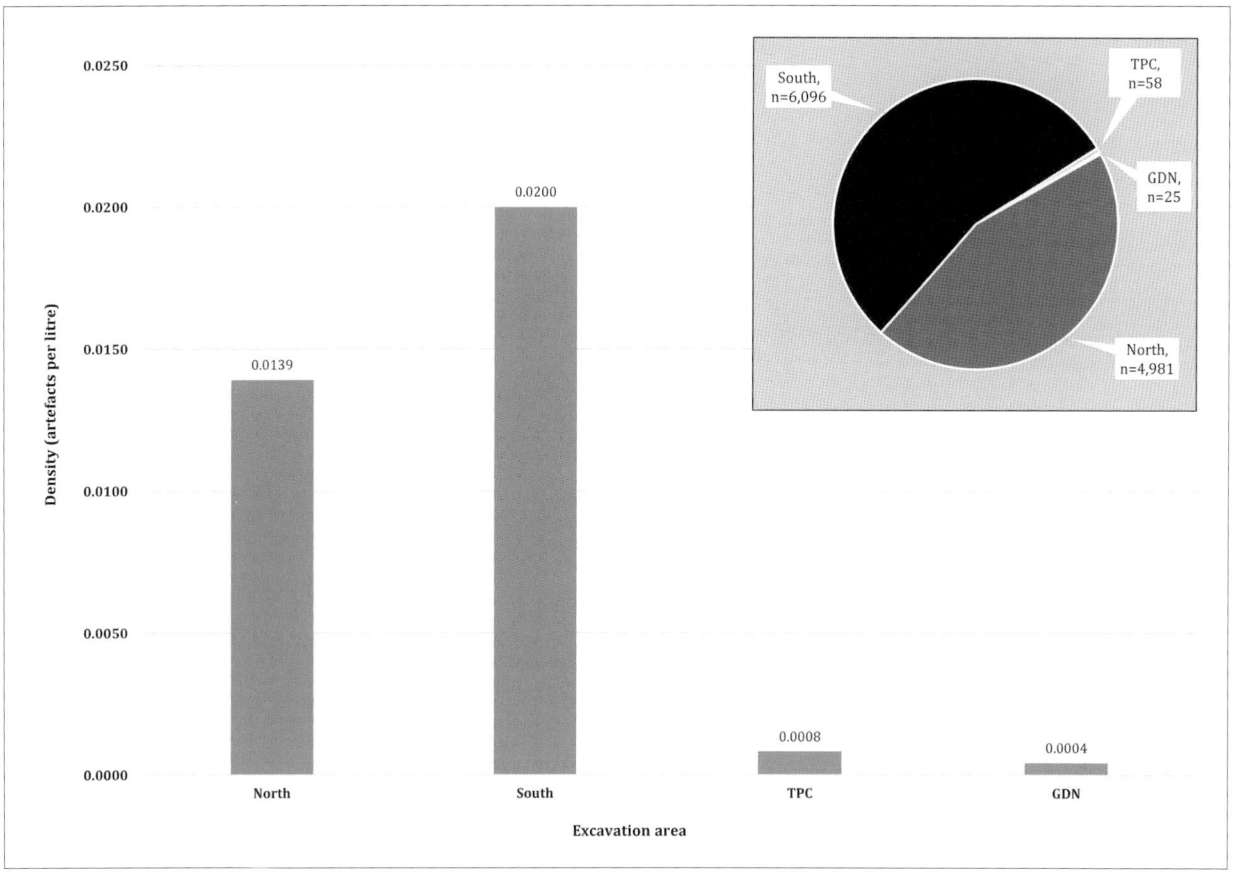

Figure 7.10. Distribution of clay balls across Çatalhöyük East's excavation areas. Main chart: density of clay balls per litre of sediment excavated per litre. Inset: counts (Level 1 and 2 data combined, n=11,190).

however, the use of clay balls is largely limited to the second half of the Early period. Eighty-six per cent (n=5,230) of clay balls of the South Area come from the Early period of settlement, the majority from a single level, South K, ca 6800–6600 BC (63%, n=3,815) (fig. 7.12 top). In contrast, no other level within the South Area contains more than 515 clay ball fragments, thus the widespread use of clay balls in the South is limited to a period of 200 years. Density analysis shows the same trend, with the greatest density of clay balls found in the South Area's Early period. There is no gradual decline in clay ball density, level by level. Levels South L (0.1650 o/l) and South K, at the end of the Early period, are both extremely dense in clay balls (fig. 7.12 bottom).

Occupied throughout Çatalhöyük's Middle and Late periods, from the North Area almost all clay balls recovered, 93%, date to the Middle period, ca 6700–6500 BC (n=4,599). This period is represented by just two stratigraphic levels, North F (n=1,992) and G (n=2,475), showing a slight increase though time within the Middle period. In contrast, the Late period in the North Area contains just 224 clay balls, almost all of which date to the start of the period, Level North H, the earliest level of the Late period (fig. 7.13 top). No clay balls have been recovered from subsequent levels within the North Area. Once more, the relative temporal distribution of clay balls is mirrored in their distribution by density. Clay balls are present at the start of the Middle period of occupation in the North Area (Level North F). They increase dramatically to 0.1009 o/l in the second half of the Middle period (Level North G) before declining in density at the start of the Late period (North H). They disappear altogether ca 6400 BC (fig. 7.13 bottom). This places the peak time of use in the North Area at ca 6700–6500 BC. Very low numbers of clay balls are in circulation during the second half of occupation at Çatalhöyük (Late and Final periods), in the South, TPC and GDN Areas. Density analysis shows an extremely limited use of clay balls towards the start of the Late period in TPC (Level M) before their use disappears almost entirely into the Final period (fig. 7.14).

The temporal and spatial patterning demonstrates that the use of clay balls was widespread during the first half of settlement at Çatalhöyük only. Clay balls were commonly used in the South and North Areas at

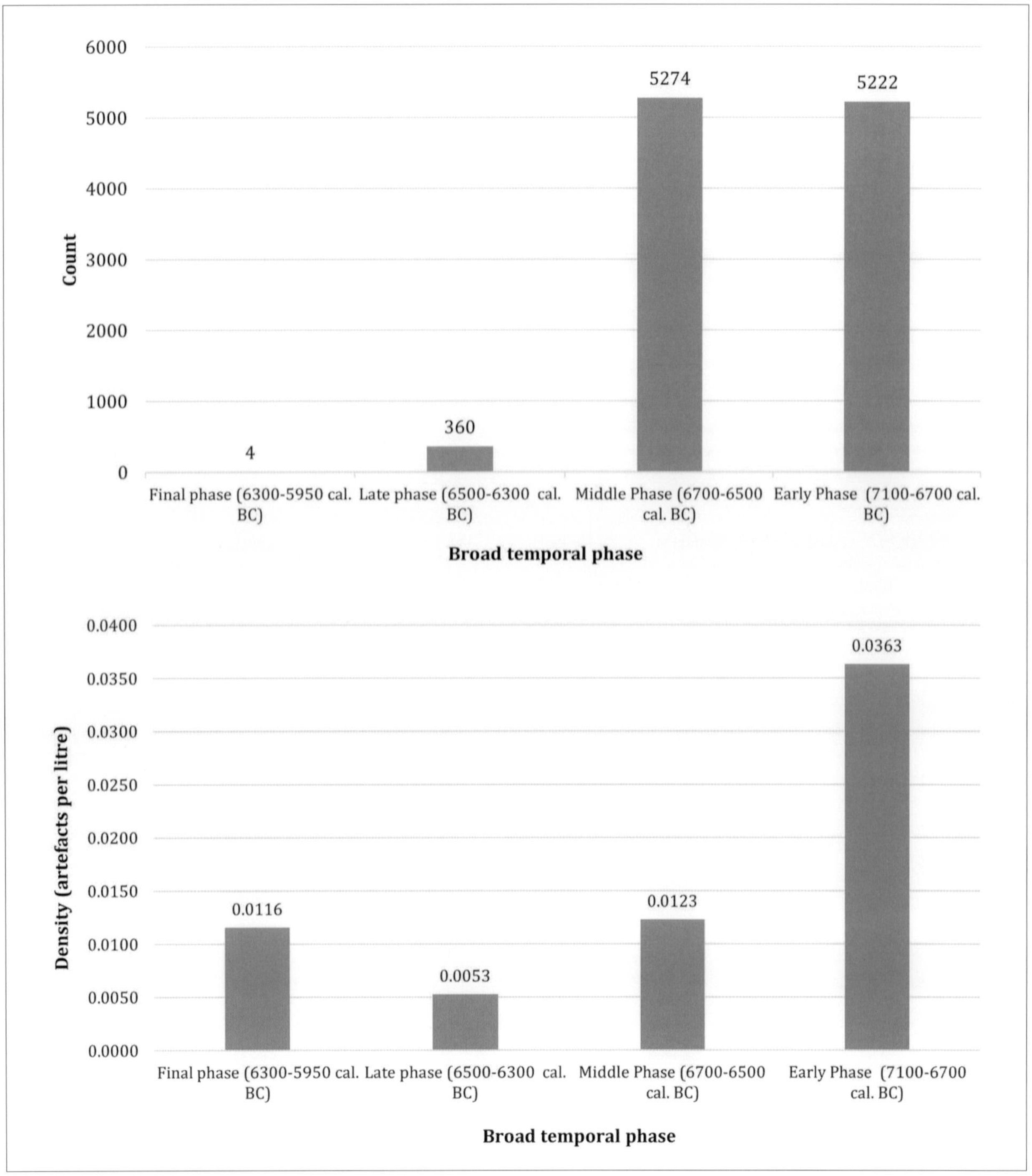

Figure 7.11. Count (top) and density (bottom) of clay balls (Level 1 and 2 data, n=11,190) across Çatalhöyük's four broad occupation periods.

this time. The South Area saw the height of clay ball use during a period of 200 years, ca 6800–6600 BC, with the peak of clay ball use in the North slightly later, ca 6700–6500 BC. Clay balls disappear in the North Area completely by ca 6400 BC and occur only in negligible quantities and densities elsewhere after this date (online fig. S7.12).

Nature of context

The distribution of clay balls according to the nature of their context shows they are most commonly recovered from refuse deposits and other secondary contexts. The Data Category context record shows a huge 86% of clay balls are recovered from the broad categories of 'fill' (n=4,503) and open area or 'midden' (n=3,448) (table

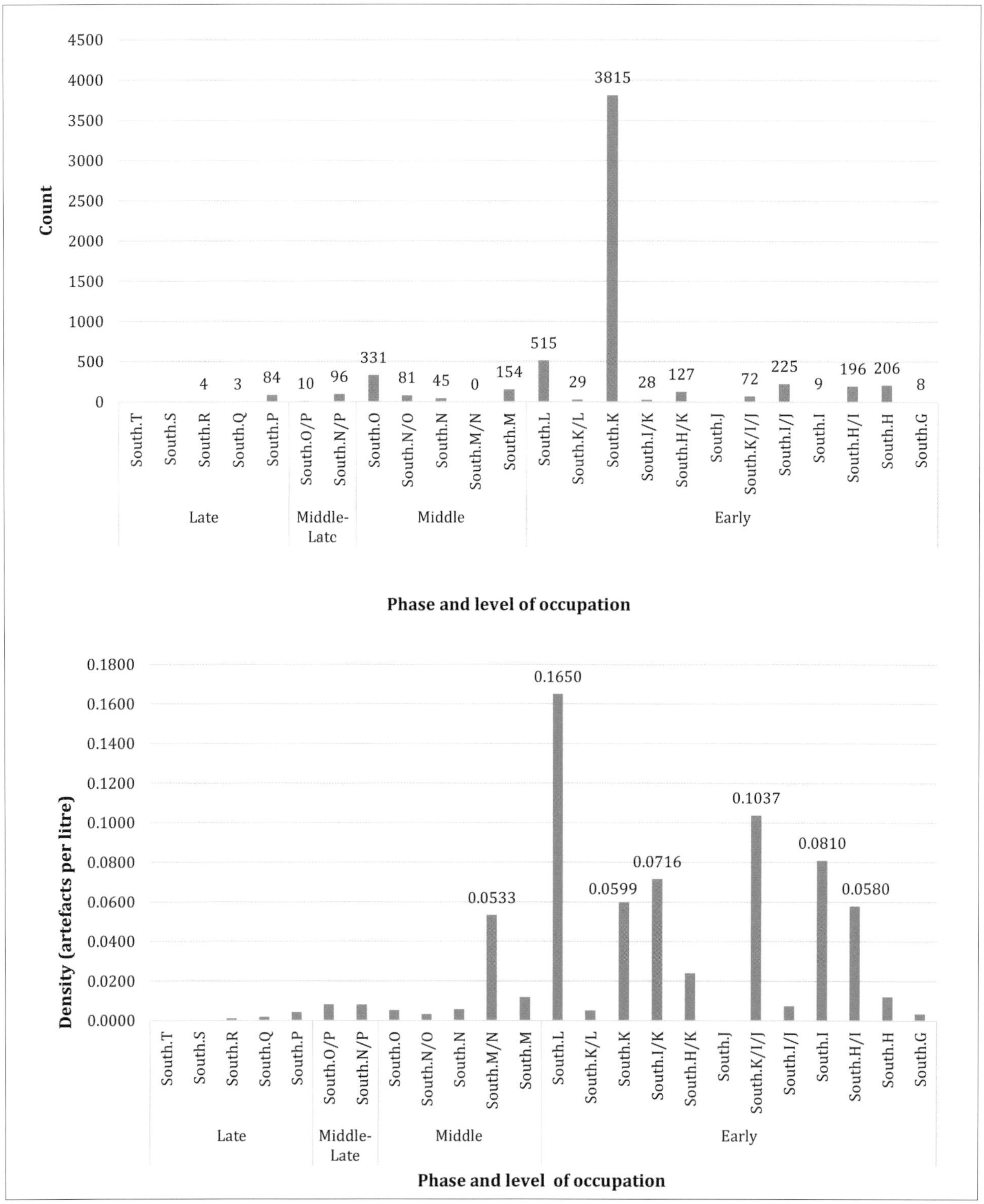

Figure 7.12. Temporal distribution of clay balls within the South Area by period and level of settlement (2009–2017 excavated artefacts, n=6,096). Top: count; bottom: density of clay balls per litre.

7.10). Midden contexts represent the accumulation and disposal of refuse in open spaces between buildings. These open areas also sometimes contained discrete surfaces upon which tasks such as craft production and food processing activities were carried out. Middens also accumulate less commonly inside disused buildings. Fill in the broadest sense represents redeposited sediment into a variety of contexts. The infilled features are almost exclusively found inside buildings, and midden material likely served as general fill, including for the infilling of abandoned buildings. Like 'fill', the broad context of 'construction/make-up/packing' also represents the

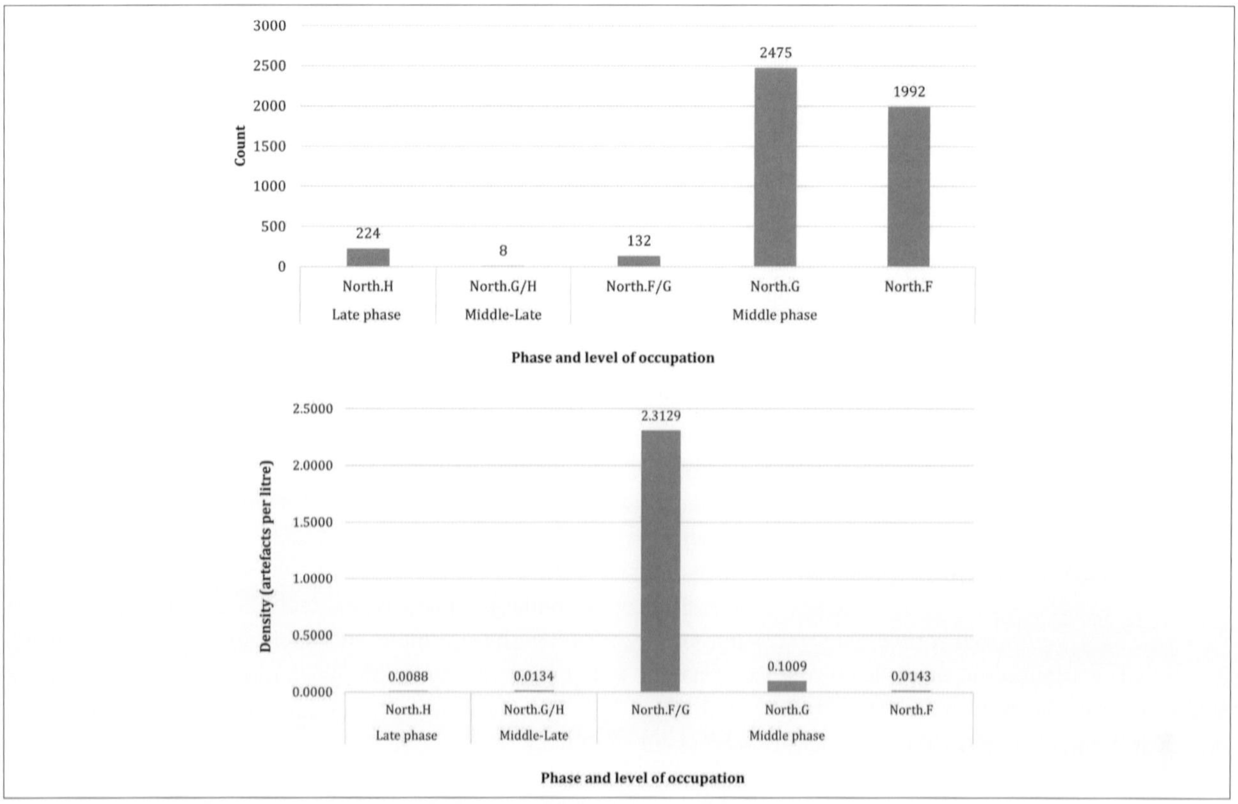

Figure 7.13. Temporal distribution of clay balls within the North Area (2009–2017 excavated clay balls, n=4,981). Top: count; bottom: density of clay balls per litre.

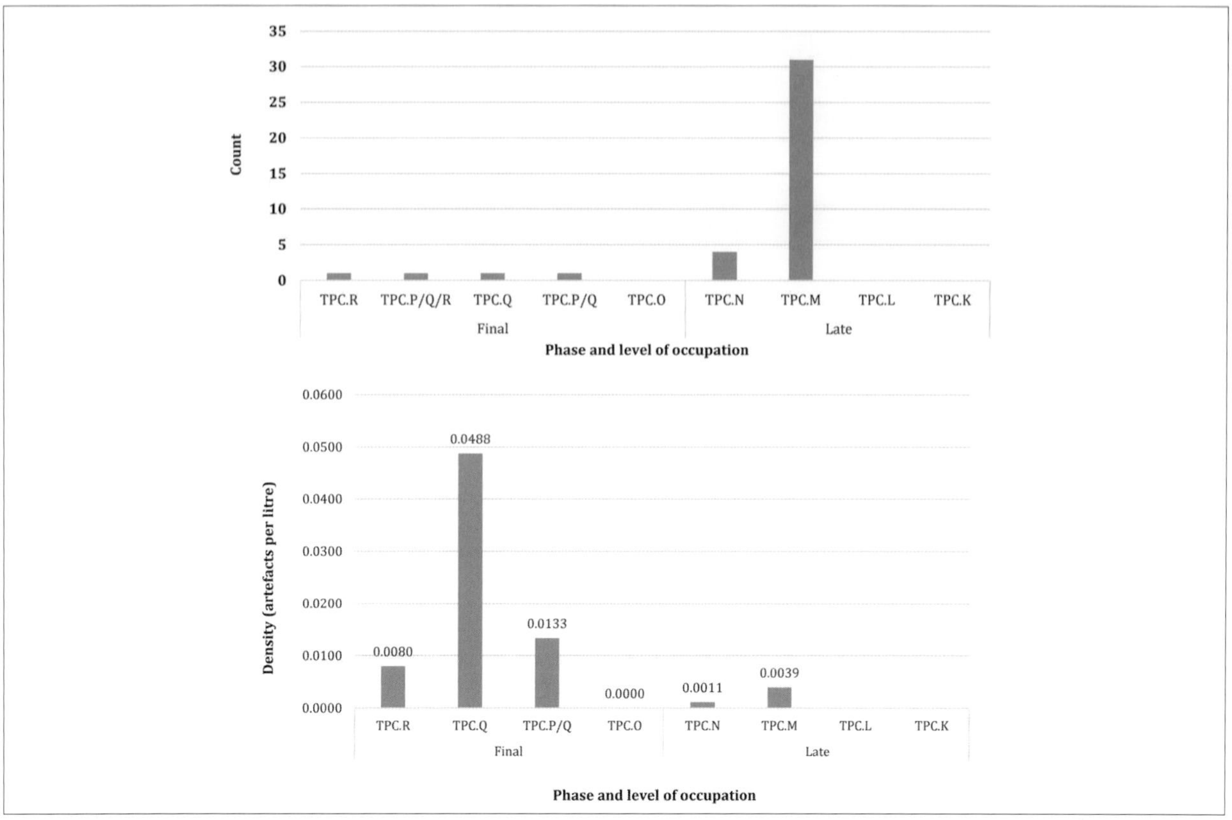

Figure 7.14. Temporal distribution of clay balls within the TPC Area (Late and Final periods). Top: count; bottom: density of artefacts per litre of sediment excavated.

incorporation of clay balls (or fragments thereof) into building material. This is the third most common broad context type after fill and midden (6.4%, n=593). All other broad (Data Category) contexts are represented by minimal quantities of clay balls. The recovery of clay balls from primary, in situ context types is therefore extremely rare. Just 4% of clay balls are recovered from the surface of floors, represented by 'floors (use)' Data Category contexts (n=378, table 7.10). Though found in small numbers in primary depositions, clay balls are most dense in contexts suggestive of primary use, including 'activity' (0.3834 o/l), 'cluster' (1.2040 o/l) and also 'skeleton' (23.8095 o/l) (table 7.10). Though when incorporated into burials, clay balls almost always appear as tiny, incidental fragments, 'cluster' and 'activity' contexts represent the intentional placement of clay balls in a primary context (see below).

General, data and interpretative category. Looking at contextual distribution in more detail, the exact nature and location of the midden and fill material (combining the most common context types as defined across all three of Çatalhöyük's context indicators: 'general', 'data' and 'interpretative', see online table S7.4), the proportion of clay balls recovered from 'midden/dump' areas remains similar, at 39% (n=3,595, table 7.11). The fill contexts are dominated by 'oven fill' (1,376, 15%), 'burial/skeleton fill' (n=1,062, 11%) and 'building/room-infill' (n=860, 9%) (table 7.11, fig. 7.15 top). Clay balls from 'construction/make-up and packing' also form a noteworthy quantity (n=593, 6%). This reflects the basic patterning seen above, demonstrating that the bulk of clay objects are recovered from contexts representing their disposal or unintentional inclusion within secondary contexts. The density of clay objects by detailed context type (all three categories combined) reveals a different picture, with primary deposits from internal spaces the densest in clay balls. Clay balls are most dense on floor surfaces (9.7881 o/l), where they are commonly recovered close to ovens. They are also dense within installations related to fire and heat ('oven fill' 1.0867 o/l, 'fire spot/installation' 0.6845 o/l, 'oven base' 0.3691 o/l) (table 7.11, fig. 7.15 bottom, online figs S7.10, S7.11). Clay balls were therefore most commonly used inside buildings in activities related to the use of heat.

Clay balls within buildings. More than half of all clay balls, 66%, are recovered from contexts associated with a building. The clay balls associated with buildings come from a variety of primary and secondary context types, including their incorporation into building material, such as walls, placements between the walls of two buildings, or underneath a building as part of a foundation deposit. Building 17 (South K, Early period) contains a disproportionately high number of clay balls, n=2,802. Most of the buildings prioritised for study by the project contain >100 clay balls, with B.77 (North G, Middle period), B.52 (North G) and B.89 (South N, Middle period) all having notably small quantities of clay balls (fig. 7.16).

Foundation deposits, the placement of items together prior to the construction of a building, incorporate a total of 157 clay balls (and fragments thereof). This practice incorporates 2% of all clay balls recovered from an internal space and represents the ritual use of clay balls (table 7.12). Such use is evidenced in B.160 (South K, Early period), where a number of foundation deposits

Context type: data category	*Count*	*Percentage*	*Density (o/l)*
Fill	4,503	48.62	0.0111
Midden	3,448	37.23	0.0360
Construction/make-up/packing	593	6.40	0.0084
Floors (use)	378	4.08	0.0291
Arbitrary	131	1.41	0.0056
Activity	55	0.59	0.3834
Cluster	36	0.4	1.2040
Skeleton	30	0.30	23.8095
Cut	6	0.06	0.0851
Natural	1	0.01	0.0003
(Blank)	80	0.86	0.0093
Total	9,261	100.00	0.0150

Table 7.10. Number and density of clay balls by context type according to the 'Data Category' (Level 1 and 2 clay balls from units with a total excavated sediment volume record, n=9,261).

Context type	*Count*	*Percentage*	*Density (o/l)*
Midden/dump	3,595	38.82	0.0269
Oven fill	1,376	14.86	1.0867
Burial fill/skeleton	1,062	11.47	0.0815
Building/room infill	860	9.29	0.0047
Construction/make-up/packing	593	6.40	0.0084
Pit-fill/infill	418	4.51	0.0140
Floor surface	328	3.54	9.7881
Oven superstructure	87	0.94	0.0867
Fire spot/fire installation	73	0.79	0.6845
Oven base/floor	67	0.72	0.3691
Post-retrieval pit fill/infill	43	0.46	0.0168
Oven superstructure	25	0.27	0.0256
Hearth base/structure	10	0.11	0.1195
Platform surface	5	0.05	0.0344
Hearth fill	2	0.02	0.1270

Table 7.11. Contextual distribution of clay balls (Levels 1 and 2 from units with a total volume of sediment excavated record, n=9,261) according to actual context type (combination of 'general', 'data' and 'interpretative' context categories). Most common contexts only.

were placed prior to building construction, many of which include clay balls (online table S7.15). Likewise, clay ball fragments are commonly recovered from a building's construction elements, representing their use during construction rather than inside a building during its main occupation phase. The large proportion of clay balls recovered from construction contexts seems to allude to the intentional incorporation of clay ball fragments as strong, study aggregate material, thus representing a secondary use of clay balls (table 7.12). Some buildings such as B.97 (South O, Middle period, n=197), record a high number of clay balls, yet closer inspection reveals they are mostly recovered from make-up and structural contexts related to the building's construction phase.

At the opposite end of a building's life history, clay balls are commonly recovered from activities related to building closure and abandonment. Over one quarter of internal space clay balls are recovered from building/room infill and midden contexts (n=1,097 and n=798 respectively) (table 7.12). Such contexts almost certainly represent the incidental inclusion of discarded clay balls. The vast majority of B.131's (North G, Middle period) n=405 clay balls are fragments, recovered from 'mixed dump' contexts dating to the building's construction phase (131.1C), 68%. Similarly, despite having over 100 clay balls, B.80 (South O, Middle period) has no evidence of their use during the main occupation phases. All recovered clay ball artefacts come from B.80's demolition material, infill or structural elements.

When recovered from a building's main occupation phase, clay balls are still most commonly incorporated into fill, unintentionally. Clay balls from such contexts are generally small fragments, suggesting non-intentional inclusion resulting from the re-deposition of midden material. Even in the case of clay balls from burials (15% of internal space clay balls), almost all examples are represented by tiny fragments placed within the loose fill, never in direct association with the skeleton or grave goods (table 7.12). Therefore, a very small number of buildings excavated during the Çatalhöyük Research Project's final phase of excavation contain significant quantities of clay balls linked to their *intentional* placement during the main *occupation phase* of a building. Contexts potentially representing in situ activities utilising clay balls include when they are found on floor surfaces, especially activity layers or dirty floors, in clusters, or recovered from fire installations and ovens. As seen above, all such contexts are dense in clay balls, suggesting these locations, despite evidencing clay balls in low quantities, represent the most common, primary locations for clay ball use. Just four buildings, B.17 (Early period, South K), B.132 (Middle period, North F), B.160 (Early period, South K) and B.161 (Early period, South J) have significant quantities of clay balls from such contexts (table 7.13, online tables S7.5, S7.6). Clay balls are dense on floor surfaces adjacent to ovens. Building 132, for example, contains 89 clay balls from 'floor surface' contexts (13% of the building's clay balls).

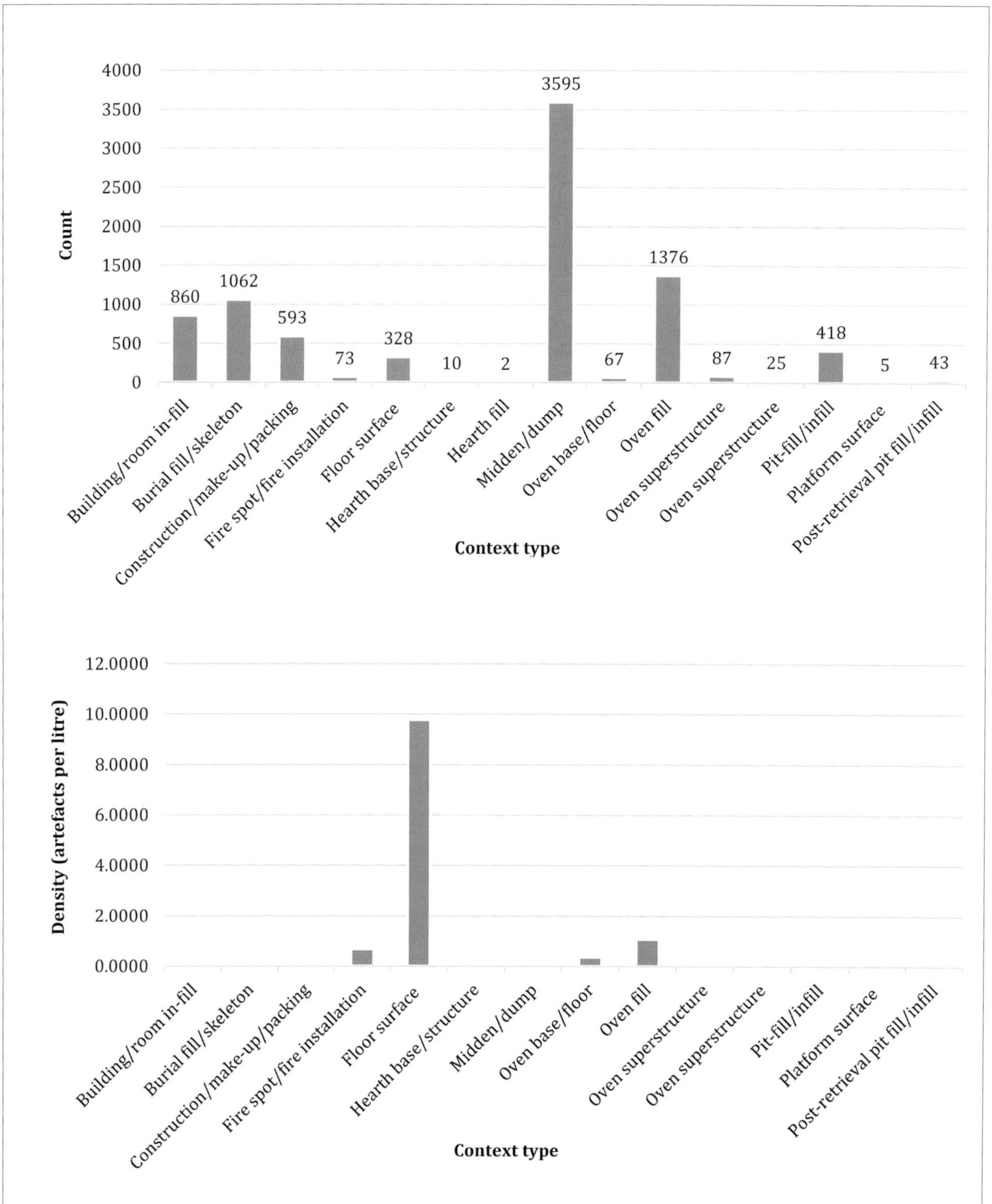

Figure 7.15. Contextual distribution of clay balls (Levels 1 and 2 from units with a total volume of sediment excavated record, n=9,261) according to actual context type (combination of 'general', 'data' and 'interpretative' context categories). Most common contexts only. Top: number of clay balls; bottom: density of clay balls.

Almost all are fragments (1–25% of the original artefact) recovered from 'dirty' floors in the main room of the building (online table S7.6). Buildings 160 and 161 also have significant quantities of clay balls recovered from their dirty floor surfaces (table 7.13). The largest proportion of clay balls recovered from buildings come from *inside* ovens. Thirteen per cent of the total assemblage, 19% of clay balls from internal spaces, are recovered from such contexts (n=1,422, table 7.12). A single oven, F. 579 in the northeast corner of B.17,

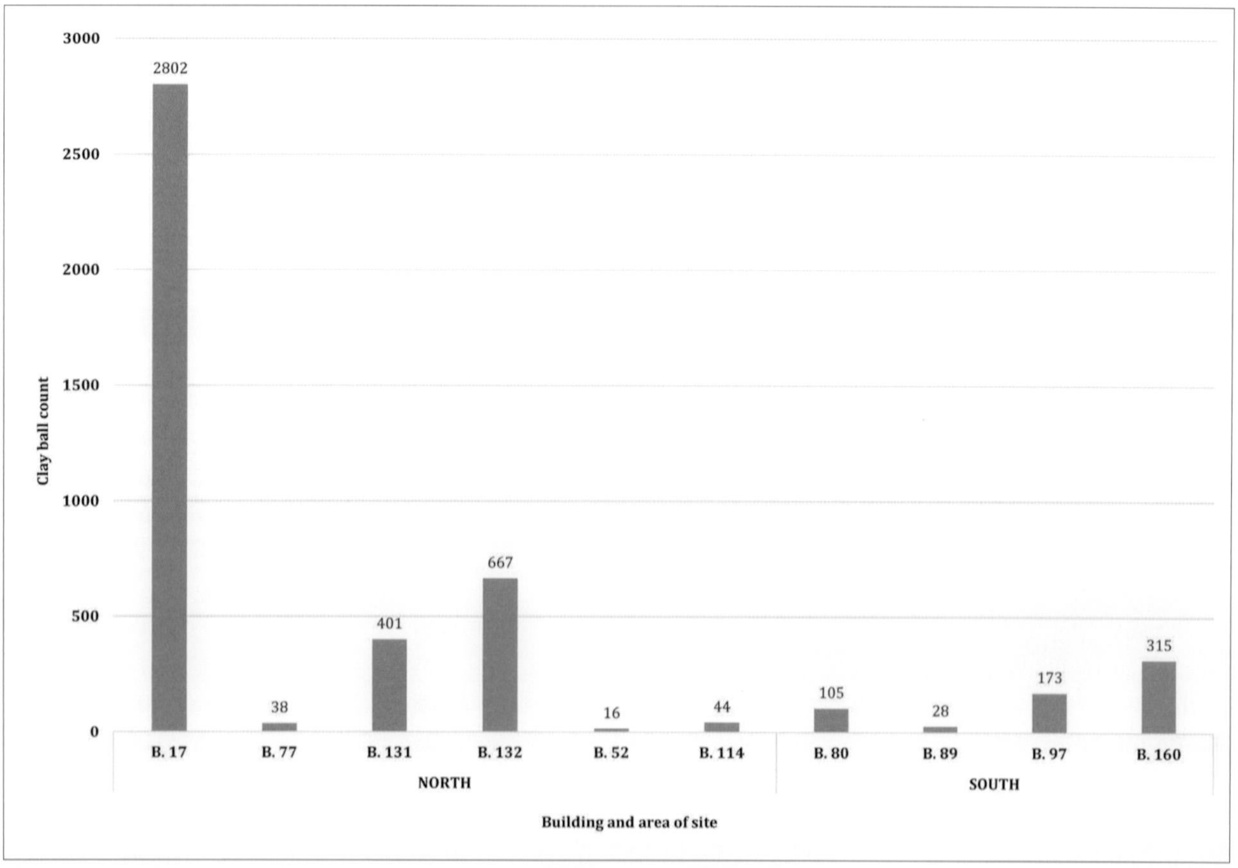

Figure 7.16. Distribution of clay balls across the 2009–2017 priority buildings.

contains 1,364 clay balls packed into its main cavity ((21849) and (21866)) (table 7.13, online fig. S7.11). Just one complete ball was recovered amongst the largely 1–24% complete fragments, which had an average weight of 30.24g. Almost all (Level 2) fragments were lightly burnt all over (online fig. S7.12). Building 132 also contains a large number of clay balls within a single oven, F.7732, a 'cluster', (32745) (table 7.14). The fragments average 109.40g, meaning they are around 25–50% complete. The most complete clay ball from this oven weighed 246.2g. Oven F.7732 is located in the main room of B.132, in the centre of the southern wall (online fig. S7.10). All other clay balls from oven fill contexts were recovered from four additional ovens, making the B.17 oven unique for the quantity of clay ball fragments recovered inside it. Though in some circumstances clay balls are used to line the base of ovens (such as oven F. 571, B.17), their use in the construction or repair of ovens remains rare.

'Cluster' as a context type represents the discrete deposition of a group of artefacts and/or cultural remains (as defined by Farid, Hodder 2014: 35). A total of 655 clay balls (6% of all 2009–2017 excavated clay balls) are recovered from a total of 23 deposits characterised as a cluster (table 7.14). The clusters range in composition: sometimes diverse, at other times consisting purely of clay balls. Clusters can therefore, contain between one and 304 clay balls. Almost all clusters containing clay balls (18 out of 23) are located inside buildings, meaning 95% of all clay balls from clusters come from internal spaces (table 7.14). Temporally, clusters incorporating clay balls are found from both the Early and Middle periods, most commonly within B.17 (Early period), B.132 (Middle period) and B.160 (Early period). Clusters incorporate a range of clay ball types from complete to highly fragmented, yet a disproportionally high quantity of complete clay balls are recovered from clusters (50% of all complete clay balls are from clusters). Many clusters appear to be ritual in function. They are placed in a variety of internal spaces, including underneath features such as walls, basins and ovens, or externally, between the walls of two buildings (table 7.14, nos. 5, 7, 9, 11 and 19). Such depositions incorporate clay balls along with faunal remains, ground stone and obsidian flakes. They likely remained present in the memories of a building's inhabitants, though they were invisible day to day. Some clusters were permanently installed as visible groups of artefacts placed on the surface of a building's floor. As buildings were cleaned as part of abandonment activities, these clusters seem to represent staged depositions, placed as part of a building closure ritual (table 7.14 nos. 12, 13, and 14).

Internal space	*Total*	*Percentage (internal space clay balls)*	*Percentage (entire assemblage)*
Oven fill	1,422	19.26	12.71
Burial	1,107	14.99	9.89
Building/room infill	1,097	14.85	9.80
Construction/make-up/packing	821	11.12	7.34
Midden/dump	798	10.81	7.13
Cluster	672	9.10	6.01
Floor surface	368	4.98	3.29
Pit-fill/infill	365	4.94	3.26
Floor layer in midden	272	3.68	2.43
Foundation deposit	157	2.13	1.40
Post-retrieval pit fill/infill	76	1.03	0.68
Fire spot/fire installation	68	0.92	0.61
Oven base/floor	64	0.87	0.57
Activity layer	59	0.80	0.53
Oven make-up	18	0.24	0.16
Hearth superstructure	12	0.16	0.11
Niche fill	4	0.05	0.04
Hearth fill	2	0.03	0.02
Platform surface	2	0.03	0.02
Ladder cut	1	0.01	0.01
Bin fill/infill	0	0.00	0.00
Total	7,385	100.00	66.00

Table 7.12. Nature of context, clay balls from buildings/internal spaces only (n=7,385, 2009–2017 excavated examples).

There is no clear correlation between the nature of a building as assessed by its elaboration index (a numerical value based upon a building's size along with the quantity and quality of additional architectural features, burials and 'artistic' elements, see this volume, Chapter 1) and the quantity of clay balls recovered from it, whether relating to the building's construction, use or abandonment phases. However, a reverse correlation is seen in *some* buildings where there is a high clay ball count. Building 160 for example, has a very low elaboration index of 5, yet has a total of 314 clay balls, most from occupational phases, including many which were definitely intentionally placed within the building. B.160 also contains six artefact 'clusters', incorporating 116 clay balls. Buildings 132 and 161 also have low elaboration index values (11 and 8 respectively) and high numbers of clay balls in clusters and floor surface deposits.

In reverse, B.131 has a higher elaboration index of 18, yet a low number of clay balls, none of which come from deposits relating to the occupation phase. Likewise, B.52 has an exceptionally high elaboration index of 42, yet just 16 clay balls in total. The reverse correlation is not true for all buildings. Building 17 is small, and plain, but other factors result in a high elaboration index of 22 (see Volume 12, Chapter 7). Yet B.17 has an extremely high number of clay balls (n=1,537), most from within an oven. Though the sample is small, the plainer, simpler buildings do appear to be those which used clay balls in larger quantities most commonly. Temporally, buildings with clay balls recovered in significant quantities from occupation phases and contexts related to intentional use come from both the Early and Middle periods, across both the North and South excavation areas. There is no grouping of evidence related to the actual *use* of clay balls limiting their distribution by phase, level or neighbourhood.

Clay balls from middens. Clay balls recovered from external spaces are almost exclusively recovered from middens. Midden clay balls are reflective of the clay ball assemblage overall. Middens contain typical clay balls of all fragment sizes, including complete examples. Non-typical shapes are also recovered from middens, as are decorated or 'marked' varieties. In fact,

Building	*Floor surface*	*Oven base/floor*	*Oven fill*	*Fire spot/installation*	*Total*
17	122	44	1,364	63	1,593
43	13	0	0	0	13
49	0	0	0	0	0
52	1	0	0	0	1
76	13	0	0	0	13
77	5	0	0	0	5
80	8	0	0	0	8
89	2	0	0	0	2
97	9	0	0	0	9
114	1	0	0	0	1
118	2	1	0	0	3
119	0	0	4	0	4
131	7	0	0	0	7
132	89	19	0	0	108
139	3	0	0	1	4
160	29	0	3	0	32
161	46	0	7	0	53
162	13	0	1	0	14
163	0	0	0	0	0
167	1	0	0	0	1
All	364	64	1,379	64	1,871

Table 7.13. The four contexts densest in clay balls (excluding 'cluster' depositions), all from within internal spaces and representing activity during a building's occupation phase. Count of clay balls per building, select buildings only (Level 1 and Level 2, 2009–2017 excavated artefacts).

the lack of distinction between the nature of clay balls recovered from inside buildings compared to middens demonstrates clay balls were used and re-used over and over again. They were not thrown into middens as soon as they fractured or became burnt and discoloured. Different dumping events can be seen in the nature of clay balls within particular layers of discrete middens. At each stage, clay ball fragments of similar sizes, levels of burning, wear and discolouration point to the use of clay ball fragments in batches, which are disposed of together at the end of their life. The midden in the North Area located in Sp.489 and Sp.490 (Middle period, North G), for example, contains discrete dumps incorporating clay balls. Unit (21116), the earliest infill of the midden, has two batches of clay balls, totalling 97 fragments. The first group, 42 clay balls (DB#s 3683–3724, (21661.m171–m212) comprises large fragments of 25–50% of the original artefact. One complete ball (208.70g, 6.1cm diameter) is also included. All have very little evidence of burning. In contrast, the second group of clay balls from the midden unit are all much smaller elements of the original artefact. None of the n=52 fragments is more than 50% complete and all are heavily burnt, on both the interior and exterior surfaces (see online figs S7.13–16). These depositions represent the clear-out of a buildings' clay balls, en masse, which were likely replaced by a fresh batch of complete balls.

Non-typical clay balls: spatial distribution

The contextual deposition of less typical clay balls mirrors that of all clay balls, suggesting that in general, the non-typical clay balls did not perform a distinct function related to their smaller size, shape or surface decoration. However, some exceptions do occur. The ovoids and oblate spheroids come from both the North and South Areas, in levels dating to the Early and Middle occupation periods. They are primarily recovered from midden and room infill contexts. The three complete ovoids all come from different deposits, ranging from middens to fire deposits and external surface dump contexts (online table S7.7). Notably, five examples of non-typically shaped clay balls (DB#s 3685/21661.m173, 3691/21661.m179, 3759/21661.m247, 3762/21661.m250 and 6280/20472.m5) come from a single midden (in Sp.489, further discussed below). Likewise, three of the

	Clay ball count	*Unit*	*Building*	*Space*	*Feature*	*Area*	*Phase*	*Level*	*Cluster description*
1	5	19343	-	459	-	South	Middle	South N	'Cluster or basket of midden material, contains...large charcoal fragments and carbonised berries/ nuts'.
2	22	19392	-	470	-	South	Middle	South M	Cluster with a wide range of artefacts. Includes one 50% and nine complete clay balls, one horn core, one antler, four ground stones. 'All were sat on the thin occupational greasy layer [19394] considered to be an upper part of the earth floor'.
3	25	20542	-	492	-	South	Early	South L	It was a cluster of a wide range of artefacts registered under one (separate) unit number. It comprised of: ten partially – and four whole preserved clay balls; number of animal bones including at least three articulated ones (!), one animal skull (X-8); dozen
4	5	23248	-	620	-	South	Early	South I	This is a cluster of two stone balls, a small chunk of plaster and two stones, embedded in a chunk of wood-charcoal.
5	1	19818	76/80	135/137	-	South	Middle	South O	Artefact cluster within a fill (u.19802) between B.80's west wall (f.5036) and B.76's east wall (f.3401)… Contained large amount and types of inclusions...ground stone, variety of animal bones, obsidian and charcoal.
6	4	32200	5	154	3810	North	Middle	North F	Cluster within burial cut.
7	38	21833	17	170	-	South	Early	South K	Cluster formed mainly by clay balls fragments, animal bones, stones fragments, shell frag. Painted small plaster frag. and obsidian flakes. It is probably related to one deposition located underneath the basin F585 and podium F574.
8	4	21843	17	170	-	South	Early	South K	Concentration of stones in the southwestern corner of Sp. 170, close to wall F8201.
9	4	23224	17	170	567	South	Early	South K	Fragments of clay balls, stone and small pieces of animal bone...under the western wall of B. 17.
10	1	17928	49	334	-	North	Middle	North G	Cluster in southwestern corner of building, dating to the construction phase. Foundation deposit?
11	11	19386	104	470/474	-	South	Middle	South M/N	Concentration of clay balls placed between walls (19835). Included three whole and five large fragments.
12	4	30630	118	510	-	South	Early	South H	Cluster of clay balls located near northwestern corner and oven of B118, Sp.510. Cluster is top of dirty floor.

Table 7.14. Detail of the location of 23 'cluster' contexts incorporating n=655 clay balls.

	Clay ball count	*Unit*	*Building*	*Space*	*Feature*	*Area*	*Phase*	*Level*	*Cluster description*
13	1	30598	132	511	-	North	Middle	North F	Contains clay ball (50%) and seven stones, including one piece of ground stone (X1). Placed in the southeastern corner of Sp.511, on the dirty floor. Stone all cracked, burnt at high temperature.
14	1	32094	132	531	-	North	Middle	North F	A cluster of artefacts and ecofacts. Set in corner of the main room, between a wall and a column. Comprised of bone, clay ball and groundstone.
15	304	32745	132	531	7732	North	Middle	North F	(blank)
16	1	32776	132	531	8330	North	Middle	North F	Located in a small niche-like depression.
17	42	32418	160	551	7832	South	Early	South K	Pit interpreted as an abandonment deposit. Pit was opened and filled in a short period, likely in late summer/early autumn. Contained large volume of faunal remains along with clay balls.
18	22	32430	160	551	7833	South	Early	South K	Pit interpreted as an abandonment deposit. Pit was opened and filled in a short period, likely in late summer/early autumn. Contained large volume of faunal remains along with clay balls.
19	36	32467	160	551	7815	South	Early	South K	Cluster of varied finds consisting mostly of clay ball fragments, baked clay, worked stone and faunal remains. The cluster was found underneath the earliest base of oven F.7815 in its southern part, a possible post-retrieval pit infill onto which the oven was later built.
20	13	32494	160	552	7847	South	Early	South K	Cluster defined in the northeastern part of Sp. 552, the side room of B. 160. This cluster is composed mostly of clay balls and stones.
21	2	32495	160	552	7847	South	Early	South K	(blank)
22	1	32600	160	552	7847	South	Early	South K	Cluster defined in the central north of Sp. 552, the side room of B. 160. This cluster is composed of five elements (two animal bones, a stone axe, a clay ball and a wood plank, on the top of which all these elements were placed).
23	108	32668	161	605/606	8175	South	Early	South J	Cluster of artefacts made up mainly of fragments of clay balls and various stones. Placed intentionally into pit F.8175 that truncated floor U.32667 which is considered to be a sequence of the earliest floors of B.161.
	Total: 655								

Table 7.14 (continued). Detail of the location of 23 'cluster' contexts incorporating n=655 clay balls.

smaller 'large' clay balls come from a distinct context, a 'cluster' of artefacts located within Sp.459 of the South Area (19392). They appear to have been specifically selected for inclusion in the deposit due to their smaller size and the fact they are completely intact. The three balls (DB# 5645, 5647, 5648/19392.m6, 8–9) were recovered from an artefact cluster, located on a thin clay occupation layer within Sp. 470 (South M, Middle period). They were placed along with a further six complete clay balls, in addition to one tiny and one larger fragment of clay ball, along with a horn core, antler and four ground stones (table 7.14).

The marked balls decorated in geometric patterns likely express personal or corporate identity, denoting ownership of a valued tool. The marked clay balls do not, however, appear to have retained any special symbolic value once their use life was complete. Almost all were disposed of in middens, with others recovered from secondary deposits of burial fill and rubble (table 7.15). An exception is DB# 5129 (32467.m127), part of a cluster of finds in B.160 (F.7815) (fig. 7.9, tables 7.14 and 7.15). The cluster contained a total of 36 clay ball fragments, along with worked stone, animal bone and obsidian flakes. It was located underneath the first oven of B.160. It is likely the highly decorated clay ball fragment was selected for inclusion in the cluster due to its symbolic value.

The midden Sp.489 (North G, Middle period) contains a number of non-typical clay balls. Along with five ovoid-shaped clay ball fragments (online table S7.7), the lowest level of the midden (21661) contains two 100 per cent intact clay balls (DB# 3683/21661.m171 and #5716/20965.m6), along with two clay balls with markings (DB# 3683 and 3684, 21661.m171–72). This level of the midden contained significant quantities of eggshell, along with clay objects, zoomorphic figurine fragments, pottery sherds, an obsidian arrowhead, mollusc shells and beads. The upper part of the midden (20965) contains a further four clay balls with markings (DB# 5681, 5725, 5823 and 5833) (table 7.15), along with faunal remains, ground stone, obsidian, shell and pottery sherds. Though this midden contains a significant quantity of clay balls (97 in the lower and 123 in the upper part), the number of non-typical examples is notable and may point to dumping activity after a ritual or other event utilising symbolic artefacts.

The contextual deposition of the n=26 re-used clay balls shaped into discs clearly differs from the normal patterning. This marks the re-used discs out as a distinct category of artefacts, as reflected by their shape and raw material. A group of seven re-used discs were recovered together as part of a foundation deposit 'cluster' underneath B.160, F.7847 (table 7.16). Most other re-used discs were deposited in groups within a building. Their distribution is limited to a small number of buildings, B.17, B.131, B.132 and B.160 (table 7.16).

Contextual distribution summary

Clay balls occur in almost equal proportions and densities in the North and South Areas, decreasing through time during Çatalhöyük's occupation. Used most in the latter levels of the Early period in the South Area and the Middle period in the North Area, ca 6500 BC, clay balls drastically decline in both quantity and density. By the Final period, where only the TPC and GDN Areas remain occupied, clay balls have completely fallen out of use. Clay balls are most commonly recovered from midden deposits, with secondary deposits (features such as ovens and side rooms which were intentionally infilled after abandonment), likely utilising midden material (fill and construction/make-up/packing), also having significant quantities of clay balls. However, in density, primary depositions representing the use of clay balls inside buildings during the main occupation phase are most dense in clay balls. They are recovered from the dirty floors of buildings close to ovens, from inside ovens and fire installations and in clusters inside the main rooms of buildings. The contextual deposition of clay balls changes little across different neighbourhoods or areas of the site. Certain buildings contain higher numbers of clay balls than others, largely due to the presence of a specific cluster or feature containing clay balls in huge quantities. The marked clay balls, though highly detailed and likely symbolic in meaning, are recovered from the same contexts as plain examples. Though they may have held a special meaning during their use life, they are disposed of into middens and as building material as readily as their plain counterparts are. The exception to this is the quantity of marked, complete and non-typically shaped clay balls recovered from the midden represented by Sp.489.

All evidence points to clay balls being used primarily inside buildings, heated and re-heated time and time again in ovens and other fire installations. This resulted in their cracking and fragmentation. Initial fragmentation resulted in distinctive, large wedge-shaped pieces, indicative of the items being exposed to high temperature, then cooled very rapidly. Only after clay balls had fragmented into very small pieces were they disposed of. Yet in some instances, complete balls in good condition were also thrown into middens. Clay balls held symbolic value, as demonstrated by their incorporation in foundation deposits and clusters. Their symbolic value was related to the function they fulfilled, as used clay balls (fragmented and burnt) rather than complete, new and decorated clay balls are most common in foundation deposits and clusters.

	DB #	Find no.	Data category	Interpretive category	General context	Building	Space	Feature	Level
1	6211	19486.m19	Midden	Arbitrary layer - midden	Midden/dump		489		North G
2	5671	19596.m8	Arbitrary	Infill	(blank)		99		North G
3	5616	19836.x1	Floors (use)	Floor	Floor surface	89	379	3492	South N
4	5681	20965.m3	Midden	Midden	Midden/dump		489		North G
5	5725	20965.m15	Midden	Midden	Midden/dump		489		North G
6	5833	20965.m123	Midden	Midden	Midden/dump		489		North G
7	7541	21634.m1	Fill	Burial infill (?)	Burial fill/skeleton		602	7634	North G
8	3683	21661.m171	Midden	Midden deposit	Midden/dump		489		North G
9	3684	21661.m172	Midden	Midden deposit	Midden/dump		489		North G
10	6113	21841.m22	Skeleton	Skeleton	Burial fill/skeleton	17	170	8017	South K
11	4304	22300.m101	Fill	Room infill	Building/room infill	160	551		South K
12	4307	22300.m104	Fill	Room infill	Building/room infill	160	551		South K
13	3840	22314.m106	Midden	Midden layer	Midden/dump		559		South L
14	4905	22343.m106	Midden	Midden layers	Midden/dump		583		South K
15	7130	23143.m29	Midden	Midden	Midden/dump	139	623, 625		North F
16	7175	23143.m74	Midden	Midden	Midden/dump	139	623, 625		North F
17	5657	30127.x1	Fill	Rubble rich layer	(blank)	119	512, 513		North F
18	4300	31594.m103	Fill	Infill-post retrieval pit	Post-retrieval pit fill/infill	132	531	7873	North F
19	4285	32106.m124	Midden	Mixed deposit	Midden/dump		85		North ?H
20	4287	32106.m126	Midden	Mixed deposit	Midden/dump		85		North ?H
21	5568	32128.m73	Midden	Mixed dumps	Midden/dump	131	500, 610		North G
22	5129	32467.m127	Cluster	Cluster of finds	Cluster	160	551	7815	South K
23	5978	32692.m45	Midden	Midden layer	Midden/dump		620, 628		South ?I/?H
24	5989	32692.m56	Midden	Midden layer	Midden/dump		620, 628		South ?I/?H

Table 7.15. Contextual distribution of the 24 'marked' clay balls.

	DB no.	Find no.	Year	Area/level	Phase	Category	Interpretive category	Data category	Building	Space	Feature
1	5690	20489.m.4	2012	North	?	Layer	Arbitrary	Midden		1000	7107
2	6925	31542.co.1	2016	North G	Middle	Layer	Wall mortar	Construction		602	7869
3	6928	21863.co.1	2016	South K	Early	Layer	Layer	Floors (use)	17	170	
4	6366	21873.m.1	2017	South K	Early	Layer	Make-up layer	Construction	17	170	
5	6926	23226.co.1	2017	South K/I	Early	Cluster	Foundation deposit	Cluster	17	170 182 620	567
6	6927	23226.co.2	2017	South K/H/I	Early	Cluster	Foundation deposit	Cluster	17	170 182 620	567
7	6588	21816.co.1	2016	South K	Early	Layer	Post fill	Fill	17	170	8016
8	5565	32128.m.70	2017	North G	Middle	Layer	Mixed dumps	Midden	131	500 610	
9	5293	22670.co.1	2015	North G	Middle	Layer	Platform surface	Construction	131	500	7952
10	5292	31583.co.1	2016	North F	Middle	Layer	Infill post-retrieval pit	Fill	132	531	7725
11	6943	32745.co.1	2017	North F	Middle	Cluster	Cluster in oven	Cluster	132	531	7732
12	5062	22301.co.1	2015	South K	Early	Layer	Room infill	Fill	160	551 552	
13	5063	22301.co.2	2015	South K	Early	Layer	Room infill	Fill	160	551 552	
14	5034	22301.m.113	2015	South K	Early	Layer	Room infill	Fill	160	551 552	
15	4249	32405.co.2	2016	South K	Early	Layer	Dirty floor	Floors (use)	160	551	
16	6592	32433.co.2	2016	South K	Early	Layer	Plaster and make up	Construction	160	551	7834
17	4950	32495.co.1	2016	South K	Early	Cluster	Foundation deposit	Cluster	160	552	7847
18	4951	32495.co.2	2016	South K	Early	Cluster	Foundation deposit	Cluster	160	552	7847
19	4952	32495.co.3	2016	South K	Early	Cluster	Foundation deposit	Cluster	160	552	7847
20	4953	32495.co.4	2016	South K	Early	Cluster	Foundation deposit	Cluster	160	552	7847
21	4954	32495.co.5	2016	South K	Early	Cluster	Foundation deposit	Cluster	160	552	7847
22	4955	32495.co.6	2016	South K	Early	Cluster	Foundation deposit	Cluster	160	552	7847
23	5144	32496.co.1	2016	South K	Early	Layer	Foundation deposit	Fill	160	552	7847
24	6924	32632.co.1	2016	South ?K	Early	Layer	Oven superstructure	Construction	161	605 606	8160
25	6942	23831.x.1	2017	North F	Middle	Layer	Floor	Floors (use)	167	635	
26	6969	31906.co.1	2015	TPC L	Middle	Layer			170	611	

Table 7.16. Context detail of the n=26 disc-shaped objects made from the re-use of large clay ball fragments.

Discussion

Clay balls are an integral element of Çatalhöyük's material culture, yet only during the first half of the period of settlement. Present across both the North and South Areas in huge quantities ca 7100–6500 BC, clay balls were manufactured in a highly regular fashion. Their primary function led to frequent breakage; however, broken clay balls remained useful. They continued to be used and re-used as fragments until a shift in lifestyle led to a rapid decline in clay ball distribution and use ca 6500 BC.

Primary function

It has long been argued that clay balls were used as cooking 'stones' in the heating of liquid foods (Atalay 2005: 156–57; Atalay, Hastorf 2005; 2006). Evidence gathered from the final phase of excavation and research at Çatalhöyük broadly supports Atalay's claims. Clearly clay balls were primarily used within buildings, heated in ovens, yet direct evidence for their placement into stews to cook them is lacking. At 9.7881 o/l, the density of clay balls found in the primary context of internal 'floor surface' is higher than the density of clay balls across all other context types combined. The fact that 78% of clay balls are burnt on both their outer and inner surfaces demonstrates that although the heating of clay balls resulted in their cracking and breakage, clay balls continued to be heated and re-heated. Their proximity to ovens when recovered from floor surfaces, in addition to their significant density as part of oven fill contexts, shows ovens inside buildings were, as Atalay identified, the primary location for the heating of clay balls.

Clay balls were clearly heated inside buildings, yet why? The Early and Middle periods of occupation at Çatalhöyük lack pottery suitable for cooking. Therefore, basketry or skins were likely used as containers in which stews and soups could have been prepared and cooked. Basketry is attested archaeologically at Çatalhöyük. Mellaart speaks of large quantities of baskets in his early research reports, stating they are far more common than other container types, those made of stone, bone or antler (Mellaart 1964: 85, fig. 34). More recent research has confirmed the presence of coiled basketry made from flax at Çatalhöyük, including examples that were waterproofed using plaster (Wendrich 2005: 334; Wendrich, Ryan 2012; Demirergi et al. 2014: 93, 95; Jørgensen et al., this volume, Chapter 11). Almost all examples of coiled basketry come from Çatalhöyük's Early and Middle periods, with a sharp decline in use evidenced at the end of the Middle period into the start of the Late period in both the North (Level G) and South (Level P) Areas (Wendrich 2005: 337, fig. 15.7; Wendrich, Ryan 2012: 61). The decline in basketry corresponds to the decline in clay balls, indirect evidence of a link between the two technologies. Yet even if baskets were used as containers for cooking, this is certainly not an exclusive function. Baskets have multiple uses and, at Çatalhöyük, could have served as vessels for transportation and storage, and as containers for bodies, as well as having a decorative role. Therefore, when interpreting changes in their temporal patterning, their multi-functional nature needs be taken into account. Likewise, the identification of basketry from plant phytoliths is subject to a number of factors. Overall, the presence of significant quantities of baskets, capable of holding liquid foods and limited to the Early and Middle periods of settlement, opens up the possibility that hot clay balls were placed inside them to cook food.

Temporal patterning in pottery forms and distributions offers additional indirect evidence to support this theory. The Early period at Çatalhöyük is characterised by very low levels of pottery use, incorporating various forms. A shift occurs ca 6700 BC (South M) with the introduction of Dark Gritty Ware (see Chapter 4). Functioning as cooking pots, Dark Gritty Wares are hole-mouth jars in form, and fired at 700–800ºC. They can be placed directly into a heat source, retain heat and do not crack, making them perfect for cooking. Dark Gritty Ware jars increase steadily during the Middle period, dominating the ceramic record of the Late and Final periods. The corresponding decline in clay balls is strong evidence that the former replaced the latter, and as Dark Gritty Ware is undoubtedly a cooking tool, clay balls likely functioned as cooking tools also (fig. 7.17).

If indeed the residents of Çatalhöyük were using heated clay balls to cook liquid foodstuffs, evidence to support this functional use of clay balls may come from food preparation and consumption evidence. Demirergi's (2015) research into food preparation and consumption at Çatalhöyük focuses on the meat element of the diet. Temporal changes in the preparation of the dominant meat animals (sheep and cow) suggest a change in the main meat-cooking practices mid occupation sequence. In the Early and Middle periods, sheep and cow-sized animal bones are dominated by those evidencing consumption cuts. These marks are produced when meat is cooked on the bone and are produced after cooking, as the meat is cut away (Russell, Martin 2005: 85). Such marks are largely restricted to baked and roasted meats; boiled and stewed meat results in meat which separates from the bone extremely easily, with no need for the use of a knife (Demirergi 2015: 133). The decrease in consumption cuts over time suggests a temporal decrease in the roasting and baking of meats at Çatalhöyük (Demirergi 2015: 144–45). There is a corresponding increase in the frequency of filleting cutmarks on sheep-

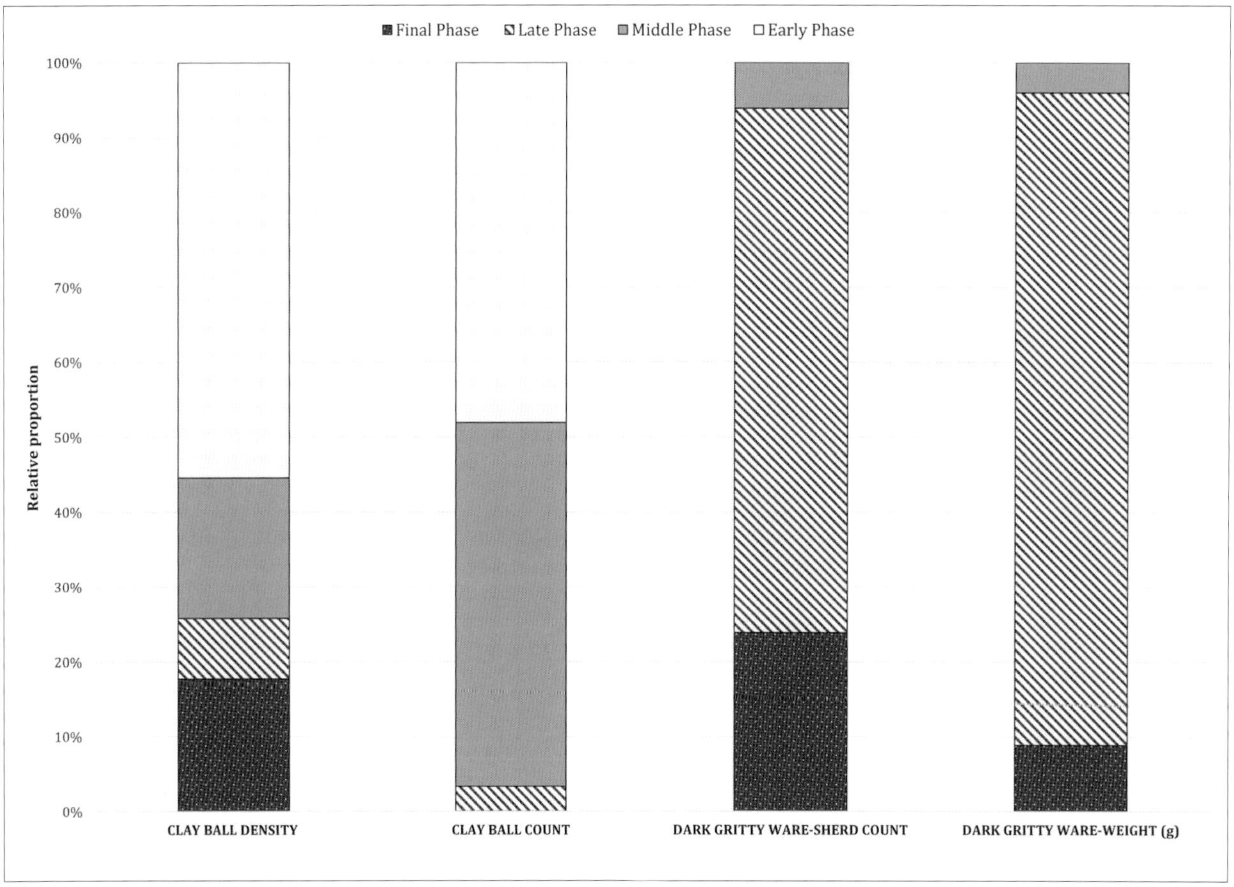

Figure 7.17. Cooking pottery (dark gritty ware) densities by period compared to clay balls density.

and cow-sized animals through time (Demirergi 2015: 143). Filleting indicates the processing of meat into small pieces, for drying and storage or to create manageable pieces of meat to be cooked and eaten off the bone (Demirergi 2015: 133, 143). The decrease in consumption cuts alongside the increase in filleting cuts implies a change in consumption practices, with meat more likely to be prepared in a liquid form, stewed or boiled, in the *latter* periods, as opposed to the roasting of large joints in the Early and Middle periods (Demirergi 2015: 143–44).

Demirergi's hypothesis is supported by wider research by Russell et al. (2013) which compared meat preparation and consumption in the South Area during the Early and Middle periods (Levels South H to M) against the Late period (Levels South P to T). Again, dismemberment and filleting patterns suggest a change from the dry roasting of sheep- and cow-sized meat fragments in the first temporal period, to the stewing of meat in the later periods (Russell et al. 2013b). The fauna data is direct evidence against the traditional theory of clay balls' use as tools to cook liquid foods, as clay balls have fallen out of use by the time stewing and boiling becomes the dominant form of meat preparation. However, it does not exclude them from a role in cooking entirely.

Evidence from the analysis of plant food remains shows a similar trend. González Carretero's research (Volume 15, Chapter 14) demonstrates a shift in cooking practices from the Early and Middle to the Late and Final settlement periods, beginning in the North Area Levels F/G, ca 6500–6700 BC. Çatalhöyük's Early and Middle periods are dominated by wheat, indicating the consumption of bread as the main source of grain at this time. With the introduction of cooking pots, a barley-based 'porridge' appears in the archaeological record. This liquid meal is prepared in pottery vessels using direct heat. The Late and Final periods of occupation see a dominance of barley, suggesting this meal formed an important element of the diet in the latter half of occupation. This evidence is interpreted as a dominance of dry baking and roasting ca 6500 BC, at which point plant-based foods are prepared in liquid form, heated via boiling and stewing, thus mirroring the faunal evidence (Volume 15, Chapter 19).

More general faunal and botanical evidence demonstrates Çatalhöyük's residents utilised a diverse array of plant and animal food sources (Jenkins 2005; Russell, Martin 2005; Russell, McGowan 2005; Bogaard et al. 2013; Hillson et al. 2013; Van Neer et al. 2013; Russell et al. 2013b; Volume 13, Chapters 3, 5, 8). Therefore, although evidence suggests the preparation of dry food cooked via roasting and baking dominated in the Early and Middle periods, then was replaced by the stewing and boiling of liquid foodstuffs in the second half of settlement, diversity of foods suggests various cooking methods co-existed throughout the duration of settlement at Çatalhöyük. A study seeking to recover traces of organic residue from the surface of various materials at Çatalhöyük (to confirm the arrival of cattle domestication and milk production) failed to recover any trace of organic residues from the clay ball study sample (Pitter 2013: 66). This further casts doubt on the conventional interpretation of clay balls at Çatalhöyük. Clay balls are large and porous. If they were placed directly into liquid foods, being completely covered for extended periods of time, residues should remain.

Cooking balls?

Clay balls undoubtedly have a connection to heat and cooking during Çatalhöyük's Early and Middle periods. Clay balls were primarily used inside buildings, being heated in the same ovens where food was cooked. The simultaneous introduction of cooking pottery and disappearance of clay balls does not seem accidental. However, there is no direct evidence that acted as cooking 'stones' as originally hypothesised (Atalay 2005, 2013; Atalay, Hastorf 2005; 2006). In opposition to the traditional view, stone is plentiful at Çatalhöyük. It would be more logical to use small stones to be heated and placed into stews rather than large clay balls, and the use of stones in this way is suggested by Tsoraki (this volume, Chapter 13). Stone has better heat retention properties than clay, and is less porous and less likely to crack and fragment when exposed to direct heat. Clay balls do have the potential, however, to be used as cooking stones, yet the ceramic and food record clearly shows a preference for liquid foods in the latter half of settlement, at the time when clay balls disappear. During the Early and Middle periods, large joints of meat were grilled or roasted, and bread was the main source of grain. During the Middle settlement period, new food preparation and cooking methods were introduced, resulting in the preference for meaty stews and plant-based 'porridges' in the Late and Final periods, ca 6500 BC onwards. With clay balls disappearing from the archaeological record at the time boiling and stewing became popular, it seems highly unlikely that they were involved in such activities during the Early and Middle periods.

Yet perhaps clay balls were still used in the heating or cooking of food. Once heated, they could have been placed on the floor, adjacent to ovens, used to grill, or more likely, to keep warm recently grilled or baked foods. This function could explain the repeated occurrence of hearths adjacent to ovens. Such a function for clay balls would also explain the continued use of fragmented clay balls, which could not have been placed into liquid foods, yet would not contaminate dry food items such as bread or cuts of meat, especially if matting, leaves or the like were placed above the heated clay balls, acting as a tray or hotplate. This interpretation seems likely considering all the evidence combined. Changes in food consumption practices beginning ca 6700 BC led to the discontinuation of clay balls by 6400 BC. The use of basketry was reduced and replaced by the appearance of cooking pottery. The reason for these changes is unclear, yet they follow a broader, site-wide pattern. The mortuary record sees a gradual shift from the dominance of primary and primary disturbed burials to secondary and tertiary during the Late and Final periods. The settlement becomes less dense in the Late and Final periods, with an increase in middens/open spaces between buildings. Charcoal analysis shows a change in tree species reliance ca 6500 BC (see Volume 13, Chapter 4). At the same point in time, there is a huge increase in sheep numbers, and isotopic analysis shows greater mobility related to sheep grazing territory in Çatalhöyük's Late and Final periods. The later periods of occupation also show a significant drop in *Unio* shell (freshwater) remains and a marked shift away from the dominance of emmer wheat in the botanical record (Asouti 2013; Pearson 2013; Russell et al. 2013b; Volume 13, Chapters 3 and 15). Whatever the impetus, the decline and disappearance of clay balls is part of the more general change in subsistence and culture at Çatalhöyük ca 6500 BC.

Alternative functions

Clearly clay balls were heated inside buildings, yet their use to cook/heat food is not the only possibility. With freezing temperatures in the winter (see Volume 13, Chapter 2), large clay balls have the potential to have acted as body warmers. They may have been wrapped in linen and held in the hands, or placed under matting or bedding, forming a warm base upon which to sit or sleep. People working on rooftops or in the fields in winter could place warmed clay balls in pockets or secure them to their body between layers of clothing, enabling them to remain outside in low temperatures. This would explain their relatively large size, as the larger they are, the more heat they retain. This theory would also account for their large numbers, as each person within a household would need at least one clay ball to heat themselves; perhaps up to ten or more to sleep on. This scenario is akin to the use of heated bricks or stones, placed at the end of children's beds among the lower classes in Victorian England up to more recent times, until the relatively recent advent of the cheap, plastic hot wattle bottle. The older generation in Greece still today heat stones and pebbles in winter, placing them, wrapped in fabric, inside babies' cots. In this context, clay balls would need to be well made, with smooth surfaces and compacted clay, in order to be durable enough to survive repeated re-heating, yet variations in size, along with the occasional variation in shape would be unimportant. It is likely that when broken, large fragments (50% or greater) would retain their original use, still being effective warmers. Once highly fragmented, clay balls were disposed of in middens or recycled (see below). This functional explanation, however, does not account for the sudden discontinuation of clay balls midway through Çatalhöyük's occupational sequence.

Secondary use

Clay balls were used and re-used, as evidenced by the heavy degree of wear and re-heating. Over 40% of clay balls are recovered from middens, yet clay ball fragments are also common in 'structural' contexts, and thus appear to have been strategically used as packing elements and in construction projects such as walls. Extremely solid and hard, clay balls and fragments would have acted like pebbles, providing useful, heat-retaining rubble for lining the base of ovens (as seen in large quantities in B.17, for example), a quick and simple way to block up ovens to be taken out of use, and an aggregate in the construction of walls.

Clay balls also held a symbolic value. Their recovery alongside other cultural materials including ground stone, chipped stone and faunal remains in structured depositions attests to this. Clay balls are commonly recovered in artefact caches or clusters placed between walls, building foundations and clusters on building floors. Their role in the heating of food, if they were used in such a way, would have made them essential and valued tools, crucial in the performance of daily household tasks, thus their inclusion as part of ritual deposits is fitting. Clay balls may have also been used as weapons in a limited capacity. It has always been surprising that such a large and densely packed community appeared to have little or no evidence of violence and warfare. Current research by Knüsel et al. (2019), however, is recovering evidence of blunt force cranial trauma (Knüsel et al. 2019; Volume 13, Chapter 16). The weight, size and shape of clay balls enables them to be easily thrown, with the potential of causing serious injury. A number of individuals exhibiting cranial trauma have been identified. Notably, this occurred in both males and females, all of whom survived their injuries. With many clay balls found easily to hand within the buildings of Çatalhöyük, at times of stress, they may have opportunistically been thrown or fired from slings.

Conclusions

Clay balls are an integral element of Çatalhöyük's material culture during the first half of its occupation. They occur frequently in both the North and South Areas, declining gradually into the Middle period, before disappearing almost completely ca 6500 BC. Their disappearance from the archaeological record corresponds to a time when many sweeping changes in lifestyle and subsistence strategies occurred. Pre-6500 BC, clay balls were tools used primarily inside buildings, in the areas close to ovens. They were heated and re-heated many times, even after they had cracked and fragmented. As such, clay balls undoubtedly functioned to heat or keep things warm.

It is highly unlikely clay balls were cooking 'stones' as originally theorised (Atalay 2005, 2013; Atalay, Hastorf 2005; 2006). Though baskets could have acted as vessels, baskets have a multitude of functions and the botanical and faunal evidence suggests that in the first half of settlement at Çatalhöyük, food was mainly baked and roasted. Furthermore, chemical analysis has failed to recover evidence of the clay balls' placement into liquid foods (Pitter 2013). In terms of cooking, all evidence points to the use of clay balls as a grill or hotplate onto which cooked foods such as bread, along with roasted meats would be placed to keep them warm. The negligible proportion of intact clay balls recovered, the high incidence of the reheating of fragmented clay balls, the density of clay balls on dirty floor surfaces and the dominance of dry cooking methods in the Early to Middle periods all point towards such a function. As time passed, more efficient methods of food preparation evolved. The introduction of Dark Gritty Ware cooking pots revolutionised cooking practices, and liquid meals in the form of stews and porridges became commonplace. Such foods were cooked and served directly from the pots, so clay balls were no longer needed. However, cooking may not have been the clay balls' sole function, as the evidence above suggests. Furthermore, the re-analysis of Çatalhöyük's clay balls has also revealed their secondary use as a construction aggregate and their symbolic value, attested by their inclusion in ritual depositions.

Supplementary material

For supplementary material related to this chapter, please visit https://doi.org/10.18866/BIAA/e-14. It comprises 16 figures: S7.1–S7.16 and eight tables: S7.1–S7.8.

8. A 'token' story? The small geometric clay objects at Neolithic Çatalhöyük

Lucy E. Bennison-Chapman

Introduction

Çatalhöyük is renowned for its range of well-preserved, finely crafted portable artefacts. Amongst its figurines, stamp seals, clay balls and the like, there are also a large number of additional items, many of which can be described as small, geometrically shaped clay objects. Study of these commenced in 2010 (Bennison-Chapman 2013) and continued until the end of the Hodder project. Similarly shaped clay objects are found at contemporary sites right across Anatolia and the wider Near East throughout the Neolithic period. They remain common into the urban settlements of the first millennium BC.

Interpreting them as accounting tokens, Schmandt-Besserat's influential publication *Before Writing* presented Neolithic clay objects as the first stage in the evolution of administration and information storage (Schmandt-Besserat 1992a; 1992b; 1996). Since then, it has largely been assumed that all small, geometric clay objects had a singular and consistent use, acting as recording tokens from the Neolithic into literate times. If true, this interpretation has profound implications for the interpretation of Neolithic society, in terms of cognitive capabilities, social organisation, and inter- and intra-site exchange networks. Yet aside from Schmandt-Besserat's work, little detailed and independent analysis considering evidence for the use of clay objects as accounting devices at Neolithic sites across the Near East has been undertaken. Research into the form and function of clay objects at Çatalhöyük seeks to address this issue. This chapter complements preliminary analysis presented in previous Çatalhöyük publications (Bennison-Chapman 2013; 2015; 2016; 2017). It reports on research carried out between 2010 and 2017, focusing on artefacts excavated during the 2009–2017 excavation seasons (n=1,879). The form and contextual distribution of clay objects are explored, allowing for a discussion on the potential uses of clay objects at Çatalhöyük East in the context of Neolithic society.

Research context

The presence of large numbers of small, geometrically shaped clay objects at sites in the Near East has long been acknowledged (de Morgan et al. 1905). The earliest study of Neolithic clay objects was published in 1958: a catalogue of objects from sixth millennium BC Jarmo, in the Zagros Mountains (Broman 1958). The initial discovery of *bullae* (hollow, spherical clay envelopes, marked with impressions on the outer surface and containing clay objects) of the proto-historic period onwards (mid-late fourth millennium BC) refocused attention to the function and meaning of the small clay objects contained inside (Oppenheim 1959; Amiet 1966). Oppenheim's research was decisive in linking administrative texts, sealing practices, *bullae* and the small objects contained inside them (Oppenheim 1959). Until relatively recently, research into early administration systems has focused on proto and historic-period clay objects. Schmandt-Besserat's (1992a; 1992b) hugely influential work theorises that clay objects operated as 'tokens', their shape and decoration representing specific units of set commodities. Carrying a meaning consistent across the entire Near East and becoming more complex through time, 'tokens' therefore represent a universal symbolic code, which developed seamlessly into cuneiform script from the eighth to the fourth millennium BC (Schmandt-Besserat 1992a; 1992b; 1996).

Though commendable for tackling such a large evidence base and difficult subject matter, a multitude of flaws can be found within the detail of Schmandt-Besserat's theory (see, for example, Damerow 1993; Michalowski 1993; Zimansky 1993; Englund 1993; 1998; Friberg 1994; Brown 1996). No clear definition of what constitutes a 'token' is given. Sweeping assumptions about the cognitive abilities and development of hunter-gatherers, early village communities and residents of early urban settlements are made. Most critically, it is claimed that from the onset, the shape, size and decorative elaboration of clay objects may have symbolised a set quantity of a specific commodity. This 'code' it is argued, was fixed and universal across all sites in the Near East, becoming more complex through time from the Neolithic into the fourth millennium BC, at which point cuneiform script was developed as a direct result of the use of 'tokens' (Schmandt-Besserat 1992a; 1996). However, the range of shapes in circulation within and across contemporary Neolithic sites was not analysed and compared to the range of commodities on-site, nor was the find context of Neolithic 'tokens' taken into account. The vast geographic expanse of the Near East and the long duration of time from the Neolithic into the historic period was also not considered (see Bennison-Chapman 2014).

Despite its downfalls, since its publication, Schmandt-Besserat's argument has dominated thought on the topic, and the word 'token' has become synonymous with small, geometric clay objects. When found at a site, clay objects are often immediately interpreted as evidence of early counting and administrative activities linked to the management of foodstuffs and other essential resources, with little independent interpretation, including the possibility of other, non-administrative functions, being incorporated (e.g., Akkermans, Duistermaat 1996; Costello 2000; 2002).

Aims and methods

Aims

The aims of this research project are:

- To identify, classify and record all excavated small geometric clay objects from Çatalhöyük East throughout all temporal periods of occupation.
- To contextualise all recorded small geometric clay objects, considering their immediate deposition context (midden, building fill, burial fill, structural context etc.) and their wider context within Çatalhöyük (area of site, occupational level and temporal grouping).
- To identify the most common forms of clay object and any changes in form according to area of site, period of occupation or the nature of their immediate find context.
- To use all of the above to assess the possible function(s) of clay objects at Çatalhöyük and any change in use over time.

Though a common artefact across the Near East from the start of the Neolithic period, little detailed research has been undertaken which considers the form and context of geometric clay objects. This project was initiated in 2010 as part of PhD research incorporating data from a number of Neolithic sites across the region (Bennison-Chapman 2014). As one of the case-study sites (alongside neighbouring ninth to eighth millennium BC Boncuklu Höyük and Tell Sabi Abyad in north Syria), a crucial element of the research was for *all* known clay objects from each site to be individually studied in detail. This was to enable a complete overview of the range of clay objects in circulation at each site, and the ability to identify any discrete variations in object appearance, manufacture, use or disposal through time or across different neighbourhoods.

Methods

Clay objects as an artefact category are defined as items intentionally manufactured into a clear geometric shape. Thus, in addition to a clear shape, evidence of human manipulation was sought (for example, fingernail impressions, fingerprints, finger-tip indentations). By definition and in order to have the possibility of having functioned as a group of tools (that is, 'tokens'), only small artefacts were included in the study (<5cm maximum dimension). Small fragments of larger objects (fragmented cones or cylinders, likely once part of a figurine) were excluded from study. Likewise, items which had two or more fragmented surfaces (and thus the original object shape and form was unknown) were not included.

Due to the uncertain function of clay objects and the often subjective nature of finds classification, all potentially relevant artefacts recovered at Çatalhöyük, regardless of the classification they were given upon excavation, were scanned and assessed against the clay object identification criteria (Bennison-Chapman 2013: 254–55). This included both newly excavated objects and archive material, thus covering all seasons of the Çatalhöyük excavation project (including some material from the 1960s excavations). All artefacts determined to be small geometric clay objects (henceforth *clay object*) and potential 'tokens' were studied in detail, regardless of whether upon excavation they were labelled as 'mini ball' (small, unfired spheres were previously separately categorised; e.g., Atalay 2001, 2005; 2012; 2013), 'figurine', 'clay object', 'clay shape' or any other determination (Bennison-Chapman 2013: 254–59; 2014; 2015). A total of 2,786 clay objects were recorded, 1,879 of which were excavated between 2009 and 2017 and thus form the basis of this paper (table 8.1).

In addition to potential 'tokens', attention was paid to the recognition of clay artefacts that may have operated as sealings. Sealings are here defined as simple lumps of clay used to close and secure the contents of a container, or a bundle of items such as a bag, basket, ceramic vessel or roll of cloth. Sealings may either be plain or bear the impression of a stamp seal (or cylinder seal in post-Neolithic periods) (see, e.g., Akkermans, Duistermaat 2004: 1–2, fig. 5; Duistermaat 1996: 19, 20, figs 2 and 3). Like small geometric clay objects, simple sealings also appear for the first time in the archaeological record during the Neolithic period in the Near East. As clay objects and sealings are recovered together from the Late Neolithic onwards (common at sites such as Tell Sabi Abyad and Tell Arpachiyah in the Upper Mesopotamian Halaf Zone), it has been argued the two formed an 'administrative tandem' with clay objects, allowing for the counting of goods which were secured and sealed with the use of clay sealings.

Detail of study	*Excavation season*	
	2009–2017	*All*
Level 1	446	758
Level 2	1,433	2,028
Total	1,879	2,786

Table 8.1. Total number of known small geometric clay objects excavated at Çatalhöyük (all of which have been recorded either in detail (Level 2) or at a basic level (Level 1: object total count and combined weight per unit).

Therefore, the identification and analysis of any potential sealings is key in assessing the likelihood of clay objects functioning as tokens at Neolithic Çatalhöyük.

Information on all the varied aspects related to clay object appearance, manufacture and use-wear was recorded as part of the detailed study process. This included clay colour, texture, the presence/type of inclusions, two- and three-dimensional shape, weight, dimensions, evidence of hardening and burning and use-wear patterning. A total of 2,028 clay objects (1,433 from the 2009–2017 season) were individually studied in this fashion (Level 2). Due to time constraints, a smaller proportion of identified clay objects (n=758) were studied at a more basic level, logging their total quantity and combined weight within an excavation unit (Level 1) (table 8.1). All studied artefacts were then analysed in terms of their form and the contexts from which they were recovered.

Research questions include:

• Are clay objects of all shapes/sizes (*types*) found in the same context types, suggesting they performed the same role at Çatalhöyük?

• Are certain types restricted or distinct in terms of their context, suggesting they held a purpose different from other clay objects?

• Is there any evidence for clay objects having been utilised as 'tokens'?

- If so, does this relate to certain clay object types, distinct time periods or neighbourhoods, or is it a universal function?

- What other possible functions might clay objects have performed, outside of the sphere of administration?

Clay object characteristics

Arguably the most distinctive aspect of clay objects is their shape. Shape was recorded for all studied objects in two dimensions (from three angles: plan, section and longitude view). Three-dimensional shape was also logged according to a series of pre-set shapes (fig. 8.1). Considering the basic three-dimensional shapes in the 2009–2017 phase of the project (n=1,433), it is significant to note that 92% of objects (n=1,323) fall into just three shape categories: sphere, disc and cone (table 8.2). Spheres overwhelmingly dominate, with two thirds of objects falling into this category (66%, n=942). Although all seven of the predefined basic shapes are represented by the clay object assemblage, some such as 'cylinder' and 'other' are represented in proportions of less than 1% (table 8.2). Comparing results of the current excavation phase with the proportion of objects of different shapes from all excavation seasons, the same three shapes dominate. However, it does appear that the assemblage overall is slightly more diverse, with cones, cylinders and cuboids comprising a more significant proportion of the larger assemblage (table 8.2).

Examination of the detailed three-dimensional shapes (16 shapes) shows similar patterning. Pure spheres comprise just over half of all of the current phase's clay objects (53%, n=760), dominating the assemblage (figure 8.2). This suggests a distinct role for at least some of Çatalhöyük's spherical clay objects. Objects of all other shapes occur in varying proportions, with rounded shapes including semi/flattened spheres (13%), disc type 2 (7%) and semi/flattened ovoids (6%) the most common shapes recovered (fig. 8.2, table 8.3). Cones are extremely underrepresented (<1% all sub-types combined), the least common shape after cube (fig. 8.3, table 8.2). In terms of three-dimensional shape, therefore, over 92% of objects fall within just three basic (table 8.2) and seven detailed (figure 8.2) three-dimensional shape categories.

Unusual shapes include a lozenge shape (classified as other), represented by four clay objects (DB# 3057, 3120, 3121 and 7556). Each is an elongated object with a curved surface (oval body in section), tapering towards either end (fig. 8.4). All are almost complete, and all are very similar in size, measuring up to 3.4cm long, 1.3–1.8cm wide and 0.9–1.35cm thick. In addition to appearance, all four examples share similarities in terms of manufacture, being well crafted from a fine clay with a very smooth outer surface. A further, hitherto unique clay object at Çatalhöyük is the exceptionally well-crafted 'pawn-shaped' clay object (DB# 6601, fig. 8.5). Relatively large, the object is circular in plan view, with a concave base and horizontal sides which taper inwards towards a flat top. The upper surface has a knob enabling the object to be held between the index finger and thumb. At 12g and measuring 2.5 x 2.5 x 2.1cm, this example is slightly larger than average.

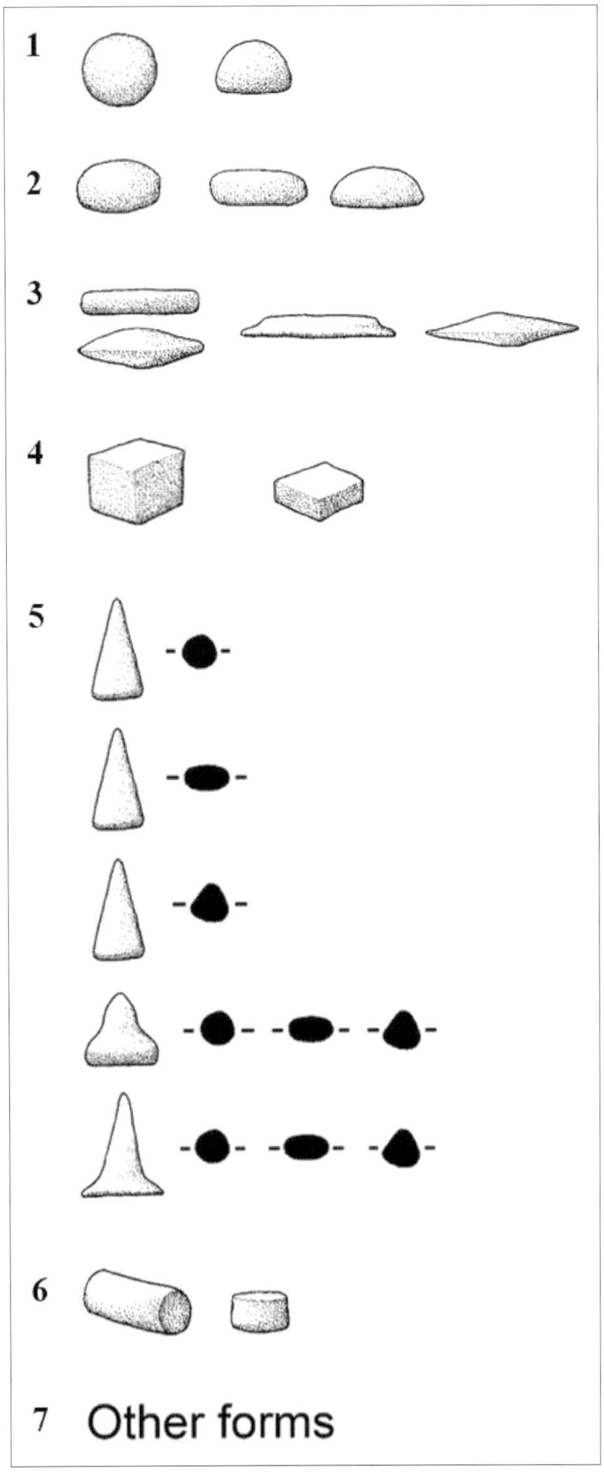

Figure 8.1. A typological diagram illustrating the three-dimensional shape categories used in the study (illustration by Mesa Schumacher).

As the clay object identification strategy dictates, all studied clay objects are small, with average length, width and height/thickness dimensions all less than 2cm (1.7cm, 1.5cm and 1.3cm respectively). However, within this range, there is modest variation in size, with the length of objects varying from between 0.35cm to 7.2cm, for example. No significant variation in size according to three-dimensional shape is evident. Spheres are not larger than discs, nor cones smaller than discs. The only variation in dimensions according to three-dimensional shape is related to the nature of each shape's detail. Discs by nature are thinner than they are wide. Cones tend to be tall and thin (fig. 8.6). Within each shape category a range of object sizes are present, covering the whole spectrum of dimensions represented within the wider assemblage. Spheres' diameters, for example. range from 0.6cm to 5.6cm, with all dimensions within this range represented. No 'small' and then 'large' groupings of spheres, groupings which could be used to convey different means or symbolise differing commodities or quantities of specific commodities, for example, are evidenced (fig. 8.7).

Weight, too, is restricted somewhat expectedly due to the small size of the objects included in the study. Objects range from 0.05g to 199.9g with an average of 4.65g. Seventy eight per cent of clay objects weigh less than 4g, rising to 93% of objects less than 10g (table 8.3). Analysis of only the *intact* objects has little influence on these figures. Ninety-five per cent of all complete clay objects (excavated 2009–2017) weigh less than 10g, the lightest being 0.09g. Complete objects record an average weight of 3.88g (table 8.3). There are slight differences in weight according to three-dimensional shape, yet these are negligible. Of the most common broad three-dimensional shapes, minor differences can be seen in the proportions of objects falling into each 2g grouping: spheres are lightest, with 44% weighing 0–2g, whilst discs are heavier, with only 27% of them weighing 0–2g. Yet discs are the only shape which deviates significantly from the overall patterning of weight distribution; 95% of all intact objects weigh <10g, but only 77% of discs weigh <10g (table 8.4, online fig. S8.1).

Çatalhöyük's clays have been investigated extensively in previous clay sourcing and clay use studies (Avis 2010; Doherty 2013; 2017). Analysis of the clay colouring, clay inclusions (range, proportion and composition), degree of smoothness and level of finish of the clay objects suggests all were created with local clays, with little preference in relation to clay type or colour (online fig. S8.2). Notably, the same is true of figurines and large clay balls (Avis 2010: 102–03). There is no evidence for the addition of temper or other inclusions for practical or stylistic purposes. Clay objects are present in a range of clay colours, yet shades of black to light grey overwhelmingly dominate (fig. 8.8). Light grey is the single most common clay colour type, with 27% of clay objects described as such. The patterning seen in clay colour changes little according to three-dimensional shape, thus clay colour was of little impor-

Basic 3D shape	*All excavation seasons*		*2009–2017 excavated*	
	Count	*Percentage*	*Count*	*Percentage*
Sphere & flattened/semi-sphere	1,144	56.41	942	65.74
Ovoid & flattened/semi-ovoid	223	11.00	177	12.35
Discs	379	18.69	204	14.24
Cube & cuboid	24	1.18	59	4.12
Cones	126	6.21	46	3.21
Cylinder	31	1.53	3	0.21
Shapeless/misc.	101	4.98	2	0.14
Total	2,028	100.00	1,433	100.00

Table 8.2. Basic three-dimensional shape of Çatalhöyük's clay objects. Left: all recorded clay objects; right: those from the present study phase (2009–2017 excavated objects).

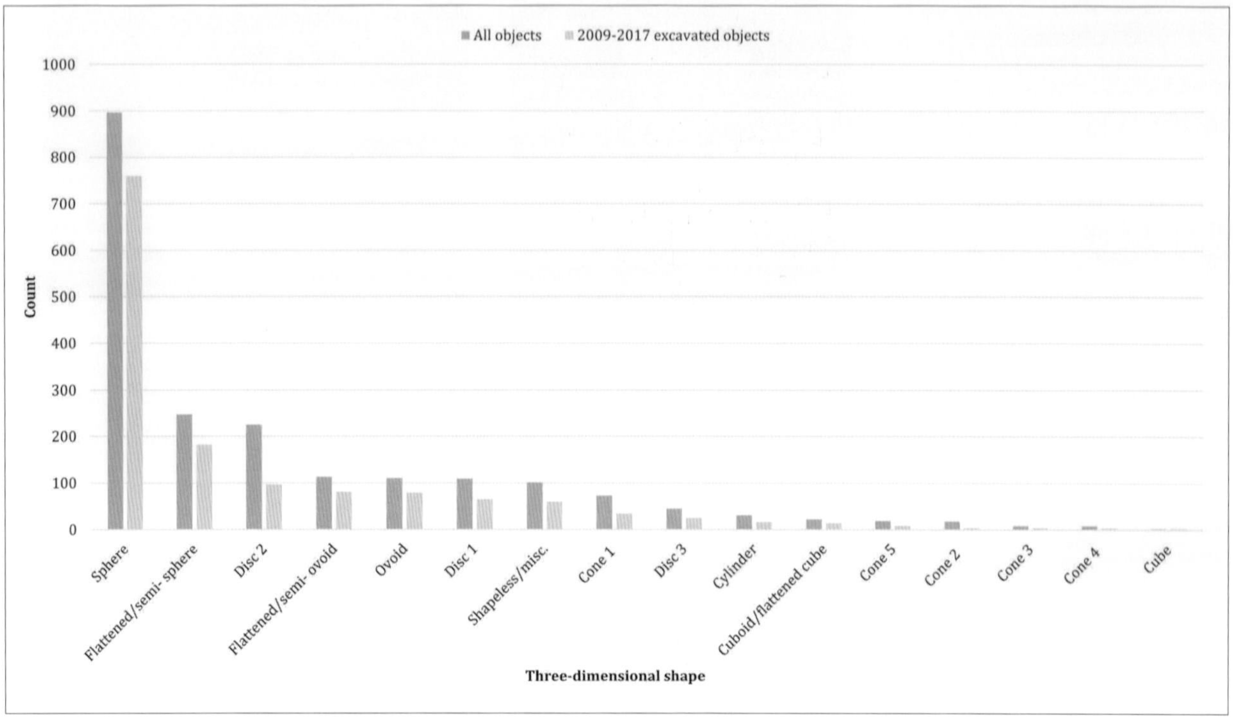

Figure 8.2. Clay objects by detailed 3D shape, all studied objects (n=2,028) compared to those from the present phase of the Çatalhöyük Research Project (excavated 2009–2017, n=1,433).

tance to the manufacturers. Despite the range of colours represented, the differences in shade are subtle for the most part. Clay appears to have been selected according to quality and availability, with colour of no importance. Likewise, though readily available and widely used at Çatalhöyük, paint was not applied to clay objects. Paint is not applied to coat the objects, nor artistically as a form of decoration. The use of paint is only evidenced on a single object – a sphere crafted from white marl-clay (DB# 5146/32606.co1, see online fig. S8.2).

Clay quality and other characteristics of the raw material must have been more important than clay colour. Clay texture, for example, was an important factor in clay selection, with 92% of clay objects exhibiting a 'fine' as opposed to 'coarse' clay texture. Accordingly, the outer surface finish of 89% of the objects is described as 'smooth' or 'very smooth' (online fig. S8.2). Though easily manufactured into a range of simply crafted basic three-dimensional shapes, care was taken to smooth the surface and remove indentations,

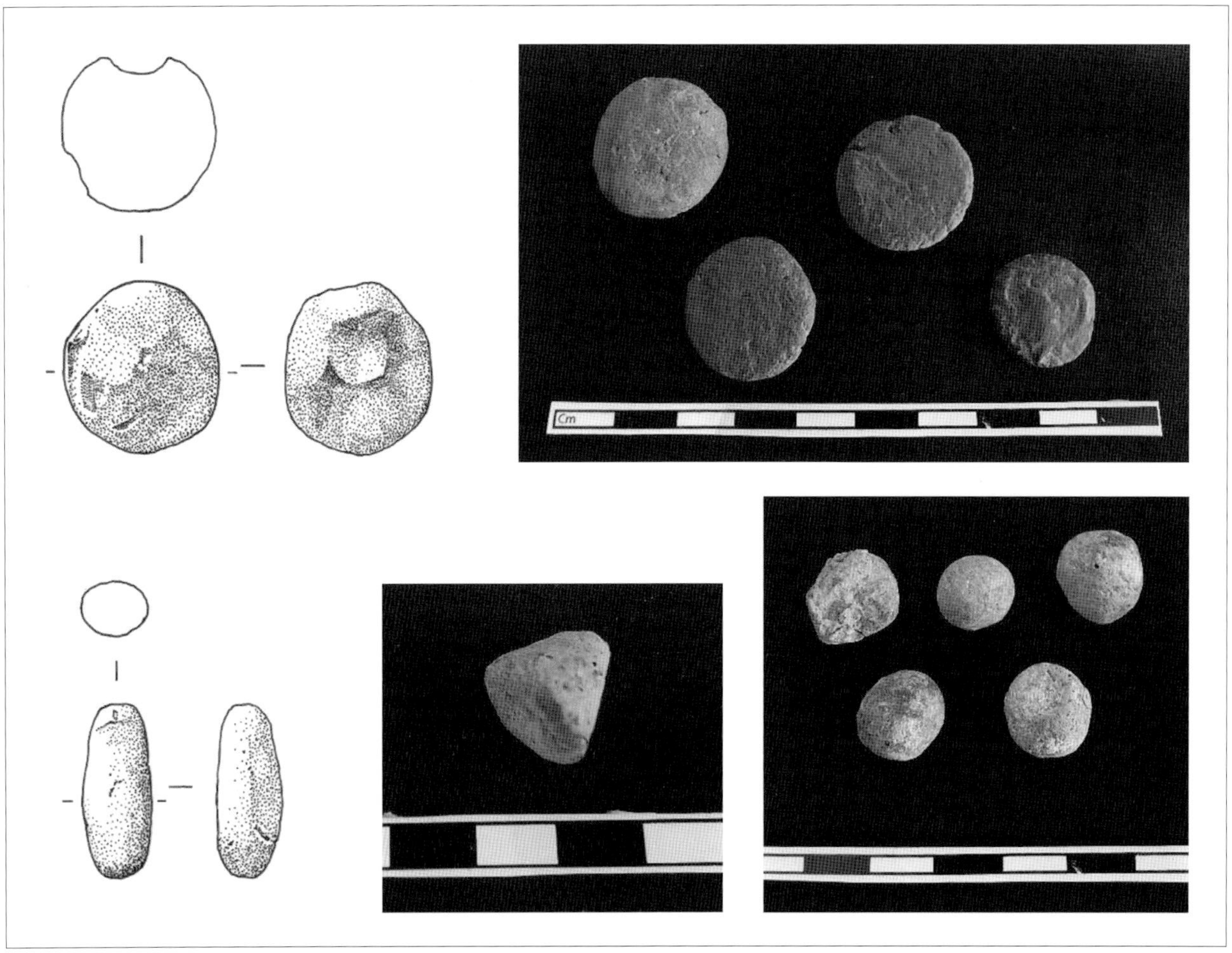

Figure 8.3. Examples of the most common shapes recovered amongst the clay objects: cone (drawing), DB# 6466 (20215.x9); discs, DB#s 3047 (20171), 3048 (20171), 3052 (20134.x10) and 3053 (20126.x22); cylinder (drawing), DB# 6587 (23733.CO4); cone (photograph), DB# 3074 (19700.H1); and spheres, DB#s 3079–83 (21128) (illustration by Jennie Anderson).

plant and matting impressions and fingerprints. Clearly some shapes were harder to perfect than others. Fingerprints (identified using a x30 magnification, 21mm loupe hand lens) can be seen on just 25 clay objects (2%), yet a huge 28% of those come from type-2 discs, suggesting that the shaping of the object between the fingertips to create an evenly contoured upper and lower surface resulted in fingerprints being left behind on objects of this particular shape most commonly (online fig. S8.3).

Just 2% of clay objects display incidental impressions (of plant or matting). Çatalhöyük's clay objects were easily manufactured, from abundant local clays, using the fingertips, palm and for certain shapes (such as cones and cubes) a flat surface. The majority of Çatalhöyük's clay objects are intentionally hardened in some way (recorded as 'baked'; 57%), exhibiting a densely compacted clay that enabled them to survive the archaeological flotation process intact. The hardening appears to be from a direct heat source, as almost half (47%) of Çatalhöyük's clay objects can be described as both baked and burnt, suggesting they were placed at the edge of an oven, fire or hearth in order to complete the manufacturing process (online figs S8.4, 8.5 and 8.6). Experimental work on the production of similar objects at the neighbouring settlement of Boncuklu Höyük has shown three of the basic shapes (spheres, cones and discs) can all be produced quickly and easily with no specialist tools (Baird 2016: 18). The objects can then be hardened to levels comparative to those seen at Çatalhöyük by placing them directly under the summer sun or close to an open fire/hearth. Thus the manufacture of clay objects was accessible to all at Çatalhöyük.

The presence of intentional, meaningful 'markings' was sought on Çatalhöyük's clay objects. It has been suggested that if the objects functioned as 'tokens', with specific shapes representing a set and consistent

Figure 8.4. Example of three of the four 'lozenge'-shaped clay objects (DB#s 3057, 3120 & 3121; find numbers 20232.x7, 30822.m101 & 30200.m101).

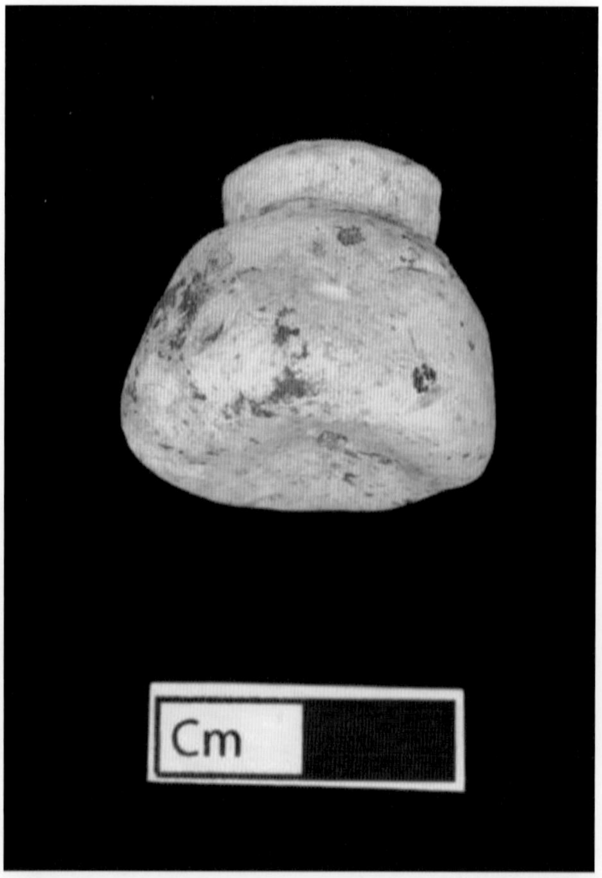

Figure 8.5. A 'pawn'-shaped clay object. Height: 2.1cm (DB# 6601/20761.co1).

range of commodities, the vocabulary was increased by the addition of markings: a line, series of dots or other conspicuous design (that is, Schmandt-Besserat's 'complex tokens': 1996; 1999. See Woods 2010: nos. 23–26 p. 62 and no. 27 p. 63 for examples). Markings are not common on Çatalhöyük's clay objects, just 2% (n=33) display them. Almost all examples involve a single, solitary marking (88%, online figs S8.7, 8.8), with straight lines (21% of all marked object designs) followed by rounded depressions (12%) (online fig. S8.9). They are clear, at 0.15–0.35mm deep, and found almost exclusively on the upper surface of the clay object. Yet the limited distribution of 'marked' clay objects suggests that even if they were used symbolically alongside a clay object's three-dimensional shape to store information, the system was neither wide-scale nor integral to the token system at Çatalhöyük.

The marked objects are almost exclusively spheres and type-1 discs, 28% and 15% of all marked objects respectively (table 8.5). As the majority of clay objects are spheres (535, n=760), this is not unexpected. However, a small proportion of clay objects overall are type-1 discs, thus the proportion of marked objects which are type-1 discs is significant. Despite this, the types of markings seen across spheres, type-1 discs and other clay objects are inconsistent. No connection between the three-dimensional shape and type of markings can be seen. Spheres incised with a cross are found alongside plain spheres, plain discs alongside discs marked with identical patterns, for example. Therefore, the addition of markings does not appear to have been used in a consistent, site-wide system to convey information at Çatalhöyük. The evidence suggests markings were added randomly, to a small number of clay objects, purely for decorative, or perhaps ad-hoc, information storage at a purely local level.

Clay objects at Çatalhöyük show little evidence of use-wear, suggesting the function they performed involved the handling and placement of the artefacts, rather than their use as rubbing tools, or in activities involving them being rolled or thrown. They show relatively little damage, with the majority of studied objects recovered 100% intact (n=911, 64% of clay objects) (fig. 8.9). There is no clear evidence of clay objects being intentionally broken at the end of their use-life (as sometimes evidenced in the Neolithic in other artefact categories such as ceramics or figurines; Oates 1978: 119; Chapman 2000: 68–79; Merpert, Munchaev 1987: 22, 26–27; Chapman, Gaydarska 2007: 138; Verhoeven 2007; Bailey 2017: 827–29), and though the level of completeness varies according to three-dimensional shape, this breakage appears to have been purely incidental, due to practical considerations, with more delicately shaped objects (thin discs and the tips of cones) more likely to be damaged than rounded shapes (fig. 8.10).

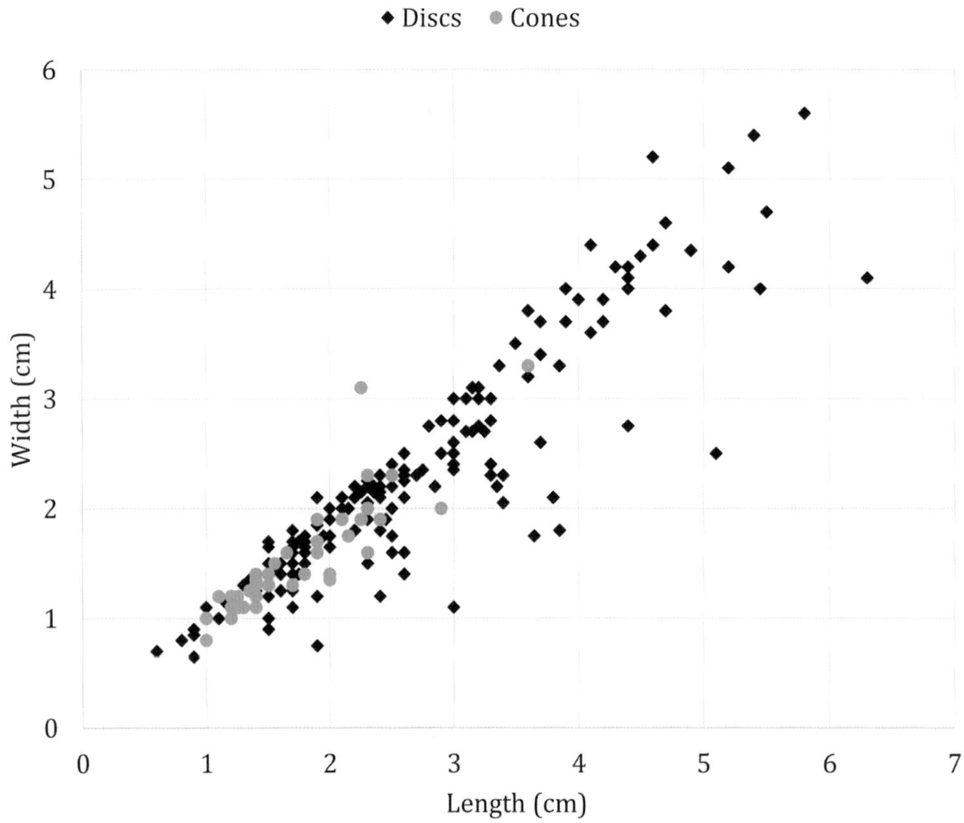

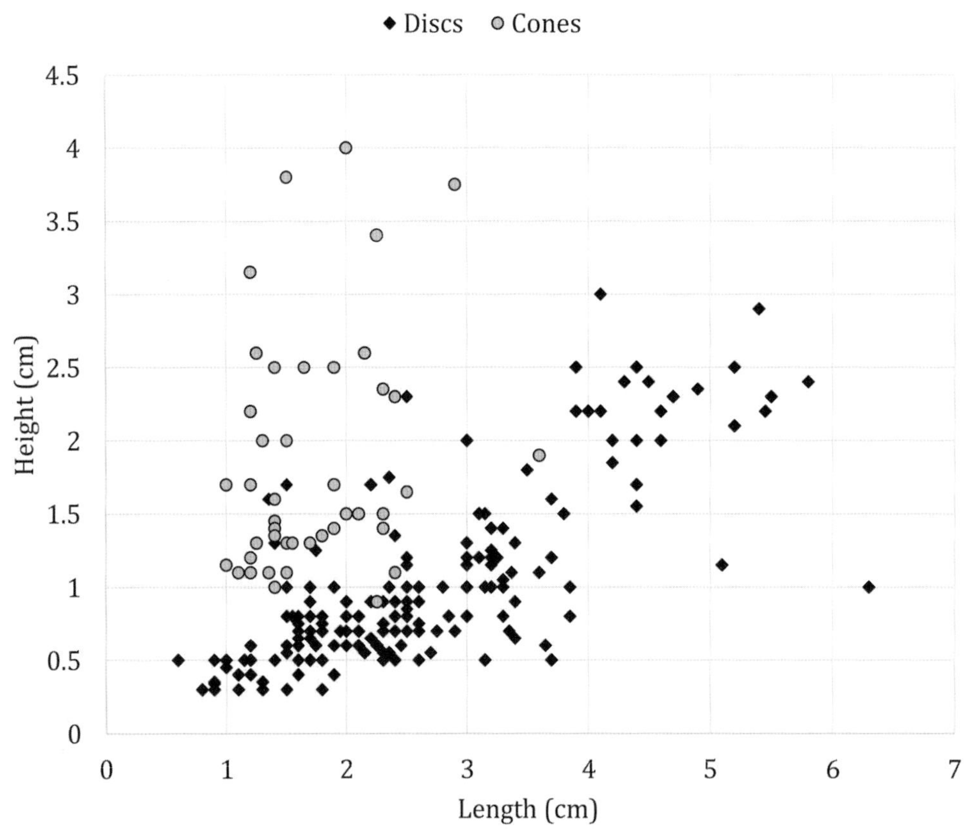

Figure 8.6. Dimensions of cones and discs (basic 3D shapes) from the present phase of the Çatalhöyük Research Project: top: length and width; bottom: length and height/thickness.

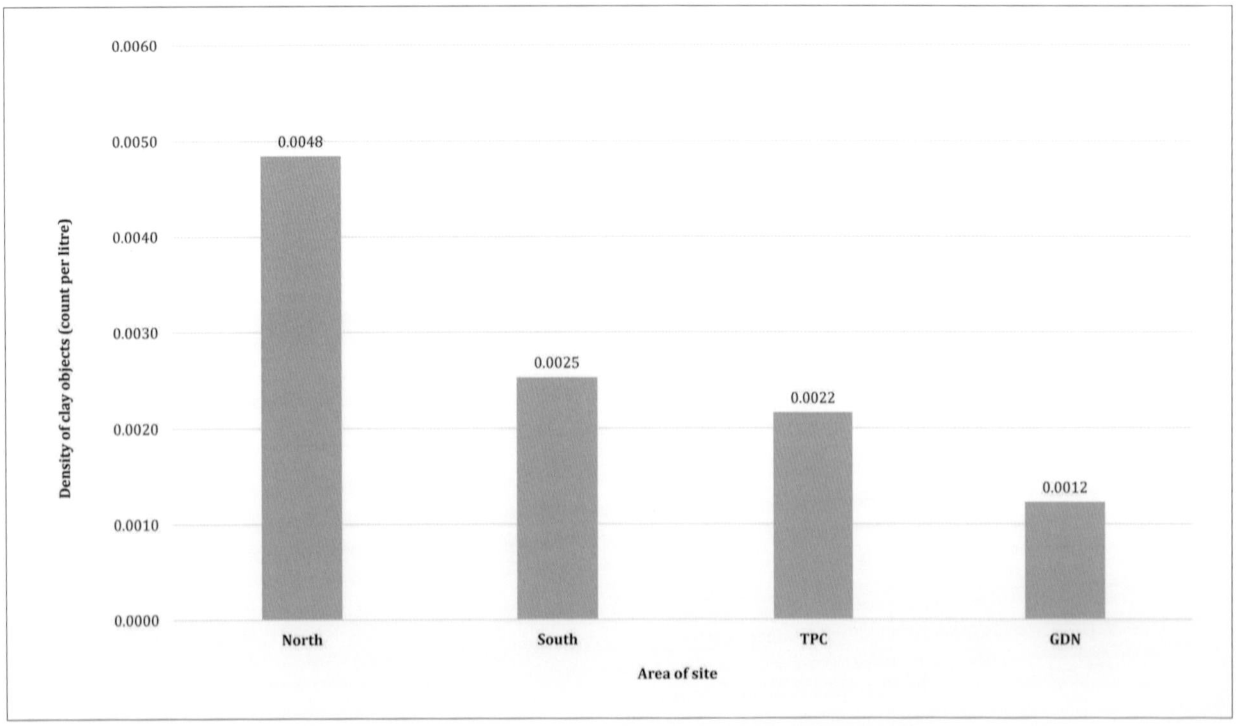

Figure 8.12. Density of clay objects by excavation area, 2009–2017 excavated objects only.

All East Mound clay objects			*2009–2017 excavated clay objects*			
Site area	*Count*	*Percentage*	*Site area*	*Count*	*Percentage*	*Density (o/l)*
North	1,685	65.7	North	1,301	69.2	0.0048
South	695	27.1	South	452	24.1	0.0025
TPC	101	3.9	TPC	0	0.0	n/a
TP	71	2.8	TP	100	5.3	0.0022
GDN	13	0.5	GDN	13	0.7	0.0012
Rec.	1	0.0	Rec.	1	0.1	n/a
Unknown	220	7.9	Unknown	12	0.6	n/a
Total	2,566	100.0	Total	1,879	100.0	0.0037

Table 8.6. Number and density of clay objects recovered and studied from each excavation area. Left: objects from all excavation seasons; right: objects excavated during the present 2009–2017 phase of work (Level 1 and Level 2 data combined).

The diversity of clay object shape varies temporally. During the Early period, 11 of the 16 three-dimensional shapes are present at Çatalhöyük. This increases to 16 during the Middle period, remaining steady into the Late period. Just eight shapes are represented during the Final period (though only 19 objects belong to this occupation period). The proportions of shapes represented also change temporally. During the Early period, spheres dominate, followed by type-3 discs and semi-spheres. The proportion of spheres increases into the Middle period, with type-3 discs almost disappearing, in favour of semi-spheres, semi-ovoids and ovoids. Spheres increase again (almost 70% of the assemblage) during the Late period, with ovoids and other-shaped objects the next most numerous. The Final period sees a dramatic change. Spheres are reduced to ca 10% of the assemblage. Type-2 discs dominate, with all other shapes represented in very low proportions. The temporal patterns suggest that if clay objects did perform a singular function at Çatalhöyük, their importance increased and remained constant during the Middle and Late periods, with the function or commodity represented by the spherical clay object increasing through time until the Final period.

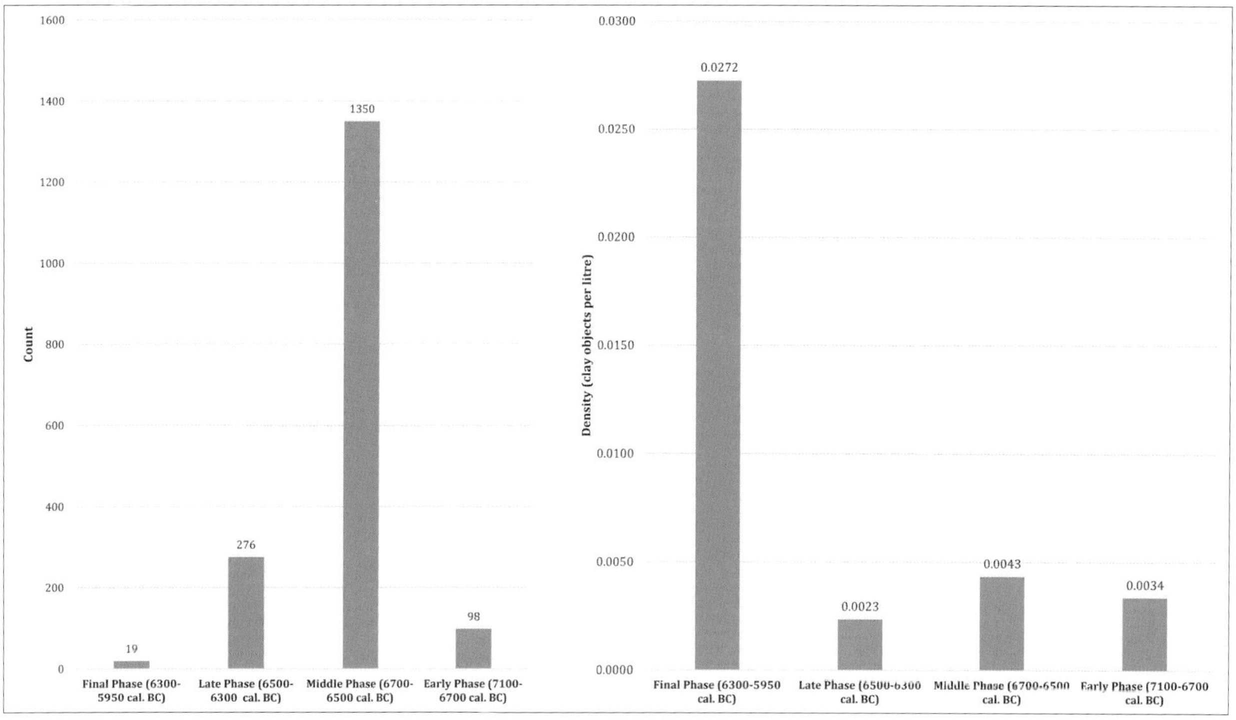

Figure 8.13. Temporal distribution of clay objects by occupation period, 2009–2017 (Levels 1 and 2) excavated objects only. Left: clay object count; right: clay object density per litre.

Nature of context

Investigation of the nature of the context from which clay objects are recovered reveals that, at the most basic level, clay objects are recovered almost exclusively from the secondary contexts of 'midden' (51.73%) or 'fill' (26.13%), 78% of all 2009–2017 excavated clay objects combined (table 8.7; see also online fig. S8.10). Middens represent the accumulation and disposal of refuse. These open areas also contain discrete surfaces upon which tasks such as craft production and food processing activities were carried out. 'Fill' in the broadest sense (as a data category) represents the unintentional inclusion of clay objects within sediments used as filler in all manner of situations. Construction contexts comprise a further 10% of clay objects. This represents the intentional inclusion of clay objects as rubble, as part of sediment collected and used in building activities.

Closer examination of the nature of specific 'fill' contexts reveals that 14% (n=254) of all clay objects come from general 'fill' contexts. Very low proportions come from contexts such as room-fill (including building-fill/infill, 4%, n=72) pit-fill (4%, n=71), burial-fill (3%, n=62), bin-fill (0.2%, n=4) and pot-fill (0.05%, n=1) (fig. 8.16, top). For most of these instances, there is little reason to suggest clay objects were intentionally or knowingly placed in these fill contexts. The evidence suggests that clay objects were readily disposed of in floor sweepings or dumping activities, later becoming incorporated into fill as a secondary context. Just two context types, 'cluster' (n=61, 3.25%, fig. 8.16, top and table 8.7) and 'floor surface' (2%, n=39) (fig. 8.16, top) stand out as potentially representing primary, in situ contexts with the potential to indicate the situation in which clay objects were actually used (see below).

Analysis of the density of clay objects within these same context types (data and interpretative category combined) shows that although middens have an above average density at 0.008 o/l, clay objects are not particularly dense in these areas (Issavi et al. forthcoming and online figs S.15–S.17 for more discussion of middens). Three context types, pot-fill (n=1, 0.286 o/l), fire spot (n=7, 0.150 o/l) and post/pillar-hole fill (0.094 o/l), are the most dense in clay objects. All three are also low in clay object numbers (table 8.8, fig. 8.16). 'Clusters' by definition (a discrete collection of artefacts and/or cultural remains) cannot be assessed in terms of density (see section below). The seven fire spot clay objects all come from the South Area. Remarkably, all but one are from a single level (South P, within the start of the Late period ca 6500 BC) and five of those were recovered from the same space. They vary in shape (two discs, a semi-ovoid and a cylinder) and are all in excellent condition. Though one is heavily burnt, the rest are all baked yet with no signs of burning.

Internal spaces. Clay objects are far more commonly recovered in external (as represented primarily by middens) as opposed to internal context types. When recovered from buildings, clay objects are most often included incidentally as part of fill or structural material, as seen above. However, in order to assess whether or not the use of clay objects was more common in particular neighbourhoods, buildings or types of buildings (as defined by their Elaboration Index (Chapter 1)), the distribution of clay objects from all context types according to building was undertaken. Thirty-seven per cent of all known clay objects (all excavation seasons, n=1,024) were recovered from contexts inside or making up the structure of buildings (see online fig. S8.11).

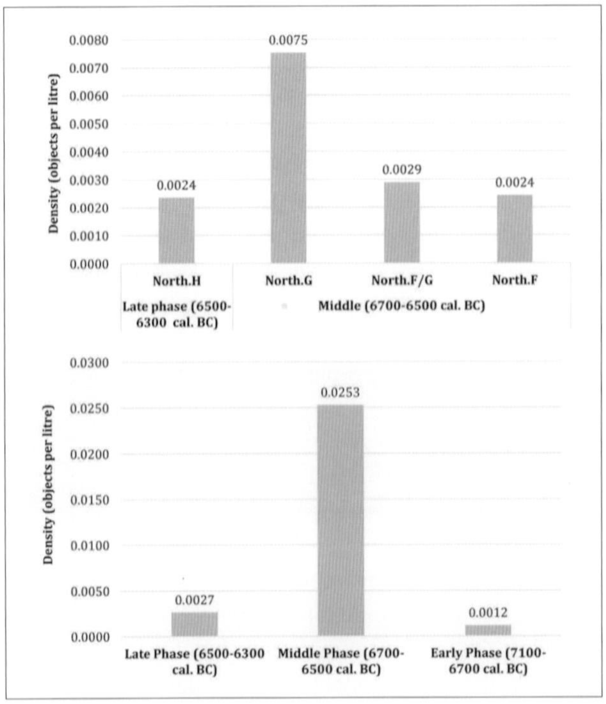

Figure 8.14. Density of clay objects within the North (top) and South (bottom) Areas by period/level (2009–2017 excavated objects only).

Twenty-four buildings had no clay objects at all, compared to 70 buildings with at least one clay object. The majority of buildings with clay objects contained them in very low numbers, with just 22 buildings containing ten or more clay objects and five buildings containing >50 clay objects each. B.131 in the North Area had an exceptionally high number of clay objects at 22 (8% of all excavated clay objects) (table 8.9; online fig. S8.11). No patterning is seen in the buildings containing significant (≥10) numbers of clay objects. They do not cluster by neighbourhood nor temporally. However, the six buildings with >40 clay objects (B.131, B.119, B.139, B.77, B.80 and B.113) are restricted to the Middle period (table 8.9). There is also some correlation between a building's size and degree of internal elaboration (its Elaboration index) and the number of clay objects it contains, with the buildings containing >40 clay objects tending to be in the higher elaboration index values (table 8.10; online fig. S8.11).

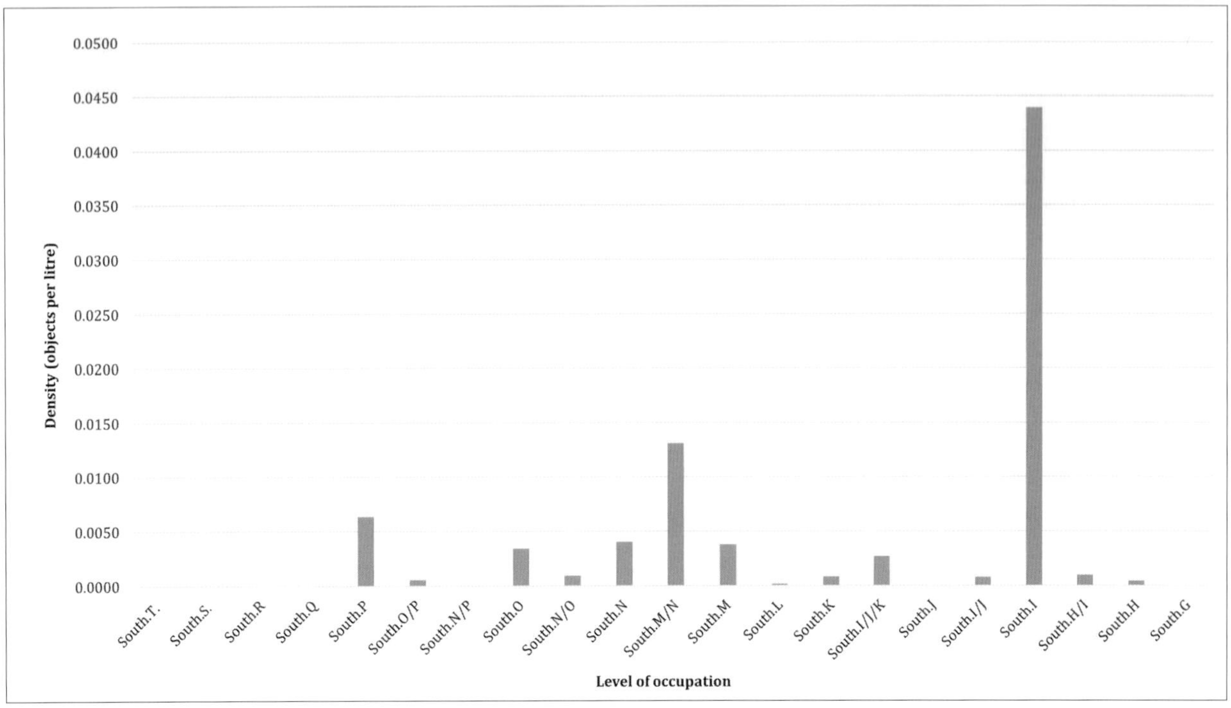

Figure 8.15. Density of clay objects within the South Area by stratigraphic level. Latest level on the left, earliest on the right (2009–2017 excavated objects only).

Data category	No. of objects	Percentage
Activity	13	0.69
Arbitrary	41	2.18
Cluster	61	3.25
Construction	189	10.06
Fill	491	26.13
Floors (use)	60	3.19
Midden	972	51.73
Other	52	2.77
Total	1,879	100.00

Table 8.7. Basic context ('data category') of the clay objects recovered during the present phase of study (2009–2017 excavation seasons, Level 1 and 2).

However, the nature of the contexts in which clay objects from within buildings are found needs to be taken into account when assessing the meaning of their presence.

Ten buildings excavated during the final phase of work at Çatalhöyük were selected for priority analysis. All contained clay objects ranging in number from 7 to 222 (fig. 8.17). In the majority of instances, the clay objects are secondary, retrieved from building infill and structural material, and do not represent the in situ use of clay objects, nor their internal deposition within the building. Building 131's (Level North G) clay objects are dominated by small complete or near complete spheres. The majority however, come from packing, make-up layers and mixed dumps such as the n=72 recovered from a single unit (32128, 'mixed dump'). Building 77 (Level North G) contains 67 clay objects, of which n=21 (all excavated 2008) were recovered from the fill of a single bin (F.3092), a probable grain store. The bin, located in the northeast corner of the western side room, is one of three and relatively large at 2.00 x 1.30 x 0.85m. It was a long-lived installation, in use from phase 2, the early occupation phase of the building, until its abandonment and destruction in phase 9 (Tung 2019). The objects were not, however, recovered in situ from the base of the bin, nor found in association with one another, representing a single, depositional event. They are incidental, secondary deposits. The clay objects were recovered from various units within the bin's fill and not associated with any of the other cultural materials recovered from the bin (which included an alabaster bracelet, a number of figurine fragments, antler, a cattle pelvis, an equid astragalus and various unworked stones) (see online fig. S8.12).

Building 160 (Level South K) contains n=38 clay objects, along with a much higher number of large clay balls (see Chapter 7). Though the building contains a number of clusters and foundational deposits, clay objects are almost absent from these primary depositions. The exception is F.7815, the cluster in Sp.551 (see Chapter 12). Here, an unusual example of a clay object (a likely rubbing tool, a cuboid measuring 7cm x 7cm x 3cm) was recovered alongside 36 fragments of large clay balls (32467). Another interesting deposition in B.160 is F.7830, a 'ladder installation pit infill'. Again located in Sp.551, this pit contained 12 small geometric clay objects in addition to 19 fragments of large clay balls (32414). However, again these objects were recovered scattered throughout the fill, not in direct association with one another, and were thus not interpreted as a single, intentional or primary deposition.

'Clusters' and in situ deposits

Clay objects recovered from primary, in situ deposits are extremely rare at Çatalhöyük East. Furthermore, when such deposits are apparently identified, one must question whether they represent the true remnants of activity, or whether cultural remains were placed in staged positions prior to building abandonment. Just 60 (3%) of the final project phases' clay objects were recovered from contexts described as 'floor (use)' (data category). This includes objects recovered from primary contexts including 'floor surface' or 'clay surface (floor)' contexts, but also from secondary deposits such as 'dump', 'infill' 'make-up' and 'structural'. The 'floor (use)' contexts cover all excavation areas and occupation periods and are found in both internal and external areas (online fig. S8.13). Only 13 of these units (found across nine buildings) represent floor surfaces from occupational phases of a building and thus can be interpreted as primary depositions (n=15 clay objects/0.8%) (table 8.11). However, the clay objects from these contexts are almost always recovered as single clay objects. Rather than representing activities related to counting, administration, gaming or the like, the 15 'floor (use)' context objects above (table 8.11) appear to represent objects accidently left behind during the floor resurfacing process, or otherwise left behind unintentionally on dirty floors alongside small fragments of bone, obsidian and shell.

In addition to these 15 objects, 61 clay objects come from units defined as 'clusters'. 'Clusters' (as defined by Farid and Hodder (2014: 35)) are variable in terms of the items incorporated (raw materials, food remains, cultural artefacts) and the terms used to describe them ('placed deposit', 'special deposit' or merely 'cluster'). They represent discrete groupings of archaeological remains,

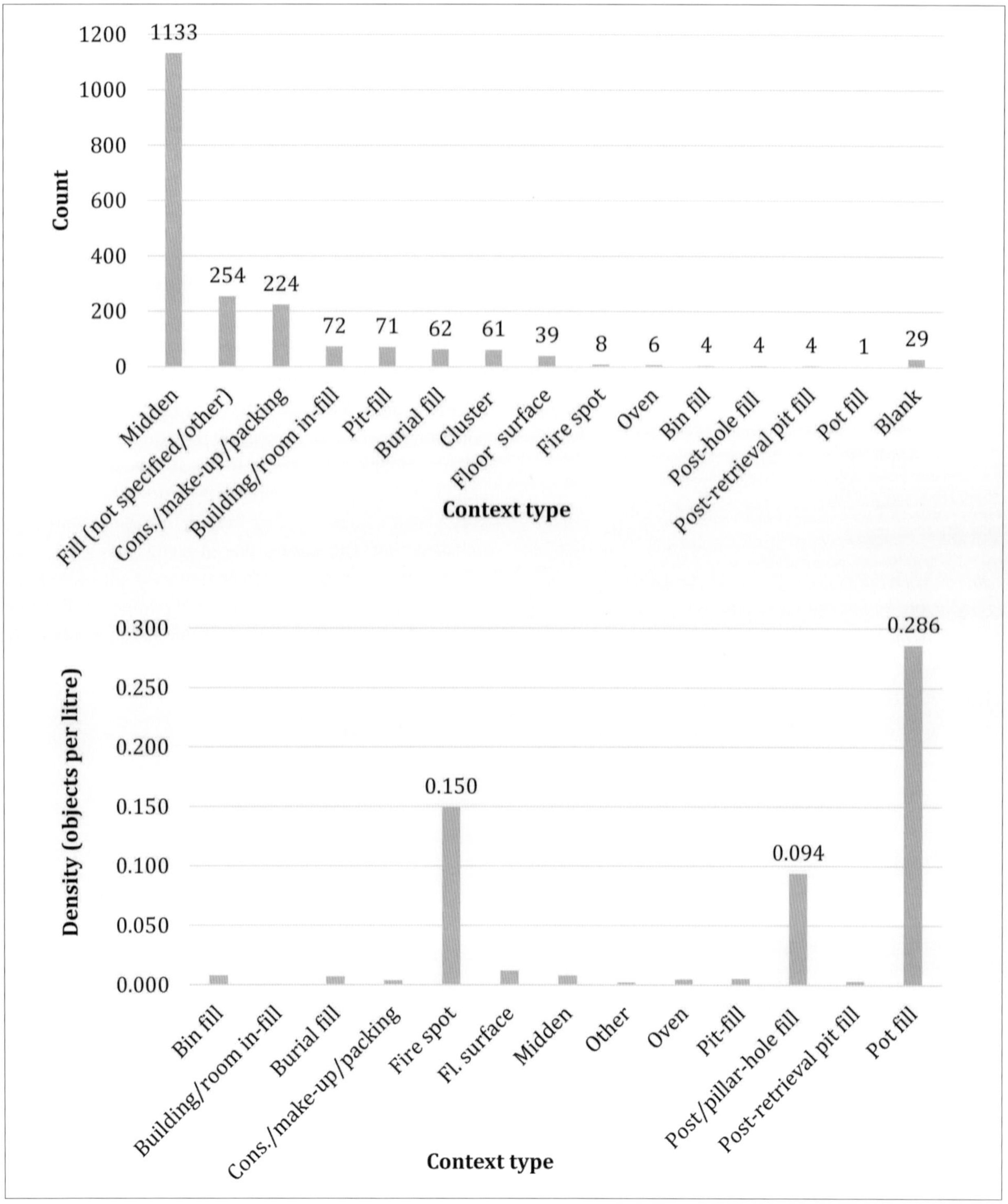

Figure 8.16. Detail of the number of clay objects recovered by context type (data and interpretative category combined, main types only), 2009–2017 season objects only (n=1,879). Top: count; bottom: density.

intentionally placed (yet again, whether they represent staged or actual use situations must be considered). Sixty-one such clay objects have been identified, coming from ten different 'clusters'. One hundred per cent of these clusters were recovered from inside a building. They occur in both the North (n=2) and South (n=8) Areas and appear to be restricted to the latter part of the Early into the Middle period (online fig. S8.14). The location of the clusters along with analysis of their make-up confirms the interpretation of them as being intentionally placed deposits or staged depositions. Just one of the clusters (18940) is high in the number of clay objects.

Context type	*Density (o/l)*
Bin fill	0.008
Building/room in-fill	0.001
Burial fill	0.007
Construction/make-up/packing	0.004
Fire spot	0.150
Floor surface	0.012
Midden	0.008
Other	0.002
Oven	0.005
Pit fill	0.005
Post/pillar-hole fill	0.094
Post-retrieval pit fill	0.003
Pot fill	0.286

Table 8.8. Detailed context of the 2009–2017 excavated clay objects by density (data and interpretative category combined, main context types).

This cluster, placed within the fill of a pillar inside B.80, contained 36 small spheres, along with broken obsidian (online fig. S8.14).

It is the only example from the current excavation phase of what in previous seasons was commonplace. Labelled as 'mini ball clusters', caches consisting of hundreds of small, similarly sized, unfired spheres were often recovered from inside buildings. These were interpreted as likely 'blanks' for the manufacture of figurines (Atalay 2001; 2005: 161; 2013: 247, 248–52; Bennison-Chapman 2013: 259–73; 2014: appendix D).

Buildings 80 and 160, both in the South Area, contain more than one cluster each, and most of the clusters are associated with high numbers of clay ball fragments and animal bone (online fig. S8.14). Discs are the most common type of clay object found within the most recent excavation phase. Cluster (21509), B.119 contains seven similar disc-shaped clay objects. All are well made with a very smooth finish, crafted from a fine, pale, marl-like, unbaked clay. These were placed along with a range of other items within a posthole retrieval pit. Likewise, cluster (32495), B.160 contains six extremely similar clay objects. All are crafted from the same fabric and crudely

Building	*Area*	*Level*	*Clay object count*	*Percentage*
17	South	South K	19	0.68
43	South	South L	19	0.68
52	North	North G	16	0.57
65	South	South Q	11	0.39
74	TP	TP N	13	0.47
75	South	South P	33	1.18
76	South	South O	11	0.39
77	North	North G	67	2.40
80	South	South O	65	2.33
96	South	South O	10	0.36
97	South	South O	22	0.79
101	North	North G	12	0.43
108	North	North H	22	0.79
110	TPC	TP N	18	0.65
113	North	North G	42	1.51
114	North	North G	13	0.47
115	TPC	TPC P–Q	16	0.57
119	North	North F	100	3.59
131	North	North G	222	7.07
132	North	North F	18	0.65
139	North	North F	95	3.41
160	South	South K	38	1.36

Table 8.9. Detail of the buildings (all seasons of excavation at Çatalhöyük East) containing ≥10 clay objects. Those with >40 objects are highlighted.

<table>
<tr><th>Building</th><th>Elaboration index (Ei)</th><th>Clay object count</th></tr>
<tr><td>52</td><td>42</td><td>16</td></tr>
<tr><td>80</td><td>25</td><td>65</td></tr>
<tr><td>17</td><td>22</td><td>19</td></tr>
<tr><td>77</td><td>20</td><td>67</td></tr>
<tr><td>131</td><td>18</td><td>222</td></tr>
<tr><td>3</td><td>17</td><td>9</td></tr>
<tr><td>2</td><td>16</td><td>1</td></tr>
<tr><td>132</td><td>15</td><td>18</td></tr>
<tr><td>59</td><td>13</td><td>5</td></tr>
<tr><td>24</td><td rowspan="3">12</td><td>1</td></tr>
<tr><td>49</td><td>2</td></tr>
<tr><td>119</td><td>100</td></tr>
<tr><td>89</td><td rowspan="2">11</td><td>1</td></tr>
<tr><td>114</td><td>13</td></tr>
<tr><td>5</td><td rowspan="4">10</td><td>2</td></tr>
<tr><td>44</td><td>4</td></tr>
<tr><td>58</td><td>1</td></tr>
<tr><td>65</td><td>11</td></tr>
<tr><td>6</td><td rowspan="2">9</td><td>1</td></tr>
<tr><td>97</td><td>22</td></tr>
<tr><td>51</td><td>8</td><td>1</td></tr>
<tr><td>56</td><td rowspan="2">7</td><td>1</td></tr>
<tr><td>57</td><td>5</td></tr>
<tr><td>23</td><td>6</td><td>1</td></tr>
<tr><td>160</td><td>5</td><td>38</td></tr>
<tr><td>64</td><td>2</td><td>1</td></tr>
</table>

Table 8.10. Clay objects in buildings according to Elaboration index. All East Mound Neolithic buildings (all excavation seasons) containing clay objects with an Elaboration index calculated.

finished. Four of the six are burnt, and two display rounded holes on their upper surface. These objects were placed with two stone balls, two further stones and four fragments of large clay balls in a cluster within the building's side-room (online fig. S8.14).

The ten clusters described above represent the deliberate accumulation and placement of a number of items together in a single location. The intention of these clusters, however, is far from clear. The composition of clusters is diverse. Rather than a group of clay objects of identical or diverse shape, along with other clay items such as figurines, stamp seals and sealings (as would be expected if these clusters represented administrative archives), clusters include stones, animal bones, obsidian flakes and so on. The location of clusters can be used to interpret their function. A number appear to be ritual deposits, being placed in between walls and under floors (for example, 18940, 'pillar fill' deposit and 19818, a cluster placed between the walls of B. 79/80) (online fig. S8.14). Others appear simply to represent piles of refuse, such as the cluster of clay objects and bones by the oven and ladder in B.80 (18964). Others such as (32495), incorporating nine identical clay objects (along with clay ball fragments, pebbles and bones) on the floor of a side room in B.160, are more likely to represent in situ activity. However the various forms of 'clusters' are interpreted, none offer support for the use of clay objects as tools which retained a numeric or mnemonic value related to the administration of goods.

Cross-contextual analysis

In most instances, little or no patterning can be observed when comparing differences in the nature of clay objects according to contextual deposition (period, stratigraphic level, context type and so on). This is also true for the context of marked objects (n=33), ruling out the possibility they had a special function or significance. Marked objects are distributed across all excavation areas in proportions representative of the general distribution of all clay objects (North 60%, South 24% and TPC 12%). Temporally, the marked objects are clustered in the Middle to the start of the Late period, again reflecting the overall temporal distribution of clay objects at Çatalhöyük. This is clear evidence against the theory of the development and increasing complexity in the use and function of clay objects tokens through time.

The four distinctive and well-crafted lozenges (detailed above, nos. 3057, 3120, 3121 and 7556) all come from different units. Notably, all were recovered from inside buildings, yet none were from primary deposits (midden, pit and room-fill contexts are represented). Temporally, these objects are limited to the Late and Final periods, yet each is from a different stratigraphic level, thus none were used together. The only clear temporal patterning evidenced in Çatalhöyük's clay objects is an increased level of finish within the assemblage through time. This is particularly evident in the objects recovered from the Final period, represented by settlement in the TPC Area. The lozenge-shaped clay objects above are an example of this, along with the well-executed pawn-shaped object (database # 6601). Distinctive not only in form, this object was recovered from the interior of a complete ceramic vessel (F.3865). The vessel was situated within B.150 (Sp.594), placed upon the floor surface. The sediment within the pot accumulated in two distinct horizons, the lower level of which contained not only the clay object, but fragments of bone, burned wood, ground stone and seeds. In this context, the pawn-shaped clay object appears to have been intentionally placed inside the pot.

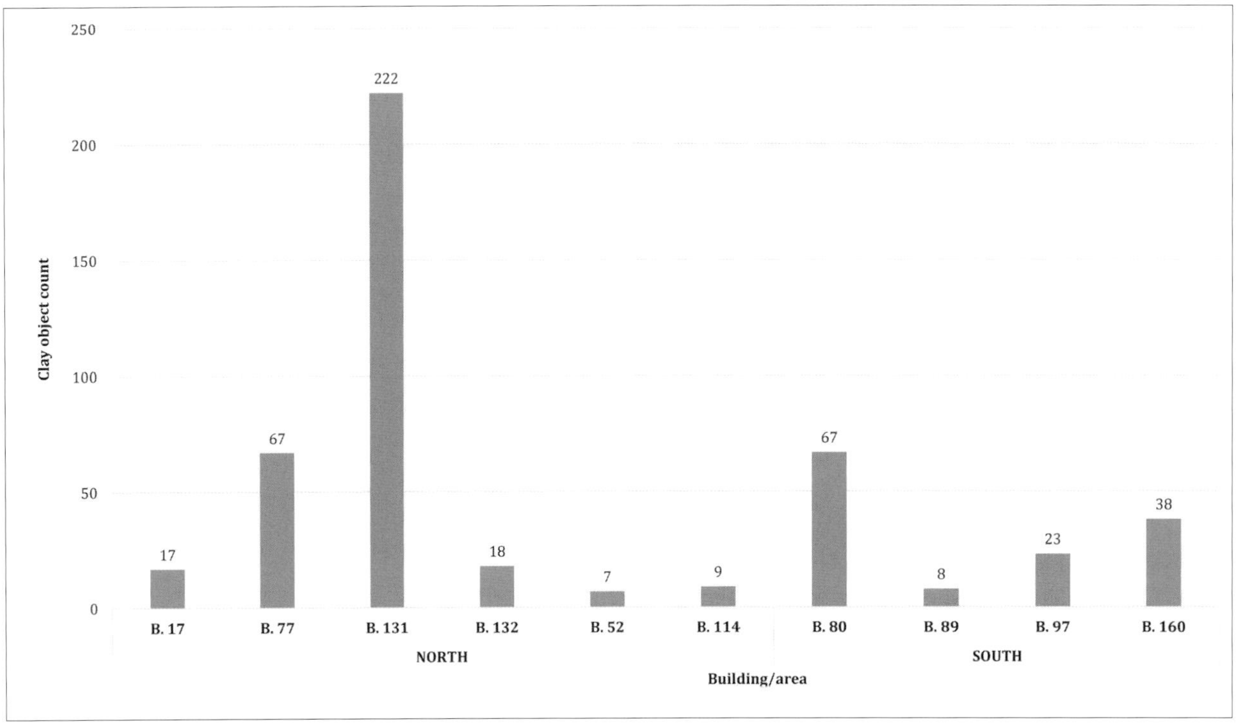

Figure 8.17. Building-by-building distribution of clay objects across the 2009–2017 season's priority buildings.

Building	*Count*	*Percentage (of 2009–2017 clay objects)*	*Elaboration index*
B. 17	2	0.11	3
B. 80	2	0.11	25
B. 97	3	0.16	9
B. 113	2	0.11	n/a
B. 114	1	0.05	11
B. 132	1	0.05	15
B. 160	2	0.11	5
B. 161	1	0.05	n/a
B. 167	1	0.05	n/a
Total	15	0.80	n/a

Table 8.11. Number of clay objects per building, objects recovered from primary, in situ floor surface deposits during the 2009–2017 phase of excavations.

Summary and discussion

The small geometric clay objects from Çatalhöyük share many basic features of appearance and manufacture. They are clearly intentionally made, with smoothed surfaces often displaying shallow finger-tip indentations and fingerprints as evidence of human manipulation. The objects were crafted into a clear range of three-dimensional shapes, with spheres by far the most common. Within the clay object assemblage, there is variety across a number of elements. Three-dimensional shapes are clear, yet they are not highly standardised, with a range of graduated sizes present within each shape category. Shapes such as discs and cones come in a variety of sub-forms, and there is a lack of correlation between the presence of markings, clay colour and three-dimensional shape, for example.

Cones vary immensely in the detail of their shape: there are straight-sided examples and others with pinched sides, some with pointed and others with rounded tips, some tall and others squat. Though such features could be intentional and a way of differentiating one set of cones from another, one set of spheres from another, each set with a different meaning, the sheer range and lack of consistency within the detail of the Çatalhöyük clay object assemblage suggests the opposite. Therefore, the high level of standardisation (clay colour, shape definition and shape range) evidenced in the later 'token' assemblages of the Late Neolithic into the historic period in Mesopotamia is lacking at Çatalhöyük. At Çatalhöyük, the nuances in object size, colour, shape definition and proportion (within three-dimensional shape categories) were unimportant and unintentional. This is supported by the contextual analysis, which shows no differential patterning of objects according to any detail of appearance or manufacture and context.

The accessibility of clay along with the ease of manufacture of clay objects at Çatalhöyük suggests these objects could be acquired by all. The sheer quantity of clay objects recovered, along with their distribution across all areas and levels of occupation at Çatalhöyük, supports this interpretation. The lack of use-wear evidence, general good condition upon recovery and lack of material value suggests clay objects had a short use-life, were created for a specific purpose and were disposed of once their purpose had been fulfilled. Clay objects were not used continuously, nor recycled, archived or stored for later use or re-use. This interpretation is supported by the contextual analysis, which shows clay objects are overwhelmingly recovered as lone clay objects, from secondary deposits or refuse. Middens and 'fill' contexts comprise the majority of clay object context types.

'Tokens'?

There are various ways in which small objects of clay could potentially have been utilised as tokens: at the most basic level, in simple one-to-one correspondence to count a number of individual items or sets of items (that is, bags of grain, bundles of reeds). A clay object would be moved from one place to another, counted side by side with each item. This could be advantageous if large numbers of items were being counted. The use of clay objects as counting aids could prevent people from losing count, which could easily happen, especially when performing large counts, undertaking counting in a busy or chaotic situation, or if there was a delay, however small, in the completion of a count. One-to-one counting with clay objects would also serve to increase the accuracy of counts, and such a system would also enable people with limited numeracy skills to perform simple counting tasks with ease. Number words, counting aloud or in the head, would be unnecessary; a clay object could simply be moved from a bag to a surface, along with each object counted. At the end, the new pile created would visually represent the number of items or units that had been counted, and the objects could then be disposed of. In this scenario, the appearance of the 'tokens' would be unimportant. A large number of small objects alike in size is all that would be necessary. The evidence from Çatalhöyük could support this interpretation for some of the clay objects in some occupation periods. Large numbers of clay objects, especially spheres, occur in the Middle period of occupation, and they are especially dense in the North Area. These objects may represent such a function. Yet this was not universal.

A more complex counting function may have used objects of different shapes to perform counts of different commodities, the resultant piles of 'tokens' being retained as a record of the count and associated transaction (trade or exchange, storage or debt). This interpretation would better explain the diversity of three-dimensional shapes within the Çatalhöyük assemblage, but little else at Çatalhöyük supports it. If such a system was used, the tokens still retained no material value and were only archived temporarily, as no artefact 'clusters' consisting exclusively of clay objects of a range of shapes have been discovered. A set and consistent range of three-dimensional clay objects is not apparent across every level of occupation period. Nor is there a steady increase in clay object diversity corresponding to an increased range of foodstuffs, raw materials and goods in circulation at the site. The caching of groups of clay objects is absent, as is storage outside of the private domestic space. Çatalhöyük is a large site with a large population size, a complex economy, and a diverse range of finished goods and raw materials, as well as processed and non-processed foodstuffs in circulation. This, in combination with the extensive evidence of storage, on-site food processing and artefact production, all combine to present an image of a settlement where a system to aid in the counting and recording of goods and products in circulation (acquired, stored and exchanged) would be advantageous. However, aside from using clay objects in the basic counting of goods before the disposal of the counters, the evidence does not point in this direction. Stamp seals have been recovered from Çatalhöyük, but in relatively low numbers (<100, Türkcan 2013). No impressions in clay or plaster have been recovered. This along with a lack of sealings has led to the hypothesis that stamps were not used as a method of securing and marking ownership of stored and sealed goods at Çatalhöyük. This leaves open the potential for clay objects to have functioned in various alternative spheres including as ritual objects (supported by some of the 'cluster' depositions such as the foundation deposits), gaming pieces, or items of divination.

Conclusions

Analysis of Çatalhöyük's small geometric clay objects highlights the complex nature of the interpretation of artefacts found and activities undertaken at the site. It seems inappropriate to attempt to assign Çatalhöyük's entire clay objects assemblage to a single functional category. The plain, schematic and varied nature of much of the assemblage would suggest the objects did not have a single function at Çatalhöyük. Clay objects are just as likely to have been used as aids to perform complex counts as they were in gaming or ritual activities. The ease of craft, accessible raw material and common midden context of most hints that their use was limited and they retained no intrinsic value. The extremely low proportion recovered from primary contexts inside buildings supports this theory. Clay objects were used yet not retained extensively for re-use, nor commonly stored or archived within the domestic space. This, along with the limited degree of clay object standardisation, opens up the possibility that their role may have been transient, with clay objects performing multiple roles within one use group, household or community.

Certainly clay objects would have been useful aids in counting tasks, yet the contextual evidence does not suggest this function was exclusive, nor that a site-wide system of accounting tokens, with objects of different shapes representing specific commodities, was understood by the residents of Çatalhöyük. The similar sets of clay objects were likely used by different people or groups of people in different ways. Aside from 'counters', residents might have also used the same set of objects for some kind of ritual or divination activity within the domestic space, as hinted at by the placement and composition of some of the 'clusters'. Society within the Neolithic Near East, and specifically at Çatalhöyük, was complex and multi-faceted. A simple, overarching interpretative function does not see appropriate for the use of clay objects here.

Supplementary material

For supplementary material related to this chapter, please visit https://doi.org/10.18866/BIAA/e-14. It comprises 18 figures (S8.1–S8.18).

9. Beads and pendants in life and death: insights into the production, use and deposition of ornamental technologies at Çatalhöyük

Milena Vasić, Matilda Siebrecht, Christina Tsoraki and Rena Veropoulidou
(with contributions by Virginia García-Díaz)

Introduction

Beads form the largest part of bodily adornment at Çatalhöyük. Over 43,000 beads have thus far been found during Mellaart's excavations and the Çatalhöyük Research Project (ÇRP) excavations, in a variety of deposits that span over 1,000 years of the occupation of the settlement. Although this chapter is primarily focused on the research conducted in the last segment of the project (2009–2017), it also provides an overview of the entire bead assemblage that was recovered during the 25 years of the ÇRP. The last nine years of excavations produced over 12,000 beads, and although patterns observed in previous studies (Bains 2012; Bains et al. 2013; Bar-Yosef Mayer 2013; Hamilton 2005; Russell 2005; 2012; Russell, Griffitts 2013; Vasić 2018; Wright 2012) have not significantly changed with the addition of these beads, we now have a better knowledge of how beads were used, as well as a better diachronic overview of their use. The studied assemblage consists of 21,546 beads from 14,294 stratified units from the Neolithic East Mound (North, South, GDN, TP and TPC Areas) excavated during the ÇRP (table 9.1). Priority units were studied in detail, and 1,165 stone and 1,272 shell beads and pendants (including finished products, roughouts and preforms) from these contexts have been subjected to a more detailed technological and use-wear analysis (see chapters 10 and 13). Beads and pendants made of animal bone and tooth were also studied in detail and the results will be published elsewhere (Garcia Diaz in preparation).

Methods

Following the methodology employed for the study of other ground stone artefacts (Chapter 13), the analysis of stone beads entailed the systematic recording of attributes that relate to production stages and manufacturing techniques, along with use-related attributes including information on the degree of wear and possible attachment methods. The latter included information on the deformation of the rim of the perforation, rounding of the perforation rim, with or without polishing in the perforation interior, and rounding on the face and margins of the beads. Microscopic observations were conducted at a low power magnification (10-60x) using a stereomicroscope (Nikon SMZ645 and Leica M80). The identification of rocks and minerals was facilitated by the use of on-site rock reference collections and identification guides (e.g., Jones 2000; Pellant 2000).

	North	*South*	*TPC*	*GDN*	*TP*	*Total*
Activity	22	125	2			149
Cluster	17	104	2		3	126
Floors (use)	231	529	63	2	16	841
Construction	338	452	50	13	42	895
Fill	748	486	85	13	95	1,427
Midden	755	1,034	5	31	40	1,865
Burial fill	11,282	737	443	76	9	12,547
Skeleton	2,567	1,034	17	9		3,627
Arbitrary	23	12			3	38
Unknown	30				1	31
Total	16,013	4,513	667	144	209	21,546

Table 9.1. Quantity of beads in different types of deposits by excavation areas.

Shell artefacts were examined under a low power magnification stereomicroscope (Nikon SMZ645; 10-40x) in order to identify taxa and skeletal parts, to take measurements and to record information on morphological characteristics of artefacts, degree of preservation and overall condition, manufacturing techniques and associated traces (facets, cuts, perforation shape and surface treatment), primary and secondary use and related use-wear traces, post-depositional variables (natural or chemical weathering and breakage patterns) and colour alterations. The recording of shell artefacts also implemented different archaeomalacological and shell artefact methodologies (see Veropoulidou, this volume, Chapter 10; Volume 13, Chapter 3).

In addition, preliminary experiments were conducted using different materials (stone, animal bone and copper) in order to provide insight into bead manufacturing processes (Siebrecht et al. in preparation). Experiments focused on the shaping, grinding and drilling of the stone and bone materials using flint, obsidian and copper drill bit tools and sandstone and schist abrading stones with different combinations of abrasive and lubricating additives such as water and sand. The experiments were designed to provide a better understanding of the properties of the tested materials in terms of workability (for example, ease of cutting and drilling), but also to create a reference collection for the subsequent stage of microwear analysis identifying distinct microwear traces that could be associated with particular motions or materials.

Assemblage overview

Materials

Over 78 different organic and inorganic materials were used for bead manufacture at Çatalhöyük. These include at least 30 different types of rocks and minerals, 24 mollusc species, copper, clay, botanical material (wood, indeterminate mineralised plant/wood material, as well as perforated hackberry seeds that potentially could have been used as beads (for the use of seeds and nuts as beads in an ethnographic context, see Falci et al. 2019)), bone and teeth of different animals and, interestingly, two examples of perforated human teeth (Haddow et al. 2019).

The analysed material confirms patterns of raw material use that were previously reported (Bains 2012; Bains et al. 2013; Vasić 2018), with sedimentary rocks, particularly limestone and tufa, being the dominant materials in the production of stone beads, followed by metamorphic rocks, most commonly fine-grained schist/phyllite and marble. Amongst minerals, there is a clear preference for carnelian, fluorapatite and turquoise.

Contrary to the previous publication (Bar-Yosef Mayer 2013), there is now more evidence to suggest that *Theodoxus heldreichi* shells were definitely used as beads, as was also suggested by Reese (2005). Together with *Unio*, these were shells of local origin that were frequently chosen for the production of beads and pendants. However, marine shells are by far more numerous in the assemblage, most commonly the *Antalis* group (Chapter 10), followed by *Nassarius gibbosulus* shells. Given the quantity of local shells on site, it is striking that marine shell beads dominate the ornamental assemblage, which clearly demonstrates a clear preference towards non-local materials.

Amongst the identified taxa, beads made of bone from large mammals (predominantly cow/aurochs/bison) and hare-sized animals, followed by sheep-sized animals, form the largest part of the assemblage. Peculiarly, this does not correspond to the actual presence of animals on site, where sheep-sized remains represent the largest proportion of the faunal remains (Russell et al. 2013b; Russell, Martin 2005).

Different forms of limestone and marble could have been procured locally within an area of 15 to 20km from the site (Bains et al. 2013), and clay and freshwater shells (*Unio* sp., *Th. heldreichi*, *Theodoxus anatolicus*, *Viviparus* sp. and *Lymnaea* sp.) were available in the immediate vicinity. The Erenler-Alacadağ volcanic formation, which is located ca 60–70km to the southwest of the site and is associated with the procurement of clays, ochre and possibly andesite, is a likely source area for carnelian (Doherty, Tarkan 2013; see also Tarkan this volume, Chapter 4; Doherty 2017a; Bains et al. 2013). On the other hand, fluorapatite, turquoise and marine shells were sourced from areas that were significantly further away from the site, with the majority of marine species originating from the Mediterranean Sea, one (*Antalis dentalis*) from the Aegean Sea and one fossil species (*Dentalium* sp. cf. *sexangulum*) from as far away as the Hatay and İskenderun basins (100–400km distant).

The vast majority of shell species employed in bead making do not appear to have had other uses. For example, the majority of marine shells were worn when procured, thus making them unsuitable for food consumption. Similarly, *Theodoxus* are rather small to be considered as a source of food. On the other hand, *Unio* species were used as food and also in the production of lime, temper and a variety of objects, pigment containers being thus far the most common type. Furthermore, copper appears to have been used exclusively for the production of ornaments. A small number of rocks and minerals were used exclusively for the production of beads (tufa, carnelian, 'galena', hematite of metallic lustre, fluorapatite and turquoise). Other stone materials, however, show simultaneous use for the production of beads and other ground stone artefacts, such as serpentinite and diabase used for axes and adzes and polishing tools, while metamorphic schist/phyllite and marble were widely used as abrasive

(32762) in B.132. In the same building, another individual (32741) was buried with a tooth-shaped limestone bead. It is interesting that both individuals were buried with these limestone beads, which do not exist elsewhere.

Unio shells were typically used for the production of type T.21, and there are also single examples of the use of *Spondylus* and *Patella* for the production of this type that were found in a burial (21802) in B.43. Interlocking beads (T.23) were typically made of animal bone, but there are also several examples in stone (see below). Peculiarly, while T.1 beads dominate across different materials, bone examples are not that frequent, and animal bone was used most commonly for teardrop and red deer canine forms. Previous studies have already noted imitation of perforated red deer canines manufactured from bone (Russell 2005; Russell, Griffitts 2013; Bains et al. 2013), whilst the recent excavations also produced a couple of examples of stone beads imitating animal teeth. Perhaps the most extraordinary examples of tooth beads are two perforated human teeth, likely intended to be worn as pendants. It is striking that although red deer canine (T.9) and tooth (T.10) forms in general seem to have been highly valued, beads made of actual animal teeth remain relatively rare on site, with a single burial (11657) thus far containing a large concentration of tooth beads. Despite the multifariousness of the bead assemblage, clear patterns have been noted. Without minimising the differences in the physical properties and limitations that various raw materials have, the observed correlations between some types and materials are suggestive of the importance of particular materials being used for certain types of beads.

Manufacturing processes and contexts of production

Bead technologies have already been discussed in previous publications (Bains 2012; Bains et al. 2013; Bar-Yosef Mayer 2013; Russell 2005; 2012; Wright 2012), and this section focuses on certain aspects of the production process that were elucidated through recent analysis of roughouts/preforms, associated debitage and finished products. Stone and *Unio* bead assemblages demonstrate that a combination of percussive (flaking, pecking) and abrasive (sawing, drilling, grinding, polishing) techniques were used to transform the raw materials into varied forms.

The manufacture of artefacts from *Unio*, *Viviparus* and *Lymnea* involved drilling, flaking and grinding and occasionally painting. In contrast, *Theodoxus* shells were naturally perforated prior to their collection. Wear traces observed at the rim of their perforations, however, demonstrate that despite their natural perforations, these shells were used as beads. Both natural and artificial perforations exist on *Nassarius* shells, with the former being created in the process of rolling on the beach and the latter through hammering and gouging techniques. Perforation of marine shells was also achieved by drilling at various sides of the shell (from the apex to the aperture, at the body whorl/surface), whilst grinding was used both as a means of creating a perforation and for modification of the natural shape of the shells.

Recent results obtained through detailed technological studies reveal similarities and differences in the *chaîne opératoire* of T.1 beads made from different materials. The vast majority of limestone, phyllite/schist (e.g., 32707.k21) and diabase (e.g., 22512.k5) disc beads have been biconically perforated and shaped mainly by abrasion all over; the occasional presence of faceted margins suggests that grinding was done in stages rather than in a rolling motion.

Both perpendicular and angled perforations (unifacial or bifacial; e.g., 22419.k9 and 18697.k1) have been noted in the assemblage, though the latter occur more frequently in limestone beads. It was previously suggested that angled perforations resulted from the use of a hand drill (Bains 2012); however, our experimental work demonstrates that angled perforations can also be achieved by using a bow drill. Given that it is more time-consuming, though, it is likely that that angled perforations on beads created with a bow drill were the result of unintentional misalignment (Siebrecht et al. in preparation).

Tufa beads show variation in the type of perforations (mostly biconical and straight, but conical perforations are also present) and the shape of the perforation rim, with both circular and ovate examples present. Based on information derived from preforms, the production sequence of tufa beads can be reconstructed as follows: after the initial shaping of the margins (e.g., 16565.k5) and faces through flaking, the faces of the beads were regularised by grinding against a rough surface, as indicated by the depth of the striations visible (e.g., 16555.k3), before being biconically drilled with conical-shaped drill bits. Most of the tufa beads appear to have been drilled using flint drill bits, as demonstrated by the ridges and bands of scoring and striations evident inside the perforation (e.g., 16565.k7), wear traces comparable to those observed on experimental beads drilled using flint. The presence of tufa beads and preforms with a more circular and regular perforation (e.g., 18137.k1, 17048.k17) suggests the use of a drill-bit material that can be more easily shaped into a cylinder, such as bone or copper. Attempts to use bone as a drill bit have been unsuccessful in previous experiments (Bains 2012). Although the copper drill bit used in the new experiments conducted by Siebrecht did not create a cylindrical perforation of the kind that occurs on archaeological pieces, the bands of scoring and small ridges visible were comparable with the traces in the archaeological pieces, thus suggesting that the use of copper is possible. The flaked margins were then ground in steps, as suggested by the presence of distinctive facets creating a polygonal/sub-square shape (e.g., 18658.k1). This clearly points away from grinding in a rolling motion (with or without a grooved abrader). It demonstrates that these beads were produced individually and not in groups (whilst being strung together) (cf. Bains 2012). Variations in the *chaîne opératoire* of tufa beads such as omission of the intermediary grinding stage and the regularisation of the face of the bead (e.g., 19773.k1) or initiation of the perforation directly on the natural weathered surface without prior treatment (e.g., 19773.k2, 21810.k1, 21810.k2 and 21810.k4) suggest that there was no strict adherence to a particular production procedure. The presence of preforms found in the same burial fill (21814) with and without evidence for the grinding stage prior to the creation of the perforation clearly suggests that such variations in the *chaîne opératoire* reflect the choices and preferences of the individual producer(s) rather than temporal differences.

On the other hand, carnelian disc beads show a different production sequence. Both conical and biconical perforations are present, but the former are more common. The faces of the beads were originally flaked, and while on one of the bead faces the negatives of the previous flake removals have been erased by subsequent grinding, they are still visible on the opposite face of the bead. The irregular finish of the carnelian beads (with the negatives of previous flake removals still visible) contrasts with the uniform appearance of the regularly shaped limestone beads (see figs 9.2 and 9.3). Considering the difficulty in shaping carnelian due to its hardness (Schumann 1992), perhaps this indicates an effort to minimise time investment in the production of these beads. Another possibility to consider, however, is whether it represents a conscious decision to preserve on the surface of the beads the different processes that brought them to life. This is reminiscent of the production processes and appearance of flint daggers at Neolithic Çatalhöyük (Nazaroff et al. 2016).

Evidence for the production of disc beads using a rod/pre-formed cylinder is provided by two marble roughouts found in the North Area. Two shallow grooves on the ground margin of roughout 32132.k11 (early stage of sawing) suggest the intention of the producer to cut the preform in half in order to create two beads. In this case, the absence of a perforation on the rod suggests that the perforations were individually drilled on the pre-existing ground surfaces. In the case of 21509.k10, however, two unfinished perforations were drilled on either side of the rod, indicating that there was variation in the manufacturing sequence of beads from pre-formed cylinders. Similarly, manufacture of T.1 beads made of shells belonging to the *Antalis* and *Dentalium* groups involved mainly sawing the naturally shaped cylinders into thin slices, with some of them exhibiting a combination of sawing and grinding at the posterior and anterior end of the shell. In contrast, bone T.1 beads exhibit different manufacturing processes, which are also evident in the ratio of the perforation and bead diameters of bone and stone discs (for more information on the production of bone beads, see Bains et al. 2013). It is likely that the production of stone and shell cylindrical beads followed similar procedures to the manufacture of T.1 beads.

Another interesting observation regarding the production of T.1 beads is a correlation between raw materials and the size of the final products, with tufa beads mostly being smaller than 3mm (Size 1) and limestone, marble, phyllite and schist beads falling mainly between 3 and 5mm (Size 2), whilst the large majority of carnelian beads are between 5 and 7mm (Size 3) and fluorapatite beads are usually between 5 and 10mm (Sizes 3 and 4). The majority of bone T.1 beads are between 5 and 10mm (Sizes 3 and 4), whilst the size of shell T.1 beads largely depends on the size of the shell chosen for the manufacture, with their never being larger than 10mm and over 58% of the shell T.1 assemblage being smaller than 3mm.

The vast majority (over 72%) of beads are smaller than 5mm. Beads that fall within the Size 3 and Size 4 groups form 14.6% of the bead assemblage, whilst beads larger than 10mm are not as common (8.4%). As disc and ring beads form the largest portion of the assemblage, consideration of size (but also colours and materials) greatly reflects the T.1 assemblage (fig. 9.4). If they are excluded, we get the results in reverse, with more than half of beads being larger than 10mm, whilst the smaller category forms only 3.9% of the assemblage and is mostly represented by smaller *Antalis* cylindrical beads.

Detailed technological studies were also conducted on interlocking (T.23) beads. Production of this type of bead made of animal bone was discussed in a previous publication (Bains et al. 2013), but as mentioned above, examples of interlocking beads rendered in stone were unearthed during recent excavations. These beads show a more demanding production sequence than that entailed in the manufacture of T.1 beads, which was also evidenced during experiments conducted by Siebrecht. Based on the study of preforms and roughouts of interlocking beads, the initial stage entailed the shaping of a long rod (average size >20mm) through grinding, occasionally performed in steps, as shown by facets created on the surface of the rod (e.g., 13127.k5); this

Figure 9.2. Carnelian beads (22623.k2) (photograph by Matilda Siebrecht) (for colour version, see online supplementary material).

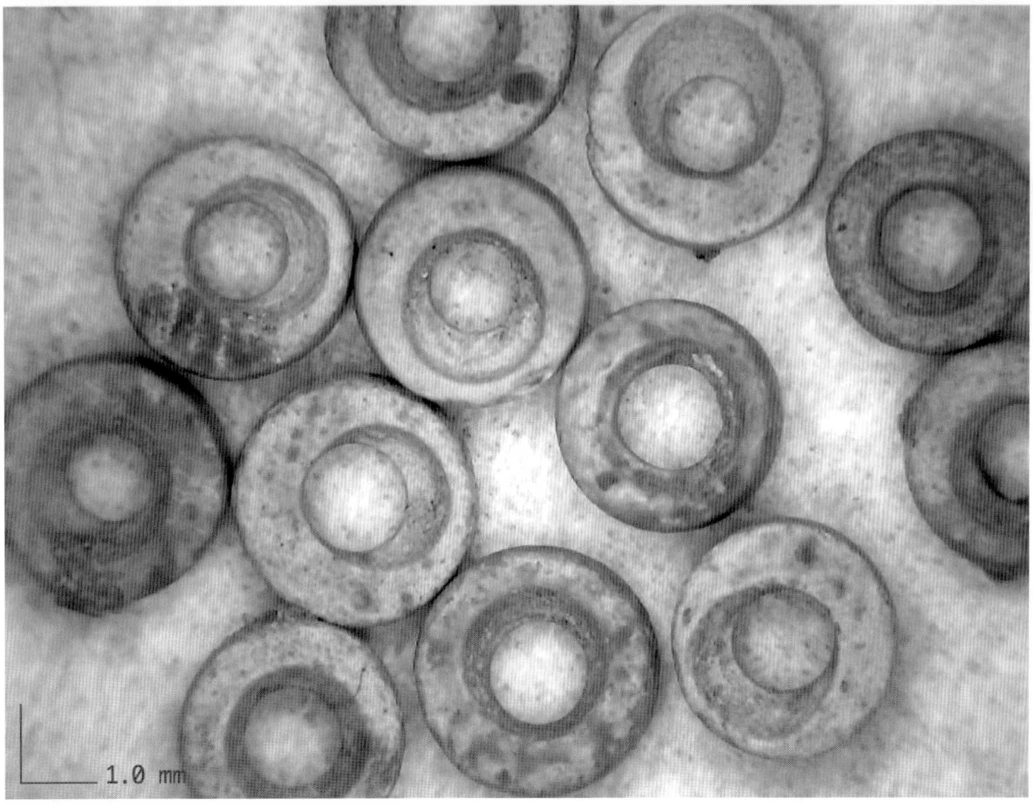

Figure 9.3. Limestone beads (30036.k2) (photograph Matilda Siebrecht) (for colour version, see online supplementary material).

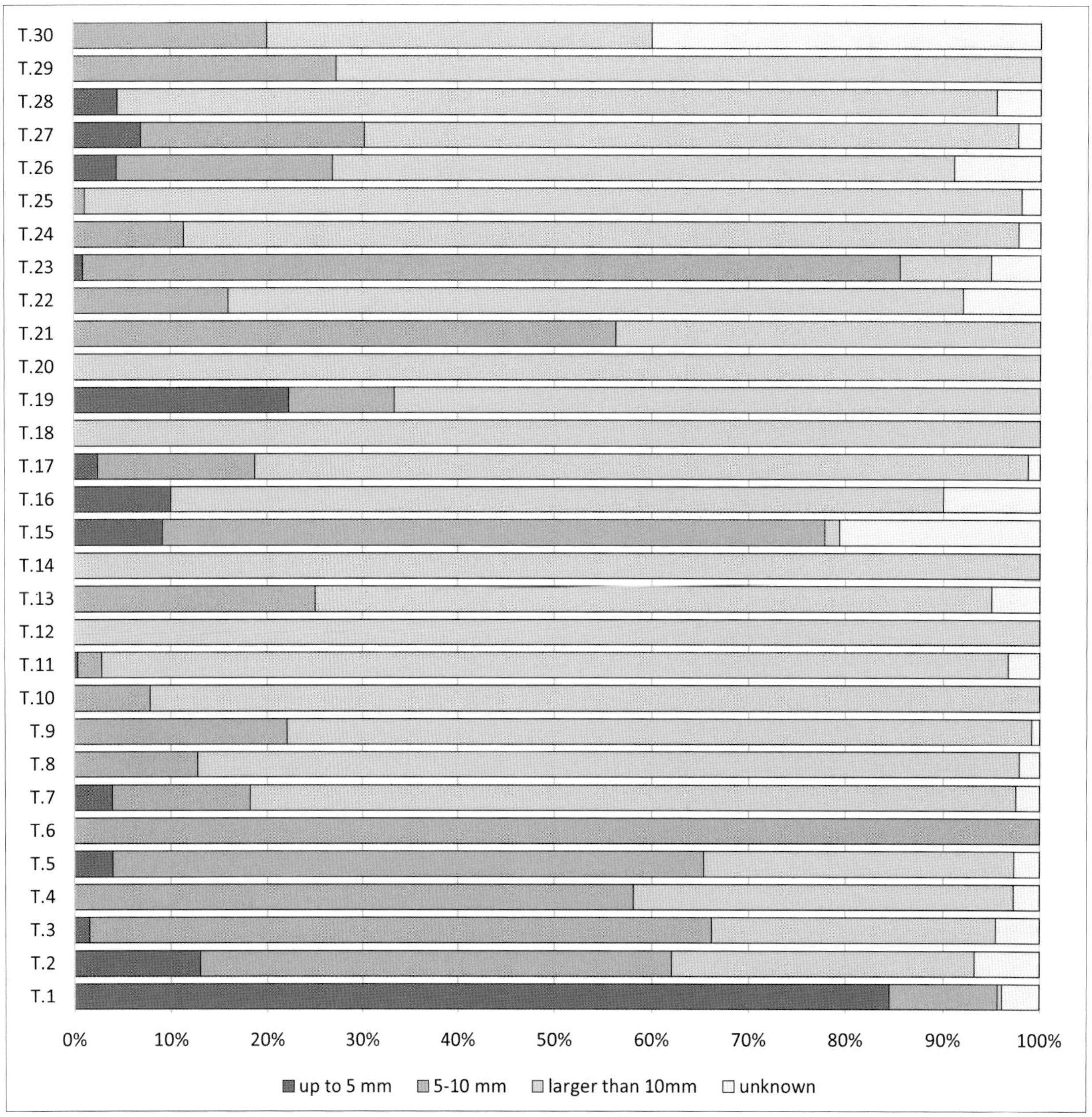

Figure 9.4. Types and sizes.

was then partly sawn bifacially into segments and biconically perforated. The next stage entailed the segmentation of the rod through snapping, and the final grinding and regularisation of the bead surface. Two finished interlocking beads made from marble (20450.k1) show scoring and fine striations on the interior of their perforation. Considering similarities with traces experimentally replicated during the use of a flint drill bit, we could suggest that the archaeological beads were drilled using flint. However, it appears that the perforations were deliberately smoothed following drilling, as there are no clear ridges within the walls of the perforation. Therefore, interlocking stone and bone beads show great similarity in production processes. The choice of marble for T.23 beads is quite interesting, due to its visual similarities to bone. Yet steatite preforms suggest that materials of other colours were also employed in the manufacture of this type.

Based on several preforms, the majority of the T.8 and T.9 pendants made of animal bone and various different stones appear to have shared a similar production process (as much as the chosen raw material allowed), which included general shaping of the raw nodule, followed by biconical perforation and final shaping and smoothing. Some stone and bone examples show that the preforms were ground extensively before a perforation was initiated (e.g., 31907.x1 and 4860.F1). In contrast, some bone pendant preforms (e.g., 4530.F22 and 4539.F35) show limited if any evidence of grinding but were instead either unworked materials or possibly flaked to a rough shape before perforation. Observed wear traces are consistent with the use of obsidian drill bits rather than flint, which creates more pronounced ridges (for example, steatite T.8 (12972.k8)).

The technological analysis of the bead assemblage reveals certain preferences, not only in terms of shapes and colours, but also in the ways beads were created. The variation in the manufacture of disc stone beads is evident, but so is the similarity in the production of shell and stone T.1 beads, the production sequence of which exhibits differences from bone beads. Different bead types were made from a variety of materials, but there is also evidence for imitation across materials. Occurrence of imitation red deer canine beads in animal bone is a recurring trait in the Neolithic (Choyke 2001; d'Errico, Vanhaeren 2002; Sidéra 2001), and there is also evidence of tooth-shaped beads being reproduced from other materials. Furthermore, a number of beads show great visual similarity with stone beads but, upon detailed inspection, were identified as non-stone materials. This includes beads found in direct association with an adolescent (23126) in B.131 that appear to have been made of wood/mineralised plant material with a very fibrous texture, but also clay beads with silty texture and highly burnished surfaces found with an adult (22620) in B.129. Similarly, some of the disc, cylindrical and barrel beads made of clay have highly burnished surfaces and look almost identical to stone beads. Considering that they were not easy to identify as non-stone materials even with the use of a microscope, it is unlikely that inhabitants of Çatalhöyük (apart from the person/people involved in their production) would have been able to differentiate (at least visually), for example, the beads that were placed with (23126) from stone beads. We cannot be certain whether this means that they were trying to replicate the stone beads, or whether their production reveals that certain aesthetic principles had to be adhered to regardless of the raw material.

Recycling has also been noted on a small number of artefacts. That is, artefacts whose use-life started as different objects at some point were turned into beads or pendants. Russell (2005) notes a ring with a pillar (5281.D1) with evident use-wear that after it had been broken was reworked and used as a pendant. Similarly, it is plausible that one T.9 bead that was found on a necklace of an adult female (10829) was created from a reworked boar tusk collar (Çatalhöyük Faunal database 10829.x10). Strikingly, both the necklace and anklet of this female contained T.9 beads made of boar incisors, whilst this individual was also adorned with five boar tusk collars. In addition, a fragment of an object that was found in an external midden deposit (32128) in the North Area appears to have started its use-life as a bangle/annulet and later was perforated, presumably with the intention of its being worn as a pendant.

Despite the large assemblage and the relatively frequent occurrence of beads, blanks, preforms, roughouts, nodules and debitage associated with bead production occur in comparatively low numbers. For example, the sampled units yielded a very small number of objects (n=23) that relate to the initial stages of stone bead production, deriving from both building interiors and external areas and mainly relating to the production of T.1 beads. While this supports the on-site production of stone beads, production contexts remain scarce. The material remains of in situ bead production that do exist in a few buildings are usually limited to one or two types. For example, T.1 preforms were recovered from B.75, whilst B.18 contained unfinished T.9 beads. There is no evidence of large-scale production of beads of a variety of types within individual units, whilst, in contrast, the existence of workshops with a range of different bead types has been noted elsewhere (for example, Kösk Höyük; see Öztan 2012).

The occurrence of unfinished beads made of fluorapatite and carnelian suggest that even these non-local materials were worked on site from the Middle period onwards. For example, carnelian roughouts and debitage were found in fills of B.119, B.97 and B.142. It is worth highlighting that B.142 is strongly associated with different types of stone-working activities and especially with the production of andesitic grinding tools (Chapter 13).

Carnelian was also found in burial fill in B.166 in the TPC Area, as well as together with tools that were found, potentially in a pouch, with an adult female (21672) in Sp.602 in the North Area. There are several other examples of unfinished shell and stone beads that were recovered from burial fills, thus providing indirect evidence of bead manufacture on site. A pouch with preforms for imitation red deer canine beads was found between the arms of an adult female (5169) in B.17, whilst another adult female (23115), buried in B.131, was directly associated with two preforms for steatite pendants, but possible use-wear traces are visible on one of the preforms. Four tufa preforms were found in burial fills associated with (21817) and (21841) in B.17, with the latter also containing waste related to the production of *Unio* beads. Two *Unio* shells (one representing an oval elongated pendant and the other, though it was shaped in a similar way, without a perforation) were placed under the cranium of an adult female (15621) in B.102. Together with other examples from non-burial contexts (for example, the floor of B.89), these attest to the manufacture of *Unio* pendants on site.

On the other hand, evidence of the on-site manufacture of beads and pendants from non-local shell species is limited. One *Cerastoderma glaucum* valve with partially finished perforations (Chapter 10) represents the best example, and, strikingly, it was also recovered from a burial fill (F.7977). Nevertheless, the occurrence of non-perforated *Columbella* examples in building fills and midden deposits, together with a few unfinished *Antalis* and *Dentalium* beads, demonstrates that marine shells were indeed worked on site, although it is possible that some were also brought in as finished products.

Currently, there is no evidence to support the specialised production of beads or the existence of specialised bead workshops at Çatalhöyük. If that was the case, we would expect to find larger quantities of unfinished products, preforms and associated debitage in at least some buildings (for example, the above-mentioned B.75 and B.18). Furthermore, the scattered nature of deposition of unfinished beads and their co-occurrence with the material remains of in situ manufacture of other types of objects (for example, axes, see Chapter 13) demonstrates that people were making different things in the same building, which is strongly suggestive of a small-scale production of different items in 'households', likely aiming at servicing individual needs as required.

The use of beads and pendants

Wear patterns and traces of use

The bead assemblage exhibits different degrees of wear. Stone beads with little or no wear account for 20.2% of the studied sample, and the remainder vary from moderately to heavily used. There is a tendency for turquoise and fluorapatite beads to exhibit a heavy degree of wear more frequently, whereas the limestone, tufa and carnelian examples more commonly tend to show light to moderate degrees of wear. The main use-related wear traces encountered across different stone materials include: partial smoothing and polishing of the perforation rims, often associated with a facet created on the edge of the perforation wall (e.g., 32453.k3) or a facet created inside the perforation (e.g., 17308.k3), resulting from tying the bead with a string; the deformation of the rims of the perforation that includes rounding with or without polish in the interior of the perforation or rounding with or without the presence of notches; and rounding and the unequal distribution of polish on the faces of the beads.

T.1 and T.2 shell beads also show similar patterns, with patchy traces of gloss on various spots of the walls of the bead coupled with slight notches/incisions at spots around the periphery of the posterior and the anterior ends that are suggestive of beads being strung (e.g., (32782), (20492), (16302) and (16131)). Smaller beads are sometimes found inside larger beads, which shows that some ornaments were composed of loosely strung beads. *Columbella* beads have a vertical perforation from the apex to the aperture that would have enabled suspension in a horizontal position. In addition, the rim of their perforations is smoothed and glossy, thus suggesting contact with a soft material (that is, thread).

There are some examples, however, that show evidence for tight attachment, most likely to clothing. These are mainly associated with the uneven deformation of the rim of the perforation and the presence of notches on opposite ends of the perforation along with thinning/flattening of the wall of the perforation on the same side. This wear pattern tends to occur in bead forms of elongated shape (e.g., T.2, T.3 and T.5). This type of wear has been observed, for instance, on a T.5 bead (22661.x2) found in direct association with an adult female (22661) in B.131, suggesting prolonged use prior to its deposition in the burial; judging from the way the perforation is worn on the two opposite ends, it appears that the bead was attached tightly, perhaps on clothing/textile, and not worn loose. A barrel-shaped bead (22678.k2) found in the fill of an adolescent female (31705) burial in the same building has rounded perforation edges, while one end has been worn through. This suggests that the attachment method was tight and was cutting through the rim of the perforation. The observed wear traces are visible only on the flat side of the bead, indicating that this was the side attached to the clothing/textile. A similar observation has been made for beads associated with an adolescent, possible female (23920), buried in B.150. This individual had a string of beads around her neck (fig. 9.5), but the use-wear traces clearly suggest that these beads had not always formed part of a necklace; rather, for a prolonged period of time prior to their deposition they had been tightly attached to another material, possibly clothing/textile. A group of black discs and black and white cylindrical stone and shell beads were found in several positions directly associated with the lower limbs. It is unclear in what form these beads were originally deposited, but as their position does not suggest anklets, it is possible that they were attached onto material as well (either used for wrapping the body, or some sort of a garment).

Similarly, evidence of beads worn with a form of tight attachment was also observed on a number of *Antalis* and *Dentalium* beads (e.g., (30038) and (19897)). Use-wear traces observed in the thinning of one part of the vertical wall of the shell and the presence of notches at the posterior and anterior ends, along with the condition of these beads, demonstrate the exhaustion of the raw material as a result of prolonged rubbing against another material (Chapter 10).

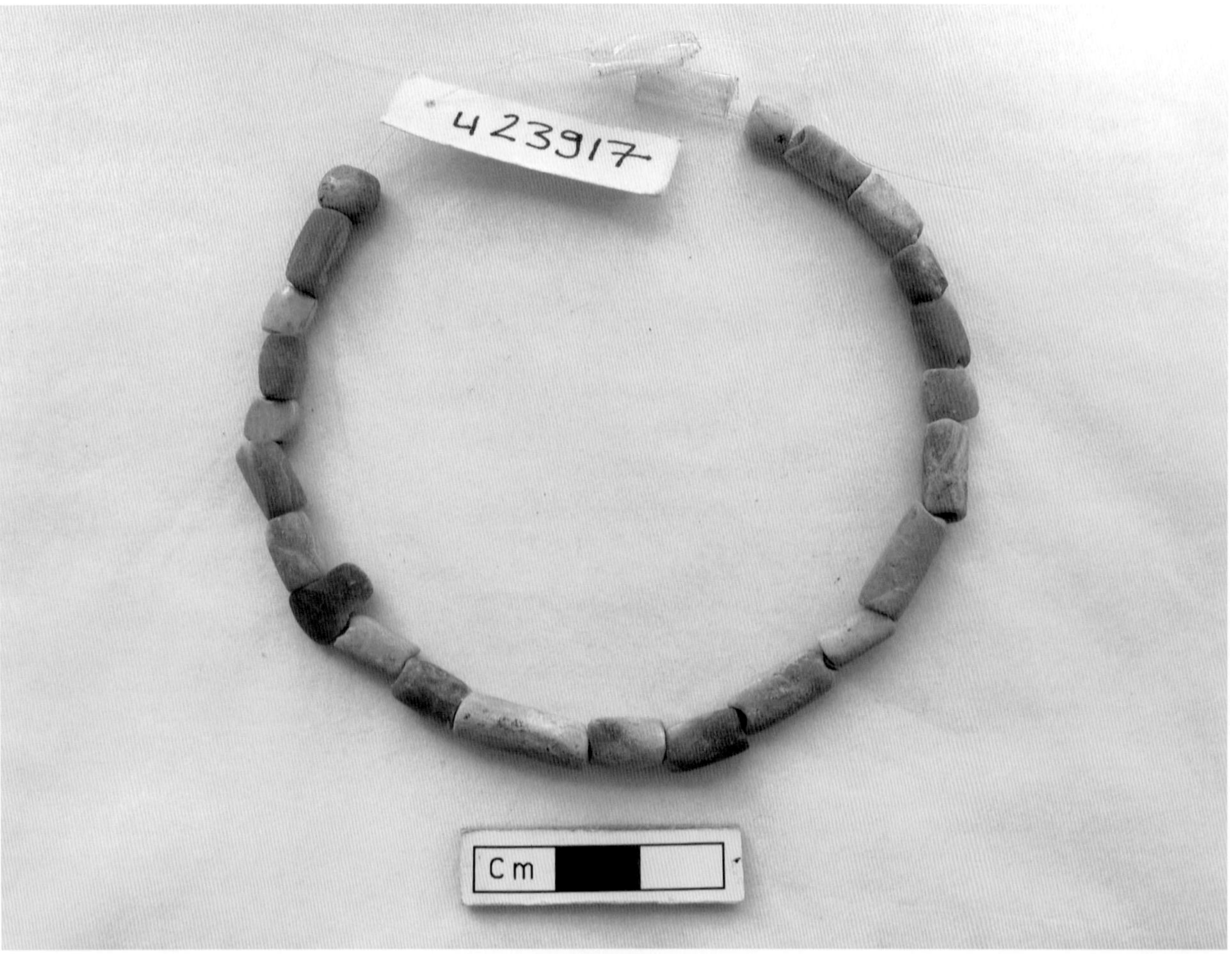

Figure 9.5. Necklace with (23920) (photograph by Ekin Ünal) (for colour version, see online supplementary material).

It is worth noting that similar wear damage patterns have been encountered in non-burial contexts as well, such as on a bead on the floor of B.89 (30977); 32806.x3, which was recovered from the building fill/burnt layer; and 20736.x5, which was found in a platform make-up layer in B.150. These examples clearly suggest that beads were being attached to materials, likely clothing, prior to their deposition in burials. Thus some of the beads originally interpreted as strings that moved due to burial disturbances might also have in fact represented clothing attachment.

Various degrees of use-wear have been noted on bone beads in an anklet found with a child (11657), ranging from very little to extensive, which suggests that some of these beads were in use for a very long time (Daly 2005). Imitation red deer canine beads found in association with an old adult individual, possible male (32762), also display different degrees of wear. Similarly, a *Nassarius* bracelet found with an old adult, possible male (20685), contained shells that had different degrees of use-wear, suggesting that these beads were not always used together. Through use, the surface of *Nassarius* shell erodes and becomes thinner. It is possible that new beads were being added to the same string in order to keep them tightly together, but it is also plausible that they had originally formed part of different ornamental compositions and at some point in their biography were assembled together to create a new string. In contrast, *Nassarius* beads found on an anklet with an adult female (23921) buried in B.166 contained beads that all displayed long use, which could suggest that this string stayed unaltered for a long time (see Chapter 10: fig. 10.7). In addition, the use-wear analysis gave an insight into how *Nassarius* beads were strung (see Chapter 10 for details). The burial assemblage demonstrates that these shells were tightly interlocked with one another, in a similar fashion to T.23 beads (see Bains et al. 2013). Both T.23 (made of stone and bone) and T.25 (*Nassarius* beads) were strung in tight clusters, thus forming very different visual patterns from strings with other bead types (fig. 9.6).

Further analysis is required, especially on strings containing beads of different materials; nevertheless, it seems that some of the bead ornaments were indeed altered at some point during their use-life. Some beads could originally have formed part of one bead string – for example, as a bracelet – but then have been restrung with other beads in order to create another ornament. While bead ornaments are 'composite artefacts through which chains of relations can be traced' (Fowler 2004), it is important to remember that each bead is an entity on its own and can therefore be used in multiple ways and form part of different ornaments. Individual beads had different paths in their use-life, and not only could they have been easily combined and recombined, strung and restrung multiple times, they could also have been used as a clothing decoration at first, but then strung into a necklace or bracelet, or vice versa.

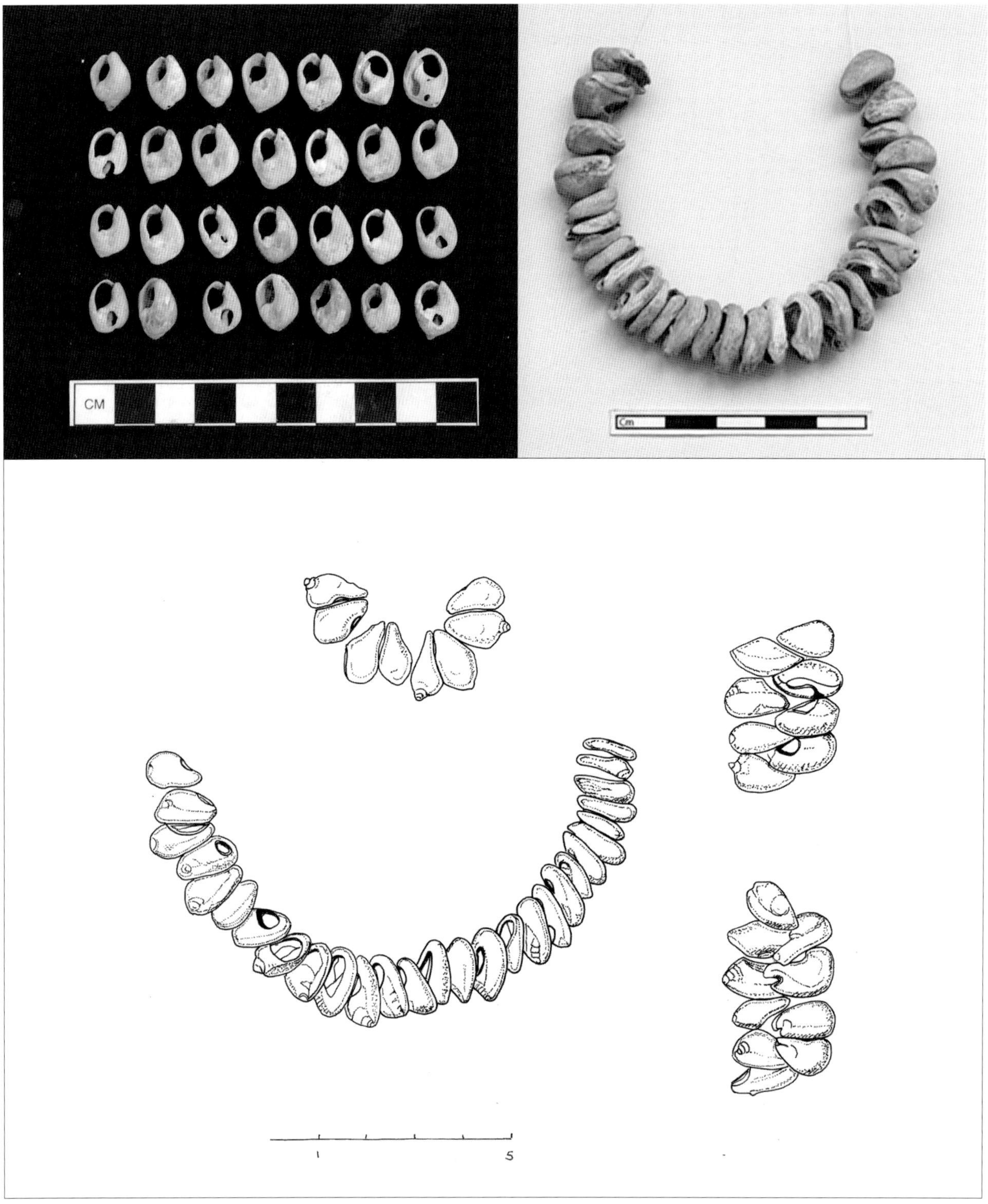

Figure 9.6. Nassarius *bracelet with (20685) (illustration by Caroline Habron; photographs by Milena Vasić and Jason Quinlan)*

Beads and pendants in context

Beads are most commonly present in burial fills, followed by middens (fig. 9.7). Although over half of the excavated midden units contained beads, beads usually occur in small numbers in these deposits and, in fact, bead densities are lowest in the midden deposits (fig. 9.8). This could partly be related to the different retrieval methods. Whilst the usual excavation and sampling procedure involved the entirety of a burial infill being sent to flotation (and the same typically applied for smaller floor deposits), due to their large volumes, midden deposits were usually sieved on site using a 5mm mesh. Given that the majority of beads are smaller than 5mm, it is likely that a lot were missed during the excavation and sieving. On the other hand, flotation ensured the retrieval of all artefacts larger than 1mm. Beads are most frequently present in burial features, where they also have the highest density. It should be noted that the density in burials is actually higher than shown in figure 9.8, as it does not take into consideration beads that were assigned to the skeleton units.

The vast majority of beads (over 75%) were recovered from burial features, some of which contained high concentrations of beads (table 9.1 and table 9.4). When they occur in other deposits, beads tend to be in smaller numbers (up to five per deposit). Primary contexts such as activity deposits, clusters and floors rarely have high concentrations of beads, although exceptions do exist. High densities of beads were noted in the dirty areas and deposits associated with fire installation rake-outs in B.18, B.80 and B.97 in the South Area and B.132 in the North Area (see also Hamilton 2005 for a similar observation regarding B.17). In some cases (B.132, B.18, B.75 and Sp.329), the higher quantity of beads can be related to bead manufacture, which can also be supported by the occurrence of preforms, drills and debitage of associated materials.

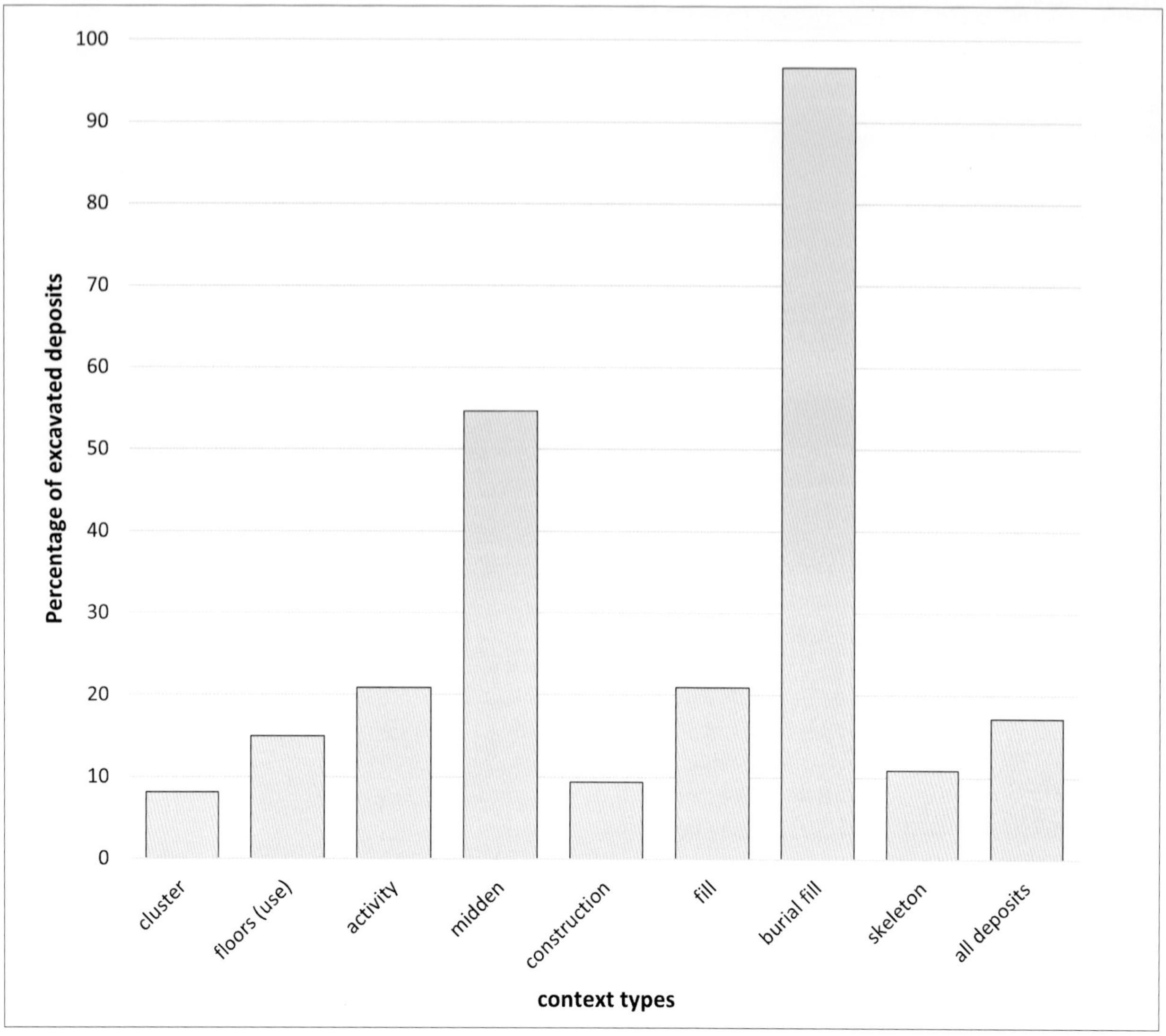

Figure 9.7. Proportion of excavated deposits with beads.

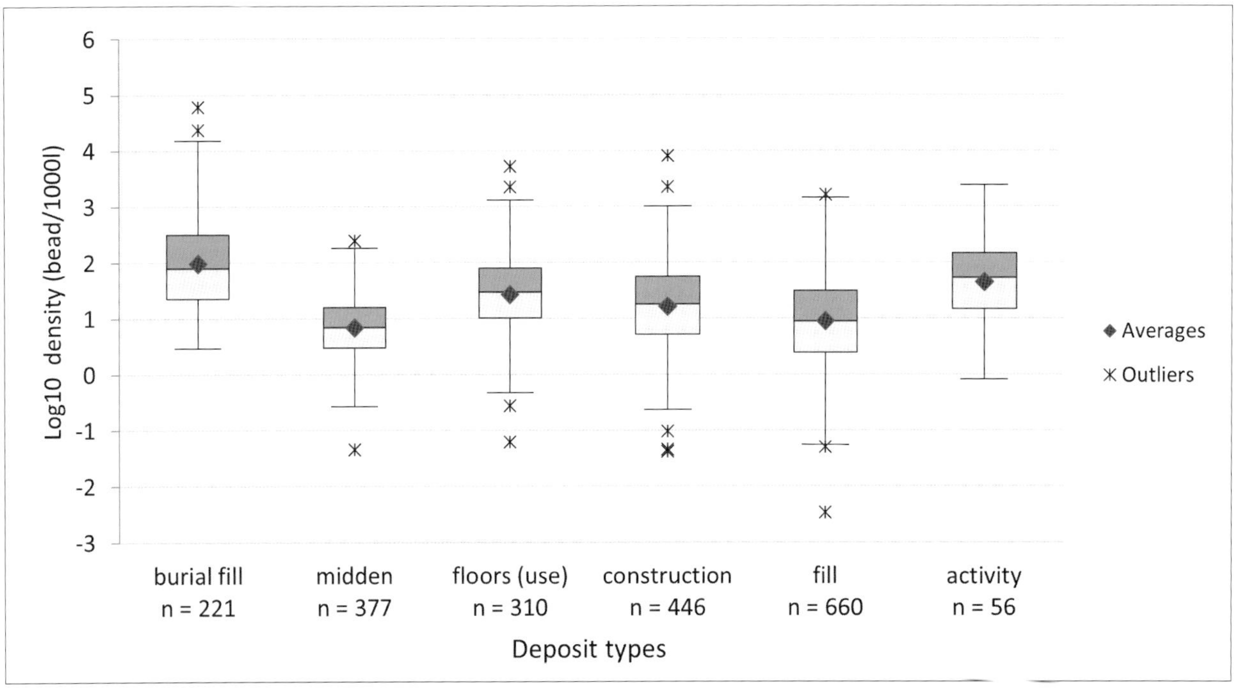

Figure 9.8. Bead density (Log10) in different deposits.

No. of beads	*Activity*	*Cluster*	*Floors (use)*	*Construction*	*Fill*	*Midden*	*Burial fill*	*Skeleton*	*Total deposits*
1 or 2	47	24	317	491	653	213	113	31	1,889
3–5	3	5	49	44	81	102	40	8	332
6–10	5	2	10	7	14	57	25	2	122
11–20	1		4	2	4	25	12	6	54
21–30	1		2	2	2	5	7	2	21
31–40				1		7	4	2	14
41–50			1		1	1	2	2	7
51–100		1	1				14	6	22
100+					1		19	10	30
Total deposits	57	32	384	547	756	410	236	69	2,491

Table 9.4. Number of deposits containing different quantities of beads.

Based on the mere count of beads per individual unit, deliberate placements of beads do stand out, as they are quite rare. For example, two clusters of beads were placed in B.56, immediately prior to its abandonment. A cluster of 69 naturally perforated *Xeropicta* shells (11691) was found in the wall blocking in the northwest part of the main room, and perforated pebbles and at least one *Viviparus* shell that was painted (13351) were found together on the floor next to the wall that was separating the main and side rooms. Similarly, a group of diverse beads was left on the floor (2798) of B.2 in the final phase of its occupation, and a group of *Nassarius* beads was found on top of a large quantity of animal bone in a packing deposit associated with an oven (8505) in B.114. Unlike other examples that represent activities related to building abandonment, the placement of these *Nassarius* beads and animal bone was attributed to an earlier (remodelling) phase of the use of this building (phase 88.2, see Stevanović 2012b).

There is also evidence of beads being occasionally buried or hidden in pits, as noted in B.64 and B.132. A group of 42 limestone disc beads were retrieved from a pit (14956) in B.64, and 107 disc beads and one imitation of a red deer canine bead, with traces of red ochre in the perforation, were found in the fill of a posthole (23638) in B.132. This posthole was located immediately next to the plastered and painted head that had obsidian in place of the eyes (21666) and was placed there during the construction phase of B.132 (see Lingle et al. 2015 for details).

Room fills tend to have a very low density of beads, and most units contain one or two beads. The most notable exception is one deposit in B.51. Building infill (11985) contained a cylindrical bead made of stone, one complete perforated *Cerastoderma*, one unperforated *Columbella* and over 20 perforated *Unio* shells. Such a large concentration of *Unio* shells has not been found thus far elsewhere (apart from a burial of a neonate (10400) with 14 *Unio* beads), and it is likely that they were deposited together. It should be noted, however, that this unit had a rather large volume, 3,680.6 litres of soil in total.

In the case of beads found in midden deposits, as these are typically not primary contexts, it is hard to discern the nature of their deposition, especially due to the fact that some of these deposits are quite large and include thousands of litres of soil. One of the rare exceptions is a midden deposit (14827) in an external area located below B.53. This unit had over 30 small stone disc beads that were all retrieved from a single flotation sample. Given the small volume of the sample and the observed uniformity of these beads, it is plausible that they were made together and likely that they were deposited together in a single event. Frequent occurrence of beads in external areas is continuous throughout the occupation of the settlement, but differences between external spaces do exist, with some of them (for example, Sp.181, Sp.279 and middens next to B.3) having a much higher quantity of beads than others.

Different types of beads, made of a variety of materials, occur in both burial and non-burial deposits. The use and discard of beads of certain shapes, colours and materials do not appear to be confined to specific contexts. With the exception of rare types and/or materials (that is, beads that appear only in a few deposits), similar beads were found left on floors, placed with the interred individuals in burials, or discarded in middens. However, some tendencies do exist, and activity and cluster deposits contain the smallest number of types (n=12), whilst the other deposits have over 20 different types. Some bead types occur across all deposits (T.1, T.2, T.7, T.11, T.13, T.24 and T.28), whilst an additional 12 types exist in all but cluster and/or activity deposits (T.3, T.4, T.5, T.8, T.9, T.13, T.15, T.22, T.23, T.25, T.26 and T.28). Types T.6 and T.31 (found in the 1960s) have been found in single burials thus far, and T.14 was recovered from two burials, whilst type T.30 exists only in burials and middens. On the other hand, a complete absence of *Viviparus* and 'bullroarer' types in burials has also been noted, whilst some beads occur only in burial fills, but not in direct association with the interred bodies (T.17, T.18 and T.19). Despite burials containing much higher quantities of beads than other deposits, a higher number of quite a few bead types was actually recovered from non-burial contexts (T.7, T.8, T.11, T.12, T.13, T.16, T.17, T.18, T.19, T.20, T.22, T.24, T.26, T.27 and T.28).

Stone beads occur relatively frequently across all deposits, but it is noteworthy that they are rather rarely present in cluster deposits (4.7%). This is quite surprising given that clusters are often associated with large numbers of a wide range of stone artefacts (Wright 2013; see also Tsoraki, this volume, Chapter 13; Tsoraki 2018). However, as demonstrated above, stone beads (as well as beads of other materials) do occur in placed deposits, and therefore their rare presence in clusters is likely to be a consequence of the ways in which cluster deposits were defined by the excavators. Nevertheless, it is quite odd that beads, regardless of their material, were not commonly combined with other types of artefacts, but instead were usually placed alone.

Bone T.1 beads were found in concentrations in two burials (4406) and (32762), but they are otherwise not very common and occur in only 15 other deposits. Similarly, larger quantities of T.23 bone beads were seen only in two burials (19460) and (15924), whilst they typically occur in smaller numbers (one or two) and were found in seven burials and 17 non-burial deposits. On the other hand, five or more beads of types T.8 and T.9 were found in eight burials, whilst an additional 16 burial features contained one or two beads. In addition, 82 non-burial deposits contained these beads, which attests to their more widespread use than the other types.

As previously mentioned, large concentrations of shell beads are found not only in burials, but also in a few placed deposits. On the other hand, outside of burials, clay beads tend to occur in smaller numbers, mainly in non-primary deposits. They are by far most common in midden deposits (18%) and burial fills (14.3%), although it should be noted that they do not occur very frequently in direct association with the interred bodies. Other types of deposits rarely contain clay beads (they are present in about 1–3.5% of different data categories).

The contextual distribution of beads demonstrates they were used in both daily life and ritualised activities such as funerary practices and practices related to building abandonment. Whilst their rare occurrence in floor deposits is likely to be due to regular sweepings, their strikingly high frequency in midden deposits cannot be simply interpreted as discard or beads being lost; rather, it could be indicative of activities taking place in the external areas. Bead production and stringing could have been one of the activities conducted outdoors.

Given the high diversity of bead strings in burials (see below), it is not surprising that some external deposits also show significant variability, and this would also explain their high occurrence in the midden deposits.

Beads and pendants in burials

Bodily decoration played a major role in funerary practices, as evident in beads being the most common type of burial associations (Vasić 2018; Vasić et al. Volume 13, Chapter 17). Due to the high level of disturbance in burials, a large number of beads moved and ended up in the burial fill rather than staying directly on the bodies of the interred individuals. Whilst some of these beads ended up in burials accidentally, the large majority seem to have been deposited deliberately. A very small number of beads (one or two) were found in burials of 82 individuals. Out of these, beads were definitely placed with 13 individuals, whilst the remaining examples are likely to have come with the soil used to fill the burial cut. For the features that contained three to ten beads, it is harder to discern the nature of their deposition, but out of 38 individuals, eight were found with beads directly associated with their bodies. Larger quantities of beads (10 or more) exist in burial features of 58 individuals, out of which only 11 were not found in direct association with beads. Nevertheless, as stated above, beads rarely occur in larger concentrations outside of burials, and it is therefore likely that most of these beads indeed belonged to the interred individuals in these features.

Burials demonstrate that beads were most commonly used in funerary practices as necklaces, bracelets or anklets, but there are also smaller numbers of beads (one or two) associated with the neck region. Although almost the same number of necklaces and bracelets was found (n=24 and n=23 respectively), at least 17 additional ornaments comprising small numbers of beads (one or two) were found in the neck region, which shows that neck adornment was favoured, whilst anklets were the least common (n=16).

Unio, fluorapatite and material of metallic lustre preliminarily identified as hematite were found in several burials under or next to crania, or in the neck region. There are also examples of beads made from other materials (for example, *Spondylus* and *Patella* were found with the adolescent (21802) in B.43), but artefacts of these three materials appear to be most commonly placed in small numbers (one or two) and associated with head or neck of the interred individuals. Hematite and *Unio* beads are almost never found on wrists or ankles, the only exception being *Unio* pendants on a bracelet of an adult male (32770) in B.132. Fluorapatite beads exhibit more diversity in their depositional patterns in burials. Apart from five individuals that had fluorapatite beads in the neck region, beads of this material were found on bracelets, necklaces and anklets with nine individuals in total.

Bead strings tend to be quite diverse and include beads made of a variety of types, colours, sizes and materials, though it should be noted that uniform strings of beads consisting of one type, size and colour do exist. Out of 67 strings, found with 37 or 38 individuals (a bracelet was found around a wrist, but it is not clear whether this wrist belongs to (19529) or (20685)), 27 strings consist solely of disc beads, whilst in an additional 26 strings, disc beads were combined with other bead forms. It appears that the strings were combined freely, although tendencies do exist. Some types (T.3, T.5, T.7, T.10, T.11, T, 15, T.21, T.22, T.23 and T.29) have thus far been found only on anklets and necklaces, and a few types (T.13, T.14, T.15 and T.24) exist only on necklaces, but it is likely that this is due to a relatively small sample. Bracelets are the most uniform in terms of colour and form, with 69.2% of them consisting of one type only and only three not including disc beads. Seven out of 17 anklets contained only one type, four of which comprised solely disc beads. Necklaces have the highest diversity, although uniform strings consisting of one bead type also exist, forming 36% (fig. 9.9). Necklaces also show high colour variability (fig. 9.10), and that is the case even for necklaces consisting solely of beads of one type.

Although the majority of strings are represented by single strands of beads, there is also evidence for multi-stranded ornaments. At least three rows of stone beads were found around the right wrist of an old adult, possible male (Sk32762), buried in B.132. Similarly, an adult male skeleton (32770), buried in the same building, had multiple strands of shell and stone disc beads around both wrists (fig. 9.11). Such firm evidence does not exist for anklets, although a child (11657) buried in B.44 might have had an anklet with two rows of beads around the left ankle (Çatalhöyük Excavation Database, Unit Sheet 11657). In the case of necklaces, evidence for multi-stranded examples comes from the burial of an adolescent (23126) whose neck was adorned with several strands of plaster and stone beads as well as beads of unidentified material of botanical origin (see below).

Strings in burials also frequently combined beads made of locally available materials with others that were harder to procure (for example, limestone and marine shells or carnelian). This is interesting, as it brings to the fore the large network of associations (people, places and origins) reflected in the creation even of a single bead string. Strings consisting of mostly beads of various

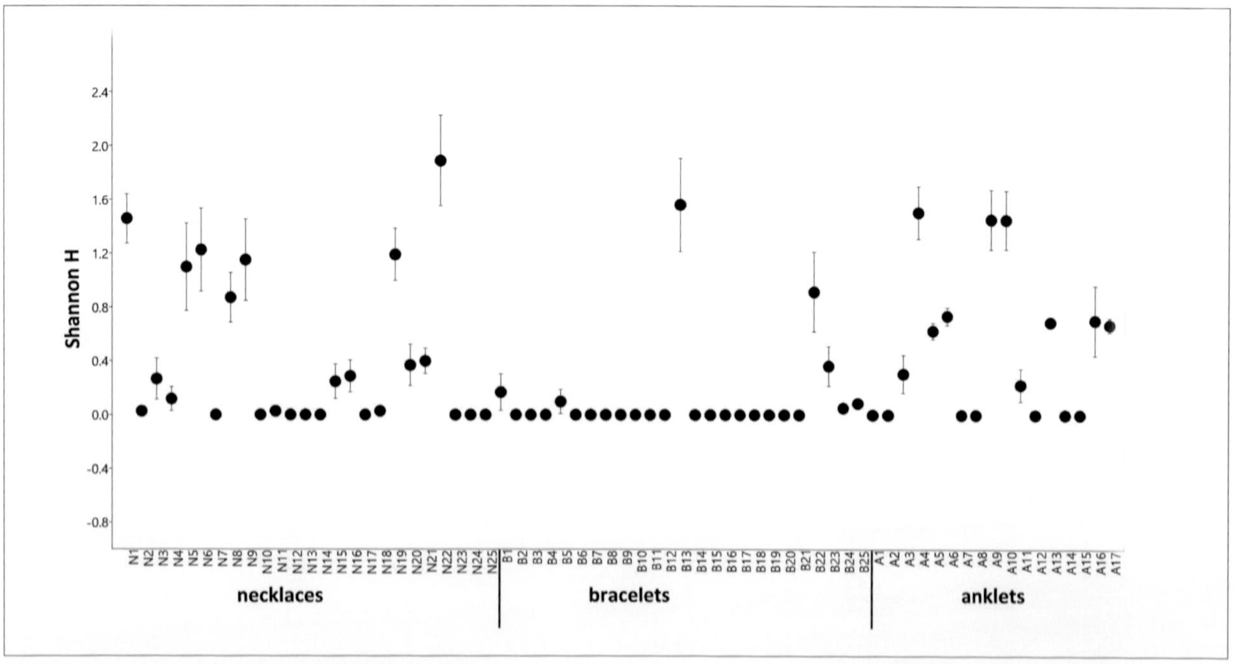

Figure 9.9. Diversity of bead strings (types).

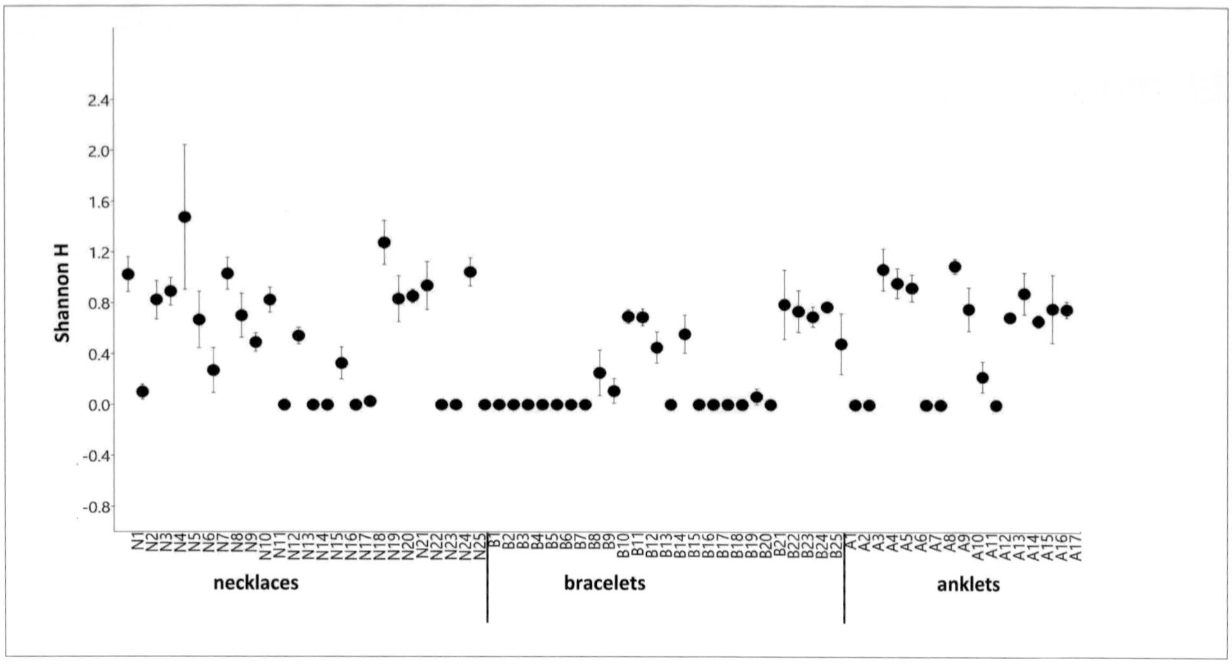

Figure 9.10. Diversity of bead strings (colours).

non-local materials (for example, a bracelet with (11306)) are quite rare, although strings consisting exclusively of beads of one marine shell species, such as *Antalis*, *Dentalium* or *Nassarius*, are relatively common.

Another interesting aspect is variability observed in cases where the interred is buried with multiple strings. This is convincingly highlighted in the burial of an adult individual (22620) in B.129. While bracelets consisted almost entirely of carnelian beads, a string of beads around the left ankle included beads made of phyllite, limestone and *Antalis*, and the right ankle was adorned with beads made from clay and phyllite. An additional 123 beads of similar materials (limestone, phyllite, clay, *Antalis* and carnelian, but also galena, fluorapatite and *Theodoxus*) were recovered from the burial fill, some, if not all, of which belonged to one of these strings. Wear analysis (see also Tsoraki 2017) suggests that all the carnelian beads, both those on bracelets and the ones recovered from the burial fill, were associated with the same production event and had been used for a similar

Figure 9.11. Multi-stranded bracelet with (32770) (for colour version, see online supplementary material).

duration prior to their deposition in the burial. Phyllite and limestone beads from the anklets show limited or no use whatsoever (Chapter 13), whilst in contrast, the *Antalis* beads display various degrees of use, with some clearly having been in use for a very long time. None of the fluorapatite beads from this burial display evidence of prolonged use; they were either used for a short period of time or not used at all.

Whilst beads in burials occur most commonly in the form of bracelets, necklaces and anklets, this was not the only manner in which beads seem to have been used/worn. For instance, in B.131 an adolescent, possible female (23126), was buried with a multi-stranded necklace and a potential *Antalis* bracelet, while a further 2,170 disc beads were found in association with the hands, left arm and lower legs, and an additional 616 beads were retrieved from the burial fill. Apart from the *Antalis* beads, phyllite, serpentinite, carnelian and plaster were identified, but the vast majority of T.1 beads were made of an unidentified material that is of wood/mineralised plant origin. They all share the same physical properties (a very distinctive fibrous texture, dark grey, almost black colour on the outer surface and light orange interior surface). These beads were found in multiple rows at various locations (lower limbs, hands, left shoulder and arm), following different directions, and it is unlikely that they formed bracelets and anklets (see Vasić et al. Volume 13, Chapter 17: fig. 17.5). Whilst no traces of cloth were recorded as having being associated with these beads during the excavation of this burial, and no use-wear suggesting attachment has been noted, soil samples taken from the area of the lower limbs contained textiles (Chapter 11). Therefore, it is plausible that these beads were indeed attached to either garments in which the adolescent was buried or the material that was used to wrap the body. Alternatively, it is possible that long strings of beads were wrapped around the body, as was previously reported to be the case with an infant (2105) buried in neighbouring B.1 (Cessford 2007b). Further evidence of multiple ways in which beads were used in funerary practices is seen in beads sometimes simply being placed on the body (for example, 15924.x2-x5 (15924)) and repeated occurrences of unfinished beads in burials, as well as the placement of beads in potential pouches (Volume 13, Chapter 17).

Only 10.5% of individuals were found buried in direct association with beads. When the indirect associations (that is, beads recovered from the burial fills) are included, the proportion of interred individuals buried with beads increases to 27.9% (Volume 13, Chapter 17: fig. 17.6). Both direct and indirect associations show that adults were more commonly buried with beads, and the older they were when they died, the more likely they were to be buried with beads (Vasić 2018; Vasić et al. Volume 13, Chapter 17). On the other hand, sex did not form the major social axis in the placement of beads in burials, although female individuals are more commonly interred with beads than males are.

Some bead types occur more frequently with adults and others with subadults. Strings with the youngest individuals (<3 years) tend to have one or two bead types, with the exception of one infant (23231) that had a necklace comprised of seven different types (fig. 9.12). Single occurrences of types T.6, T.13, T.14 and T.19 have thus far been noted only with subadults. On the other hand, types T.7, T.24, T.26, T.28 and T.30 have been found only with adult individuals. Females are associated with more bead types, and so far the direct associations of T.3, T.5, T.7, T.21, T.23, T.24, T.26, T.27 and T.30 have been observed only with females. Colour does not seem to have played an important role in the differentiation of funerary practices, as beads of different colours were found with both subadults and adults, and with males and females.

A similar diversity of necklaces and anklets has been noted with both subadults and adults; that is, there are no significant differences in terms of how beads were combined. On the other hand, 50% of bracelets found with adults contained at least two bead types, with the bracelets placed with an old adult female (11306) and possibe male (32762) being the most diverse. As previously suggested, bracelets generally display the highest level of uniformity and typically contain one bead type, and whilst this is the case for all bracelets found with subadults, it seems that adults were buried with bracelets of a higher variability. The sample is quite small, consisting of 14 bracelets that were found with eight (or potentially nine) adult individuals, and further excavations will show if this pattern holds true.

Spatial distribution

Buildings display a remarkable variability in bead assemblages that does not seem to be a result of either their location in the settlement or the period in which they were used. This variability has been observed in the occurrence of beads in different types of deposits, as well as in their densities in floor deposits (fig. 9.13).

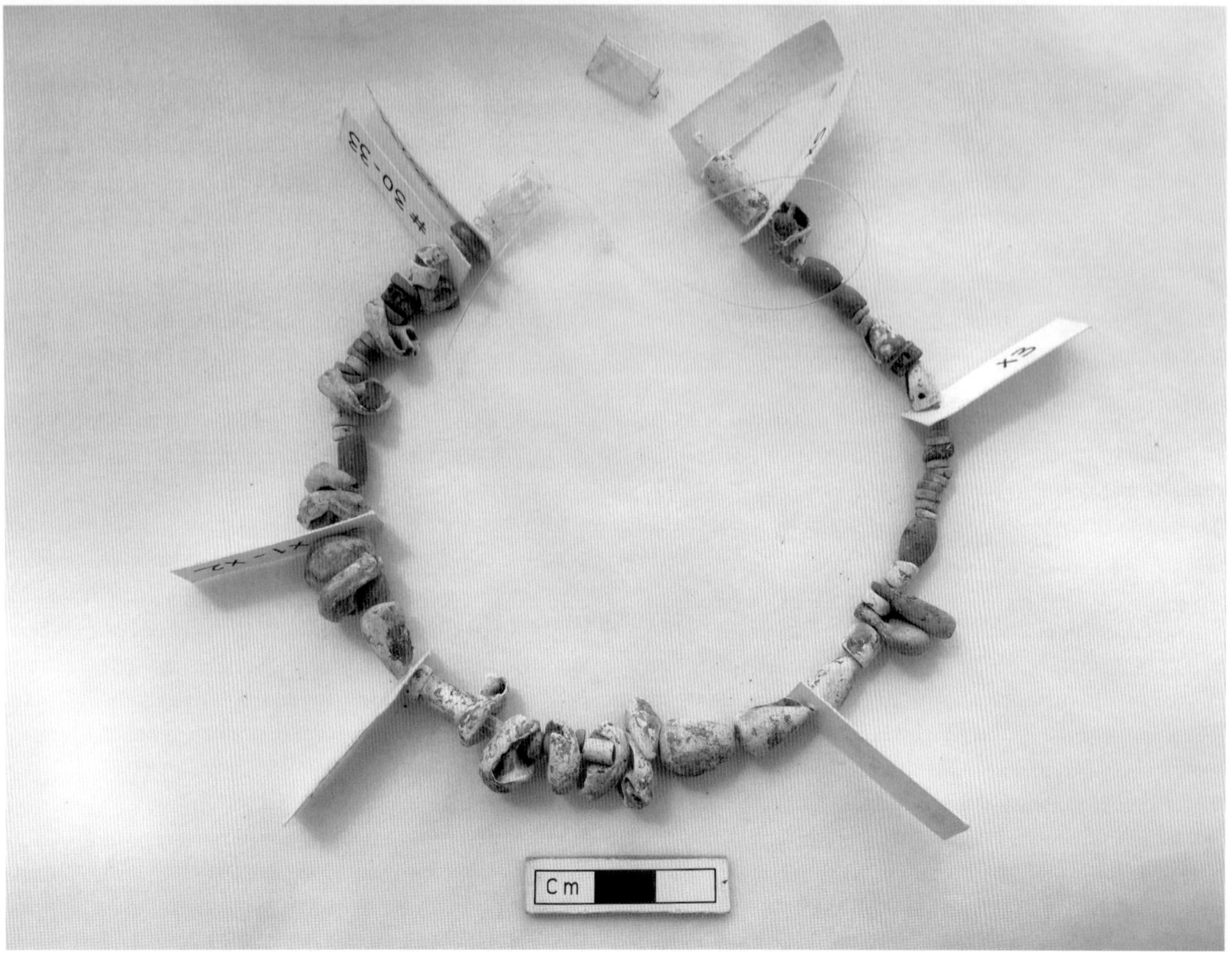

Figure 9.12. Necklace with (23231) (photograph by Ekin Ünal) (for colour version, see online supplementary material).

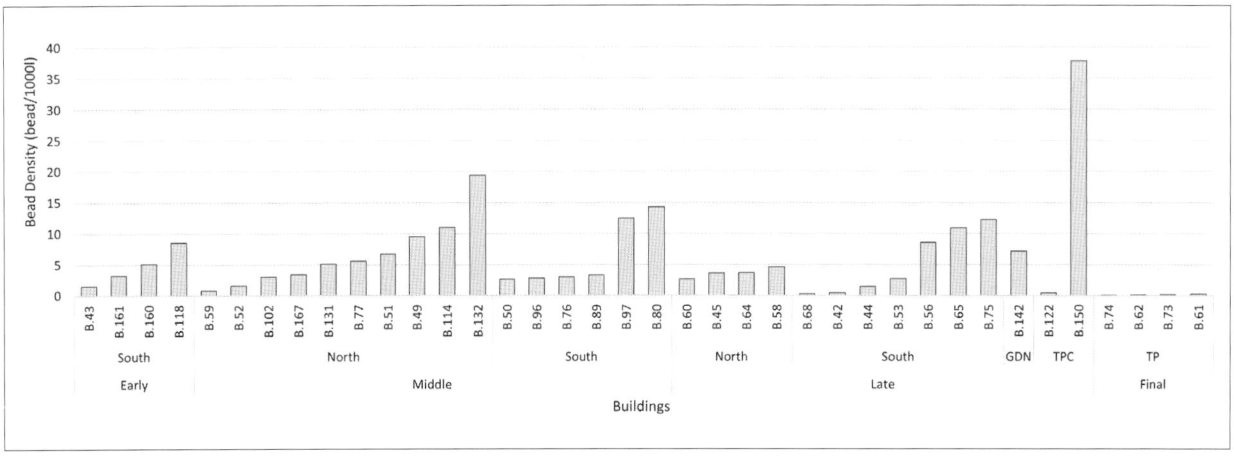

Figure 9.13. Estimated bead densities in floor deposits.

Individuals buried in the same building rarely have the same bead combinations, and neighbouring buildings do not necessarily contain similar beads. Although types do occasionally repeat (especially the disc beads and other common types), individuals are rarely adorned with similar items, although cases of similar beads being placed in several burials have been noted. For example, two neonates buried in B.42 had hematite pendants associated with the neck region, and two infants were buried in B.52 with two blue and blue/green beads, one on each side of the crania, but one of the infants had a large concentration of limestone discs in the abdominal region, whilst the other one had no other direct associations. Similarly, several individuals in B.77 were adorned with strings composed solely of *Nassarius* beads. Over 70% of the entire *Nassarius* assemblage has been recovered from these burials and one placed deposit in the adjacent B.114. Strikingly, no other contexts in these two buildings contained *Nassarius* beads.

Three adjacent and roughly contemporary buildings – B.1, B.131 and B.77 – can be taken as a good example to illustrate both the variability of the bead assemblage and the observed repetition of bead ornaments in burials. These buildings contained a large number of inhumations, some of which were heavily disturbed due to being located under the same platforms. For that reason, only direct associations were taken into consideration.

Only four out of 61 individuals interred in B.1 were found directly associated with beads. A blue stone bead and a deliberately burnt bead made of bone were found on the neck of a male (1924). In addition, this individual had a bracelet composed solely of *Dentalium* beads on his upper left arm. Lying between the femur and the pelvis of a female individual (1955) was a large green pendant with two perforations. The body of an infant (2105) was wrapped with hundreds of disc beads, whilst similar-looking beads were found together with a single *Dentalium* bead on the neck of a child (1913). Outside of burials, a bone T.2 and a stone T.1 were recovered from two deposits interpreted as clusters, whilst 31 beads of a variety of materials (*Dentalium*, animal bone, phyllite, marble, limestone and clay) were found in the floor deposits.

The neighbouring B.131 contained 41 inhumations, out of which at least six were buried with beads. Interestingly, all individuals adorned with beads seem to have been females. As previously discussed, an adolescent (23126) was found with hundreds of beads associated with different parts of the body, an adult (23115) was found with two unfinished steatite preforms, and an adolescent (31705) had four shell T.1 beads and a carnelian T.3 bead associated with the lower left limb (however, a large quantity of beads was recovered from the burial fill). Two females were found buried in a mirrored position (see Volume 13, Chapter 17: fig. 17.15); one of them (22661) had a single pendant on her neck, whilst the other female (32324) was buried with a necklace consisting of T.1, T.2 and T.20 beads and two bracelets, one of which was located on the right mid lower arm (T.1, T.9, T.28 and T.29/T.11) and the other on the right wrist (T.1, T.8, T.9 and T.28). In addition, 12 concentrations containing over 600 beads (see above) were recovered from the fills of three burial features and are likely to have been deposited with either (30044) or (30040). In contrast to the large quantity of beads found in burials, no beads were recovered from clusters in this building, and the floor deposits contained a very small number of beads (n=10).

A similar number of inhumations (n=39) was found in the adjacent B.77. Peculiarly, individuals interred under the north and east platforms were not found directly associated with any beads, whilst at least seven out of 15 individuals buried under the northeast platform (F.6051) were adorned with beads. In total, 1,270 beads were recovered from these burials. The earliest inhumation belongs to a child (21681), buried with a large number of beads (n=269) of various types and materials, the large majority of which were limestone T.1 beads. During the next occupation phase (B.77.2C), four individuals were interred, with only one of them (19529) potentially being adorned with a *Dentalium* bracelet (alternatively, this bracelet could have belonged to (20685)). A single individual, an adult male (20685), was buried in the following phase (B.77.2D) and was adorned with a *Nassarius* bracelet. In addition, 22 beads made of a variety of materials (galena, turquoise, greenstone, copper etc.) were recovered from the burial fill. A single individual (19535) was buried during the phase B.77.5, and a group of *Nassarius* beads were found under her hip. The final occupation phase, B.77.7, included interment of four individuals. Over 700 beads of different types (T.1, T.3, T.5, T.9, T.21, T.23 and T.25) were found with a female (20683) and appear to have formed at least two ornaments, one of which was likely to have been a necklace. In addition, long bones of two possible males ((19541) and (19554)) were placed in a pile together with a large concentration of *Nassarius* beads. Similarly to the adjacent B.131, a very small number of beads were recovered from the floor deposits (n=9), and two beads were recovered from clusters. A very large pendant (19532.x1) was found incorporated between two layers of plaster on the northern platform (F.3617), which is surprising, given that burials below this platform did not contain beads.

In contrast, a block of buildings B.80, B.76, B.96 and B.97 in the South Area that are roughly contemporaneous with the northern buildings discussed above did not have large quantities of beads. Out of 44 individuals interred in these buildings, it appears that only one male (18701) buried in B.76 was adorned with beads. A group of galena disc beads were found on his left shoulder, whilst imitation red deer canine beads were found in the waist area, potentially in a pouch. Although burials in B.80 did not have direct associations of beads with the interred individuals, 13 beads were recovered from a clay surface (22421) in the dirty area of B.80, 25 beads were found in a packing layer below the floor (22425) and, similarly to B.77, a large pendant was incorporated between two plaster layers (20013) in the eastern platform.

The disparity between the occurrence of beads in burials and in other primary contexts in individual buildings is quite remarkable, as it appears that buildings with large numbers of beads in burials rarely contain a high quantity of beads in other deposits. This has been noted in the buildings discussed above, but also in other buildings across the site. For example, only four limestone disc beads were recovered from the floor deposits of large and elaborate B.52, which is assumed to be contemporary with B.1, B.131 and B.77. Two clusters found in bins contained a small number of beads (five in total) made from different materials (naturally perforated pebble, marble and marine shells), whilst one bin infill also had a complete 'bullroarer' type made of antler. On the other hand, several individuals interred in this building were adorned with beads. One burial feature (F.7127) contained an adult, possible male, and five subadults. The adult (30514) had a black bead with metallic lustre under his cranium, and one of the subadults, an infant (30511), was buried with a green bead on each side of the cranium. Similarly, as previously discussed, another infant (30523), buried in another feature had two green/blue beads in the same position, but this infant also had a string of beads around the left ankle and at least three rows, with over 100 limestone disc beads, found in the abdominal region. Finally, an infant (23805) was buried with a necklace largely consisting of pink limestone T.1 beads but also a few T.1 beads of other materials and highly damaged roughly sub-spherical plaster beads whose type could not be determined. As no stratigraphical link could be determined between this burial and the platform's surface, it is unclear whether this burial represents the first inhumation in the northwest platform of B.52 or if it belongs to the building below, B.167 (Barański, personal communication). In addition, this infant had grey/black T.1 beads around the right wrist and a string of alternating black and white T.4 beads around the left ankle (fig. 9.14). Strikingly, this combination of black and white T.4 beads also exists in a bracelet in a burial in nearby B.132, as well as in a necklace in one burial in B. 43 in the South Area.

Although beads are most commonly present in burials, a complete absence of beads has been observed in burials of B.59 and, strikingly, B.75 and B.18, buildings that contained clear evidence of continual bead production, as witnessed in the presence of preforms and roughouts throughout their occupation. It should be noted, however, that B.75 contained only three inhumations, and the other two buildings had one burial each. The absence of beads in burials in these buildings could therefore be a result of a small number of inhumations, as well as the age-at-death categories of the buried individuals. Their absence could also be related to the fact that B.18 was partially excavated by Mellaart, and B.75 was a heavily eroded building with only one part preserved. A very small number of beads (five or less) were recovered from burials in B.54 and B.58 (see Volume 13, Chapter 17 for further discussion) in the North Area and B.80, B.160 and B.97 in the South Area, likely reflecting accidental deposition. In addition, only eight beads were recovered from 16 burials in B.114, and

Figure 9.14. Anklet with (23805).

16 burials in B.65 contained 16 beads in total. Only two beads were recovered from four burial features in the subsequent B.56, whereas two strings of beads were left on the floor prior to its abandonment.

Some buildings, however, contain a high quantity of beads in both burial and non-burial deposits. For example, 216 beads were recovered from three burial features in B.150. At the same time, a number of beads were found in the floor deposits, with 51 beads originating from a single 30 litre flotation sample from one unit (23739). Both burials and floor deposits contained limestone and fluorapatite beads.

Four out of ten individuals interred in B.132 were adorned with beads. Three adult individuals ((21685), (32741) and (32762)) had beads that are likely to have represented necklaces, and (32741) also had a multi-stranded bracelet comprised of limestone T.1 and limestone and serpentinite T.4 beads. Another individual, a young adult male (32770), who was also adorned with a belt hook and eye and four rings, had bracelets on both wrists with T.1, T.2, T.9 and T.22 beads. Unlike the buildings discussed above, this building also contained a large concentration of beads in a posthole pit (see (23638) above), and 46 beads were recovered from the floor deposits, out of which 37 were stone T.1 beads. These beads were made from different materials (limestone, tufa, serpentinite, marble, phyllite, steatite, quartz). Overall, the presence of finished beads with fresh appearance and no obvious use-related traces, and debitage from equivalent materials (limestone, tufa, marble, quartz) strongly suggests that B.132 was associated with bead production along with other stone working activities (Chapter 13), which is reminiscent of the picture we get in the later B.75 in the South Area (Bains 2012; Bains et al. 2013) (see discussion above).

Occurrence of beads does not seem related to either the size or the elaboration of buildings. This is evident in the lack of correlation between the count of beads in burials, clusters and floors or other deposits and size and elaboration index. As a result, some of the large and elaborate buildings such as B.77, B.131 and B.52 have lower bead density in the floor deposits in comparison to not-so-elaborate and smaller houses (figs 9.15 and 9.16).

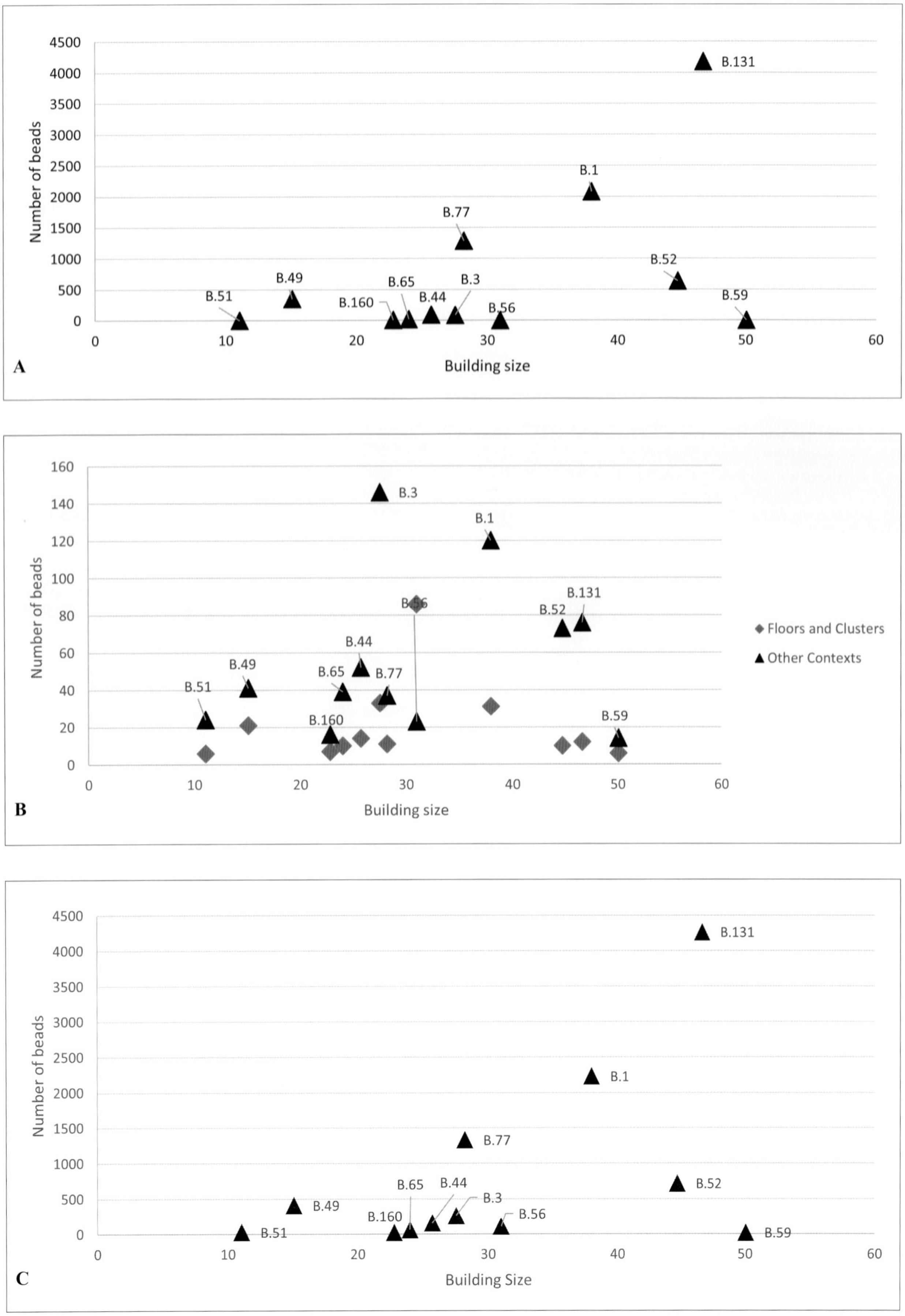

Figure 9.15. Beads and building size: (A) burials; (B) non-burial contexts; (C) all contexts.

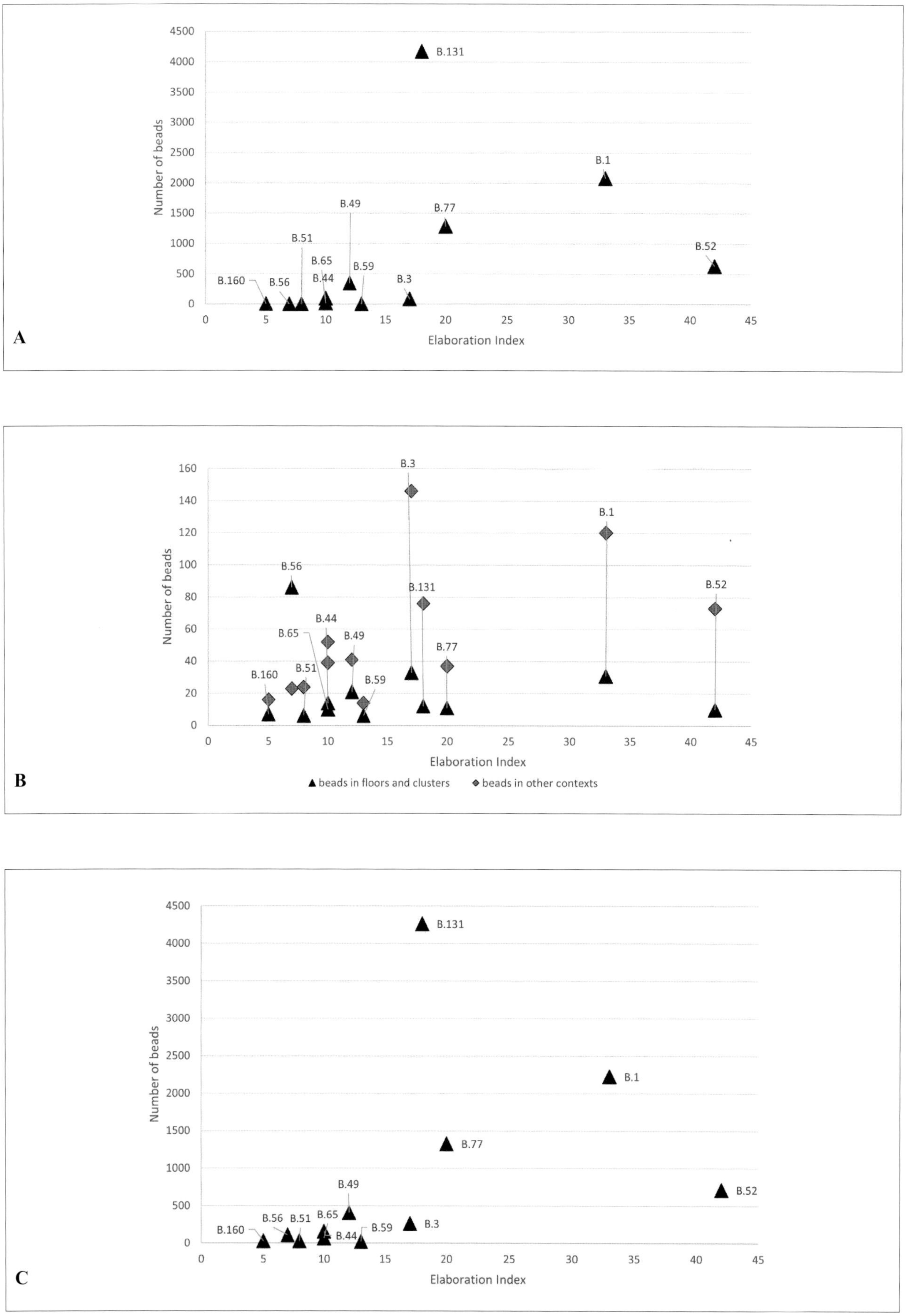

Figure 9.16. Beads and elaboration index: (A) burials; (B) non-burial contexts; (C) all contexts.

Given the variability observed in individual houses, the lack of clear differentiation between the areas in the settlement is not surprising. Similar beads occur in the same types of contexts in the North and South Areas, which does not suggest clear-cut differences between people occupying different parts of the settlement. Nevertheless, it is clear that use of beads, items that are one of the most potent mediums for the expression of identity, does not reflect any restricted access to resources, nor does it suggest that (groups of people) occupying and/or being buried in individual buildings in the various areas of the settlement had an exclusive 'right' to adorn themselves with specific beads.

Temporal distribution

The earliest deposits at Çatalhöyük (Sp.181) demonstrate that the vast majority of bead types were already in use from the beginning (T.1, T.2, T.7, T.8, T.9, T.12, T.13, T.15, T.16, T.17, T.20, T.22, T.24, T.25, T.26, T.28 and T.29), and several types seem to have been introduced towards the end of the Early period, in Levels South K and/or South L (T.4, T.5, T.10, T.11, T.14, T.21 and T.27). The Middle period also brought a few new types (T.12, T.18, T.19, T.23 and T.29), whilst only three new types (T.6, T.30 and T.31) appeared in the Late period. Strikingly, despite the clear distinction in the material culture of the Final period (for example, the burial chamber in the TP Area), no novelties in the bead assemblage have been noted, and people continued using and creating the same bead types in the Final levels of the occupation as they did before.

Despite the existence of most of the bead types from the beginnings of life at the settlement, the diversity of beads increases through time; that is, the assemblages from the Late and Final periods have a much higher diversity than those of the Early and Middle periods (fig. 9.17). Although disc beads are the most common type in every period, their relative quantities significantly drop after the end of the Middle period.

Similarly, an increased variability has been observed in the expansion of raw materials used in bead production. For example, carnelian beads, along with fluorapatite and turquoise, first occur in the Middle period, and the use of these materials continues throughout the Late and Final periods. The earliest evidence of copper beads has been recorded in Levels South K and South L, although it should be noted that the ÇRP excavations did not produce copper beads from levels earlier than South M/North G. Copper beads are quite rare; they have thus far been found in a small number of burials, but they exist in a few non-burial contexts as well.

T.3 beads in the Early period were made from clay and bone only, whilst the use of stone for the manufacture of this bead type is attested in the Middle period. Similarly, T.7 beads exist only in clay in the Early period, whilst stone and wooden T.7 beads appear in the Middle period. The material selection for the production of T.8 beads expands as well, as witnessed in the use of only animal bone for the manufacture of these beads in the Early period and the addition of clay and stone materials in the Middle period. On the other hand, whilst T.2 beads were made from bone (Level South G), stone (Level South L) and shell (Level South H) from the Early period onwards, the use of clay for the manufacture of cylindrical beads seems to have started in the Middle period. Interlocking beads (T.23) made of animal bone start appearing in the Middle period (Level North G), but a few Late period contexts demonstrate that the selection of material expands to include stone as well.

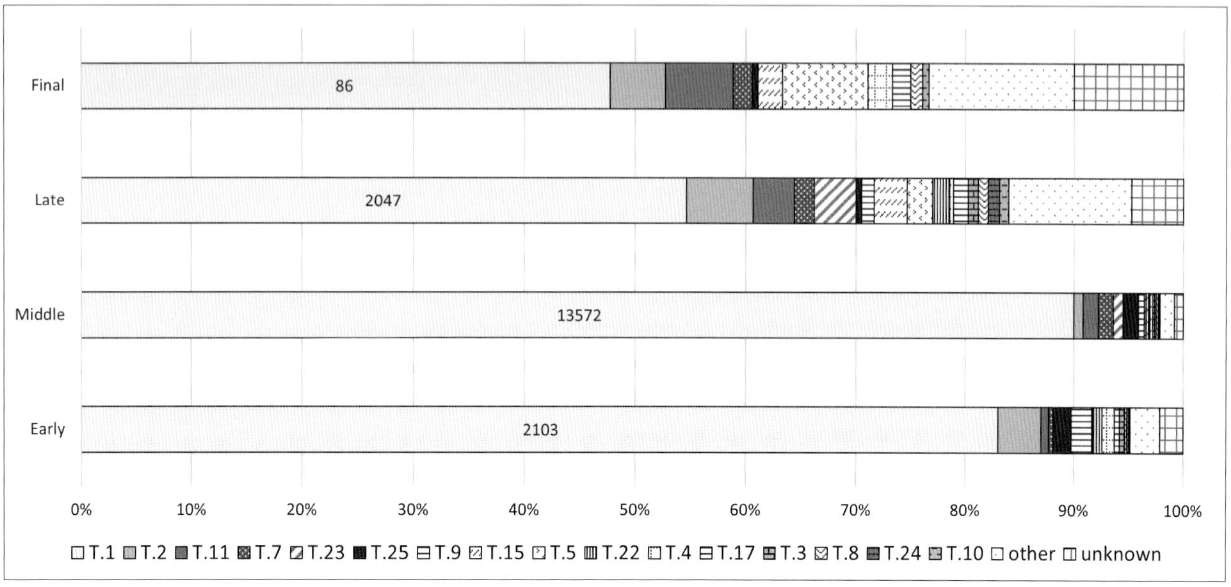

Figure 9.17. Types through time.

In addition, an increase in the relative quantities of shell and clay beads in the Late period, in which they form almost 30% of the assemblage, has been observed. Almost all shell species were used from the earliest levels and throughout the sequence, with marine shells being prevalent in the assemblage of each period. Nevertheless, similarly to observations from previous publications (Bains et al. 2013; Bar-Yosef Mayer 2013; Vasić 2018; see also Veropoulidou, this volume, Chapter 10), the studied assemblage suggests that the proportion of local shells (*Unio*, *Xeropicta* and *Viviparus*) significantly increases in the Late period, but an increase in *Columbella* beads has also been noted. At the same time, fossils such as *Anadara turoniensis*, *Clavatula calcarata* and *Athleta ficulina* occur only in Late period deposits, albeit in small numbers.

Therefore, there is an evident continuity with beads slowly diversifying through time. Stone disc beads remain dominant, but the use of other materials and types becomes more common. Similarly, the use of larger beads becomes more frequent in the Late and Final periods (fig. 9.18), and consequently the average number of beads on strings drops through time.

Whilst the increase in overall diversity is evident, different types of strings do not individually follow the pattern, as there are still uniform bracelets, necklaces and anklets in the Late period. The sample is small, but the occurrence of uniform strings demonstrates that the increased diversity did not necessarily mean that strings consisting of one type of bead disappeared. For example, only one bracelet from the Late period (found with the adult (11306) with a plastered skull), which is also the most diverse bracelet thus far, contained multiple types of beads, whilst the other five bracelets included one type of bead only. On the other hand, increasing diversity has been noted on anklets, with three out of six from the Middle period containing one type, whereas only two out of eight Late period anklets were composed of one type.

The presence of beads in different kinds of deposits seems pretty consistent through time, with an evident decrease in their presence in burial features over time (fig. 9.19). Bead density, on the other hand, demonstrates a gradual decrease through time in the most types of deposits, with the largest difference seen in the Final period (fig. 9.20). As discussed above, buildings are highly variable, regardless of the period in which they existed, although it should be noted that in the Middle period, northern buildings tend to have higher bead density in the floor deposits, whilst the floors in the South Area buildings have a higher density in the Late period (see fig. 9.13).

Discussion

Although bodily ornamentation at Çatalhöyük included a variety of objects, such as bangles, finger rings and boar tusk collars, and probably other organic materials such as feathers that do not survive in the archaeological record, beads formed its main component. This is evident in their relatively frequent occurrence throughout the entire occupation of the settlement, but also their prevalence in burials in comparison to other artefacts. Judging by their frequent occurrence, bodily decoration with beads also represented one of the main aspects of funerary practices.

This chapter builds upon previous studies, with the purpose of exploring how beads were made and used at Çatalhöyük. The recent analyses shed more light on manufacturing techniques, with perhaps the T.1 assemblage providing the most substantial evidence and producing interesting results regarding its manufacture. We are still missing important elements of the *chaîne opératoire*, however, mainly in relation to the number of people involved in the procurement of raw materials and the production of individual beads, as well as the beads' exchange and use.

Some strings in burials included beads of a variety of local and non-local materials, and it is unlikely that a single individual would have travelled to each of the sources to procure the raw materials and make each of those beads themselves for personal use. Therefore, it is more reasonable to assume that a group of people were involved in the creation of these strings. This is also evident in different types of technological know-how that were observed on some bead groups. Raw materials were probably traded within the site, as well as with individuals from other settlements, and finished products as well.

As discussed in the previous sections, there is substantial evidence to suggest that the use of beads was quite fluid, and that individual beads served multiple purposes and formed part of different objects; this is evident in individual beads on the same string exhibiting various degrees of use-wear, but also in the possibility of beads being used for different purposes (clothing attachments and necklaces, bracelets etc.) throughout their use-life.

Combinations of beads on strings found in situ in burials do not exhibit any strict patterning. Beads seem to have been combined freely and their order appears to have been a matter of personal preference. Similarly, individual choices and preferences may explain the lack of strong correlations between bead types, colours of materials and types of strings (that is, necklaces, bracelets, anklets), as well as their occurrence with individuals of specific age and/or sex, although these choices are likely to have depended at least to some

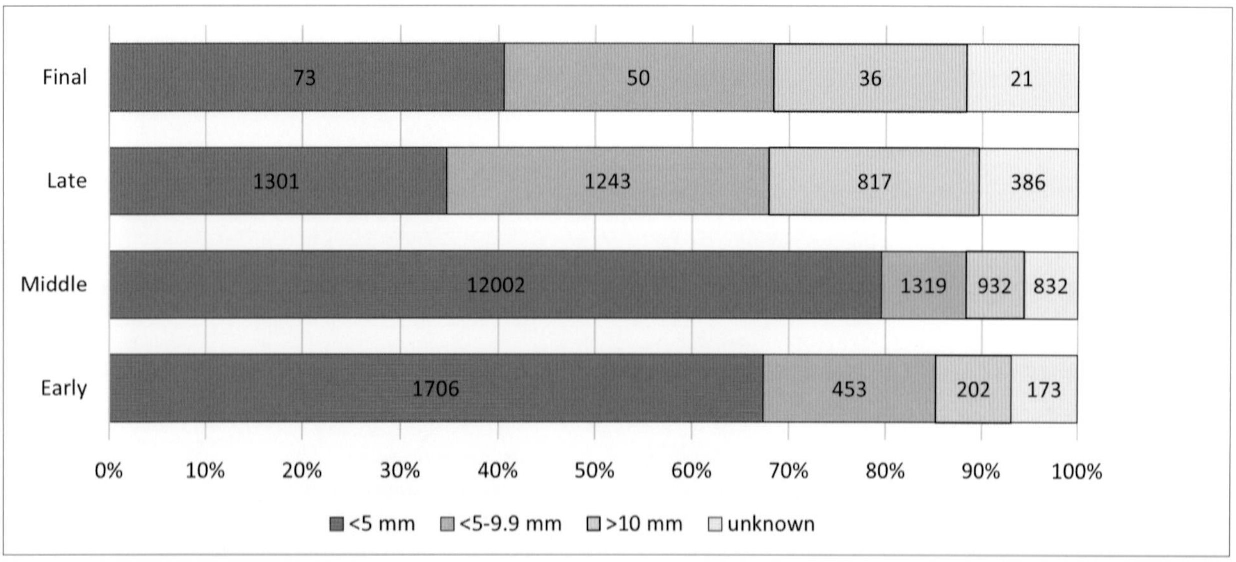

Figure 9.18. Bead size through time.

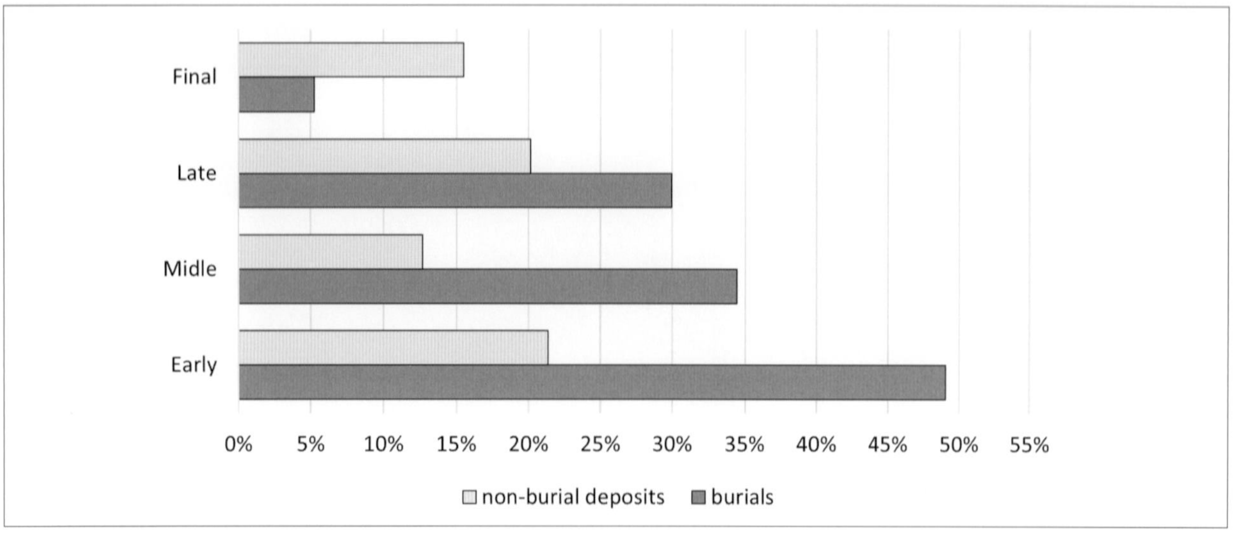

Figure 9.19. Proportion of excavated units with beads in each period.

extent on the availability of beads at the given moment. It is important to bear in mind that strings in burials are just palimpsests; that is, they only represent the final combination of individual beads before they were taken out of circulation through their placement with the deceased. As previously suggested, rather than personal possessions, some of the bead strings in burials represented newly created objects that were possibly formed by a group of people (Vasić 2018).

Due to their transformative properties, bead strings could have been continually strung and restrung, and this could be the reason why there are so many complete and still usable beads in a variety of contexts, especially in the external areas. Whilst some of them might have been lost, it is plausible that a relatively frequent occurrence of a small number of beads (one or two) in, for example, midden deposits could in some cases have been a result of them being superfluous after the stringing of an ornament (for example, a bracelet) was completed. Although the bead assemblage displays a high variability, there are not that many unique beads and pendants that would be easily recognisable when worn. Some shapes are quite distinguishable (for example, pyramid or bird beads), but generally speaking, it would be hard for people to recognise individual beads on a string as something they created, or to associate particular bead(s) with a specific individual. In that sense, apart from a few beads that are unique, the gestalt of the created multi-composite strings would have played a bigger role in the negotiation of identities than individual beads ever could, and bead making and the creation of strings might have

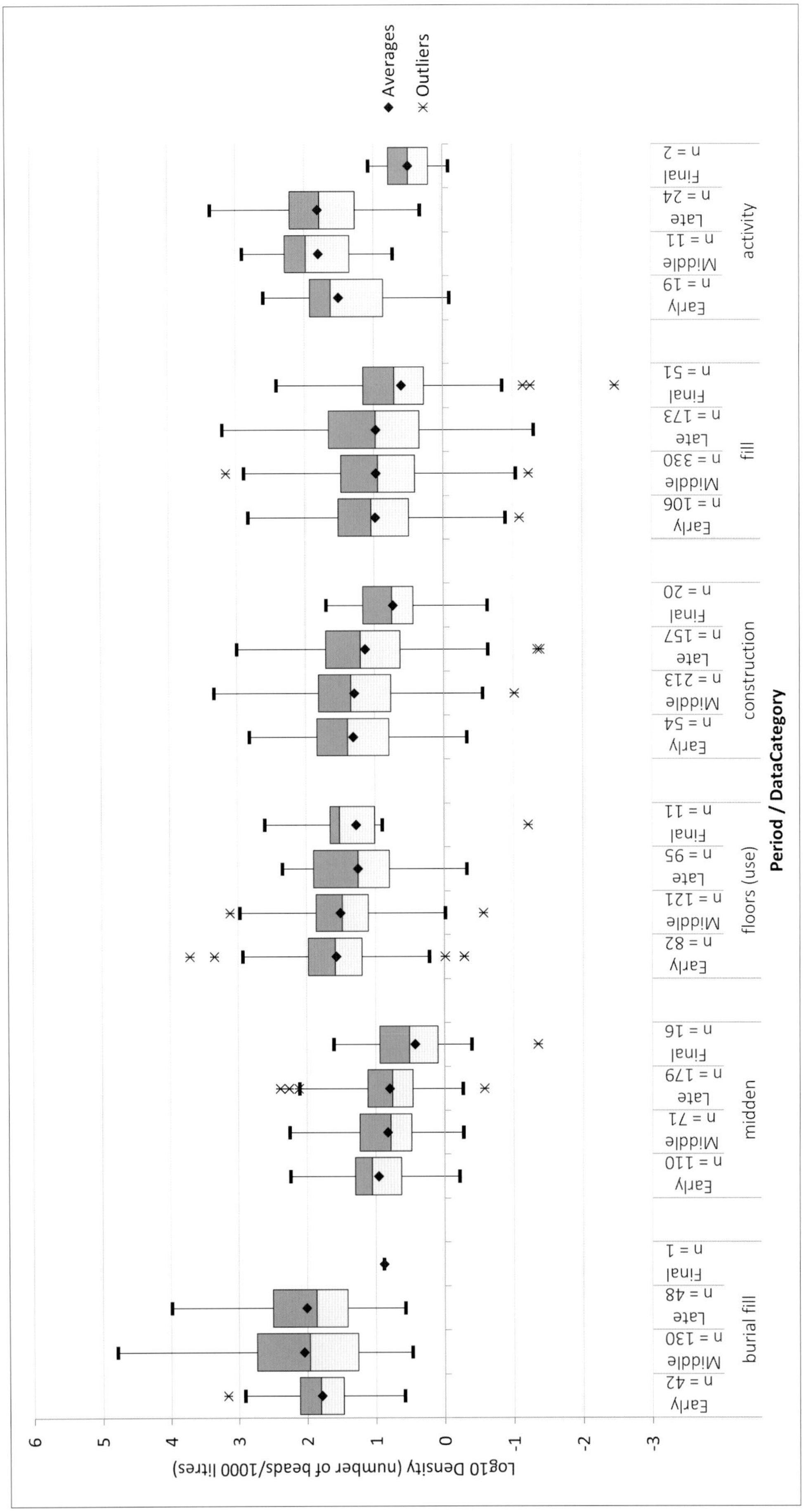

Figure 9.20. Bead density in different deposits through time.

been as significant as the final products themselves. Stringing beads might have been a communal affair, with a group of people coming together, exchanging beads and creating individual strings.

Bodily decoration is a powerful medium for propagating individual and social identities and an active tool in the creation of social meaning. Beads – as the main type of adornment at Çatalhöyük – actively participated in the formation, expression and renegotiation of identities. The production and use of beads, and possibly the procurement of raw materials, were highly social processes. If beads were used to create and/or strengthen relations within the settlement, as well as with other communities, strings with multiple beads made of a variety of materials would reflect complex networking and a high level of sociality at Çatalhöyük (Vasić 2018).

Whilst sequences of buildings built on the same spot (that is, 'history houses') exhibit a remarkable continuity in various aspects of material culture (Hodder, Pels 2010), the burial associations and beads specifically stand out due to their high variability. Some individuals were adorned with beads, others had none. To take the B.65-B.56-B.44 sequence as an example, out of 36 individuals buried in these buildings, only one child (11657) in B.44 was found in direct association with beads. (In total, 19 additional beads were recovered from seven burial features, likely reflecting accidental inclusion, as no beads were found in direct association with these individuals.) In cases in which subsequent buildings do have beads in burials (for example, B.77 and B.132), beads are quite diverse. It is clear that people were buried with different things in these buildings, which is in contrast with the overarching sense of continuity observed in these houses. Therefore, people using and/or being buried in these buildings did not have distinct adornment, something that would distinguish them from other groups, and they were buried with beads similar to those that occur across the settlement. It is possible, however, that the diversity of beads in these buildings reflects complex networks and affiliations of individuals buried there with other people in the settlement, a suggestion also supported by the rich ground stone assemblage from B.77 (Tsoraki 2018).

Without downplaying the increased diversity of beads throughout the occupation of the settlement, the bead assemblage displays an overall sense of continuity, in terms of both manufacture and the ways in which beads were used. For example, the production of stone beads was carried out alongside other stone-working activities, and this is a recurrent pattern for the Middle period (B.132) and Late period (B.142), and perhaps for the Early period too (B.17). Common production locales of bone and stone beads were also noted previously in B.18 (Bains et al. 2013). In other words, our sample of bead materials and the study of debitage and nodules from a wide range of buildings does not support the existence of specialised bead workshops operating at Çatalhöyük.

The spatial distribution of beads and the lack of clear-cut differentiation in the bead assemblages of each of the areas of the settlement go along with the other lines of evidence that suggest a high complexity of social networks that were crisscrossing the settlement (Hodder 2014c). The use of beads reflecting complex social relations and networks would be a reason why B.75, which had clear evidence of tufa T.1 bead manufacture, did not contain any finished products nor beads placed in burials, or why buildings with large groups of beads in burials did not necessarily contain a high quantity of beads on floors and/or in placed deposits. Bead strings as composite artefacts depended on a group (or groups) of people that were involved in different steps of the *chaîne opératoire*; from procurement of raw materials to their manufacture and use, beads were part of socially entangled practices (Hodder 2012) that brought a network of people together. Beads were a fundamental part of life and death at Çatalhöyük and were socially embedded in both daily activities and ritualised practices. They could have been used to propagate individual identities, but it is clear that they were also the means of proliferation of social networks. Given that the assemblage exhibits the high level of sociality in their use, beads represented social adornment as much as they were items of personal adornment (Tsoraki, Vasić in preparation). Beads were able to mediate complex notions of personhood and identity, through relations within and outside the site, as well as between the living and the dead.

Acknowledgements

We would like to thank Ian Hodder and all members of the Çatalhöyük Research Project. This chapter would not have been possible without the meticulous recording of the excavators, the human remains team, the finds officers, particularly Lisa Guerre, and the various specialists that worked on the material over the last 25 years, especially Nerissa Russell, Roseleen Bains, Daniella Bar-Yosef Mayer, Karen Wright and Emanuela Christiani. Thanks are also due to Kathryn Killackey and Caroline Hebron for their illustrations and reconstructions and to Ekin Ünal for photographing the artefacts. We are grateful to all the heavy residue sorters who thoroughly went through the samples and pulled out even the smallest fragments of beads, and to all the workmen for all the hours they spent sieving the material. Christina Tsoraki is indebted to Markéta Šťovíčková for her contribution during the on-site analysis of the ground stone material. Funding for Tsoraki's analysis was provided by the European Commission FP7-People Marie-Curie Actions Program (grant agreement 328862-Project CRAFTS) and the Çatalhöyük Research Project. The analysis undertaken by Matilda Siebrecht was only possible thanks to funding provided by the Catherine van Tussenbroek Foundation Grant, the Prehistoric Society Research Grant and the Leids Universiteits Fonds International Study Fund.

Supplementary material

For supplementary material related to this chapter, please visit https://doi.org/10.18866/BIAA/e-14. It comprises colour versions of figures 9.2, 9.3, 9.5, 9.11 and 9.12.

10. The shell artefact assemblage at Neolithic Çatalhöyük

Rena Veropoulidou

The shell artefact assemblage of Çatalhöyük is one of the richest unearthed in Anatolia, even taking into account the large scale of excavations, thus demonstrating their importance in the lives of the Neolithic people. Previous research has recognised the great potential of shell artefacts at Çatalhöyük, in particular shell ornaments, in discussing issues of technology and trade, social and individual identities, and ritual and symbolism (Hamilton 1996; Reese 2005; Bar-Yosef Mayer 2013; Bains et al. 2013; Vasić et al. 2014).

The results from the study of 1,300 shell artefacts from the 2009–2017 excavations are presented here. The term 'shell artefact' denotes shells that bear traces of human modification resulting from the transformation of the raw material or the use of the shell. The assemblage includes beads, pendants, 'serrated' and painted artefacts, expedient tools and unworked shells that were used as containers/palettes for pigments, as well as items in various stages of manufacture (blanks, semi-worked) and extent of use (not or little used, exhausted). The current approach builds on previous work, but also differs by drawing attention to the properties of shell raw materials and the contextual approach of the life cycle of artefacts, including procurement, production, consumption and discard. Furthermore, in the previous excavation seasons (2000–2008), the majority of shell artefacts were found in midden and room deposits (Bar-Yosef Mayer 2013: 330), but in the current excavation seasons, large quantities of shell artefacts were associated with burials and human skeletons, thus allowing discussion of their social, economic and symbolic attributes contextually.

On this basis, the present study seeks to explore the origins, qualities and histories of shell artefacts to open up insights into the appropriation of raw materials by craftspeople and consumers and thus discuss their economic, social and sensorial properties. In addition, the presence of local and non-local shells through the sequence is investigated to shed further light on networks of procurement and exchange, trade and mobility. Another aim is to address artefact production, including methods of manufacture and contexts of production, to obtain an understanding of the social and economic implications of this activity and contribute to the discussion of craft technologies in the settlement. Moreover, emphasis is given to the study of use-wear traces coupled with the investigation of contexts of consumption, deposition and discard of artefacts, especially ornaments, to elucidate the ways and extent of their use in everyday and more special contexts. The fact that the majority of artefacts were recovered from burial contexts, indirectly or directly related to the human body, benefits the discussion of social and individual identities, rituals and symbolism. The ultimate aim is to look at shell ornaments as an integral part of the whole ornament assemblage so as to allow a holistic approach to the organisation of production and consumption of these artefacts at household and community levels.

Materials and methods

Sample size and previous studies

The shell artefact assemblage from the 2009–2017 excavations comprises 1,231 items (including items discussed in Chapter 9). The present report also includes 69 artefacts studied by Bar-Yosef Mayer (2013), which were re-examined for manufacture and use-wear traces. Therefore, 1,300 artefacts are presented here.

To obtain a better understanding of the quantity of shell artefacts recovered from Çatalhöyük and the proportion of this assemblage in comparison to those previously studied, research on publications and the Excavation Database was conducted (table 10.1). Despite the difficulties in dealing with the 1960s excavations and findings, a detailed report on the grave goods is given by Hamilton (1996: 247, table 12.7), who notes 2,666 shell beads (1,344 stratified), the vast majority of which were made from non-local shells. She also mentions some shell pendants and a few pigmented shells, most possibly local shells. This material comes from burial contexts and usually in groups, but it is highly possible that a number of shell artefacts, and especially the minuscule beads, might have been overlooked during excavation, considering the lack of sampling for flotation.

In addition, Reese (2005; personal communication September 2018) discusses ca 1,700 shell artefacts that come from both the 1960s (ca 1,465) and the 1995–1999 surface scrapings and excavations (ca 235). This number does not include some hundreds of scaphopod beads and other artefacts (ca 1,322). More detailed reports come from the 2000–2009 studies, as Bar-Yosef Mayer (2013: 329) discusses 734 artefacts (though 957 artefacts are

Study (East Mound)	Shell artefacts	Comments
Hamilton (1961–1965)	1,201	Unstratified=1,323; plus 142 (unknown); in total 2,666
Reese (1961–1965 and 1995–1999)	1,700	1961–1965 excavated=1,465; 1995–1999 excavated=ca 235
Bar-Yosef Mayer (2000–2008) Bains et al. (2000–2008)	734 (956)	Unstratified=421; including some from Reese's study
Veropoulidou (2009–2017)	1,231	Plus 69 from Bar-Yosef Mayer's study; in total 1,300
Total	4,866 (5,088)	Plus some hundreds of scaphopods not studied

*Table 10.1. Shell artefacts from the East Mound excavations (Hamilton 1996; Reese 2005; Bar-Yosef Mayer 2013; Bains et al. 2014; current study).**

mentioned in Bar-Yosef Mayer 2013: tables 16.1, 16.2), among which a small number had already been studied by Reese. The 2010 database includes, in total, 956 shell artefacts (Bains et al. 2013: 357), of which 421 come from unstratified contexts (Bar-Yosef Mayer 2013: tables 16.1, 16.2). As stated above, the 2009–2017 excavations brought to light at least 1,231 artefacts (see also Veropoulidou 2017); they were recovered in the trenches, by sieving and flotation. Despite the problematic sums and some inconsistencies in the reports, one can assume that the shell artefacts unearthed at Çatalhöyük number ca 4,860 to 5,500–6,000 (table 10.1). The fact that at least 3,665 artefacts (75%) have been studied by a shell specialist provides reliability in taxa identifications, while Bar-Yosef Mayer also provided information on technology, size of beads and contexts of production and discard (Bains et al. 2013: 355–59).

Methods

All shell artefacts discussed here were examined individually under a low magnification stereoscope combining a specific archaeomalacological methodology (Volume 13, Chapter 3) with additional methodologies of shell artefacts studies (e.g., Francis 1982; 1989; Hayes 1989; Miller 1997; Veropoulidou 2011), the aim being to obtain as much archaeomalacological and material culture data as possible. Basic archaeomalacological information includes:

- Taxon identification on the basis of atlases (Poppe, Goto 1991; 1993) and web databases (Worms Editorial Board 2019; AnimalBase Project Group 2005–2019).
- Recording of skeleton part (right, left or undetermined valve or umbo for bivalves; intact, apex, aperture for gastropods and scaphopods) and, in the case of fragments, which part or zone of the shell was used.
- Assessment of the state of shell during collection; that is, whether it is a recent fresh, a recent beach-/water-worn or a fossil shell.

Additional attributes recorded include:

- Typology and morphological characteristics of artefacts.
- Measurement of all dimensions in terms of plan and longitudinal sections for all objects, such as length, width, thickness, and internal and external diameter.
- Measurement of any additional feature (perforations, incisions or projections).
- Manufacturing techniques, such as facets, cuts, shape of perforation, projections, incisions, surface treatment.
- Use history of artefacts, such as primary and secondary use, repairs, types of use-wear.
- Post-depositional variables, such as degree of wear, breakage patterns, trampling, chemical weathering.
- Qualitative attributes; in particular, colour, burning, degree of finish.

All this information was recorded in the Shell Artefact Database, created by the author. All raw data are available online in the Çatalhöyük database (www.catalhoyuk.com).

Raw material selection: local and non-local shells

A large variety of mollusc species (n=19) were selected for the production of shell artefacts (table 10.2). These raw materials are characterised by similarities and differences in terms of sensorial and mechanical properties, but most importantly in terms of origin (table 10.3). The latter variable is selected here as a tangible property through which to build discussion and explore issues related to differential access to raw materials, as well as networks of exchange within and beyond the site. Hence, on the basis of their origin, the shell raw materials are divided into two large groups: local and non-local (exotica) shells.

Local shells. The local shell group comprises five different freshwater species: the *Unio* sp. bivalve, two large gastropods (*Lymnaea* sp., *Viviparus* sp.) and two gastropods of small size and rather similar appearance (*Theodoxus heldreichi*, *Theodoxus anatolicus*) (table

Species	*Beads*	*Pendants*	*Other types*	*Production*	*Total*	*%*
Antalis dentalis	204			2	206	15.8
Ostrea edulis			1		1	0.1
Antalis inaequicostata	37			1	38	2.9
Antalis vulgaris	104			1	105	8.1
Antalis sp.	171				171	13.2
Cerastoderma glaucum	1			1	2	0.2
Columbella rustica	5				5	0.4
Conus mediterraneus			1		1	0.1
Fustiaria sp.	267				267	20.5
Luria lurida			1	1	2	0.2
Nassarius gibbosulus	182			1	183	14.1
Patella caerulea		1			1	0.1
Ranella olearium			1		1	0.1
Spondylus gaederopus		1			1	0.1
Dentalium sp.	187			1	188	14.5
Unio sp.	13	13	22	15	63	4.8
Theodoxus heldreichi	56				56	4.3
Theodoxus anatolicus	4				4	0.3
Viviparus sp.	2	1			3	0.2
Lymnaea sp.		2			2	0.2
Total	1,233	18	26	23	1,300	100.0

Table 10.2. Shell raw materials and types of artefacts.

Origin	*Species*	*Beads*	*Pendant*	*Other types*	*Production*	*Total*	*%*
Aegean	2	204		1	2	207	15.9
Aegean and Mediterranean	12	767	2	3	5	777	59.8
Hatay and İskenderun	1	187			1	188	14.5
Local	5	75	16	22	15	128	9.8
Total		1,233	18	26	23	1,300	100.0

Table 10.3. Origin of shell raw materials and types of artefacts.

10.2; fig. 10.1). Among local species, the most commonly used are *Unio* and *Th. heldreichi*. These are also among the most common species encountered in large quantities, especially the *Unio*, in the archaeomalacological assemblage of the site and in the natural environment (Volume 13, Chapter 3). Despite, however, the proximity to readily available freshwater raw materials, the number of artefacts made of local shells is particularly low (9.8% of total amount, n=128). A similar taxa representation and frequency of artefacts made of local shells was observed by both Reese (2005) and Bar-Yosef Mayer (2013).

Unio, *Viviparus* and *Lymnaea* were turned into artefacts by drilling, flaking, grinding and painting. On the other hand, the majority of *Theodoxus* were collected with a natural opening (a naturally abraded apex due to rolling on sand), but the traces identified at the periphery of the perforation (gloss, wear or notches) point towards wear from the string used for suspension. A number of local shell artefacts were deposited in burials, indicating that this material had some symbolic connotations and could accompany the dead. Nevertheless, the level of exploitation of the abundant local shells for artefact manufacture and consumption is particularly low, thus

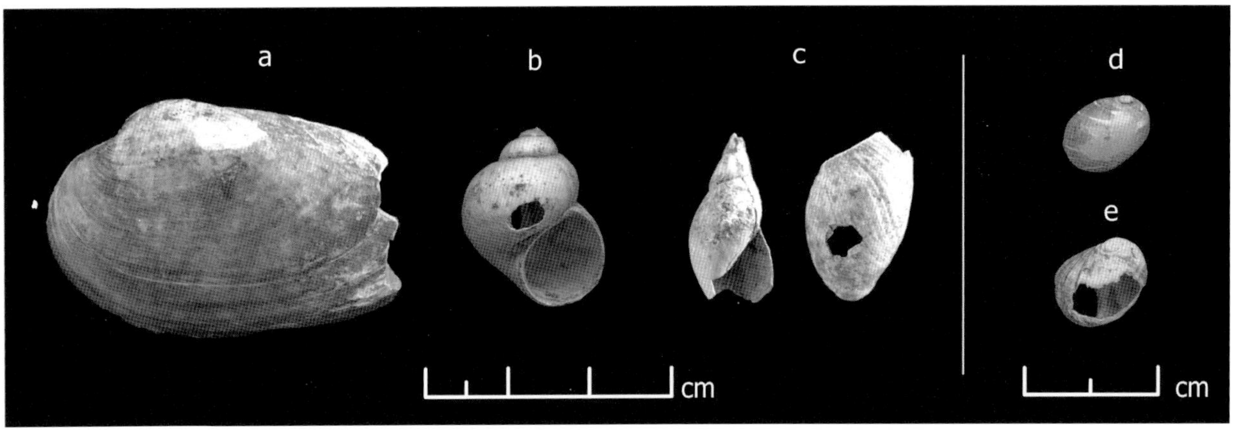

Figure 10.1. A selection of local shell artefacts: (a) Unio *sp., (b)* Viviparus *sp., (c)* Lymnaea *sp., (d)* Th. heldreichi, *(e)* Th. anatolicus *(photographs b, d and e by Jason Quinlan; a and c by Rena Veropoulidou) (for colour version, see online supplementary material).*

demonstrating the differences in the perceptions and appropriateness of raw materials. A similar pattern has been observed in other Neolithic societies (e.g., Baysal 2016a; Veropoulidou 2014; in press) and could be explained on functional and non-functional (aesthetic, economic and symbolic) grounds.

Non-local shells. The non-local shell artefact group is the richest in terms of number (n=1,172) and variety. It comprises recent marine and fossil shells, indicating three different places of origin and thus participation in different networks of communication and exchange.

Recent marine shells. There is a large variety of recent marine shells (n=14), of which the vast majority originate from the Aegean and the Mediterranean seas of Turkey (n=12), followed by two species, *Antalis dentalis* and *Ostrea edulis* which live only on the Aegean coasts (tables 10.2, 10.3). According to studies of the biogeography of modern molluscs (Demir 2003; Öztürk 2011), the marine shells in this assemblage can be found ca 150–400km distant in a straight line from Çatalhöyük to the Mediterranean and the Aegean respectively. There are no Red Sea mollusc species in this assemblage, though Reese (2005) reports six items from the 1961–1965 and 1995–1999 excavations. A similar species variety is reported in Natufian sites and in Epipalaeolithic and Neolithic sites in Turkey (Bar-Yosef Mayer 2005; Baysal 2017a; Baysal, Erek 2018; Yelozer 2018). This seems to reflect a continuation of a Palaeolithic tradition, and according to Bar-Yosef Mayer (2013: 331) these species and artefact types may indicate connections with the Levant.

The preferred marine molluscs are four congeneric species of scaphopods, the *Antalis* group (tables 10.2, 10.3; fig. 10.2a–b): *Antalis dentalis*, *A. vulgaris*, *A. ineaquicostata*, *Antalis* sp. (for those not able to be identified to species level) and *Fustiaria* sp. cf. *rubescens*. The *Antalis* species are small and medium in size (10 30mm in height), with a conical shell open at both sides and slightly curved. They have a Mediterranean origin (apart from *A. dentalis*, which lives only in the Aegean Sea) and similar functional properties, but also some minor visual differences, for instance, in the diameter of the intact shell, the colours (whitish, yellowish, pinkish, brownish) and the presence of sculpture or not (ribs and striations or a smooth surface). The subtle differences, especially in artefacts, and the fact that they are usually found mixed in the same composite ornaments or in the same burial fill suggests that people from Çatalhöyük did not discriminate between these species.

Completely different in both visual and mechanical properties is the second most common group of marine shells, the *Nassarius gibbosulus* (current name: *Tritia gibbosula*) (table 10.2; fig. 10.2d). This is a small marine gastropod with a thick, ovate and gibbous last whorl of orange, pink, brown or olive colour. Shells from the *Nassarius* genus are first used as beads from the Upper Palaeolithic in Africa (D'Errico et al. 2009), a practice that continued in the Palaeolithic and Early Neolithic of the Levant and Turkey (Bar-Yosef Mayer 2005; Baysal 2017a; Yelozer 2018).

The visual and mechanical properties of the remaining 14 artefacts, made from eight different marine species, have similarities (*Columbella rustica*, *Conus mediterraneus*) and differences (*Cerastoderma glaucum*, *Ostrea edulis*, *Spondylus gaederopus*, *Ranella olearium*, *Patella caerulea*, *Luria lurida*) (table 10.2; fig. 10.2e–l). Despite the fact that they originate from the same sources as the *Antalis* and *Nassarius* groups, the variety of raw materials and the rarity or uniqueness of these artefacts may have been important for the consumers.

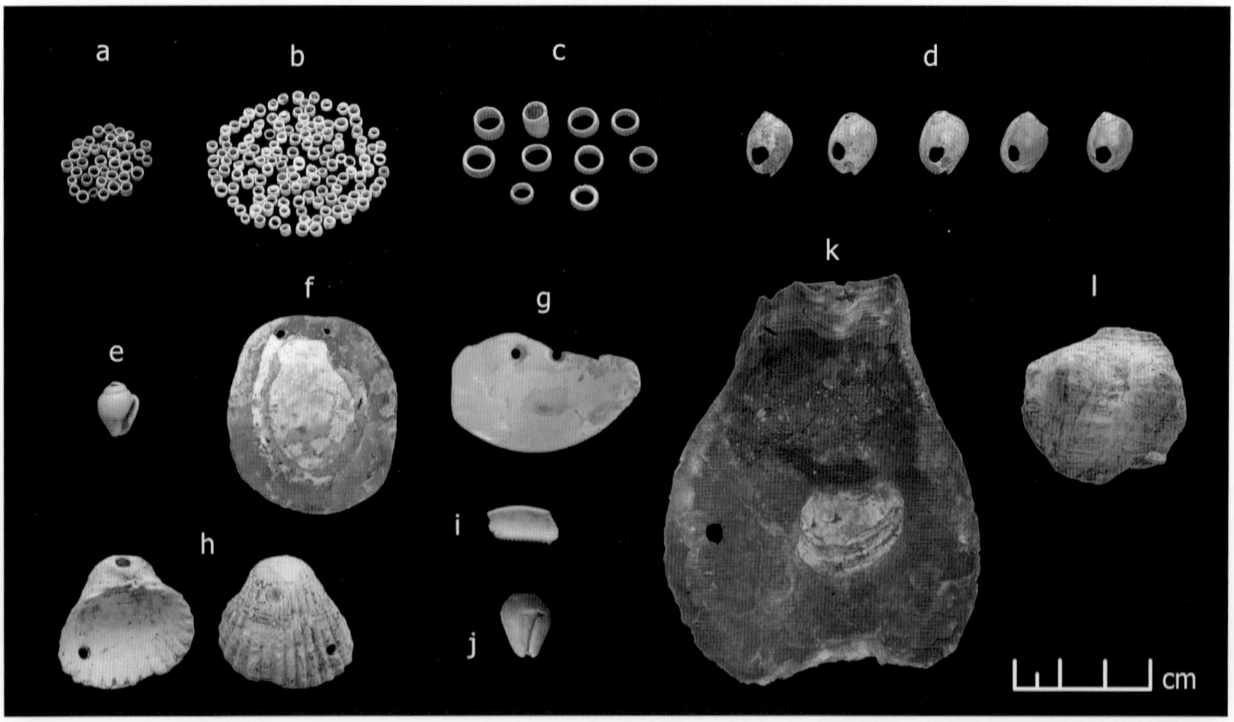

Figure 10.2. Examples of artefacts made from marine shells: (a) Antalis, *(b)* Fustiaria, *(c)* Dentalium, *(d)* Nassarius, *(e)* C. rustica, *(f)* P. caerulea, *(g)* Sp. gaederopus, *(h)* C. glaucum, *(i)* L. lurida, *(j)* C. mediterraneus, *(k)* O. edulis, *(l)* R. olearium *(photographs a, b, c, d, e, h, i, k and l by Jason Quinlan; f, g and j by Didem Turan) (for colour version, see online supplementary material).*

As regards molluscan ecology, these species live at rocky (*Sp. gaederopus*, *P. caerulea*, *R. olearium*) and sandy substrata (*Antalis* group, *C. glaucum*, *L. lurida*), or in between (*C. rustica*, *N. gibbosulus*, *C. mediterraneus*, *O. edulis*), from full marine to estuarine aquatic conditions. The assemblage includes species that live near the coast but also species from deeper water zones, like, for instance, the scaphopods (in depths from 4 to 100m). The range of ecological niches represented implies the familiarity of collectors with the seascapes and especially their intention to select a particular variety of mollusc species out of the large variety available in the natural habitats.

Two questions arise: whether the procurement of these species was an easy task or required experience and labour input, and whether the state of shell during collection was important for the visual and mechanical properties of raw material and subsequently of the artefact (Miller 1997; Veropoulidou 2011: 77–84). The vast majority of shells (91.2%, n=897) bear evidence of post-mortem wear; in particular, loss of gloss, some bio-erosion and smoothed surfaces due to rolling on sand. In other words, these shells were collected empty from the coast and the beach, an activity that could be performed by any individual with no equipment at all. However, the fact that most of these shells still bear their natural colours and decoration implies that they were carefully selected from the coast and the beach according to some standards.

There are only a small number of items (7.3%, n=72) made from species that live in deep water zones and do not bear any, at least visible, evidence of post-mortem wear. If indeed these shells were collected fresh from their original habitat, this task would require some equipment (boats, tools), experienced divers and different organisation in terms of time and labour input compared to the 'easy' task of beachcombing, notwithstanding possible territorial and resource rights (Veropoulidou 2011: 465). The knowledge and danger that such maritime activities entailed, coupled with the significance of the desired raw materials, are the main reasons put forward to explain the performance of rituals required in such expeditions (Malinowski 1922). Apart from the enhanced functional properties of the freshly collected shell material – for instance, moisture that prevents breakage, brighter colours and gloss – elaborate maritime activities may have added to the value and symbolism of shells (e.g., Safer, Gill 1982).

Taking into consideration the long distance of the site from the Aegean and Mediterranean coasts, coupled with the evidence for exploitation of different ecological niches for the selection of particular shells, it seems plausible to suggest that people from Çatalhöyük acquired these raw materials (or finished artefacts) through various types of networks of communication and exchange, rather than by visiting the sea coast. Meanwhile, the taxa represented (especially *Antalis* and *Nassarius*, less so *Columbella*) and types of shell artefacts that have affinities with Palaeolithic traditions in Turkey and the Levant (Bar-Yosef Mayer 2005: 2013) may be taken to suggest the participation of Çatalhöyük people in effective and extended networks (*sensu* Gamble 1998) encompassing shorter and longer distances in various directions. The evidence for participation in these networks first dates from Level South G (Sp. 181), with a small number of shell artefacts, and steadily increases in the later levels.

Fossil shells. The last group of non-local shells comprises the fossil *Dentalium* sp. cf. *sexangulum*, which is represented by a significant number of items (tables 10.2, 10.3; fig. 10.2c). This is a scaphopod, bearing many similarities to the recent marine *Antalis* (shape), but differing in colour (grey) and size (larger). It must also have had different mechanical properties, as fossil shells, having lost moisture and elasticity, are more difficult to work (Miller 1997). There is also a *Turritela* sp. and another fragment of an undetermined gastropod in the assemblage, but they show no evidence of human modification and are not further discussed in this chapter. Bar-Yosef Mayer and colleagues (2010) have investigated the origin of fossil shells at Çatalhöyük and convincingly suggested that *Turritela* fossils (as well as other items found in the previous excavations) must originate from the Taurus Mountains (at least ca 50km away), while *Dentalium* fossils come from the Hatay and İskenderun basins (more than 300–400km away). The evidence for participation in these networks begins in Level South I (Sp.198) and South J (B.23), with individual finds, and steadily increases in the late levels of habitation.

Typology, production and use of shell artefacts

This section discusses the types of shell artefacts encountered, providing information about their characteristics, methods of manufacture and use-wear traces. For consistency with the previous publications of the Çatalhöyük shell artefacts and other sites in Turkey (Baysal 2017a; Yelozer 2018), the present study follows a similar typology.

The category 'beads' includes all small artefacts bearing a natural or human made perforation, regardless of whether the shell preserves its natural shape and characteristics (*Nassarius*, *Columbella*, *Conus*) or if it is completely transformed (*Antalis*, *Dentalium*). In terms of bead typology, the seminal works of Francis (1982; 1989) and Hayes (1989) have been followed. The term 'pendants' refers to large-in-size artefacts (preserving the original shell shape or not) with single or double perforations that could have been used either individually or in the centre of a composite artefact. The category 'palettes' refers to the *Unio* and *Ostrea* unworked valves that contain traces of pigment, while 'painted' shells refer to examples with traces of coloured decoration. 'Serrated' artefacts are parts of *Unio* valves with a rectangular, trapezoid or triangular shape and bearing serrations on one edge. There are also some expedient tools made from *Unio* shells, with traces resulting from use.

The discovery of large numbers of shell ornaments in burials, and in many cases on the skeletons, gives information on how shell beads were used, individually or in strings, in various arrangements on the human body, as body or cloth decoration. It also allows insights into the patterns of ornament use in relation to gender and age (see also Chapter 9).

Antalis *beads*

Antalis beads comprise the largest group of shell artefacts (60.5%, n=787). Apart from four items that represent preforms and manufacturing waste, all others are finished artefacts (table 10.2; figs 10.2a, 10.2b, 10.3). The main manufacturing technique was to saw the long cylindrical shell into thin slices (89.5%, n=701). Evidence for the combination of sawing and grinding at the posterior and anterior end of the shell or its wall was found in a few artefacts (10.5%, n=82). Sawing resulted in the 'ring-like' bead type; that is, a short thin-walled bead. The vast majority of the *Antalis* group (99.3%, n=777) measures on average 1–2mm in height, 0.4–2mm in internal diameter and 1–3mm in external diameter (table 10.4; fig. 10.4). Only a few examples belong to the 'short barrel' (n=8) and 'long barrel' types (n=2), measuring in length on average 3.8–7.1mm and 8.7–19.6mm respectively. Therefore, beads were created to be similar in size and appearance.

A question that arises concerns the amount of raw material used for the manufacturing of 787 beads. Considering the average height of an intact *Antalis* shell (10–30mm) and the average height of a 'ring-like' *Antalis* bead (1–2mm), then 60–100 shells would have been needed. Taking into consideration the long time span of habitation, it seems plausible to suggest that the amount of raw material used was not very large, and definitely smaller than that used for *Nassarius* beads. This could be taken to imply a limited procurement and consumption of *Antalis* beads or an economising behaviour towards this raw material, which in turn may have added more value to these artefacts.

The *Antalis* beads are usually well preserved and complete (84.4%, n=661). However, there is a moderate quantity (15.6%, n=122) preserved to various degrees (39 items are >50%; 29 items are 25–50%; 54 items are 5–25%) that allows a better understanding of the different types of use-wear traces (fig. 10.5). The first type relates

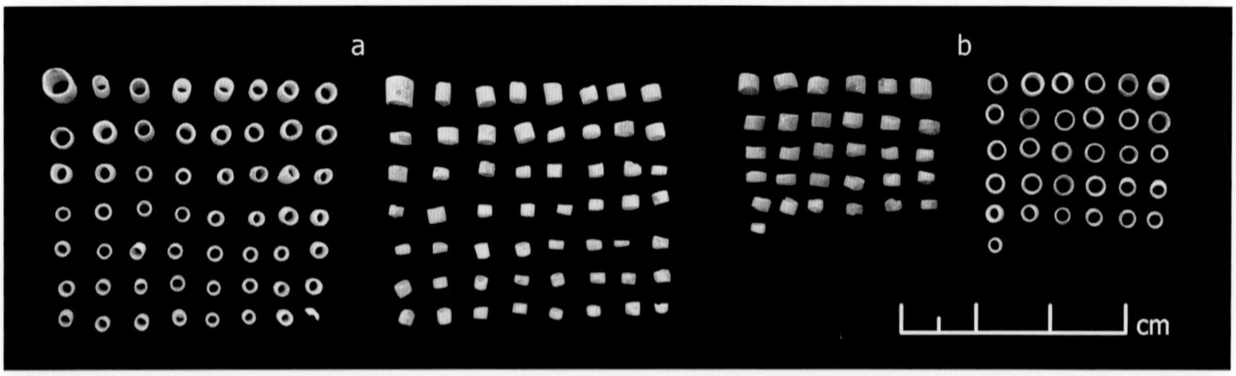

Figure 10.3. Examples of Antalis *beads: (a) F.7713 (22619), fill from infant burial; (b) F.7962 (22678), Sk31705 (female and other adolescent?) (photographs by Didem Turan) (for colour version, see online supplementary material).*

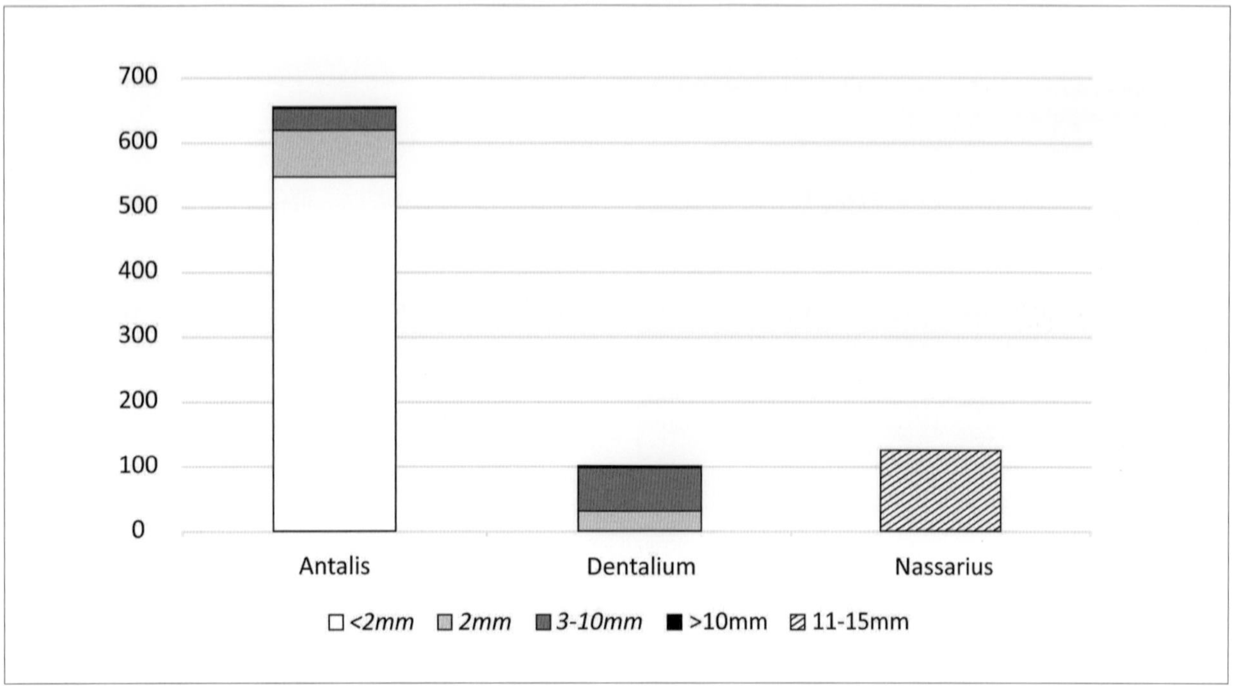

Figure 10.4. Height distribution of Antalis, Dentalium *and* Nassarius *beads.*

to the thinning of one part of the vertical wall of the shell, usually coupled with the presence of notches at the posterior and anterior ends. These traces must have resulted from the pressure and rubbing of a string and the continuous contact of one wall of the bead with a surface. Therefore, it seems that these beads were possibly sewn onto a textile, as evidenced also by the finding of groups of beads in relation to human bodies. Notably, there are items where part of their wall has been completely ground down (recorded as partially preserved beads), indicating exhaustion of the raw material due to prolonged rubbing against a surface. A second common type of use-wear is the patchy traces of gloss on various spots of the walls of the bead, coupled with slight notches/incisions at spots around the periphery of the posterior and the anterior end of the bead. These traces suggest the use of a string for multiple beads to form a composite ornament, as evidenced also from the finding of strings of beads in relation to ankles, wrists and other parts of the body. There are even beads with a very small diameter that were found interlocked with larger beads, corroborating the suggestion that some beads were loosely strung together.

Despite the fact that there are only four items that can be related to the manufacture of beads, there are beads with incisions and particular breakage patterns that can be related to pre-forms and manufacturing waste. One example was recovered from a domestic context (floor in B.89, Level South N), thus providing one of the very few pieces of evidence for on-site production, or at least repair, of beads. Interestingly, the remaining three examples of manufacturing waste were found in burials (skeleton (21841), B.17, South K; F.7977, B.131, Level North G).

	Height			Diameter - internal		Diameter - external	
Measurements	Antalis	Dentalium	Nassarius	Antalis	Dentalium	Antalis	Dentalium
>2mm	548	4		381	1	226	
2.1–3mm	72	14		151	3	306	2
3.1–4mm	20	15		16	24	8	4
4.1–5mm	9	15		6	64	29	16
5.1–6mm	8	11					62
6.1–10mm	3	36					
10.1–11mm		1					
11.1–12mm		2	4				
12.1–13mm		1	23				
13.1–14mm			48				
>14mm	1		51				
Total	661	99	126	554	92	569	84

Table 10.4. Measurements of height, internal diameter and external diameter of a large sample of the Antalis *and* Dentalium *beads, as well of the height of a large sample of the* Nassarius *beads.*

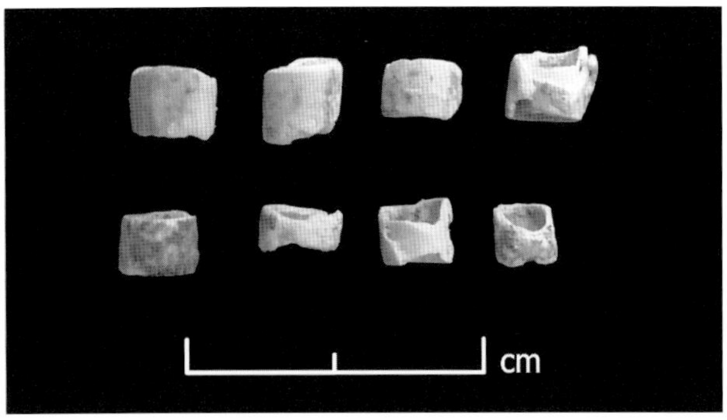

Figure 10.5. Use-traces on A. vulgaris *beads (22623) (photographs by Didem Turan) (for colour version, see online supplementary material).*

Therefore, the study of use-traces on *Antalis* beads suggests, on the one hand, different ways of using and displaying these artefacts and, on the other hand, prolonged use that leads to their exhaustion. Of particular interest is the co-existence of beads bearing different degrees and types of wear (new, little used, exhausted) in the same composite ornament decorating the dead human body. On this basis, the complex life histories of *Antalis* beads before being discarded or placed into burials are implied. Even exhausted material and manufacturing waste was deposited in burials, presumably being highly significant.

Dentalium *beads*

The *Dentalium* bead group is the second most popular (14.5%) in the shell artefact assemblage, with 187 beads and one item with evidence of slicing (incision) on its wall (table 10.2; figs 10.2c, 10.6). The state of preservation of *Dentalium* beads is moderate to good, as most (n=143) are 100% complete. Those remaining are preserved in various percentages (n=14 are 60%; n=31 are 25–50%). Almost all *Dentalium* beads belong to the 'ring-like' type, measuring on average 4.5–7mm in height, 3.6–8.8mm in internal diameter and 3.5–9mm in external diameter (table 10.4; fig. 10.4). There are a few examples that belong to short (n=4) and long barrel types (n=1), measuring in length on average 4–13mm. Therefore, the *Dentalium* bead group is of a significantly larger size than the *Antalis* group (figs 10.3, 10.4, 10.6). Other visual differences between the two groups include colour, as *Dentalium* is greyish, and sculpture, with ribbed *Antalis* and smooth *Antalis* and *Dentalium*.

Dentalium beads are also the result of slicing the elongate shell (n=139) and of combined slicing and grinding (n=49). An incision on one item provides clear evidence for sawing (fig. 10.6a), indicating that some production or repairs took place on site. As with the *Antalis* bead group, at least half (n=94) of the fossil beads bear similar use-wear traces due to prolonged contact with a surface and possibly loose stitching. These examples have thinner walls and grooves on the posterior and anterior ends of beads.

Therefore, *Dentalium* beads were manufactured and displayed in similar ways as *Antalis* beads. However, the distribution of *Dentalium* beads in space differs significantly from the distribution of the *Antalis* group (see below). Taking into account the similarities in the ways they were used, it seems plausible to suggest that each material was categorised on the basis of origin, and thus different networks of procurement. This differentiation could also imply different perceptions and symbolisms for each group of scaphopod shells.

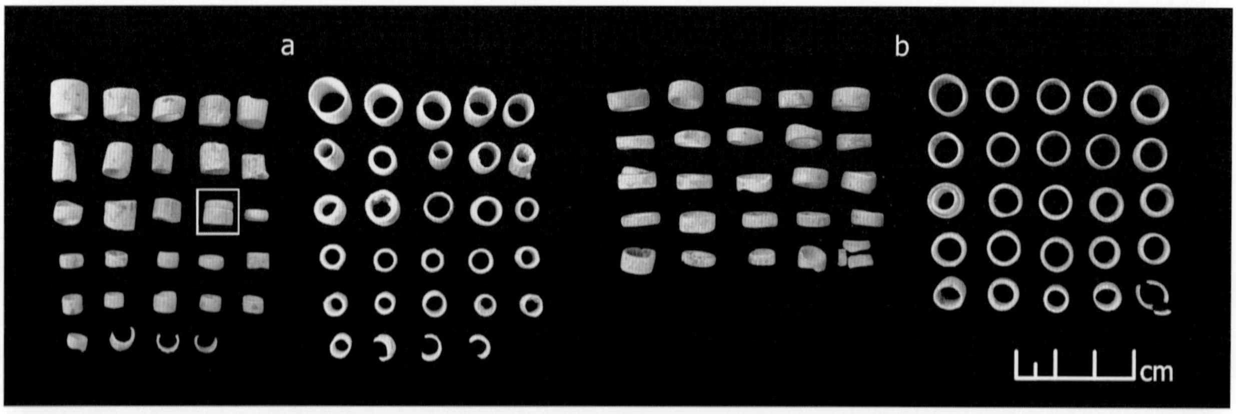

Figure 10.6. Examples of Dentalium *beads: (a) F.3479 (19879), burial fill (one infant, one older infant, one adult), bead with incision in white rectangle; (b) F.7133 (20989), fill of adult male skeleton (20685) (photographs by Didem Turan) (for colour version, see online supplementary material).*

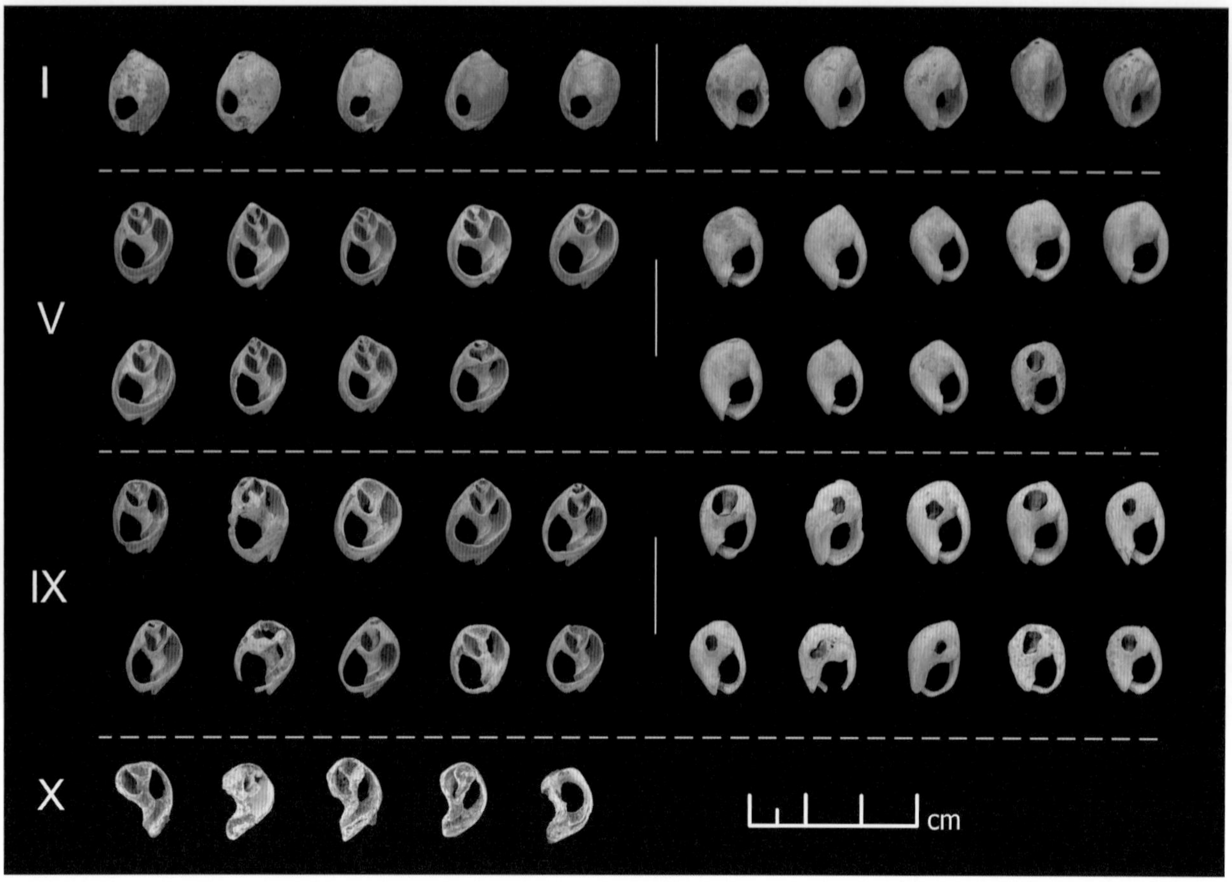

Figure 10.7. Examples of different use-traces of Nassarius *beads: categories I, V, IX and X (photographs by Jason Quinlan) (for colour version, see online supplementary material).*

Nassarius *beads*

The *Nassarius* beads belong to the third most popular group (14.1%) in the shell artefact assemblage (table 10.2; fig. 10.2d). A large number are preserved in good condition, complete (n=126) or nearly complete (n=6), followed by highly worn items that only partially preserve the outer lip and the siphonal canal (n=51) (fig. 10.7). In contrast to scaphopod beads, *Nassarius* are large in size (table 10.4; fig. 10.4) and retain their natural characteristics.

Though the vast majority of *Nassarius* shells were collected empty from the beach, they were in good condition overall and retained their colour and sometimes their gloss. The shell was transformed into a bead though a natural or human-made perforation either at the upper or the gibbous last whorl, created due to rolling on the beach, and with hammering or gouging respectively. The removal of the gibbous whorl in various degrees, the rubbing of the slightly protruding ventral area that results in unpatterned striations, notches, furrows and ultimately perforations and extensive wear on the outer lip are among the most commonly recorded wear traces. The openings on the last whorl bear scratches and notches at several spots, thus implying the use of a thread to facilitate suspension and stitching of shells together. To obtain a better idea of how these traces were created, of the sequence of events and temporalities entailed, the traces were categorised into ten different categories according to the extent of wear (fig. 10.7) (for similar traces, see D'Errico et al. 2009). In this scheme, Category I comprises items with the lowest degree of wear; that is, bearing a natural or artificial perforation, but no other wear. Categories II to IX comprise items with increasing removal of the last whorl and rubbing of the ventral area that results in smaller or larger perforations; in other words, artefacts with different extents of use. Lastly, Category X comprises partially preserved beads, showing total exhaustion of the raw material.

Therefore, it seems that *Nassarius* were used as interlocking beads for extended periods of time, a practice that led to the rubbing of all protruding spots that were in direct contact with each other. The prolonged use of interlocking shell beads resulted in flattening the shape of the shell and exposing the columella, the axis of the shell which is normally invisible. Interestingly, there are a few *Nassarius* beads (n=8) in pristine condition, but flattened artificially from both the lateral and the ventral side through grinding, as vertical and horizontal striations are visible on both surfaces. In other words, shells in pristine condition and without evidence of being used (new) were ground down to imitate those beads that were flattened and worn due to extensive use (old).

In a nutshell, the study of use-wear indicates shorter and longer periods of use and maybe also different ways of assembling the beads in composite ornaments. The use of *Nassarius* beads for various periods of time points towards the creation of biographies and complex histories for these artefacts. The importance of long-used artefacts is demonstrated by the production of imitations, indicating cognitive acts of staged use/deposition of beads.

Other marine shell artefacts

The remaining 14 marine shell artefacts were made from eight different species (table 10.2; fig. 10.2e–l). Shells were used intact with minimum human intervention (*C. rustica*, *P. caerulea*, *C. glaucum* and *O. edulis*), or parts of shells were transformed through various techniques (*Sp. gaederopus, L. lurida*, *C. mediterraneus* and *R. olearium*).

There are some *C. rustica* and *C. mediterraneus* beads (figs 10.2e, 10.2j) that visually resemble (shape, colour, dimensions) the *Nassarius* shells. However, they bear completely different perforations and use-wear traces. Most *C. rustica* beads (n=4) bear a perforation created from the apex down to the columella. The conical shape of the perforation and the concentric striations on its walls indicate that it was made with a drill. Another item bears evidence for grinding the whole upper whorl and then drilling through the columella. The periphery of all perforations is slightly smoothed and glossy, thus suggesting contact with a soft tissue. A vertical perforation from the apex to the aperture would facilitate suspension of the shell in a horizontal position.

There are three more pendants in the assemblage (*Sp. gaederopus*, *P. caerulea*, *C. glaucum*). The *C. glaucum* pendant bears a single square perforation on the centre of the shell, made by gouging; there is no transformation of its natural characteristics. The *P. caerulea* (fig. 10.2f) pendant bears two perforations near the lip, made with a drill. Its outer surface has been ground down but preserves its natural brown colour and original conical shape. In contrast, the *Sp. gaederopus* valve (fig. 10.2g) has been ground down to acquire an oval shape, eliminating its natural characteristics (pink colour, spines and sculpture). Two perforations were drilled from the internal and the external surfaces, as suggested by their bi-conical shape, and are located at the long side of the artefact. Gloss and some notches on the upper part of the perforations indicate the use of a thread for suspension.

The only evidence for the production of perforations on non-local shells on site is a *C. glaucum* valve (fig. 10.2h) that bears a partially opened perforation (drilling), striations on the concave part of the umbo (grinding), a finished perforation created by drilling at the posterior ventral part and a finished perforation created using a combination of grinding and gouging at the umbo. The latter is a good example of on-site experimentation with the handling and working of non-local shell material.

Moreover, there are a few 'elongate' artefacts made from the lip of *L. lurida* (fig. 10.2i) and *C. mediterraneus* that resemble the enigmatic 'cassid' lips; that is, the outer curved lips of helmet shells (*Cassidae*) that bear evidence of being carefully removed from the rest of the shell and are a common find throughout the Eastern Mediterranean and Levant from the Epipalaeolithic to the Iron Age (Reese 1989). The items from Çatalhöyük were cut from the last whorl of the shell and slightly ground at their ends. There are no perforations, incisions or other traces on their surfaces to allow insights into how they were used, while their contexts of finding (Sp.489 and Sp.610, middens) do not allow further suggestions.

The last two marine species are an *Ostrea edulis* valve (fig. 10.2k) and a part of the gastropod *Ranella olearium* (fig. 10.2l) that is encountered at Çatalhöyük for the first time. The former bears patches of a dark red colour all over its internal surface, as well as a neat perforation (created with drilling) at the ventral side that hardly bears any evidence of use. Considering the patches and the traces of pigment, it is suggested that it was used as a 'palette' for pigments. The *R. olearium* is part of the last whorl of the shell bearing two rounded tubercles of natural decoration, which, however, have been enhanced and given a slight conical shape by grinding and polishing. In addition, traces of a dark yellow pigment were recognised on the tips of the projections, as well as extremely thin vertical and horizontal black lines present all over the surface.

Unio *artefacts*

This assemblage includes worked shells and items in several stages of manufacture, such as beads (n=13) and pendants (n=12), 'serrated' items (n=8), 'palettes' or containers for pigments (n=14), an expedient tool (burnisher?) and items in various stages of manufacture and use (n=15) (table 10.2; figs 10.1a, 10.8). It must be noted that 50 items are reported here for the first time, while 13 more were previously reported by Bar-Yosef Mayer (2013: 333), but re-examined for evidence of manufacture and use-wear.

Unio *beads and pendants.* The *Unio* ornament assemblage comprises beads (n=13) (fig. 10.8a) and beads/pendants (n=12) (fig 10.8b, c, d), as well as a number of blanks, semi-worked items and production waste (fig. 10.8e), thus providing solid evidence for the manufacture of *Unio* artefacts on site.

Unio shells were used in two ways for the manufacture of ornaments: as valves with a perforation and as raw material that was transformed in several ways. In the first case, the valve was perforated by drilling, grinding or gouging, usually near the umbo (fig. 10.8c, d) or at the centre of the dorsal part. In the second case, the valve was ground and sometimes polished to give the shell the desired shape (triangular, rectangular, oval) and then it was drilled from the nacreous inside towards the calcite outside (fig. 10.8a, b). The conical shape of perforations and the concentric striations around their walls point towards the use of a hand, pump or bow drill. There are examples bearing concentric striations, notches and furrows on the periphery of the perforation, thus indicating the combination of both gouging and drilling. Drilling is a more elaborate and time-consuming technique, producing circular and neat perforations. On the other hand, grinding down the concave part near the umbo is the easiest and least time-consuming method to acquire a perforation for suspension. The occurrence of different techniques indicates good knowledge of the mechanical properties of *Unio* shell and conscious choices in the process of working this raw material, but maybe also the co-existence of craft specialisation and expedient production.

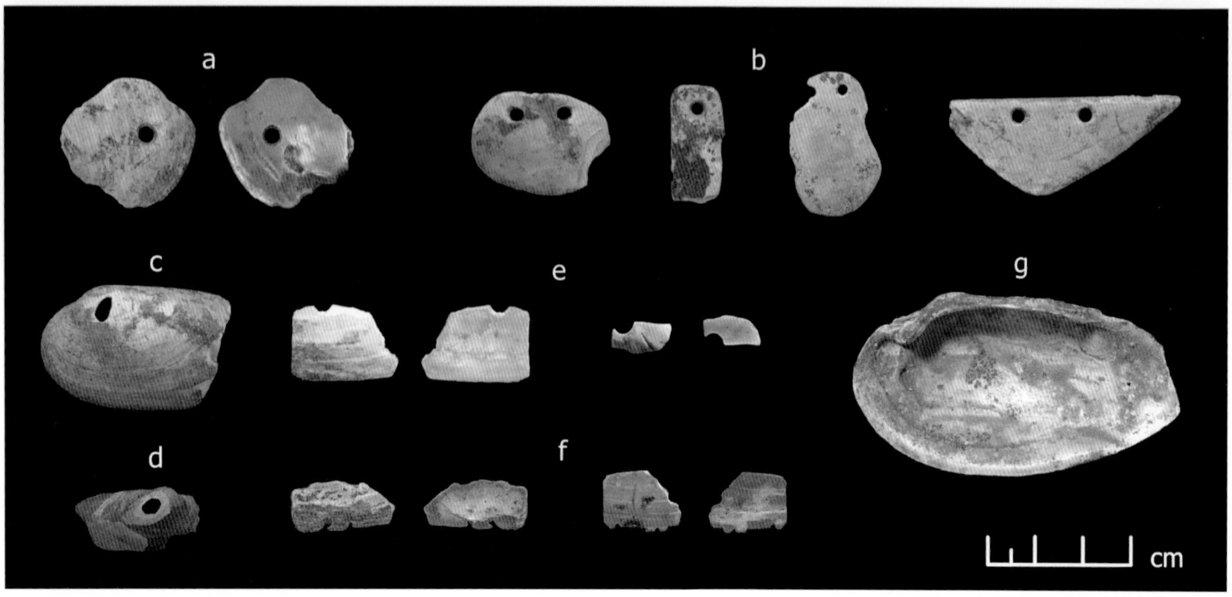

Figure 10.8. Examples of Unio *artefacts: (a) bead; (b) beads/pendants of various shapes; (c) valve perforated with gouging; (d) valve perforated with grinding; (e) production waste; (f) 'serrated' artefacts; (g) 'palette' for pigments (photographs a–e, h–l by Jason Quinlan; photograph f by Didem Turan; photograph g by Rena Veropoulidou) (for colour version, see online supplementary material).*

The ways *Unio* beads and beads/pendants were used is envisaged on the basis of ornaments found in direct association with a skeleton. One significant example comes from child skeleton (21681) (B.77, Level North G) with five *Unio* ornaments, of which one was near the neck and one under the left shoulder. These shell beads, along with a larger quantity of stone beads and one copper bead, were suspended either from several strings that were placed on particular parts of the body or from one single thread that was bound around the body. Similar positions of *Unio* beads/pendants (ankles, wrists and neck) are reported for an adult skeleton (22620) (B.77, Level North G).

In cases where *Unio* shells were not directly associated with a skeleton, the study of use traces proved particularly useful. For instance, there is one item (22676) with intense gloss at the outer periphery of the perforation, possibly due to a soft thread for suspension, and two items (21622, 31585) with scratches at the outer periphery of the perforation that can be related to a hard thread for suspension. These *Unio* artefacts were probably strung free, like pendants. There is also another item (21655) with a rather deep (diameter: 2mm; width: 2mm) incision, which indicates that the thread was either bound around the perforation to keep the shell steady and in position or used to sew the shell onto a textile. Based on these examples, it seems plausible to suggest that *Unio* beads/pendants were both suspended loosely and sewn onto panels.

Unio *serrated artefacts*. The term 'serrated' is used to characterise the saw-toothed pattern at one side of the valve (fig. 10.8f). The current study reports seven new serrated artefacts, thus increasing the number on site to 14 (Bar-Yosef Mayer 2013: 336). One previously reported (17457) was re-studied for use-wear traces. These artefacts are triangular, trapezoid or rectangular in shape, measuring 14.1–21.3mm in height and 10.2–20mm in width. They are cut-offs from the ventral area of the valve and the lip was sawed. The serrations have a semi-rounded end and are located 1–2.6mm apart. The length and width of the serrations is uniform in most artefacts, ranging from 0.8 to 2mm, but it is not always standardised. The study of the serrations under the stereoscope revealed the presence of directional striations and scratch marks on the sides of serrations, suggesting the use of a tool with a sharp edge to saw them. Therefore, the manufacture of these artefacts included a combination of techniques, namely cutting, grinding and finally sawing.

Determining how these artefacts were used, if at all, is not straightforward. Use-wear was observed in three artefacts and comprises mainly gloss, but also slight scratch marks and very small flakes at the pointed end of each serration. Considering the hardness of *Unio* shell material (3-4 in Mohs Scale) and the small size of the serrations, these artefacts may have been used as tools in the processing of leather, straps, cord, flax and other soft materials. In this vein, Bar-Yosef Mayer (2013: 336) suggested that they could have been used in pottery decoration or as 'fish-scalers', but the latter interpretation does not agree with the fish bone analysis, which suggested that the small size of fish at Çatalhöyük meant de-scaling was not required (Van Neer et al. 2013). There are also some (n=4) artefacts that do not bear any evidence of wear on the serrations, indicating that they must have had a decorative function. Similar examples in some Aegean Neolithic sites have been interpreted as imitations of the spines of fish (Kyparrisi-Apostolika 2001: 62, 113). An interesting example is a 'serrated' item with no use-wear, but with slight traces of a red pigment on its inner side and on the serrations, that was accompanied by three *Unio* 'palettes' (infant skeleton (17457), B.49, Level North G). The association of these artefacts may be taken to imply the use of 'serrated' artefacts as stamps for creating 'serrated' patterns on the body or clothes.

As regards the contexts of finds, four serrated artefacts without any use-wear were found in middens: three (32692) in Sp.620 (Level South ?I) and one (22314) in Sp.559 (Level South L). In addition, two serrated artefacts with use traces accompanied a young adult male Sk21636 (Sp.602, Level North G) and one the infant skeleton (17457) mentioned previously. There is a change in the context of deposition between the Early and Middle periods, which could imply a change in the use and perception of these artefacts.

Unio *palettes for pigments*. A moderate number (n=14) of *Unio* valves are categorised as containers or palettes for pigments (fig. 10.8g). The term 'palette' denotes a tool used to lay, mix and hold pigments. The nacreous and nonporous surface of the *Unio* shell and its shallow concave shape make it a perfect container and surface for this task.

The basic characteristic of these artefacts is the presence of stains, coating or lumps of different pigments with yellow, orange and vivid red hues (ochre and cinnabar) in the concave, inner side of the valve. These traces are usually lighter and thinner at the central part of the valve, but thicker and more substantial near the concave part, while in some examples there are brush strokes on the shell. All these traces suggest the use of liquefied pigments. It is also worth mentioning that there are two *Unio* palettes that bear a perforation; this was probably aimed at furnishing a handle to facilitate the use of the palette or intended as a way to strap it from the belt or the wrist or suspend it as a bead/pendant.

The use of *Unio* shells as palettes for pigments is also supported by the context of finds, as, for instance, in burial F.8759 (B.150, TP-M). There, a *Unio* valve coated with a red pigment was lying on top of the right shoulder of an adult male skeleton (32818) whose frontal bone had a straight 'stroke' of a red pigment. PXRF analysis of both red pigments identified them as cinnabar (HgS) (Haddow et al. 2017: 134). Another example was found placed on the feet/ankles of a two-year-old child's skeleton ((17939), B.49, Level North G) (Hager, Haddow 2009: 80–81). Apart from the intense traces of pigment on the inner side and the scratches of a tool on the inner surface, it had been ground into a triangular, pointed shape. Then a perforation was drilled near the angle of the triangle. Also taking into account its place in the burial, it seems plausible to suggest that this was a palette that was secondarily used as a painted 'bead/pendant' for the decoration of the infant's ankle.

Other freshwater shell beads and pendants

This category includes four freshwater gastropod species that were used as beads and pendants, of which the most commonly encountered are the *Th. heldreichi* shells (n=56). The remaining species were used rarely, as there are only a few *Th. anatolicus* (n=4), *Viviparus* sp. (n=3) and *Lymnaea* sp. (n=2) artefacts (table 10.2; fig. 10.1).

The majority of *Th. heldreichi* (n=46) and all *Th. anatolicus* bear a naturally abraded apex with evidence for use-wear around the periphery of the perforation, in particular, gloss, wear and notches. These items were collected intentionally as ready-to-use artefacts and did not require extra labour. The remaining *Th. heldreichi* beads (n=10) bear traces of manufacture, in particular, scratches and furrows that seem to have been the result of gouging or scratching the area of the apex with a pointed tool. In some items, the naturally abraded apex was enlarged with the use of a tool to give it an oval shape, while there is also one item that preserves traces of a red pigment on the inner side of the last whorl.

Moreover, there are two *Viviparus* intact shells that were perforated by gouging or hammering at the last whorl, and the examination of perforations showed wear at the periphery of the hole (fig. 10.1b). Thus, they were also used as pendants. In addition, there is another *Viviparus* and one *Lymnaea* (fig. 10.1c) that are painted and pierced for suspension. The colour is partially preserved, but close examination showed that each was decorated with a striped pattern (a narrow black line and a wider red stripe), the width of which increases relative to the incremental growth of whorls. In other words, stripes are thinner in the short upper whorls and gradually widen as the width of the lower and last whorl widens. The last whorl of both the *Viviparus* and the *Lymnaea* has been pierced with a pointed tool (gouging). Perforations are not neat, but examination under the stereoscope indicates wear (incision) from suspension, at least in the *Lymnaea* example. Similar patterns and coloured decorations are common in nature (marine gastropods from the Mediterranean and the Red Sea). In this framework, it would be tempting to suggest that this pattern imitated the design of fresh marine shells that were more difficult to acquire than the locally abundant species with dull colours and no decorative patterns.

Temporal and spatial distribution

This section discusses the temporal and spatial distribution of shell artefacts unearthed during the 2009–2017 excavations (table 10.5; fig. 10.9), but also adds the results from the 2000–2008 excavations (Bar-Yosef Mayer 2013: tables 16.1, 16.2) to allow a better appreciation of presence and distribution (fig. 10.10).

In general, the use of local and non-local shells for the manufacture of artefacts is documented from the Early habitation levels (3.3%, n=43) in low frequencies. Their use becomes widespread during the Middle period (78.1%, n=1,012), when the variety of raw materials and the quantity of artefacts proliferates. During the Late period (18.5%, n=240), there is a significant decrease, but the spectrum of raw materials and types of artefacts is still large (table 10.5; fig. 10.9). The results from previous studies show a different temporal trend (fig. 10.10), with an already significant quantity in the Early period (19%, n=100) and a continuous increase from the Middle (28%, n=147) to the Late period (53%, n=278), as well as considerable spatial differences (Bar-Yosef Mayer 2013: tables 16.1, 16.2).

Starting with the local shell artefacts, their use is documented in the earliest levels of habitation (Level South G) (Bar-Yosef Mayer 2013: 333) and shows a slight increase from Level South H to South L (table 10.5; figs 10.9a and 10.10). The number of local shell artefacts is larger in non-burial contexts. A substantial increase is evidenced in the Middle period, in all types of contexts, though local shell artefacts in non-burial contexts still outnumber those in burials. The combined results from the current and the previous study show that the quantity of non-local shell artefacts is large in the Late period, and they are usually found in non-burial contexts. The inclusion of *Unio* shell artefacts in burials demonstrates their symbolic value for the inhabitants of Çatalhöyük; however, the low number of these artefacts, despite the abundant raw material, coupled with the different sensorial properties compared to non-local shells, is intriguing. Taking all this into consideration, it may be suggested that local shells served particular needs related to ornamentation and tools for the treatment of the dead – for instance, large pendants and palettes for pigments – that were not met by non-local shells.

Groups	*Levels*	Antalis	Nassarius	*Marine*	Dentalium	Unio	*Freshwater*	*Total*	*%*
	South H						3	3	0.2
	South I	2				5	1	8	0.6
Early	South K	7	4	2		1	7	21	1.6
	South J					2		2	0.2
	South L	2		2		4	1	9	0.7
	North F	5	1		1	7	1	15	1.2
	North G	541	153	6	93	29	10	832	64.0
Middle	North ?G	46	3					49	3.8
	South ?M					1		1	0.1
	South N	9			73			82	6.3
	South O	7			1		25	33	2.5
	North H	164			3	2	12	181	13.9
	South P				1			1	0.1
	South R	1		1				2	0.2
Late	South S					1	3	4	0.3
	TP-M		22	2		2		26	2.0
	TP-N						1	1	0.1
	GDN			1	16	6		23	1.8
	IST	3				2		5	0.4
Final	TP-Q					1	1	2	0.2
	Total	787	183	14	188	63	65	1,300	100.0
	%	60.5	14.1	1.08	14	5	5	100	100.0

Table 10.5. Temporal distribution of shell raw materials.

A completely different picture emerges from the non-local shell artefacts. The moderate variety (n=10) and low quantity (n=21; 8.5 artefacts per number of units) of marine raw materials in the Early period (table 10.5, fig. 10.9) is reversed if we add the large quantities of marine shells (n=101) and the few fossils (n=6) recorded by Bar-Yosef Mayer (2013: table 9.16.1) (fig. 10.10). Though they seem to be represented in similar numbers in burial and non-burial contexts (fig. 10.9), this is a sampling bias, as according to the previous study, the vast majority (106 out of 107) were found in middens (n=73) and various domestic contexts (n=33) of mundane and more special character, such as foundation deposits, bins and postholes. Therefore, in the Early period of habitation, non-local shell artefacts were deposited in middens and various domestic contexts, such as floors (32453, 32455, 21851), a foundation deposit (32496) and fills (22355, 32605, 32606), suggesting that at least some of them were used and displayed by the living community, possibly being highly significant as evidence for their participation in networks of communication and exchange. However, at some point they were put out of circulation (to increase their value?) and were deposited (intentionally or thrown away?) in areas of communal character, maybe during feasting activities.

The variety of non-local materials and of artefact types increases considerably in Level South K, but they proliferate in absolute (n=939) and relative (33 artefacts per number of units) terms in the Middle period (table 10.5; fig. 10.9). Interestingly, the vast majority are concentrated in burial contexts (n=903: marine=741, fossils=162), sometimes in groups, followed by a particularly low quantity in domestic contexts (n=36). This picture is slightly reversed by adding the results from the previous study (Bar-Yosef Mayer 2013: table 9.16.1, 16.2), according to which the quantity of artefacts is quite similar in burial (56.3%) and non-burial contexts (43.7%). Therefore, there is a shift in the contexts of deposition of non-local shell artefacts in the Middle period that allows inferences to be drawn about the shift in meanings and value of non-local shell artefacts for the living community, while also implying shifts in emphasis from the communal to the individual and household identity.

Worth noting are the concentrations in some burials. In B.77 (North G), 86 *Nassarius* beads were found with the remains of three individuals (F.3620), and 32 *Nassarius* beads in unison with 53 *Dentalium* beads were found with the skeleton of a young male (20685). In B.131 (North G), in burial F.7961, there are 30

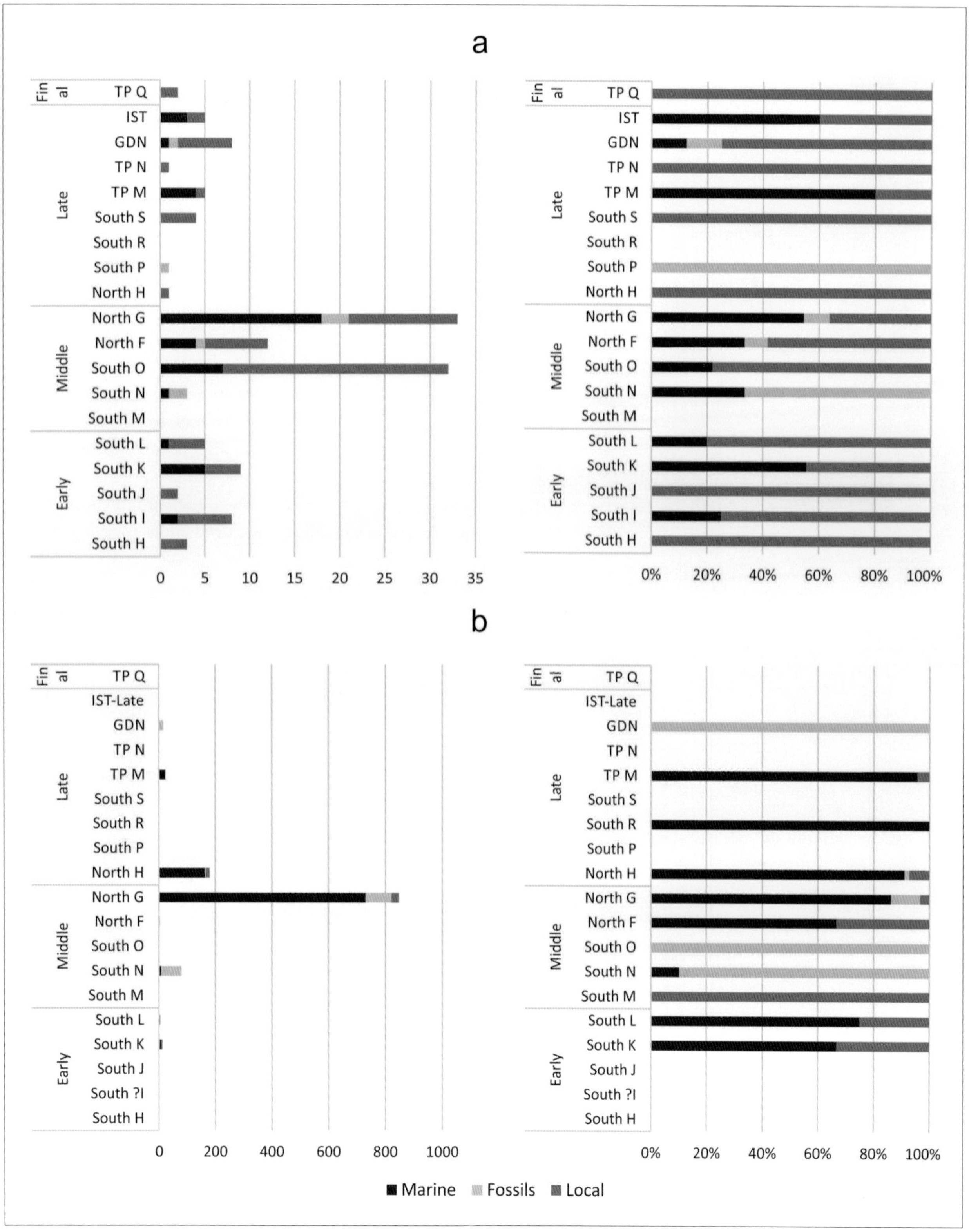

Figure 10.9. Temporal distribution and frequency of shell artefacts of different origin in (a) non-burial and (b) burial contexts.

Dentalium beads and two *Antalis* beads adorning the skeleton of a young adult, as well as the 329 *Antalis* beads with the skeleton of an adolescent female (31705). In B.129 (North F), 106 *Antalis* beads were associated with the burial of an infant (F.7713), 35 *Antalis* beads were found along with a young adult skeleton (22620) and 22 *Antalis* beads were found with an adolescent skeleton (19481).

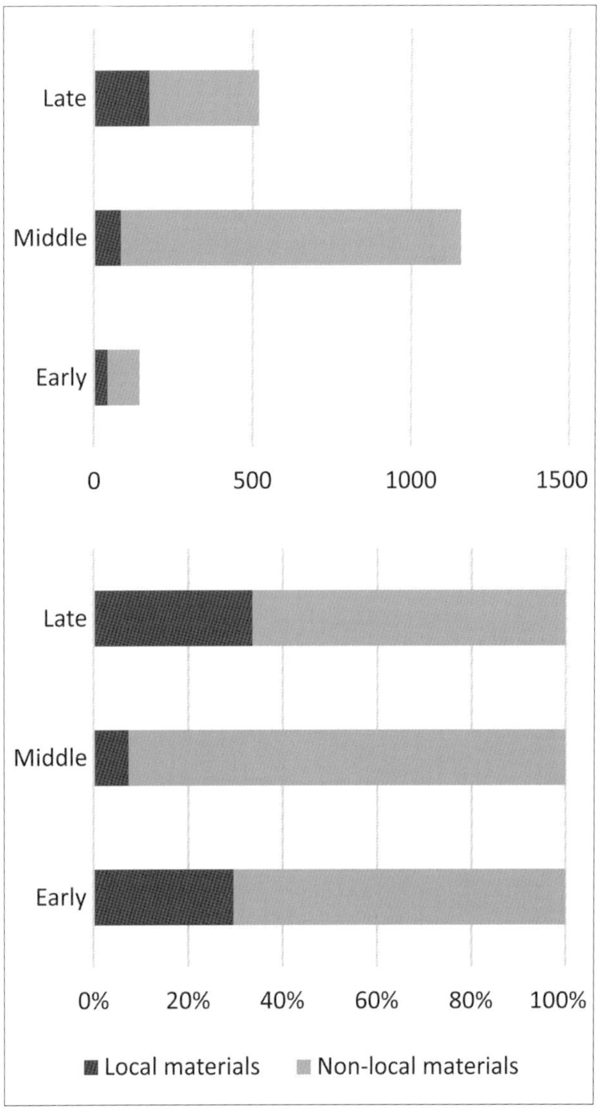

Figure 10.10. Combined plotting of the temporal distribution of local and non-local shell artefacts from the 2000–2008 (Bar-Yosef Mayer 2013) and 2009–2017 (current study) excavations.

Moving to the Late period, the total number of marine (n=10) and fossil shells decreases (n=214; 23 artefacts per number of units) (table 10.5; fig. 10.9). There is still a significant quantity of marine shells (n=194), of which the majority belong to the *Antalis* group (n=168), followed by a small number of fossils (n=20). As also observed in the Middle period, the vast majority come from burial contexts (n=206) and only a few (n=8) from non-burial contexts. Despite the fact that the material studied by Bar-Yosef Mayer was mainly recovered from non-burial contexts (n=148), in particular middens (n=28), non-local shell artefacts in burial contexts still outnumber (60%) those recovered from non-burial contexts (40%). The largest quantity of marine shells was recovered in Level North H (n=164), of which the majority are of a group of *Antalis* beads (n=105) that were associated with an infant burial (F.7713) in B.129. Another concentration worth mentioning is the 22 *Nassarius* beads in burial F.3896 (B.166, TPC-M), where the raw material is highly exhausted and only part of the shell is preserved (fig. 10.7, last row).

In summary, the relative frequency of different raw materials in the Early period levels indicates a balance between the use of local and non-local sources, a distribution that is completely reversed in the Middle and Late periods. The presence of a large variety and quantity of non-local raw materials indicates participation in different networks of communication and exchange that increase and expand over time. This also suggests increased mobility of individuals or groups of people to or from Çatalhöyük, as has also been proposed on the basis of other materials, like obsidian and stone. In addition, the use of non-local shell artefacts as burial goods and ornaments for the dead underlines their significance throughout the sequence. However, their meanings and values change during the Middle period, as indicated by the shift in the contexts of deposition and the increased elaboration of types. Interestingly, there are significant differences in the variety and quantity of shell artefacts associated with different burial contexts/individuals that highlight the differences between these individuals in life or after death and surely also between their social groups in displaying, using and depositing artefacts of exotic origin. In this vein, a pattern related to age seems to emerge, as young people and infants receive the vast majority of shell ornaments, while older people have the least. This seems to be in contrast with the discussion of all burial goods provided by Nakamura and Meskell (2013). Therefore, further analysis and integration of all different types of burial goods is needed.

Conclusions

The above represents the first attempt to quantify and synthesise the shell artefact assemblage from the 2009–2017 excavations, while also drawing from previous studies to allow a comprehensive appreciation of the role of shell in the act of bodily adornment.

The variety of shells used as raw materials for artefact production by the inhabitants of Çatalhöyük is large, with substantial differences in sensorial and mechanical properties, as well as in origins. These shells are characterised by different sizes, shapes, textures, decorations, colours and lustre, but also produce different sounds when one artefact hits the other in cases of artefacts in unison. Even when similar items were created and used together, such as the *Antalis* and

Dentalium beads or the *Nassarius* beads that changed shape according to the extent of use, there were substantial differences in visual appearance, especially when they were combined with artefacts made from stone and bone, let alone perishable materials like plants and feathers.

Some shells were transformed into artefacts with the least possible intervention, thus preserving their characteristics and acting as references to the natural materials, places and people involved in their procurement, production and use. The majority, however, were used as raw materials for the production of various types of artefacts. Their different mechanical properties, related to shape, hardness and thickness, in combination with the desired final forms, required the employment of different manufacturing techniques, such as sawing, gouging, grinding and drilling, individually or in combination. All these techniques were already employed from the earliest levels of habitation, while the effort needed to drill neat perforations or to gouge a perforation in a thin-walled shell without breaking it or to saw the tiny beads underlines the skills and time needed for these tasks. Despite the lack of substantial evidence for the production of non-local shell artefacts on site, local shell artefacts and some of the non-local shell artefacts were manufactured on site, as indicated by the drilling experiments on a cockle and the sawing traces on some scaphopods. In addition, the presence of standardised techniques and types of artefacts, together with the employment of different techniques for the production of similar artefacts, like the *Unio* pendants, may be taken to suggest the co-occurrence of specialised and expedient production of non-local and local shell artefacts respectively.

The shells originate from different places, including local freshwater bodies, the Aegean and Mediterranean coasts of Turkey, and the Hatay and İskenderun basins, ca 100 to 400km distant in a straight line from the site. Evidence for participation in shorter- and longer-distance networks of communication and exchange is already found from the earliest levels. However, it is during the Middle temporal group that an expansion of different materials and origins is witnessed, as also seems to be the case with a range of materials, such as obsidian, various stones and pigments.

The presence of shell artefacts throughout the sequence of habitation and in all types of contexts, though in varying proportions, indicates that they were used in different ways, but mostly as items of bodily adornment for the living and the dead, underlining their importance. Moreover, many items were used extensively, hence implying that they were socially significant for their owners. The hard shell materials could have been used throughout and beyond the human lifespan and by different individuals, thus acquiring complex life histories and subsequently more value. The intentional transformation of 'new' *Nassarius* shells to imitate 'old' (extensively used) beads further supports this argument. In addition, the shift in the contexts of deposition of shell artefacts (from middens and other everyday or more special contexts in the Early period to burials in the Middle and Late periods) may be taken to suggest a shift in the perception and value of these artefacts to the living, as well as a shift from the importance of communal practices in the earlier levels to the importance of individual and household identities in the later levels.

To conclude, both local and non-local shells were procured and used for the production of artefacts and especially items of bodily adornment. Raw materials, manufacturing techniques and artefact types became more elaborate in the course of habitation, pointing towards participation in more distant networks, specialisation in production and the existence of complex means of adorning the body. Meanwhile, the presence of shell artefacts in various deposits provides compelling proof of their importance for the Neolithic people at Çatalhöyük, while the shift in the context of deposition of exotic raw materials from the Early to the Middle and Late periods underlines the shift from communality to the individual and the household. Overall, it is clear that the changes in shell artefact consumption and deposition practices through time can be linked to the important social and economic transformations that took place at the settlement.

Acknowledgements

I would like to thank Ian Hodder for inviting me to participate in the Çatalhöyük Research Project and Christina Tsoraki for making the first arrangements. This chapter would not have been possible without the work of the excavators, the meticulous work of the Heavy Residue team and the stimulating discussion during the Catania meeting. Particular thanks go to Milena Vasić for discussions and help, Gesualdo Busacca and Eline M.J. Schotsmans for help with 'pigmented' shells and Lisa Guerre for bringing the shell artefacts for study. I also thank Jason Quinlan and Didem Turan for the photographs and Nikos Valasiadis for editing the figures.

Supplementary material

For supplementary material related to this chapter, please visit https://doi.org/10.18866/BIAA/e-14. It comprises colour versions of figures 10.1–10.3, 10.5–10.8.

11. Textiles, cordage and basketry from Çatalhöyük

Lise Bender Jørgensen, Antoinette Rast-Eicher and Willeke Wendrich

Introduction

Perishables like textiles and basketry are the most fragile objects, and rather than being discarded, they make suitable kindling or fire starters. They survive only if the conditions of preservation are ideal, such as a very dry or very wet environment. That is why we get only glimpses of what was used in the past. This makes materials of this kind particularly valuable, as they represent a significant but little-known part of the material culture of the past. This certainly applies to the textiles from Çatalhöyük, which include the earliest preserved woven textiles currently known. These combine weaving with the technique of weft-twining, which precedes weaving as a way to make flexible fabrics that could be used for clothing, wrapping, carrying and a multitude of other purposes. The textiles from Çatalhöyük thus reflect the emergence of weaving as a new technology to produce objects made from various fibres. Textiles made by looping, netting techniques and weft- or warp-twining are known from earlier sites and are likely to go back to the Palaeolithic and certainly to the Mesolithic (Bender Jørgensen 1992: 114–16; 2013; Rast-Eicher, Dietrich 2015).

Cordage is essential for binding and tying, for netting, fishing lines, et cetera. Such items are found in abundance in the Neolithic Alpine lake dwellings (Rast-Eicher, Dietrich 2015) and in caves, lakes and desert sites in West Asia and Egypt (Nadel et al. 1994). Cordage found at Çatalhöyük will have served similar purposes.

Baskets served as containers, including burial receptacles, while small circular platters were used to cover or support household items or foods. Mats were used as space dividers, floor coverings and possibly sleeping surfaces (Wendrich 2005; Wendrich, Ryan 2012).

Preservation

At Çatalhöyük, tiny remains of textiles, basketry and cordage survived as charred objects due to burnt houses, as phytoliths, or in a few cases through oxidation of metal beads. In cases where the building above burials had been burnt, very fragile remains were found, described as 'burnt' or 'charred' in the record. Other remains, especially of baskets and mats, are preserved as phytoliths and impressions in clay. During Mellaart's excavations in the 1960s the groundwater was high, later lowered by irrigation canals (Hodder 1996: 2); this probably influenced preservation negatively, as organics preserved in wet conditions decay when drying out, although phytolith remains of floor matting and textiles were still found. Phytolith remains are sometimes difficult to interpret, because they represent remains of the fibres but not of the structure. Upon closer examination, some of the large patches of phytoliths turned out to be from wood, even if catalogued in the Çatalhöyük database as 'basket', while the presence of layers of unworked grasses might represent bedding (Nadel et al. 2004).

Definitions (figures 11.1 and 11.2)

Textile. The term 'textile' is sometimes used of basketry and cordage as well as of woven textiles; some scholars, like Seiler-Baldinger, include this in their technical descriptions (Seiler-Baldinger 1994). This may easily lead others to conflate the techniques (e.g., Gilligan 2019). It is important to distinguish between these techniques, as they have different histories, are used for different artefacts and require different raw materials, tools and techniques (Wendrich 1999). In this paper, 'textile' is used of flexible fabrics made by weaving or twining.

Weaving requires a loom, that is, a construction that allows parallel threads (warp) to be held taut and often has a mechanical system (heddles) that divides the warp into two parts. This way a shed is formed, into which weft threads can be inserted. Woven textiles can be made in different weaves, such as tabby, twill and a variety of patterned weaves. There are no preserved woven textiles dated before the Neolithic period.

Tabby or plain weave is the simplest form of woven textiles (Fig. 11.1e). Tabby occurs in varieties such as balanced tabby, which has approximately the same number of warp and weft threads per cm^2, weft- or warp-faced tabby, where one system partly covers the other, or repp, where one system covers the other completely. Further variations of tabby are half-basket and basket weave, where threads are paired in one or both systems. Several of these varieties may appear in the same fabric, as edges (transverse borders and selvedges) are often more densely woven than the main web. All woven textiles from Çatalhöyük are in tabby.

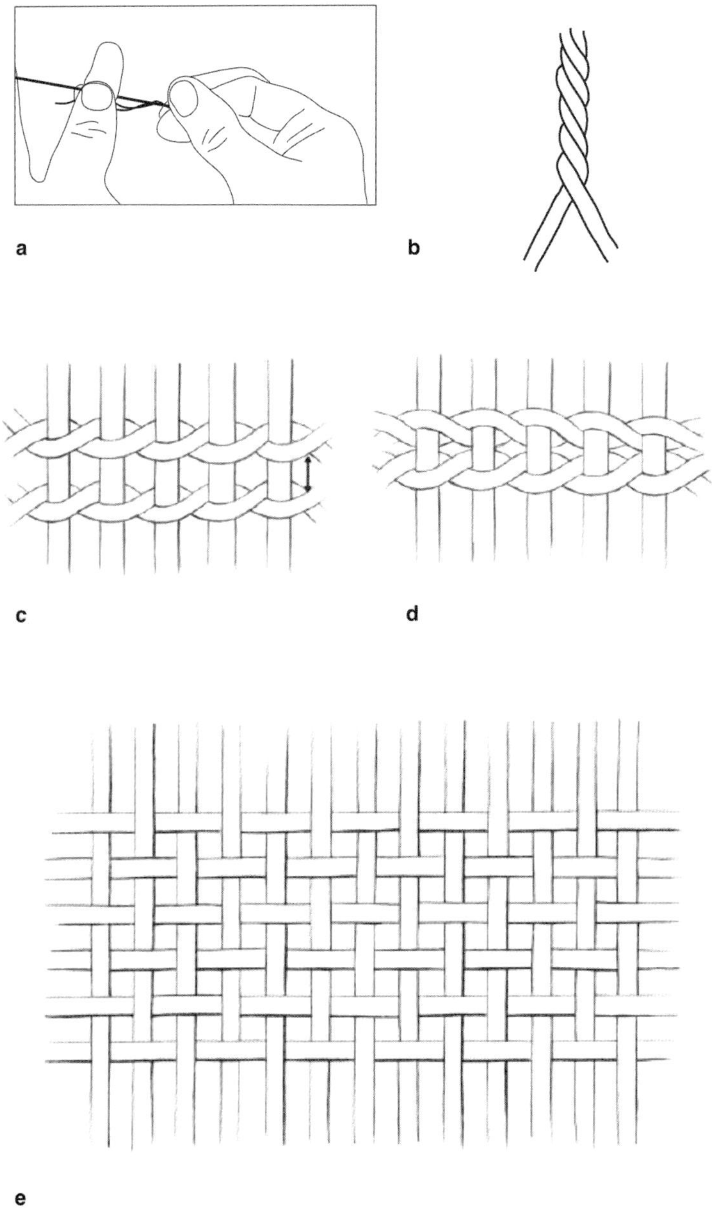

*Figure 11.1. Techniques: textiles and twinings: (a) splicing of two fibre strips; (b) plied yarn (S2*i); (c) twining in same direction (S); (d) twining in two different directions, S and Z or herringbone; (e) tabby weave (illustrations a and b by A. Rast-Eicher; c–e by Atelier Oculus, Zürich).*

Soumak is a technique in which the warp is wrapped with a continuous thread, with or without having weft threads in between. Soumak was found in some textiles from the Mellaart excavations at Çatalhöyük.

Twining (fig. 11.1c, d) is a basketry technique but can also be used for fine, flexible fabrics that resemble woven textiles if made of finely prepared fibres. It is made of untwisted or plied yarns, with an active system 'weft' that binds with two threads the passive system 'warp'. The 'weft' can be introduced densely or with spaces of different widths. Twining at Çatalhöyük occurs as weft-twining.

Basketry is defined as objects made of plant or animal materials of limited length that often retain their natural shape and are worked into a fabric or structure (Wendrich 1999). The basketry found at Çatalhöyük consists of four categories: coiled baskets and platters, as well as tabby and twill plaited matting (Wendrich 2005). Coiled basketry is constructed from a flexible strand, such as twigs or a bundle of grasses (passive system), fastened as a coil that spirals out from the centre and is fastened by an active system that wraps around the bundle and stitches through the previous coil, often with the aid of an awl or needle. These active elements, the winders, often consist of tough grass leaves, thin strips of broader leaves, or bands of other materials. Matting is made from reeds, leaf strips or grasses, using two active systems. Depending on the raw materials, the results may appear similar to the techniques of tabby and twill used in woven textiles. To distinguish between matting and weaving, matting techniques are designated 'basketry tabby' or 'basketry twill'. The difference is mostly that in basketry, multiple strips are inlayed to arrive at the required length and width of the mat.

Splicing. Threads and cordage are usually described as s- or z-twisted, Sz- or Zs-plied (Emery 1994: 11; Bender Jørgensen 1992: 15, fig. 2). This is not the case at Çatalhöyük. Here threads were *spliced* and then plied (fig. 11.1a). This is rendered as S2*i; that is, S-plied from two elements of spliced (*) fibres, untwisted/without discernible twist (i). Splicing – that is, adding lengths of fibres to each other by tying them at the ends – is a technique first noted in Pharaonic Egyptian textiles (Barber 1991: 47–48; Granger-Taylor 1998; 2003) but also since found in Neolithic Europe (Leuzinger, Rast-Eicher 2011; Rast-Eicher 2016; Rast-Eicher, Dietrich 2015: 34–39). The fibre strips have still-connected fibres and parallel nodes. Splicing is now found in virtually all prehistoric textiles made of plant fibre in Europe, the Mediterranean and the Near East until at least 600 BC (Gleba, Harris 2018).

Textiles, cordage and basketry at Çatalhöyük: a short history

Textiles and basketry from Çatalhöyük were first found during the excavations by James Mellaart in the 1960s (Mellaart 1962; 1963; 1964; 1966; 1967) and were investigated by Harold B. Burnham (1965). They are now in the Anadolu Medeniyetleri Müzesi in Ankara. A small box of fragments is in the collections of the Textile Research Centre in Leiden.

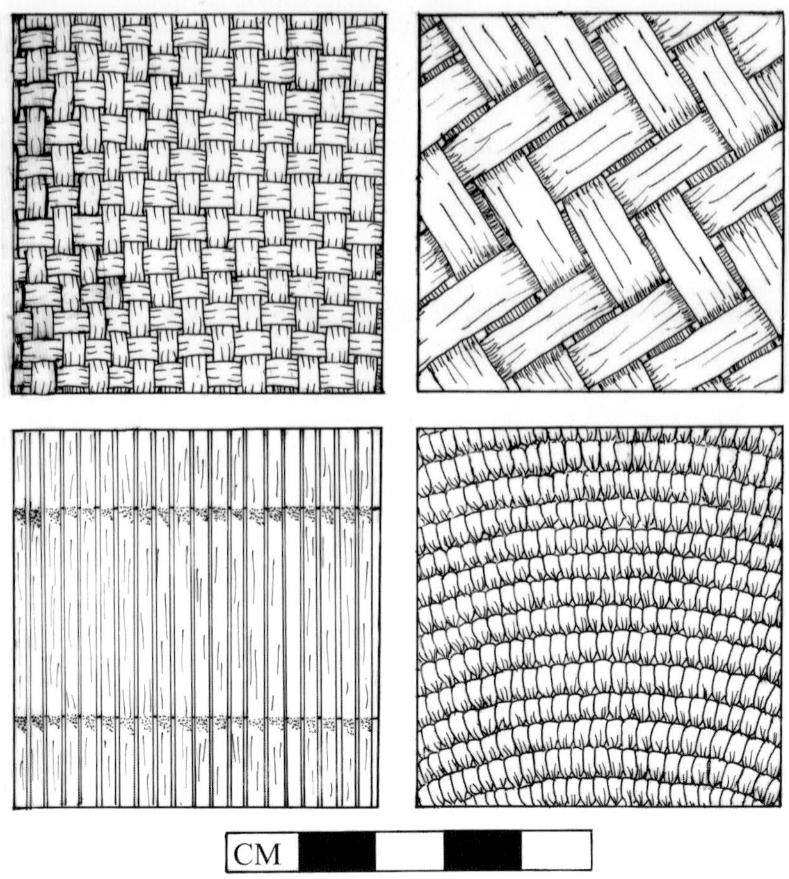

Figure 11.2. Basketry techniques found at Çatalhöyük. Top: fine plaited tabby matting (left); coarse plaited twill matting (right). Bottom: sewn or stitched matting (left); fine coiled basketry (right).

During the excavations directed by Ian Hodder in 1993–2017, further textiles and basketry have been recovered. W. Z. Wendrich's report on Çatalhöyük during the 2000 season includes a catalogue of 31 items of basketry (Wendrich 2005; Wendrich, Ryan 2012). Textile specialists L. Bender Jørgensen and A. Rast-Eicher came to the site in the 2017 season to examine textiles and cordage, and items of basketry found since Wendrich's visit. This resulted in the recording of 19 items designated as 'textile', 28 as 'cordage' and 30 as 'basketry' (Bender Jørgensen, Rast-Eicher 2017). The basketry samples examined by Bender Jørgensen and Rast-Eicher were excavated between 2006 and 2017; basketry finds discovered from 2000–2006 are not catalogued. The selection of finds to investigate was carried out by searching the project database for finds labelled 'textile', 'baskets' or 'cord'. During our work, further items turned up, such as imprints of textile or basketry on clay balls, or strings in beads. Some skin and hide items were also discovered by looking through the boxes with textiles. During the excavations, further finds had been observed, but only a few samples were recovered. This means that the finds described below only represent a part of what was in fact found but not recognised. Still, the recovered remains are far from reflecting the full range of items made from fibres. No remains of hunting or fishing equipment such as nets or fishing lines have been found.

Methods

The boxes with textiles and cordage were examined on site with magnifying glasses and with a stereo microscope available from the project. Photographs were made with a Nikon, augmented with macro objective (1:1) so that they could be taken quite close to the object.

Preservation of many items in the form of phytoliths or burnt layers precluded fibre analysis. As preserved fibres are mostly burnt, or even carbonised, samples were taken for analysis with SEM (scanning electron microscope) in Switzerland by Rast-Eicher. Samples were also taken from hides. One sample was taken for chemical analysis. We used a SEM at the Institute of Geography of the University of Bern (Oeschger Centre), Switzerland (Zeiss EVO 50). The samples were sputtered with gold (20nm) to be able to work with 10 to 15KV in the vacuum column. The aim of having SEM samples was to have a better picture for fibre analysis and also a picture for discussion of fibre processing. The fibres were measured with a measurement program on SEM pictures. The basketry was recorded as described in detail in (Wendrich 1991; 2005).

Cross-disciplinary questions were discussed with other members of the team, in particular the botany, human remains, conservation, zooarchaeology and use-wear analysis teams.

Areas and dating

Cordage, textiles and basketry have mainly been found inside buildings. They were found in Buildings 49, 52 and 131. Cordage was also recovered from Building 77; any potential textile remains from this building were too badly preserved to be confidently identified as such. These four buildings are all in the North Area and are assigned to Level North G. This means they date to between 6700 and 6500 cal BC. Textiles and cordage from Mellaart's excavations derive from his Levels VI A/B. They were then C14 dated to 6200–5800 BC (Mellaart 1964: 116). They are now calibrated and assigned to the Middle period of Çatalhöyük, aligned with Level North G and thus contemporary with the textile finds from the Hodder excavations. It means that all Neolithic textiles and cordage from Çatalhöyük are dated from 6700–6500 cal BC.

Basketry has been found in Buildings 4, 5, 6, 17, 18, 23, 56, 80, 89, 97, 108, 123, 132, 150, 160, 161 and 162. Further basketry was recovered from Spaces 489, 583 and 610.

Buildings 4, 6, 17, 18, 23, 160, 161 and 162, and Space 583 belongs to the Early period (7100–6700 cal BC); Buildings 5, 49, 52, 77, 80, 89, 97, 131 and 132, and Spaces 489 and 610 date to the Middle period (6700–6500 cal BC); and Buildings 56, 108 and 150 date to the Late period (6500–6300 cal BC).

Regarding post-Chalcolithic and later textiles and basketry: textile fragment (8770.s3) is Iron Age or later; the basket from (12634.s20) and textile from (19586.s3) are Post-Chalcolithic.

Textiles from the Mellaart excavations

The textiles from the Mellaart excavations were examined by Harold B. Burnham (1965). Previously, they were briefly described by botanist H. Helbæk (1963), who had excavated most of them himself. The textiles are now in the collections of the Anadolu Medeniyetleri Müzesi in Ankara, except for a small sample. We did not have occasion to study the textiles in Ankara, and our catalogue of these pieces, below, has been put together based on Burnham's. As Burnham was a textile specialist, we assume his technical descriptions are accurate, although his descriptions of fibres and yarns should be revised according to our findings. We did, however, examine a sample of one of the textiles now in the Textile Research Centre in Leiden.

Leiden sample: 29 small fragments in a box labelled 'Textile from skull, lower layer VI' (table 11.1 and fig. 11.3). As the sample is dated 'Summer 1963', it may derive from a textile that had been used to replace the extracted brain of a skull found under Room E.VI.1 (Burnham 1965: 172; Mellaart 1964: 93, pl. XXIV a, b; 1967: pl. 94).

Leiden no.	Size of largest piece	Yarn type	Yarn diam.	Technique	Threads/cm	Borders/selvedges
1–5, 10	2.3 x 0.8cm	S2*i/S2*i	0.7/0.7 mm	Tabby	14/14	Reinforced selvedge/starting border
6–9, 12–18, 20–26	1.1 x 1.3cm	S2*i/S2*i	0.5mm	Tabby	20/14	-
11, 28	1 x 0.5cm	S2*i	0.7–0.8mm	-	-	Tassel or end piece?
19	1.1 x 0.3cm	S2(S2*i)	0.5–0.7mm	Twining	-	-
27	0.3 x 0.4cm	S2*i	0.7mm	Twining	-	Lump of fibres
29	-			Single threads		Sample taken

Table 11.1. Textile fragments in the Leiden collection.

Figure 11.3. Textile fragments in the Leiden collection (Mellaart excavation).

Textiles and basketry from the Hodder excavations

Raw materials (table 11.2)

Cordage and basketry. According to the phytolith identifications, Çatalhöyük basketry is made from wild grasses, sedges and cereal straw, and matting from sedges and reed (Ryan 2011; Wendrich, Ryan 2012). One piece of cordage which has been analysed appears to be made of tree bast (oak bast, see fig. 11.7; Rast-Eicher et al. forthcoming).

Twining and woven textiles. Generally, the fibres used for twining and woven textiles are bast fibres, which includes tree basts and flax. Animal fibres would melt with heat and could not be charred or burnt. Fibres for the Çatalhöyük textiles have previously been identified as flax (Ryder 1965; Vogelsang-Eastwood 1988; Fuller et al. 2014). The question of the first domesticated flax has been very much discussed (latest articles: Bar-Yosef 2020; Shamir, Rast-Eicher 2020). Fibre analysis of a few samples has shown that it was not very easy to differentiate between the basts, but that tree bast has been used for woven textiles as well (Rast-Eicher et al. forthcoming). In two cases, flax could not be excluded, but flax fibres can be obtained from several species of wild and domesticated flax. The very low number of flax seeds in the settlement – according to their size, wild flax – leads to the assumption that wild flax was more important than domesticated flax for fibre use. The presence of domesticated flax would have required enough seeds for sowing out. The use of tree basts even for woven textiles is well known in the Neolithic lake dwellings of Switzerland (Rast-Eicher, Dietrich 2015; Rast-Eicher 2016).

Animal fibres. Hides found in Çatalhöyük clearly show animal fibres but are so badly preserved that the species could not be identified. Fibre diameters and pores with large fibres of about 100µ and fine undercoat (wool) of about 15–20µ might, however, fit with ancient sheep types (fig. 11.4).

Unit no.	*Technique*	*Building/Space*	*Date (cal BC)*	*Fibre*	*SEM samples*
20465	coiled basketry	Space 489	6700–6500	stitch: gramineae leaf	sample 2
3051.s4	cordage	Building 52	6700–6500	plant fibre, oak bast	sample 4
17457.x10	woven; tabby	Building 49	6700–6500	badly preserved, with conservation product, dm. 7–15um, plant fibre	sample 9
22661.s5	net? tassel?	Building 131	6700–6500	charred; plant fibre with large lumen, dense fibre bundles, ev. rays, tree bast fibre? flax?	sample 15 sample 16
30503.s8	woven; tabby	Building 52	6700–6500	burnt; badly preserved, cross-section with hollow fibres, remain of ray, oak bast probable	sample 11

Table 11.2. Çatalhöyük: samples analysed from the Hodder excavation.

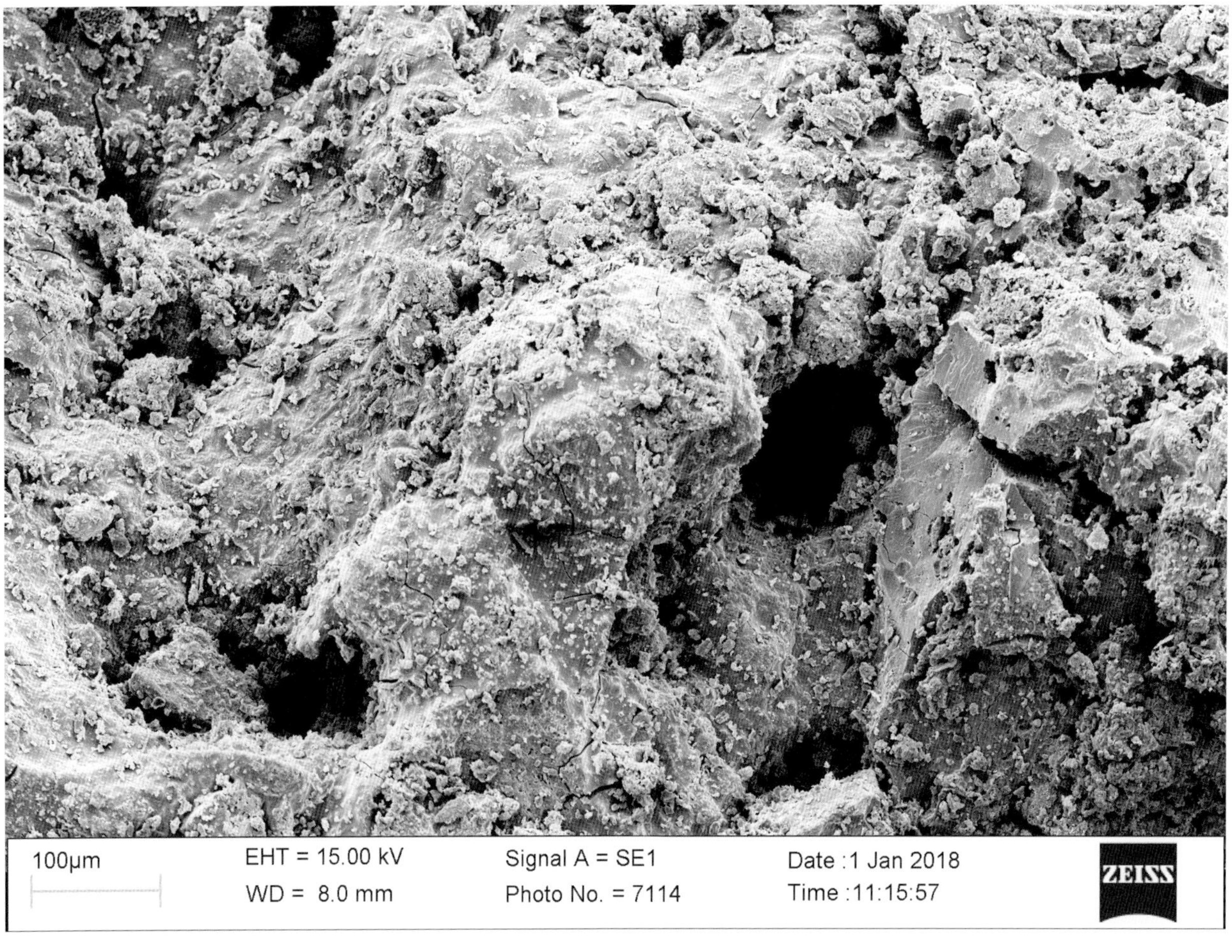

Figure 11.4. Surface of hide with pores.

Presentation of the material

Threads and cordage (table 11.3)

Twenty-eight fine threads and cordage were catalogued, with diameters varying between 0.5 and 10mm. Several items – especially when preserved as phytolith cord – were so deteriorated that the diameter could not be measured. Three items were found inside beads: one is a plied thread (7580) from unstratified burial F.1202 in Sp. 1003; the others are fine strings composed of a multiple-two-ply thread – (17457) from infant burial F.4023 in B.49 and (21682) from burial F.8060, B.77. The diameter of these plant fibre strings indicates that they were used for stringing the beads rather than for sewing them onto a textile or hide (figs 11.5 and 11.6).

Building 52, multiple burial F.7127. A thread with a diameter of 2mm (30510.x1) was found around the mandible of a child. Medium strings of 5–7mm were used to tie the deceased in a foetal position. Cordage (30511.s4) was found wrapped around the legs of infant skeleton (30511), also holding the body in flexed position.

Building 77, burial F.7611, skeleton (21606). (21605.s5) was wrapped around the skeleton with orange-brown material. In the same building, cordage preserved as white phytoliths (30154) was connected to burial F.7309. Cord remains from skeleton (21606) (an infant, 2–4 years old) are thicker, with diameters up to 10mm. They are preserved as phytoliths. One piece (21606.s1) was found wrapped several times around the head. One cord went over the brow ridge and into the eye socket, showing that when the wrapping took place the body must have been decomposed (fig. 11.7).

Unit no.	Find or sample no.	Spin direction	Thread diameter (mm)
7580	X1	S2i	1.30
17457	X3	S6(S2i)	4.00
21605	s5	S	5.00
21605	s8	S2i	10.00
21606	s1	S2i	10.00
21682	X7	S2i	3.00
22661	s17	S	5.00
22661	s18	S	5.00
22661	s19	S	5.00
22661	s20	S2i	0.50
30154	s3	S2i	7.00
30510	X1	S	2.00
30510	X1	S2i	1.50
30511	s4	Z	4.00

Table 11.3. Strings and cords with visible twist.

Building 131, skeleton (22661). Three thread samples (22661.s17–19) were found in connection with the skeleton. They are burnt and very brittle and probably made of bast fibres.

Figure 11.5. Fine string found in bead (7580.x1–4).

Figure 11.6. Multi-ply string in copper bead (17457.x3–5).

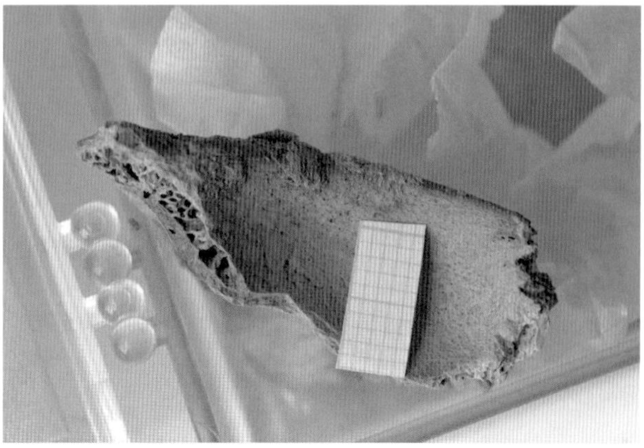

Figure 11.7. Multi-ply string. Cordage over the eyelid in burial F.7611 (no. 21606).

Basketry

In his report on the 1963 season, Mellaart showed different forms of baskets found in his Level VI (Mellaart 1964: pls XXIIb–XXIIIa). The baskets and mats recorded in 2017 are listed in table 11.4. There are two main techniques: coiled baskets and matting (see fig. 11.2). The coiled baskets are made of a coil of grasses, sewn with grass leaves or bast. The sewing is always stitched – there are no examples of wrapping around the bundle without stitching through the bundle of the previous coil. The stiches are irregular and dense, or V-shaped. In some cases, they run into the last stitch or the stitch of the lower row (no 32673; 32803). The mats are all made in basketry tabby.

Coiled baskets. Most of the coiled baskets seen in 2017 by Lise Bender Jørgensen and Antoinette Rast-Eicher are preserved as phytoliths; two are imprints on clay balls. They are quite variable in form, coil diameter and stitching (see table 11.4). One of the new finds, (32610.x1), found in Building 160 as an infill in a burial, was preserved in several big fragments that allow the calculation of its

Unit no.	*Excavated*	*Type*	*Found as/preservation*	*Building/Space*
12634.s20	2015	Coiled basketry	Imprint on phytolith	B.41
12806	2006	Coiled basketry	Phytolith	B.56
18924.x1	2010	Coiled basketry	Phytolith	B.80
20386.s3	2014	Coiled basketry	Imprint	B.97
20395.s3	2014	Coiled basketry	Phytolith	B.97
20413.s2	2013	Coiled basketry?	Fibrous material	B.77
20465.ch2 12	2012	Coiled basketry	Carbonised	Sp.489
20938.m1-4	2013	Coiled basketry	Imprint on clay balls	Sp.489
20938.m 1-4	2013	Basketry tabby	Imprint on clay ball	Sp.489
21634.s6	2015	Basketry tabby	Burnt material	B.132
21688.s8	2015	Wood?	Phytolith, burnt	B.132
21859.m1	2016	Coiled basketry	Imprint on clay ball	B.17
22051.x1	2014	Basketry tabby	Phytolith	B.77
22351.m113	2016	Basketry tabby	Imprint on clay ball	Sp.583
22501.x2	2015	Coiled basketry	Phytolith	B.6
22676.s15	2017	Bark? Wood?	Phytolith	B.131
23804.s1	2017	Strings/basketry	Phytolith	B.52
30043.s5	2017	Basketry tabby	Organic material	B.131
30129.m1-2	2013	Coiled basketry	Imprint on clay ball	B.123
30154.s4	2014	Coiled basketry	Phytolith	B.77
30197.s3	2014	Basketry tabby	Carbonised and phytolith	B.108
32055.s5	2016	Basketry tabby	Carbonised	B.132
32128.s153	2017	Basketry tabby	Carbonised	Sp.610
32373.x1	2016	Basketry tabby	Burnt	B.131
32610.x1	2016	Coiled basketry	Phytolith	B.160
32646 s2	2017	Basketry tabby	Burnt	B.161
32673.m1	2017	Coiled basketry	Imprint on clay ball	B.162
32696 sk	2017	Coiled basketry	Phytolith	B.161
32803.x5	2017	Coiled basketry	Carbonised	B.150

Table 11.4. Basketry found in Hodder excavations, recorded 2017.

diameter. One of the fragments, where both bottom and remains of the lid were preserved, had a height of 13.5cm. Over the lid, a fragment of a cover could be observed; this fragment had larger coils (fig. 11.8).

Mats

In Mellaart's reports, impressions of mats are often mentioned. One example is particularly interesting, as it was found on pottery from Shrine E, Level IX. This was the earliest pottery found (Mellaart 1964: 82). Mellaart further observed matting on the floor in several rooms, 'carpeted with fine reed matting' (Mellaart 1964: 55). Mats were also found in burial contexts; the skeletons were laid on mats or placed in baskets (Mellaart 1964: 94).

We found eight mats, one in the form of an imprint on a clay ball. All are in basketry tabby (see table 11.4); in most cases the fibre strips of both systems have approximately the same width. Strip widths vary between 3 and 11mm. Some look like grasses, others like bast. The imprint on the clay ball (22351.m113) is particularly interesting, as the imprint of the mat changes its direction, partly superposing itself (fig. 11.9). This means that the ball was rolled on the mat in order to obtain its round shape.

Figure 11.8. Coiled basket with cover (32610.x1).

Textiles (table 11.5)

Neolithic textiles were found in Buildings 49, 52 and 131, all of which had burnt, causing textiles in recent burials to be charred. One item (19586.s3) proved to be post-Chalcolithic, while (8770.s3) dates to the Iron Age or later.

As mentioned above, 'textiles' is used of flexible fabrics made by two different techniques, weft-twining and tabby weaving (see figs 11.11 and 11.12). The two techniques appear linked, directly or because they are made from the same yarns within the same unit. This corresponds with Burnham's records of the textiles from the Mellaart excavations (Burnham 1965). One of the pieces he recorded (1965: 172) had a twined heading cord with 13 warp threads and 16 weft threads per cm; another fragment, probably of the same piece, had a plain rolled hem sewn in coarse running stitch. He also described two types of narrow tapes (1965: 171–72); one is 7–8mm wide, strongly warp-faced and composed of 20 threads with about six wefts/cm. The other is wider, ca 1.5cm, and has on average nine warp and eight weft threads/cm. Textiles from the current project are similar, with between eight and 14 threads/cm.

Textiles from Building 49, Feature 4023. The textiles from Building 49 were found in 2008. They are labelled (17457.x3 and x10) and derive from F.4023, an infant burial wrapped in matting. Grave goods included a copper bead and a twine necklace and textiles. The burial was the earliest in Building 49. The copper beads were on a string, described above (17457.x3, s10; fig. 11.6). One textile fragment was found on a copper bead (17457.x3, x10); three further fragments, (17457.x10), were also found with copper bead(s) (fig. 11.10). All are mineralised and have been preserved by oxidation of the copper.

Textiles from Building 52. The textiles from Building 52 were found in 2013 and were first examined in 2014 by D. Fuller (Fuller et al. 2014). They derive from Space 94, F.7127, a multiple burial that contained an adult skeleton (30514) and four infants/children, (30510), (30511), (30513) and (30524). The textile remains were found between (30511) and (30513) and on the cranium of (30510). They are labelled (30503.x9), (30503.s5–9, .s10) box 1 and 2. Two samples labelled (30510.x1) were described as 'burial bandage from skull and cord sample'; both were found to be cordage. Another sample labelled (30511.s4) described as 'textile from legs (cords)' was also found to be cordage. Textile samples (30503.s2) and (30503.s10) subsample were exported by D. Fuller in 2013. These, as well as (30503.s5), stored in the Konya museum, were not examined by the present authors.

Unit no.	*Found*	*Size*	*Yarn*	*Yarn dm (mm)*	*Technique*	*Threads/cm*	*Comments*
17457.x3 s10	Building 49, Sp.100. Matting over/under body, together with textile, beads etc. earliest burial in sequence.	1.8x0.7cm	S2*i/S2*i	0.5/0.5	Tabby	10/12	Mineralised textile fragment on copper bead
17457.x10		3 textile fragments, 1.4x0.8cm, 0.7x0.8cm, 0.7x0.5cm	S2*i/S2*i	0.7/0.7	Tabby and/or basket weave	10/-	As x3. One fragment is not always plain tabby; some threads in both directions over two, a kind of basket weave. Possibly border/selvedge.
22661.s2	Building 131, Sp.500, F.7956, burial fill	1x0.8cm	S2*i	0.5/0.5	Twining and maybe knots	-	Fringe? Small textile and thread remains; threads end in loop.
22661.s3		Largest piece 2x1.2cm	S2*i/S2*i	0.5	Twining	-	Bundles of parallel threads, like bobbin. One piece clearly twining.
22661.s5		Largest piece 3.2x1cm	-	0.7	Net or tassel	-	Bundles of parallel threads with small knots.
22661.s5 (fill)		1x0.3cm	S2*i	0.5	Threads	-	Bundle of parallel threads
22661.s11-13		Not seen. Exported 2015 by D. Fuller					
22676.s4	Building 131, Sp.500, F.7961 and F.7963, burial fill. With skeleton (30045/30040), Sp.91-92	1x0.7cm	S2*i/S2*i	0.8/0.8	Tabby	14/10	Bone-mineralised textile fragment. Grasses on back.
22676.s12, subsample of s5		Largest piece 1.3x3cm	S2*i/S2*i	0.8/0.8	Tabby	8/12	Small fragments, largest piece folded. Floating threads on surface of two fragments.
30039.s10	Building 131, F.7961, skeleton (23126). Organic material on leg	Tiny	S2*i?	-	-	-	Possible textile.
30503.x9	Building 52, Sp.94, F.7127. Infill of 30502, skeletons (30511) and (30513)	7x4cm, 2.5x1cm, 1.2x1.7cm	S2*i/S2*i	1.0/1.0	Tabby	10/8	One piece with unclear threads, full of conservation products. Possible starting border.
30503.s2		Not seen. Exported 2013 by D. Fuller					

Table 11.5. Neolithic textiles of the Hodder excavations.

Unit no.	Found	Size	Yarn	Yarn dm (mm)	Technique	Threads/cm	Comments
30503.s5		Not seen. In Konya Museum					
30503.s6 Thorax region of sk (30511)	Building 52, Sp.94, F.7127. Infill of 30502, skeleton (30511)	Largest piece 3.3x1.5cm; others 1.6x1cm, 1.6x1.1cm, 1.6x1.2cm and less	S2*i/S2*i	1.0/1.0	Tabby	10/8	Ten fragments of similar fabric, one fragment with fringes. One piece has a knot. One tine piece inside brown fragment looks red.
30503.s7 Neck region of sk (30511)		1.9x1.5cm	S2*i/S2*i	1.0/1.0	Tabby	10/8	Possible end fringe. Clearly visible on side 1. Folded around edge. On side 2, only one thread visible, before turning into fringe.
30503.s8 Abdominal region of sk (30511)		1.9x1cm, 1.2x0.5cm	S2*i/S2*i	1.0/1.0	Tabby	10/8	Two small fragments, one in several layers but folding not visible.
30503.s9 Upper abdominal region of sk (30511)		4.5x3.2cm, 1.1x1cm	S2*i/S2*i	0.8/1.0	Tabby	10/8	Several layers, one of them folded. Some loose fragments.
30503.s10, box 1 From head region of sk (30513)	Building 52, Sp.94, F.7127. Infill of (30502), skeleton (30513)	Largest pieces 1.3x0.5cm	S2*i/S2*i	1.0/1.0	Tabby and weft twining	10/8	Tiny fragments, two ending in weft twining. One has two rows of herringbone weft twining, one only one row. Likely to be end of web.
30503.s10, box 2 Textile from above sk (30513)		Largest piece 2.1x0.5cm	S2*i/S2*i	1.0/1.0	Tabby and weft twine	10/8	Several small fragments, full of conservation products. One has single row of weft-twining, likely an end border/fringe.
30503.s10 subsample		Not seen. Exported 2013 by D. Fuller					
32373.x	Building 131, Sp.500, F.7997 fill basket	7x2cm	-	1.0/1.0	Tabby	-	Faint textile impressions under mat.

Table 11.5 (continued). Neolithic textiles of the Hodder excavations.

Figure 11.9. Mat impression on clay ball (22351.m113).

Figure 11.10. Textile found with beads (17457.x10).

Textiles (30503.s6–9) were found with an infant, (30511.s6) in the thorax region, (30511.s7) in the neck region and (30511.s8–9) in the abdominal region. Textile (30503.s10) box 1 was found in the head region of infant (30513) and box 2 above (30513). The textile fragments from (30503) are small; the largest piece is 7 x 4cm (figs 11.11, 11.12).

Textile remains from Building 77, Feature 7611. Possible textile remains and cordage were found in 2014. They are labelled (21605.s3 fill, .s5 and .s8), and 21606.s1, .s2, .s3, .s4, .s5, .s6, .s8, .s9 (associated with .s5), .s10 and .s11 (continuation of .s1, .s10). They derive from Space 336, F.7611, skeleton (21606). This was a juvenile primary burial, flexed and bound. As the fibres were badly preserved, none of the samples could be identified unequivocally as textiles but may be remains of them. Several were strings and cordage. Some of the cordage (21606.s1) was found attached to the skull (fig. 11.7). Hide remains (non-human) were found underneath the cordage (see fig. 11.16).

Textiles from Building 131. The textiles from Building 131 were found in 2015, 2016 and 2017. They derive from Space 500, Features 7956, 7977, 7997 and 7961.

Feature 7956, skeleton (22661). Skeleton (22661) is a female, 20–30 years old, tightly flexed and wrapped in cords and textile. The textile remains from F.7956 are labelled (22661.s2, .s3 and .s5); cordage from the same feature is labelled (22661.s4, .s17, .s18, .s19 and .s20). Samples labelled (22661.s11–.s13) were exported by Fuller in 2015 and were not seen by the current authors.

Feature 7977, skeleton (30040/30045). Skeleton (30040/30045) is a young adult possible female from F.7977. The textiles are labelled (22676.s4 and .s12) (subsample of .s5). A basketry impression, (22676.s15), was initially described as a textile impression.

Figure 11.11. Textile in tabby (30503.s9).

Feature 7997. Feature 7997 is designated a pit that contained the remnants of a carbonised basket (32373). Textile fragment (32373.s1) derives from this basket.

Feature 7961, Skeleton 23126. Feature 7961 is a multiple burial. The primary inhumation is skeleton (23126), an adolescent female found with objects such as an obsidian mirror with ochre, beads, pigment, a wooden bowl and a large pottery fragment placed underneath the cranium. A tiny possible textile fragment labelled (30039.s10) was found among other organic material on the leg of skeleton (23126).

Summing up

The textiles from Çatalhöyük are very homogeneous. They are all made from spliced threads. Yarn diameters vary slightly between find spots: in Building 49 and Building 131, F.7956 they are 0.5–0.7mm; in Building 131 F.7961 and F.7963 they are 0.8mm; and in Building 52 they are 1.0mm. Most are woven in tabby, in Building 49 possibly with an edge in basket weave. One of the fragments from Building 52 has remains of what may be a transverse (starting) border, others appear to end in fringes, and two appear to end with one or two rows of weft-twining. This also applies to one of the fragments from Building 131,

Figure 11.12. Weft twining (30503.s10).

F.7961 and F.7963 (22661.s2 and .s3, 30503.s10). The twining in (30503.s10) consists of both a single and a double row of weft-twining, the latter creating a herringbone effect (see fig. 11.12). Organic remains from Building 77 were too badly preserved to identify as textiles.

Some pieces (22661.s2 and .s5) from Building 131, F.7956 had fine knots or loop ends and could derive from a net or a tassel. They may represent a form of decoration at the end of fringes or, perhaps, have become detached from the textiles (fig. 11.13). The woven fabrics have thread counts between eight and 14 threads/cm, but they appear almost balanced in each individual piece. As regards the weft-twining, distances between 'weft-rows' are 7–10mm.

This corresponds well with the findings of Harold Burnham (1965). He documented twining and soumak along with the tabby-woven textiles. The woven tabbies of Burnham had thread counts of 10–16 threads/cm. Like those from the current project, they are nearly balanced. Burnham also found narrow tapes in warp-faced tabby with simple selvedges that were used to tie the deceased into a foetal position. Burnham describes the weft yarns of these as z-twisted, but photos (Burnham 1965: pl. XXXIIb) show that, like all other fine threads, they were spliced. Burnham also found various edges. One textile, from EVI,1 reportedly has a heading or starting cord, made by inserting what was to become warp threads into each turn of a plied cord. Three to four threads follow in the same shed to strengthen the starting border before the regular tabby weave begins (fig. 11.14). The same textile is also reported to have a rolled hem, sewn with a quite coarse thread. Burnham mentions another rolled hem but does not supply details of the textile. Another tabby (without specified technical details) is described as having been mended; it was sewn with overcast stitches onto a different fabric.

Non-Neolithic textiles

One textile, (19587.s3) proved to be post-Chalcolithic (table 11.6). It is a very fine textile imprint on clay, tabby or possibly twining. Another, (8770.s3), is a fine tabby with fine z-twisted threads that proved to be cotton. The single yarns and cotton fibres indicate that the textile is not Neolithic, corroborated by the fact that it derives from a burial in one of the top layers of the site.

Hides

A number of hides were recognised while looking through boxes with organic remains from burials. Samples for the SEM were taken to try to find out more: if it was hide and which kind of hide (table 11.7). These remains are especially interesting as they reflect inhumation practices. Exposure to high temperatures in a burning building will cause animal fibre to melt; hide will be charred and obtain a glassy surface.

Results from examination by SEM demonstrated that some samples were indeed remains of hide; others were botanical ((20413) grass, (32330) wood). The burial of child skeleton (21606) F.7611 in Building 77 was especially interesting, as remains of hide (21606.s3) were visible on parts of the cranium (figs 11.15 and 11.16). If the pores are of different size, it cannot be human skin. Such differences are especially well visible in the hide remains in burial F.7056 in Building 131 (22661). The SEM photo shows some clearly visible pores, large ones of nearly 100µ and, next to them, fine ones of 15–20µ (see fig. 11.4). The 'wall' on the right of the large hole (pore) has a glassy shape from the heat. In the same grave, a hide could be documented on a textile. The results show that the dead in some instances must have been wrapped in hide(s).

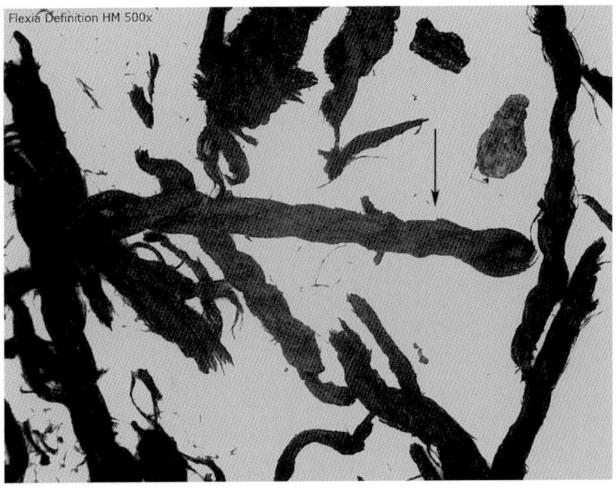

Figure 11.13 Fragment of unit no. 22661.s2 with loop ends.

Figure 11.14. Technical drawing of starting border (illustration by Atelier Oculus, Zürich).

Unit no.	*Found*	*Size*	*Yarn type*	*Yarn dm (mm)*	*Technique*	*Threads/ cm*	*Comments*
8770.s3		Largest fragment 1x1cm	z/z	0.5/0.5	Tabby	10/12	Iron Age
19586.s3	Burial fill in Sp.1002, F.3689	Largest fragment 4x1.2cm	-/-	0.5/0.8	Twining or tabby	18/12	Post-Chalcolithic Several lumps with textile impression, very fine fabric

Table 11.6. Non-Neolithic textiles.

Unit No.	*Context*	*Stereomicroscope*	*SEM*
20413	Infant skeleton (19494)	fibrous material over base of skull/vertebras, ev. pores	plant remains, stomata visible, grass-types, oats?
21606.s3	F.7611, primary disturbed burial from construction phase of Building 77	distinct remains of cordage over a larger piece of skin on left temporal	hide, surface with pores, animal indet., not human skin
22661.s6	Building 131, Sp.500, F.7956, burial of young female, around right foot	remains of skin/hide	melted/glass-like surface, badly preserved lumps of fibres, dm. about 20-100μ, indet.
22661.s21	Id., wrapping under legs?	remains of skin/hide	surface with pores, large and fine, dm. up to 100um, sheep?
23765	Burial	remain of possible skin or hide, yellowish, with little bobbles at the edges; under this layer phytolith plant remains (wood or bast). Bloc heavily treated with conservation product	no sample
30039	Building 131, burial with 4,000 beads	several fragments of hide	not well preserved, indet.
31705	Burial, organic layer under neck	fragment of orange layer, probably hide, badly preserved	no sample
32330	Building 131, Sp.500, F.7963, burial fill	orange layer, hide?	wood

Table 11.7. Hides.

Production processes

The *chaîne opératoire* of textile production consists basically of the procurement and processing of raw materials (fibres), the transformation of these fibres into threads and then the creation of flexible, flat materials by weaving or twining. This is followed by finishing processes such as wetting, and perhaps further shaping and/or decoration. Cord-making also consists of the harvesting and processing of fibres and turning them into threads that are then plied and re-plied until the desired type of string or cord has been obtained.

Most time and effort for basketry production is spent on the collection and selection of grasses, leaves, culms and other raw materials. Most materials would have been found in the swampy environs of the settlement. The best period for harvesting was the late summer. Typically, basketry materials are cut and then left to dry, a procedure needed both to avoid shrinkage of the elements in the fabric and to allow storage for later use in periods when time- or season-sensitive tasks are done. Depending on the basketry technique, the dried plant parts are briefly soaked or only wetted before final use to restore flexibility and are then intertwined in different ways as outlined below.

Figure 11.15. Hide on fragment of skull, unit no 21606.s3.

Figure 11.16. SEM-photo of the hide (see fig. 11.15) (21606.s3).

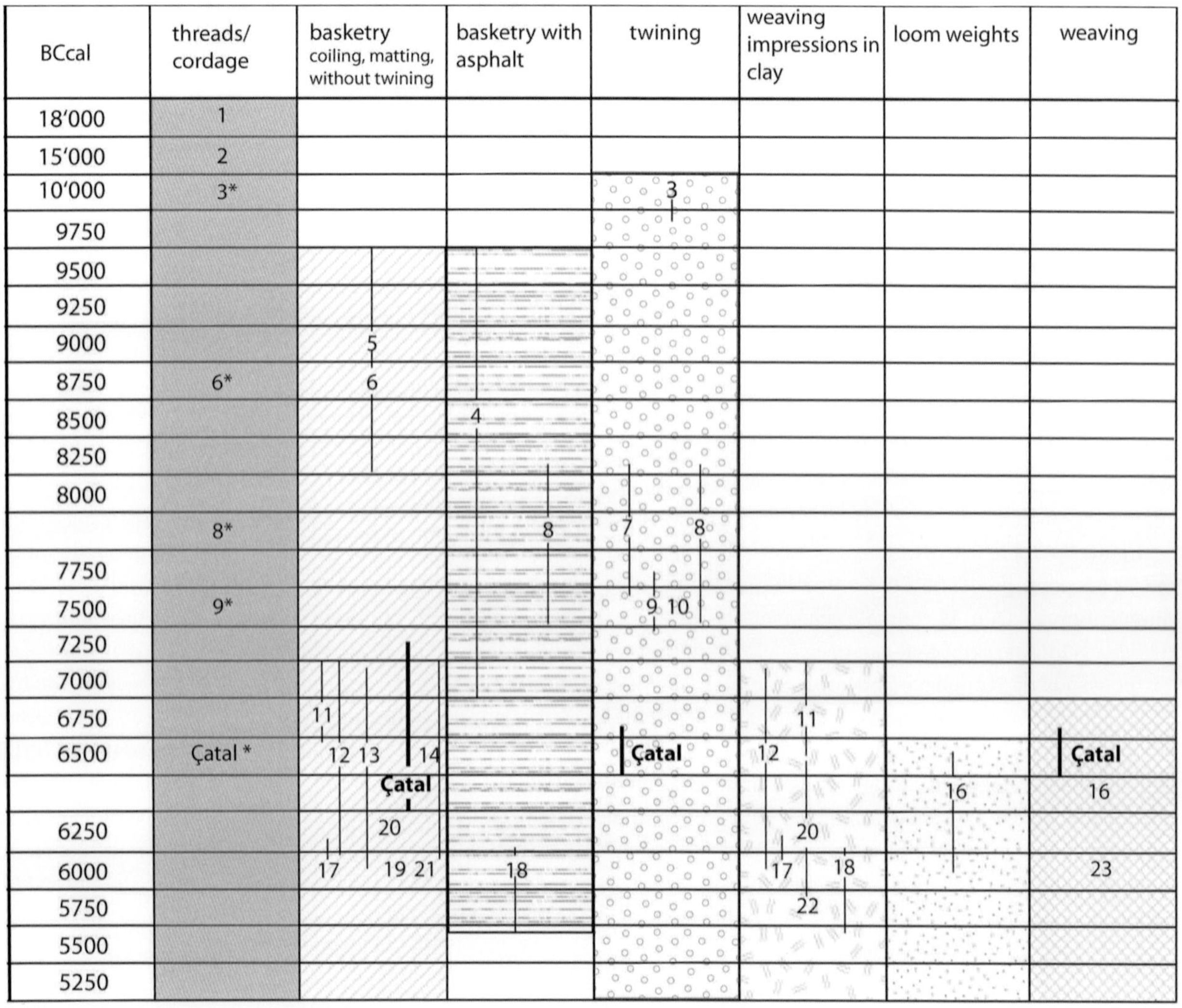

Figure 11.17. Development of techniques.

during the third quarter of the seventh millennium BC (Rooijakkers 2012: 101–03), and loom weights seem to begin about the same time. The loom weights found in levels dated to 6200–5800 cal BC at Ulucak Höyük in western Anatolia (Çilingiroğlu 2009; Gleser 2016) are currently the oldest from this area, but it is uncertain whether they were used for weaving. It is quite possible that they were weights for large items of twining, as has been the case in Europe. Another argument is the lack of starting borders, which are typical for a warp-weighted loom.

The woven textiles from Çatalhöyük were used as cloth to wrap bodies. Before, in everyday life, these textiles were probably used as clothing, as suggested by the early figurines found at Jarmo wearing long tunics with fringes (Braidwood et al. 1983: fig. 161). The early depictions in Çatalhöyük, such as the figurine published by Mellaart (phase II; Mellaart 1962, pl. 23b), suggest that skirts with fringes were worn, but otherwise most garments were presumably made of hides.

Hide

The fragments of hides from Çatalhöyük were all found in connection with inhumations. The dead were obviously wrapped in or covered with hide. The preservation of hide is difficult, and finding and recording it is even more difficult. Hide fragments were for a very long time not really a research subject, but it is now beginning to emerge (Van Driel-Murray 2000; Hatziminaoğlou, Boyazoğlu 2004; Rast-Eicher 2010; Spangenberg et al. 2010; Ruß-Popa 2018).

As wall paintings show (e.g., Mellaart 1967: fig. 61), hides were a main component of clothing at Çatalhöyük, but they would also have been used for bags or other items. The hides depicted seem to be from leopards, as they are spotted. But hides from other animals such as sheep, goat or ox are very well possible, too. Throughout history, hides have been important, from the Palaeolithic period to modern times – as protection against a cold climate, as well as to display affluence and social prestige. Hides were certainly the first materials used for garments.

Evidence from Palaeolithic graves shows beads in very structured compositions, indicating that they were sewn onto hide. In Sunghir (Russia), two individuals from the Gravettien Period (29,000–26,000 bp) were buried in clothing decorated with thousands of mammoth ivory beads (Trinkaus, Buzhilova 2012). Two children in the 'Grotta dei Franciulli' (prov. Imperia/Italy), dated to the late Upper Palaeolithic (13,000 BC), also had thousands of beads (Mussi 2002); the grave of a child in the cave of 'Arene Candide' (prov. Savona/Italy) contained about 80 squirrel tails, very probably the remains of a fur mantle with tail decoration (Mussi 2002). In Scandinavia, late Mesolithic burials excavated at Vedbæk (Denmark) also display groups of beads, indicating decorated skin garments (Albrethsen, Brinch Petersen 1976).

In the Neolithic lake dwellings of Central Europe, the soil is alkaline, preventing the preservation of hides or animal fibres. Due to climate change and melting glaciers, Neolithic skins that had been enclosed in the ice for thousands of years have appeared. The most famous of them were with the 'Iceman' found in the glacier of the Tisenjoch in Northern Italy and dated to 3300 BC (Egg, Spindler 2008). He was wearing arctic clothing, a combination of hide garments and a grass mantle to prevent the hides from becoming wet. The trousers and the upper garment, a sort of long jacket, were made of red deer and goat hide (very solid and smooth), the girdle is from cowhide and the head covering from bear fur. Similar objects of the same period were found on the Schnidejoch, a passage at 2700m altitude in the Swiss Alps (Hafner 2015). One item was a large piece from trousers of goat hide – interestingly, according to DNA analysis, an Asian type (Schlumbaum et al. 2010) – and a cape of willow bast with a fine leather strap to close it under the chin (Rast-Eicher 2015). As soon as metals appear, hides can be documented on metal objects added into graves, often pointing to capes or wrappings. In northern Europe, many hides have been found in bog burials or oak coffins dated to the Bronze or Iron Age (Hald 1980). At Çatalhöyük, beads (no 17457, see fig. 11.6) found with hide may well represent similar clothing (see Chapter 10).

Due to difficult preservation of the organic material at Çatalhöyük – textiles and hides – and the lack of specialist analyses during the excavations, interpretation is difficult. As the textiles appear to have been made from wild plants, we find large-scale textile production for clothing unlikely. It follows that hides remained an important part of clothing.

Conclusion

The body at Çatalhöyük was covered in life and in death, but most of the evidence we have speaks to the latter. Small textile fragments, mostly preserved in contact with metals or as charred remains, are the only evidence for what was probably clothing. The dead, buried under the feet of the living, were separated from the earth either by wrapping the body in hides or putting it in a contracted position in a coiled basket. Matting does not seem to have been used to wrap the dead, but only as floor covering; this speaks to the stiffness of the matting materials, but perhaps also to an understanding of the need to separate the living from the earthen floor. No continuous matting was found on the platforms under which the dead were resting, but it is on these platforms that plate-sized coiled mats were found. These similarly form a separation of objects, or maybe foodstuffs, from the ground surfaces. From the little evidence that was found, we can conclude that there is consistency in techniques, which points to local craft production. Detailed study of these coherent technical traditions shows, however, quite some variation in those aspects, which is indicative of several non-specialised households being involved in the production.

Acknowledgements

Our thanks to Ian Hodder, director of the Çatalhöyük Research Project, for inviting us to join the project, and to members of the project for sharing their knowledge. A generous grant from the Agnes Geijer Foundation for Nordic Textile Research made it possible to carry out fibre analysis of selected samples of textiles, cordage and basketry from Çatalhöyük. We also owe thanks to Gillian Vogelsang-Eastwood of the Textile Research Centre in Leiden for allowing us to examine the textile sample from the 1963 season at Çatalhöyük now in the collections of the Textile Research Centre. We thank the Institute of Geography of the University of Bern (Oeschger Centre), Switzerland for use of their scanning electron microscope.

12. Spatial contexts, temporalities and experiences of architectural paintings at Çatalhöyük

Gesualdo Busacca

Introduction

Since the early days of excavation at Çatalhöyük, architectural paintings have been in the spotlight of both scholarly interpretations and public perceptions of the site. This is due not only to the legacy of Mellaart's publications, which provided an impactful visual corpus and relied heavily on the paintings' iconography in his interpretation of the site, but also to the special importance accorded to iconographic artefacts within archaeological assemblages, in particular in the field of prehistory (Conkey 2010). Because of their perceived exceptionality, scholarly approaches have tended to treat Çatalhöyük paintings as an isolated category of artefacts to be interpreted solely based on their visual content (De Jesus 1985; Gimbutas 1990; Forest 1993; Czeszewska 2014). Since the beginning of the Çatalhöyük Research Project (ÇRP), there has been an increase in studies dealing with the contexts (Last 1998; 2005b), material composition and technical aspects of paintings (Çamurcuoğlu 2015; St. George 2012), along with a number of studies investigating the geochemical composition and application of plasters (Koppelson 1996; Matero 2000; Wiles 2008; Anderson et al. 2014; Doherty 2017).

Building upon these important contributions, recent research has investigated specific themes related to the contexts and materiality of architectural paintings at Çatalhöyük, based on the enriched archaeological dataset resulting from 25 years of excavation and research at the site. The main aim of this research has been to fully situate painting practices within their spatial and temporal contexts, further investigating aspects such as their temporality and duration of exposure, as well as their experience in relation to lighting conditions within houses (Busacca forthcoming). Besides presenting this recent research, another important aim of this chapter is to provide a synthetic overview of the painting corpus at the end of the Çatalhöyük Research Project, with an emphasis on temporal changes in painting practices across the site's occupation. For this reason, data are often broken down according to the four-period chronology used elsewhere in this volume: Early (7100–6700 BC), Middle (6700–6500 BC), Late (6500–6300 BC) and Final (6300–5900 BC).

Given the scope of this volume, this chapter focuses primarily on the paintings uncovered in the main North (including BACH) and South excavation areas of the East Mound, with occasional references to Mellaart's corpus and to the paintings from the IST, TPC, TP and West Mound excavation areas.

The architectural painting corpus

Quantifying Çatalhöyük's painting corpus is not an easy task, given the heterogeneity and variable state of preservation of the archaeological assemblage. A general challenge of studying earthen materials from a horizontally multi-stratified site such as Çatalhöyük is that, except for the few extant specimens selected for preservation, sampling or under excavation, most of the data comes from the documentation of excavated paintings (textual information, photography, drawings, 3D recording and so on) contained in the excavation database. The site's excavation database includes a total of 256 excavation units containing paintings, spanning from scanty remnants of paint to better-preserved paintings (out of a total of 1,106 units categorised as plaster). Given the multi-layered nature of plasters at the site, some of the painted units include multiple layers of painting, bringing the minimum total number of painted layers to 331. Most of the paintings recorded in the database belong to Level North G and, in general, to the stratigraphic levels attributed to the Middle period (fig. 12.1). This distributional bias is certainly due in part to the more extensive archaeological exposure of the middle levels, especially in the North Area, in comparison with the early and late ones, but it also reflects a peak in painting activity during the Middle period, as suggested by the percentage of painted units and that of painted layers in relation to the total number of plaster units per period, as shown in figure 12.1.

Paintings occur on a variety of plaster-lined house interior features, such as walls (42.3%), platforms (24.7%), niches, benches, posts/pillars (each 6.6%), floors (4.9%) and others (fig. 12.2). Given this variable distribution, the umbrella term 'architectural paintings' is preferred here to previously used terms such as 'wall paintings' (e.g., Czeszewska 2014; Çamurcuoğlu 2015) or 'murals' (St. George 2012).

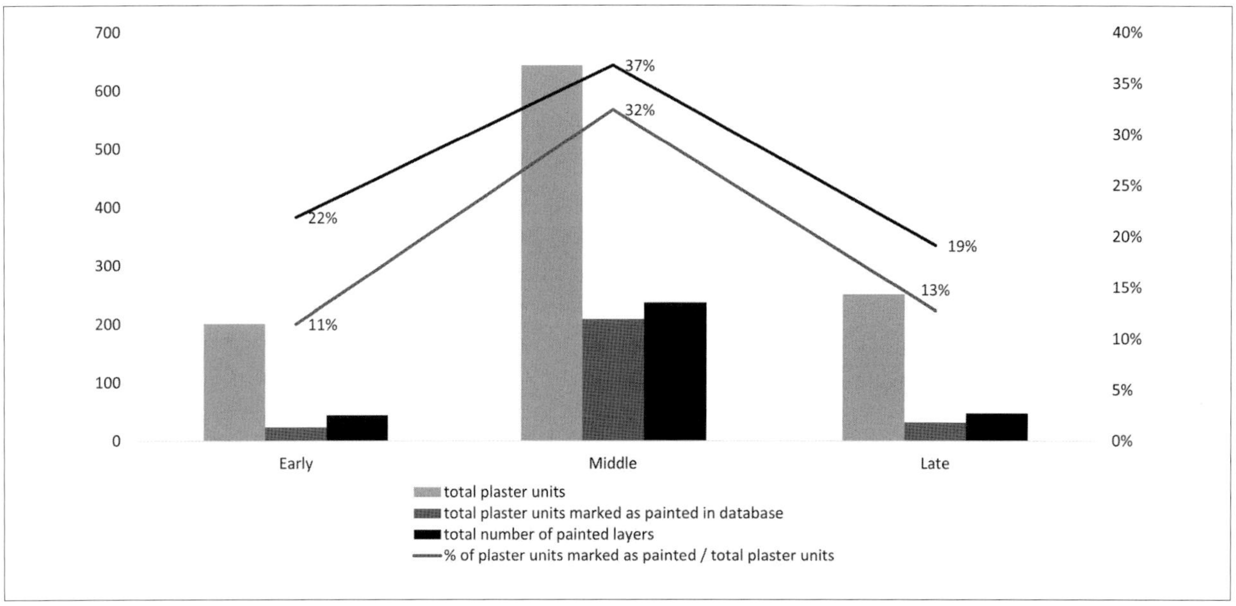

Figure 12.1. Temporal distribution of paintings at Çatalhöyük. The first column (light grey) indicates the total number of excavation units categorised as plaster, whereas the second (medium grey) and third (black) show the number of paintings, counted respectively as painted units and as singular painted layers. The trendlines show the ratio of painted units to total plaster units (medium grey) and of painted layers to total plaster units (black).

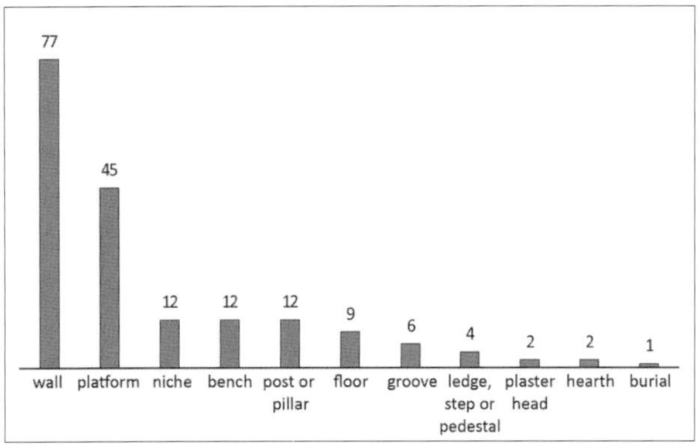

Figure 12.2. Distribution of painted units by feature type.

In terms of painted motifs, the vast majority of architectural paintings from Çatalhöyük are composed of monochromatic red layers (58.6%), followed by a significant portion of paintings whose motif cannot be identified due to poor state of preservation or insufficient exposure (23.1%). Geometric motifs make up more than 15% of the painting corpus, while hand motifs (2.1%) and combinations of geometric and hand motifs (0.6%) are less frequent. Figure 12.3 shows the motif distribution through the three main periods of site occupation considered in this study. The most surprising feature of the ÇRP corpus as a whole, especially when compared with Mellaart's one, is the almost total lack of anthropomorphic or zoomorphic motifs, with the only exception being the human depiction found in a burial context in the TP area (Czeszewska 2014: 187; Marciniak et al. 2015a: 170). Zoomorphic motifs, however, occur in the related category of plaster reliefs, which sometimes show painted decoration (Lingle et al. 2015; Busacca, Lingle 2017).

Unlike anthropomorphic or zoomorphic motifs, the geometric and hand motifs uncovered by the ÇRP show parallels with some of the paintings excavated in the 1960s. Hand motifs are not attested at Çatalhöyük before the mid-levels of occupation and continue during the Late period (see fig. 12.3). Direct examination of hand paintings reveals evidence of brush strokes, indicating that the design was obtained by paint brushing and not simply by hand printing; it is possible that brushing was performed after hand printing in order to retouch and finish the hand impression (fig. 12.4). Hand motifs are most often red-painted on a white background, but sometimes appear as white negative spaces on a red background, as in the case of the platform in Building 49 (fig. 12.4).

Geometric motifs show a wide range of designs and combinations, but some recurring groups of motifs can be identified. Among the most often to appear are rhomb motifs, honeycomb motifs, ladder motifs and combinations of vertical lines, triangles and brick motifs. Rhomb motifs are usually formed by juxtaposed concentric rhomboidal shapes and are sometimes both painted and incised, as in the case of Building 119 (fig. 12.5). Rhomb motifs seem to occur mainly in the Early period (Mellaart Levels VIII and VII; Building 2, Level South K) and in the earlier levels of the Middle period (Building 119, Level North F).

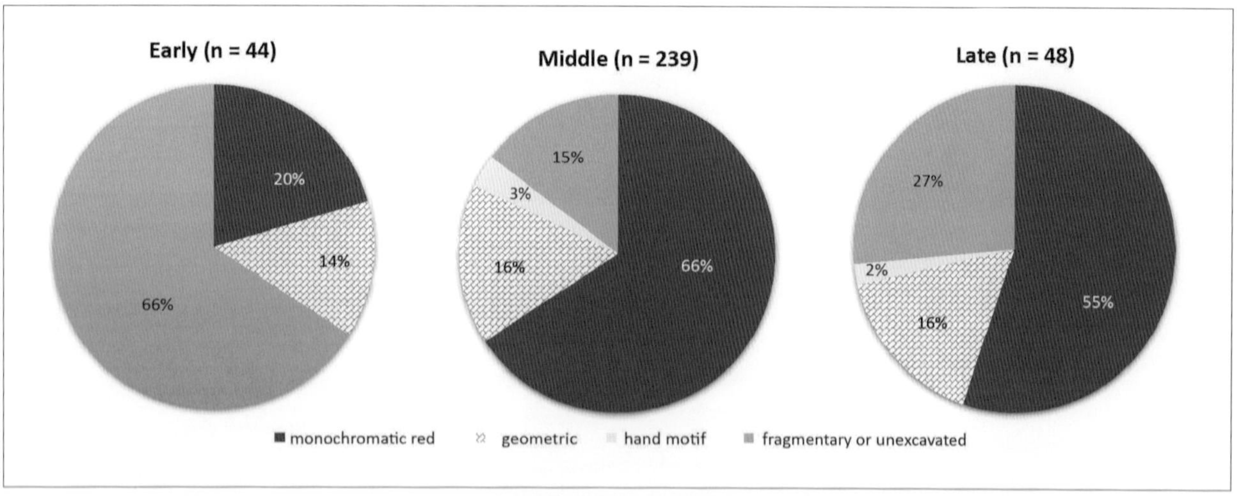

Figure 12.3. Distribution of painted layers by motifs during the main occupation periods of the site.

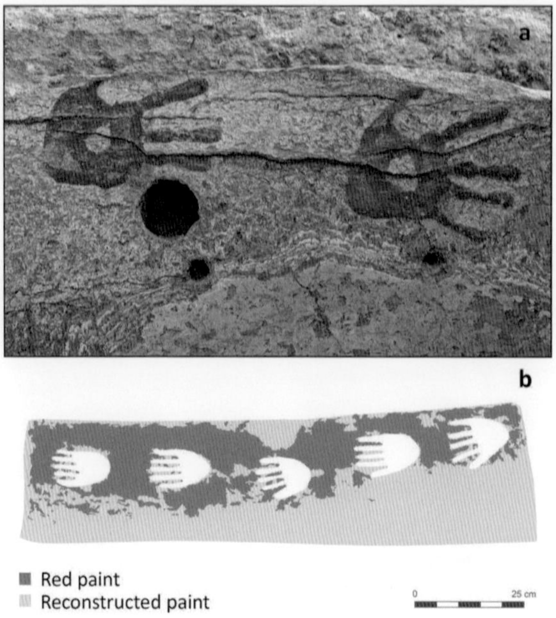

Figure 12.4. Hand motifs: (a) from B.77, F.3094; (b) from B.49, F.1651 (illustration by Kathryn Killackey) (for colour version, see online supplementary material).

Honeycomb motifs are broadly identified as round or round-edged rectangular lines dividing the painted space into regular partitions (fig. 12.6). Like rhomb motifs, these designs occur mainly in the early levels (Mellaart Levels VIII and VII; Building 2, Level South K) and in the early levels of the Middle period (Building 8, Level South M). Ladder motifs are groups of three (or more) red-painted or black-painted vertical lines that are intersected at regular intervals by oblique lines, with the corners of intersection sometimes marked by additional red paint (fig. 12.7). Their occurrence is strongly concentrated in the Middle period, as they are mainly documented in Level North G (B.77, B.114), South O (B.96) and in Mellaart's Level VI.B (fig. 12.7).

A complex geometric motif that occurs multiple times in variable configurations is a combination of two or more vertical lines (sometimes showing zigzag patterns), red-lined or red-filled triangles (sometimes containing rosette motifs) arranged along said lines, and brick patterns completing the design (fig. 12.8). Like the ladder motif, this complex combination of motifs appears concentrated in the Middle period, being attested exclusively in Levels North G (Buildings 1 and 49), South O (Building 80) and in Mellaart Levels VI.A and VI.B. To this loose categorisation of geometric painted motifs, one should add a number of paintings showing parallel lines, bands, checkerboards, spirals, triangles, brick motifs and various combinations of two or more of these motifs.

Spatial distribution and contextual associations

Given the highly structured nature of the use of space at Çatalhöyük, on both the short-term scale of daily and seasonal activities (Boivin 2000; Hodder, Cessford 2004; Matthews 2005a; 2005b) and the longer-term scale of house renovation and rebuilding (Hodder, Pels 2010; Lercari 2018), an investigation of the spatial patterning of paintings is vital for the understanding of their contextual associations with other archaeologically documented activities and interior features of the house. Indeed, during the early years of the ÇRP, the spatial distribution of wall paintings was investigated by Last (2005b), who noticed a frequent spatial association between paintings and burials, and Düring (2006: 194), who argued that paintings were concentrated on the eastern and northern walls of buildings. These observations, however, relied on a small dataset of paintings (besides legacy data from Mellaart's publications) and need to be revisited in the light of more recent data.

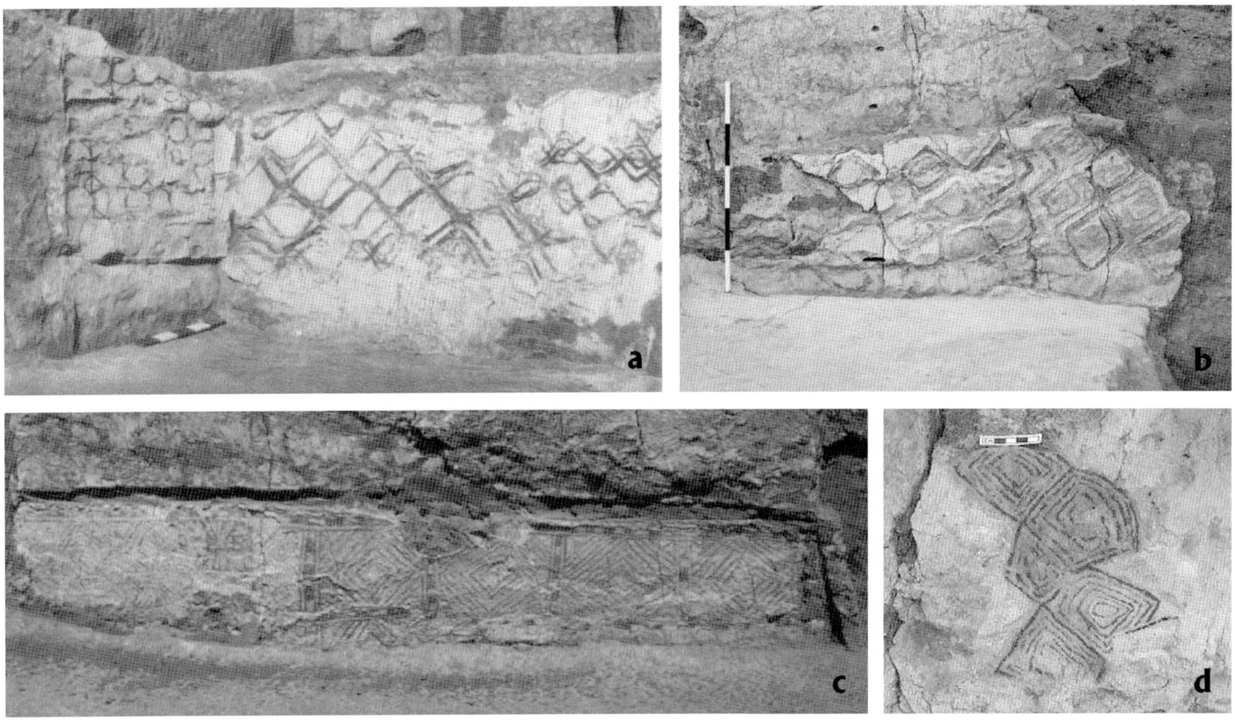

Figure 12.5. Rhomb motifs: (a) from Shrine VIII.25 (reproduced from Mellaart 1966: pl. 44); b) from B.119, (21501) (photograph by Jason Quinlan); (c) from Shrine E.VII.1 (reproduced from Mellaart 1989: pl. 6.1); (d) from B.2, (4223) (for colour version, see online supplementary material).

Figure 12.6. Honeycomb motifs: (a) from E.VI.8 (reproduced from Mellaart 1963: pl. XIb); (b) from Shrine VII.14 (reproduced from Mellaart 1964: pl. Vb); (c) from Shrine VIII.14 (reproduced from Mellaart 1966: pl. XLIIIb); (d) from B.8, F.415 (reproduced from Farid 2007: fig. 8.34); (e) from B.2, (3978) (for colour version, see online supplementary material).

A renewed investigation of the spatial patterning of architectural paintings at Çatalhöyük can now benefit from a larger dataset comprising not only paintings but also other kinds of archaeological data included in the Çatalhöyük geodatabase (Mazzucato 2013). Using the site's geodatabase and the absolute coordinates of paintings entered by excavators, as well as additional information such as photographic documentation and sketches, a general mapping of architectural paintings unearthed at the site since the beginning of the ÇRP has

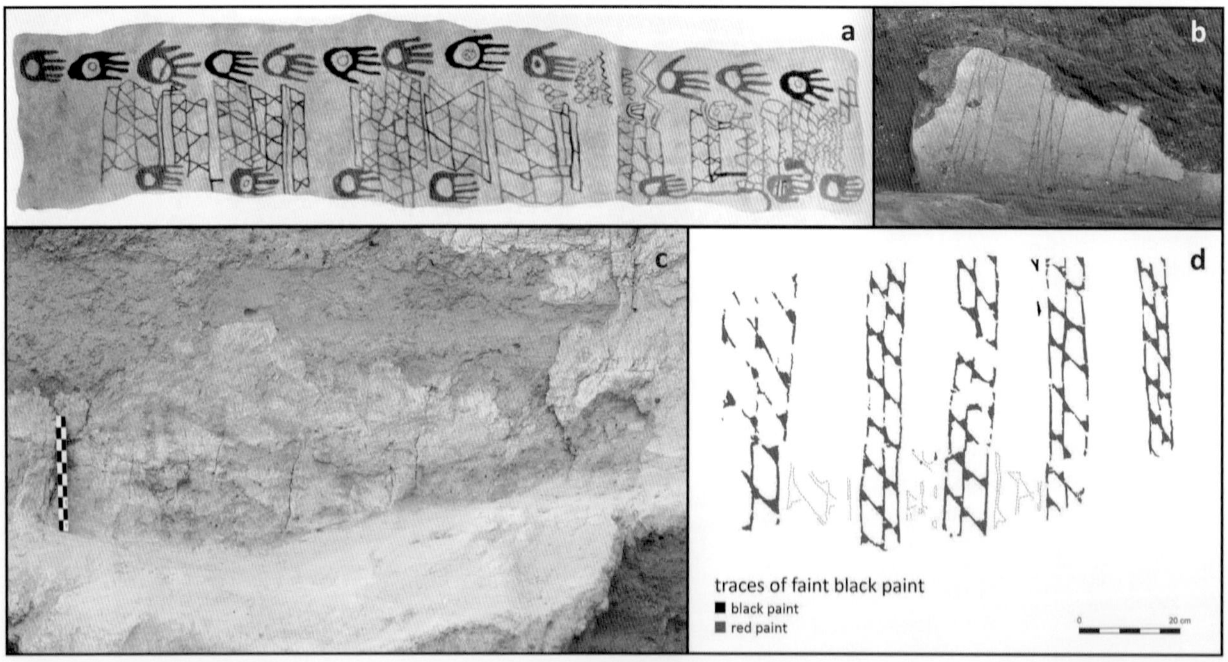

Figure 12.7. Ladder motifs: (a) from Shrine VII.8 (reproduced from Mellaart 1967: pl. 43); (b) from Shrine VI.B.50 (reproduced from Mellaart 1966: pl. 34b); (c) from B.114, (30031) (photograph by Jason Quinlan); (d) from B.77, (19051) and (19469) (illustration by Kathryn Killackey) (for colour version, see online supplementary material).

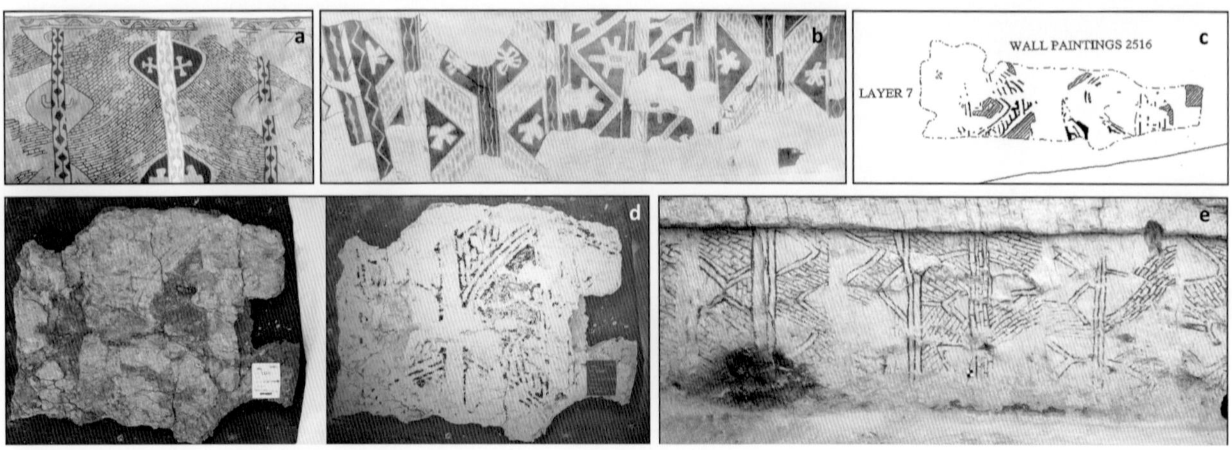

Figure 12.8. Combinations of vertical lines, triangles and brick patterns: (a) from VI.A.50 (reproduced from Mellaart 1967: pl. 37); (b) from Shrine VI.B.1 (reproduced from Mellaart 1967: pl. 29); (c) from B.1, (2516) (reproduced from Cessford 2007: 430); (d) from B.49, (13669) (the version on the right is colour enhanced for design detection); (e) from B.80, (18918) (photograph by Jason Quinlan) (for colour version, see online supplementary material).

been created. The first outcome of this mapping project has been the creation of phase maps showing the location of paintings in association with broadly contemporary buildings, following the current level system (for an example of a phase map, see figure 12.9; for the other phase maps, see Busacca forthcoming).

Besides this site-wide mapping, spatial analysis has also included an exploration of statistical patterns in the distribution of paintings across and within houses at the site. For this study, the main room (which is where, with almost no exceptions, paintings are located) of each house has been partitioned into nine sectors using an ideal grid dividing each wall into three sectors, following where possible the observed partitioning of indoor spaces by means of architectural features such as platforms, benches, and posts (fig. 12.10). In comparison to earlier spatial analyses, this method allows for more fine-grained statistical analysis, also enabling the inclusion of non-wall features and spaces such as platforms, benches and floors.

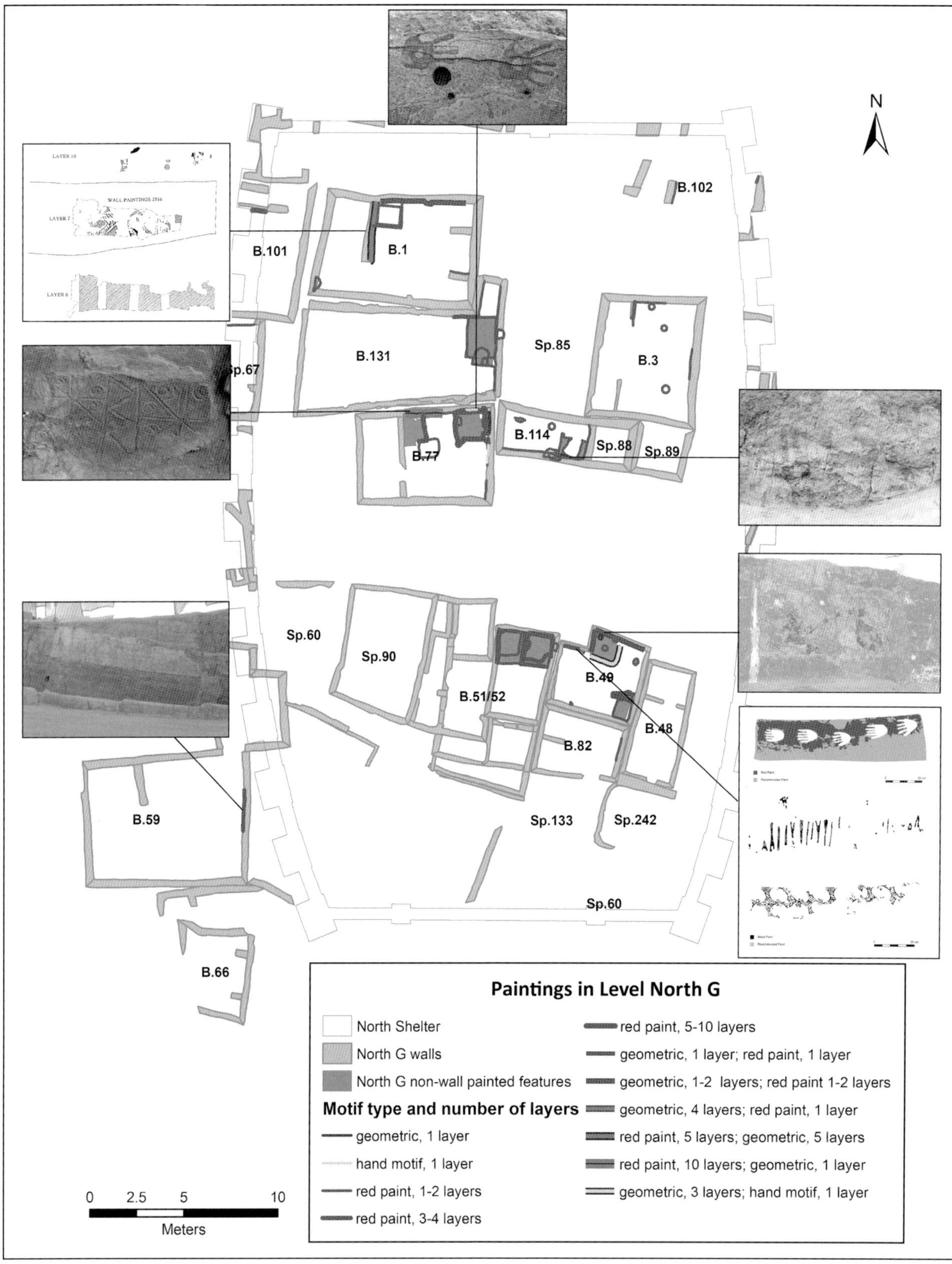

Figure 12.9. Building outlines and paintings attributed to Level North G (for colour version, see online supplementary material).

This analysis reveals significant concentrations of paintings in certain areas of houses, but also relevant changes in painting locations across different periods of site occupation (see fig. 12.11). If we consider data from all periods, paintings appear heavily concentrated in the northern and eastern sectors of houses. This reflects a

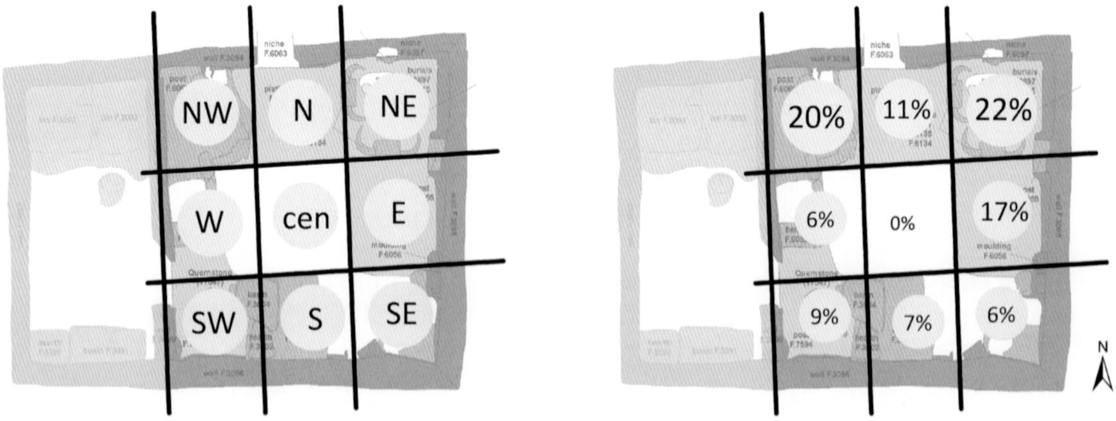

Figure 12.10. Left: ideal grid dividing the main room into nine sectors; right: statistical distribution of architectural paintings from all periods and houses (sketches based on map of B.77 produced by Camilla Mazzucato).

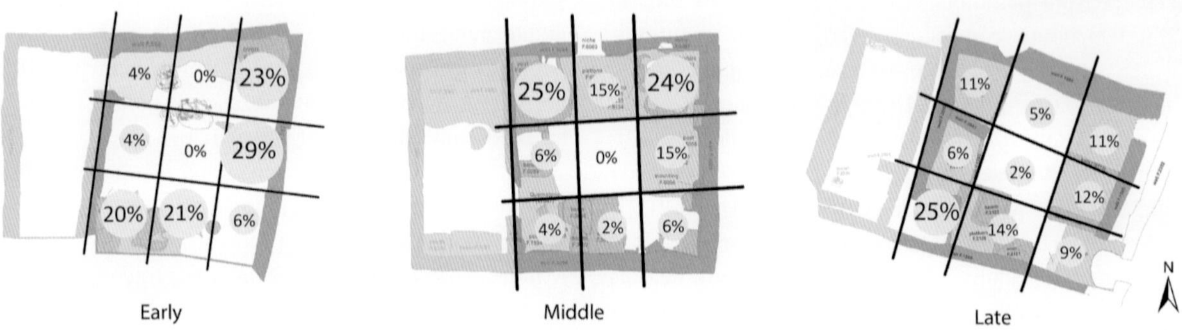

Figure 12.11. Statistical distribution of paintings within the houses of the Early, Middle and Late periods (illustrations based on phase maps of B.17 (Early), B.77 (Middle) and B.58 (Late) produced by Camilla Mazzucato).

standard organisation of Çatalhöyük houses where white-plaster-surfaced platforms, benches, burials and installations tend to be located in the northern areas of houses, while 'dirty' floors, activity areas, ovens, hearths and roof entrances generally occupy the southern areas and side rooms. In this general framework, paintings are usually associated with the northern, 'clean' areas of houses. However, significant variations from this general pattern occur when the data are plotted according to different occupational periods, indicating relevant changes in painting locations through time.

During the Early period, paintings appear strongly clustered in the eastern and northeastern sectors – reflecting in particular a preference for eastern walls – and also in the southern and southwestern sectors of the house (fig. 12.11). This bimodal distribution corresponds to general observations regarding architecture and use of space during this period, characterised by significant variability in the location of features such as ovens and hearths and by the lack of the clear spatial demarcations that become visible during the following period.

As house layouts and use of space became more neatly partitioned and standardised across the site during the Middle period, the location of paintings underwent an analogous clustering in the northern and eastern sectors of houses, with peaks in the northwestern and northeastern sectors, usually occupied by platforms. This strong pattern of concentration indicates an association of paintings with the 'clean' area of houses and with interior features such as white-plastered platforms, benches, plaster or bone installations and, in particular, intramural burials. During the Late period, the practice of architectural painting at Çatalhöyük underwent significant changes, reflected not only in the observed decrease in painting activity (see fig. 12.1), but also in different patterns of spatial location and distribution. Unlike in the previous period, paintings appear now spread out across different areas of the house, with a concentration in the southwestern sector. As suggested below, this remarkable change in location reflects a new association of paintings with fire installations and/or food-processing or production-related activities.

Feature-based spatial analysis was conducted in order to explore spatial associations of proximity between paintings and other house-based features such as burials, bone and/or plaster installations, benches, platforms, ovens, hearths and bins. Seven buildings were selected for this analysis, being completely or at least 75% excavated, complete in plan, and containing paintings and at least some of the other features considered in this analysis. Having mapped paintings and other features as points within building phase maps in a GIS platform, point distance analysis was performed for each phase of each building, calculating the average distance between paintings and other categories of features for each phase.

The results of this analysis are shown in figure 12.12. Buildings from the Early period (B.160, B.17 and B.43) do not show clear patterns of spatial association, as different features are located on average at similar distances from paintings, with the exception of bins in B.43, which are considerably farther from paintings. Clearer patterns emerge in some of the buildings of the Middle period (B.80, B.131 and B.77), where paintings are most closely associated with bone and plaster installations – when present, as in the cases of B.80 and B.77 – and burials, especially in the cases of B.131 and B.77. Platforms and benches also seem to occur relatively close to paintings, especially when compared to ovens, hearths and particularly bins, which tend to occur at the farthest distance from paintings. This confirms the observed pattern of painting concentration in the northern and eastern sectors of the house and away from the activity and food-processing areas, generally located in the southern areas.

Again, this analysis shows very significant changes during the Late period (B.58 and B.44). In an almost complete inversion of the previous period, burials are now on average at the farthest distance from paintings compared to other features, while ovens and hearths are the nearest, the latter sometimes showing traces of paint on their surfaces. A relevant case in this sense is that of an elaborated set comprising hearth and oven, the latter incised in its upper part with linear decoration, found in Building 57 (Sadarangani 2014), further indicating the emergence of unprecedented forms of elaboration associated with features such as ovens and hearths. As a whole, these significant shifts in painting frequency, location, and spatial and contextual associations are to be related to widespread socioeconomic transitions observed throughout the site during Levels South P and North H (Marciniak, Czerniak 2007; Hodder 2014d). In terms of visual culture and symbolic elaboration, this transition involved a decrease in architectural forms of visual elaboration (for example, paintings or plaster reliefs) and a parallel increase of more portable forms of visual culture, such as pottery (Yalman et al. 2013) and stamp seals (Türkcan 2013).

The relationship between painting and burial practices needs further elaboration. Besides a strong spatial association observed primarily during the Middle period – a pattern that was noticed already by Last (1998) – there also seems to be a correlation between the frequency of burials and of paintings across houses. A comparison between the minimum number of buried individuals and minimum number of painted layers across houses – considering only buildings excavated to at least 75% of their extent –

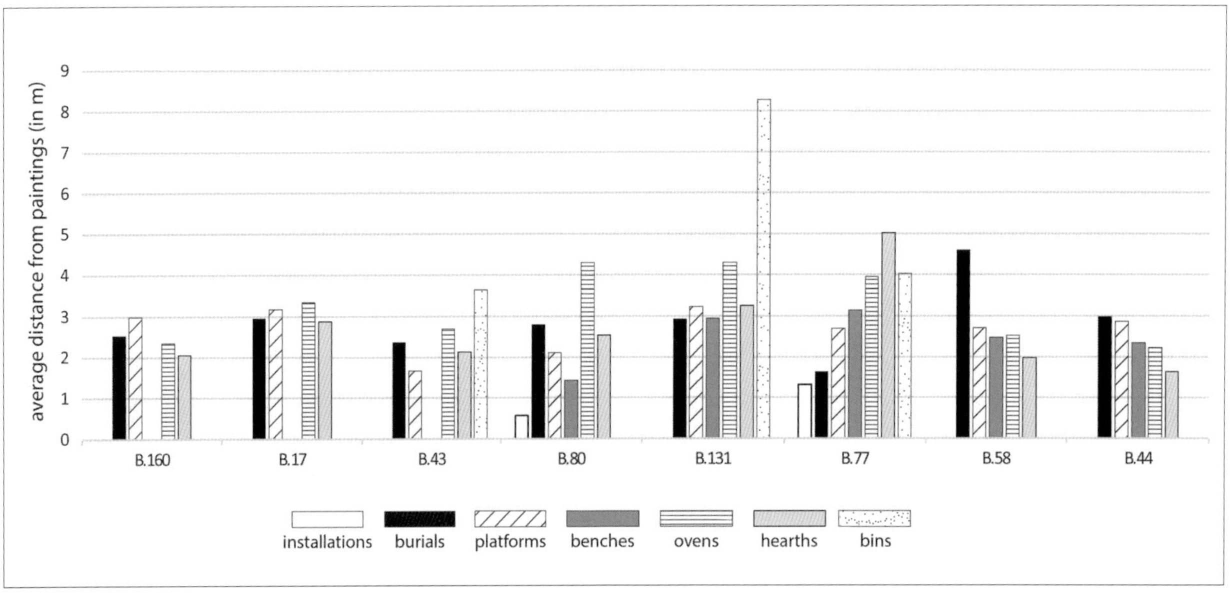

Figure 12.12. Feature-based spatial analysis of paintings in relation to other archaeologically identified features in Çatalhöyük houses.

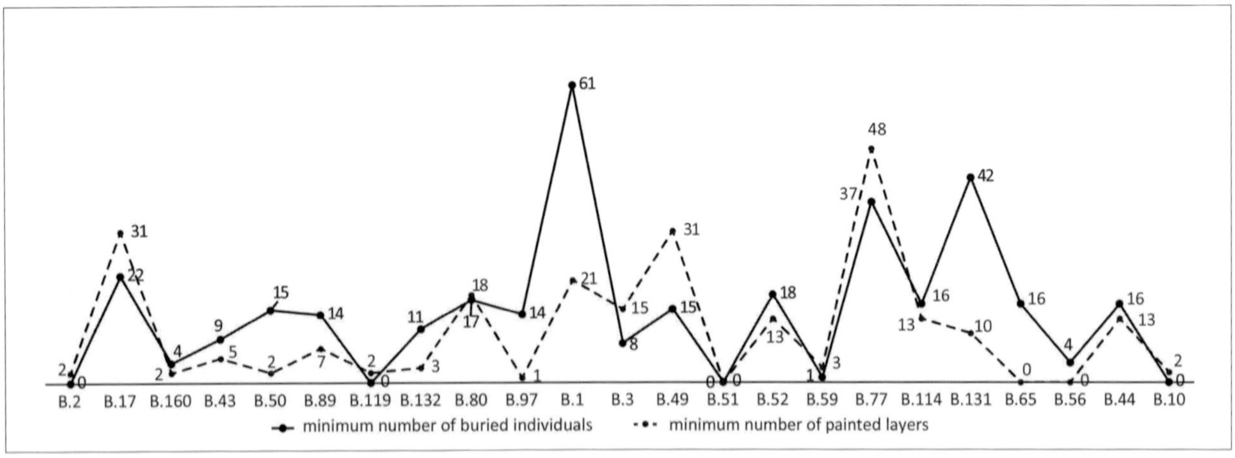

Figure 12.13. Minimum number of painted layers (dotted line) and burial MNIs (black line) in buildings. Only 75–100% excavated buildings were considered.

shows that buildings containing multiple burials often show multiple paintings (see fig. 12.13). Of the nine buildings showing a number of painted layers above the overall average of ten (Buildings 17, 80, 1, 3, 49, 52, 77, 114 and 44), eight of them (all except B.3) also show an above-average number of buried individuals (more than 14). In these buildings, therefore, heightened painting activity is accompanied by heightened burial activity, suggesting a direct association between painting and burial practices.

Another interesting question regarding the relationship between paintings and burials concerns their stratigraphic correlation. Although the conditions of preservation and almost complete lack of stratigraphic continuity between single paintings and burial events makes it difficult to link specific paintings with specific burials, some insight can be gained at the broader level of stratigraphic phasing. The attribution of paintings and burials to buildings' phases and subphases shows that, in some cases – for example, Building 1 – burials and paintings from the building sequence are concentrated in the same subphases (B1.2b, B1.2c, see table 12.1), indicating a strong temporal correlation between painting and burial episodes. In other buildings, however, most clearly in the case of Building 80, burial and painting events largely occur in different phases of the building's lifespan, although a clear spatial association exists between them (for example, between geometric painting F.5014 and the three burials dug in the adjacent platform, which belong to a later phase). Other buildings show more variability in the stratigraphic association between paintings and burials. This variability suggests that while in some cases paintings may have been executed during phases of burial depositions, in others, the two actions may have occurred at very different times – albeit in close spatial proximity – perhaps suggesting a link between painting activity and the commemoration of past burial events.

Temporalities of plastering and painting

As part of the architectural fabric of Çatalhöyük's built environment, paintings were enmeshed in processes of frequent house renovation and transformation, which characterise the site's earthen architecture, as observed both archaeologically (Barański et al. 2015) and ethnographically (Boivin 2000; Matthews et al. 2000; Stevanović 2012b). Far away from the idea of permanence and stability that defines the modern idea of 'retinal' architecture (Pallasmaa 2005), Çatalhöyük buildings, in particular their interior surfaces, were in constant transformation, as revealed by the multi-layered sequences resulting from repetitive plastering and painting. Micromorphological analyses have shown that paintings may have been left exposed for as little as a few months or up to one or two years, according to the annual and inter-annual rhythms of re-plastering observed at different locations within houses (Matthews et al. 2013).

Virtual reconstructions of fully excavated buildings with multiple painted phases enable an effective display of the mutability of painted interior surfaces throughout the lifespan of the house, also challenging an idea of Çatalhöyük's paintings as durable surfaces intended for prolonged visual contemplation (fig. 12.19). Recent studies of painted plaster sequences enabled further explorations of the frequency and temporal distribution of paintings within single sequences, as well as long-term transitions in plastering and painting practices throughout the entire occupation span of the site (Busacca forthcoming).

Building	Subphase	Burials	Paintings
1	B1.1b	5	0
	B1.2a	0	0
	B1.2b	14	3
	B1.2c	7	3
	B1.3	0	0
	B1.4	3	0
49	B49.2a-b	0	0
	B49.2c	0	0
	B49.3-5a	3	0
	B49.3-5b	3	0
	B49.3-5c	1	0
	B49.3-5d	0	3
	B49.3-5e	0	1
	B49.3-5f	0	3
	B49.3-5g	1	0
	B49.4	0	0
	B49.5a	3	1
	B49.5b	2	0
	B49.5c	1	1
	B49.6	0	0
	B49.7	1	4
77	B77.2	7	2
	B77.3	7	3
	B77.4	0	5
	B77.5	0	4
	B77.6	2	2
	B77.7	2	6
	B77.8	0	0
80	B80.1/2.1	0	2
	B80.2.2	2	0
	B80.2.3	1	1
	B80.2.4	5	0
	B80.2.6	0	3
	B80.2.7	0	0

Table 12.1. Number of burials and paintings in multiple-painting buildings of the Middle period.

In order to provide a general overview of different levels and areas at the site, cross-sectional sampling of painted plaster sequences involved areas excavated during the last years of the ÇRP (primarily the early levels in the South Area) as well as previously excavated materials, including plaster samples from the TP and Çatalhöyük West areas, whose investigation enabled an extension of this analysis into the final stages – Neolithic and Early Chalcolithic – of the site (table 12.2). Samples were analysed and photographed through optical stereo-microscopy in the Archaeology Laboratory at Koç University, Istanbul. Building on existing knowledge of the elemental composition of plasters (Wiles 2008; Anderson et al. 2014; Doherty 2017) and pigments (Çamurcuoğlu 2015) and of their multi-stratified application (Matthews et al. 2013), this analysis aimed to characterise entire plaster sequences and to identify the frequency and relative locations of painted layers within long-lived sequences. Following Doherty (2017), marl and softlime were identified as the two essential raw materials of Çatalhöyük plasters. Following Matthews and colleagues (2013), a plastering event is here considered a couplet of a foundation coat of marl and a thin, white softlime finishing coat, followed, in the case of a painted layer, by the application of pigment. However, since this research has documented that the marl-softlime couplet scheme is not largely attested beyond the Middle and Late levels, the notion of plastering event has been enlarged to include single marl applications, sometimes painted over, which were the norm during the Early and Final periods. Another important aim of this recent study was to extend the scope of previous investigations beyond a focus on the better-documented Middle period at the site, in order to understand broader dynamics of long-term changes in plastering and painting practices at Çatalhöyük. While the complete results of this study will be published elsewhere (Busacca forthcoming), an overview of the main results is presented here.

During the Early period, plaster sequences are composed of single applications of thin marl-based plaster layers, with rare applications of softlime coatings – unlike the regular couplets of marl and softlime coatings that will become common during the Middle period. Paint layers are extremely thin (0.02–0.04mm) and often show irregular sections, suggesting non-flat plaster surfaces. The frequencies and locations of paintings within sequences greatly vary across buildings: in Building 160, single layers of paint are located at the very end of the two plaster sequences examined; in Building 161, a single layer of paint is located toward the end of the sequence; in Building 17, several painted layers – reaching a peak of 11 painted layers on the eastern wall of the building (see fig. 12.14c) – are interspersed throughout the plaster sequences. In Building 17 sequences, painted layers appear sometimes in isolation, preceded and followed by unpainted layers; more often, however, they are clustered in groups of two or more consecutive layers, sometimes with the intermission of one white layer (see fig. 12.14). This pattern suggests that, in the life history of Building 17, phases of intense

Period	*No. of samples*	*Buildings and Spaces sampled*	*Average no. of plastering events*	*Average no. of painted layers*	*Highest no. of painted layers recorded*
Early	6	B.17, B160, B.161	21.8	3.4	11
Middle	15	B.3, B.5, B.40, B.82, B.89, B.96, B.101, B.114, Sp.455	32.6	2.6	10
Late	4	B.55, B.75, Sp.370, Sp.562	14.67	1.5	3
Final	3	Sp.410, Sp.412	2.3	1	1
Chalcolithic (Çatalhöyük West)	3	B.105, Sp.189, Sp.197	1	1	1
All periods	31		24.06	2.27	11

Table 12.2. Overview of cross-sectional analysis of painted plaster sequences.

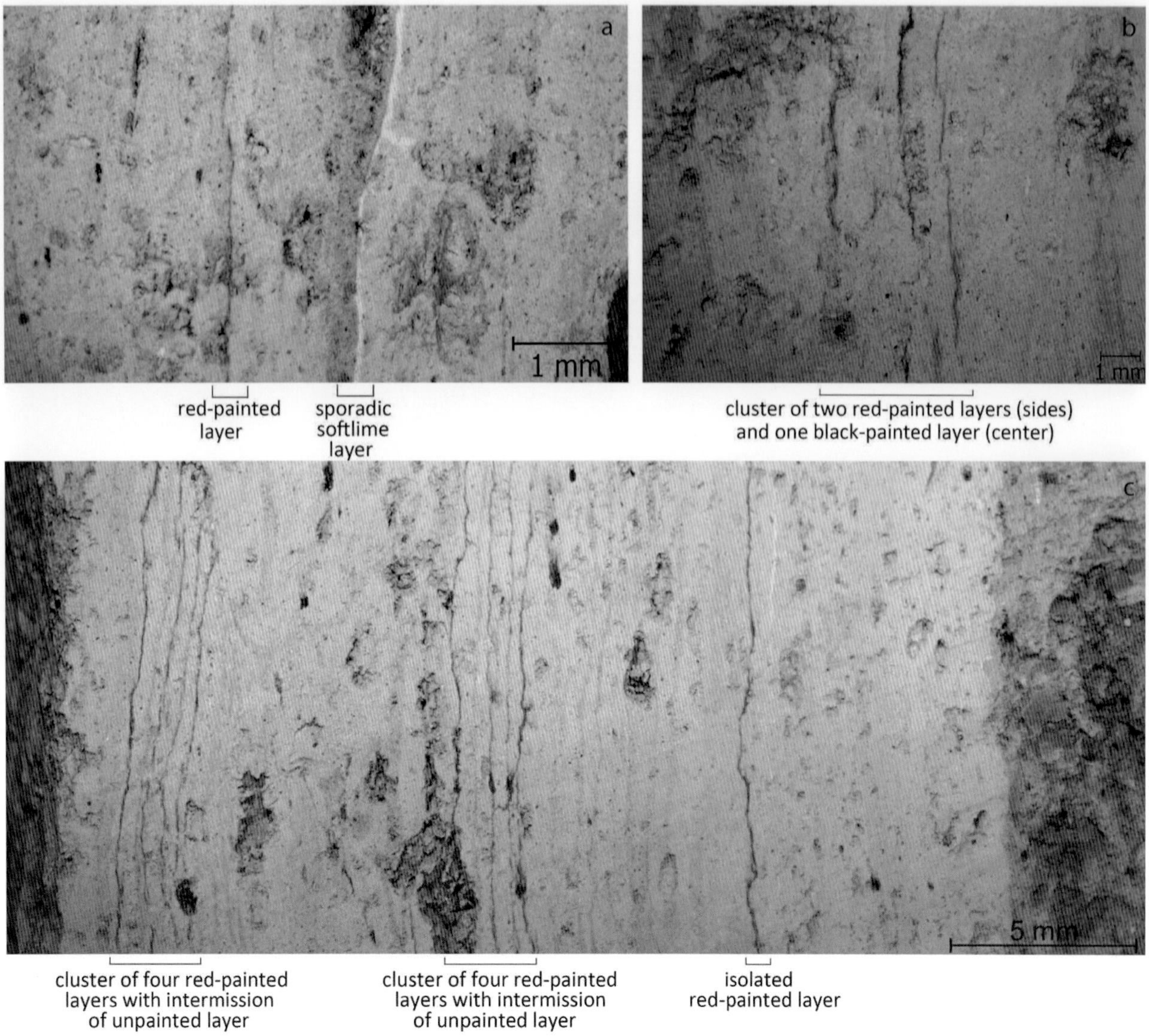

Figure 12.14. Cross-sections of painted plaster sequences from the Early period: (a) B.161, (32637); (b) B.17, (17348); (c) B.17, (20567) (for colour version, see online supplementary material).

painting activity occurred along with potentially longer periods of absence of paintings. The repetition of consecutive plaster layers may also be interpreted as the renewal or repair of a painting after one or more episodes of re-plastering.

The larger sample from the Middle period comprises a range of painting and plastering techniques, including single-layered painted plasters, multi-layered plasters composed of marl and softlime coating couplets, and mixed sequences with inconsistent use of softlime, especially in the earlier levels of the period (South M, North F). As already observed during the Early period, painted layers may appear in groups of two to six with no more than one unpainted layer in between, as observed in Buildings 40, 101, 114 and 96 (see fig. 12.15c). Another interesting characteristic of this period is that, in several cases, painted layer(s) occur immediately after a major episode of plastering, revealed by the application of a thicker marl plaster layer. In Building 96 (20871), for example, the first of a series of painted layers is laid onto a thick layer of marl plaster covered by a thin softlime coating (see fig. 12.15c); in Building 101, the first of a series of painted layers begins immediately after a major episode of remodelling consisting in the application of a thick marl plaster horizontal ridge on the wall surface (see fig. 12.16b). One possible interpretation for this observed practice is that a thicker make-up layer was laid in order to level the plaster surface in preparation for a painting. Another possibility is that painting activity itself marks a phase of extraordinary remodelling that may be associated with a specific event in the life history of the house.

The Late period is characterised by a decrease in painting activity (see fig. 12.1), which also results in a lower average number of painted layers per plaster sequence (table 12.2), with a maximum of three painted layers per sequence. In terms of plastering technique, there is continuity with the previous period, with the traditional alternation of marl and softlime coatings characterising the plastering technique across the sample (fig. 12.17). A dramatic change in plastering technique is observed during the Final Neolithic period, represented by the TP area, where the use of a much thicker and inclusion-rich marl plaster coated with a thinner white layer (presumably softlime) becomes commonplace (fig. 12.18). Plasters are single-layered or composed of a few layers, and no more than one painting per plaster sequence has been detected. This trend continues into the Chalcolithic period in the West Mound. The painted plasters examined for this period are all single-event sequences, composed of one thick, inclusion-rich plaster and a thinner layer on top of which paint is applied.

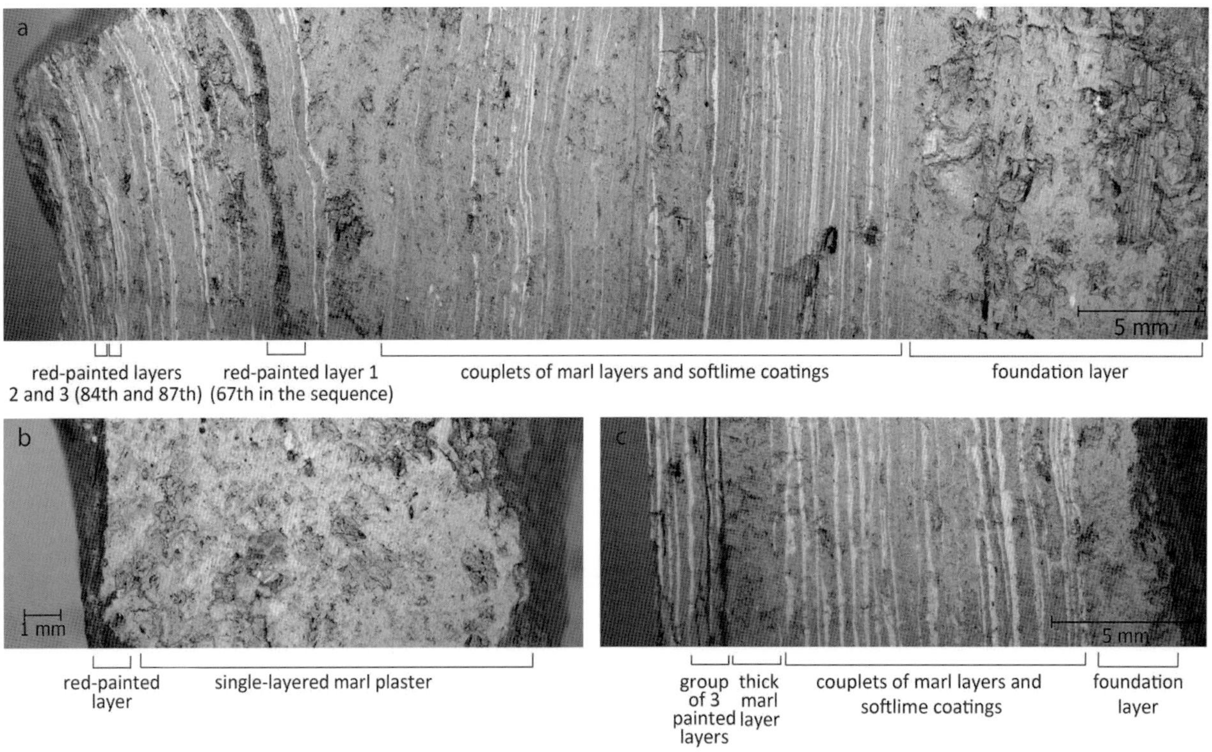

Figure 12.15. Cross-sections of painted plaster sequences from the Middle period: (a) B.89, (20564); (b) B.82, (17750); (c) B.96, (20871) (for colour version, see online supplementary material).

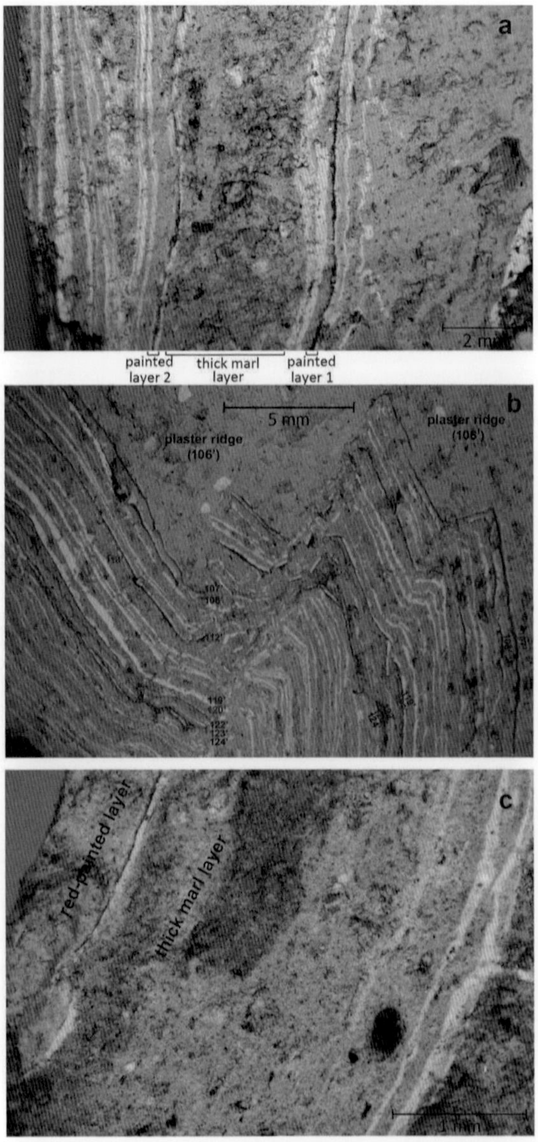

Figure 12.16. Cross-sections of plaster sequences with painted layers following thicker marl plaster layers: (a) B.89, (19830); (b) B.101, (16136); B.114, (20568) (for colour version, see online supplementary material).

Visibility and experience

Besides the limited temporal exposure of paintings due to frequent re-plastering, another observation challenges an understanding of architectural paintings as durable surfaces intended for visual contemplation and display. As the study of the spatial distribution of paintings within houses has shown, paintings occurred preferentially in the northern half of houses (see fig. 12.10), often relatively distant from fire installations – except for in the Late period, which also witnessed a decrease in painting activity (see fig. 12.11). Paintings were also relatively distant from the roof openings, whose locations corresponded with fire installations, given their use as chimneys as well as entrances and sources of sunlight. These observations suggest that paintings occupied rather dimly illuminated areas of the house, and that their placement did not reflect an attempt to maximise the availability of light.

Light is a central actor in the performance of living spaces (Bille, Sorensen 2007), affecting the experience of built environments as well as the execution of tasks (Dawson et al. 2007). Starting from this acknowledgement, archaeologists have begun to explore lighting conditions in excavated spaces, often with the use of virtual simulations (Gutierrez et al. 2007; Papadopoulos, Sakellarakis 2010; Earl et al. 2013; Papadopoulos et al. 2015; Papadopoulos, Holley 2017). A similar approach, combined with experiments conducted on physically reconstructed replica houses at the site, has been employed to further explore lighting conditions within Çatalhöyük houses in relation to the experience and visibility of paintings. One caveat that needs to be taken into consideration in undertaking this approach regards the gaps in the archaeological reconstruction of the site's architecture, especially in relation to the upper-wall and roof areas of the house. Some of the major uncertainties related to the lighting of Çatalhöyük's houses regard the exact size and dimensions of roof openings and the possible use of light sources additional to those documented archaeologically.

Experimental lighting analyses were conducted in Çatalhöyük's experimental house (Stevanović 2012b), as well as in a recently built replica of Building 77. A daylight lighting analysis was conducted in both replica buildings. The experiment consisted in multiple illuminance measurements being taken at a number of set points within each building at three different times of the day (the experiment was carried out in August). A similar night-time experiment was conducted in the experimental house, using alternatively the reconstructed oven and an improvised hearth, and combining the archaeological knowledge on fuel use in internal fire installations (Asouti 2005; 2013) with local knowledge on fire ignition and smoke management (fig. 12.20). In addition to these experiments, a virtual lighting simulation was conducted based on a virtual reconstruction of Building 49 (see fig. 12.19) and involved illuminance mapping and photorealistic rendering with both flame light and sunlight at different times of the day and the year.

The daylight experiment yielded wide differences in illuminance levels across different areas of the house, with broad variations according to the time of day. A stark difference in illuminance levels separates the southern half of the house, more exposed to the

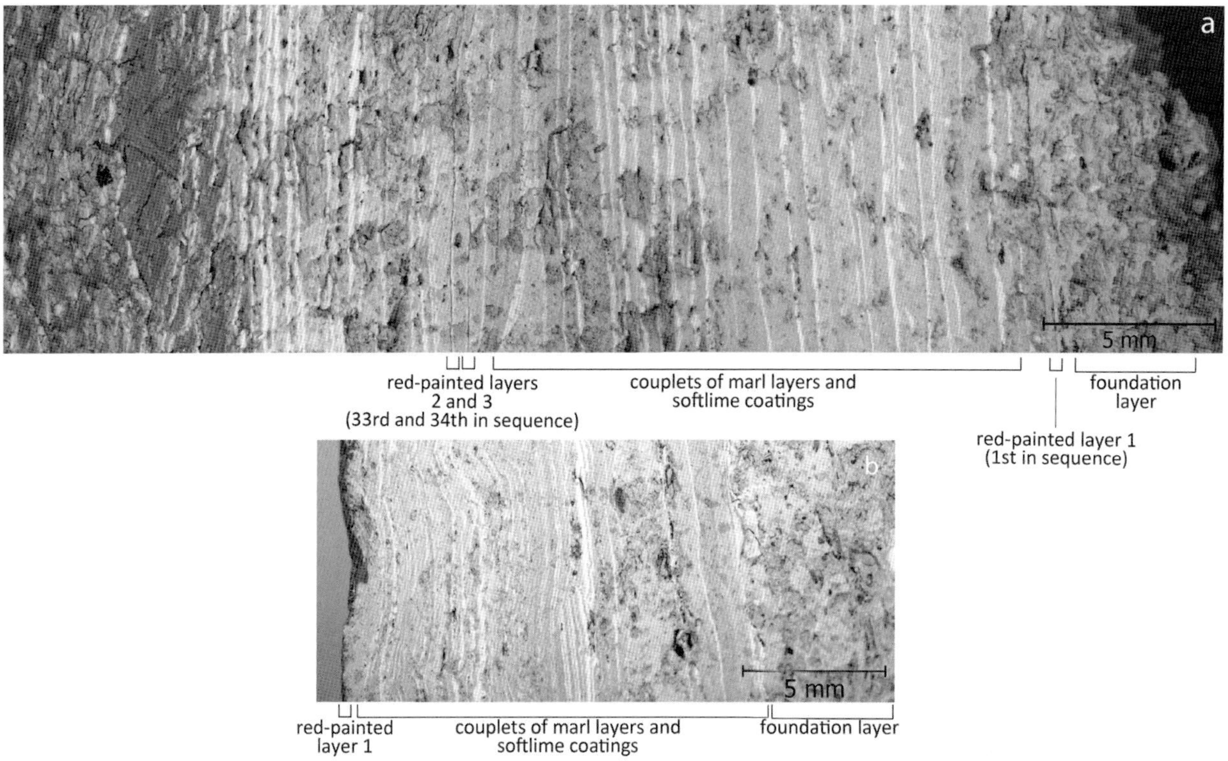

Figure 12.17. Cross-sections of painted plaster sequences from the Late period: (a) B.75, (17037); (b) Sp.370, (19307) (for colour version, see online supplementary material).

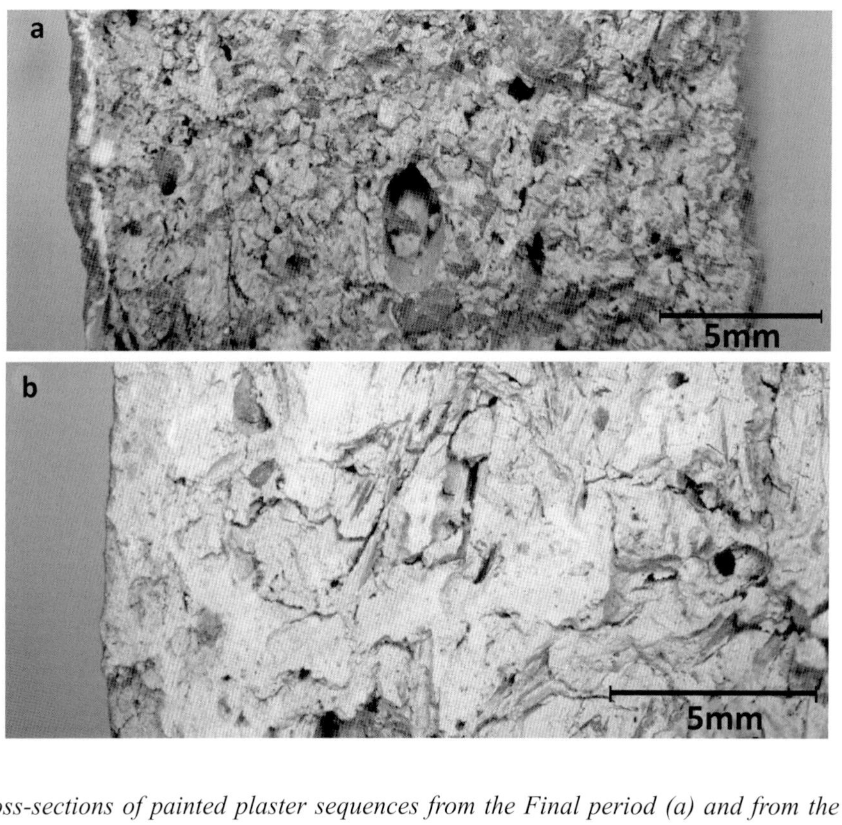

Figure 12.18. Cross-sections of painted plaster sequences from the Final period (a) and from the Chalcolithic West Mound (b): (a) Sp.410, (7886); (b) Sp.197, (6570) (for colour version, see online supplementary material).

Figure 12.19. Changing painted surfaces throughout the occupational sequence of B.49. Phasing and reconstruction is based on Eddisford (2014). 3D modelling by Grant Cox: (a) During subphase 2a (not long after the construction of the building) large areas of the northern, eastern and western walls, as well as the northeastern platform are red-painted; (b) during the later subphase 3–5b, the dogleg walls and the eastern sector of the northern wall are decorated with geometric motifs; (c) to the following subphase 3–5d are attributed a hand design on the southern side of northeast platforms and red-painted panels on the northern wall and posts; (d) the hand motif is soon after covered by two superimposed black-painted geometric designs; during phase 5 (e), limited red-panel painting occurs in the southeastern area of the house, to disappear completely in phase 6 (f) (for colour version, see online supplementary material).

light coming from the roof opening, from its northern area, both in the experimental house (fig. 12.21) and in the Building 77 replica (fig. 12.22). The side rooms yielded the lowest illuminance values in both buildings.

The night-time experiment was primarily aimed at evaluating the performance of ovens and hearths as light sources in the absence of sunlight. It became clear almost immediately that the oven, given its domed shape with a small frontal opening, contributed very little to the lighting of the house during the night, as confirmed by the very low illuminance levels detected both in the lighting experiment (see fig. 12.21) and in the lighting simulation (fig. 12.23). The hearth, on the other hand, provided some lighting, but mostly localised in the southern area of the house (see fig. 12.20). However, the flame had to be kept to a low level, in order to minimise the smoke produced by the interior fire installation, thus dramatically downsizing the role of the hearth as a light source. The experiment thus suggested that ovens and hearths played a very limited role in the lighting of the building and almost no role in the illumination of most of the paintings, given their preferential locations. This observation was also supported by the photorealistic lighting simulation based on the virtual reconstruction of Building 49 (fig. 12.23). Given these results, the use of additional flame light sources such as torches or lamps cannot be ruled out altogether, even though they have not been documented archaeologically.

Despite the uncertainties related to the architectural reconstruction of Çatalhöyük houses, the results of these lighting simulations suggest generally low levels of lighting within houses, especially beyond the southern sectors of houses where all the light sources – as well as most of the archaeological traces of food-processing or manufacturing activities – were located (fig. 12.23).

Figure 12.20. Lighting experimental simulations: (a) daytime experiment in B.77 replica; (b) night-time experiment in the experimental house with hearth; (c) illuminance measurement in B.77 replica; (d) daytime experiment in the experimental house; (e) Küçükköy resident Emine Bülüç ignites the oven in the experimental house in preparation for the night-time experiment; (f) photorealistic rendering of B.49 based on virtual reconstruction (3D modelling by Grant Cox) (for colour version, see online supplementary material).

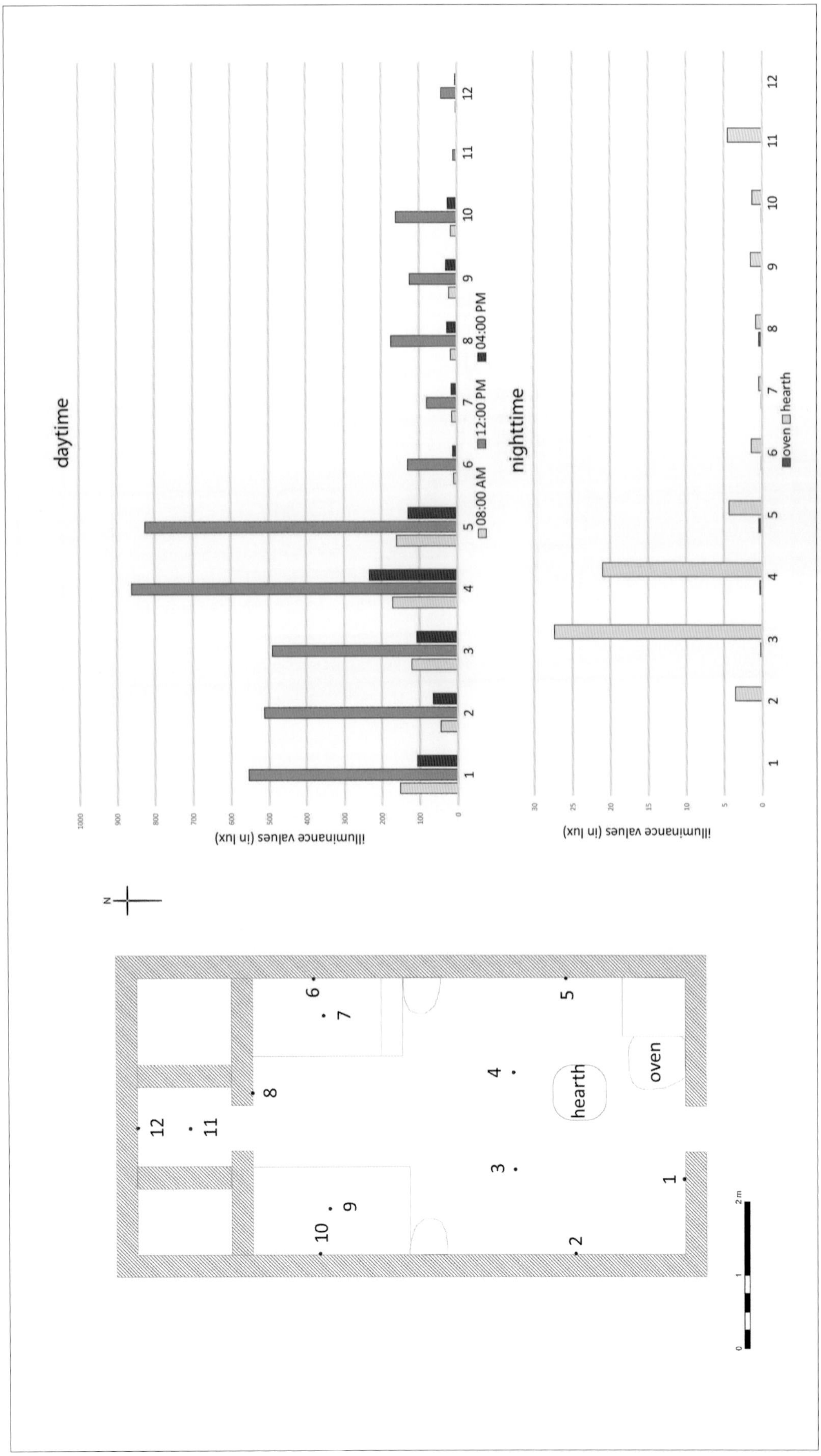

Figure 12.21. Results of lighting experiment in the experimental house (schematic map of experimental house by Camilla Mazzucato).

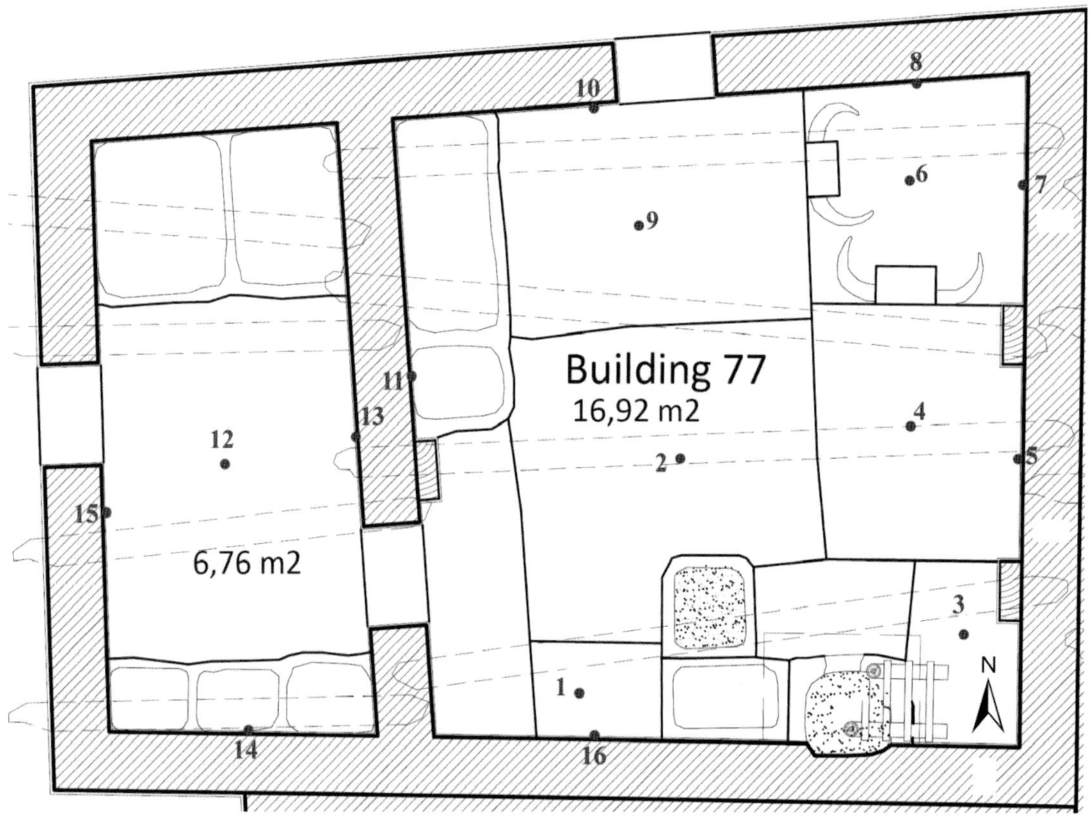

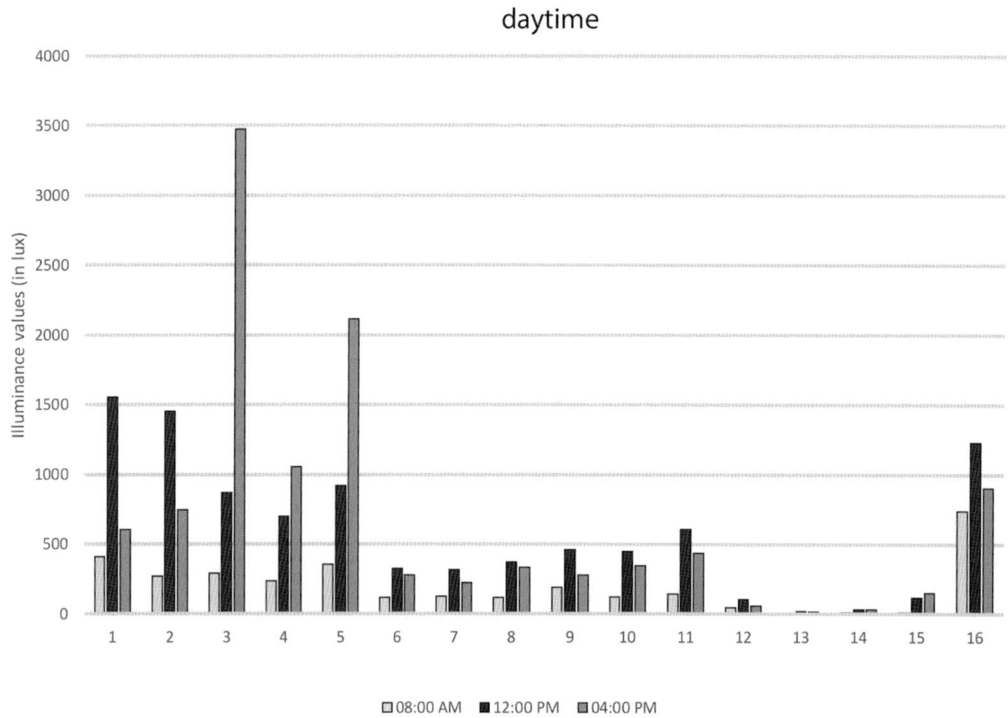

Figure 12.22. Results of lighting experiment in B.77 replica (map reproduced from building plan by Marek Barański).

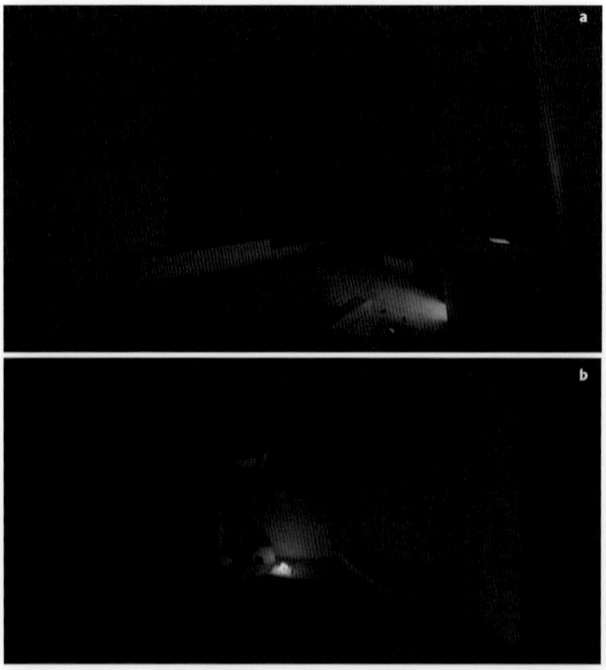

Figure 12.23. Photorealistic lighting simulation based on the virtual reconstruction of B.49: (a) view from southwest with lit oven; (b) view from north with lit hearth (3D modelling by Grant Cox) (for colour version, see online supplementary material).

Paintings were generally located in areas that were among the most poorly lit of the house, which strongly undermined their prolonged visibility over a significant part of the day and of the year (fig. 12.24). This does not fully undermine the importance of their visual experience, which could even have been enhanced by their variable exposure due to daily and annual fluctuations of light (Bosch 2017). However, if combined with the limited exposure of painted surfaces due to frequent re-plastering, a lack of visibility due to poor lighting conditions seems to suggest that the act of painting itself – likely in association with a particular event; for example, a burial or an episode of remodelling – was more important than the visual experience of painted surfaces within the house.

Conclusion

By investigating spatial and contextual associations, painting practices and lighting conditions, recent research on Çatalhöyük's architectural painting corpus has attempted to move beyond the iconographic focus that characterises most of the scholarship on prehistoric art and visual culture (Busacca forthcoming). This does not mean to disregard the visual and iconographic aspects of paintings altogether but, as Jonathan Last put it in the early days of the ÇRP, to link 'content to context' (Last 1998: 362); that is, the iconic and visual content of the paintings to their material existence in the dimension of daily, household life. Indeed, treating the painting corpus as a material assemblage – rather than as a repertoire of images and motifs – helps us better understand not only how paintings were practically made, but also the role they played within the houses and communities at Neolithic Çatalhöyük.

Sampling and analysis of painted plasters' cross-sections, for example, reinforces our understanding of paintings as surfaces of limited exposure and durability, but also as events marking specific moments in the history of a house, such as episodes of burial or architectural renovation. Close investigation of the spatial patterning of paintings reveals their contextual associations with other house features and activities, especially in the highly repetitive and demarcated space of Çatalhöyük's built environment. The experimental simulation of lighting conditions within houses enables explorations into the daily experience of paintings, further reinforcing an interpretation of paintings as episodic practices rather than durable surfaces intended primarily for visualisation and display.

Temporal variation is another key aspect to consider in any synthetic study of an archaeological assemblage from a site with an occupation spanning more than a millennium. Like many other aspects of the site, paintings are attested throughout the entire site occupation, but their contexts, associations, technical execution and painted motifs underwent very significant changes through time. During the Early period, for example, paintings were laid directly on top of thin marl layers, were often multi-layered (with examples of multiple consecutive layers) and their preferential locations were the eastern walls and the southern areas of houses, although with no clear spatial association with other house features. Despite the arduous identification of most of the depicted motifs due to poor preservation, we know that geometric motifs were common during this period, with several examples of rhombs and honeycomb patterns.

During the Middle period, paintings were usually laid onto bright white softlime coatings, often as part of multi-layered plaster sequences with multiple paintings (sometimes in groups of consecutive layers), and – unlike in the previous period – were spatially clustered in the northern area of houses, in spatial association with features such as burials, bone or plaster installations, platforms and benches. Although most of the paintings from this period were simple solid red panels, a number of other designs are also attested, including hand motifs and geometric designs. Some of these designs, including ladder designs and combinations of triangles, vertical lines and brick motifs, are only attested during this period.

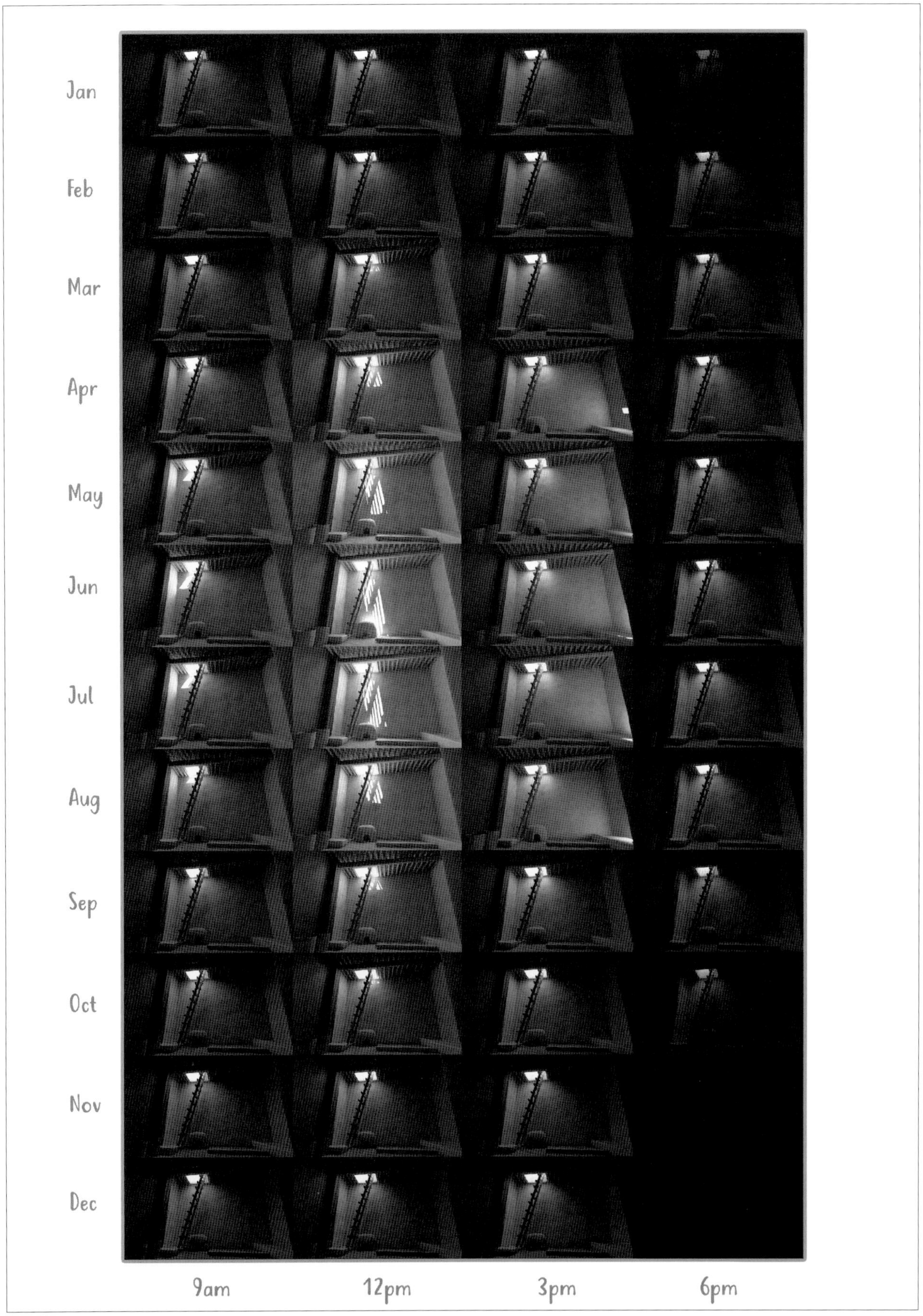

Figure 12.24. Yearly cycle of a virtually reconstructed house (B.49), considering the roof opening as the only source of sunlight (3D modelling, rendering and illustration by Grant Cox) (for colour version, see online supplementary material).

A series of substantial changes in painting practices occurred at the beginning of the Late period (ca 6500 BC). The frequency of paintings decreased dramatically, as did the number of paintings within the same plaster sequence. In striking contrast with the previous period, paintings were now spread throughout the house, with peaks in the southern area of houses, suggesting an unprecedented association with fire installations and other production-related activities. Very few paintings are known from the latest periods of occupation of the site, but they all indicate drastic changes in the material composition and in the practices of plastering and painting, with a generalised trend toward single-layered, red-painted surfaces that continues during the Chalcolithic occupation of the West Mound.

Acknowledgements
The research presented in this chapter is part of the author's doctoral research at Stanford University, Department of Anthropology. Research in the field has been greatly facilitated by Ashley Lingle, Emine Bülüç, and the conservation, excavation and support teams of the 2015, 2016 and 2017 seasons. Post-field research has been supported by Koç University's Research Center for Anatolian Civilizations. I am thankful to Sıla Mangaloğlu-Votruba and Dalila Alberghina for their help during laboratory analyses at the Koç University Archaeology Laboratory in Istanbul. Virtual reconstructions were created with support from the Institute of Research in the Social Sciences at Stanford University.

Supplementary material
For supplementary material related to this chapter, please visit https://doi.org/10.18866/BIAA/e-14. It comprises colour versions of figures 12.4–9, 14–20 and 23–24.

13. The ground stone technologies at Neolithic Çatalhöyük: issues of production, use and deposition

Christina Tsoraki

Introduction

Ground stone technologies received attention from the early stages of the Çatalhöyük Research Project (ÇRP) (Wright 2005; Baysal 2009; Wright, Baysal 2012; Wright 2013; 2014). The final phase of research on the ground stone assemblage unearthed during the ÇRP excavations aimed to improve our understanding of food and craft technologies and the social practices surrounding the use and discard of ground stone artefacts at Neolithic Çatalhöyük. Among the issues investigated were the organisation of stone production, the scale of subsistence activities and the structuring principles of different depositional practices.

Methodology

The classification of the material follows a broad activity-based classification system (Adams 2002) that was developed for the analysis of other prehistoric assemblages (Tsoraki 2008; 2012); it draws upon previous work (Wright 2013), with some adjustments and simplifications (see also Tsoraki 2013). Emphasis was placed on the systematic recording of use-related attributes including information on maintenance strategies, fragmentation patterns, fracture weathering and the overall state of artefact preservation. A refitting study was undertaken, while detailed analysis of flotation samples allowed further insight into questions about production locales and the scale of stone production. The identification of rocks and minerals was facilitated by the use of on-site rock reference collections and identification guides (Schumann 1992; Jones 2000; Pellant 2000).

In addition, objects were sampled for microwear analysis, an analytical technique that involves the identification of microscopic wear traces that develop on the surface of objects during manufacture and use, but also through handling and post-depositional processes (Van Gijn 2010; 2014; Dubreuil et al. 2015; Hayes et al. 2018). Microscopic observations were conducted at both low (10-60x) and high magnifications (up to 500x) using a stereomicroscope (Leica M80), an incident light microscope (Leica DM6000M equipped with LeicaDFC 450 digital camera) and a stereomicroscope with a coaxial illumination unit (Leica M80 LED5000 CXI); the latter provides a 3D view of the surface microtopography revealing the microstratigraphic relations and the spatial extent of wear traces (Tsoraki forthcoming). A sampling strategy that takes into consideration the overall variability encountered in the assemblage in terms of object types, type of contexts and temporal patterns was implemented. Objects sampled include grinding tools, grooved abraders, polishers, palettes, edge tools, 'maceheads', bracelets and beads. Questions addressed through the microwear analysis so far include the relation of tool type and function, identification of activity areas, object biographies and the understanding of depositional processes. The results of the microwear analysis will be presented in more detail elsewhere. Furthermore, questions about food-processing techniques and, more broadly, foodways at Neolithic Çatalhöyük were investigated through the integrated study of microwear traces and micro-botanical residues on grinding tools (phytoliths and starch) (Tsoraki et al. 2015a; 2015b).

Assemblage composition

The analysed material, which includes x-finds (that is, 3D plotted artefacts in the field), bulk material and flotation samples, derives from contexts excavated during the 2009–2017 seasons in the North, South, TPC and GDN Areas and from levels that span the entire occupation sequence at Neolithic Çatalhöyük. Material from both priority units selected for the final project publication volumes and non-priority units from contexts rich in ground stone material were subjected to detailed study. Material excavated prior to 2009 was revisited for additional data collection; materials targeted for analysis include a) all edge tools (axes, adzes, chisels), polishing tools and palettes, and complete or almost complete grinding tools, b) clusters from the B.65-B.56-B.44-B.10 sequence, c) the detailed analysis of all material, including material from selected burial units, from B.77, the excavation of which was completed in 2015.

In total, 20,899 objects were recorded that derive from 84 buildings and 55 open spaces, and from different types of contexts (for example, floors, clusters, fills). Natural, waterworn small-sized pebbles, mostly from flotation samples, make up a large proportion of the recorded material (ca 9,700 objects). As these have no apparent use in ground stone technologies, they are excluded from the following discussion. The repertoire of stone objects used by the Neolithic community of Çatalhöyük includes different types of grinding and abrading tools (for example, querns, abraders, pestles, palettes and polishers), stone axes and adzes, hammerstones, centrally perforated spheres ('maceheads'), abraded pigment nodules, stone beads (see also Chapter 9) and bracelets, along with roughouts/preforms and debitage from the production of ground stone artefacts. Certain categories reported in previous publications have limited or no occurrence in our sample, and this may be due to sampling strategies of contexts or occupation levels (for example, stone vessels that mainly appear in the upper levels). The following discussion focuses on material from stratified Neolithic contexts (n=11,066).

Grinding/abrading tools, which correspond to Wright's categories of Class B (coarse grinding), Class C (fine abrading) and Class D (polishing tools) (Wright 2013), and edge tools occur more commonly in the ground stone assemblage, and their production and use are discussed in more detail in the following sections (table 13.1). Within the miscellaneous category, fire-cracked, natural cobbles, mostly of limestone, predominate (n=1,687). Based on their size (ca 5–10cm) and depositional patterns (often found near hearths and/or ovens together with clay balls), these have been interpreted as elements of heated stone technology used for different cooking practices (stone boiling-convection cooking, roasting and baking) (Tsoraki 2016).

Among the object types with limited representation are pestles, with 25 Neolithic examples known to date (see also Wright 2013: table 20.6). In our sample, they occur in the Middle period (North G/South O) and mainly in the Late period (TP M/TP N). On average, complete examples are 147.06mm long and their average weight is 685.4g. They are mostly conical in shape and rarely bell-shaped (fig. 13.1). No pestles with stylised upper ends like those encountered in earlier Aceramic Neolithic sites such as Hallan Çemi (Rosenberg, Redding 2002) and Çayönü (Davis 1982) occur at Çatalhöyük.

Another type with limited presence are grooved abraders/polishers (see also Wright 2013: table 20.26). These tend to have between one and four U-shaped grooves along their surface, while occasionally both V-shaped and U-shaped grooves are encountered on the same object (e.g., 21103.x1). A wide range of materials have been used, including basalt, gabbro, sandstone and schist. Two examples have incised decorative motifs on their surfaces (fig. 13.2): a steatite/serpentinite example (22300.k5) found in B.160, the earliest example from the ÇRP excavations, has four parallel incised grooves creating a linear pattern on the surface opposite the grooved surface. A more elaborate example (22858.x2), with incised decoration on all surfaces, was found on the floor of Sp.555 in B.81, attributed to the Late period (Level TP M). It is made from well-cemented sandstone. It survives intact, though burnt, with overall dimensions of 30.30 x 30.14 x 19.13mm and weighing 34g. The surface is decorated with an incised geometric pattern that consists mostly of cross-hatched triangles interspersed between parallel lines. Short incised lines are also present on the edges of the groove, while two incised lines, representing the beginning of an unfinished pattern, are visible on the bottom surface of the object. Overall, the pattern shows inconsistencies in its execution (for example, abandoned lines, irregularly executed cross-hatching pattern inside the triangles). The object shows extensive traces of use in the interior of the groove, associated with a motion executed parallel to the length of the groove. While no microwear analysis was conducted on this specific object, the results of the analysis of other grooved abraders found at Çatalhöyük mostly suggest use with a linear motion and contact with multiple materials and mainly bone (e.g., 22300.k5) and wood. A further fragment from a decorated grooved polisher (287.k1) with a cross-hatching incised pattern on the body was unearthed, but it derives from an unstratified context.

Grooved abraders (also known as 'shaft straighteners'), including decorated examples, occur in other Pre-Pottery Neolithic sites including Nevalı Çori, the later levels at Cafer Höyük, Çayönü and Mureybet III (Cauvin et al. 1999; Davis 1982; Hauptmann 1999) and are also reported from Pınarbaşı and Boncuklu in the Konya plain (Baird et al. 2013; 2017). The grooved abrader (22300.k5) from B.160 evokes in material and technological terms earlier examples such as those from Çayönü (Davis 1982: 110–11), yet the presence of the incised example from B.81 is interesting, as it carries many similarities with motifs in wall paintings reported by Mellaart and in particular with an earlier wall painting with the 'groove and painted kilim pattern' from Shrine VII.21 (Mellaart 1967: 72, 76, Pl. VIII).

Bracelets/annulets form a component of the ornamental repertoires at Çatalhöyük. Bracelets appear in Near Eastern sites from the Neolithic onwards (Baysal 2017b), with marble examples noted at Nevali Çori and marble and diabase ones at Cafer Höyük (Cauvin et al. 1999; Hauptmann 1999). Nineteen complete and fragmented specimens have been recovered during the ÇRP excavations, 16 of which were recovered between 2009 and 2017 (fig. 13.3). In addition, another three fragments of bracelets have been unearthed from unstratified contexts, while Vasić et al. 2013 reported a further fragment from the North Area (14121.x2). The vast

Object type	*no.*	*%*	*Object type*	*no.*	*%*
Edge tool - indeterminate	81	0.8	Misc. - pigment abraded nodule	8	0.1
Edge tool - axe	46	0.4	Misc. - pigment raw nodule	16	0.2
Edge tool - adze	35	0.3	Misc. - rocks with natural perforations	19	0.2
Edge tool - chisel	18	0.2	Misc. - fire-cracked rocks	1,687	16.2
Edge tool - roughout/preform	4	0.0	**Miscellaneous objects total**	1,778	17.0
Edge tools total	184	1.8	Mult. - indeterminate	5	0.0
Percussive - indeterminate	5	0.0	Mult. - polisher/hammerstone	36	0.3
Percussive - hammerstone	26	0.2	Mult. - grooved abrader/grinding slab	1	0.0
Percussive - pounder	4	0.0	Mult. - grinder/hammer	6	0.1
Percussive tools total	35	0.3	Mult. - abrader/hammer	5	0.0
Perforated objects - macehead	12	0.1	Mult. - grooved abrader/polisher	2	0.0
Perforated objects - weight	1	0.0	**Multiple-use tools total**	55	0.5
Perforated objects total	13	0.1	Ornaments - indeterminate	13	0.1
Grind/abras - indeterminate	234	2.2	Ornaments - beads	1,201	11.5
Grind/abras - abrader	150	1.4	Ornaments - pendants	2	0.0
Grind/abras - grooved abrader	2	0.0	Ornaments - bracelets/annulets	18	0.2
Grind/abras - grooved abrader U-shaped	15	0.1	Ornaments - bead roughout	4	0.0
Grind/abras - abrading slab	30	0.3	Ornaments - bead preform	16	0.2
Grind/abras - palette	160	1.5	Ornaments - pendant preform	4	0.0
Grind/abras - abrader-knife	2	0.0	Ornaments - bead preform cylinder	1	0.0
Grind/abras - sander	7	0.1	**Ornaments total**	1,259	12.1
Grind/abras - polisher/smoothed stone	199	1.9	Debitage - indeterminate	2,234	21.4
Grind/abras - polishing tool roughout	2	0.0	Waste by-products/shatter	169	1.6
Grind/abras - polishing slab	6	0.1	Flake	359	3.4
Grind/abras - pestle	11	0.1	Debitage - primary flake	9	0.1
Grind/abras - bell muller	1	0.0	Debitage - secondary flake	15	0.1
Grind/abras - grinder	77	0.7	Debitage - tertiary flake	93	0.9
Grind/abras - grinder one hand mano	37	0.4	Debitage - rejuvenation flake	96	0.09
Grind/abras - grinder two hand mano	3	0.0	Blade/bladelet	3	0.0
Grind/abras - grinder roughout	1	0.0	Debris/fragments	1,487	14.3
Grind/abras - grind. slab/quern	200	1.9	Possible debitage	32	0.3
Grind/abras - grind. slab roughout	37	0.4	Ground stone core	2	0.0
Grind/abras - grind. slab or grinder frag.	342	3.3	Possible nodule	46	0.4
Grind/abras - mortar	3	0.0	Unworked nodule	300	2.9
Grinding/abrasive tools total	1,519	14.6	Quarried material	1	0.0
Misc. - indeterminate	10	0.1	Natural pebble	566	5.4
Misc. - incised stone	1	0.0	Natural cobble	143	1.4
Misc. - retouched tool	1	0.0	**Debitage, cores & nodules total**	5,555	53.3
Misc. - stone ball	19	0.2	Indeterminate worked	33	0.3
Misc. - pigment	17	0.2	**Total**	10,431	100.0

Table 13.1. Frequency of object categories from stratified Neolithic contexts (excluding indeterminate cases, n=635).

majority derive from the Middle period (n=17), one from the Late period, and one from an unstratified Neolithic burial. Only simple circular forms are present in the assemblage and no occurrences of moulded designs have been noted at Çatalhöyük (for example, Canhasan I, Baysal 2017b; Cafer Höyük upper Levels III–I, Cauvin et al. 1999: 98). There is a clear preference for white/off-white marble, but minerals (gypsum, possibly calcite and an indeterminate mineral with translucent texture) of white/off-white hues were used occasionally. Their internal diameters vary from ca 50 to 68mm and the only example found in direct association with an adult female skeleton has an outer diameter of 100.74mm and was worn on the arm.

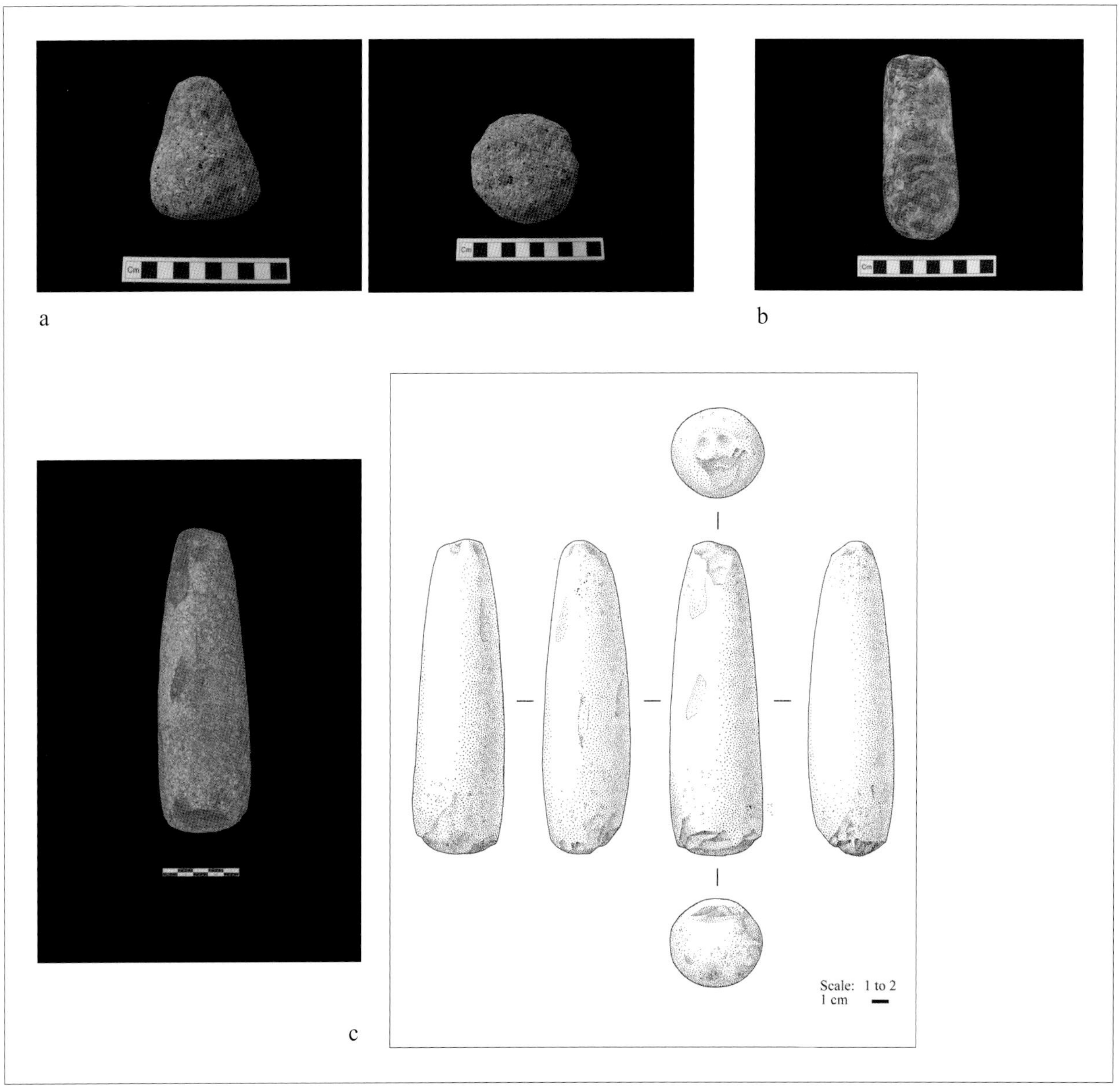

Figure 13.1. Pestles: (a) 23756.x11 plan view and view of the use face; (b) 32860.k33; (c) 32860.k34 plan view and illustration (photographs by Christina Tsoraki and Jason Quinlan; illustration by Kathryn Killackey).

Raw material use

Analysis of the 2009–2017 ground stone assemblage confirms the previous results regarding raw material use at Çatalhöyük (Wright 2013). In total, more than 30 types of rocks and minerals have been employed for stone technologies in varying relative frequencies (table 13.2). Igneous rocks are represented mainly by different varieties of volcanic porphyritic andesite and basalt, followed by diabase. Andesites have been widely used for grinding tools that show, however, high fragmentation rates (see below, also Wright 2013). Yet, the high relative frequencies of andesites are retained even if fragments of grinding tools are excluded from raw material analysis. Among sedimentary rocks, mainly limestone, often a recrystallised variety, predominates, along with sandstones and sedimentary orthoquartzite and tufa. Metamorphic rocks are represented mainly by schists/phyllite and marble, followed by greenstone. While in previous publications green-coloured rocks were invariably grouped under the term 'diabase', during this phase of the analysis a distinction was drawn between 'diabase' and 'greenstone' mainly in terms of macroscopic textural features. The term 'diabase', which corresponds to Wright's 'dark greenish black diabase variant' (2013: 370), was employed for rocks with a fine to medium-grained massive texture usually of dark green or greenish black colour. The term 'greenstone' refers to fine-grained green-coloured rocks with homogenous, crystalline texture that often exhibit some foliation, and corresponds with the 'bright green variant' reported by

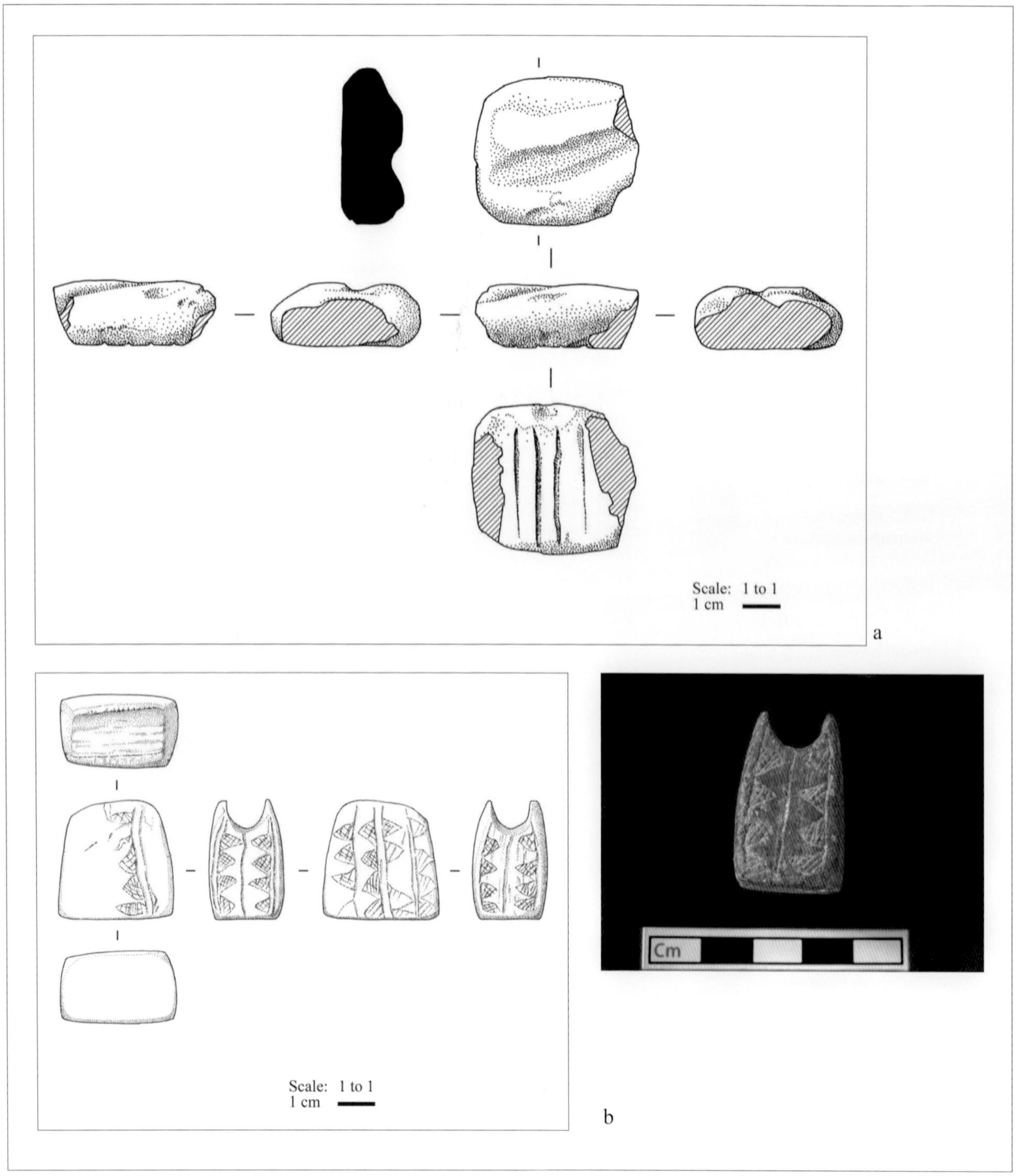

Figure 13.2. Incised grooved polishers: (a) 22300.k5; (b) 22858.x2 (illustration by Kathryn Killackey).

Wright (2013: 370). Preliminary results of p-XRF analysis of the two macroscopically identified groups indicated that these belong to two distinct geochemical groups, possibly associated with at least two different material sources (Nazaroff 2014).

Raw material use suggests clear selectivity patterns (table 13.3): tufa, carnelian, hematite of metallic lustre, fluorapatite and turquoise were used exclusively for the production of beads. Andesites are preferentially associated with grinding tools, a material choice widely encountered in different communities with stone milling technology (e.g., Hayden 1987; Wright 1998; Schneider 2002). Occasionally, grinders were fashioned from quarzitic sandstone/orthoquartzite and metaquartzite rocks. Pestles are more commonly made from andesite and basalt followed by sedimentary quartzite, while one example (32860.k33) was made from possible gneiss. Limestone and marble are widely used for polishing

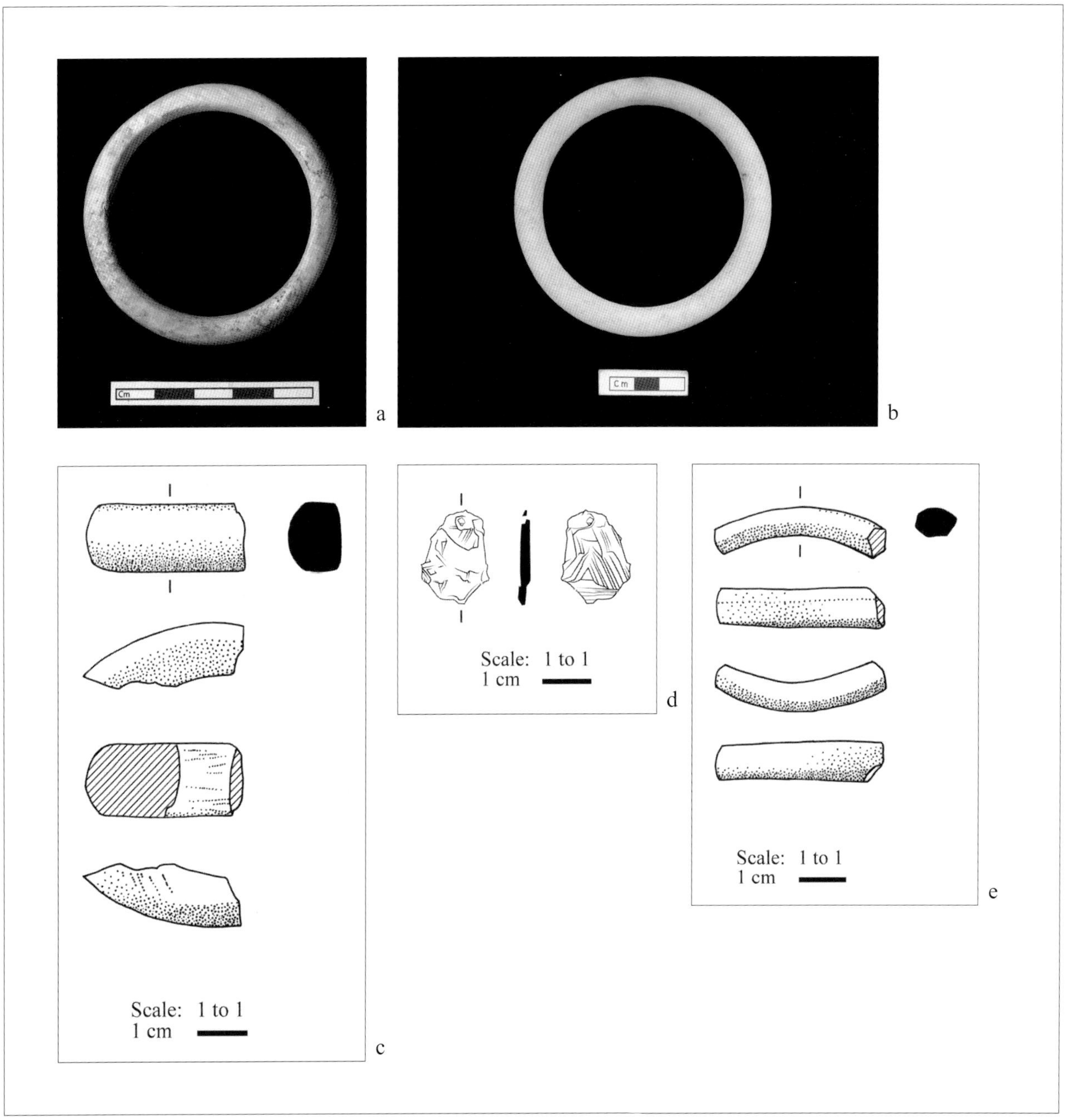

Figure 13.3. Bracelets/annulets and pendants: (a) 30503.x8; (b) 7580.x2; (c) 23155.k1; (d) 30503.k1; (e) 23143.k7 (photographs by Jason Quinlan; illustration by Kathryn Killackey).

tools, 'maceheads' and disc beads. Metamorphic rocks of fine-grained texture and in particular schist were favoured for the production of palettes and other abrading tools, a material choice that remained constant throughout the habitation of the site.

In the case of edge tools, there is a clear preference for metamorphic rocks and in particular greenstone and serpentinite followed by igneous diabase. Less common is the use of steatite, basalt and well-cemented sandstone. All materials occur across the three variant types (axes, adzes and chisels) and no variation between the North and South Area of the site was noted. The preference over metamorphic rocks for the production of edge tools is well documented in other Neolithic sites in Turkey (e.g., Özbek 2011), while the use of green-coloured stones for polished axes is reported throughout the sequence at Cafer Höyük (Cauvin et al. 1999).

Interestingly, basalt, a rock that has the appropriate mechanical properties (durable, tough material with ability to withstand impact force (Attewell, Farmer 1976; Dickson 1981)) and was used relatively

Raw materials	*no.*	*%*
Sedimentary rocks		
Sandstone	81	0.8
Quartzite/Orthoquartzite	93	0.9
Flint	5	0.1
Chalk	3	0.0
Conglomerate	7	0.1
Limestone	2,093	21.4
Lake limestone/Marl	24	0.2
Tufa	247	2.5
Indeterminate	354	3.6
Total	2,907	29.7
Metamorphic rocks		
Serpentinite	45	0.5
Schist	449	4.6
Gneiss	1	0.0
Marble	436	4.4
Slate	5	0.1
Phyllite	113	1.2
(meta)Quartzite	37	0.4
Steatite	12	0.1
Greenstone	156	1.6
Indeterminate	65	0.7
Total	1,319	13.5
Igneous rocks		
Pumice	3	0.0
Gabbro	1	0.0
Diabase	74	0.8
Basalt	134	1.4
Andesite	4,546	46.4
Andesite hornblende	86	0.9
Andesite hornblende-biotite	130	1.3
Andesite basalt	7	0.1
Indeterminate	58	0.6
Total	5,039	51.4
Minerals		
Carnelian	142	1.4
Quartz	25	0.3
Calcite	204	2.1
Hematite	1	0.0
Pigment indeterminate	8	0.1
Ochre	32	0.3
Gypsum	2	0.0
Hematite with metallic lustre	5	0.1
Indeterminate	69	0.7
Fluorapatite	28	0.3
Turquoise	16	0.2
Biotite	1	0.0
Total	533	5.4
Total	9,798	100.0

Table 13.2. Frequency of rocks and minerals (excluding indeterminate raw materials, n=1,268).

frequently for other percussive tools at Çatalhöyük, has been used only rarely for the production of edge tools. Similarly, while greenstone, serpentinite, diabase and steatite have been fashioned into beads and less frequently polishing tools, they were never transformed into bracelets, while widely occurring volcanics were never fashioned into ornaments. This would suggest that raw material selection and use at Çatalhöyük did not rely strictly on practical considerations such as ease of working, mechanical properties and raw material availability, but choices reflected culturally specific understandings of different material categories that determined which materials were appropriate for the creation of different objects (cf. Tsoraki 2011). Overall, material choices for the production of edge tools seem rather conservative over the 1,000 years of habitation of the site, with a clear preference for the use of colourful rocks with greenish hues and distinctive textures.

The location of Çatalhöyük in an alluvial deposit (Baysal 2009; Wright 2013) suggests that ground stone technologies at Çatalhöyük relied on the acquisition of rocks and minerals from sources located further away from the site and possibly dispersed across the landscape (Baysal, Wright 2005; Türkmenoğlu et al. 2005; Bains et al. 2013; Wright 2013). While some andesitic materials may have been procured from the Karadağ complex, located 35 kilometres to the southeast of Çatalhöyük (Türkmenoğlu et al. 2005), more recent work on clay provenance and pottery fabrics indicates that andesites used for grinding tools resemble macroscopically those encountered within the gritty pottery fabrics that derive from the volcanic Erenler-Alacadağ area ca 60–70km to the southwest of Çatalhöyük (Doherty, Tarkan 2013; Doherty 2017; Tarkan, this volume, Chapter 4). Overall, the andesitic-rhyodacite activity associated with the Erenler-Alacadağ formation is associated with a range of resources found on site, including red ochre, gritty pottery fabrics, pumice and carnelian. Stone resources possibly originating from this upland volcanic region were used from early on (Level South J) and prior to the appearance of the gritty pottery fabrics on site in South M. While procurement of waterworn material from the Çarşamba and May rivers is highly likely, the presence of large-size angular andesite blocks on site suggests that these do not represent materials that have been transported within the river system over long distance from the original source. This in tandem with the on-site presence of rectangular blocs of quarried material suggests that at least on some occasions, the Neolithic inhabitants of Çatalhöyük would have travelled closer to the sources of materials for raw material acquisition.

Raw material		*Object category*									*Total*
		Edge tools	*Percussive tools*	*Perforated tools*	*Grinding/abrasive tools*	*Misc.*	*Multiple-use tools*	*Ornaments*	*Debitage/cores/ nodules*	*Indeterminate-worked*	
Sandstone	no.	1	0	0	58	1	5	1	5	0	71
	%	0.6	0.0	0.0	4.0	0.1	9.4	0.1	0.1	0.0	0.8
Sedimentary quartzite/orthoquartzite	no.	0	1	0	45	5	1	0	31	1	84
	%	0.0	2.9	0.0	3.1	0.4	1.9	0.0	0.6	3.1	0.9
Flint	no.	0	1	0	0	1	0	2	0	0	4
	%	0.0	2.9	0.0	0.0	0.1	0.0	0.2	0.0	0.0	0.0
Chalk	no.	0	0	0	0	0	0	0	2	0	2
	%	0.0	0.0	0.0	0.0	0.0	0.0	0.0	0.0	0.0	0.0
Conglomerate	no.	0	0	0	3	0	0	0	1	0	4
	%	0.0	0.0	0.0	0.2	0.0	0.0	0.0	0.0	0.0	0.0
Limestone	no.	0	10	3	116	1,020	15	458	211	6	1,839
	%	0.0	29.4	33.3	8.0	78.1	28.3	39.5	4.1	18.8	19.7
Lake limestone/Marl	no.	0	0	0	1	3	0	11	6	1	22
	%	0.0	0.0	0.0	0.1	0.2	0.0	0.9	0.1	3.1	0.2
Tufa	no.	0	0	0	0	0	0	243	4	0	247
	%	0.0	0.0	0.0	0.0	0.0	0.0	20.9	0.1	0.0	2.6
Indeterminate sedimentary	no.	0	1	1	18	155	1	9	141	2	328
	%	0.0	2.9	11.1	1.2	11.9	1.9	0.8	2.7	6.3	3.5
Serpentinite	no.	11	0	0	1	0	0	28	5	0	45
	%	6.1	0.0	0.0	0.1	0.0	0.0	2.4	0.1	0.0	0.5
Schist	no.	0	0	0	315	1	0	4	68	4	392
	%	0.0	0.0	0.0	21.9	0.1	0.0	0.3	1.3	12.5	4.2
Gneiss	no.	0	0	0	1	0	0	0	0	0	1
	%	0.0	0.0	0.0	0.1	0.0	0.0	0.0	0.0	0.0	0.0
Marble	no.	0	0	3	58	0	16	78	273	1	429
	%	0.0	0.0	33.3	4.0	0.0	30.2	6.7	5.3	3.1	4.6
Slate	no.	0	0	0	3	0	0	1	1	0	5
	%	0.0	0.0	0.0	0.2	0.0	0.0	0.1	0.0	0.0	0.1
Phyllite	no.	0	0	0	14	0	0	97	1	0	112
	%	0.0	0.0	0.0	1.0	0.0	0.0	8.4	0.0	0.0	1.2

Table 13.3. The correlation of raw material and object category (% within Object Category, excluding indeterminate raw materials and objects, n=1,713).

Raw material		*Object category*									*Total*
		Edge tools	*Percussive tools*	*Perforated tools*	*Grinding/abrasive tools*	*Misc.*	*Multiple-use tools*	*Ornaments*	*Debitage/cores/ nodules*	*Indeterminate worked*	
(meta)Quartzite	no.	0	2	0	11	7	0	0	11	0	31
	%	0.0	5.9	0.0	0.8	0.5	0.0	0.0	0.2	0.0	0.3
Steatite	no.	2	0	0	0	0	0	10	0	0	12
	%	1.1	0.0	0.0	0.0	0.0	0.0	0.9	0.0	0.0	0.1
Greenstone	no.	110	0	0	2	0	0	11	27	6	156
	%	61.1	0.0	0.0	0.1	0.0	0.0	0.9	0.5	18.8	1.7
Indeterminate metamorphic	no.	11	0	1	12	1	3	13	15	2	58
	%	6.1	0.0	11.1	0.8	0.1	5.7	1.1	0.3	6.3	0.6
Pumice	no.	0	0	0	2	0	0	0	0	0	2
	%	0.0	0.0	0.0	0.1	0.0	0.0	0.0	0.0	0.0	0.0
Gabbro	no.	0	0	0	1	0	0	0	0	0	1
	%	0.0	0.0	0.0	0.1	0.0	0.0	0.0	0.0	0.0	0.0
Diabase	no.	44	0	0	5	1	0	6	12	3	71
	%	24.4	0.0	0.0	0.3	0.1	0.0	0.5	0.2	9.4	0.8
Basalt	no.	1	17	1	11	58	8	0	24	0	120
	%	0.6	50.0	11.1	0.8	4.4	15.1	0.0	0.5	0.0	1.3
Andesite	no.	0	0	0	624	4	1	0	3,893	1	4,523
	%	0.0	0.0	0.0	43.3	0.3	1.9	0.0	75.8	3.1	48.4
Andesite hornblende	no.	0	0	0	34	0	0	0	51	0	85
	%	0.0	0.0	0.0	2.4	0.0	0.0	0.0	1.0	0.0	0.9
Andesite hornblende-biotite	no.	0	0	0	85	0	0	0	41	0	126
	%	0.0	0.0	0.0	5.9	0.0	0.0	0.0	0.8	0.0	1.3
Andesite / Basalt	no.	0	0	0	1	1	3	0	2	0	7
	%	0.0	0.0	0.0	0.1	0.1	5.7	0.0	0.0	0.0	0.1
Indeterminate igneous	no.	0	1	0	18	4	0	0	28	0	51
	%	0.0	2.9	0.0	1.2	0.3	0.0	0.0	0.5	0.0	0.5
Carnelian	no.	0	0	0	0	0	0	137	5	0	142
	%	0.0	0.0	0.0	0.0	0.0	0.0	11.8	0.1	0.0	1.5
Quartz	no.	0	1	0	2	1	0	2	18	0	24
	%	0.0	2.9	0.0	0.1	0.1	0.0	0.2	0.4	0.0	0.3

Table 13.3 (continued). The correlation of raw material and object category (% within Object Category, excluding indeterminate raw materials and objects, n=1,713).

Raw material		Object category									Total
		Edge tools	Percussive tools	Perforated tools	Grinding/abrasive tools	Misc.	Multiple-use tools	Ornaments	Debitage/cores/ nodules	Indeterminate worked	
Calcite	no.	0	0	0	0	0	0	0	202	1	203
	%	0.0	0.0	0.0	0.0	0.0	0.0	0.0	3.9	3.1	2.2
Hematite	no.	0	0	0	0	1	0	0	0	0	1
	%	0.0	0.0	0.0	0.0	0.1	0.0	0.0	0.0	0.0	0.0
Pigment indeterminate	no.	0	0	0	0	8	0	0	0	0	8
	%	0.0	0.0	0.0	0.0	0.6	0.0	0.0	0.0	0.0	0.1
Ochre	no.	0	0	0	0	32	0	0	0	0	32
	%	0.0	0.0	0.0	0.0	2.5	0.0	0.0	0.0	0.0	0.3
Gypsum	no.	0	0	0	0	0	0	1	1	0	2
	%	0.0	0.0	0.0	0.0	0.0	0.0	0.1	0.0	0.0	0.0
Hematite (metallic lustre)	no.	0	0	0	0	0	0	2	1	2	5
	%	0.0	0.0	0.0	0.0	0.0	0.0	0.2	0.0	6.3	0.1
Fluorapatite	no.	0	0	0	0	0	0	28	0	0	28
	%	0.0	0.0	0.0	0.0	0.0	0.0	2.4	0.0	0.0	0.3
Turquoise	no.	0	0	0	0	0	0	16	0	0	16
	%	0.0	0.0	0.0	0.0	0.0	0.0	1.4	0.0	0.0	0.2
Biotite	no.	0	0	0	0	0	0	0	1	0	1
	%	0.0	0.0	0.0	0.0	0.0	0.0	0.0	0.0	0.0	0.0
Mineral indeterminate	no.	0	0	0	0	2	0	2	57	2	63
	%	0.0	0.0	0.0	0.0	0.2	0.0	0.2	1.1	6.3	0.7
Total	no.	180	34	9	1,441	1,306	53	1,160	5,138	32	9,353
	%	100.0	100.0	100.0	100.0	100.0	100.0	100.0	100.0	100.0	100.0

Table 13.3 (continued). The correlation of raw material and object category (% within Object Category, excluding indeterminate raw materials and objects, n=1,713).

The transportation of material back to the site would have required the negotiation of considerable distances, entailing most likely the coordination and cooperation of a larger social group, especially with regard to the transportation of the large andesitic boulders used for quern production.

Temporal patterns

In the sample studied, andesites along with greenstone and diabase occur from early on, though raw material use during the Early period seems to have relied on a smaller range of materials (table 13.4). Material repertoires expand during the Middle period to include basalt, steatite and minerals such as carnelian, fluorapatite and turquoise. Object types show consistent presence throughout the habitation of the site, with a few exceptions such as bracelets and pestles that appear in the Middle period (table 13.5). The relative frequencies of fire-cracked rocks, comprising the vast majority of the miscellaneous category, decrease over time in both the North and the South Areas, a pattern mirroring the decreasing relative frequencies of clay balls over time (Chapter 7) (table S13.1 and online fig. S13.1). Yet, as discussed in more detail in the following sections, the main difference encountered in the Late period is an increase in the quantities of ground stone artefacts used by the inhabitants of Neolithic Çatalhöyük.

Production processes

The production of grinding tools

Three hundred and sixty-three objects have been attributed to the category of upper (n=124) and lower (n=239) grinding tools (grinding slabs/querns), including finished tools and roughouts (fig. 13.4). This includes 209 upper and lower grinding tools that were unearthed prior to 2009 and survive complete or retain more than half of their original size, all grinding tools from B.77, and grinding tools from B.44 which were revisited for further data collection. A further 464 fragments of grinding tools were also studied, but due to their small size they cannot be securely attributed to upper or lower tools. The high level of fragmentation of grinding tools (cf. Wright 2013) over-exaggerates the actual number of grinding tools in use at Neolithic Çatalhöyük. The number of complete specimens unearthed during the Çatalhöyük Research Project excavations is rather low: 27 lower grinding tools and 53 grinders.

The analysis of material from the 2009–2017 seasons reaffirms certain trends in the production and use of grinding tools at Çatalhöyük that have been highlighted by previous work (Baysal, Wright 2005; Baysal 2009; Wright, Baysal 2012; Wright 2013). The production sequence of querns entailed the on-site modification of large-size waterworn cobbles and boulders acquired from secondary sources such as riverbeds. Emphasis was placed mainly on the preparation of the use-face through pecking, while the margins and the rest of the body surface were shaped through the application of percussive techniques including hammering, flaking and pecking (for a detailed description of the *chaîne opératoire* of querns, see Wright 2013).

While the production of grinding tools shows consistency in technological choices (raw material use, production strategies) throughout the habitation of the site, some variation is, however, evident over time. A new technological form, resulting from a more formalised production sequence that exhibits an increased level of investment in the production stage, appears during the Middle period in the North and South Areas and continues throughout the later part of the occupation sequence (Levels North G, South N, O, Q, TP M, TP N) (Tsoraki 2018; see also Wright 2013). Rendered only in pink-coloured porphyritic andesite that contrasts with the grey-coloured variant used more commonly at Çatalhöyük, this new form stands out. Querns of this type exhibit margins and often body surfaces shaped by gracile pecking followed by grinding. The margins were often shaped at right angles to the use-face. In addition, some examples exhibit an incised groove across the margins, possibly an additional form of decoration (fig. 13.5). Querns with grooved decoration occur in B.80 and in B.77. The presence of roughouts of this type of quern and unfinished examples demonstrates their on-site production alongside the production of regular querns and other grinding tools. Furthermore, an example from B.140 (31911.x1) was subjected to this type of modification (for example, regularisation of the margins) following a long period of use, as indicated by its concave use-face. The tool was intended to be shaped in a similar fashion to the rest of the carefully manufactured querns, but the final form was never completed. This suggests that the rendering of this particular form was on some occasions associated with subsequent modifications of the tools and was not always part of the early stages of the production sequence. Elaborate querns occur in limited numbers (n=28) and are encountered in single numbers in buildings with different degrees of elaboration, and in both burnt and unburnt buildings (B.77, B.102 and B.49 in the North Area, and B.89, B.80, B.65, B.68, B.140, B.150 and B.63 in the IST Area). B.77, however, stands out, as the largest concentration (n=17, 60.7% of all known examples) of elaborate querns and roughouts were deposited there during the abandonment of the building.

Raw Material		Temporal groupings of levels				Total
		Early	Middle	Late	Final	
Sandstone	no.	2	31	47	1	81
	%	0.1	1.0	0.9	1.4	0.8
Sedimentary quartzite/orthoquartzite	no.	13	46	32	2	93
	%	1.0	1.5	0.6	2.7	0.9
Flint	no.	1	0	4	0	5
	%	0.1	0.0	0.1	0.0	0.1
Chalk	no.	2	1	0	0	3
	%	0.1	0.0	0.0	0.0	0.0
Conglomerate	no.	1	2	4	0	7
	%	0.1	0.1	0.1	0.0	0.1
Limestone	no.	693	1,059	312	29	2,093
	%	51.2	34.1	5.9	39.7	21.4
Lake limestone/Marl	no.	2	16	6	0	24
	%	0.1	0.5	0.1	0.0	0.2
Tufa	no.	18	224	5	0	247
	%	1.3	7.2	0.1	0.0	2.5
Indeterminate sedimentary	no.	164	152	37	1	354
	%	12.1	4.9	0.7	1.4	3.6
Serpentinite	no.	2	32	10	1	45
	%	0.1	1.0	0.2	1.4	0.5
Schist	no.	29	135	280	5	449
	%	2.1	4.3	5.3	6.8	4.6
Gneiss	no.	0	0	1	0	1
	%	0.0	0.0	0.0	0.0	0.0
Marble	no.	20	98	315	3	436
	%	1.5	3.2	6.0	4.1	4.4
Slate	no.	0	2	3	0	5
	%	0.0	0.1	0.1	0.0	0.1
Phyllite	no.	12	78	23	0	113
	%	0.9	2.5	0.4	0.0	1.2
(meta)Quartzite	no.	7	17	12	1	37
	%	0.5	0.5	0.2	1.4	0.4
Steatite	no.	0	4	8	0	12
	%	0.0	0.1	0.2	0.0	0.1
Greenstone	no.	20	64	67	5	156
	%	1.5	2.1	1.3	6.8	1.6
Indeterminate metamorphic	no.	12	24	27	2	65
	%	0.9	0.8	0.5	2.7	0.7
Pumice	no.	0	1	2	0	3
	%	0.0	0.0	0.0	0.0	0.0
Gabbro	no.	0	1	0	0	1
	%	0.0	0.0	0.0	0.0	0.0

Table 13.4. The temporal distribution of raw materials (% within Temporal groupings of levels, excluding indeterminate raw materials, n=1,268).

Raw Material		*Temporal groupings of levels*				*Total*
		Early	*Middle*	*Late*	*Final*	
Diabase	no.	13	29	32	0	74
	%	1.0	0.9	0.6	0.0	0.8
Basalt	no.	0	121	13	0	134
	%	0.0	3.9	0.2	0.0	1.4
Andesite	no.	272	692	3,567	15	4,546
	%	20.1	22.3	67.8	20.5	46.4
Andesite hornblende	no.	18	58	10	0	86
	%	1.3	1.9	0.2	0.0	0.9
Andesite hornblende-biotite	no.	26	81	23	0	130
	%	1.9	2.6	0.4	0.0	1.3
Andesite / Basalt	no.	0	7	0	0	7
	%	0.0	0.2	0.0	0.0	0.1
Indeterminate igneous	no.	4	35	17	2	58
	%	0.3	1.1	0.3	2.7	0.6
Carnelian	no.	0	6	136	0	142
	%	0.0	0.2	2.6	0.0	1.4
Quartz	no.	2	20	3	0	25
	%	0.1	0.6	0.1	0.0	0.3
Calcite	no.	3	31	167	3	204
	%	0.2	1.0	3.2	4.1	2.1
Hematite	no.	0	0	1	0	1
	%	0.0	0.0	0.0	0.0	0.0
Pigment indeterminate	no.	0	7	1	0	8
	%	0.0	0.2	0.0	0.0	0.1
Ochre	no.	16	7	8	1	32
	%	1.2	0.2	0.2	1.4	0.3
Gypsum	no.	0	2	0	0	2
	%	0.0	0.1	0.0	0.0	0.0
Hematite (metallic lustre)	no.	0	2	3	0	5
	%	0.0	0.1	0.1	0.0	0.1
Fluorapatite	no.	0	10	18	0	28
	%	0.0	0.3	0.3	0.0	0.3
Turquoise	no.	0	1	14	1	16
	%	0.0	0.0	0.3	1.4	0.2
Biotite	no.	1	0	0	0	1
	%	0.1	0.0	0.0	0.0	0.0
Mineral indeterminate	no.	0	12	56	2	69
	%	0.0	0.4	1.1	1.4	0.7
Total	no.	1,353	3,108	5,264	73	9,798
	%	100.0	100.0	100.0	100.0	100.0

Table 13.4 (continued). The temporal distribution of raw materials (% within Temporal groupings of levels, excluding indeterminate raw materials, n=1,268).

Object type		*Temporal groupings of levels*				*Total*
		Early	*Middle*	*Late*	*Final*	
Edge tool - indeterminate	no.	12	28	37	4	81
	%	0.8	0.8	0.7	5.3	0.8
Edge tool - axe	no.	8	20	18	0	46
	%	0.6	0.6	0.3	0.0	0.4
Edge tool - adze	no.	3	14	16	2	35
	%	0.2	0.4	0.3	2.6	0.3
Edge tool - chisel	no.	1	11	6	0	18
	%	0.1	0.3	0.1	0.0	0.2
Edge tool - roughout/preform	no.	1	3	0	0	4
	%	0.1	0.1	0.0	0.0	0.0
Percussive - indeterminate	no.	0	4	1	0	5
	%	0.0	0.1	0.0	0.0	0.0
Percussive - hammerstone	no.	0	18	8	0	26
	%	0.0	0.5	0.2	0.0	0.2
Percussive - pounder	no.	0	1	3	0	4
	%	0.0	0.0	0.1	0.0	0.0
Perforated objects - macehead	no.	1	5	6	0	12
	%	0.1	0.1	0.1	0.0	0.1
Perforated objects - weight	no.	0	1	0	0	1
	%	0.0	0.0	0.0	0.0	0.0
Grind/abras - indeterminate	no.	17	86	130	1	234
	%	1.2	2.4	2.4	1.3	2.2
Grind/abras - abrader	no.	4	51	93	2	150
	%	0.3	1.4	1.7	2.6	1.4
Grind/abras - grooved abrader U-shaped	no.	1	2	12	0	15
	%	0.1	0.1	0.2	0.0	0.1
Grind/abras - grooved abrader	no.	1	1	0	0	2
	%	0.1	0.0	0.0	0.0	0.0
Grind/abras - abrading slab	no.	4	7	17	2	30
	%	0.3	0.2	0.3	2.6	0.3
Grind/abras - palette	no.	9	37	113	1	160
	%	0.6	1.0	2.1	1.3	1.5
Grind/abras - abrader-knife	no.	0	2	0	0	2
	%	0.0	0.1	0.0	0.0	0.0
Grind/abras - sander	no.	0	2	5	0	7
	%	0.0	0.1	0.1	0.0	0.1
Grind/abras - polisher/smoothed stone	no.	15	65	113	6	199
	%	1.1	1.8	2.1	7.9	1.9
Grind/abras - polishing tool roughout	no.	0	0	2	0	2
	%	0.0	0.0	0.0	0.0	0.0
Grind/abras - polishing slab	no.	0	4	2	0	6
	%	0.0	0.1	0.0	0.0	0.1
Grind/abras - pestle	no.	0	4	7	0	11
	%	0.0	0.1	0.1	0.0	0.1
Grind/abras - bell muller	no.	0	0	1	0	1
	%	0.0	0.0	0.0	0.0	0.0
Grind/abras - grinder	no.	9	34	33	1	77
	%	0.6	0.9	0.6	1.3	0.7

Table 13.5. The temporal distribution of object types (% within Temporal groupings of levels, excluding indeterminate objects, n= 635).

Object type		*Temporal groupings of levels*				*Total*
		Early	*Middle*	*Late*	*Final*	
Grind/abras - grinder one hand mano	no.	1	9	27	0	37
	%	0.1	0.2	0.5	0.0	0.4
Grind/abras - grinder two hand mano	no.	0	1	2	0	3
	%	0.0	0.0	0.0	0.0	0.0
Grind/abras - grinder roughout	no.	0	0	1	0	1
	%	0.0	0.0	0.0	0.0	0.0
Grind/abras - grind. slab/quern	no.	10	124	63	3	200
	%	0.7	3.4	1.2	3.9	1.9
Grind/abras - slab or grinder fragment	no.	46	158	134	4	342
	%	3.2	4.4	2.5	5.3	3.3
Grind/abras - grind. slab roughout	no.	0	23	13	1	37
	%	0.0	0.6	0.2	1.3	0.4
Grind/abras - mortar	no.	1	1	1	0	3
	%	0.1	0.0	0.0	0.0	0.0
Misc. - indeterminate	no.	0	0	10	0	10
	%	0.0	0.0	0.2	0.0	0.1
Misc. - incised stone	no.	0	0	1	0	1
	%	0.0	0.0	0.0	0.0	0.0
Misc. - retouched tool	no.	1	0	0	0	1
	%	0.1	0.0	0.0	0.0	0.0
Misc. - stone ball	no.	1	6	11	1	19
	%	0.1	0.2	0.2	1.3	0.2
Misc. - pigment	no.	14	2	1	0	17
	%	1.0	0.1	0.0	0.0	0.2
Misc. - pigment abraded nodule	no.	0	7	1	0	8
	%	0.0	0.2	0.0	0.0	0.1
Misc. - pigment raw nodule	no.	2	5	8	1	16
	%	0.1	0.1	0.2	1.3	0.2
Misc. - rocks with natural perforations	no.	5	8	5	1	19
	%	0.4	0.2	0.1	1.3	0.2
Misc. - fire-cracked rocks	no.	568	984	133	2	1,687
	%	40.0	27.2	2.5	2.6	16.2
Mult. - indeterminate	no.	0	5	0	0	5
	%	0.0	0.1	0.0	0.0	0.0
Mult. - polisher/hammerstone	no.	1	12	21	2	36
	%	0.1	0.3	0.4	2.6	0.3
Mult. - grooved abrader/grinding slab	no.	0	1	0	0	1
	%	0.0	0.0	0.0	0.0	0.0
Mult. - grinder/hammer	no.	0	5	1	0	6
	%	0.0	0.1	0.0	0.0	0.1
Mult. - abrader/hammer	no.	0	1	4	0	5
	%	0.0	0.0	0.1	0.0	0.0
Mult. - grooved abrader/polisher	no.	0	2	0	0	2
	%	0.0	0.1	0.0	0.0	0.0
Ornaments - indeterminate	no.	2	2	9	0	13
	%	0.1	0.1	0.2	0.0	0.1
Ornaments - beads	no.	99	819	282	1	1,201
	%	7.0	22.6	5.3	1.3	11.5

Table 13.5 (continued). The temporal distribution of object types (% within Temporal groupings of levels, excluding indeterminate objects, n= 635).

Object type		*Temporal groupings of levels*				*Total*
		Early	*Middle*	*Late*	*Final*	
Ornaments - pendants	no.	0	1	1	0	2
	%	0.0	0.0	0.0	0.0	0.0
Ornaments - bracelets/annulets	no.	0	17	1	0	18
	%	0.0	0.5	0.0	0.0	0.2
Ornaments - bead roughout	no.	0	1	3	0	4
	%	0.0	0.0	0.1	0.0	0.0
Ornaments - bead preform	no.	8	2	6	0	16
	%	0.6	0.1	0.1	0.0	0.2
Ornaments - pendant preform	no.	1	2	1	0	4
	%	0.1	0.1	0.0	0.0	0.0
Ornaments - bead preformed cylinder	no.	0	1	0	0	1
	%	0.0	0.0	0.0	0.0	0.0
Debitage - indeterminate	no.	40	175	2,018	1	2,234
	%	2.8	4.8	38.0	1.3	21.4
Waste by-products/shatter	no.	39	76	30	24	169
	%	2.7	2.1	0.6	31.6	1.6
Flake	no.	142	149	65	3	359
	%	10.0	4.1	1.2	3.9	3.4
Debitage - primary flake	no.	0	9	0	0	9
	%	0.0	0.2	0.0	0.0	0.1
Debitage - secondary flake	no.	2	9	4	0	15
	%	0.1	0.2	0.1	0.0	0.1
Debitage - tertiary flake	no.	11	72	10	0	93
	%	0.8	2.0	0.2	0.0	0.9
Debitage - rejuvenation flake	no.	7	57	31	1	96
	%	0.5	1.6	0.6	1.3	0.9
Blade/bladelet	no.	0	0	3	0	3
	%	0.0	0.0	0.1	0.0	0.0
Debris/fragments	no.	21	103	1,360	3	1,487
	%	1.5	2.8	25.6	3.9	14.3
Possible debitage	no.	0	27	5	0	32
	%	0.0	0.7	0.1	0.0	0.3
Ground stone core	no.	0	2	0	0	2
	%	0.0	0.1	0.0	0.0	0.0
Unworked nodule	no.	125	73	93	9	300
	%	8.8	2.0	1.7	11.8	2.9
Quarried material	no.	1	0	0	0	1
	%	0.1	0.0	0.0	0.0	0.0
Nodule?	no.	2	24	20	0	46
	%	0.1	0.7	0.4	0.0	0.4
Natural pebble	no.	131	169	266	0	566
	%	9.2	4.7	5.0	0.0	5.4
Natural cobble	no.	42	69	32	0	143
	%	3.0	1.9	0.6	0.0	1.4
Indeterminate worked	no.	11	7	15	0	33
	%	0.8	0.2	0.3	0.0	0.3
Total	no.	1,420	3,620	5,315	76	10,431
	%	100.0	100.0	100.0	100.0	100.0

Table 13.5 (continued). The temporal distribution of object types (% within Temporal groupings of levels, excluding indeterminate objects, n= 635).

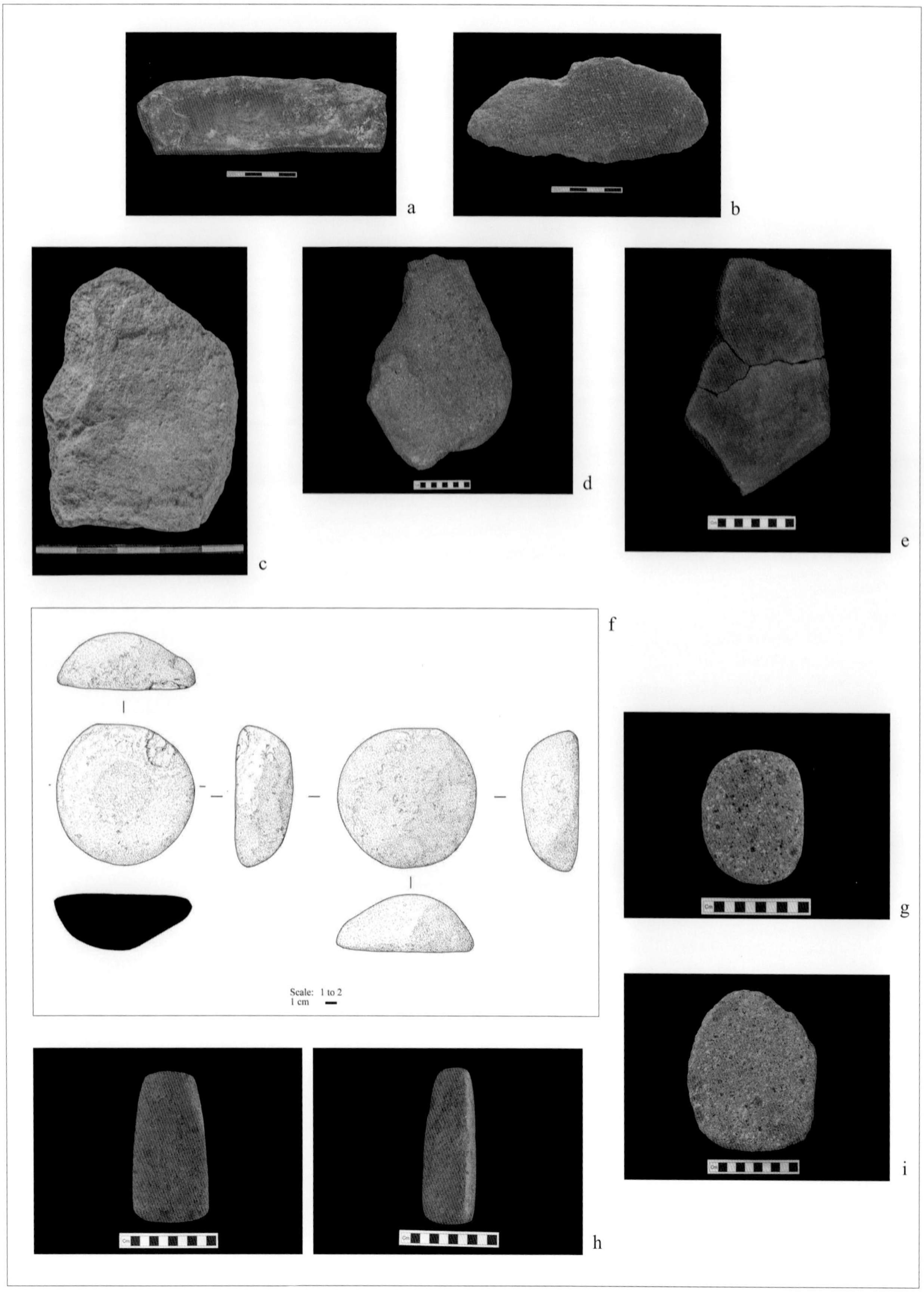

Figure 13.4. Upper and lower grinding tools: (a) 16492.x14; (b) 18523.x2; (c) 23765.x9; (d) 18545.x2; (e) 23765.x4; (f) 18595.x3; (g) 30928.x1; (h) 18596.x1 (front and side view); (i) 18421.k1 (photographs by Jason Quinlan and Christina Tsoraki; illustration by Kathryn Killackey).

Figure 13.5. Querns with increased investment in their production: (a) 16454.x6 plan and side view from B.77; (b) 30945.x1 from B.89; (c) 21767.x2 from B.80; (d) 14019.x36 from B.65 (photographs by Jason Quinlan).

The production of edge tools

In total, 237 axes, adzes and chisels (collectively, edge tools) were recovered during the 1993–2017 field seasons, but only 184 derive from stratified Neolithic contexts. All three common morphological variants (axes, adzes and chisels, a distinction based on the cutting edge profile and the relationship of the cutting edge to the haft) are present in the assemblage, though chisels occur less frequently (fig. 13.6). A large number of tools, however, could not be attributed to any of the three categories because they did not retain a significant part of their cutting edge. Axes and adzes are often trapezoidal or sub-rectangular in plan view, while chisels tend to be rectangular/sub-rectangular. Greater variability is encountered in the morphology of the transverse section; axes tend to be lenticular, ovate/spherical or tapered more frequently, while adzes and chisels tend to have rectangular/flat or irregular sections more often. The vast majority of edge tools have mostly convex cutting edges (68.6%), while lopsided (25.6%) and straight (4.7%) ones are less common.

Mostly material from secondary deposits (that is, waterworn cobbles from stream beds), often retaining the convexity of the natural cobbles, was used for the production of edge tools, but the presence of quarried material was also noted on site (e.g., 21661.k18) (fig. 13.7). The concurrent use of quarried material and material collected from river contexts is first encountered during the Early period in association with a foundation deposit placed during the construction of B.160 (South K) (32496.k4 and 32496.k8). Interestingly, the quarried material resembles macroscopically a stone axe found in B.17 (South K).

The presence of roughouts, tools under modification, debitage and cores of associated materials allows us to reconstruct the technological processes entailed in the production and maintenance of edge tools. Following initial shaping mainly by flaking, hammering and pecking and occasionally sawing, the final form and appearance of the Çatalhöyük edge tools was effected by grinding and polishing. No variation is evident in the production sequence of greenstone and diabase edge tools. Of all edge tools, excluding indeterminate cases, 67.4% have been well polished and a further 24.2% exhibit highly polished surfaces. It is worth noting that the degree of polishing does not correlate with the hardness of the rock type used, but an interest in creating smooth and nicely polished surfaces is evident across different rocks. The presence of very few fine striations on the surface of the edge tools suggests that the final stage of polishing entailed the use of a very fine abrasive frequently combined with a non-abrasive soft material such as leather.

An interesting element in the production of the Çatalhöyük edge tools is the use of the sawing technique (fig. 13.7), a complex and labour-intensive technique that requires great understanding of raw material properties

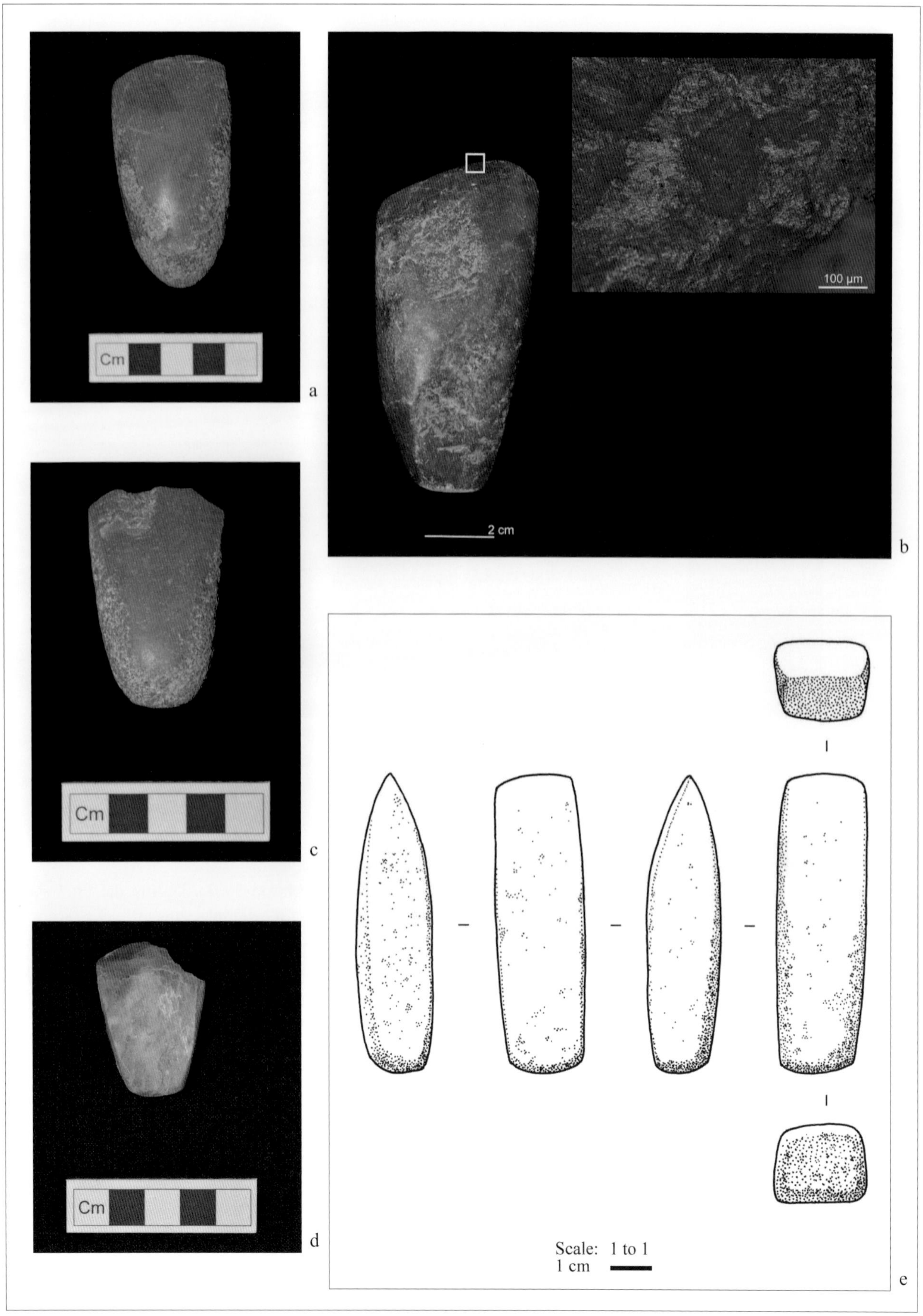

Figure 13.6. Edge tools: morphological variability and microwear traces: a) 18135.x3; b) 21509.k1; c) 18605.x1; d) 30443.x1; e) 13405.x1 (photographs by Christina Tsoraki; illustration by Kathryn Killackey).

Figure 13.7. Edge tool production: (a) core 21661.k18 plan and side view and (b) debitage (32496.k4 and 32128.k3) (photographs by Sophie Vullings and Christina Tsoraki).

(Moundrea-Agrafiòti 1981: 181). Introduced during the Middle period of occupation, from Levels North F and South N onwards, sawing was applied to materials with varied physical properties and hardness (for example, diabase, steatite), suggesting that the difficulty of application did not preclude its use (cf. Tsoraki 2011). It was employed at different stages in the production sequence for shaping nodules, during the initial stages of manufacture (e.g., 21509.k1) and for subsequent modifications of the tools, mainly to correct the overall dimensions of the reworked tools (for example, sawing of the margins) and to repair faults (e.g. 20456.x1). A reconstruction of the sawing technique based on ethnographic, experimental and archaeological evidence suggests that precision and control were particularly valued elements during this process (Kelterborn 1991; Beek, Maika Mason 2002; Croutsch 2005; Tsoraki 2008; 2011). Microwear analysis of sawing grooves suggests that the Çatalhöyük edge tools were sawn using stone materials (cf. Beek, Maika Mason 2002). During the final stage, traces of sawing (for example, lips of sawn grooves) were erased by re-polishing the surface (e.g., 19359.x2), indicating the importance attached to the visual appearance of the tool (cf. Tsoraki 2011).

The production of edge tools quite often would finish with the secure adjustment of the tool into a haft. Hafting devices made from antler and possibly wood have been reported previously at Çatalhöyük (Asouti 2013; Russell, Griffitts 2013). Interestingly, the antler hafts in B.75 and B.56 were found in close association with edge tools left on the floor, while a large number of edge tools were deposited in the same room as the wooden haft during the abandonment of B.77. Indirect hafting evidence is provided by pecked surfaces on margins and lower body surfaces often exhibiting smoothed rounded grains, most likely the result of contact with a haft. Edge tools were also used as non-hafted implements, with a hammerstone often combined with an intermediary material to cushion the butt area as the presence of percussive traces on the butt area show. In this case, the presence of extensive percussive traces and damaged or completely destroyed cutting edges suggests the use of the tools as wedges for activities that required heavy impact force.

The production of palettes

Palettes are small-sized abrading tools that are on average 125mm long and 84mm wide. In total, 160 of these implements can be securely attributed to Neolithic levels. Manufacture entailed mainly the coarse shaping of the margins by flaking (53.2%), occasionally followed by grinding, while often the natural surfaces remain (32.3%). An important element in their morphology is the angled transverse sections and their sub-rectangular/irregular plan (fig. 13.8), a deliberate technological choice that facilitated the comfortable handling of the tool during use. This in tandem with their consistent small size indicates that these tools were used as handheld implements.

The production of polishing tools

Among the polishing tools identified at Çatalhöyük there is a type made from metamorphosed limestone/marble that shows great consistency in its overall morphology, with mostly discoidal, ovate and sub-rectangular forms, and usually two or more flat use-faces that are well to highly polished (fig. 13.8). Wright (2013: 382) noted an association between the appearance of similar tools and the development of plaster technology during the PPNB period in Near Eastern sites. Previous hypotheses regarding the use of these tools for burnishing plastered surfaces at Çatalhöyük (Mellaart 1962: 48; Baysal, Wright 2005; Wright, Baysal 2012) were tested through microwear analysis (see below).

The production processes of these objects entailed a multi-stage production sequence, as suggested by roughouts and finished objects. The initial shaping entailed the severing of the original nodule in half using percussion and the shaping of the whole surface of the nodule through flaking and pecking (22813.x24 and 22813.x10). Following this, the surface of the object was ground with the aid of abrasive agents, removing any irregularities present and creating the flat surfaces seen on the finished examples. Microwear traces and their microstratigraphic relations show the presence of two types of micropolish and associated linear features that can be attributed to the stage of manufacture and use of the tools (fig. 13.8). This confirms that the smooth finish and flat surfaces encountered on these tools was an intentional manufacturing choice and not the result of use.

The production of 'maceheads'

During the Hodder excavations only 16 'maceheads' have been recovered from stratified and unstratified Neolithic contexts, and together with the 15 examples from Mellaart's excavations stored at the Konya Museum (Wright 2013), the total known number is still relatively low. Previous suggestions about their late appearance in the temporal sequence at Neolithic Çatalhöyük (from the Middle period onwards) and an increased presence in the Chalcolithic levels at Çatalhöyük West (Wright 2013; 2014) are broadly supported by the more recent findings. A recently recovered example, however, dated to the Early period (South K), suggests an earlier introduction of this elaborate form of ground stone technology than previously thought (fig. 13.9).

Both complete spheres and fragments have been found (Tsoraki 2015; 2016; 2017; for 'maceheads' excavated prior to 2009, see Wright 2013). On average, complete examples measure ca 63mm in diameter and weigh 303g, and their size distribution does not show standardisation in their dimensions (fig. 13.10). They are almost spherical/sub-spherical in shape and often exhibit flattening around the area of the perforation (e.g., 30008.x1). The perforations are biconically drilled, with visible bands of concentric striations. The drilling was not performed with great accuracy, and the interior of the perforations is irregular and often off-centre. The body surface was shaped by grinding and polishing executed in a continuous motion, as the lack of faceting on the body surfaces indicates. The degree of polishing ranges from well to highly polished, and the lack of visible striations indicates the use of a soft non-abrasive material such as leather for the final surface finishing. Although all known examples from Mellaart's and Hodder's excavations are finished objects and no identified preforms have been found to date, the on-site production of these objects cannot be disregarded.

An example that stands out is a complete 'macehead' (22194.x1) with an intricately carved design, a product of fine craftsmanship with no known parallels in other Near Eastern and Aegean Neolithic sites (Tsoraki 2016). It was made of high-quality white marble and placed in a burial (22196) in B.5 together with obsidian projectile points and a possible shell container with pigment traces (fig. 13.9). It measures 66mm in diameter and weighs 360g. It has a biconically drilled perforation exhibiting concentric striations. The lack of wear traces in the interior of the perforation argues against the use of the 'macehead' with a shaft (handle/haft). This together with the overall surface condition (for example, the sharp lines

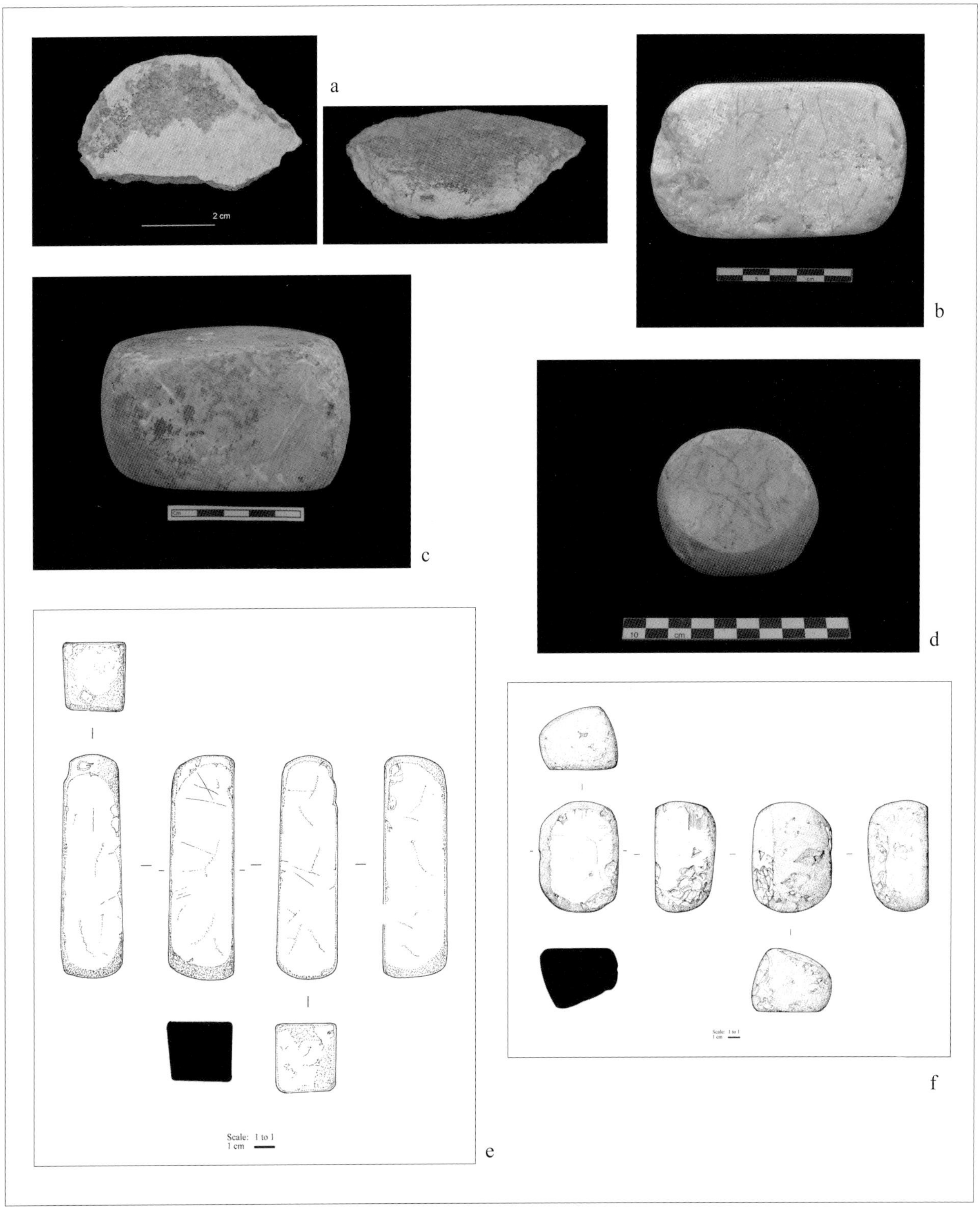

Figure 13.8a–f. Tools used for plastering activities: (a) applicator 21509.k3 plan view and view of the margin that has plaster residues; polishing tools (burnishers) (b) 19818.x2; (c) 18593.x1; (d) 11648.x2; (e) 22813.x27; (f) 22813 (photographs by Jason Quinlan and Christina Tsoraki; illustration by Kathryn Killackey and Caroline Habron).

of the incised pattern and lack of rounding of the ridges from contact with wrapping material/handling) suggest that the 'macehead' was deposited in the burial not long after it was made. The execution of the intricate motif, consistency in the spacing between the incisions, and the lack of unintentional faceting of the surface or other manufacturing errors during the carving suggests that this was the work of a highly skilled craftsperson.

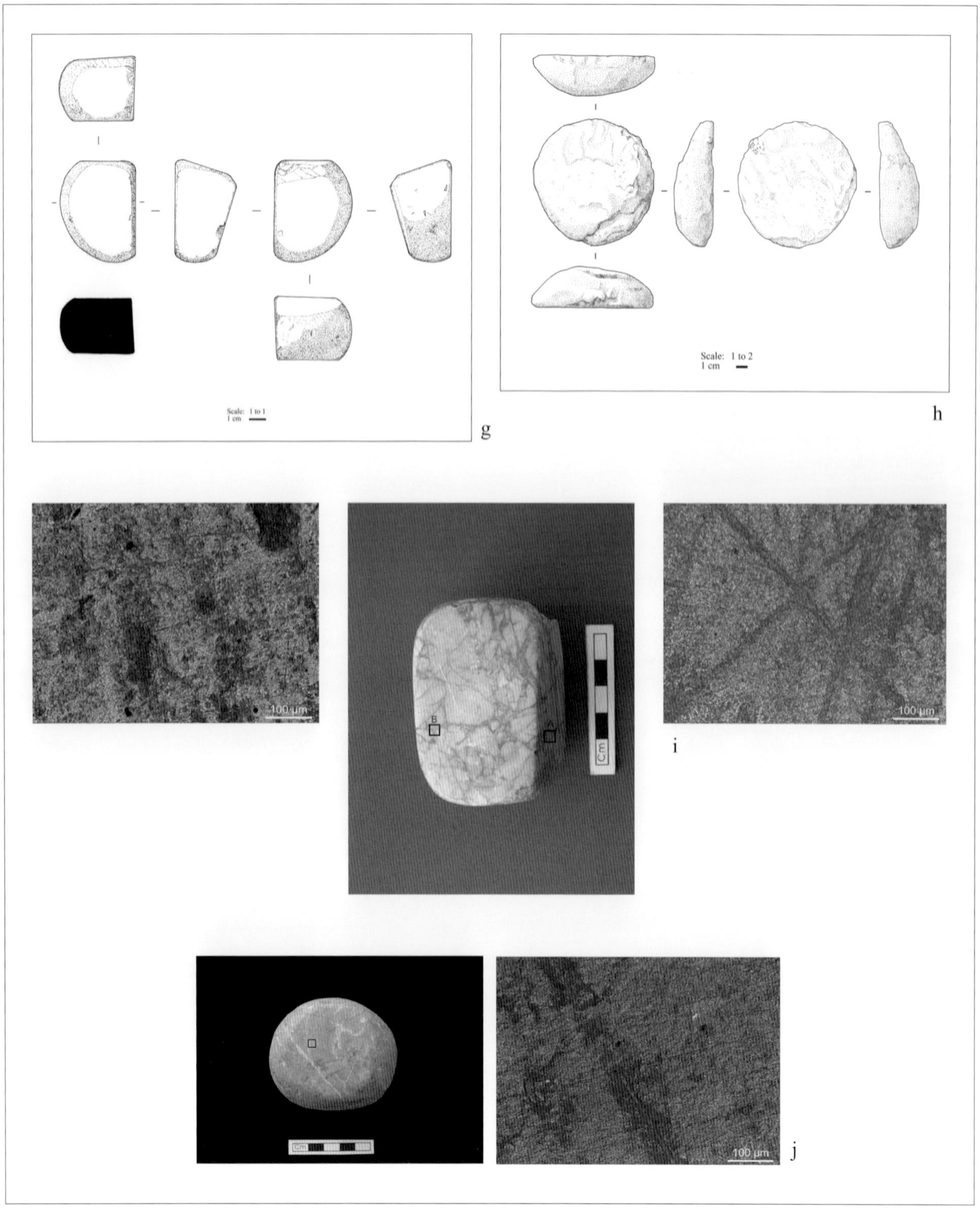

Figure 13.8g–j. Tools used for plastering activities: (g) 20761.x1; (h) polishing tool roughout 22813.x10; (i) polisher 16483.x30 and microwear traces associated with the production (Location A) and use of the tool (Location B); (j) polisher 16480.x2 and microwear traces associated with the use of the tool (photographs by Jason Quinlan and Christina Tsoraki; illustration by Kathryn Killackey and Caroline Habron).

During the execution of such an elaborate and unique design on the actual stone material, a mock-up made of clay or an organic material such as wood may have been used. While incised decorative motifs have been rendered on the surface of other materials (for example, decorated grooved abraders and stone vessels), this is the only example of a decorated 'macehead' found on site to date.

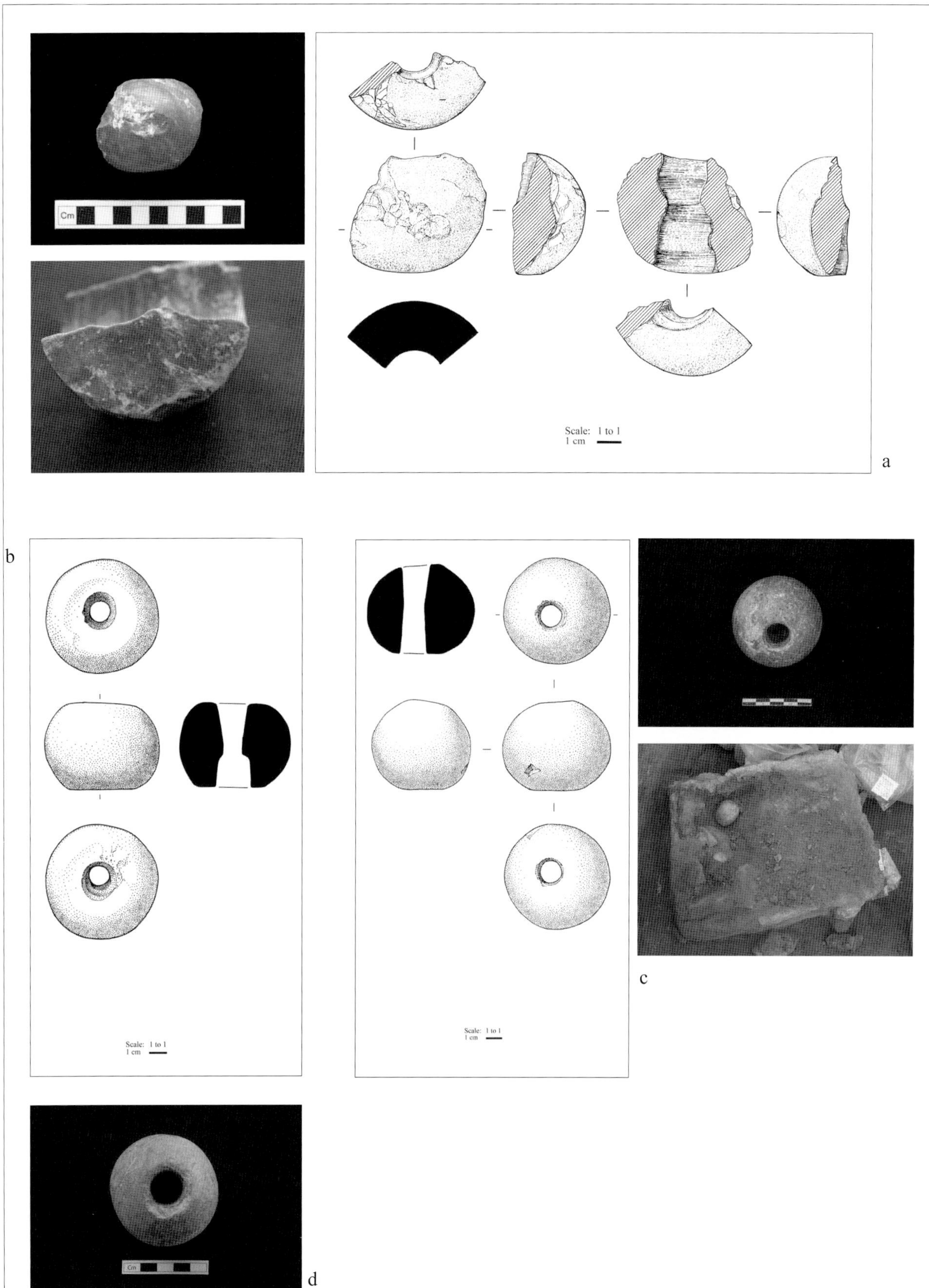

Figure 13.9a–d. 'Maceheads': (a) 21872.k1 from B.17, plan view and view of the crushed margin from re-use; (b) 22813.x1; (c) 32860.k35 from B.150 and the location of the macehead in bin; (d) 12343.x1 (photographs by Jason Quinlan and Christina Tsoraki; illustration by Kathryn Killackey and Caroline Habron).

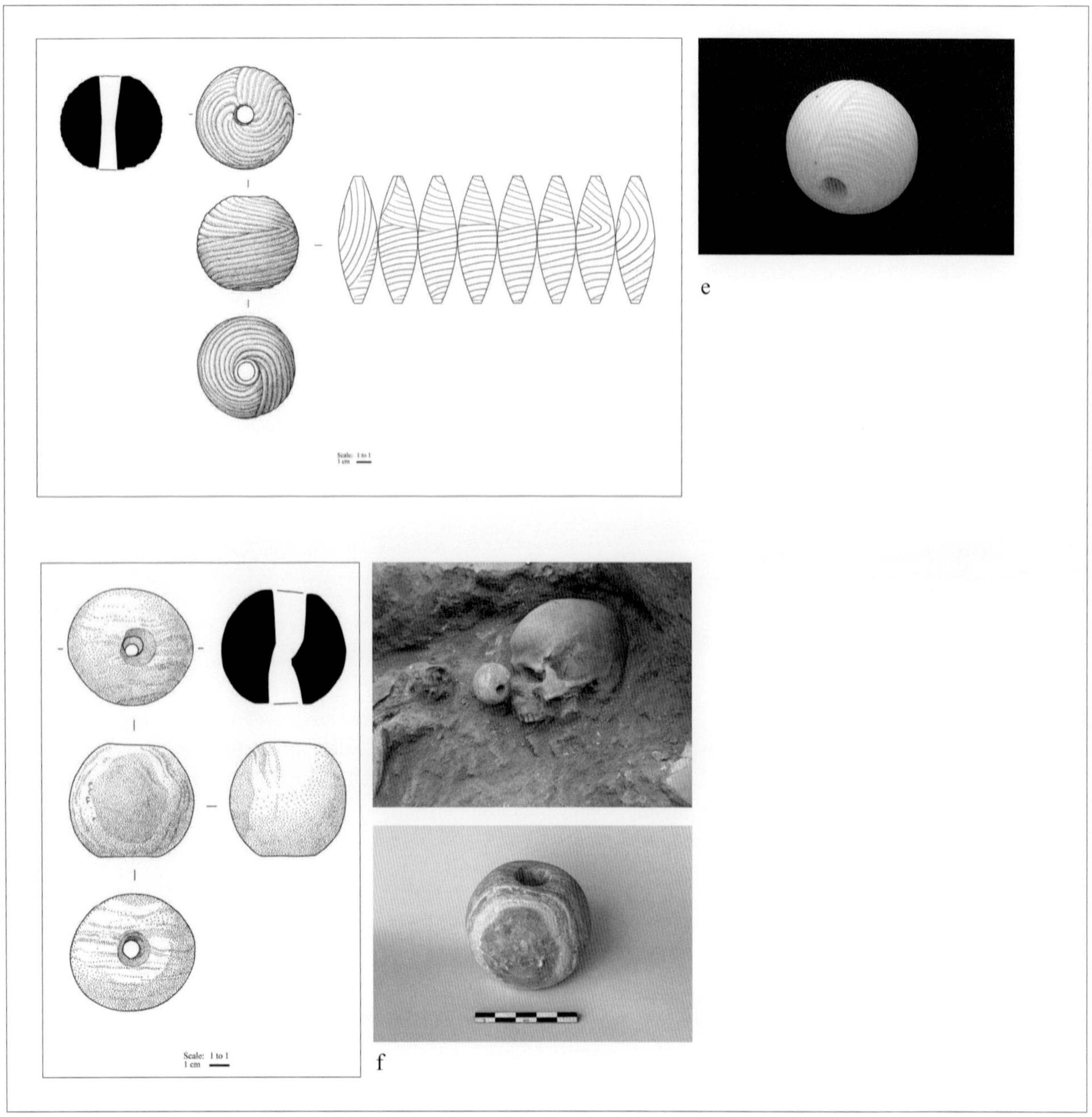

Figure 13.9e–f. 'Maceheads': (e) incised marble 'macehead' 22194.x1 from burial context in B.5; (f) 30008.x1 from burial context in B.114 (photographs by Jason Quinlan and Christina Tsoraki; illustration by Kathryn Killackey and Caroline Habron).

Depositional practices

Ground stone artefacts occur across all types of depositional contexts, with varying relative frequencies (online fig. S13.2). Construction/make-up deposits are the most common category, with more than 3,500 artefacts, followed by middens, fills and burial fills. While floors, fills and activity deposits are broadly similar in object variability, middens tend to have a higher relative frequency of miscellaneous artefacts and mostly fire-cracked rocks, while ornaments out of all ground stone categories are most commonly found in burials (fill and skeleton) (see also Chapter 9).

All complete lower grinding tools or tools that retain more than half of their original size (n=38) derive from building interiors, a pattern also noted by Wright (2013). This is also the case for complete grinders, of which all except one were found inside buildings. While most buildings tend to have fewer than three querns each, B.77 (North G) and B.150 (TP M) stand out with 12 and six querns respectively. No complete querns derive from the Early period. In terms of depositional patterns, complete querns tend to be associated with clusters/placed deposits more frequently.

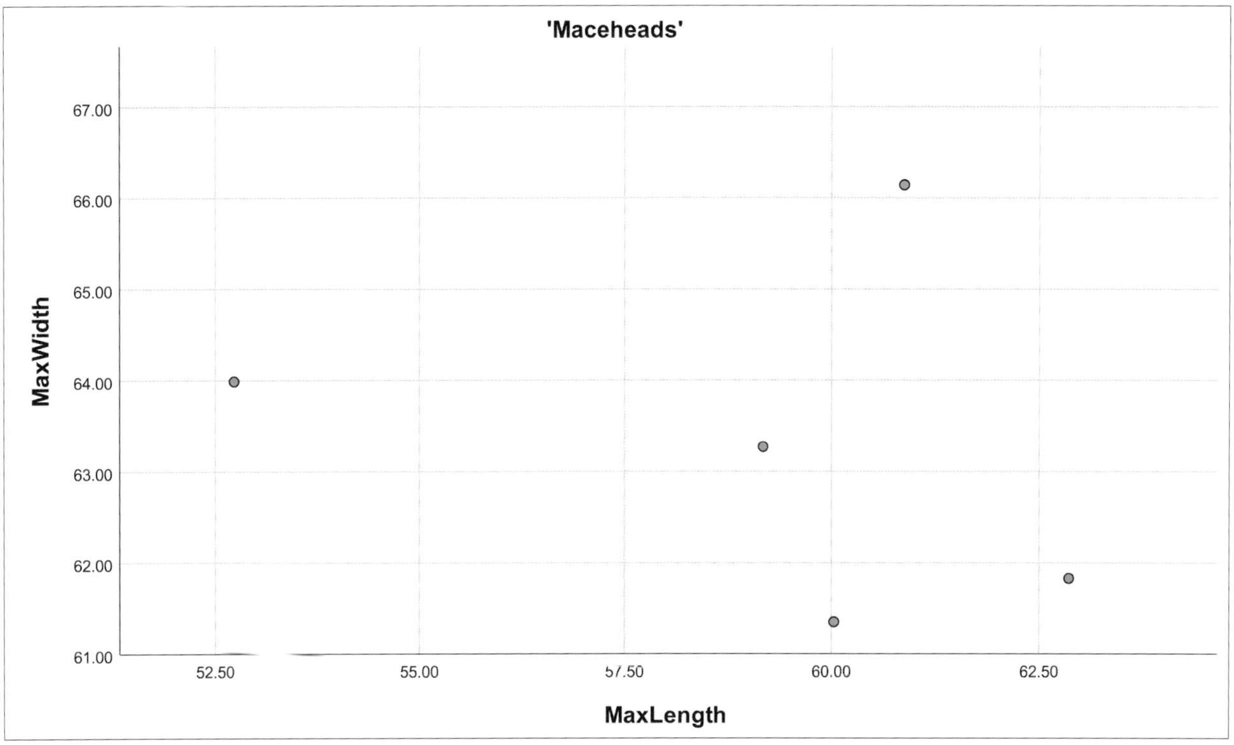

Figure 13.10. The size distribution of complete 'maceheads' (measurements were available for five specimens).

Edge tools occur throughout the occupation sequence at Neolithic Çatalhöyük, but higher densities are encountered in the Late period for both the North and the South Areas (fig. 13.11). Edge tools occur more frequently within buildings than in open areas (70.6% vs 29.4%), but during the Late period deposition in open areas is more common, though it is not a statistically significant association. Edge tools are distributed across different buildings, and there is no strong correlation between the density of edge tools and buildings with high elaboration index (Ei) (see Introduction, this volume). Among buildings that are completely or more than 75% excavated, B.65 (Late period) has the highest density, followed by B.56 and B.77 (Middle period) (fig. 13.12). Edge tools are most commonly found in fill (38.1%), followed by midden layers (26.7%), and clusters/placed deposits (14.2%). When deposited in clusters/placed deposits, edge tools appear in single numbers. The clustered depositions in B.77, however, stand out, as ten edge tools formed elements of clustered depositions, seven of which were found in two discrete concentrations [(16491) n=3, (16492) n=4]. In total, 13 edge tools were found in B.77 – the largest concentration from a single building – and they are all associated with the burning event and abandonment of the building (Tsoraki 2018). No significant correlations are evident between degree of wear or fragmentation patterns and the place and form of deposition (floor, fill, cluster etc.), with lightly used and heavily used edge tools encountered across all depositional contexts. Edge tools that survive in a more complete state are more commonly found inside buildings (82.1%) and are mostly associated with fills (47.2%) and clusters/placed deposits (26.4%). The presence of tools with light wear in open areas may suggest, however, that tools were not always carefully curated inside the buildings and were discarded at the place of use.

Plastering tools (applicators and polishers/burnishers) occur throughout the habitation sequence at Çatalhöyük. There is no significant variation in their density in the Middle and Late period in the North Area, while in the South Area there is an increase in their densities over time, with the highest attested in the Late period and more specifically in level TP.M (0.00033 artefacts per litre of soil) (fig. 13.13). Plastering tools, and especially complete tools, are more commonly found inside buildings (63.5% of 115 objects), but there is no statistically significant association between location and condition of tools. High peaks in the density of plastering tools are encountered in B.142, and B.150 (TP M) and B.140 (TP N) (fig. 13.14). Of the 23 buildings that are fully or more than 75% excavated, 11 had plastering tools in their inventories, while no correlation between buildings with high elaboration indices and number of tools was established.

All except one 'macehead' from the ÇRP excavations derive from building interiors, but no strong correlation can be discerned with burnt/unburnt buildings or buildings with high elaboration. The earliest 'macehead'

(21827.k1) was found on the floor of B.17 (South K). During the Middle period (North F and North G), 'maceheads' occur in burials and one complete example (1306.x1) was found in an external area next to B.1. During the Late period, 'maceheads' are found only in non-burial contexts and show more variability in their depositional contexts, including floor deposits, clusters and fills.

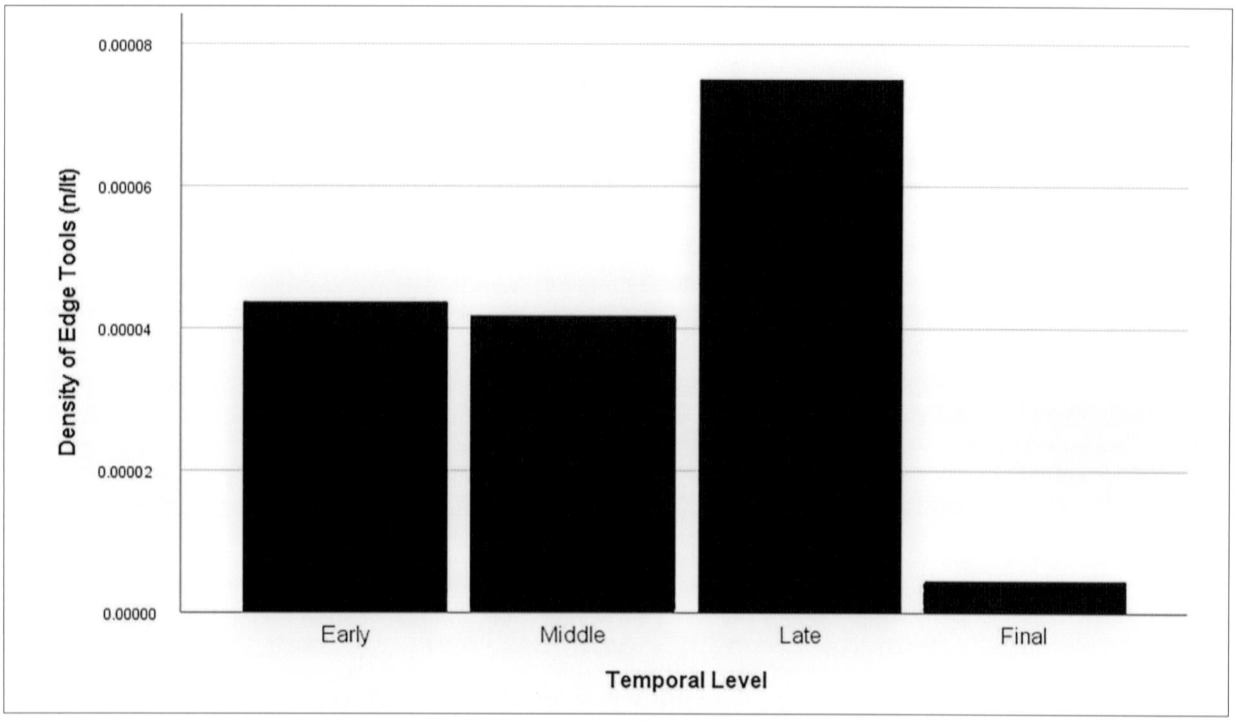

Figure 13.11. The density of edge tools across the temporal groupings of levels.

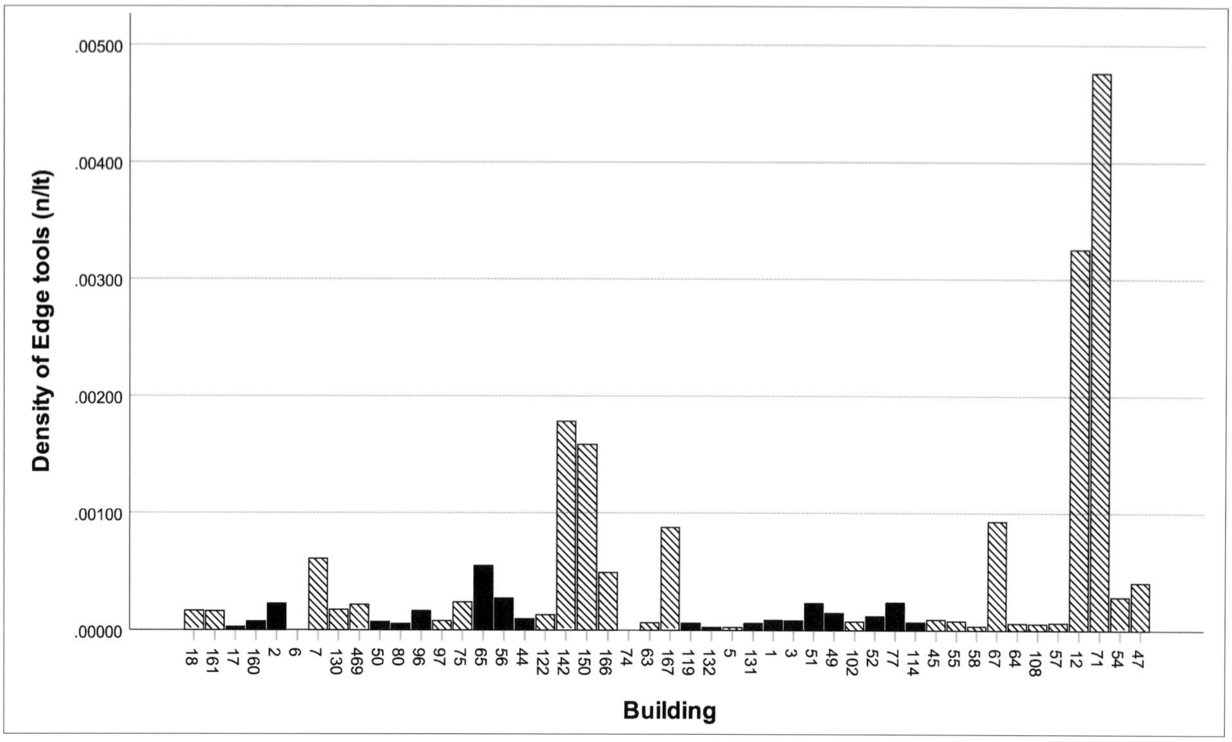

Figure 13.12. The density of edge tools across buildings (all areas and periods, black bars indicate buildings that are completely or more than 75% excavated).

Pestles follow previous patterns in their distribution, with more frequent occurrence inside buildings. Interestingly, pestles tend to concentrate in specific buildings: two examples were found in a bin in B.77, three examples, also in a bin, in B.150, and three examples in B.122. In the case of annulets/bracelets, they occur across different depositional contexts but are more frequently found in middens (10 out of 19) and fills (5 out of 19). A further fragment of

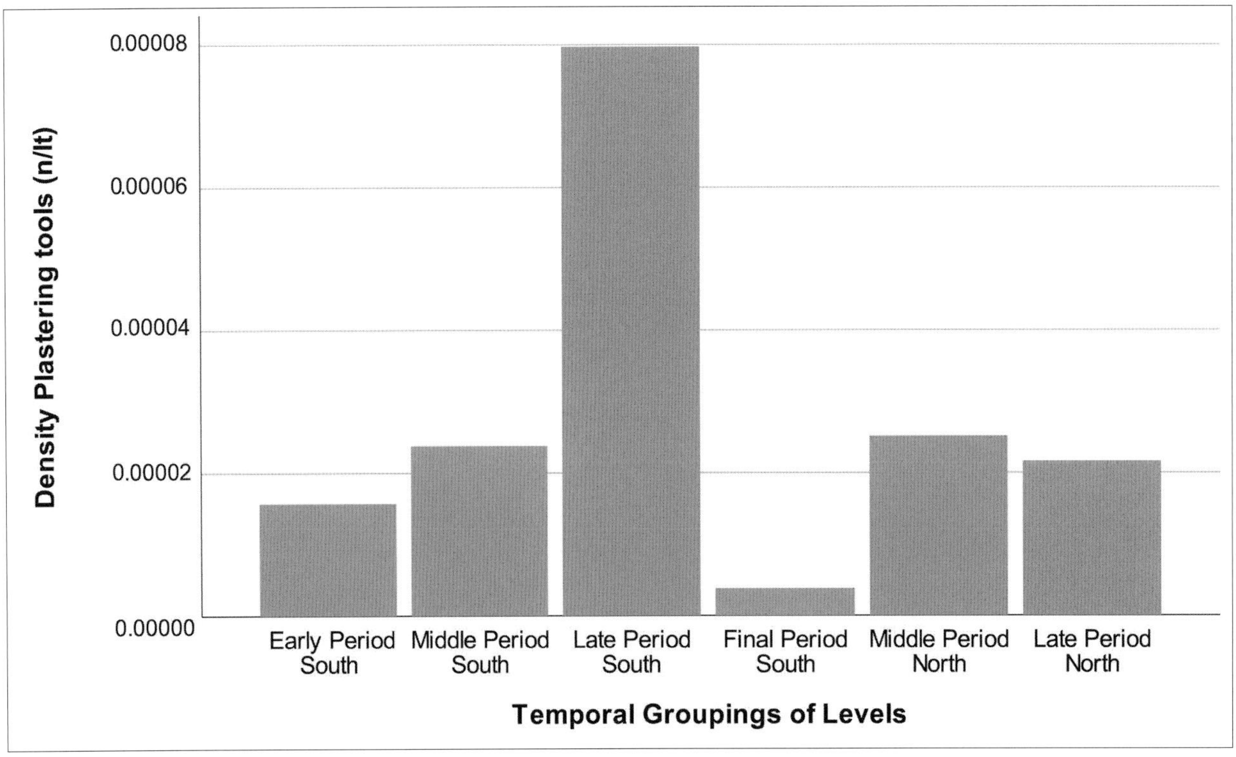

Figure 13.13. The densities of plastering tools across the temporal groupings of levels.

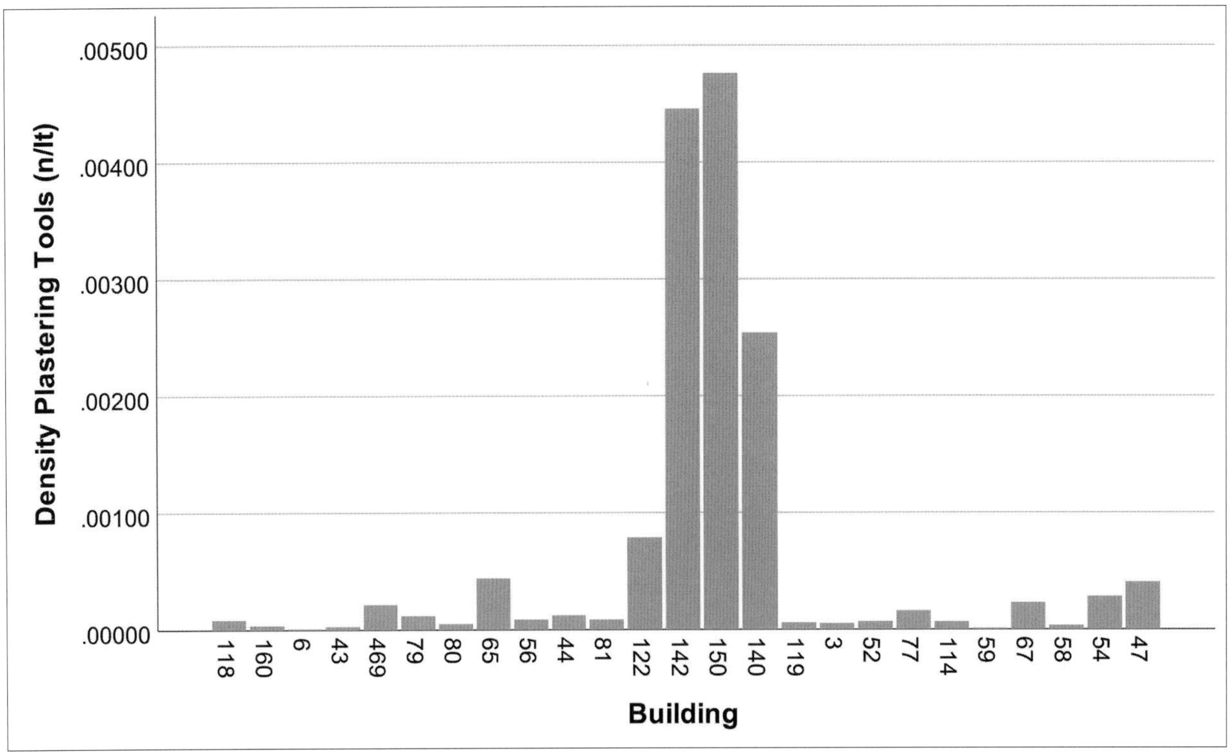

Figure 13.14. The density of plastering tools per building, including all areas and all periods.

a marble annulet/bracelet that was in the process of being modified into a pendant was found in an external midden deposit (32128) in the North Area (Chapter 9); as this was found at the very end of the final excavation season, it was not recorded in the ground stone database and therefore it does not appear in the overall counts mentioned in the text. Six annulets/bracelets were found in Sp.489 and two of the fragments refit (19486.x2, 19486.x5). When found in building interiors they tend to derive from buildings with high Ei (>18).

Construction/make-up layers

A wide range of mostly small-sized broken tools (less than 50cm in size) along with debitage from stone-working activities have been regularly incorporated in the construction of structural elements (for example, platforms) throughout the occupation of the site. Yet the deposition of a small number of artefacts in the walls of various buildings is intriguing and raises questions about the intentionality behind this act. During excavations in the South Area, a large andesitic quern (20892.x1) that was broken in half was found embedded in the western foundation wall (F.8003) of Sp.558 (Middle period) (fig. 13.15) (Tsoraki 2015). Following the original construction of the wall, the implement was intentionally embedded in it, cutting through several courses of mortar and brick (ÇRP Excavation Database). Furthermore, the heavily concave use-face of the tool – a rather rare occurrence at Çatalhöyük – attests to a tool with a prolonged biography prior to its deposition. The re-use of querns in the wall construction of stone-built structures is a practice that has been reported from other ethnographic and archaeological contexts in the Near Eastern and the Neolithic Aegean (e.g., Evans et al. 1964; Wright 2008; Rosenberg 2013; Tsoraki 2014; Nixon-Darcus, D'Andrea 2017). A common assumption is that the incorporation of seemingly mundane objects such as grinding tools in walls and other construction elements is guided by practical considerations and notions of efficiency (for example, availability and re-use of materials). In the case of 20892.x1, however, an argument such as this cannot be supported, as the wall was made up of bricks very uniform in size and shape, and the embedded quern reduced, therefore, the stability of the wall. Instead, the act of placing seemingly mundane objects in structural elements needs to be approached as a practice embedded in a wider frame of social meanings. At Çatalhöyük, quern fragments have received special treatment on numerous occasions. For example, during the Late period a large quern fragment was embedded in an upright position in the northeastern platform of B.150. This is reminiscent of similar practices encountered in the B.65-B.56-B.44-B.10 sequence (for example, intentionally squared quern fragment placed on a hearth upon abandonment of B.65 (Regan 2014)).

Figure 13.15. Quern (20892.x1) found embedded in foundation wall F.8003 in Sp.558 (South M, Middle period) (photographs by Jason Quinlan and Christina Tsoraki).

Similarly, during the Late period three edge tools, all broken, were found embedded inside the mortar of walls in the North Area in B.12 (16460.k1-F.3087) and the South Area in B.150 (32823.x1-F.3899) and B.65 (15727.x1-F.2514). This is reminiscent of similar practices at Neolithic Çayönü that entailed the incorporation of complete and broken edge tools in the walls of the Grill Buildings (Özdoğan 1999: 56).

Clusters

A depositional practice commonly encountered at Çatalhöyük is clustered depositions of ground stone objects, often found in association with other materials (Nakamura 2010; Carter et al. 2015; Nakamura, Pels 2014; Russell et al. 2014; Tsoraki 2018). This type of deposition displays great variability in composition and structural principles, ranging from clusters with a single type of object, to discrete concentrations of natural pebbles, to clustered depositions of increased variability. The complexities of this form of depositional practice intensify in the Middle and Late periods of occupation (online fig. S13.3). We see a shift from clusters that seem to relate to more 'functional assemblages', mostly consisting of fire-cracked rocks and clay balls (B.17, B.160, B.132) (see also Tsoraki 2016), to clusters of a more diverse composition that incorporate objects of varied forms and biographies dated to the Middle and Late periods. Clustered depositions in B.77 in the North Area (Tsoraki 2018), along with depositions in the B.65-B.56-B.44-B.10 sequence, B.80 in the South Area and B.150 in the TPC Area, stand out in terms of material 'richness' and divergence in object itineraries.

Recurrent components of these clusters are complete and fragmented grinding tools. Detailed analysis of fragmentation patterns concurs with previous suggestions about the deliberate destruction of grinding tools (Wright 2013). Variation, however, in the degree of weathering of the fractured edges, which ranges from fresh-looking, sharp edges to heavily worn ones, and the lack of refits among fragments suggest that this practice was not associated with the in situ breakage of grinding tools; instead, these fragments have varied depositional histories associated with different primary contexts and were purposefully assembled at the moment of deposition during particular events such as the establishment or closure of a house (Tsoraki 2018).

Interestingly, depositional practices associated with B.150 (TP M/Late period) are reminiscent of similar events encountered in the B.65-B.56-B.44-B.10 sequence (see also above), but also in B.49 ((7957)/(14460)), attributed to Level North G. These include a clustered deposition (23765) placed in the southwestern part of Sp.637, as well as a distinct concentration of 259 natural waterworn pebbles (32821.x1) of a similar size (ca 12mm), shape and colour that were placed in the construction layer of platform (F.8693) in Sp.639 (fig. 13.16). Similar caches of natural pebbles were encountered in B.65 deposited during the construction of a small bench (Regan 2014: 188). Unit (23765), on the other hand, furnished an exceptional concentration of materials and objects, the vast majority of which are different types of ground stone objects (n=129), including grinding implements, schist palettes and other abrading and polishing tools (Tsoraki 2017). Variability in object types is also reflected in the type of activities represented, including plant processing, pigment processing, stone polishing and possible bone working (fig. 13.17). Similarly, in B.44 and B.49 a large number of ground stone artefacts along with obsidian, faunal remains, bone tools, pottery sherds, stone beads, pigment, quartz crystal and figurines were deliberately placed during the construction of the southwestern platforms (Wright 2013: 406; Eddisford 2014: 321–26; Nakamura, Pels 2014; Regan 2014: 169; Carter et al. 2015). While previous interpretations approached these depositions as the result of cleaning up activities of household toolkits (Eddisford 2014; Regan 2014), re-evaluation of the material from (11648) in B.44 by the current team suggests that this represents an intentional deposition that entailed the concerted actions of multiple social groups (multiple houses/households) (Tsoraki in preparation).

Finally, a distinct form of clustered deposition is the placement of stone objects inside bins. Similar practices have been noted previously (for example, B.52, B.59, B.1 and B.77: Wright 2013; Baysal and Wright 2005). Recent examples have been found in B.150 (F. 8674) (fig. 13.16), B.122 (F.7198) and B.81 (F.8081) in the South Area and B.102 (F.3698) in the North. All known occurrences are dated to the Middle (North G) and Late periods. Some of these deposits are associated with infilling processes during abandonment (for example, B.81) (Barański et al. Volume 15, Chapter 5). Based on the variability and the condition of the material, some of these deposits, however, point towards the possible storage of still usable tools with functional correlations, which may represent household toolkits (for example, B.150 (32860), which contained 35 objects; Tsoraki 2017; cf. Wright 2013).

Ground stone in burial contexts

Stone beads form the main type of inclusion found in burial contexts, either in the burial fill or in direct association with interred individuals (for detailed discussion see Chapter 9) (fig. 13.18 and table 13.6).

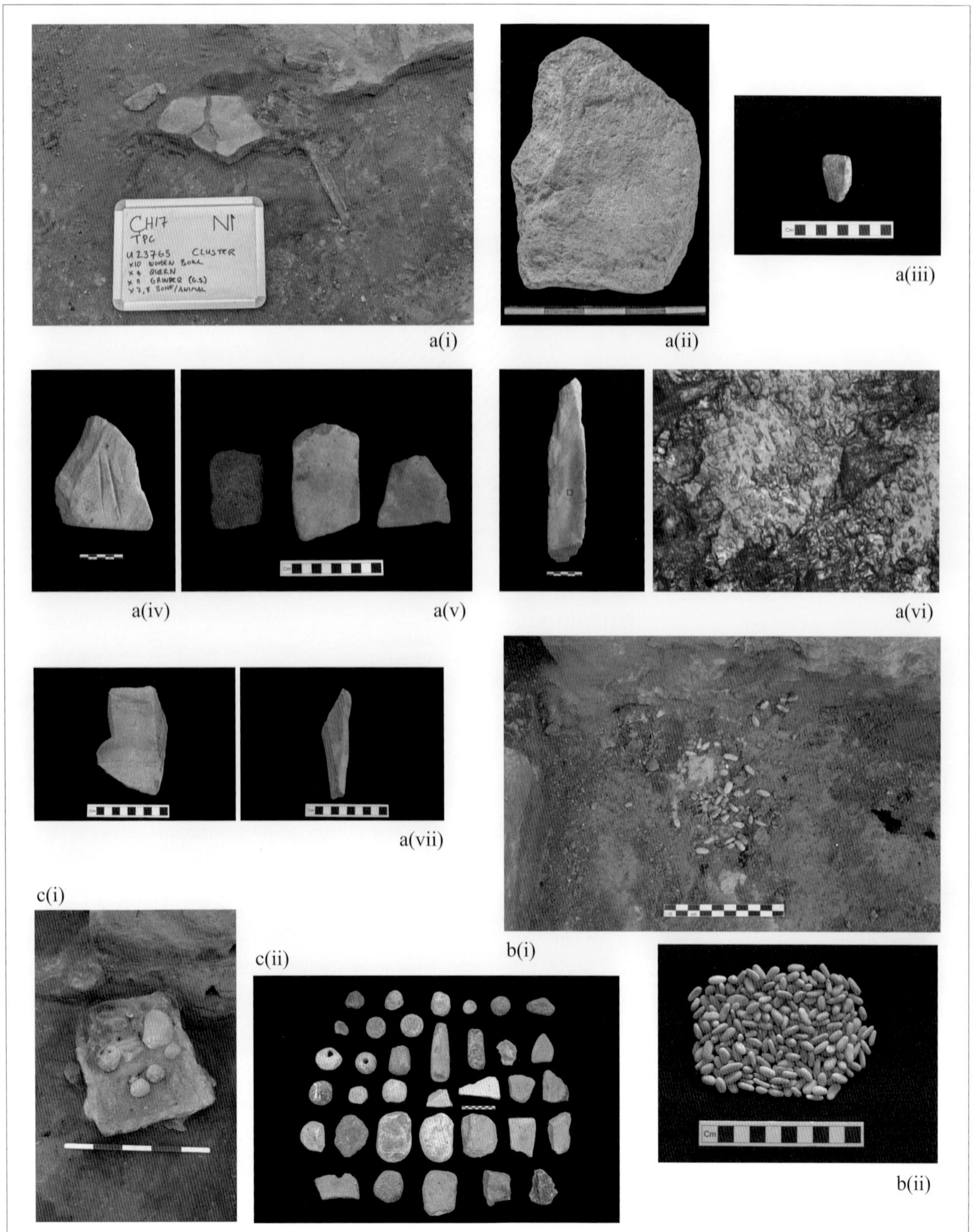

Figure 13.16. Clustered depositions in B.150, TPC Area: (a) cluster (23765) in Sp.637: (i) view of the excavation context of (23765), (ii) large quern 23765.x9, (iii) edge tool 23765.x102, (iv) grooved abrader 23765.x5, (v) abrading tools, (vi) polishing tool 23765.x100-x.78 and microwear traces associated with hard mineral contact (micrograph taken at 200x magnification), (vii) palette 23765.x99 plan and side view; (b) cluster (32821) of natural pebbles associated with the construction of F.8693 in Sp.639: (i) the pebbles in context, (ii) the complete cluster; (c) cluster (32860) in Sp.639: (i) view of excavation of bin F. 8674, (ii) the complete assemblage of stone tools stored in the bin (photographs by Jason Quinlan and Christina Tsoraki).

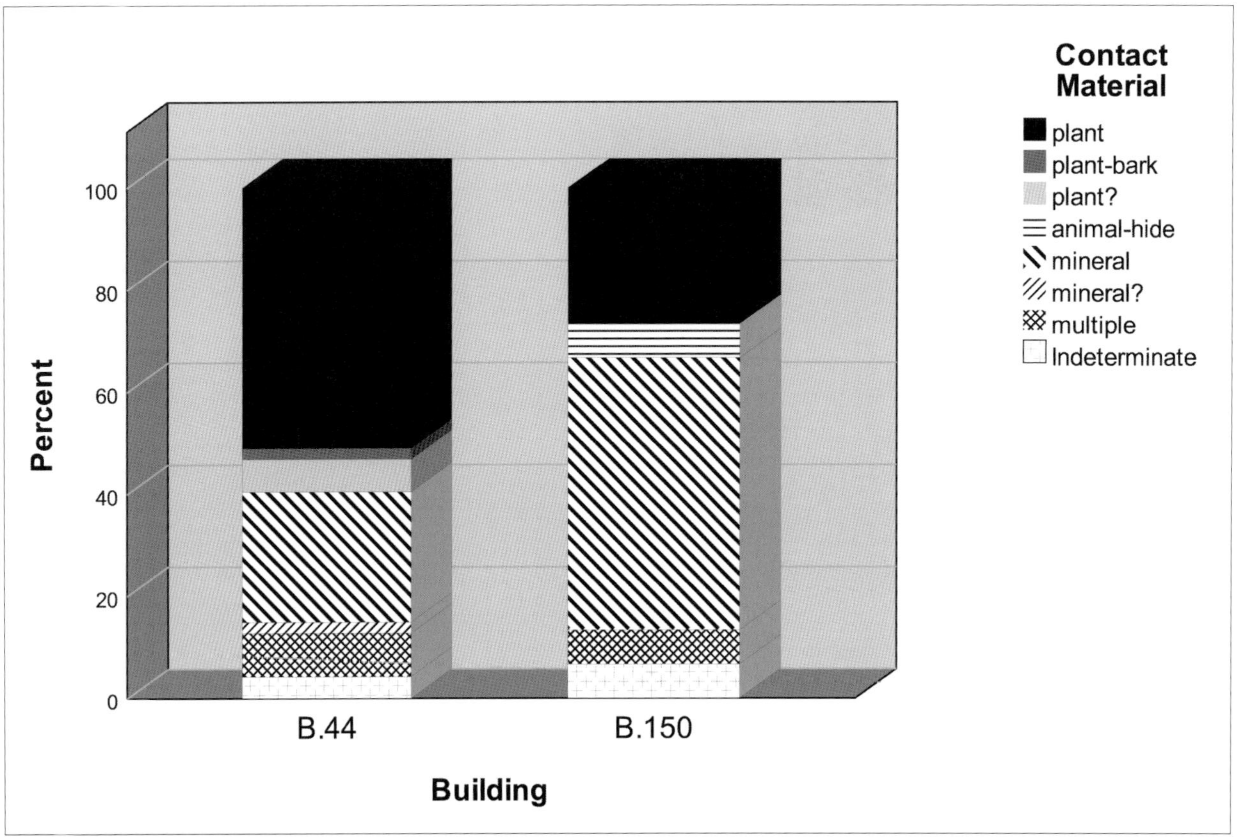

Figure 13.17. Contact materials identified through the microwear analysis of ground stone artefacts in clusters B.44 (11648) and B.150 (23765).

While beads in burial contexts present greater variation in their degree of wear, ranging from examples with no use or limited use to heavily used ones, the tools found in the fill and in direct association with individuals tend to be tools with extensive wear traces, indicating that they had been used for a considerable time prior to their placement in the burial, though none are worn out.

While some materials may have entered burial fills accidentally (for example, in floor sweepings), in many cases, beads found loose in the fill may originally have formed components of ornaments (for example, strings of beads used as bracelets) adorning the interred individuals that were disturbed during subsequent burial events. This is convincingly showcased by bead materials found in association with adult skeleton (22620) buried in B.129. A large number of stone beads forming bracelets and anklets were found together with the skeleton, along with loose beads in the burial fill. Detailed analysis of technological and use-wear traces suggests that the carnelian beads that formed the bracelets and those found loose in the burial fill were associated with the same production event and had been used for a similar duration prior to their deposition in the burial. Moreover, this burial assemblage highlights nicely variations in the microhistories of the different objects selected to furnish the dead. In this case, while the left anklet (22623.x5) consists of beads with a fresh appearance suggesting no use or limited use prior to deposition, the right anklet (22623.x3) as well as the two carnelian bracelets exhibit a moderate degree of wear. Since both phyllite and limestone beads that form the left anklet exhibit the same degree of wear, we can attribute this with confidence to the duration of use the anklet saw before it was placed in the burial and not to the effect weathering processes have on different stone materials. The fresh appearance of the anklet may, therefore, suggest that it did not see much use between the moment of its manufacture and its placement in the burial, raising the possibility that its production may have been deliberately commissioned for the event of the burial. In that respect, the bracelets and anklets accompanying the individual interred in F.7714 can be perceived as an assemblage of materials of distinct origin and appearance but also technological skill and temporalities (Tsoraki 2017).

Other forms of ornaments that have been found in burials include two marble annulets/bracelets, the only complete examples found to date. (7580.x2) was found on the left arm of a mature adult female skeleton, in a disturbed burial (F.1202) attributed to unstratified Neolithic deposits in the North Area (tables 13.7 and 13.8),

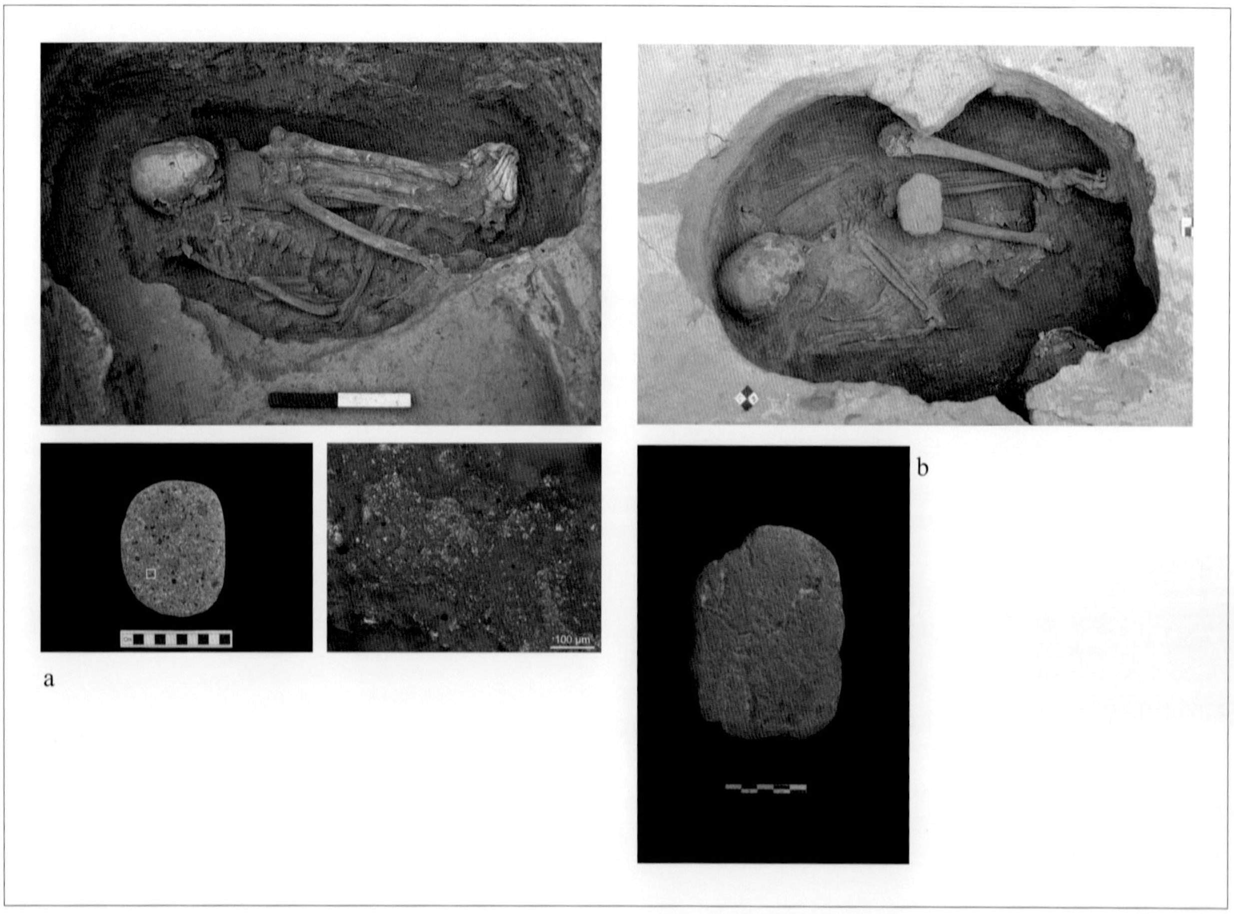

Figure 13.18. The deposition of grinding tools in direct association with interred individuals: (a) grinder with skeleton (30928) and microwear traces associated with plant processing; (b) grinder with skeleton (23827) (photographs by Jason Quinlan and Christina Tsoraki).

and (30503.x8) was placed at the bottom of the fill of burial cut (30502), which housed a multiple burial (F.7127) in B.52 (fig. 13.3). (7580.x2) has an outer diameter of 100.74mm and is 10.51mm thick; in places the surface shows a low degree of polishing, the result of use. (30503.x8) has an outer diameter of 63.52mm and inner diameter of 50.09mm, is ca 8mm thick and weighs 19g (Tsoraki 2013). The bracelet, of rather irregular shape, does not exhibit macroscopically visible use-related wear traces, suggesting limited or no use prior to its deposition. Overall, when compared with bracelets from non-burial contexts (for example, 20965, 19486.x2, 19486.x5, 19486.x2 from Sp.489), the bracelets found in burials do not stand out in terms of quality of working or the overall level of technical skill invested in their production. In addition, an unused pendant (30503.k1) possibly made from hematite (Tsoraki 2013) was found under the cranium of a mature adult individual, possibly male (30514) (fig. 13.3), while two pendant preforms were found with a mature female adult individual (23115) buried in B.131.

Grinding tools also featured among burial goods. A palette was found with a disturbed burial (F.1231) excavated in 2003, while two complete grinders were placed directly onto the bodies of two mature adult individuals, one female buried in B.167 (23827.x1, F.8521, North F) and one possible female buried in B.89 (30928.x1, F.3484, South N) (fig. 13.18). In the case of B.89, a further complete grinder was found in the fill (19897.x1) that sealed this burial. All three grinders had been used extensively prior to their placement in the burial. Edge tools are found in burial fills (6.3% of all edge tools), and in a single occurrence, a greenstone edge tool (16133.x2) was placed along with other burial goods under the chin of a mature adult female individual (16309) in Sp.17 (North G). As mentioned previously, during the Middle period of occupation 'maceheads' tended to be deposited in association with burials of mostly young male and female adults. The incised 'macehead' (22194.x1) was found in association with a young male. None of the 'maceheads' found in burial contexts show use-related traces (for example, percussive traces). Similarly, 'maceheads' were among the objects selected to furnish the dead at PPNA Körtik Tepe (Özkaya 2009: 6).

Object Type		*Data Category*		*Total*
		Burial fill	*Skeleton*	
Edge tool - indeterminate	no.	6	1	7
	%	0.4	1.7	0.5
Edge tool - axe	no.	1	0	1
	%	0.1	0.0	0.1
Edge tool - adze	no.	3	0	3
	%	0.2	0.0	0.2
Edge tool - chisel	no.	1	0	1
	%	0.1	0.0	0.1
Perforated objects - macehead	no.	4	0	4
	%	0.3	0.0	0.3
Grind/abras - indeterminate	no.	6	0	6
	%	0.4	0.0	0.4
Grind/abras - abrader	no.	4	0	4
	%	0.3	0.0	0.3
Grind/abras - polisher/smoothed stone	no.	4	0	4
	%	0.3	0.0	0.3
Grind/abras - grinder	no.	1	1	2
	%	0.1	1.7	0.1
Grind/abras - grind. slab/quern	no.	1	0	1
	%	0.1	0.0	0.1
Misc. - pigment	no.	1	0	1
	%	0.1	0.0	0.1
Misc. - rocks with natural perforations	no.	1	0	1
	%	0.1	0.0	0.1
Misc. - fire-cracked rocks	no.	133	0	133
	%	10.0	0.0	9.5
Ornaments - indeterminate	no.	9	0	9
	%	0.7	0.0	0.6
Ornaments - beads	no.	965	51	1,016
	%	72.2	86.4	72.8
Ornaments - pendants	no.	2	0	2
	%	0.1	0.0	0.1
Ornaments - bracelets/annulets	no.	1	1	2
	%	0.1	1.7	0.1
Ornaments - bead preform	no.	4	0	4
	%	0.3	0.0	0.3
Ornaments - pendant preform	no.	0	2	2
	%	0.0	3.4	0.1
Debitage - indeterminate	no.	11	0	11
	%	0.8	0	0.8
Waste by-products/shatter	no.	19	2	21
	%	1.4	3.4	1.5
Flake	no.	29	0	29
	%	2.2	0.0	2.1
Debris/fragments	no.	59	0	59
	%	4.4	0.0	4.2

Table 13.6. The distribution of ground stone artefacts in burial contexts (% within Data Category).

Object Type		*Data Category*		*Total*
		Burial fill	*Skeleton*	
Unworked nodule	no.	30	0	30
	%	2.2	0.0	2.2
Possible debitage	no.	1	0	1
	%	0.1	0.0	0.1
Grind/abras - slab or grinder fragment	no.	8	0	8
	%	0.6	0.0	0.6
Grind/abras - palette	no.	1	0	1
	%	0.1	0.0	0.1
Grind/abras - grinder one hand mano	no.	1	1	2
	%	0.1	1.7	0.1
Misc. - pigment raw nodule	no.	2	0	2
	%	0.1	0.0	0.1
Indeterminate worked	no.	3	0	3
	%	0.2	0.0	0.2
Debitage - rejuvenation flake	no.	2	0	2
	%	0.1	0.0	0.1
Debitage - tertiary flake	no.	2	0	2
	%	0.1	0.0	0.1
Natural pebble	no.	21	0	21
	%	1.6	0.0	1.5
Total	no.	1,336	59	1,395
	%	100.0	100.0	100.0

Table 13.6 (continued). The distribution of ground stone artefacts in burial contexts (% within Data Category).

Object Type		*Burial Sex*					*Total*
		adult female	*adult female?*	*adolescent*	*child juvenile*	*adult*	
Edge tool - indeterminate	no.	1	0	0	0	0	1
	%	14.3	0.0	0.0	0.0	0.0	1.7
Grind/abras - grinder	no.	1	0	0	0	0	1
	%	14.3	0.0	0.0	0.0	0.0	1.7
Grind/abras - grinder one hand mano	no.	0	1	0	0	0	1
	%	0.0	50.0	0.0	0.0	0.0	1.7
Ornaments - beads	no.	2	1	0	28	20	51
	%	28.6	50.0	0.0	100.0	100.0	86.4
Ornaments - bracelets/annulets	no.	1	0	0	0	0	1
	%	14.3	0.0	0.0	0.0	0.0	1.7
Ornaments - pendant preform	no.	2	0	0	0	0	2
	%	28.6	0.0	0.0	0.0	0.0	3.4
Waste by-products/shatter	no.	0	0	2	0	0	2
	%	0.0	0.0	100.0	0.0	0.0	3.4
Total	no.	7	2	2	28	20	59
	%	100.0	100.0	100.0	100.0	100.0	100.0

Table 13.7. The correlation of ground stone type and sex of interred individuals (% within Burial Sex).

Object type		*Burial age group*						*Total*
		Child (3-12)	*YA (20-35)*	*MA (35-50)*	*OA (50+)*	*A (20+)*	*Adolescent (12-20)*	
Edge tool - indeterminate	no.	0	0	0	1	0	0	1
	%	0.0	0.0	0.0	50.0	0.0	0.0	1.7
Grind/abras - grinder	no.	0	0	1	0	0	0	1
	%	0.0	0.0	16.7	0.0	0.0	0.0	1.7
Grind/abras - grinder one hand mano	no.	0	0	1	0	0	0	1
	%	0.0	0.0	16.7	0.0	0.0	0.0	1.7
Ornaments - beads	no.	28	1	1	1	20	0	51
	%	100.0	100.0	16.7	50.0	100.0	0.0	86.4
Ornaments - bracelets/annulets	no.	0	0	1	0	0	0	1
	%	0.0	0.0	16.7	0.0	0.0	0.0	1.7
Ornaments - pendant preform	no.	0	0	2	0	0	0	2
	%	0.0	0.0	33.3	0.0	0.0	0.0	3.4
Waste by-products/shatter	no.	0	0	0	0	0	2	2
	%	0.0	0.0	0.0	0.0	0.0	100.0	3.4
Total	no.	28	1	6	2	20	2	59
	%	100.0	100.0	100.0	100.0	100.0	100.0	100.0

Table 13.8. The correlation of ground stone type and age group of interred individuals (% within Burial age group).

Building biographies and indoor activities

Analysis of material from buildings excavated during the 2009–2017 seasons corroborates previous results regarding the nature of activities taking place indoors (Hodder 2013, and chapters therein). Our findings confirm the consistent use of the southern parts of buildings (dirty floors) for the production and maintenance of andesitic grinding tools and other forms of stone production; this pattern is documented from the earlier excavated layers (for example, cluster of flakes in (30615), B.118, Level South H) and throughout the occupation sequence. The southern part is also associated with grinding activities and plant processing (see below). Ground stone objects have been found in both occupation and abandonment phases of buildings and observed patterning suggests variation in quantitative and/or qualitative terms between the different phases of the buildings. This is very convincingly showcased in B.77, a fully excavated structure that contained large quantities of stone materials deposited on the floors and platforms (Wright 2014). As I argued elsewhere, material richness evidenced in the abandonment levels does not correlate with the stone assemblages from the pre-burning levels, which are characterised by a low number of tools and limited variability. The specificities of the depositional practices in B.77 (that is, deposition of all quern roughouts, used tools and fragments of previously used grinding tools with varying degrees of weathering during the abandonment and infilling process of the building) suggest an emphasis on scale and performativity (that is, deposition of large functional querns on the northern platform, an area of increased symbolism) (House 2014; Tung, Volume 12, Chapter 22). Interestingly, while beads occur in large quantities in burial fills (e.g., F.7860), they are rare in all other phases and depositional contexts (for a discussion of the staging of depositional practices in B.77 see Tsoraki 2018; see also Taylor et al. 2015).

A different picture emerges, however, when we consider depositional practices in B.132. Apparent material correspondence between occupation and abandonment phases, but also between the different occupation phases, suggests continuity in material depositions during the life cycle of the building (online fig. S13.4). The presence of debitage of various materials (for example, andesite, tufa, marble, greenstone) suggests that the production of grinding and edge tools and stone beads was an activity regularly performed in B.132.

Similarly to B.77, grinding tools were found in the abandonment phase of B.79 in the South Area, also attributed to the Middle period. Yet, there are distinct differences in the depositional practices between the two buildings, not only in quantities, but also in fragmentation patterns and object variability: B.79 contained a much smaller assemblage of complete tools (n=26), including two querns and three grinders that show functional variability (for example, plant processing, wall plastering). In sharp contrast to B.77, quern fragments are almost absent from the abandonment deposits of B.79. Based on this, we could suggest that the tool assemblage from B.79, which seems to have been stored in the upper storey of the building (Taylor, Volume 12, Chapter 10), is more consistent with an actual household toolkit that was abandoned during the burning event.

Finally, an interesting set of practices has been encountered in the B.43-B.160-B.161-B.162 sequence in the South Area dated to the Early period. During the construction of B.160, a complete diabase edge tool (32600.k1) was placed together with animal bones and a clay ball in the foundation of wall F.7847 (foundation deposit). Evidence for different episodes of modification and re-use of the tool in a grinding/abrasive activity suggest that the tool had a prolonged biography prior to its deposition. During the occupation of the building (Phase B.160.2.3) an interesting collection of artefacts (32453), which included a diabase stone adze and an axe, stone beads and debitage from stone-working activities along with bone, flint and obsidian materials, was deposited on the floor around the ladder (F.7839). The placement of the cluster by the ladder evokes a comparable deposition in B.77 (16491). Similar to the edge tool from the foundation deposition, the axe and adze carried complex and prolonged biographies prior to their deposition. Wear traces on the surface of the adze suggest that the tool was re-used for ochre grinding. In the case of the axe, following its breakage the fractured edge was once again re-used for grinding/polishing activities. The two edge tools seem to have had parallel lives, sharing distinct similarities in the raw material used for their production (both made of the same type of diabase and possibly the same nodule), their re-use in grinding activities and their subsequent placement by the ladder. A fourth diabase edge tool (22301.k10) was retrieved broken from the room fill of the building and similarly to all other examples from this building, it exhibits different episodes of modification and re-use for ochre grinding. In addition, four greenstone edge tools including a preform were found in the abandonment fill of the preceding B.161. On the whole, repetition characterises the depositional patterns in the B.43-B.160-B.161-B.162 sequence, which seem to evoke the repetitive nature of depositional events encountered later on in the B.65-B.56.-B44-B.10 sequence (Regan 2014), during which practices associated with the intentional placement of stone materials intensify.

The use of open spaces: temporal trends

Ground stone artefacts, debitage and natural cobbles also occur in external spaces, which seem to have hosted a range of activities along within the regular disposal of waste from individual buildings. More than 2,300 objects derive from open areas dated from the Early to the Late period, mirroring variability seen in building interiors. Variations of statistical significance (χ^2=0.000, df=14) are evident in the distribution of material in open areas between the Early and Late periods of occupation (table 13.9 and online fig. S13.5). While fire-cracked cobbles form the majority of the material found in exterior spaces attributed to the Early and Middle periods (54.9% and 59.8% respectively), their relative frequencies decrease in the Late period (22.7%), while there is simultaneously a significant increase in the number of grinding/abrasive tools present in this period. As there is no variation in the degree of damage/preservation among grinding/abrasive tools from all three periods, observed variation in their relative frequencies is not the effect of preservation.

Comparative analysis of external areas excavated between 2009 and 2017 provides further insight into the compositional differences and temporal variation in the ground stone artefacts found in these spaces across different areas of the site. All studied spaces, including Sp.620, a possible penning area, present broadly similar variability in the range of objects found (online fig. S13.6). Moreover, debitage from the production of different stone objects including andesitic grinding tools, edge tools and ornaments are present in all spaces. A closer look at the density of andesitic debitage across these areas shows higher densities in open areas attributed to the Middle period, and in particular Sp.631, than in the Late period. When considering only material deriving from flotation samples it becomes apparent that material densities for Sp.489, Sp.631 and Sp.610 are not too dissimilar (fig. 13.19), while densities in the Late period are lower. In the case of Sp.489 (see fig. 13.26), the consistent presence of flakes that derive from the same reduction episodes and refit strongly suggests in situ production and maintenance of andesitic grinding tools, while the presence of debitage from other materials, including a complete serpentinite ground stone core that was in the process of being sawn in half, points to other stone-working activities as well.

While debitage from the working of different stone materials was found in Sp.420, no refits have been noted during the analysis, possibly indicating that this external space was not used for in situ activities. This external area, however, contained a wide range of abrasive and polishing tools, including U-shaped undecorated grooved abraders, schist palettes, a fragment from a sanding tool and 18 polishing tools, along with numerous fragments of grinding tools that do not refit.

Overall, observed differences between the open areas seem to reflect variations in their formation processes and ultimately in the way they were appropriated in different areas of the settlement and/or different phases in the life cycle of the settlement.

Object type		*Temporal groupings of levels*			*Total*
		Early	*Middle*	*Late*	
Edge tools	no.	5	16	31	52
	%	0.9	1.2	6.2	2.2
Percussive tools	no.	0	2	0	2
	%	0.0	0.2	0.0	0.1
Perforated tools	no.	0	1	0	1
	%	0.0	0.1	0.0	0.0
Grinding/abrasive tools	no.	45	105	194	344
	%	8.5	8.0	38.6	14.6
Miscellaneous	no.	293	795	120	1,208
	%	55.3	60.3	23.9	51.4
Ornaments	no.	4	31	21	56
	%	0.8	2.4	4.2	2.4
Debitage, cores, nodules	no.	183	366	134	683
	%	34.5	27.8	26.6	29.1
Indeterminate worked	no.	0	2	3	5
	%	0.0	0.2	0.6	0.2
Total	no.	530	1,318	503	2,351
	%	100.0	100.0	100.0	100.0

	Value	*df*	*Asymptotic significance (2-sided)*	*Monte Carlo significance (2-sided) Significance*
Pearson Chi-Square	419.261*	14	0.000	0.000
N of valid cases	2,351			

*Table 13.9. The temporal distribution of ground stone objects in open spaces (excluding indeterminate cases). *Nine cells (37.5%) have expected count less than five. The minimum expected count is 21.*

Comparative analysis of the ground stone material of internal and external areas highlighted variability in assemblage composition and degree of preservation of material. In the Middle period, middens are strongly associated with fire-cracked rocks, whereas building interiors present greater variability, with a clear tendency towards the deposition of tools and ornaments. This is convincingly shown by analysis of material from the contemporary Sp.489 and Sp.490 and B.77 and B.52. There is a statistically significant difference in the assemblage composition between B.77 and the midden area, with all tool categories, including edge tools, grinding tools and fragments, multiple-use tools and ornaments, being more common in the building, whereas fire-cracked rocks and debitage occurred more frequently in the midden (table 13.10). In the Late period, however, material depositions in open areas show greater variability, with grinding/abrasive and edge tools found in the external areas and debitage being more common in buildings (table 13.11).

Activity patterns and wear traces: towards an understanding of the nature of activities at Neolithic Çatalhöyük

The use of edge tools

The Çatalhöyük edge tools, with an average length of 53mm, tend to be small in size. Tools of different sizes occur throughout the occupation sequence, but tools attributed to the Middle period tend to be larger than those attributed to the Late period (fig. 13.20). Tools made from diabase tend to be larger in size than their greenstone counterparts (fig. 13.21). Despite their small size, edge tools show evidence for intensive use. The majority of edge tools, irrespective of their edge morphology, show a moderate to heavy degree of wear, while 40.1% (=51) have been used so heavily that their use faces have been worn out. The intensity of use of the edge tools is also evidenced by the high number of tools (54.4%) that are broken in half or into smaller fragments.

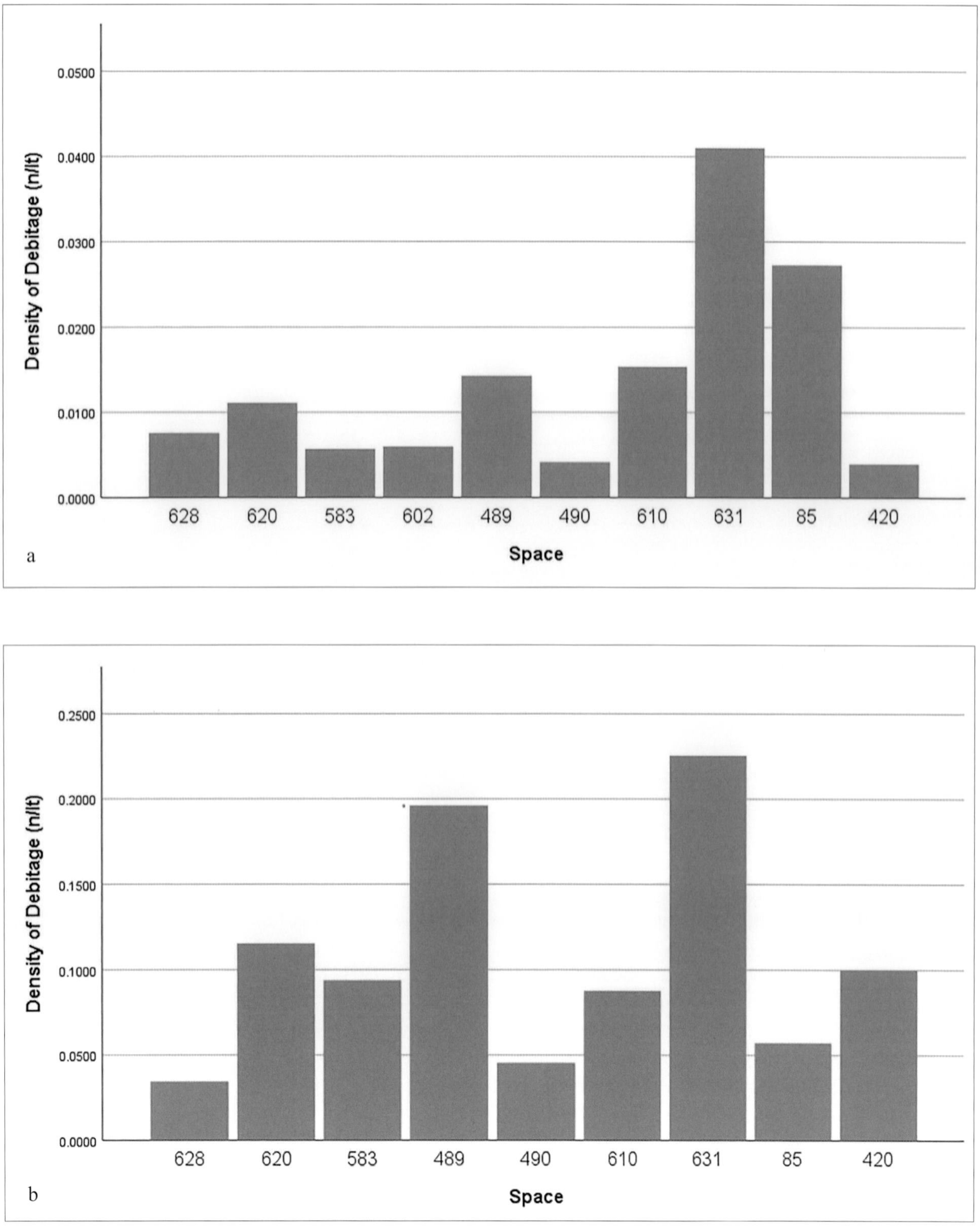

Figure 13.19. The density of andesitic debitage in open spaces: (a) all debitage; (b) debitage from flotation samples only (densities for Sp.628, Sp.620, Sp.583, Sp.490 and Sp.420 are based on one flotation sample per space).

In terms of wear patterns, unifacial or bifacial edge removals represent the most common type of wear encountered on all types of edge tools (70.3%, 71 out of 101), which ranges from micro-chipping to large-size removals, the size being indicative of the hardness of the contact material and the force applied during the task. Edge removals are often associated with striations that are often positioned diagonally along the cutting edge,

Object category		*Location*		*Total*
		Building	*Open area*	
Edge tools	no.	13	2	15
	%	2.1	0.3	1.2
Percussive tools	no.	14	2	16
	%	2.3	0.3	1.3
Perforated tools	no.	1	0	1
	%	0.2	0.0	0.1
Grinding/abrasive tools	no.	246	73	319
	%	40.4	12.0	26.2
Miscellaneous	no.	1	324	325
	%	0.2	53.3	26.7
Multiple-use tools	no.	20	0	20
	%	3.3	0.0	1.6
Ornaments	no.	161	15	176
	%	26.4	2.5	14.5
Debitage, cores, nodules	no.	153	192	345
	%	25.1	31.6	28.3
Total	no.	609	608	1,217
	%	100.0	100.0	100.0

	Value	*df*	*Asymptotic significance (2-sided)*	*Monte Carlo significance (2-sided) Significance*
Pearson Chi-Square	578.422*	7	0.000	0.000
N of valid cases	1,217			

*Table 13.10. The distribution of ground stone artefacts in B.77 and Sp.489/Sp.490 (midden area). *Two cells (12.5%) have expected count less than five. The minimum expected count is .50.*

suggesting contact with a relatively hard material of abrasive nature. Other common wear traces include rounding of the cutting edges associated with a band of micropolish. Often this overlays the topography of the edge removals, indicating a secondary use and contact with materials of different hardness. Variation in the wear traces is indicative of variation in contact materials and tasks performed with these tools. While microwear traces are consistent with the use of these tools for woodworking activities including debarking, wear patterns resulting from contact with materials of pliable nature such as skin/leather were also observed.

Edge tools were subjected to numerous episodes of resharpening and subsequent modifications throughout their use-life, as also demonstrated by numerous examples that were abandoned while under modification (e.g., 30189.k1). The biography of these tools was frequently extended after their breakage by their re-use in other activities, mainly for tasks associated with grinding/abrasive activities, including pigment-processing activities and in the case of 22676.x12, re-use as a U-shaped grooved polisher for the shaping of pointy objects/shafts. The re-use of edge tools in grinding/abrasive tasks, which is recurrent throughout the habitation of the site, is intriguing, especially when juxtaposed with grinding tools and the abundant quern fragments, which, on the other hand, were rarely re-used.

The function of 'maceheads'

'Maceheads' show certain trends in the way they were used during the habitation of the site. 'Maceheads' (22813.x1, 32860.k35) dated to the Late period bear evidence of use in the form of edge removals around the rim of the perforation, possibly the result of use in an activity requiring impact force (cf. Wright 2013), but also of highly polished perforation interiors, suggesting hafting (fig. 13.9). The majority, however, do not exhibit any visible traces of use (30008.x1, 22676.x1, 22194.x1, all associated with burial contexts). This is also the case for the 'macehead' (21827.k1) from B.17 that was found broken in half. Following its breakage, however, the 'macehead' was re-used as a hammerstone against a hard contact material, as flake scars on the body and crushing of the fractured edge clearly demonstrate (fig. 13.9).

Object category		*Location*		*Total*
		Building	*Open area*	
Edge tools	no.	46	31	77
	%	1.0	6.2	1.5
Percussive tools	no.	12	0	12
	%	0.3	0.0	0.2
Perforated tools	no.	6	0	6
	%	0.1	0.0	0.1
Grinding/abrasive tools	no.	559	194	753
	%	11.7	38.6	14.2
Miscellaneous	no.	48	120	168
	%	1.0	23.9	3.2
Multiple-use tools	no.	26	0	26
	%	0.5	0.0	0.5
Ornaments	no.	282	21	303
	%	5.9	4.2	5.7
Debitage, cores, nodules	no.	3,798	134	3,932
	%	79.3	26.6	74.3
Indeterminate worked	no.	12	3	15
	%	0.3	0.6	0.3
Total	no.	4,789	503	5,292
	%	100.0	100.0	100.0

Table 13.11. The distribution of ground stone artefacts in buildings and open spaces attributed to the Late period.

Grinding tools and the technology of plant-processing activities

Grinding tools form part of a suite of stone implements that constituted the regular household toolkit at Çatalhöyük. A morphometrical analysis of grinding tools and their use-faces provides insights into processing techniques and motor habits adopted during the execution of grinding tasks, while material choice and varying surface textures are indicative of the processing of crops of different size and hardness (Horsfall 1987; Nixon-Darcus, D'Andrea 2017).

The majority of querns and grinders have flat use-faces (74.5% and 73.9% respectively) while concave use-faces for querns (17.2%) and convex ones for grinders are less common (15.9%). Querns and grinders most often have one use-face (86.4% and 56.8% respectively), though grinders with multiple use-faces are also common (40.8%). Querns exhibiting a light to moderate degree of wear predominate (66.3%). The fact that grinding tools at Çatalhöyük were not subjected to intensive use can also be extrapolated from the comparative analysis of the thickness of roughouts and used tools, the thickness being the dimension most affected during the use and maintenance of the tools. Roughouts have an average thickness of 91.61mm, while the average thickness of used tools is 74.92mm (fig. 13.22). This in tandem with the high frequency of tools with one flat use-face and the rarity of tools with heavily concave use-faces clearly suggests that grinding tools had not gone through extensive episodes of use and maintenance by the time of abandonment. Morphological correlation of upper and lower tools (cf. Risch 2002) along with microwear traces is consistent with the simultaneous use of grinding slabs/querns and stone grinders for the processing of intermediary substances.

Plant processing at Çatalhöyük relied heavily on andesitic querns and grinders, mainly for the processing of large-sized grains. A prerequisite for the processing of large-size grains, such as wheat and barley, is the abrasive capacity of grinding surfaces, as the roughness of the use-face enables the shearing and pulverising of the grains (Stone 1994). The importance placed on sharp texture and material roughness for efficient grinding tools (Horsfall 1987; Risch 2002) is further indicated by the production sequence and maintenance practices (that is, modification of use-faces through pecking) aimed at retaining or increasing the abrasive capacity of the grinding surface. The processing of small-sized grains, on the other hand, requires a combination of crushing and (mainly rotary) grinding to reduce the grains into smaller particles. This is best achieved on non-vesicular surfaces which reduce the risk of seeds getting trapped in the vesicles (that is, seed loss) (Stone 1994: 682).

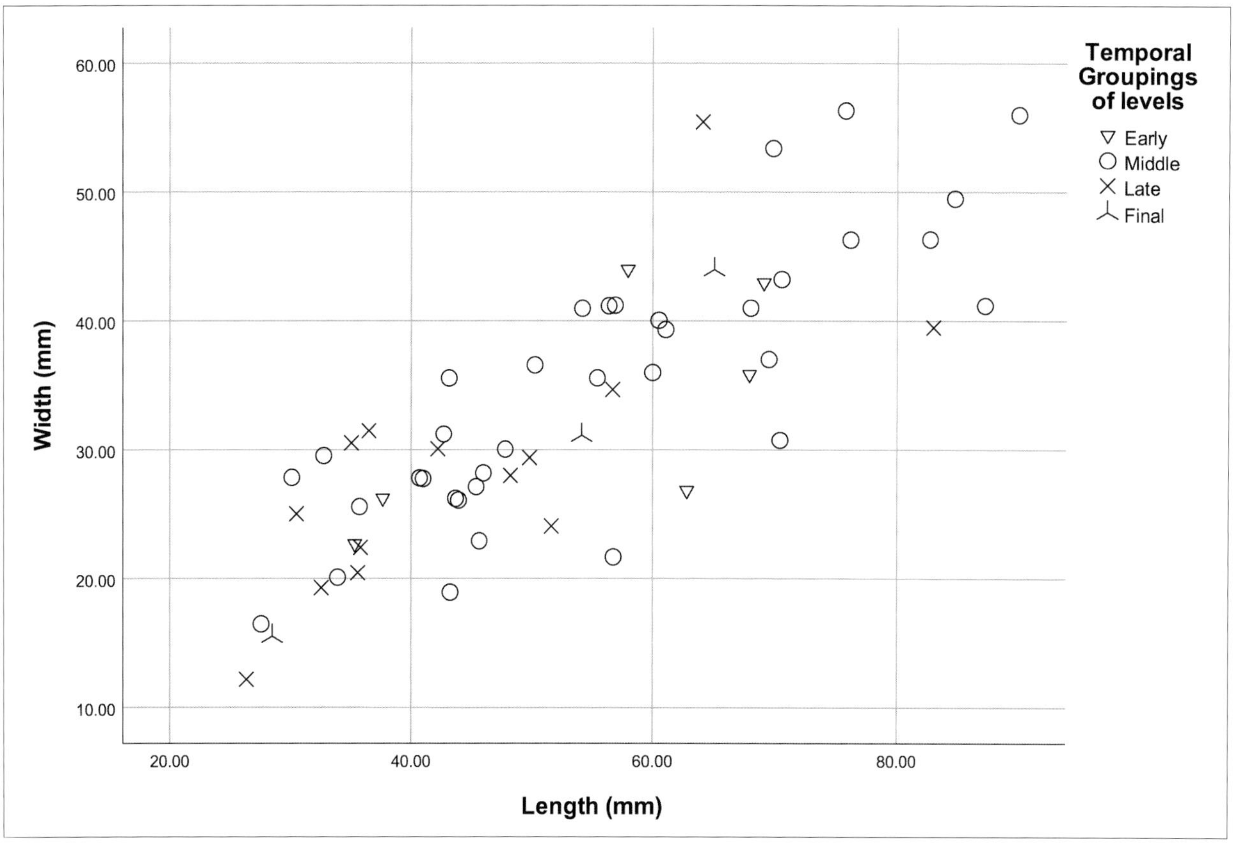

Figure 13.20. The size distribution of edge tools across temporal groupings of levels (only examples with complete length and complete width are included).

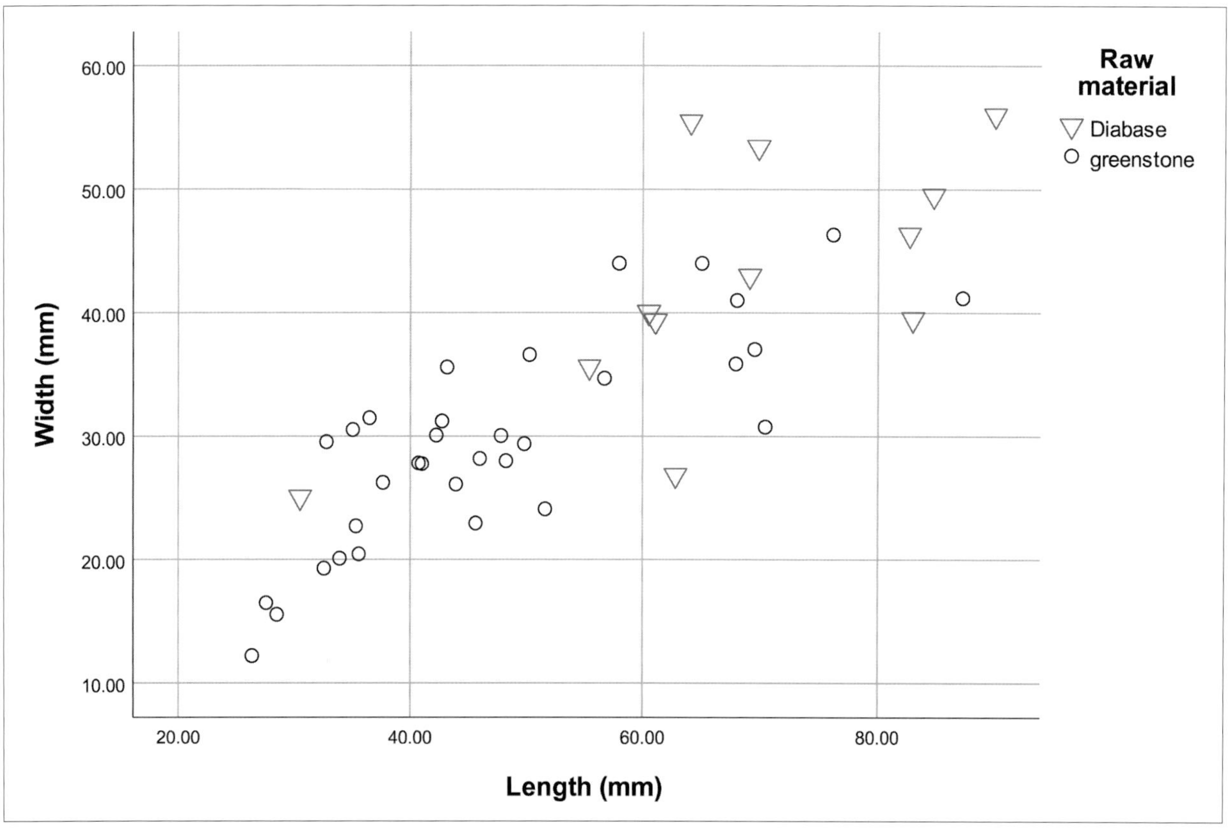

Figure 13.21. The size distribution of diabase and greenstone edge tools (only examples with complete length and complete width are included).

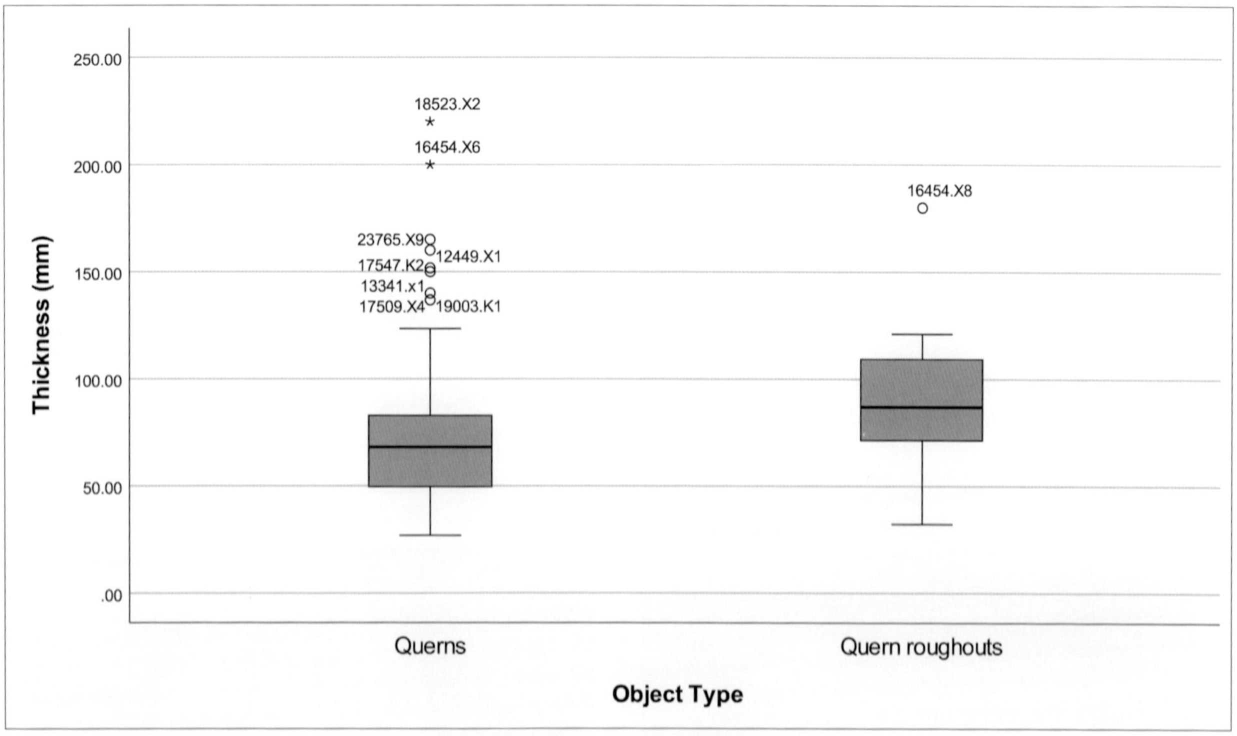

Figure 13.22. Correlation of the thickness of quern roughouts and used querns.

Microwear analysis conducted on grinding tools confirms their use for plant-processing activities and mostly cereal processing (fig. 13.23), an observation corroborated by starch analysis (Volume 13, Chapter 7). The results of the archaeobotanical analysis point to regular dehusking of grains stored in the chaff and processed piecemeal (Bogaard et al. 2013: 120–21), an activity that was more commonly associated with building interiors (Ryan 2013). Traditionally, dehusking activities have been associated with the use of mortars and pestles, which could be made of wood or of a combination of a wooden pestle and a stone mortar (cf. Hamon, Le Gall 2013). Although stone mortars with rather shallow hollows and pestles have been found during the ÇRP excavations, albeit in limited numbers, there is no clear indication that these were employed for dehusking or other plant-related activities. Pestles subjected to microwear analysis exhibit wear traces consistent with the processing of minerals (e.g., 32860.k33), while the use-face of a pestle found in B.77 has pigment residues, most likely from the use of the tool for pigment processing. Ethnographic studies, however, show that the dehusking of hulled cereals does not always entail pounding on mortars. Instead, the grains can be freed from the chaff through other processes such as roasting and/or grinding on querns followed by sieving, or grinding and rubbing in shallow baskets or pits in the ground (Peña-Chocarro, Zapata 2014; Procopiou 2014). The initial results of residue analysis of the Çatalhöyük grinding tools indicate that these did not partake in cereal dehusking activities (Tsoraki et al. 2015a). Drawing upon the ethnographic studies, however, it would be interesting to consider whether at least some of the small-sized pits regularly found in the southern areas of the buildings at Çatalhöyük may have functioned for similar purposes (for example, pit F.7740 in B.132, which contained a high-density deposit associated with cereal dehusking; see Bogaard et al., Volume 13, Chapter 5).

Moreover, contextual associations between plant-processing tools and botanical remains have been noted during the ÇRP excavations. For instance, in B.79 a layer of charred grain (18596) deposited along the northern platform (F.5073) directly above the floor (Taylor, Volume 12, Chapter 10) was found in direct association with a complete grinder (fig. 13.4). Macroscopic and microscopic wear traces on the grinding surface of the tool are consistent with its use for plant-processing activities. Three further complete grinders were found in the burnt layer (18595) that sealed (18596).

Other forms of plant processing are suggested by a limited range of implements that, based on their wear traces, raw material choice and morphological characteristics, can be linked to the processing of non-cereal plants (for example, a sandstone grinding slab with a shallow basin (16492.x2) from B.77) (fig. 13.4).

Figure 13.23. Microwear traces associated with plant processing on grinding tools: (a) 23765.x4; (b) 18595.x2 (micrographs taken at 100x magnification).

Location of grinding activities. As noted previously, grinding tools that are complete or retain more than half of their original size are found inside buildings. The association of lower grinding tools and building interiors is further supported by the presence of grinding tools found embedded in the floors and platforms of buildings, an observation also made by Mellaart (1967: 63). Fixed grinding installations occur across different buildings with varying elaboration indices; these include B.77 (Ei 20, F.6058) and B.114, Sp.88 (Ei 11, F.610) in the North Area, and B.89 (Ei 11, F. 8154) and B. 80 (Ei 25, F.3440, Sp.135) in the South Area (Stevanović 2012a; House 2014; Tsoraki 2018) (fig. 13.24). All known occurrences, including possible examples from B.55 (Ei n/a) (Yeomans 2014), B.1 (F.27) (Ei 33) (Cessford 2007; but see Baysal, Wright 2005) and B.44 (F.1329) (Ei 10) (Regan 2014; but see Wright 2013) are dated to the Middle and Late periods, suggesting that during this period grinding activities became more structured, signalling therefore a change in the performance of grinding tasks. Microwear analysis of the B.77, B.80 and B.89 specimens revealed wear traces consistent with plant processing.

The grinding installations tend to be found in the southern areas of the main rooms of the buildings, where there is more evidence for productive activities (dirty floors, dehusking, knapping residues) (Hodder 2013 and chapters therein) and sometimes in proximity to the side room, as seen in B.77. Similarly, in B.140, a quern (31911.x1), which survives almost complete, was found in relation to a plastered platform (F.7656) located in the southern part of Sp.541. It may have been originally set

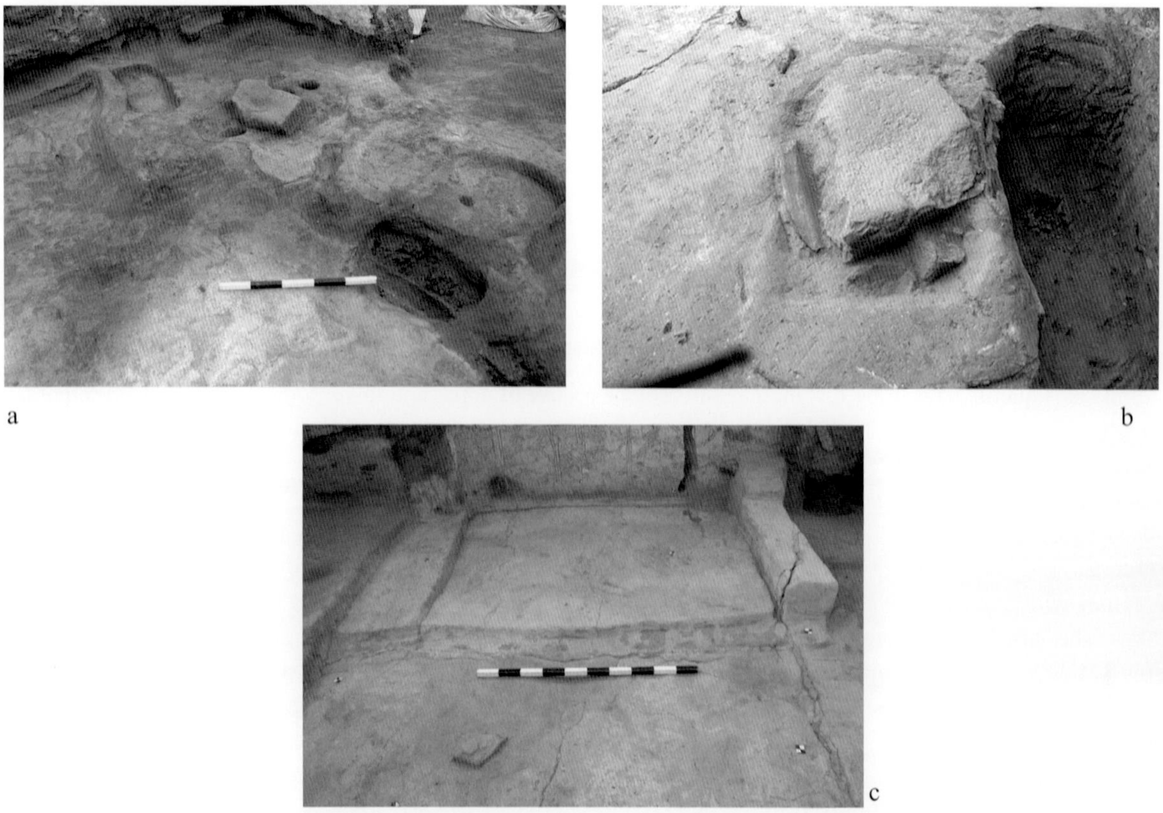
a b c

Figure 13.24. Fixed grinding installations: (a) 17547.k2 in B.77; (b) 30945.x1 in B.89; (c) 21767.x2 in B.80 (photographs by Jason Quinlan).

into the platform, but the heavily eroded nature of the platform does not allow for more concrete associations (Barański, personal communication). Yet, its location follows broader conventions for the use of the southern areas of the main rooms for grinding activities. B.80, on the other hand, stands out, as the location of the fixed quern in the northern part of the main room – an area closely associated with burial practices – does not conform to the conventional 'North-South' divide usually seen in the organisation of internal space at Çatalhöyük. Interestingly, a bin with a rich plant assemblage (18945) was located in the northeastern platform of B.80 in phase B.80.2.7, though during this phase the quern goes out of use (Taylor, Volume 12, Chapter 11). Despite the location of the quern in the northern area of the main room, activity areas were identified towards the southern part of the main room, along with ovens and hearths (Taylor, Volume 12, Chapter 11).

A further point worth highlighting is that the placement of the querns in their fixed locales was not part of the initial construction and planning of the buildings; instead, all fixed installations were placed there during the occupation of the structures and went out of use just before their abandonment (for example, B.80: B.80.2.3–B.80.2.6; B.77: B.77.4–until the end; B.89: early occupation?–B89.B.6; B.114: Sp.88.3–Sp.88.5). Moreover, cycles of maintenance, use and modification evident on the surface of the querns, along with the fact that the plastering of the adjacent surfaces respected their position, suggests that the tools were actively used during the habitation of the building and people tended to them regularly.

Scale of plant processing. Ethnographic research suggests a correlation between the morphology and overall tool design of grinding tools with a large grinding surface area and increased grinding capacity (Mauldin 1993; Stone 1994; Hard et al. 1996). Grinding capacity is measured as output of flour per hour of time invested (Hard et al. 1996). Based on various experimental and ethnographic studies, an average estimate of three hours per 1–2kg of cereal grains (barley and wheat) has been suggested (Runnels 1981; Wright 1992; Samuel 2010). Different studies have formulated models for grinding intensity and reliance on plant processing based on expectations of average grinding surface area for intensive grinding (Hard et al. 1996; Risch 2008). Risch (2008) has suggested that an adequate use-face of a quern used for efficient grain processing should measure more than 250cm^2.

Metrical data were collected for the overall size of the tools, and the dimensions and area of their use-faces. Based on the shape of the use-face, the area for rectangular/sub-rectangular use-faces was calculated as length x width, while for ovate use-faces the surface area was calculated as π x L/2 x W/2. In morphological terms, grinding slabs/querns tend to be rectangular/sub-rectangular (46.8%) and ovate/obovate (31.9%) in plan view, while grinders have spherical/discoidal (30.4%), ovate/obovate (26.8%) and rectangular/sub-rectangular forms (19.7%).

Querns are on average 342.06mm long and 223.41mm wide, and their weight ranges from 2kg to 54.6kg. The grinding surface area of the complete querns varies between 166cm^2 and 2,196cm^2, with an average grinding surface measuring 713.85cm^2. The vast majority of querns (82.1%) have a grinding surface area that is less than 1,221cm^2. Their size is therefore consistent with implements employed in the processing of plants and in particular cereals (see Ivanova 2018, table 10.4 for comparative ethnographic data on grinding surface area). Furthermore, during the Middle and Late periods tools of different sizes were in use, but there is no statistically significant difference in the size of the tools (Independent-samples Mann-Whitney U test, $p=0.306$) or in the size of the grinding surface area between the two periods of occupation (Independent-samples Mann-Whitney U test, $p=0.243$), indicating, therefore, no significant change in the scale of plant processing over time (fig. 13.25). Moreover, in both periods, grinding technologies entailed the simultaneous use of upper and lower tools of different sizes that correspond to two distinct modes of operation mainly in terms of portability. Analysis so far has shown no functional differences between grinding tools of different size. Variations in the size of upper grinding tools have implications for the kinetics and suggest that these were operated either with one hand, possibly using a rotary grinding motion, or with two hands in reciprocal grinding. Furthermore, the lack of raised platforms suggests that a kneeling posture was adopted during grinding activities, which also seems to be supported by activity markers identified on skeletal remains at Çatalhöyük (Sadvari et al. 2015).

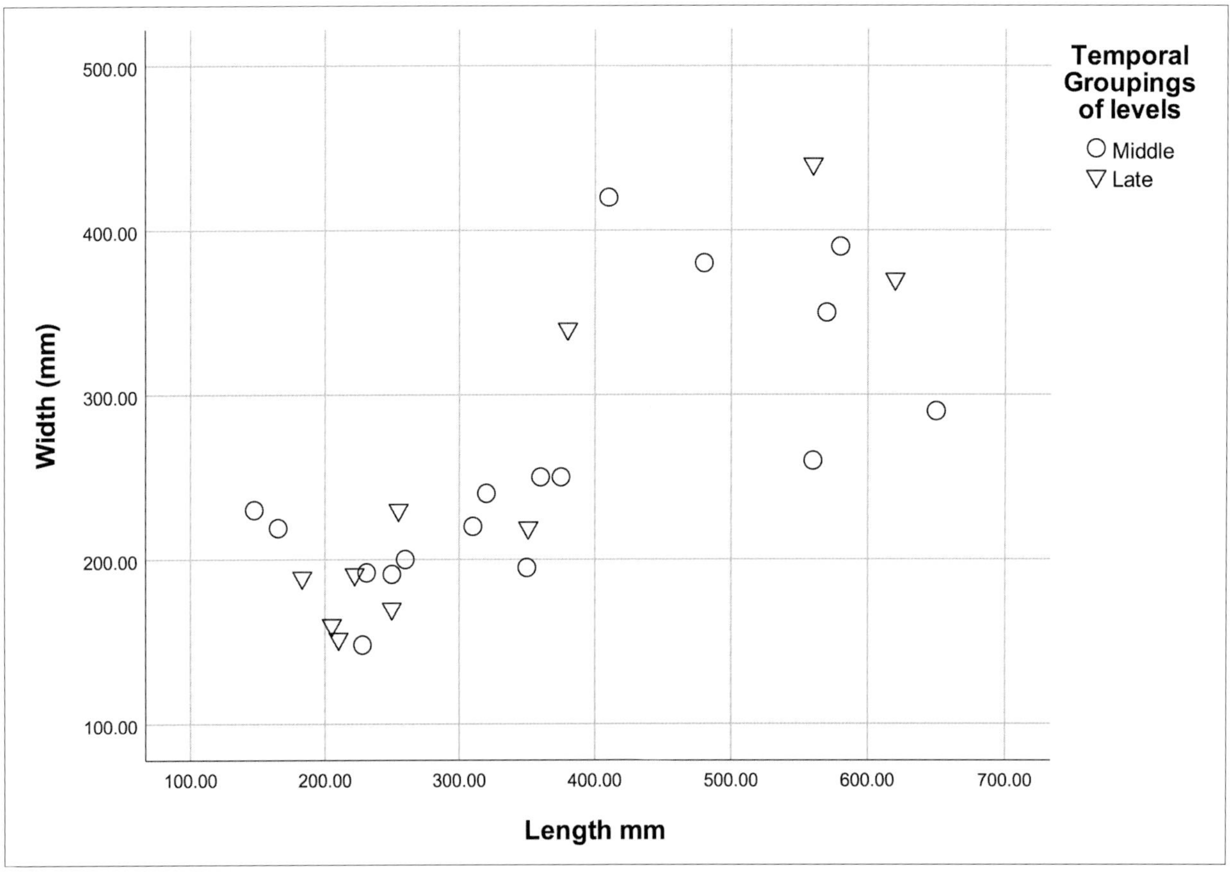

Figure 13.25. The size distribution of querns across the temporal groupings of levels.

Furthermore, there is no statistically significant correlation between elaborate buildings and increased grinding intensity and productivity (table 13.12). Large-size tools occur in buildings with high elaboration, but not exclusively, while buildings with high elaboration have a range of tools of all sizes. Buildings associated with querns that exhibit amongst the largest grinding surface areas include B.77, B.52 and B.150, as well as B.63 from the IST Area. While B.77 contained a high number of large-sized querns (n=12), the largest tools are found in abandonment deposits and in the building infill.

Fixed grinding installations, on the other hand, provide an indication of grinding productivity more directly associated with the actual occupation and use of a building. A fixed grinding installation found in association with the occupation phase of B.77 had a surface area of 627cm^2, which is significantly smaller than that indicated by implements associated with the abandonment of the building and more in agreement with the scale of processing indicated by tools from other buildings. Overall, the average grinding surface of fixed installations is 440cm^2, while the surface area of the grinding tool from B.140 is 416cm^2 and is consistent with plant processing organised on a small-scale, a suggestion also supported by the analysis of archaeobotanical remains (Bogaard et al. 2013; Volume 13, Chapter 5). Increased grinding capacity, however, is expected in buildings with multiple grinding toolsets, if used simultaneously. All identified fixed grinding installations include single implements, and no sets of grinding tools positioned together that could be interpreted as 'cooperative grinding' (Nixon-Darcus, D'Andrea 2017: 211) have been found in interior or exterior spaces at Neolithic Çatalhöyük. As mentioned above, B.79, however, contained two querns and with a combined grinding surface area of 2,025cm^2 – if indeed these relate to tools stored together on the upper floor of the building – this would indicate intensive crop-processing activities taking place in B.79.

Stone-working activities: specialisation of production?

A key question for understanding the socio-economic development of the Neolithic society at Çatalhöyük concerns an investigation of the organisation of stone-working activities and an assessment of the intensification and increase in the specialisation of production over time (e.g., Hodder 2013).

Within the recorded assemblage, more than 5,600 objects are associated with the production and maintenance of grinding tools, edge tools, stone ornaments (see also Chapter 9) and polishing tools. These include preforms and roughouts, different forms of debitage including flakes, microdebris and rejuvenation flakes (fig. 13.26), a limited number of ground stone cores and approximately 1,000 natural unworked materials that based on size and raw material characterisation seem to be associated with ground stone technologies. Different rocks and minerals were worked on site across different buildings from the earlier phases of habitation onwards, as suggested by the distribution of debitage and unworked nodules. The widespread use of rock materials points away from unequal access to resources and their strict regulation by a particular sector of the community.

The organisation of grinding tool production: questions of scale

The vast majority of the debitage (ca 4,000 pieces) is associated with the production of andesitic grinding tools. All identified varieties of andesite (porphyritic andesite with hornblende and biotite crystals, porphyritic andesite with elongated hornblende and/or large size plagioclase mineral crystals) of both grey- and pink-coloured variants (the latter of dacitic origin) appear from the Early period (Level South H) onwards and were simultaneously worked throughout the habitation of the settlement, often within the same building spaces. Furthermore, the presence of debitage on floors throughout the occupation of particular buildings (for example, B.160, B.132) and often throughout the sequence of superimposed structures (for example, B.43-B.160-B.161-B.162) suggests that the production of grinding tools was one of the activities the occupants of the buildings engaged in regularly throughout the lifetime of the structure. This is in agreement with the results of the analysis of buildings excavated prior to 2009 (e.g., Baysal, Wright 2005; Cessford 2007; Wright 2013).

It was previously suggested that during the later periods of habitation at Çatalhöyük East, an increased level of specialisation in stone working activities started to appear, and a case has been put forward for the emergence of specialised production of grinding tools during the Middle period of occupation, possibly in association with seemingly elaborate buildings such as B.77 (Wright 2014). Grinding tool roughouts occur across different buildings dated to the Middle, Late and Final periods (fig. 13.27). B.77 has the largest number of complete and broken roughouts (n=23), all of which, however, derive from the abandonment phase of the building and do not relate to productive activities (see above). The question of specialised production was further investigated through a consideration of the distribution of andesitic microdebitage across different buildings in the North and South

Area of use-face (cm^2)		*Elaboration index (Ei)*							*Total*
		0	*7*	*10*	*11*	*20*	*25*	*42*	
166	no.	0	0	0	0	0	0	1	1
	%	0.0	0.0	0.0	0.0	0.0	0.0	33.3	2.6
179	no.	0	0	0	1	0	0	0	1
	%	0.0	0.0	0.0	33.3	0.0	0.0	0.0	2.6
230	no.	1	0	0	0	0	0	0	1
	%	8.3	0.0	0.0	0.0	0.0	0.0	0.0	2.6
269	no.	1	0	0	0	0	0	0	1
	%	8.3	0.0	0.0	0.0	0.0	0.0	0.0	2.6
280	no.	0	0	1	0	0	0	0	1
	%	0.0	0.0	16.7	0.0	0.0	0.0	0.0	2.6
288	no.	0	0	1	0	0	0	0	1
	%	0.0	0.0	16.7	0.0	0.0	0.0	0.0	2.6
308	no.	0	0	0	0	1	0	0	1
	%	0.0	0.0	0.0	0.0	8.3	0.0	0.0	2.6
316	no.	0	0	0	1	0	0	0	1
	%	0.0	0.0	0.0	33.3	0.0	0.0	0.0	2.6
323	no.	0	0	0	0	0	1	0	1
	%	0.0	0.0	0.0	0.0	0.0	100.0	0.0	2.6
324	no.	0	0	1	0	0	0	0	1
	%	0.0	0.0	16.7	0.0	0.0	0.0	0.0	2.6
349	no.	0	0	0	0	1	0	0	1
	%	0.0	0.0	0.0	0.0	8.3	0.0	0.0	2.6
353	no.	0	0	0	0	1	0	0	1
	%	0.0	0.0	0.0	0.0	8.3	0.0	0.0	2.6
382	no.	0	0	1	0	0	0	0	1
	%	0.0	0.0	16.7	0.0	0.0	0.0	0.0	2.6
407	no.	1	0	0	0	0	0	0	1
	%	8.3	0.0	0.0	0.0	0.0	0.0	0.0	2.6
416	no.	1	0	0	0	0	0	0	1
	%	8.3	0.0	0.0	0.0	0.0	0.0	0.0	2.6
442	no.	1	0	0	0	0	0	0	1
	%	8.3	0.0	0.0	0.0	0.0	0.0	0.0	2.6
450	no.	1	0	0	0	0	0	0	1
	%	8.3	0.0	0.0	0.0	0.0	0.0	0.0	2.6
451	no.	0	0	0	0	1	0	0	1
	%	0.0	0.0	0.0	0.0	8.3	0.0	0.0	2.6
456	no.	1	0	0	0	0	0	0	1
	%	8.3	0.0	0.0	0.0	0.0	0.0	0.0	2.6
459	no.	0	0	0	0	1	0	0	1
	%	0.0	0.0	0.0	0.0	8.3	0.0	0.0	2.6
481	no.	1	0	0	0	0	0	0	1
	%	8.3	0.0	0.0	0.0	0.0	0.0	0.0	2.6

Table 13.12. The correlation of grinding surface area of querns and the Elaboration index (Ei) of buildings.

Area of use-face (cm^2)		*Elaboration Index (Ei)*							*Total*
		0	*7*	*10*	*11*	*20*	*25*	*42*	
493	no.	0	0	0	1	0	0	0	1
	%	0.0	0.0	0.0	33.3	0.0	0.0	0.0	2.6
544	no.	0	0	1	0	0	0	0	1
	%	0.0	0.0	16.7	0.0	0.0	0.0	0.0	2.6
558	no.	0	0	0	0	1	0	0	1
	%	0.0	0.0	0.0	0.0	8.3	0.0	0.0	2.6
618	no.	1	0	0	0	0	0	0	1
	%	8.3	0.0	0.0	0.0	0.0	0.0	0.0	2.6
627	no.	0	0	0	0	1	0	0	1
	%	0.0	0.0	0.0	0.0	8.3	0.0	0.0	2.6
680	no.	0	0	0	0	1	0	0	1
	%	0.0	0.0	0.0	0.0	8.3	0.0	0.0	2.6
713	no.	0	0	1	0	0	0	0	1
	%	0.0	0.0	16.7	0.0	0.0	0.0	0.0	2.6
740	no.	0	0	0	0	1	0	0	1
	%	0.0	0.0	0.0	0.0	8.3	0.0	0.0	2.6
1060	no.	0	0	0	0	1	0	0	1
	%	0.0	0.0	0.0	0.0	8.3	0.0	0.0	2.6
1221	no.	0	1	0	0	0	0	0	1
	%	0.0	100.0	0.0	0.0	0.0	0.0	0.0	2.6
1407	no.	1	0	0	0	0	0	0	1
	%	8.3	0.0	0.0	0.0	0.0	0.0	0.0	2.6
1640	no.	0	0	0	0	0	0	1	1
	%	0.0	0.0	0.0	0.0	0.0	0.0	33.3	2.6
1702	no.	0	0	0	0	1	0	0	1
	%	0.0	0.0	0.0	0.0	8.3	0.0	0.0	2.6
1734	no.	1	0	0	0	0	0	0	1
	%	8.3	0.0	0.0	0.0	0.0	0.0	0.0	2.6
1904	no.	0	0	0	0	0	0	1	1
	%	0.0	0.0	0.0	0.0	0.0	0.0	33.3	2.6
2166	no.	0	0	0	0	1	0	0	1
	%	0.0	0.0	0.0	0.0	8.3	0.0	0.0	2.6
2196	no.	1	0	0	0	0	0	0	1
	%	8.3	0.0	0.0	0.0	0.0	0.0	0.0	2.6
Total	no.	12	1	6	3	12	1	3	38
	%	100.0	100.0	100.0	100.0	100.0	100.0	100.0	100.0

	Value	*df*	*Asymptotic significance (2-sided)*	*Monte Carlo significance (2-sided) Significance*
Pearson Chi-Square	228.000*	222	0.377	1.000
Likelihood Ratio	122.497	222	1.000	1.000
Fisher's Exact Test	330.138			1.000
N of valid cases	38			

*Table 13.12 (continued). The correlation of grinding surface area of querns and the Elaboration Index of buildings (Ei). *266 cells (100.0%) have expected count less than 5. The minimum expected count is .03.*

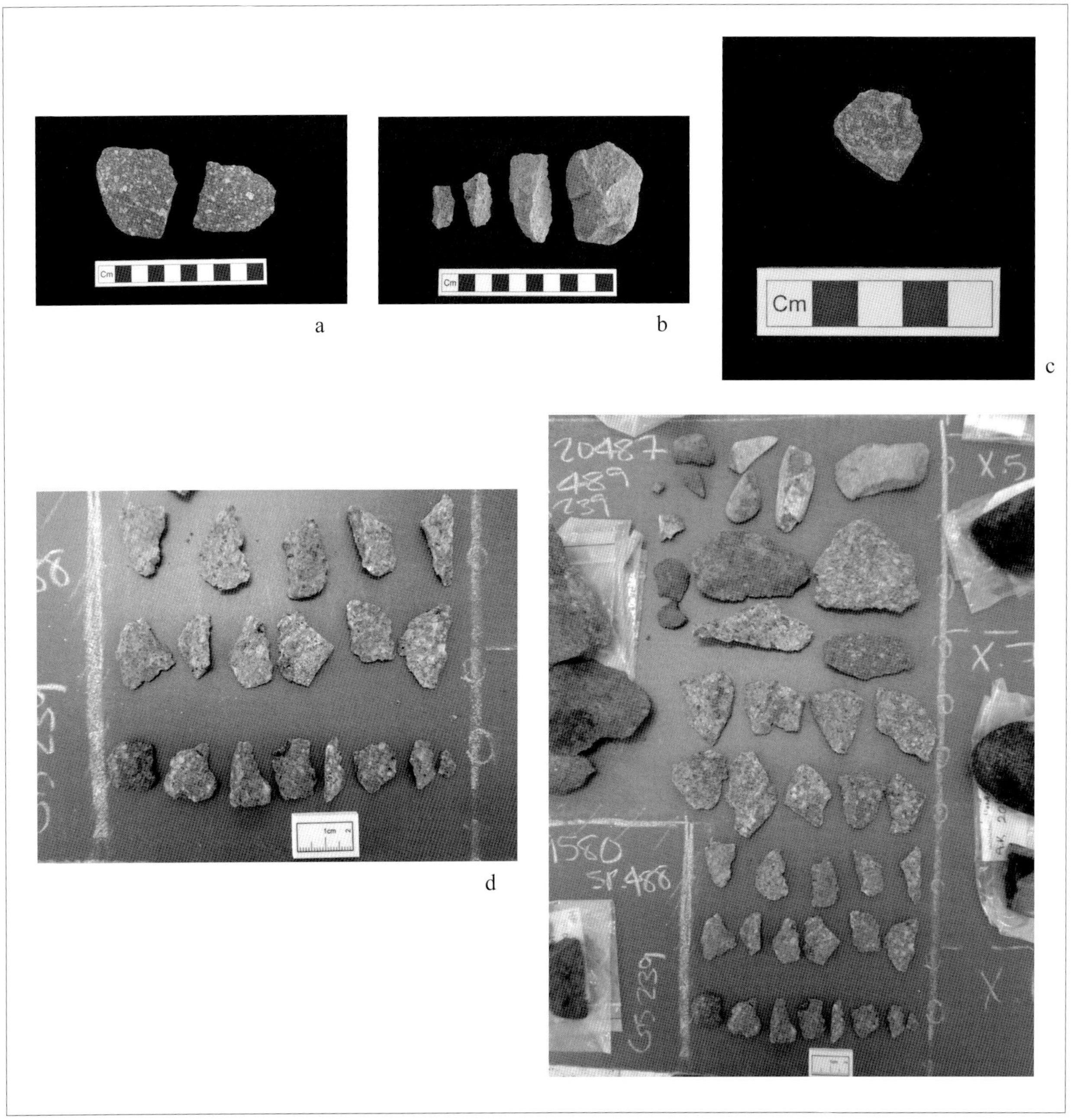

Figure 13.26. Debitage from the production of ground stone artefacts: (a) andesitic flakes from (21850.k9); (b) andesitic flakes from (18928 and 18543); (c) greenstone flake from (18578); (d) andesitic flakes from (20487) in Sp.489.

Areas. Only microdebitage from units associated with floors or floor/platform make-up layers deriving from flotation samples was selected, as this is more likely to reflect actual activity associated with the use of the building. When considering the distribution of microdebitage across different periods, the Late period of occupation exhibits higher densities, with 15.49 artefacts per litre of soil, while no significant variation is present between the Early and Middle periods of occupation (fig. 13.28). It should be noted that the distribution of material in the Final period is not representative of the actual population, but instead the low frequencies are the effect of unequal excavation. As seen in figure 13.29a, while evidence for the production of grinding tools is present across different buildings in all areas of the site, the highest peak in the densities of microdebitage, at 30.171 artefacts per litre of soil, is registered for B.142 (GDN Area, Late period). No significant variations are evident between the remaining buildings when considering the density of microdebitage on floors, as illustrated in fig. 13.29b. In the case of B.77, densities of microdebitage are on average 0.221 artefacts per litre of soil, and in that respect they do not differ significantly from other buildings. This in tandem with the specificities of the depositional practices in B.77 does not support the association of these deposits with in situ production areas.

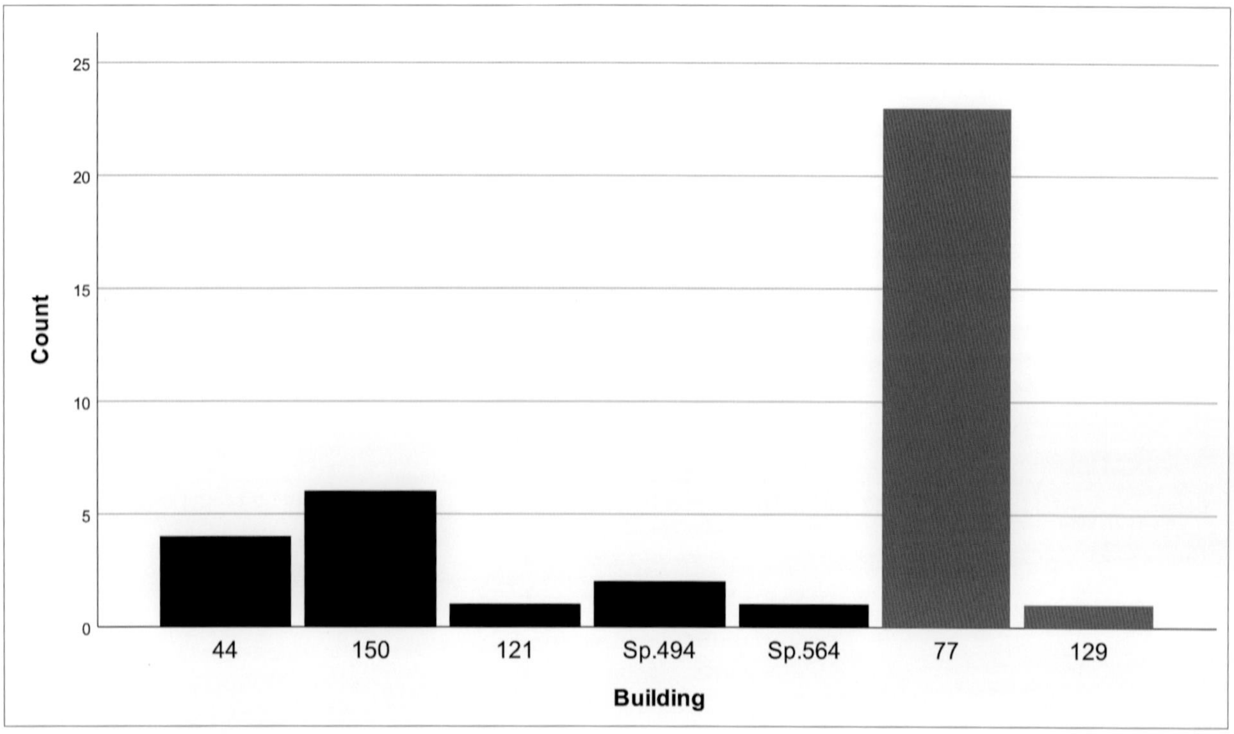

Figure 13.27. The distribution of grinding tool roughouts (n=38) (grey-shaded bars indicate buildings located in the North Area and black-coloured ones, in the South Area).

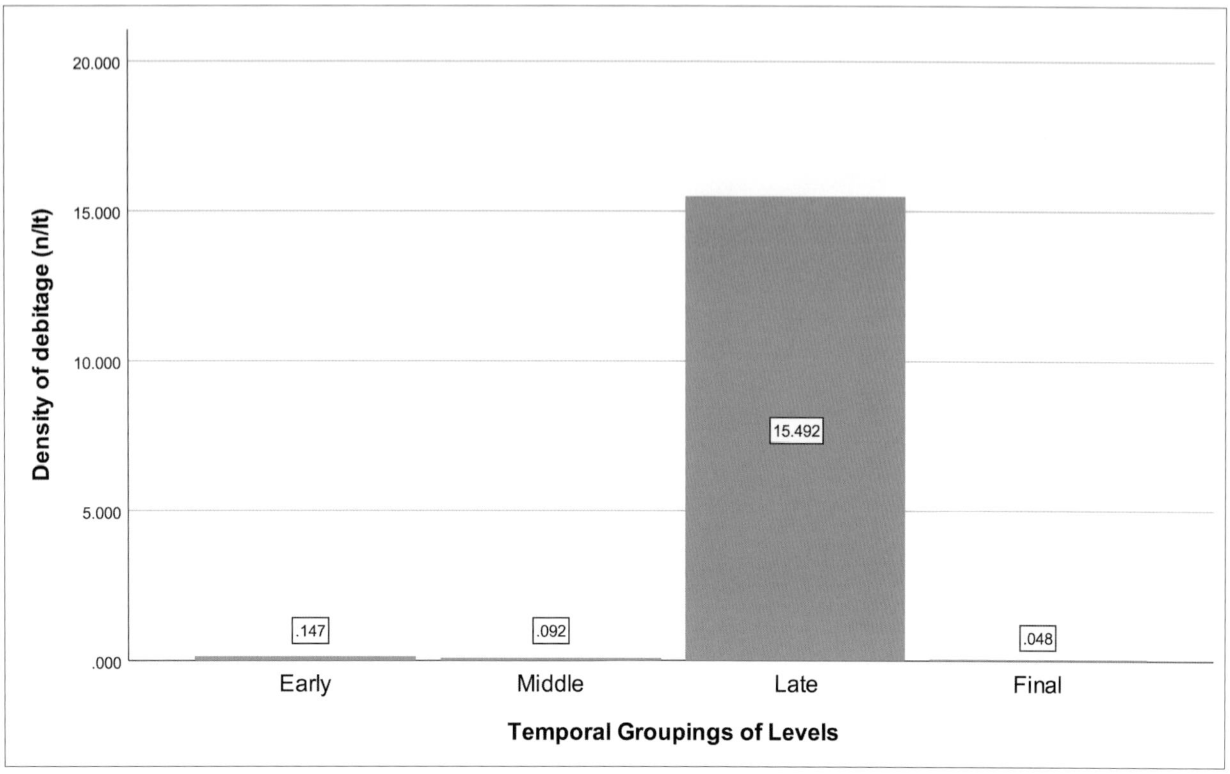

Figure 13.28. The density of debitage across the temporal groupings of levels.

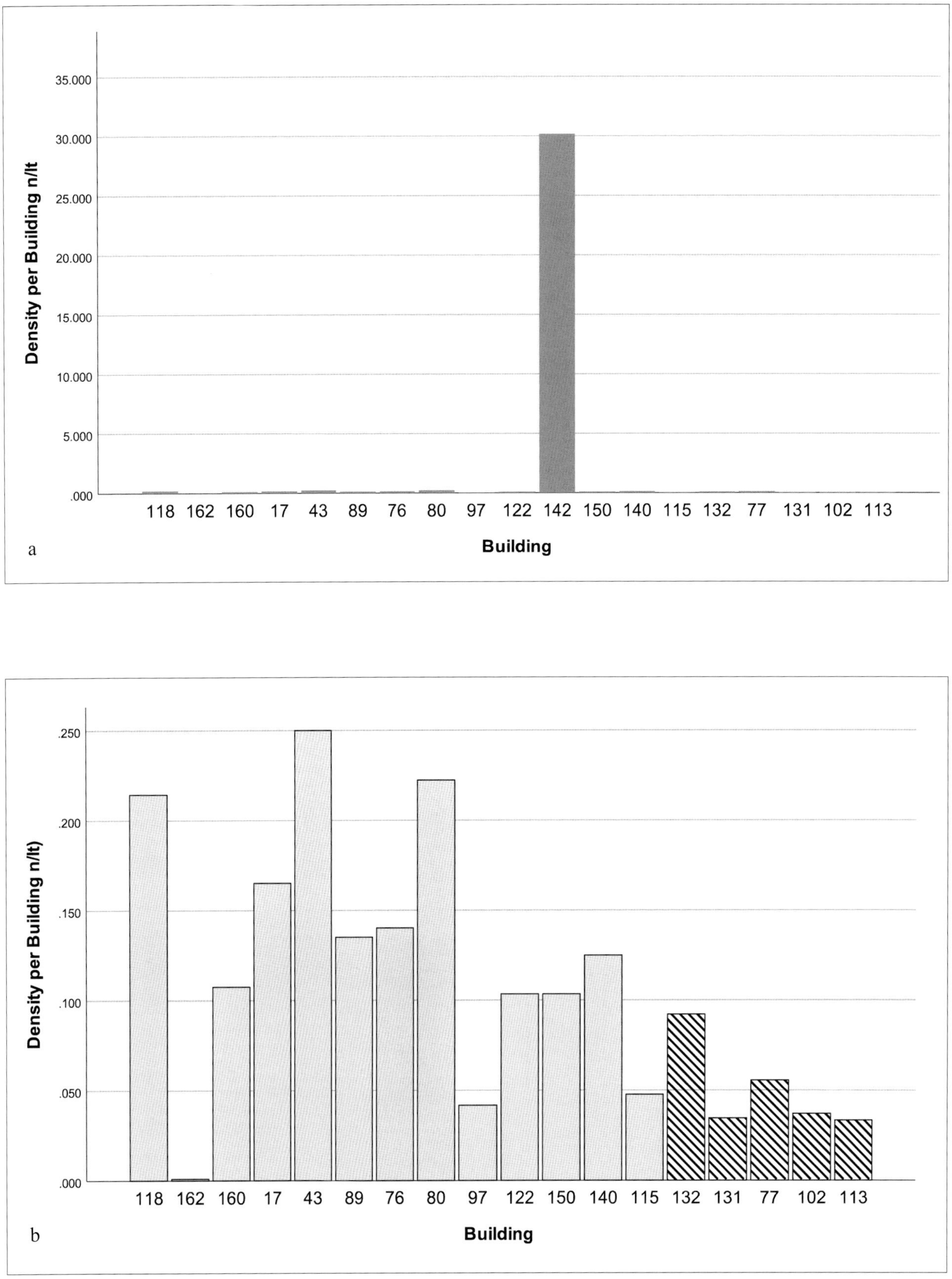

Figure 13.29. The density of andesitic debitage per building, all areas/periods: (a) only flotation samples from units associated with floors or floor/platform make-up layers (density for B.162 was calculated based on the total volume of a single unit as the material (n=1) did not derive from the flotation sample and was included here for indicative purposes); (b) only flotation samples from floor units considered and excluding B.142 (patterned bars indicate buildings located in the North Area).

The picture we gather from B.142 is, however, intriguing. Excavation of a sequence of floor make-up layers in the southeastern platform (F.8077) of Sp.537 and more specifically of (22891) revealed a particularly rich deposit of andesitic debitage; with more than 3,100 flakes and mostly angular pieces of debitage of varying sizes deriving from different types of andesite, this is the highest concentration of andesitic debitage ever noted during the ÇRP excavations. The average density of debitage per floor unit for other buildings, at 0.232 artefacts per litre of soil, is significantly lower than the 107.1429 artefacts per litre of soil registered for (22891). This deposit represents the residue of high-intensity activity undertaken within this space and, in all probability, in this particular locale. Indeed, the scale of production indicated by B.142 seems significantly different to what we have observed so far for the earlier periods of occupation. Whether B.142 represents a building engaging in the specialised production of grinding tools is an issue that with the current state of excavation and knowledge we cannot investigate further. Interestingly, B.142 is adjacent to B.140, which corresponds with Shrine A.III.2, which Mellaart described as a building that 'produced a large number of stone tools as well as raw material and might have been a stoneworker's shop' (Mellaart 1962: 55). While the recent reinvestigation of the building (Barański et al. Volume 15, Chapter 5) could not shed more light onto this interpretation, debitage from different types of stone-working activities and stone tools were found in association with the southwestern platform of B.140. In addition, B.81, contemporaneous with B.142 (attributed to Level TP M), also has evidence for stone-working activities, as the presence of two preforms of polishing tools typically used for burnishing plastered surfaces (fig. 13.8) and other pieces of debitage indicates. All this taken together, we may wish to consider whether this particular area of the settlement was associated with extensive stone-working activities over an extended period of time, if indeed B.140 and B.142 were not in use simultaneously at any point in their history (Barański et al., Volume 15, Chapter 5). Yet, the densities of debitage from Late period external areas in the South (Sp.420) and North Areas (Sp.85) are not particularly high, as might have been expected if production at an increased scale took place during the later occupation levels at Neolithic Çatalhöyük.

The production of edge tools: a small-scale affair

Although specialised axe production areas may have existed during the Neolithic, as suggested by a probable Neolithic axe workshop in northwest Turkey (Hamaylitaria, Özbek 2000; 2011), at Neolithic Çatalhöyük there is consistent evidence in the form of nodules, preforms, debitage and a related abrading toolkit for the on-site production and maintenance of edge tools in all occupation periods, with the earliest occurrences noted in Levels South H and South K (fig. 13.30). Following a

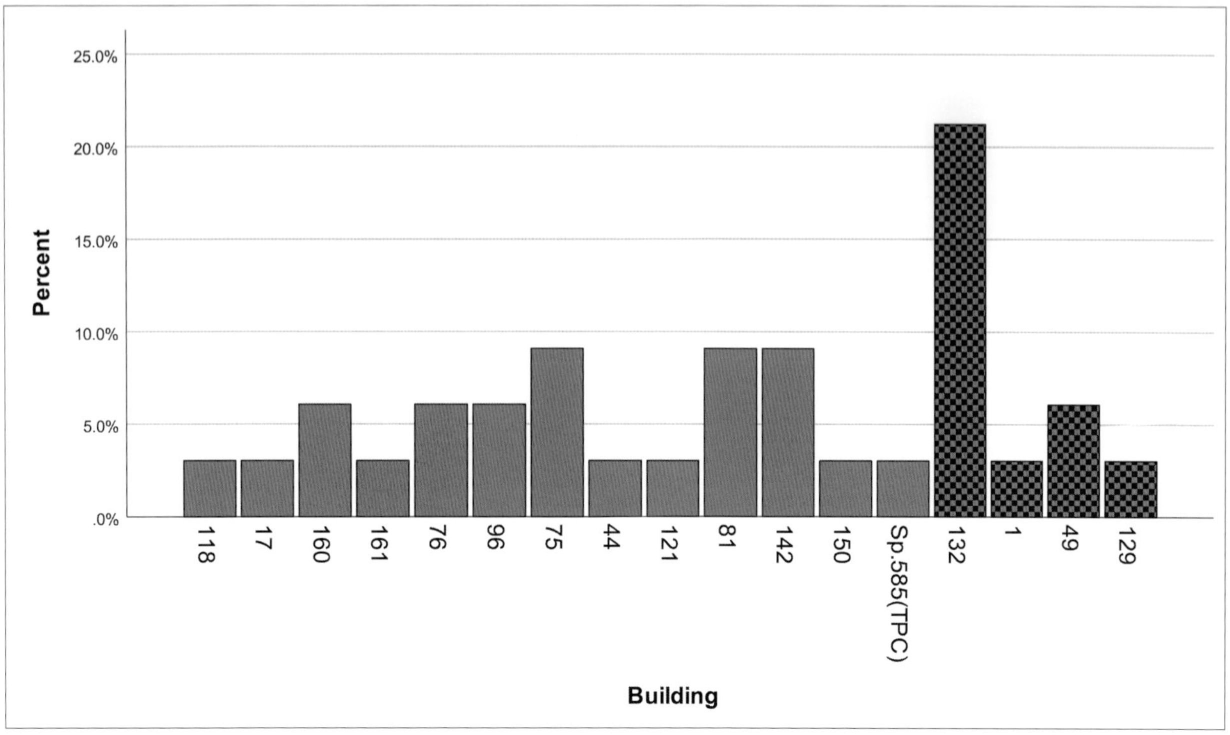

Figure 13.30. The distribution of debitage and preforms associated with the production of edge tools (patterned bars indicate buildings located in the North Area, grey-coloured bars indicate buildings in the South Area).

reassessment of all edge tools found between 1993 and 2017, only a limited number of preforms/roughouts were identified in the North and South Areas. Three examples were found in association with buildings B.1, B.96 and B.161. In addition, a diabase roughout derives from the open area Sp.610, while a quarried serpentinite nodule was found in contemporary Sp.489. A possible sub-rectangular greenstone preform or tool that was under complete modification when discarded (19345.x23) was found in Sp.459, an open area in the South Area also dated to the Middle period.

Flakes and related microdebitage of diabase, greenstone and serpentinite are found in building interiors, mainly in association with dirty floors and fills, and in open areas, but overall no significant concentrations of debitage are encountered in any particular space. Similarly, B.75 provided clear evidence for the in situ working of diabase edge tools where refitting flakes from a single rejuvenation episode were found on the floor of Sp.332. Interestingly, the diabase debitage was found in association with a layer of obsidian debitage indicative of in situ obsidian knapping activities but also bead production (Bains et al. 2013).

Overall, intense use and frequent modification episodes of edge tools have resulted in objects of an irregular morphology; this in association with the high degree of variability encountered in the production sequences and maintenance practices encountered across the assemblage strongly suggests that the production of these tools was a small-scale affair at the hands of individual people rather than work by a specialised group from the community, a suggestion also supported by the distribution of roughouts and associated debitage across different buildings and spaces.

Ochre-processing activities

Palettes tend to have flat to slightly concave use-faces, with one (53.7%) or two use-faces on opposite sides of the body (42.7%). The vast majority (83.2%) exhibit a light to moderate degree of wear. Microwear traces consistent with mineral contact material and the frequent presence of light red-coloured ochre on the use-faces (42.2%), found superimposed with wear traces, demonstrate that these implements were used for pigment-processing activities. Their morphometrical characteristics, raw material choice and wear patterns strongly indicate that they were employed for small-scale processing of pigments that had already been reduced to small particles, producing a fine-grained powder as the final product. This finding is in agreement with Çamurcuoğlu's suggestion that pigments used for the creation of wall paintings were finely ground, ensuring their smooth application while achieving a brighter colour (Çamurcuoğlu 2013: 327). Apart from the decoration of house surfaces, pigments were used widely at Çatalhöyük for different purposes including the decoration of wooden posts (Mellaart 1967: 58; Kabukcu, Asouti 2014) and pots (Chapter 4), and for bodily adornment, as suggested by burial practices (Boz, Hager 2013; Nakamura, Meskell 2013a; Vasić et al. Volume 13, Chapter 17). Yet, there is a significant correlation between buildings that have palettes and wall paintings (χ^2=0.000, df= 28), though the lack of palettes from B.77 – a building with a large ground stone assemblage and extensive wall paintings – is intriguing.

Palettes appear in the Early period, with the earliest example attributed to Level South K, and they remain in use throughout the habitation of the site, but higher densities are noted in the Late period. This change is paralleled in both the North and the South sequence, but the difference is more pronounced in the South Area, and more specifically in Levels TP M and South T (0.00051 and 0.00053 artefacts per litre of soil respectively) (fig. 13.31a). The distribution among different buildings is unequal, however, and B.150, with 23 examples in total, has the highest density, at 0.1217 object per litre of soil (fig. 13.31b). Caution is required, however, about the observed variation in densities between B.150 and other buildings, as depositional practices in the former suggest that at certain points in its history, large quantities of ground stone artefacts, along with other materials (for example, cluster (23765) in Sp.639 discussed above), were deposited in the building that do not necessarily relate to actual activities taking place there, a situation not that dissimilar to B.77 (Tsoraki 2018). Yet, the consistent presence of ground stone tools throughout the life history of the building may be indicative of the execution of mineral processing and other stone-related activities at a significant scale within this building. Detailed analysis of the stratigraphy of B.150 will allow us to assess the significance of this pattern in more detail.

Regarding the scale of ochre processing, it is worth noting that there is no significant difference in the size of palettes between the different periods (fig. 13.32); an increase in the number of palettes, however, during the Late period suggests an intensification of ochre-processing activities in relation to earlier periods, an intensification that did not, however, result in significant changes in the technological choices and other elements of the production of these tools.

Plastering toolkits and the organisation of plastering activities

The smoothing and burnishing of house interiors, including the careful coating of wall surfaces with plaster prior to the execution of the wall paintings, was one of the activities performed on a regular basis at

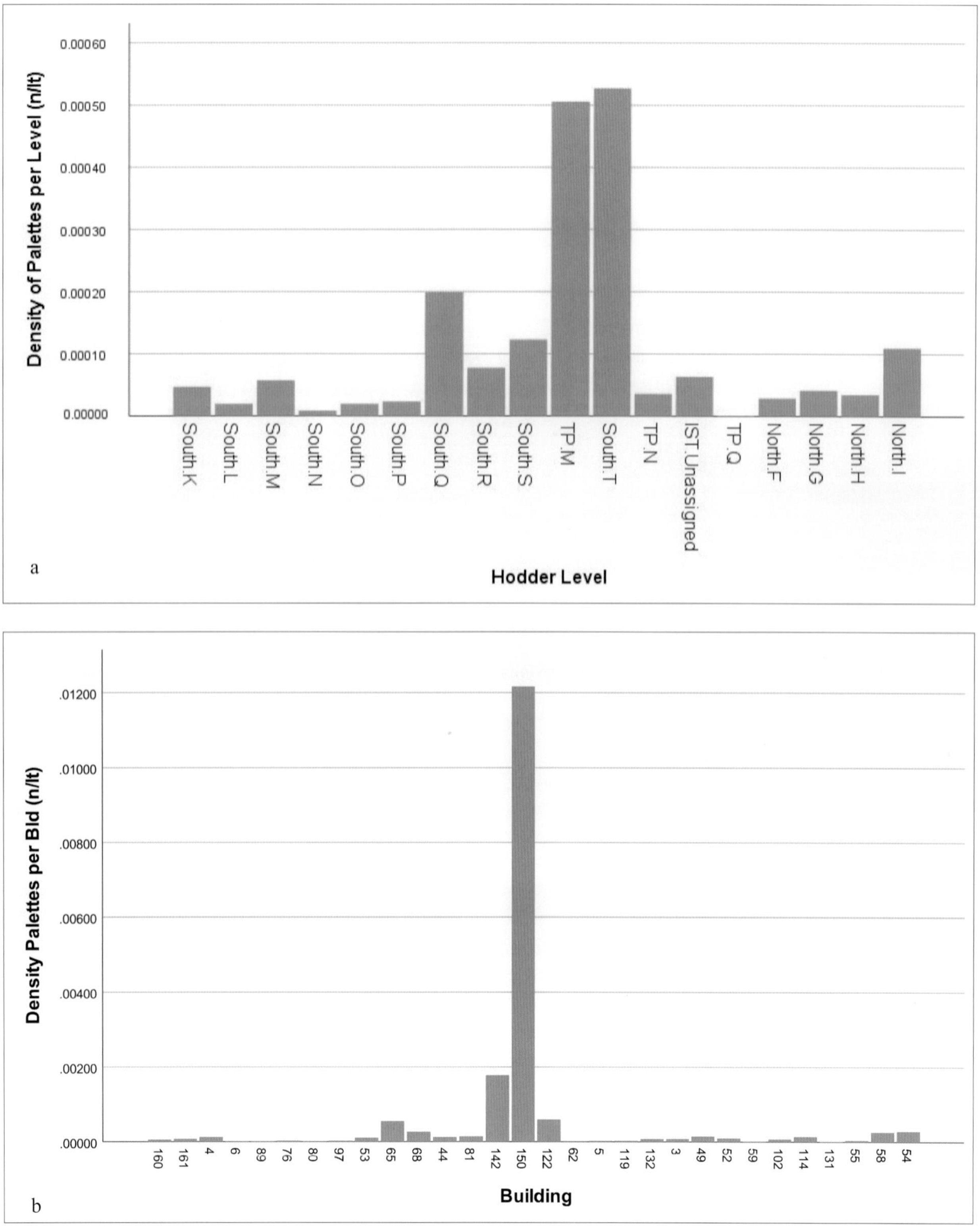

Figure 13.31. The distribution of palettes: (a) density of palettes across Hodder Levels; (b) density per building.

Neolithic Çatalhöyük. The house surfaces consist of multiple layers of plaster made from soft-lime and mainly white marls – a highly calcareous clay – while the finishing coat is very thin, and has a shiny appearance and smooth texture (Matthews 2005b; Doherty 2013; Matthews et al. 2013).

At Neolithic Çatalhöyük, plastering toolkits consisted of multi-media components including stone materials along with bone scapulae (Russell, Griffitts 2013) and shells (Bar-Yosef Mayer 2013). Based on detailed techno-logical and microwear analysis of the ground stone assemblage, 125 hand-sized tools with different textural

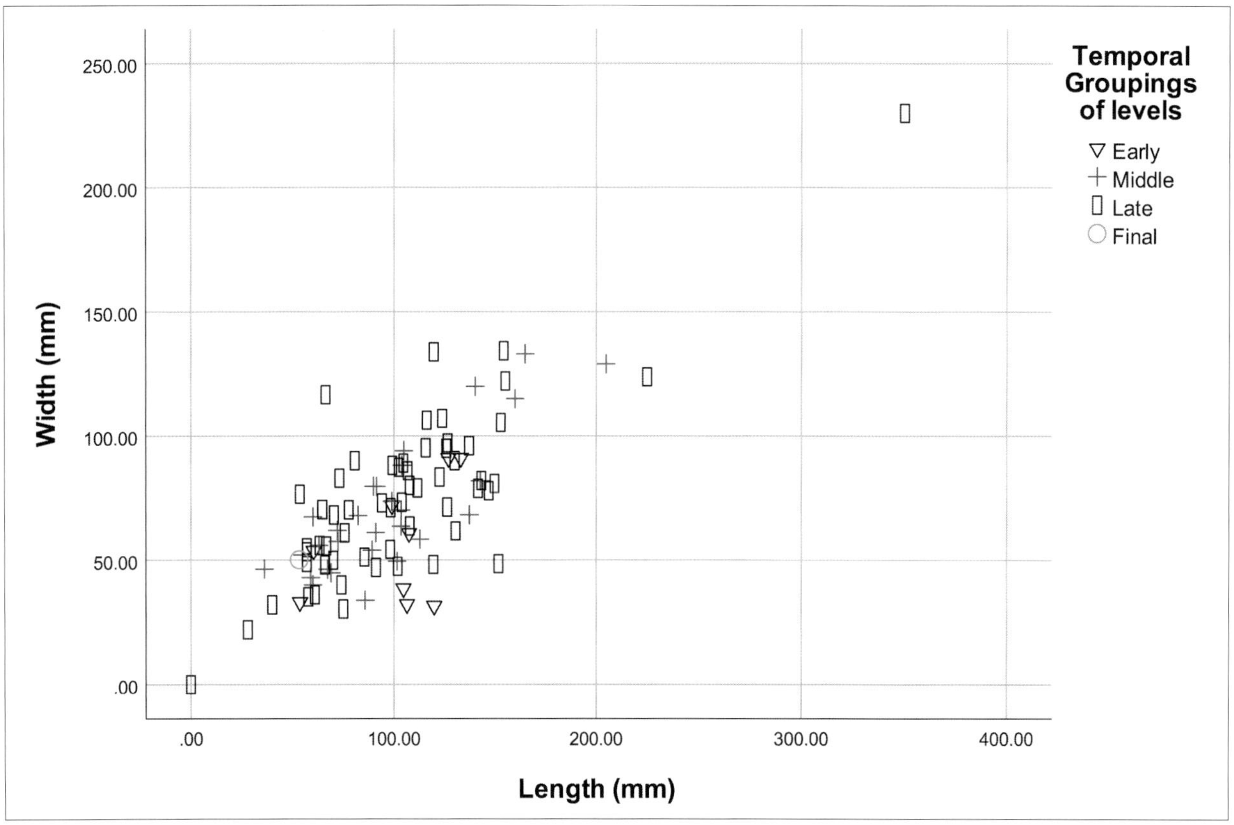

Figure 13.32. The size distribution of palettes across the periods.

characteristics can be associated with different stages of the plastering sequence. Stone tools with a relatively rough texture, usually made of schist and limestone, were used during the earlier stages of plaster application (applicators). They exhibit smoothed faceted surfaces along their narrow margins which are often coated extensively with layered plaster residues (fig. 13.8). Due to their relatively rough and uneven surface texture, these tools could not have been used in the final stages of the plastering process that resulted in the smooth and shiny finish seen on the walls of the Çatalhöyük houses.

Intentionally modified metamorphosed limestone/marble cobbles were employed for the final burnishing of the plastered surfaces (burnishers). Microwear analysis has established the presence of consistent microwear traces on their surfaces (fig. 13.8). This type of micro-polish resembles that developed on experimental tools used for burnishing leather-hard clay (Tsoraki forthcoming) and has strong similarities with the microwear traces observed on archaeological plastering tools (Dubreuil, Grosman 2013). Results of the microwear analysis along with the presence of plaster residues on the surface of the tools suggest that the burnishing of the surfaces took place before the plaster was completely dry, an observation that supports previous suggestions (Çamurcuoğlu 2013: 327).

Burnishers are on average 85mm long and 390g in weight. Variation in their size most likely reflects variation in the size of the surfaces being plastered. Greater size variability is evident during the Late period, with three size groups being present (fig. 13.33). The lack of size standardisation along with increased diversity in raw material use (table 13.13) and morphology (table 13.14) observed during this period suggests that these tools were not the products of specialist producers. The wide distribution of plastering tools across different buildings (fig. 13.14) indicates that access to these tools was not restricted and most likely the burnishing of house surfaces was an activity performed by different members of the local community. It is worth highlighting, though, that the higher densities of plastering tools in B.142 and B.150 are in broad agreement with the picture we get from the distribution of palettes.

Final discussion and conclusions

Ground stone technologies formed an important component of the Çatalhöyük material worlds from the early levels of the occupation of the site. Detailed analysis of technological processes, production patterns and wear traces bring to the fore the diversity in object forms and associated activities. Consideration of

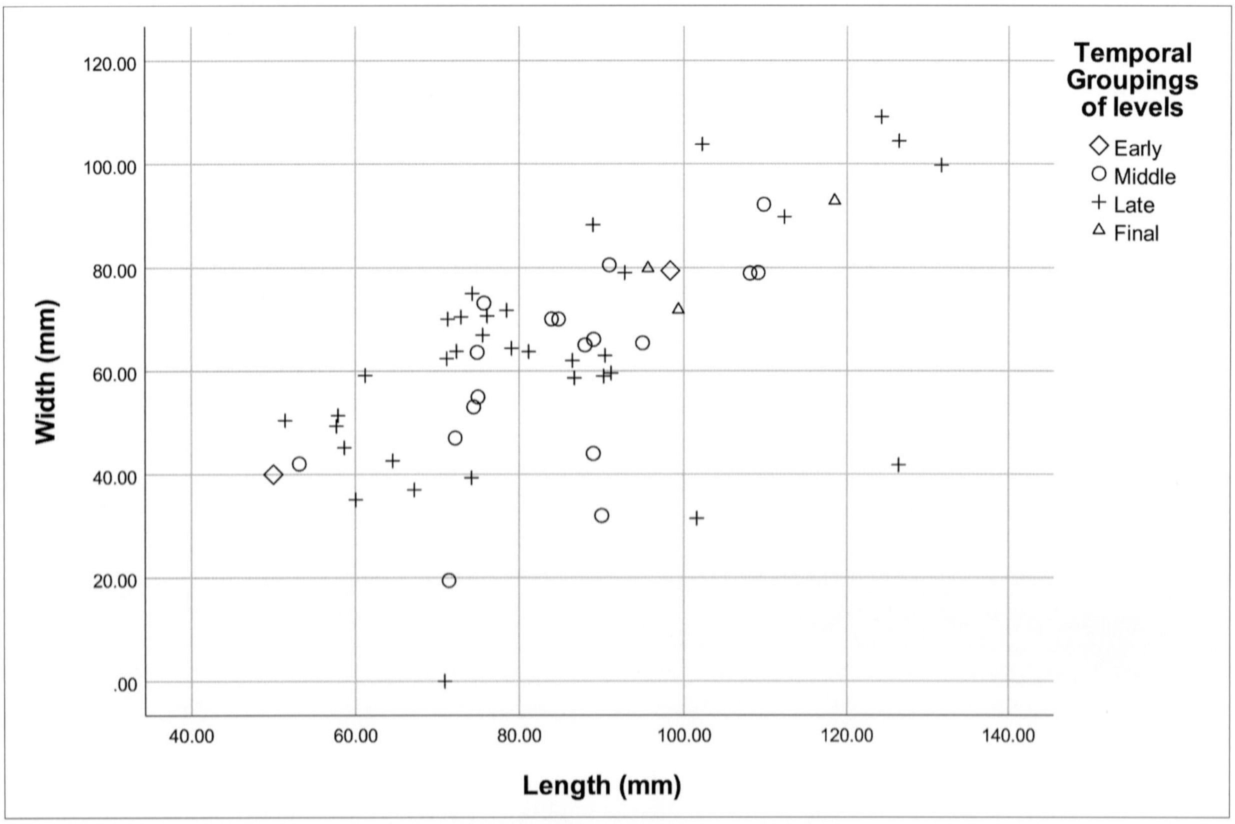

Figure 13.33. The size distribution of burnishers/plastering tools across periods.

production sequences, the organisation of food and craft production and depositional practices gives insight into the multiple ways ground stone artefacts were incorporated into the everyday life of the Çatalhöyük community.

Towards an evaluation of the scale and organisation of stone production at Çatalhöyük

Questions about the specialised nature of productive activities were investigated through detailed analysis of products associated with the early stages of the *chaîne opératoire* of mainly grinding and edge tools. As discussed above, the high debitage densities in B.142 and its location in close proximity to B.140, which Mellaart interpreted as 'a stoneworker's shop', and B.81, which also produced manufacturing evidence, may indicate specialised stone production in the Late period, associated with specific buildings/areas of the settlement.

There are, however, two further points that require consideration. If indeed the production of grinding tools (as is the case for all stone categories) became a specialised affair during the Late period, then it is rather surprising that the principles of stone production did not undergo significant changes. Technological choices evident in the earlier and later phases of the production sequence of querns exhibit consistency, with no apparent change in raw material use, most likely suggesting continuous use of the same sources and procurement networks, technological practice and final form of the produced objects. Lack of high morphological standardisation of the finished querns (cf. Van Pool, Leonard 2002) – even when the more elaborate quern type is considered – does not support the suggestion of the specialised productions of querns. Moreover, the fact that B.142 contained debitage from the working of a range of stone materials (serpentinite, schist, calcite, carnelian and possible hematite) suggests that the occupants of the building engaged simultaneously in the production of different stone categories including grinding and edge tools, stone ornaments and possibly other abrasive tools. This clearly suggests no strict specialisation in stone-working activities. The concurrent production of different types of stone objects within the same spaces is a recurrent pattern encountered in other Late period buildings (for example, B.75), but more importantly, it is also noted in the Early (for example, B.17) and Middle (for example, B.132) periods. Therefore, based on current data, the existence of 'specialised workshops' in a strict sense cannot be substantiated for any periods at Neolithic Çatalhöyük.

Raw material		*Temporal groupings of levels*				*Total*
		Early	*Middle*	*Late*	*Final*	
Conglomerate	no.	0	1	1	0	2
	%	0.0	3.1	1.9	0.0	2.0
Sedimentary quartzite/ orthoquartzite	no.	0	1	1	0	2
	%	0.0	3.1	1.9	0.0	2.0
Limestone	no.	3	8	13	1	25
	%	37.5	25.0	24.1	25.0	25.5
Indeterminate sedimentary	no.	0	0	0	1	1
	%	0.0	0.0	0.0	25.0	1.0
Schist	no.	0	0	2	0	2
	%	0.0	0.0	3.7	0.0	2.0
Marble	no.	4	22	24	1	51
	%	50.0	68.8	44.4	25.0	52.0
(meta)Quartzite	no.	0	0	2	1	3
	%	0.0	0.0	3.7	25.0	3.1
Indeterminate metamorphic	no.	1	0	5	0	6
	%	12.5	0.0	9.3	0.0	6.1
Diabase	no.	0	0	4	0	4
	%	0.0	0.0	7.4	0.0	4.1
Basalt	no.	0	0	1	0	1
	%	0.0	0.0	1.9	0.0	1.0
Indeterminate igneous	no.	0	0	1	0	1
	%	0.0	0.0	1.9	0.0	1.0
Total	no.	8	32	54	4	98
	%	100.0	100.0	100.0	100.0	100.0

Table 13.13. The temporal distribution of raw materials used for plastering tools.

Shape in plan		*Temporal groupings of levels*				*Total*
		Early	*Middle*	*Late*	*Final*	
Irregular	no.	0	0	5	0	5
	%	0.0	0.0	17.9	0.0	11.1
Ovate	no.	2	7	2	2	13
	%	100.0	63.6	7.1	50.0	28.9
Sub-rectangular	no.	0	4	9	1	14
	%	0.0	36.4	32.1	25.0	31.1
Discoidal	no.	0	0	12	1	13
	%	0.0	0.0	42.9	25.0	28.9
Total	no.	2	11	28	4	45
	%	100.0	100.0	100.0	100.0	100.0

Table 13.14. The morphology of plastering tools and its correlation to main occupation periods.

The lack of specialised activities is also supported by the study of edge tools, plastering tools and possibly ochre-processing activities (see also Wright 2013). Despite plaster burnishing not being a specialised activity, the level of consistency evident in technological and material choices surrounding the use of this technology suggests that enduring community-based traditions, social rules and conventions regulated the performance of plastering activities at Çatalhöyük. These social rules seem to have been adhered to most strictly during the Middle period and to have weakened in the Late period.

Finally, an object type linked to the emergence of specialised production of prestige goods is 'maceheads' (Wright 2013: 408). Yet, when the production of these objects is considered within their particular technological and cultural context, it becomes apparent that their transformation into objects of a distinctive appearance relied on a repertoire of techniques (abrasive techniques, drilling and most likely percussive techniques for the initial shaping of the original nodule) that were widely used at Çatalhöyük since the early levels. Their producers, therefore, would have drawn upon layers of existing knowledge of rocks and their material properties, appropriate technological gestures and toolkits, a know-how that was repeatedly applied to other material categories, such as edge tools, polishing tools/burnishers, bracelets/annulets and other forms of ornamentation, throughout the history of the site. Moreover, the fact that 'maceheads' appear in the Early period (Level South K) further demonstrates that the creation of these objects was part of an existing technological system, and in that sense, their production does not represent some form of technological advancement in stone-working activities that crystalises in the later part of the occupation sequence. In fact, no considerable difference in object variability is evident over time, as most of the ground stone types – with a few exceptions such as bracelets/annulets, pestles and vessels – appear already in the Early period. Differences, however, are materialising in the later periods in terms of material richness, expressed not necessarily in qualitative terms but in quantitative terms, that support an argument for the intensification of craft activities (for example, mineral/ochre processing) during the Late period, but also indicate an increase in human-thing entanglements (Hodder 2012).

Ground stone objects as personal possessions and indicators of status

One of the recurrent themes highlighted during the analysis was the differential treatment various stone objects received. In the case of edge tools, the cycle of production, use and maintenance transformed them into objects of a rather individual appearance. At first glance, the time and effort invested in their production, maintenance and subsequent re-use in other activities may indicate attempts to conserve a non-local raw material. Yet, broader cultural attitudes towards raw material use suggest that the curation of stone materials acquired over long distances was not a salient feature of the technological traditions at work at Neolithic Çatalhöyük. This is particularly evident in the treatment of non-local andesitic materials. Particular choices in the making of these objects (for example, polishing on all surfaces, along with the time-consuming technique of sawing) suggest that people at Çatalhöyük were willing to spend time and effort enhancing the appearance of seemingly mundane axes and adzes, and instead of throwing away broken, heavily used tools, they chose to prolong their biography. Their repeated use in seemingly prosaic tasks, however, may have facilitated a strong sense of attachment, individual ownership and curation (e.g., Toth et al. 1992; Pétrequin, Pétrequin 1993). Through repeated modification, which attests to their enduring value for their owners, they became rather highly personalised objects.

'Maceheads' have been invariably interpreted as weapons and symbols of social power and political authority (e.g., Rosenberg 2010; Wright 2014) and have been linked to 'rising social tensions' (Wright 2014: 28). While their production and the raw material selection show clear intent to create visually distinctive and potent objects (Tsoraki 2008; cf. Taçon 2004), consideration of wear traces and depositional practices does not substantiate their interpretation as weapons at Çatalhöyük. What becomes clear, however, is that static labels such as 'weapons', 'prestige goods' or 'symbols of power' do not capture persuasively the fluctuations evident in the perception and treatment of these objects throughout the habitation of Çatalhöyük: from objects that were re-used following breakage and therefore received no differential treatment from other materials in the Early period, to objects that were deliberately chosen to furnish the dead (but not always) in the Middle period, to hafted objects used in percussive activities and deposited in various non-burial contexts in the Late period. It is worth highlighting that when 'maceheads' appear in burial contexts, they are associated with both young female and male adults.

The sociality of grinding activities

Ethnographic studies invariably show that grinding activities are among the practices that contribute to the socialisation of small-scale communities and enable the construction of social categories based on age and gender roles (e.g., Searcy 2011; Hamon, Le Gall 2013; Robitaille 2016; Nixon-Darcus, D'Andrea 2017). Grinding activities were among the tasks that structured the perception and use of physical and social space. During the Middle period, elements in the use and deposition of grinding tools potentially indicate changes in their treatment and social significance. Grinding activities become more noticeably connected with particular buildings through the installation of fixed grinding facilities in interior spaces. While the presence of portable toolkits may have allowed for greater social interaction and flexibility during the performance of grinding tasks, the choice to create fixed grinding locations inside buildings signals a shift in the focus of grinding activities performed in restricted spaces, but also highlights the close associations between the occupants of certain buildings and grinding implements. The fact that these querns were left in situ at the end of the use-life of a building, and were therefore taken out of circulation despite being still functional tools, suggests that their use was not regulated strictly by notions of utility, but other factors also influenced their perception and treatment. This along with the lack of heavily used tools strongly suggests that grinding tools at Çatalhöyük do not represent heavily curated implements that were passed down from one generation to the next, an observation also made by Wright (2013). Yet, the absence of querns from fully excavated buildings (for example, B.97) despite the presence of complete upper grinding tools indicates that in some cases the occupants of the buildings did remove their grinding tools prior to the abandonment of the building.

This close association between people and grinding implements is further reflected in wider contemporary practices. Grinding tools were among the objects chosen to furnish the burials of two mature female adult individuals attributed to the Middle period. The gender associations referenced through these burial practices are further corroborated by an integrated study of activity markers on skeletal remains, projectile points and grinding tools at Çatalhöyük (Sadvari et al. 2015), a correlation also highlighted by multiple ethnographic studies of small-scale agricultural societies (e.g., Searcy 2011; Hamon, Le Gall 2013; Robitaille 2016). At Çatalhöyük grinding tools, fragments and complete objects, constituted an integral component of clustered depositions that seem to have formed through the concerted actions of multiple social groups and were deliberately placed in the buildings during their construction or abandonment (for example, B.77, B.150 and B.44). In that respect, they acted as material symbols chosen to commemorate the contribution of different households to socially significant and symbolically charged events (Tsoraki 2018). In ethnographic studies, the gifting of grinding tools features as an important element during the creation of a new household (e.g., Searcy 2011: 72, 74; Hamon, Le Gall 2013). Among the highland Maya communities, grinding tools are commonly offered as housewarming or wedding gifts because they are regarded as the symbolic representation of the unity of men and women, a unity required for the success of a household unit (Searcy 2011). Perhaps such cultural ideas may offer an alternative explanation for the reasoning behind the incorporation of a heavily used quern in the foundation wall in Sp.558, also dated to the Middle period.

Overall, at Çatalhöyük, there is ample evidence to suggest that plant- and ochre-processing activities were organised on a small scale. As noted above, while there is no physical evidence for 'cooperative grinding' (that is, multiple grinding toolkits positioned in the same space), the presence of small-sized grinding toolsets would have allowed for a certain degree of flexibility during grinding tasks, facilitating different forms of social interaction. This is particularly pertinent in the way ochre-processing tasks seem to have been organised. We can envisage the contribution of multiple actors in the processing of ochre, especially for tasks that required significant quantities of ground ochre, such as the creation of large-scale wall paintings or the treatment of dead bodies. This element of cooperation most likely also characterised the plastering and burnishing activities of house interiors, as became apparent during the construction of the experimental house at Çatalhöyük, during which the input of a group of people was required (Stevanović 2012b). Importantly, the performance of seemingly mundane activities such as these and the opportunities for socialisation they offered would have strengthened social bonds while facilitating the development of different community support networks.

Entrenched ideas about the function and seemingly mundane character of grinding tools and other ground stone categories have only rarely allowed for their consideration in practices beyond the realm of utility and the everyday. By shifting attention, however, to aspects of production, use and deposition, the aim of this chapter was to move away from static typologies and to consider ground stone technology as a socially embedded practice that is structured by particular technological traditions and culturally specific frames of reference.

Acknowledgements

I would like to express my sincere thanks to Ian Hodder for facilitating the analysis of the Çatalhöyük ground stone assemblage. I am indebted to Markéta Šťovíčková who has been instrumental during the on-site analysis of the material. The 2010–2017 excavation teams (North, South, GDN, TPC) are thanked for providing important information about the contextual associations of the ground stone material. I am grateful to all members of the Çatalhöyük Research Project team for creating a stimulating environment and in particular to Arek Klimowicz, Marek Barański, James Taylor, Onur Yüksel, Numan Arslan, Burcu Tung, Justine Issavi, Cristina Belmonte Santisteban, Christopher Knüsel, Marco Milella, Scott Haddow, Josh Sadvari, Duygu Tarkan, Dominik Lukas, Sean Doyle, Cristina Lemorini, Davide d'Errico, Alice Vinet, Levent Özer, Lilian Dogiama, Matilda Siebrecht, Lynn Meskell, Carrie

Nakamura, Gesualdo Busacca, Camilla Mazzucato, Ceren Kabukcu, Eleni Asouti, Ashley Lingle, Milena Vasić, Karen Wright, Shahina Farid, Amy Bogaard, Sara Perry, Marco Madella, Lisa Guerre, Jason Quinlan and Katy Killackey. I would like to thank all the undergraduate and masters students from the universities of Sheffield, Aberdeen and Leiden who took part in the recording of the material between 2013 and 2017, with special thanks to Juliette Hemelaar and Sophie Vullings from Leiden University. Funding was provided by the European Union's Seventh Framework Programme (FP7/2007–2013) under grant agreement no. PIEF-GA-2012-328862-Project CRAFTS and the Çatalhöyük Research Project.

Supplementary material

For supplementary material related to this chapter, please visit https://doi.org/10.18866/BIAA/e-14. It comprises table S13.1 (distribution of object types) and six figures (S13.1–S13.6).

14. The chipped stone of Çatalhöyük

Sean Doyle

Introduction

This chapter provides a synopsis on most aspects of the chipped stone assemblage from Neolithic Çatalhöyük, revitalises past discussions with new data from the material excavated between 2009 and 2017, outlines results of the current study period and establishes future directions of research. A number of objectives include disseminating the quantified data, discussing the major themes and patterns that have appeared and setting the stage for further exploration of specific topics. Such a task could not be accomplished without the work of those who came before, thus the chapter begins with an overview of the research completed on the obsidian and flint artefacts since the site was first investigated archaeologically. What follows are a number of thematic discussions on the trends evident in the data that highlight several dramatic changes through time. The focus throughout is the multi-faceted quality of chipped stone production and consumption at Çatalhöyük as the media through which socio-economic questions can be confronted. A parting aim of this chapter is to lay the quantitative framework on which future discussions will be based, in particular how chipped stone production is entangled within the socio-economic mesh of the immediate community and the wider regional networks into which they are in myriad ways intertwined (see Hodder 2011a; 2011b; 2012; Nazaroff 2015).

The final study period of the Çatalhöyük Research Project witnessed the excavation of buildings from nearly all levels of the site. This provided a large and varied sample that permits most aspects of previous conversations to be updated. Of great import was the horizontal expansion and intensive excavation under the new permanent shelter of the North Area, where a number of buildings and external spaces were excavated. These data provide critical support for arguments surrounding relationships between the North and South Areas, and a wider spatial net in which to capture the nature and pace of numerous changes that occurred throughout the occupation of the site. This chapter is largely organised thematically and follows certain trends and patterns evident in the data. The themes are discussed chronologically, with illustrations and graphics that stress differences between the Early, Middle, Late and Final periods. More detailed investigations of trends by level will occur in the future, once the radiocarbon dating program results have been achieved (for preliminary results see Volume 12, Chapters 3 and 19).

Methodology

The new data presented in this chapter were primarily obtained through a technological lens, with few assumptions made regarding the types and intended functions of tools. Traceology was performed microscopically, thus in this volume only the use-wear specialists Cristina Lemorini and Davide D'Errico discuss specific uses for tools (Chapter 15). The only aids used to produce the following new data were a hand lens and a thorough consideration of chipped stone reduction sequences and technological parameters. As such, I report primarily on the formality of tool type based on macroscopic analyses of morphology, type and degree of retouch, and intensity of use-related micro-flaking. In this scheme, informal tools can be retouched, for which I would prefer to report on the angles and retouch types of their edges, but in this chapter the terms 'cutter' and 'scraper' are used for brevity. Generally speaking, a low angle has a sharp, acute edge that is well suited for cutting, and an abrupt angle has a strong, steep edge that works better for scraping. In reality, this is not a strict dichotomy, as a low retouched edge can also be used for scraping activities, while an abrupt retouched edge can also be used for cutting activities. This reflects the necessity of categorisation for the purposes of this general report and minimises assumptions made about tool function based solely on retouch and micro-flaking characteristics. The term 'retouch' is reserved only to describe pressure-flaked edges and 'micro-flaking' strictly for the edge damage that occurs during a use activity. Some of the more formal tool types stand apart as tradition dictates, such as diagnostic projectiles (though hereafter referred to as 'retouched points') and large ovoid scrapers, thus they are assigned official typologies and given more attention during the discussions in this chapter. These types of objects tend to weigh more heavily in a wider range of theoretical and functional arguments. In future publications, however, attention will shift more towards the many characteristics of chipped stone production, in an attempt to better understand the nuanced socio-economic conditions responsible for changes in production strategies across the site and through time.

The bulk of new analyses outlined in this chapter were conducted on the 385 prioritised units spanning the North and South Area sequences (see Chapters 1 and 16; the 71 units from TPC will be reported in a future volume (Marciniak et al. forthcoming)). Material from a number of significant or otherwise interesting units not included on that list were also studied to varying degrees. Due to time and energy constraints, not all units were analysed to the same level of detail. All were subjected to Level 1 analysis (count, weight and material type per bag of artefacts), and most were assigned a basic technological classification, but only the more significant or informative units could be studied most intensively. For these reasons, there are what might appear to be discrepancies in numbers between the types of analyses in any of the subject populations. For example, B.160 had 2,420 artefacts, but only 287 obsidian artefacts were visually attributed to a specific source, 511 artefacts were given a technological classification, and a state of use was determined for 243 of them. All used, retouched and formal tools were assigned a typological classification. With such a massive assemblage it was virtually impossible to record every attribute of every piece of chipped stone. Doing so would surely provide more information, but that task will have to wait until another day.

The illustrations in this chapter focus on the Early and Middle periods for two reasons: first, because most of the units excavated in this final study period belonged to those periods, and second, because the chipped stone chapter in the previous volume targeted the Late period for study and illustrations (Carter, Milić 2013a), a decision that reflected the younger age of many of the buildings excavated between 2000 and 2008. The illustrations here are organised in two ways. Some are arranged contextually, like the four retouched points recovered from burial fill (22194) in B.5 (fig. 14.8), or the contents of an obsidian cache from B.131 (fig. 14.6). Others are organised typo-technologically, showing groups of related items such as retouched points on bidirectional blades (fig. 14.7), or obsidian mirrors (fig. 14.10). These rather rigid groupings of artefacts do not immediately demonstrate the diversity of Neolithic toolkits, but they do give a general sense of the chipped stone implements that were made and used at Çatalhöyük. In future collaborative work, illustrations and discussions will include the ground stone, bone, wood and other materials that comprised different toolkits. The knappers and other members of a building or household had lives and material assemblages that were multi-stranded cables, entangled with one another in numerous ways (see Hodder 2011: 164; 2012; Nazaroff 2015).

In a previous volume, the chipped stone chapter breathed life into the assemblage from the 2000–2008 study period through a window 'into the daily lives and cultural traditions of Çatalhöyük's inhabitants during the later stages of the Pottery Neolithic (PN)' (Carter, Milić 2013a: 417). Using a variety of scientific and contextual approaches aimed at the consumption of chipped stone, including the trace-elemental and visual characterisation of obsidian sources and artefacts (see Carter et al. 2006b; Carter, Shackley 2007; Carter et al. 2008; Milić et al. 2013) and further exploration of the change in blade production from percussion to pressure (Carter 2011; Carter, Milić 2013a; 2013b), a number of socio-economic aspects were identified and confronted. As a result, the chemical compositions of the obsidian sources in Anatolia have been well established, thus the sources for obsidian artefacts newly reported in this chapter were exclusively determined visually. This was only possible due to the significant investment by a number of researchers who have found and elementally characterised Anatolia's many obsidian sources (fig. 14.1) in the years since Renfrew's first study (Renfrew et al. 1966; Gratuze 1999; Gratuze et al. 2005; Carter et al. 2006b; 2008; Carter, Shackley 2007; Milić 2014). These studies and others have resulted in the development of a visual identification protocol for Çatalhöyük's obsidian that is up to 94–97% accurate, as determined by a rigorous replicable training program and a series of blind tests (Milić et al. 2013: 4).

My own research input brings the focus squarely back to technology, especially the traces of on-site production and a strict consideration of the depositional context. Following Conolly (1996; 1999) and others (Grace 1989; Hurcombe 1992) I present informal tool types based strictly on retouched-edge morphologies; thus cutter, scraper and perforator are the three main categories, with a multi-tool category used for anything that exhibits any combination of those three tool types. Formal and generally recognised types such as bifaces, retouched points and backed perforators were identified using a suite of technological parameters beyond simply edge-retouch morphology. In the past this formal/informal dichotomy had largely been discussed as 'curated' versus 'expedient' tool types (Binford 1973; Shott 1996, *inter alia*), a concept that has come to encompass several definitions with no universal agreement on its meaning and complex theoretical discussions that are too lengthy for this chapter. Without the right type of data to accurately calculate reduction intensity, as was done for the assemblage at Boncuklu (Muller et al. 2018), modern arguments surrounding the concept of curation are not yet confronted, so in this chapter the simple formal/informal distinction of tools

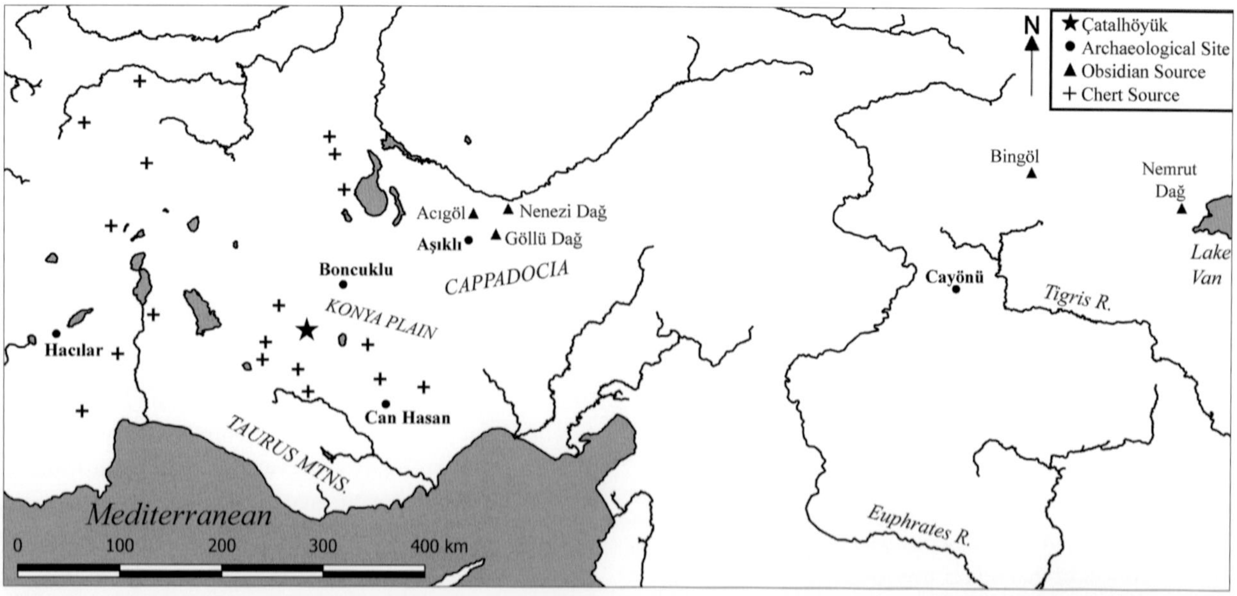

Figure 14.1. Revised map of all known obsidian and flint sources in central and eastern Anatolia, with sites and regions mentioned in the text (built upon Milić, Carter 2013a; Nazaroff 2015).

is preferred, because it is more generalised and less problematic. This does inevitably impart some researcher bias into the process of identifying the intended function of a tool, but arguably much less so than other options; therefore, for the intents and purposes of this chapter the simplest option is best. Future publications will further explore the notion of curation as a dynamic and fluid process, highlighting the versatile functionality and frequently recycled nature of chipped stone tools at Çatalhöyük (see Chapter 15). Other future studies will quantify the trace elements of manufacturing debris to further elucidate the nature of craft specialisation and labour investment at the site.

A history of research

The 'broken obsidian arrowheads' James Mellaart found on the mound in 1958 allowed him to first identify Çatalhöyük as an Early Neolithic occupation (Mellaart 1964: 94) and, as such, they have been one of the longest-studied aspects of the site's material culture. When one thinks of Çatalhöyük, they must also think of obsidian, the two being inextricably linked among the archaeological community for over 60 years. Surprisingly, however, there was only one dedicated chipped stone report that came out during Mellaart's tenure in the 1960s, and that was after just the first season of excavation. In that report, Perry Bialor stated that the chipped stone industry was homogenous from bottom to top (Mellaart level VIII to II at the time), with no changes in tool proportions, sizes or production techniques (1962: 67). This belief held sway for a long time, and though Mellaart did mention some of the more significant chipped stone finds in his four preliminary reports on the excavation seasons (Mellaart 1961; 1962; 1963; 1965), his main arguments regarding the chipped stone would remain relatively unchanged from Bialor's initial assessments. A more comprehensive report on the 1961–1963 excavated material was in fact completed by Mortenson (1964). This was supposedly a more traditional culture-historical account of the tool typologies, raw materials and chronological distributions, but unfortunately it was never published, remains hard to access and, generally speaking, has not contributed anything to the wider conversation.

Several important studies occurred contemporaneously with the original excavations, including the use of elemental characterisation techniques conducted on obsidian artefacts from Çatalhöyük and samples from several known Anatolian sources (Renfrew et al. 1966; Wright 1969). Mellaart himself made the following evocative statement regarding the socio-economic, magico-religious significance of obsidian to the site's residents:

> It may be surmised that it was not only for its great cutting power, its transparency, reflective power and its jet black appearance that this material was so highly prized. Its volcanic and thus chthonic origin would have linked it to the underworld, the place of the dead, and it was a true gift of mother earth, and therefore imbued with magical potency (Mellaart 1967: 177).

Even this mythically inspired and romanticised account of obsidian's significance could not divert attention from the marvellous wall paintings, numerous other forms of artistic expression and elaborate architecture that attracted most interest in the site from the public and the scientific community. For a time, the chipped stone carried on in the background and did not garner any new academic attention in the intervening years between excavations.

It was not until the Çatalhöyük Research Project came into existence that things really began to gain momentum. The first chipped stone report of the new project came from Nur Balkan-Atlı (1994), who devised a modified typological system to reanalyse material from the earlier levels of Mellaart's excavations as part of a larger dissertation on the Neolithisation of Anatolia. One relevant result of that report was the conclusion that up to 7% of the retouched implements were microliths (Balkan-Atlı 1994: 128), a chipped stone technology that generally characterises Epipalaeolithic or early PPNA sites in Anatolia. This contradicted earlier claims that genuine microliths were not present at Çatalhöyük (Mortenson 1964: 4). Balkan-Atlı also firmly established that projectiles were primarily made on bidirectional percussion blades, a tradition that was present in the earliest Neolithic at Aşıklı Höyük (Todd 1966; Balkan-Atlı 1994) and whose production virtually defined the Pre-Pottery Neolithic B (PPNB) period of the Levant (Borrell 2011; Barzilai 2013).

Next, James Conolly presented the results of the 1993–1995 surface scrapes and outlined his research objectives and methodology moving forward (1996). Delving deep into both the theoretical and practical applications of chipped stone analyses, he noted that traditional typological, technological and functional methods conducted in isolation were not very conducive to the study of a site such as Çatalhöyük, whose lengthy occupation and well-preserved contexts offered much more varied information than was usually available. He envisaged 'a contextual approach that emphasizes the interpretation of typological and technological data' (Conolly 1996: 173). Conjuring the French school of cultural technology, a tenet of which states that technical actions are both social and cultural phenomena (Pelegrin 1990; Inizan et al. 1992: 12), he proposed that technological analyses could in fact be considered contextual if they followed a methodology that adhered to the *chaîne opératoire* (Conolly 1996: 173). This was also the first time several important patterns in the data were truly discussed in a publication, including a possible shift in production from flake-based to blade-based industries around Level V, and certain differences in the depositional contexts of bifaces, projectiles, cores and thinning flakes. This was made possible via the exhaustive comparison and integration of all previous reports (Conolly 1996: 194). These initial hypotheses still ring true and have since evolved into more complex themes that are integral to the understanding of the Çatalhöyük chipped stone assemblage.

Conolly went on to complete his PhD dissertation on the material using his fully established techno-contextual approach and later published a condensed version (1999). He had collated the information in the first three reports (Bialor 1962; Mortensen 1964; Balkan-Atlı 1994) and Mellaart's various publications and recognised several issues that required attention. He made the following three primary observations, which confirmed a need for more systematic analyses: 1) there was a classification problem in defining non-standardised assemblages, with a reliance on broad inclusive categories; 2) a change in primary technology over time had been tentatively suggested by the three report writers, but none of them had studied all levels themselves and as such they could not quantify this change; and 3) there were several hints, but nothing conclusive, regarding the possibility that some production and deposition of certain tools, like projectiles and bifaces, was not strictly based on functionality (Conolly 1999: 6–7). Conolly set out to analyse as much material as possible, ultimately including the chipped stone assemblages from all major excavations at Çatalhöyük in his study. He recorded attributes for over 15,000 pieces in total and obtained a 'detailed reconstruction of the Çatalhöyük knapped-stone industry, as well as its temporal, spatial and contextual relationships' (Conolly 1999: 16).

Conolly's results forever changed the way we understand the chipped stone of Çatalhöyük and its significance in the wider socio-economic system that encapsulated not only the site, but also the surrounding regions and beyond. Not only did he produce the first multi-attribute statistical classification of projectile points for any Neolithic site in Central Anatolia (Conolly 1999a: 39–41), he also brought questions regarding craft specialisation and symbolic deposition to the forefront of scientific discussion. He studied the obsidian caches on a contextual scale and established their regularity of deposition within the dirty area of buildings, most often in the southeastern corner near a hearth or oven. He also began to quantify and contextualise the nature of change within the production industry, specifically from flake-based to blade-based and from percussion to pressure techniques (Conolly 1999a; 2003).

More recently, Tristan Carter and Marina Milić led the charge during the second study period and adopted a rigorous approach towards investigating the social dimension of chipped stone consumption. They made great strides in the elemental characterisation of Çatalhöyük's obsidian, which allowed for comparisons between the North and South Areas based on ratios of the major obsidian sources and contributed much information to larger discussions on regional zones of interaction (Carter et al. 2005a; 2008; Carter, Milić 2013a; 2013b). This was accomplished using a wide array of scientific instrumentation, including ICP-AES, ICP-MS and LA-ICP-MS (Carter et al. 2005b; 2006b), EDXRF (Carter, Shackley 2007; Carter et al. 2008; Poupeau et al. 2010), pXRF (Carter 2009) and PIXE (Carter et al. 2008; Poupeau et al. 2010). There was also an extensive visual characterisation program established for the Cappadocian obsidian sources, resulting in a 94–97% accuracy rating for artefacts that originated from the Göllü Dağ, Nenezi Dağ and Acıgöl sources (Milić et al. 2013), the three of which provide over 99.9% of Çatalhöyük's obsidian in all levels. These studies collectively revealed an interesting trend in the primary obsidian source used through time, from almost 90% Göllü Dağ to almost 90% Nenezi Dağ, a process that took approximately a century to develop toward the end of the Middle period (fig. 14.3). This phenomenon is discussed further in the 'Raw materials' section below.

Carter also investigated the presence of 'characterful objects', injecting some much-needed agency into the assemblage, and sought to delineate the types of human-thing relationships that might have occurred with particular items like mirrors, daggers and weapons of the hunt (Carter 2011; Carter, Milić 2013a; 2013b). Drawing inspiration from numerous studies on the influential *Kula* ring of Melanesia (e.g., Malinowski 1922; Munn 1986; Mauss 1990; Weiner 1992), he envisioned the movement of certain items between the people of Çatalhöyük and also between individuals or groups from neighbouring communities (Carter 2011: 5). He explored this concept through the lens produced from larger conversations around continuous gifting behaviours that 'ultimately tied people together in a materially mediated social entanglement/enchainment through space and time' (Carter 2011: 5; see also Appadurai 1986; Fowler 2004: 53–78; Lazarri 2005; Hoskins 2006).

Building on discussions of the figurines and their roles as potent media that could manipulate social learning through their production, portability and whether or not they were on display or concealed within a building (Meskell et al. 2008), Carter added certain aspects of the chipped stone assemblage to that list of social media, specifically 'characterful' objects that appear to have been more meaningful than others (2013: 451–52). This followed assessments that some items could be more active than others in the maintenance of important social relationships (Gero 1989: 103). Many of these 'characterful objects' were relatively common tools like cutters, scrapers and retouched points (Carter, Milić 2013a: 472–74) and would likely not have been identified as such had they not been set apart by their depositional context. These discussions continue below, in the 'Caches' section.

As part of his PhD research, Adam Nazaroff co-directed the Anatolian Archaeological Raw Materials Survey (AARMS) with Adnan Baysal (Nazaroff et al. 2013; Nazaroff 2015). This survey scoured large parts of Central Anatolia for sources of chert, the term they used for a wide variety of silicified materials that could have been used for chipped stone production, synonymous with the term 'flint' used in this chapter and more generally for Çatalhöyük's non-obsidian chipped stone assemblage. Before the commencement of this project, just five sources of flint had been regularly discussed in the literature, of which only two were located in Central Anatolia; neither had been located or sampled, but simply hypothesised based on geological maps and reports (Nazaroff et al. 2013; Nazaroff 2015: 301–02). Mellaart himself suggested that Çatalhöyük's flint could have formed in the Taurus Mountains and Gaziantep region of southeastern Turkey (1967: 213), but it would later be discovered that suitable stone could be found much closer to the site than predicted. AARMS located 20 sources of chert in four years of work surveying five major regions around Çatalhöyük: the Western Konya Basin, the Southern Konya Basin, the Antalya region, the Afyon-Eskişehir region and the Cihanbeyli and Haymana Plateaus (Nazaroff 2015: 302–03). Many of these sources were gravels or secondary deposits, generally unsuitable as a knapping medium, but a number of them appear to have been of large enough size and high enough quality to produce the types of flint artefacts found at Çatalhöyük. In the future, an experimental knapping study conducted on material from these identified sources could provide valuable insight into why some sources were used over others, what constraints each had in terms of their production potential and the roles each played in the socio-economic milieu of the settlement.

Having proved there is indeed no shortage of flint in Central Anatolia and that the variability among deposits is great, Nazaroff suggested that the notion of provenance could be conceptualised in a different way than it traditionally had been (2015: 299). He argued that material provenance should be perceived as a thing that is itself entangled within the wide networks of engagement that comprised the social and physical realities of Neolithic life (Nazaroff 2015: 299–300).

Lamenting how provenance has generally been considered a discrete phenomenon, completely at the whim of its active social agents, he successfully 'demonstrated that provenance, as *a thing*, is in fact a diverse and far-reaching collection of materials, environments, peoples, and temporalities, and that the provenance thing emerges as unique combinations of these relations at different points in time' (Nazaroff 2015: 677). I would honour the notion of entanglement in a similar vein and propose that the act of chipped stone production is itself another thing that is enmeshed within this network, a thing that shapes and is shaped by the materials, environments, peoples and temporalities that it is connected with. During all the life stages of any unit of raw knappable stone material – from procurement, to transport, production, storage, use, maintenance and final deposition – there are an exponentially greater number of social and physical relationships that are built and maintained to ensure the system remains reliable and sustainable. This approach lies at the heart of future research on chipped stone production and the development of craft specialisation at Çatalhöyük and beyond.

Lilian Dogiama conducted PhD research on the projectile point assemblage of Çatalhöyük, aiming to examine the role of hunting and why it remained such a popular activity in an agropastoral economy, and the symbolic roles projectile points played in the Neolithic community (2017: 2). Using the hunter's weapons as the medium, Dogiama studied the projectiles and bifaces of Çatalhöyük, recording technological, morphological and contextual attributes to situate them within the social fabric of the larger community. She separated these types of chipped stone products into two groups, Group A Points and Group B Bifaces, and asserted that they belonged to different technological strategies and followed different operational sequences (Dogiama 2017; in press). I agree with the assessment that Group A points were often multi-purpose tools, exhibiting both impact and micro-flaking damage resulting from use as projectiles and for a number of other activities (Dogiama 2017: 174–75; see also Chapter 15). As such, I cannot strictly classify them all as projectiles; rather, I simply call them 'retouched points' as an objective claim to signify production technology and geometric shape, with nothing implied about intended or actual function. Thus the classifications of projectile, arrowhead, spearhead, dart point and dagger used in this chapter fall under the umbrella term 'retouched point'. See Dogiama (2017; in press) for a calculable morphometric distinction between arrowheads and spearheads, and the socio-economic implications therein based on regional comparisons of traditional projectile point typologies.

All retouched points exhibit some degree of pressure retouch, usually along most if not all of their edges, but even the most minimally modified point was subjected to at least some shaping of the stem and/or tip. These features distinguish this group from a less common technological class, the true bifaces, which are made not on unidirectional or bidirectional percussion blades like the majority of retouched points, but are instead bifaced down from larger spalling flakes and often found in varying stages of production. These are equivalent to Dogiama's Group B Bifaces, which are generally unused and unfinished and belonged to a distinct technological tradition that followed a different path from source to site. From procurement to final deposition, they underwent a very different operational sequence than their more utilitarian counterparts, the retouched points, which are most often found in midden deposits and frequently broken and/or repurposed.

Another major contribution by Dogiama was a regional approach to traditional projectile typologies, combined with an intensive attribute analysis that differentiated arrowheads from spearheads and related certain point types to certain types of weapons (Dogiama 2017; in press). Of the points that could be typified were five variants, the Byblos (n=125, 59%), Amuq (n=38, 18%), Herzeliya (n=17, 8%), Jericho (n=14, 7%) and Nizzanim (n=13, 6%), along with a small number of Abu Maadi points (n=4, 2%), which are likely to be local versions of the Amuq point (Dogiama 2017: 154) and thus are not considered as a primary point type in this chapter. Although one must always be wary of relying too heavily on typologies, it is interesting that many of the Levantine PPNB point types appear to have endured together all through the occupation of Çatalhöyük, perhaps yet another indication that the community was made up of multiple distinct groups from different places, who managed to coexist while all maintaining their own nuanced traditions from their lineal ancestors. This interpretation is qualitative and not yet based on any conclusive confirmation, but there is a growing body of evidence that indicates social cohesion of the site's large population, with communal investment in certain rituals and actions, but also small-scale variation. Perhaps retouched point type was just one of many deliberate acts of agency individuals were able to perform that visibly connected them to their particular group, an acceptable act meant to maintain a social identity that helped to keep order within such a populous settlement, alongside beads and other forms of ornamental technologies (see Chapter 9).

This research left us with three reasonable possibilities as to why hunting remained a viable activity in a growing agropastoral community: 1) to create or maintain group identity; 2) to acquire social status in an otherwise egalitarian society; or 3) as a leisure activity to inject some excitement and danger into an otherwise mundane agricultural life (Dogiama 2017: 235). I would add some nuance to these scenarios and acknowledge that it is possible for all, or none, of them to have been the true case. It is possible that the hunting of wild animals was a symbolically charged, socially stratified activity that only a select few were privy to, or skilled enough to partake in, and which was used to forge bonds in the community and stimulate competitive display. Conversely, hunting of wild game could have provided a dietary supplement with a strictly functional or utilitarian role in daily life, no longer the symbolically charged and dangerous act that it was before domesticated meat and plants became the primary sources of sustenance. Nevertheless, it is tempting to view the existence of hunting alongside the magnificence of the wall paintings and elaborate architecture and jump to possible symbolic or social explanations.

Some initial new data

A number of databases have been implemented and used by the chipped stone team for different purposes over the last 25 years, with most of them now imported into the current primary database. The final number of individual artefacts stands at an even 367,000, split between 32,438 bags, with a weight of 213.1kg. Of that, 97.9% (n=359,210) was obsidian, while only 2.1% (n=7,790) was flint. The term 'flint' has always been employed at Çatalhöyük, but in reality this category is composed of a variety of silicified materials that fall under the umbrella term 'flint', even though geologically speaking flint is a specific type of chert that forms in beds of chalk. For ease and comparability, use of the term 'flint' is continued in this chapter. Interestingly, the proportion of flint by weight is over six times higher than it is by frequency. Table 14.1 displays these totals and breaks down the numbers by the three major study periods. It reveals how during this last study period the same proportion of flint objects were recovered, but they had a much higher average weight than those from previous years. This phenomenon is partially explained by the excavation of a higher number of Early and Middle period buildings after 2008, where the proportion and average size of flint artefacts was higher than in Late and Final levels. Furthermore, very little knapping of flint seems to have occurred on site in any level, beyond the maintenance of tools likely brought into the settlement already made. In terms of production debris quantities, a dearth of flint and a plethora of obsidian are an obvious way to explain the discrepancy in weights, but the trend is further explored below.

This chapter incorporates material from all study periods, as it relates to the larger themes expressed in the data, but the majority of what is covered was excavated between 2009 and 2017. Aside from the units prioritised for study in this volume, a number of others that contained interesting material and spanned a range of contexts were also subjected to analyses. Some were sampled and studied to different degrees of detail, depending on the form and size of a unit's assemblage. I attempted to select these extra units as objectively as possible, but due to the sheer volume of the assemblage and the study time constraints it should be assumed that some researcher bias exists towards units that contained more technologically informative material, as analyses had to be prioritised.

The buildings listed in table 14.2 are all those that were excavated to at least 75% completion in this final study period. I discuss some finds from spaces and buildings not on that list, but it has been suggested that site-wide patterns should not be investigated using buildings with less than 75% of their total volume excavated, because many of the more significant deposits are often found in sub-floor features, which are generally the final units excavated in any building. For this reason, I chose to limit my larger pattern identification to using the nine buildings listed, which together provide sufficient detail to illuminate the key themes from both the North and South Areas. Table 14.3 displays these results in bulk, with the analysed assemblages from these nine buildings condensed into three tables, one each for the technological, use-related (functional) and typological parameters. Table 14.4 further explores these data by categorising everything by temporal grouping, which demonstrates the striking differences between the Middle and Late levels that always seem to exist in every respect.

Study period	*Bags*	*Artefacts (no.)*	*Weight (kg)*	*Obsidian no. %*	*Flint no. %*	*Obsidian weight %*	*Flint weight %*
1993–1999	11,254	98,074	33.1	97.7	2.3	91.8	8.2
2000–2008	10,202	142,058	105.6	98.5	1.5	90.3	9.7
2009–2017	10,982	126,868	74.4	97.5	2.5	79.0	21.0
Total	32,438	367,000	213.1	97.9	2.1	86.6	13.4

Table 14.1. Level 1 chipped stone data organised by study period.

Building	Hodder Level	% Excavated	Obsidian (n)	Flint (n)	HR Density (n/L)
77	North G	100	3,436	66	0.96
80	South O	75	3,850	124	1.16
89	South N	75	2,722	45	1.25
97	South O	100	3,347	51	1.28
114	North G	75	2,538	47	1.52
119	North F	75	1,022	18	0.93
131	North G	100	4,419	83	1.14
132	North F	75	6,134	214	3.31
160	South K	100	2,326	94	2.78

Table 14.2. Number of chipped stone artefacts recovered from buildings targeted for primary study.

Middens and external spaces have also become an important focus of research at Çatalhöyük, in the past overlooked as inconsequential refuse pits but more recently understood to have been much more complex. They certainly do contain a wealth of chipped stone data in all forms, and some of them are discussed at length further below, but they remain a point of contention due to their apparent secondary and tertiary depositional contexts (see Volume 12, Chapter 33). In most situations it is impossible to say with ultimate confidence where these items originally came from, or what their exact relationship with the surrounding buildings was. Furthermore, they generally do not contain anything that would contribute to the larger spatio-temporal discussions pertaining to particular contextual phenomena like obsidian caches, other forms of symbolic deposition, intensity of in-house domestic production and so on. Several of these external areas did, however, contain some significant and potentially in situ artefacts, so they were analysed and are used in some of the broader discussions of the site below, especially the early, flint-rich sequence in Space 628 in the South Area and the gridded, well-documented sequences in Spaces 610, 625 and 631 in the North Area.

The raw materials

Mellaart bestowed a near-mythical status on the obsidian of Çatalhöyük and believed it must have been accorded a major socio-economic and symbolic importance within the community. He went so far as to say Çatalhöyük's occupants controlled the procurement and trade of it in the wider region, with a well-organised and extensive commercial system that was responsible for the apparent wealth of the settlement (Mellaart 1964b: 101). As already noted, the primary component of Çatalhöyük's chipped stone assemblage is, of course, obsidian, in all occupation levels and nearly every individual context. In terms of numbers, obsidian makes up 97.9% of the entire collection but only 86.6% of its total weight (table 14.1, fig. 14.2). This notable difference is primarily due to the fact that most flint artefacts are larger, heavily retouched tools, with longer use-lives and higher average weights than their obsidian counterparts.

Although some flint production did undoubtedly occur on site, as evidenced by several blade cores and associated production debris (Conolly 1996: 178), in most instances flint is found as either a retouched tool or the pressure flakes and small fragments that are removed during maintenance. The majority of flint tool manufacture ostensibly occurred off site, with a system of procurement that may or may not have included methods of trade and exchange as we might traditionally recognise them. This procurement strategy had a focus on the more functional and utilitarian, bringing ready-made blanks and/or more formal tools that were highly valued and enjoyed long use-lives at Çatalhöyük. Of course, some exquisite flint objects, such as large 'daggers' that sometimes ended up alongside individuals in burials, made their way to the site as well, but overall, flint was regarded as a functional and durable material that could handle heavy jobs for long periods of time. The high concentration of long-lived highly retouched flint tools in Sp.628 of Level South H could indicate that people had travelled from many distant locations across the landscape with their toolkits made of diverse materials to become some of the earliest inhabitants of the site. In any case, these early occupants exploited a larger proportion of the stone resources available in the wider region than those in later levels (e.g., Carter et al. 2005a: 278–79; Carter 2011: 5).

Technology	*n*	*%*
Pressure bladelet	140	3.8
Pressure blade	220	5.9
Pressure core	25	0.7
Percussion bladelet	44	1.2
Percussion blade	217	5.8
Percussion core	41	1.1
Percussion flake	436	11.7
Biface thinning flake	209	5.6
Biface	52	1.4
Core preparation	147	3.9
Pressure retouch flake	240	6.4
Burin spall	9	0.2
Fragment	1,761	47.4
Chunk	120	3.2
Unknown	63	1.7
Total	3,724	100.0
Cortical	114	3.5

Use	*n*	*%*
Unused	2,656	78.2
Used	379	11.2
Used/retouched	185	5.4
Formal tool	176	5.2
Total	3,396	100.0

Typology	*n*	*%*
Informal tool	358	54.4
Backed tool	13	2.0
Pièce esquillée	97	14.8
Retouched point	141	21.4
Perforator	4	0.6
Burin	1	0.2
Denticulated	2	0.3
Hooked	6	0.9
Notched	11	1.7
Knife	6	0.9
Dagger	1	0.2
Formal scraper	17	2.6
Total	657	100.0

Table 14.3. Results of technological, functional and typological analyses presented in bulk.

Technology	*Early (%)*	*Middle (%)*	*Late (%)*
Pressure bladelet	1.4	3.0	38.6
Pressure blade	2.2	1.6	23.2
Pressure core	0.7	0.5	2.0
Percussion bladelet	6.1	5.9	4.6
Percussion blade	1.4	1.3	0.0
Percussion core	1.0	1.3	0.0
Percussion flake	5.6	12.6	19.4
Biface thinning flake	0.4	1.9	0.0
Biface	4.6	6.6	0.3
Core preparation	2.9	4.3	4.1
Pressure retouch flake	8.8	6.6	0.0
Burin spall	0.2	0.3	0.0
Fragment	60.4	48.4	7.2
Chunk	2.4	3.8	0.6
Unknown	1.9	1.9	0.0
Total	100.0	100.0	100.0
Cortical	4.3	3.1	0.9

Use	*Early (%)*	*Middle (%)*	*Late (%)*
Unused	79.5	83.2	41.7
Used	8.4	7.6	39.8
Used/retouched	7.5	3.4	16.6
Formal tool	4.6	5.8	1.9
Total	100.0	100.0	100.0

Typology	*Early (%)*	*Middle (%)*	*Late (%)*
Informal cutter	37.8	32.8	76.6
Informal scraper	11.3	5.7	4.7
Informal cutter/scraper	5.7	3.8	0.8
Informal cutter/perforator	4.7	2.4	0.8
Backed cutter/perforator	3.8	2.1	0.0
Pièce esquillée	5.7	11.1	4.7
Tool on broken point	2.8	0.5	0.0
Pièce esquillée recycled	4.7	6.2	5.4
Retouched point complete	3.8	18.0	2.3
Retouched point broken	13.2	8.5	2.3
Burin	0.0	0.2	0.0
Perforator	0.9	0.7	0.0
Dagger	0.0	0.2	0.0
Knife	1.9	0.7	0.8
Formal scraper	0.9	3.8	0.0
Denticulated	0.0	0.0	1.6
Hooked	0.9	1.2	0.0
Notched	1.9	2.1	0.0
Total	100.0	100.0	100.0

Table 14.4. Results of technological, functional and typological analyses presented by temporal group.

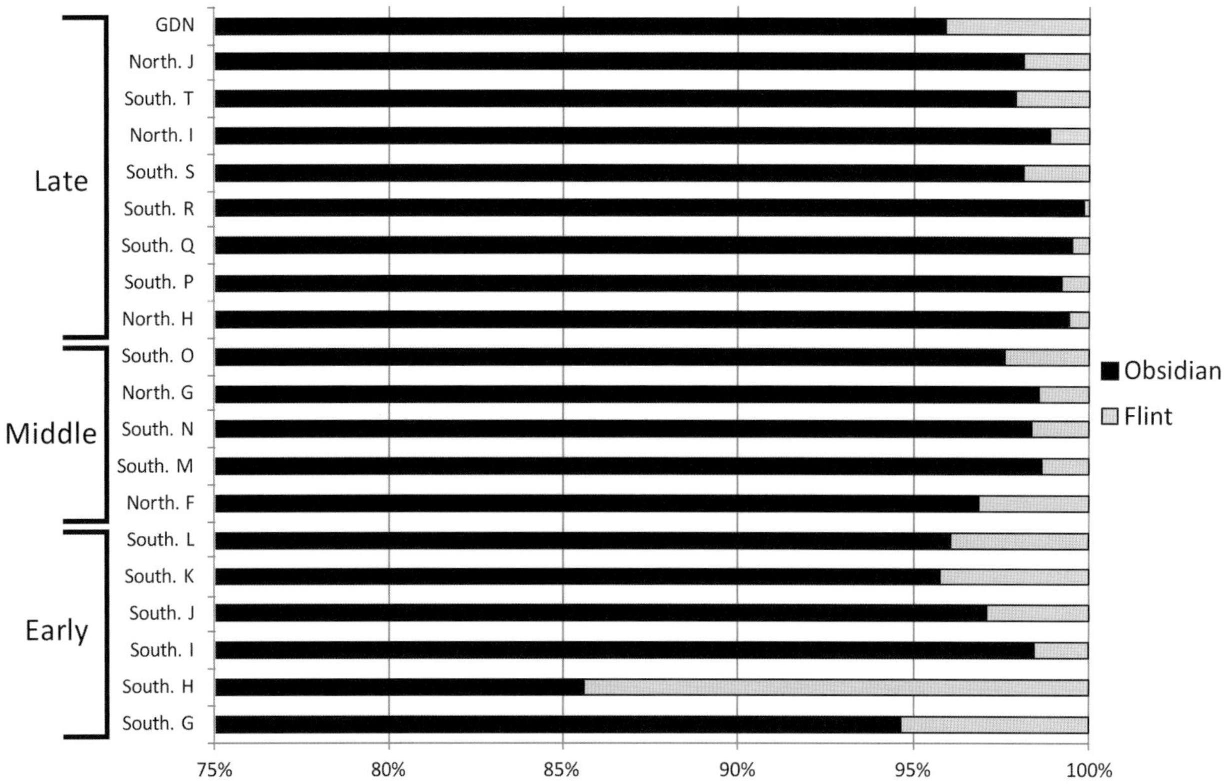

Figure 14.2. Updated obsidian versus flint percentages through time.

Obsidian as a resource was regarded altogether differently and associated with a wider array of behaviours. In the Early and Middle periods, there existed seemingly inefficient production methods with high amounts of obsidian waste occurring all over the site; it was simply an abundant raw material that was used for domestic-scale production and use. At the same time, it was turned into exquisite mirrors, tucked away in impractical caches, and involved in various forms of ritualised deposition that indicate there was a symbolic or spiritual quality attached to it. Given these varied behaviours associated with obsidian, I would argue that it was not the raw material itself that was most important, but rather the amount of energy that went into the production of an object, or what the object represented or could potentially become as a tangible or intangible part of an individual, a household or the settlement.

With so many varieties of flint in the near vicinity, it begs the question, why so much obsidian? One might not blame Mellaart too harshly for presuming that the people of Çatalhöyük held a near-mythical adherence to obsidian, even though it had to be carried over 250km from source to site, assuming the longer route around the southern edge of the Konya plain was taken to follow water sources (Carter 2011: 5). In this case, the practical argument is perhaps most likely, wherein obsidian from the Cappadocian sources could be obtained in much larger, fracture-free pieces of raw material than any of the flint outcrops in the area could provide. Obsidian is also more homogenous, and as long as it is tested properly at the sources, it is more likely to be free of faults than the local flint types and thus less risky to transport over long distances, even though it is more brittle. Arguments could also be taken in more symbolic directions, as explored below, but neither approach should be taken in isolation. Obsidian was entangled with the daily life of a Neolithic inhabitant of Çatalhöyük in a number of ways, and throughout the site's occupation it was procured in different forms, along different routes, with different meanings and value systems attached to it (e.g., Carter 2011: 5). It is quite possible, even likely, that a number of established procurement strategies and exchange networks existed simultaneously and operated independently of each other, resulting in a complex process by which obsidian as a material in various stages of production entered into the site.

Whether there were distinct and autonomous trade routes or not, some significant social event or chain of events occurred during the Middle period that caused, or was affected by, a change in the mechanisms of procurement and distribution from the major obsidian sources in Cappadocia. These changes are reflected in the relative quantities of obsidian procured from the

Late period pressure blade technology

One of the primary chipped stone characteristics of the Late period is the specialised unipolar pressure blade(let) technique that dominates the production strategies (see Carter, Milić 2013a: 430 for illustrated examples of these conical and bullet type cores). In this chapter, anything under 12mm wide is classified as a 'bladelet' (for brevity the term 'microblade' to characterise those under 8mm is not used here), and anything wider than 12mm is called a 'blade' (Pelegrin 2012: 466). The pressure blade technique in general has become a major interpretive tool for chipped stone research and is significant in the context of Çatalhöyük for two interrelated reasons. The first is that there are a number of changes in production strategy and intensity, and thus the second is a potential change in the socio-economic organisation of the community. Pressure blades and bladelets existed in every level of occupation, but starting in the Late period levels (South P/North H), their production was intensified, streamlined and prioritised. They gained a monopoly over the production industries that supplied blanks for all activities around the site and could be modified into a number of different shapes for different functions. In the Early and Middle periods, the blade component of the assemblage contained fewer than 40% pressure products, but in the Late period, the diversity of tool blanks and production methods dropped significantly, while the proportion of blades produced via pressure reached a crescendo at 93% (fig. 14.5a).

Perhaps even more striking is the change in the blade component as a percentage of the overall assemblage. In the Early and Middle periods, blade(let)s made via any of the production techniques represented only 12% of the total assemblage, while in the Late period this number jumped to 67%, with pressure blade(let)s comprising the vast majority (fig. 14.5b). This is significant for several reasons, not least because it resulted in a massive decrease in the amount of waste produced during chipped stone production and a corresponding increase in the efficiency of labour output. These numbers are in compliance with a drastic change in the organisation of blade production between the Middle and Late levels.

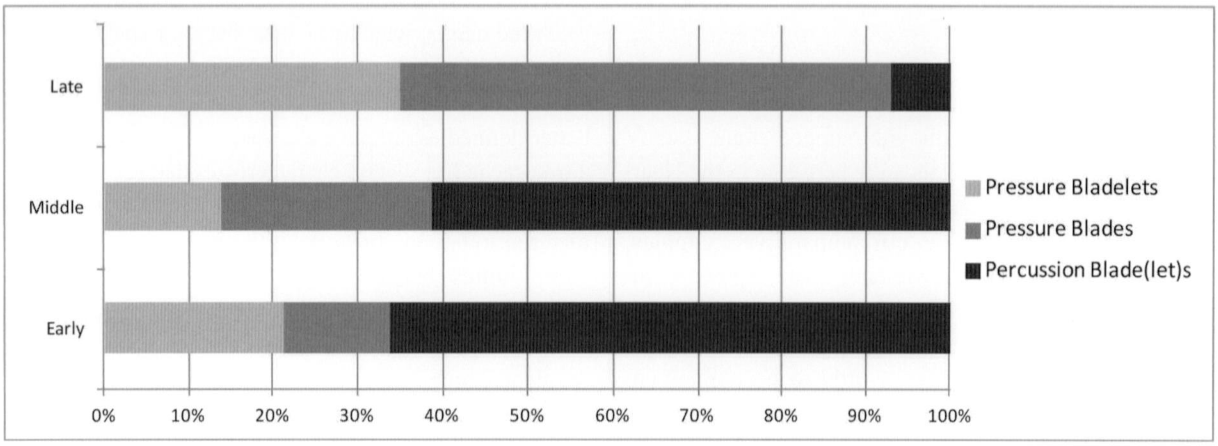

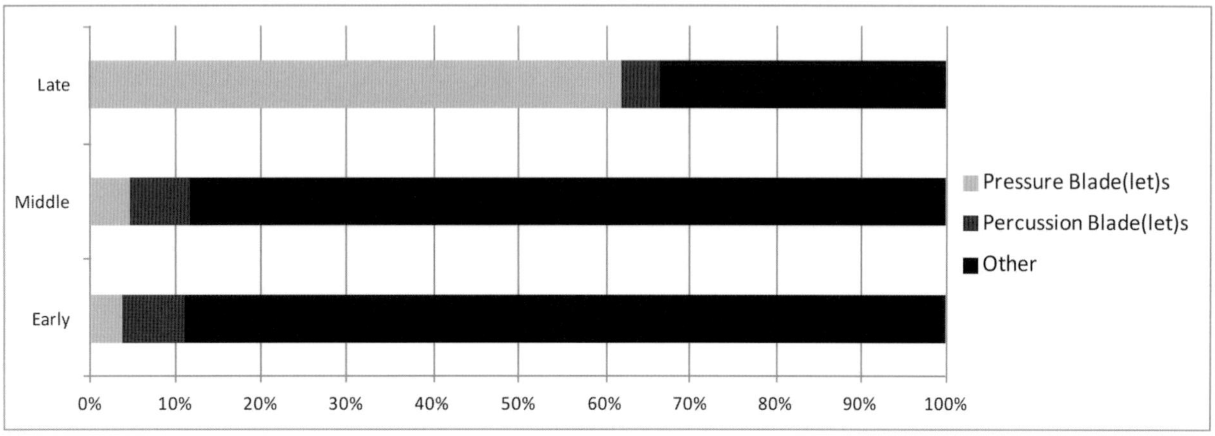

Figure 14.5. (a) Blade assemblage categorised by production technique and period; (b) blades as percentage of overall assemblage, categorised by production technique and period (building on Carter, Milić 2013a: 436, fig. 21.13).

Inefficient methods of production such as bifacial reduction and opportunistic type cores were replaced with a highly efficient use of raw material in the form of unipolar conical pressure blade cores. Cores in the Late period likely entered the site expertly preformed, with several initiation blades already removed, as we still do not see many crested blades in the assemblage, leaving the experienced knappers on site a perfect blank canvas to work with. Blades were removed sequentially and the core did not require much maintenance during the process, as indicated by the scarcity of platform or face rejuvenation flakes. For these reasons, we find much lower quantities of unused waste material, and we see the monopolisation of the pressure technique for the production of versatile, easily interchangeable blades that were used for a number of functions by everyone in the community.

Thematic depositional discussions

The following sections expand on a number of categories and themes that have appeared in the data. They are organised by either depositional context or artefact type, and their spatio-temporal specificities are discussed when possible.

Caches

The obsidian caches have become one of the most iconic aspects of the Çatalhöyük chipped stone assemblage, with no close parallels anywhere else in the Near Eastern Neolithic (Conolly 2003: 55), a fact that remains true to this day, as no comparative examples have yet been announced. Mellaart's site reports from the 1960s excavations mentioned several groups of large blades and bifaces found together, sometimes referred to as caches, but with little to no contextual information attached to them (Conolly 2003: 58). This makes it difficult to compare those original caches with the ones found in the last 25 years. Nevertheless, for some time we have been aware that there are several major characteristic patterns associated with the caching phenomenon. Caches were not present in every building, they generally contained relatively unworked large flakes and rough bifaces, and there was some depositional spatial patterning evident in the majority of them, particularly that they were placed in the southern part of a building in the 'domestic' areas that typically contained the ovens, hearths and food preparation areas (Conolly 2003: 59). Only two buildings produced caches during this final study period, B.97 from the South Area and B.131 from the North, but each of them contained two separate examples, each with different characteristics (see fig. 14.6 for the B.131 example).

After 25 years of excavations and a growing number of caches discovered, some variability in their form, size and context has come to light, and thus I made the decision to slightly modify the definition of an obsidian cache at Çatalhöyük and to quantify how many of these 'true' caches have been unearthed. Table 14.5 displays a list of caches that have been excavated by the Çatalhöyük Research Project since 1993, a grand total of 24. However, several caveats are required. Empty scoops or potentially retrieved caches are no longer used in this schema as they were in the past (Carter, Milić 2013a: 419, table 21.1). It appears that scoops had a number of different purposes, and many of them were eventually filled with at least some knapping debris. Several of these had previously been misinterpreted as the remnants of a cache that had been retrieved some time after initial deposition and worked on the spot, with the debris put back in the pit and covered up again. This should not be eliminated as a possibility in some cases, but because intramural knapping activities were seemingly commonplace events, it is impossible to differentiate between the original form of the worked raw material and whether it represents a retrieved cache. It seems likely that knapping often occurred inside the buildings, probably on reed mats or something else to catch the debris, which was then swept or shaken into an open pit in the domestic or 'dirty' area. Some of the concentrations in these scoops or larger pits would be better defined as simple waste disposal deposits, whereas a true cache has several characteristics that set it apart.

In a rough sense, these caches contained units of raw material, relatively unworked, that were highly valuable commodities because they had the potential to be converted into many different things, be it an opportunistic flake core or a well-made thin bifacially reduced retouched point destined for burial with an important person. Generally, these caches contained large pieces related to the biface industry, mostly in the rough preform stage of production, but sometimes they contained raw quarry blanks or retouched points on blades, with two even containing a large flint knife. In order to make the list of caches, a deposit had to comprise at least two artefacts in direct association with each other in a sub-floor pit that was later plastered over or otherwise intentionally sealed (although two examples exist with just one artefact, they did meet all the other requirements). Another criterion required the average weight of artefacts within a cache to be at least 10g (though one example exists where 11 artefacts had an average weight of 7.5g, but they were all still useable tools that defied any other classification). The final, and perhaps most important, criterion is that a cache must contain some form of raw material whose potential value was often much higher than its value in its cached state.

Figure 14.6. Obsidian cache (23000) and (23034) from B.131, containing 23 large percussion blades with extensive bifacial modification (photograph by Jason Quinlan).

In a community where small cores were regularly milked to their last bladelet and then often reused as a wedge until they were smashed to bits, it is quite odd that large amounts of raw material that could have been turned into any number of useful things were instead buried under floors, seemingly never to be seen again. This is the allure of this caching phenomenon and why it remains so enigmatic.

Not only do the majority of caches represent different technological forms than the site's more utilitarian tools, they also represent an entirely different *chaîne opératoire* from beginning to end and a different role in the complex social, economic and symbolic world of Çatalhöyük. They are entirely unique entities, with their own agency and purpose. It is difficult to imagine exactly what that purpose was, but it seems unlikely that these caches were ever meant to be retrieved at a later date, or that they even represent projectile preforms that were ever meant to be functional at all, as it had seemed in the past (Carter, Milić 2013a: 441, 445). In fact, it appears that the majority of retouched points with traces of use were made on bidirectional percussion blades, while true bifaces rarely exhibited any traces of use (see Chapter 15) and were largely connected to the symbolic or ritualised behaviours inherent in their deposition as caches or in burials (cf. Dogiama 2017; in press). These caches were almost always contained within the construction or first occupation phase of a building, always in the southern or south-central area of a space and sometimes sealed up by hearths and ovens that were constructed at a later date. Furthermore, we do not have any concrete evidence of a cache that was retrieved. Thus my final interpretation asserts they were deposited as part of a symbolic foundation ritual, or perhaps a social 'housewarming party', during the earliest levels of a building's occupation and intended to rest there forever.

	Building	Space	Phase	Hodder Level	Mellaart Level	Cache	Quantity	Avg. weight (g)	Source	Location
Late	60	278	60.2.a	North.H	IV–V	13111	5	40.1	1ND, 4GD	Southeast
Middle	97	365	97.C.9	South.O	VIA	20337	8	114	8GD	Centre
	97	365	97.C.9	South.O	VIA	20369	13	45.6	13ND	South Centre
	1	71	1.1.B	North.G	VIA–B	1451	1	23.1	1ND	Southeast
	1	71	1.2.B	North.G	VIA–B	1461	12	18	12ND	Southeast
	52	255/290	52.E(iii)	North.G	VIA–B	10342	7	23.9	6ND, 1Flint	Southeast
	131	504	131.1.D	North.G	VIA–B	23034	24	65.6	24ND	Southwest
	131	500	131.1.D	North.G	VIA–B	32536	2	69.7	2GD	Centre
		113	113.2	South.M	VII	1836	11	7.5	8ND, 3GD	South Centre
		113	113.2	South.M	VII	2038	34	45.8	n/a	South Centre
		113	113.2	South.M	VII	2039	15	49.8	n/a	South Centre
		109	109.2	South.M	VII	2810	2	n/a	n/a	Centre
		109	109.2	South.M	VII	2812	7	n/a	n/a	Centre
	50	231	n/a	South.M	VII	10819	4	33.1	4GD	Northwest
	92	208	n/a	South.M	VII	5665	77	116.1	77GD	Southeast
Early	6	163	6.1	South.L	VIII	4276	2	94.6	2GD	Southeast
	6	163	6.2	South.L	VIII	4280	18	70.6	1ND, 14GD, 3ACG	Southeast
	2	117	2.2	South.K	IX	4138	11	13.9	1ND, 9GD, 1Flint	n/a
	2	117	n/a	South.K	IX	4209	39	66.4	12ND, 27GD	n/a
	2	117	n/a	South.K	IX	4210	6	18	2ND, 4GD	n/a
	16	164	16.2	South.K	IX	4317	3	32.2	3GD	Southeast
	9	167	n/a	South.J	X	4205	1	79	1ND	n/a
	23	178	23.2	South.J	X	4989	4	17.3	n/a	n/a
	23	178	23.2	South.J	X	5096	3	20.3	n/a	n/a

Table 14.5. All caches listed by temporal group (expanding on Carter et al. 2013).

They became part of the fabric of that building, both physically and symbolically, and remained entangled with those living inside or using that structure, just as the ancestors buried beneath the floors had done.

Retouched points

A lengthy discussion of retouched points is not possible in this chapter, nor is it necessary, as several authors have already covered the topic extensively (Conolly 1999; Carter, Milić 2013a; 2013b; Dogiama 2017; in press). Some of this discussion began with Mellaart's original claims pertaining to hunting iconography, wherein the hunt was integrally and actively involved in the religion and daily lives of the community. Dogiama explored this topic and argued that hunting was more than just a part of the subsistence strategy and that it served a higher purpose, one in which symbolic expression and social identities could be performed and negotiated (2017).

With a straight, narrow face from which to consistently remove regular blades, bidirectional blade core technologies like Naviform had a distinct advantage over the unipolar or multidirectional cores that the rest of the Çatalhöyük sharp-edge technology was made from, at least when it came to making standardised, strong, straight blade blanks for retouched points (fig. 14.7). This technology was especially important to the community throughout the occupation and is emblematic of the hunting weapons that were revered by its inhabitants, as represented in the hunting scene wall paintings depicting hunters with their weapons facing off against a large dangerous animal, and the ritualised deposits that often included a retouched point on one of these bidirectional blades. This type of blade technology, however, was not limited to Anatolia. Similarities in hunting symbolism and chipped stone production techniques suggest a close relationship to communities in the Levant that strengthened in the Late period. This occurred in contrast to the earliest levels, where there were higher proportions of flint and a multitude of chipped stone production methods, which is best explained by people and or resources coming into Çatalhöyük from all different directions, particularly the west and north.

While the majority of retouched points were derived from bidirectional blade blanks, a small number of exquisitely manufactured points on true bifaces with further bifacial parallel ribbon pressure flaking have been recovered. Four of these came from one burial fill, (22194) in B.5, that was particularly rich in grave goods (fig. 14.8).

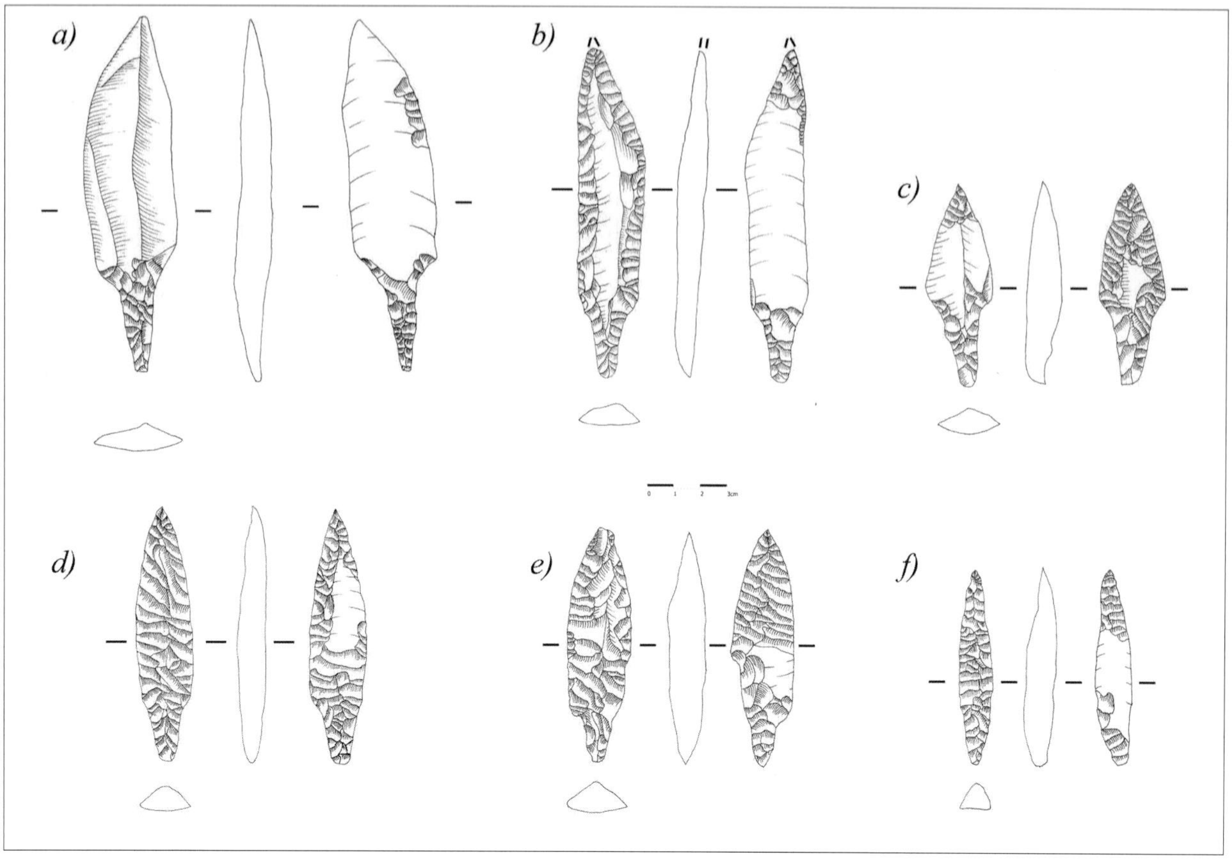

Figure 14.7. Retouched points on bidirectional blades: a) 18982.x1; b) 20447.x1; c) 18578.A1; d) 19498.x1; e) 18578.A2; f) 11648.x12 (illustrations by Danica Mihailović).

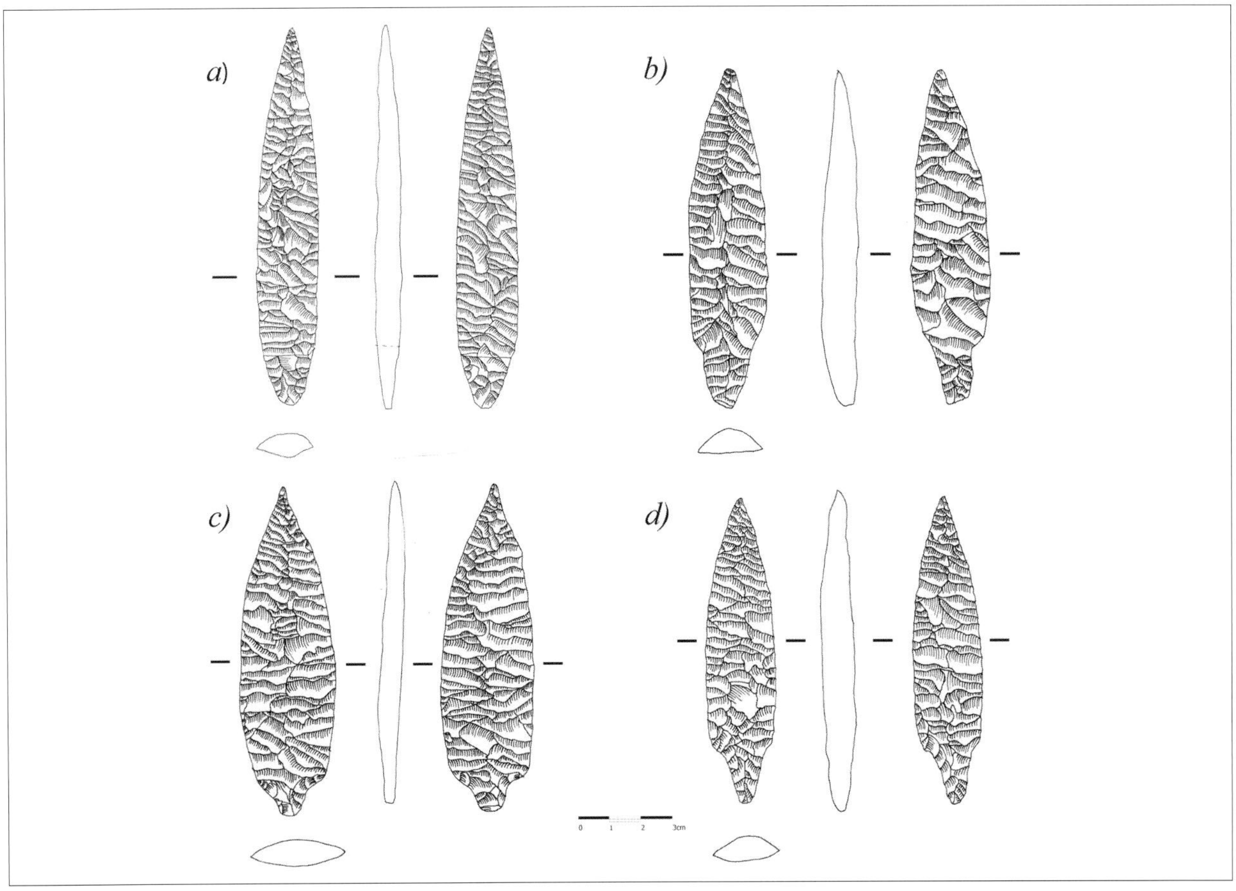

Figure 14.8. Retouched points from B.5 burial fill (22194): (a) 22194.x5; (b) 22194.x2; (c) 22194.x3; (d) 22194.x4 (illustrations by Danica Mihailović).

Within this burial was an unparalleled marble macehead with a spiral pattern carved into it, a painted skull on a wooden platter, a number of pigments and, among other things, two 'daggers' made on large flint blades, one produced via indirect percussion and the other via a lever pressure technique, that were symmetrical and minimally retouched (Doyle 2016: 162). The entire assemblage sets this burial apart, but the four retouched points in particular are something entirely unique across the whole site, at least in this quantity and in a burial context. They still have delicate edges and show no evidence of any use, as if they were knapped and immediately placed within the burial. To further cement their special status, they were first carefully wrapped in a bundle or mat made of *Phragmites* reed (see Volume 12, Chapter 20).

Can Hasan III inscribed points

Another peculiar aspect of the assemblage are the five unifacially retouched points with symbols or shapes carved into their unmodified ventral faces, akin to the inscribed points of Can Hasan III (Ataman 1986). While quite similar in many aspects, the Çatalhöyük examples are not as intricately carved as the Can Hasan III points and come from different levels and areas of the site, some of which are much later than the occupation of their type-site. The first two examples came from Level South G middens (Carter et al. 2005a: 277), a third from a Level North H midden and a fourth from Trench 7 on the West Mound (figs 14.9b, 14.9c; Carter, Milić 2013a: 454). A fifth example with a clear 'X' inscribed into the ventral face was recovered in this last study period from floor surface (32492) in B.160, Level South K (fig. 14.9a; Doyle 2016: 165, fig. 5). All five examples were made from Göllü Dağ obsidian on bidirectional percussion blades that had extensive dorsal retouch but minimal modification to the ventral face. All were broken to some degree and, with the exception of the most recent, were deposited without fanfare into middens with the rest of the trash.

It has been argued that the recovery of two of these points in Late Neolithic contexts suggests they had been 'carefully curated heirlooms, storied weapons that had been passed down through the generations' (Carter, Milić 2013a: 454). However, with over 1,000 years separating the earliest and latest examples and a lack of wear associated with being handled for long periods of time, that theory does not seem very likely. While it is tempting to make the obvious comparison with Can

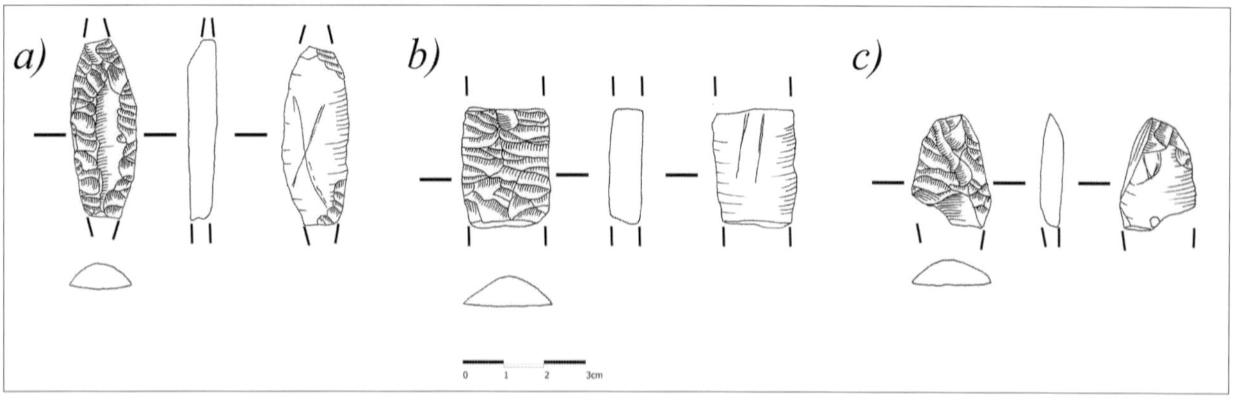

Figure 14.9. Three of the five Can Hasan III Inscribed Points from Çatalhöyük: (a) 32492.x1; (b) 14931.A1; (c) 15101.A28 (illustrations by Danica Mihailović). The other two are illustrated in Carter, Milić 2013a: 452, fig. 21.24.

Hasan III due to the many similarities, especially as the earliest three from Çatalhöyük were roughly contemporary with the Can Hasan III examples, it becomes much more difficult to do so with the later two. In reality, they probably do not have any direct correlation with Can Hasan III, but rather were coincidental markings that somehow survived a number of retouch episodes and the eventual snapping of both tips. One might wonder how many similar examples existed but were lost to these processes. It does seem likely, however, that these inscriptions were some sort of symbol. Whether it represented a maker's or hunter's personal mark, or a more socially charged mnemonic device used to convey personal, lineage and community histories (Carter, Milić 2013a: 454) remains difficult to clarify with so few examples. At the very least they present yet another form in which the inhabitants of Çatalhöyük used chipped stone materials and tools to express themselves as individuals.

Obsidian mirrors

The obsidian mirrors of Çatalhöyük are some of the most emblematic and enigmatic objects from the site, and from the Neolithic period in general, as they remain the earliest mirrors found anywhere in the world (Mellaart 1963: pl. XXVb; 1967: pl. XII; Todd 1976: 85; Vedder 2005: 597). A total of eight complete examples and four possible fragments of obsidian mirrors were found during Mellaart's excavations in the 1960s (Vedder 2005: 597), while another three have been recovered during the 25 years of the Çatalhöyük Research Project (Carter 2012: 195–96; Doyle 2017: 199). The three most recent mirrors were all unearthed during this final excavation period, two in 2012 and one in 2017, and in a remarkable example of memory construction and retention, they were found in two separate burial fills from two buildings connected in a vertical sequence, in the same horizontal locations within those buildings. From opposite ends of the fill associated with burials F.3630 and F.3684 in B.129 came two obsidian mirrors (19447.x3 and 19447.x4; figs 14.10b, 14.10c), both produced on Nenezi Dağ obsidian, subconical in shape, unipolar percussion flake cores that had their platforms ground and polished completely smooth (Carter 2012: 195). Both reflective surfaces exhibit scratches that look deliberate, as if the mirrors had been intentionally 'killed' or rendered inert before their deposition in the burial.

In contrast, the mirror below in B.131 (30039.x2; fig. 14.10a and 14.11) was associated with burial F.7961 of an adolescent female skeleton (23126), which also contained ochre, beads, pigments, a wooden bowl and a large pottery fragment placed underneath the cranium, practices that were remarkably similar to those found with the two mirrors in the building above (Barański et al. 2012: 44). Some differences in the mirrors themselves are notable, with (30039.x2) standing out as a unique specimen, a large blade core platform rejuvenation tablet that had been ground and polished smooth on the dorsal face of the platform. A thick layer of white plaster was attached to the ventral face that must have acted as a handle or grip to make the mirror easier to hold and to limit the risk of cutting one's fingers on the still sharp edges of the ventral face where the tablet was originally removed from the core. The dorsal surface was still remarkably free of scratches or any blemishes, unlike those of the B.129 mirrors, and still reflects a clear image after almost nine millennia in the ground. Ultimately the similarities, yet at the same time stark and surely intentional differences, between the objects and their depositional contexts exhibit rare characteristics of social memory maintenance. This speaks to the common theme of continuity through change that binds together so many patterns in the chipped stone data. It is noteworthy that the B.131 mirror came from

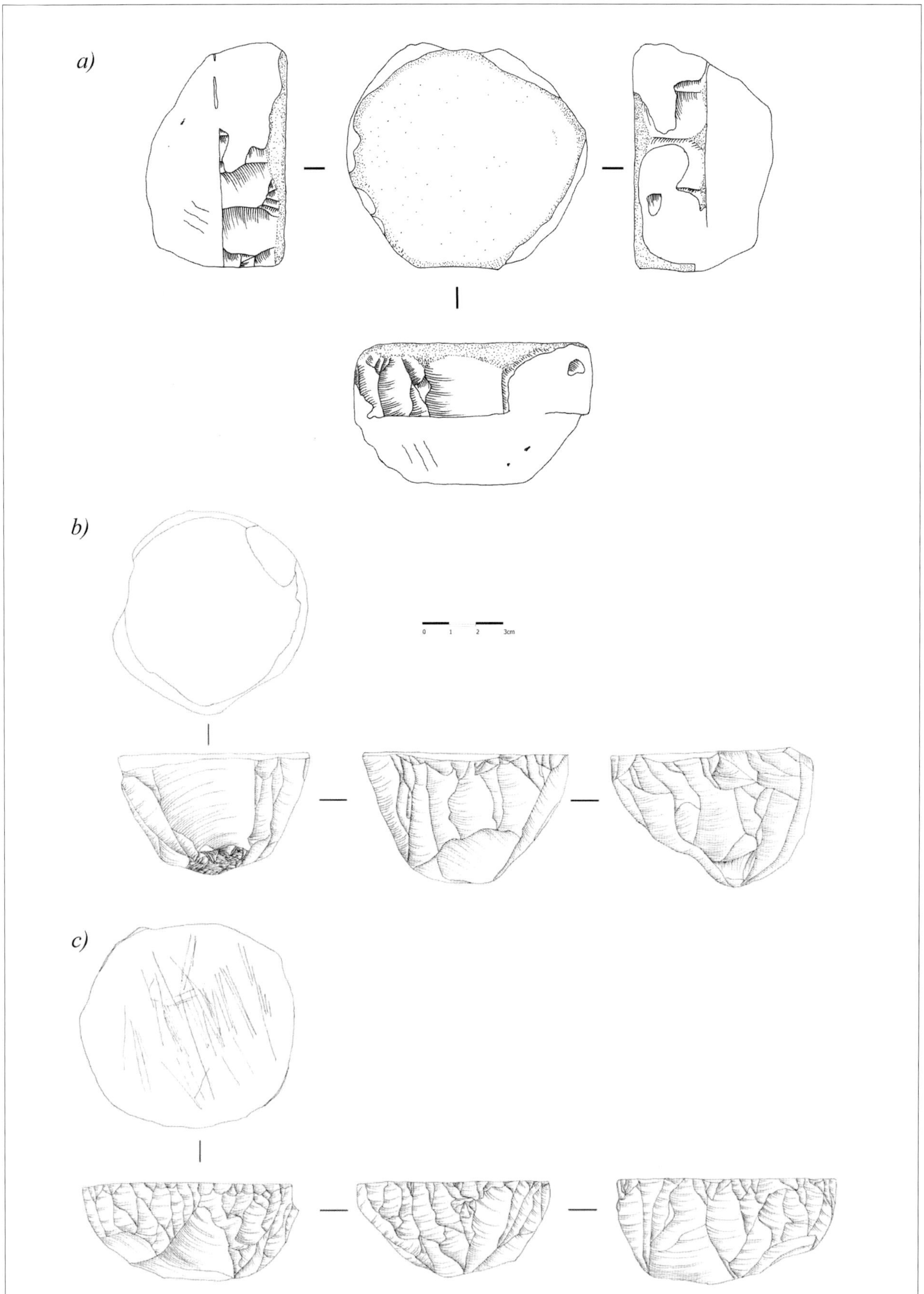

Figure 14.10. The Çatalhöyük Research Project's obsidian mirrors: (a) 30039.x2; (b) 19447.x3; (c) 19447.x4 (illustrations by Danica Mihailović).

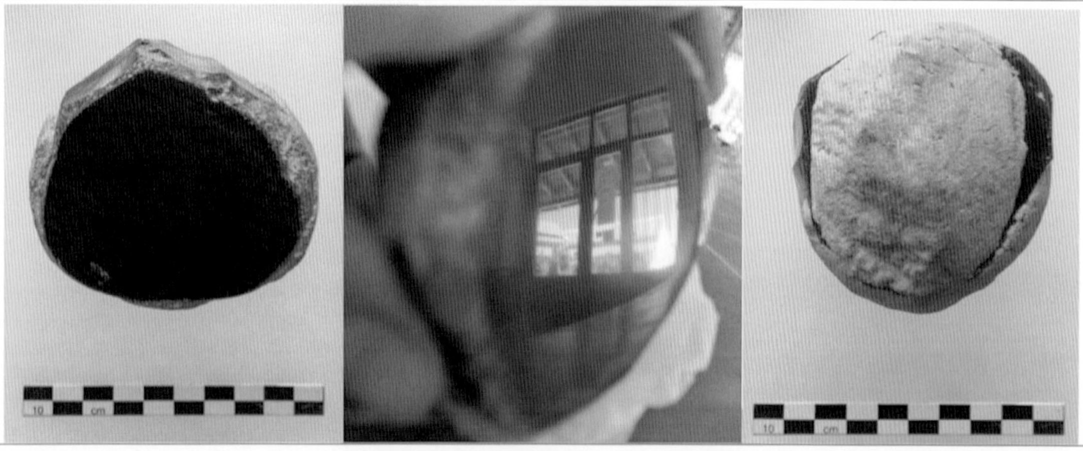

Figure 14.11. Obsidian mirror from B.131 burial fill (30039.x2). Left: dorsal view; right: ventral view (photographs by Jason Quinlan); middle: front door of lithics lab clearly shown in reflection off mirror's surface (photograph by Sean Doyle).

Level North G, the last level of the Middle period in the North Area, while the two mirrors that came from B.129 belonged to North H, the first level of the Late period. This transition was a turbulent period for the settlement, during which time a number of buildings were seemingly intentionally burned down and after which a number of drastic changes were surely caused by social and/or economic factors and manifested within the material culture of its inhabitants (see Chapter 1 for an overview of these changes).

Discussion

Connections between Central Anatolia and the Levant during the Holocene period have been defined by a number of authors (e.g., Balkan-Atlı et al. 1999; Bellot-Gurlet et al. 1999; Binder, Balkan-Atlı 2001; Cauvin et al. 1998, *inter alia*), and with the use of obsidian distribution patterns it has also been noted that these links might have been strongest in the first stages of the Neolithic (Binder 2002: 79). In the context of Çatalhöyük, these claims have been tested more rigorously over the previous 25 years of study and a great deal has been added to the conversation.

Perhaps the most significant trend evident in the data and the themes discussed above is the number of large-scale, site-wide changes that occurred in relatively rapid succession, effectively separating the Middle and Late periods. These changes manifested in the chipped stone in a number of ways: 1) the major source of obsidian switched from Göllü Dağ to Nenezi Dağ; 2) production of pressure blades became more refined and standardised, which suggests early craft specialisation; 3) the practice of caching obsidian disappeared; 4) obsidian appeared from Eastern Anatolian sources over 650km away (Carter et al. 2008); and 5) there was an apparent decline in the frequency of symbolic deposition and a more utilitarian, functional relationship with chipped stone, though retouched points did remain a valued commodity. These changes were a major focus of Carter and Milić in the chipped stone chapter in the last volume (2013), where much work was done to quantify and interrogate them.

Some questions still plague this researcher, which only a deliberate approach toward them will answer. The first is just how much blade production via core reduction was actually done on site. Previous specialists have claimed it was minimal, especially in the production of desired bidirectional blanks that were primarily used in the manufacture of retouched points, but some evidence put forth in this chapter suggests that a lot more happened within the confines of Çatalhöyük than was formerly believed. Future analyses of the assemblage will highlight the core shaping, preparation and rejuvenation flakes that are so characteristic of the relevant operational sequences, in order to determine how self-sufficient Çatalhöyük's Neolithic inhabitants actually were. This type of new information can then be used to reassess the nature of trade and exchange, the chronological trajectory of craft specialisation and the socio-economic status of Çatalhöyük in the wider region.

Conclusions

Contrary to the beliefs of Bialor and Mellaart, the chipped stone of Çatalhöyük was a dynamic and fluid system that underwent many changes through at least 1,150 years of occupation on the East Mound. Now that these big picture trends and patterns have been identified and discussed at length, the focus of future research will turn inwards to quantify on spatial and temporal scales the traces of production within the settlement. There is much evidence to suggest that in the Early and Middle periods the settlement imported most of its formal tools and blanks from elsewhere, after they were likely made at workshops located at or near the sources, and that in the Late and Final periods, the majority of equipment was produced on site. The increasing standardisation and regularity of the pressure blades produced on site hints at the existence of craft specialisation and towards a reorganisation of the social systems that controlled the procurement of raw material and the production of chipped stone tools at Çatalhöyük. This is not to presume that some form of craft specialisation began at this site, in fact, there is much data from earlier Anatolian sites to prove otherwise, but there does seem to be a shift in standardised production away from the quarry workshops and into this large settlement. A fine-grained analytical investigation and quantification of this change in production will lend insights to the wider discussions of socio-economic considerations and political organisation in the Late Pottery Neolithic.

Acknowledgements

First, I would like to give the biggest thank you to my wife, Duygu Ertemin, who I met at Çatalhöyük and who motivated me to complete this chapter in a way that I could be proud of. Thank you to Tristan Carter for first bringing me to the site and to Ian Hodder for letting me return for so many years. Thank you to Christina Tsoraki, who made the lithics lab a fun and exciting place to work, as long as we had enough table space! To Lisa Guerre, who held everything together, you helped me out so much. A big thank you to everyone else I worked and lived with for the last six field seasons and to all those who came before. It was, and remains, a massive team effort that continues to benefit from each and every one of us who passes through. Throw some water on that taxi, until we meet again.

15. Chipped stone use-wear

Cristina Lemorini and Davide D'Errico

Introduction

This chapter presents and discusses the results of the use-wear analysis of the chipped stone tools from Çatalhöyük. During our work at the site, we did not have time to analyse the chipped stone tools coming from the entire excavation of the East Mound. We decided to fully analyse some well-preserved buildings that could exemplify living units through the Neolithic sequence. We applied this sampling program to Buildings 65, 56, 44 and related middens of the South Area, whose excavation preceded 2009 (and is not reported on here), and Buildings 77, 131, 132 and related middens of the North Area (Spaces 439, 610, 631). We also analysed two deposits (the infill corresponding to Space 623 and the midden corresponding to (23143)) that preceded the construction of Building 131. Regarding the oldest building of the sequence, Building 139, we analysed only the chipped stone tools that came from its infillings (Sp.623). Thus, these chipped stone tools testify to activities not directly connected with B.139.

Before presenting our data, we want to underline that the activities inferred from the use-wear analysis may give a general picture of (a) the functions carried out with the chipped stone tools, (b) the possible links between the morphology of the chipped stone tools and their use, and (c) the possible strategies of curation and re-use of the chipped stone tools.

The reconstruction of activities carried out in the living spaces is arduous. The living spaces of the buildings of Çatalhöyük were systematically cleaned; thus, we cannot expect to find many chipped stone tools inside the rooms. However, it is possible to find chipped stone tools in situ when they are recovered in association with burials or foundation deposits in the buildings, or in caches or storage deposits. Clearly, the chipped stone tools deposited as goods in foundation deposits or burials or in caches do not have direct connections with the activities carried out in the buildings and around them. Burials, caches and other deposits pertaining to the building during its utilisation may, however, give tangential information about the daily practices of the inhabitants. Middens, as well, give a general picture of the possible activities linked to a household. In fact, in their deposits we can expect the rubbish cleaned off from the working spaces inside and outside the related buildings. Moreover, if distinct stratigraphic units that pertain to the middens contain chipped stone tools that show the processing of the same materials, as an example stone sawing and abrading, it is possible to suggest that distinct sections of stone working were carried out by the inhabitants of a nearby household. Different reasoning must be applied to the same evidence found in the infilling of buildings. At Çatalhöyük, it was common to reconstruct new buildings on top of the remains of the abandoned ones. The abandoned areas were usually infilled and levelled to prepare the new space on which to rebuild. Thus, the chipped stone tools found in these infillings are possibly related to activities slightly more recent than the abandoned buildings and slightly older than the new buildings. In other cases, the infillings contain clusters consisting of various objects, some of them clearly no longer able to function as tools, and of organic or mineral remains. These clusters are interpreted as deliberate accumulations, representing common ritual practices connected with the abandonment of the household (Hodder, Cessford 2004: 17–40; Tsoraki 2018: 238–62).

In accordance with this general explanation, after a description of the protocol adopted for the analysis, we will start with a general description of the activities inferred for each building and midden and, thus, we will present possible significant functional associations observed in specific stratigraphic units. We present middens and buildings following the stratigraphic sequence and starting from the oldest to the most recent.

Methods

The use-wear analysis was carried out with two types of optical light microscopes (OLM): a stereomicroscope with a reflecting light source to observe macro-traces (edge-removals, edge-rounding, striations, abrasions) at magnifications up to 63X and a metallographic microscope with a reflecting light source to observe micro-traces (micro edge-rounding, polishes on surface, striations, abrasions, polishes on abrasion) at magnifications of 100X, 200X and 400X. The observation of macro-traces with a stereomicroscope is known as the Low-Power approach; the observation of micro-traces is known as the High-Power approach (Lemorini 2000: 7–8 and references therein). Our protocol of analysis consists of an integrated approach, Low Power plus High Power, to observe, define and interpret the use-wear identified on the chipped stone tools.

At Çatalhöyük, we used the Low-Power approach to make a first survey of the entire assemblage of each building or midden. We discarded forthwith the by-products of the knapping activities. The survey of the remaining flakes, blades, bladelets and retouched tools allowed us to identify for each tool possible alterations of their surface (for example, the presence of edge-damages, of thermal stress, of patination; Hurcombe, 1992: 71–78; Lemorini, 2000: 35–43). Thus, from the entire assemblage, we selected all the chipped stone tools in a good state of preservation that suggested the possible retention of use-wear.

With the Low-Power approach we detected, as well, the presence of macro-traces of use that allowed us to reach a first general evaluation of the activity carried out. Then we refined this evaluation using the High-Power approach. The observation of micro-traces makes it possible, in fact, to infer in better detail the actions carried out and the materials worked with the chipped stone tools.

This protocol was carried out at Çatalhöyük with a stereomicroscope Nikon SMZ-T with oculars 10X, objective 1X and a magnification range of 0.63x to 6.3X, and with a metallographic microscope Nikon M-T with oculars 10X and objectives 10X and 20X. A ToupView CMOS® Camera was applied to the microscopes for the documentation of the macro and micro use-wear.

We did not have time to complete the interpretation of the micro-traces at Çatalhöyük. For this reason, we decided to mould the areas showing micro-traces with a high-definition two-component silicon, Provil Novo Light Fast and Provil Novo Light Medium Heraeus®. The moulds were subsequently analysed and documented at the LTFAPA laboratory of the Department of Classics of Sapienza University. In the LTFAPA laboratory we used a metallographic microscope Nikon Optiphot with oculars 15X and objectives 10X, 20X and 40X and a Nikon digital camera DMX 1200 and TV Lens C-0.6X connection, or a metallographic microscope Nikon Eclipse with oculars 10X and objectives 10X and 20X and an Olympus SC100 digital camera. Helicon Focus software was used for picture stacking in order to obtain images that were entirely in focus.

At Çatalhöyük, before the metallographic observations, we cleaned each item, first in a bath of tap water and soap and then by gentle rinsing with tap water. Before the moulding session, we rinsed again to clean each item. The observations under the metallographic microscope and the moulding procedure were carried out with un-powdered gloves to prevent the deposition of residues that might compromise the observation of the micro-traces.

A reference collection of replicas of chipped stone tools made of obsidian was built and stored in the LTFAPA laboratory to allow the authors of this chapter to improve their knowledge on the use-wear formation and morphology that develop on this type of raw material. The replicas were used on various materials, with a variety of motions. As a reference collection for the chipped stone tools made of chert, we used the collection that already existed at the LTFAPA laboratory

We show in figures 15.1 and 15.2 use-wear developed on replicas in our reference collection. In the captions we briefly describe the use-wear, following the terminology proposed by Hurcombe (1992: 128, table 5) for micro use-wear. We, however, use a different terminology for Hurcombe's variables 'relief on edge', which we define as 'edge-rounding', and 'attrition', which we define as 'abrasion'. We added a description of the topography of the polish (Lemorini 2000: 133) that is not present in Hurcombe's terminology.

The analysis of the archaeological assemblage

The state of preservation of the assemblage analysed is, in general, very good. Chemical alteration generating a 'corrosion-like' appearance on the surface of the obsidian is very rare. The decay of the organic matter that, we can suppose, was a major source of chemical percolation in a densely occupied site did not produce visible effects on the chipped stone tools. As is suggested in Volume 13, Chapter 14, the organic rubbish in external deposits was often covered with wood ash or clay that, with their hygroscopic properties, could have attenuated the decay effects of the organic matters. The only type of alteration that was rarely observed, especially in external deposits, was mechanical alteration attested by edge damages at the macro-scale and chaotic stations at the micro-scale. These mechanical alterations were possibly caused by trampling, or by erosion and re-deposition in unsealed deposits (see also Hurcombe 1992: 74–75).

In the following paragraphs we will define as 'formal tools' the retouched flakes, blades or bladelets that have a shape well defined by retouches; we will define as the 'active edge' the area of the tool showing use-wear.

Space 623 Unit (12343) (midden), Space 623 Units (23139), (23153), (23159) (infillings) and Space 631

Unit 12343 of Space 623 represents a large midden partly found directly below Building 131 and assigned to Level North F and thus to an earlier phase in the North Area sequence. This deposit consists of many distinct layers of ash, objects and organic remains that correspond to various episodes of common activity and rubbish accumulation.

We selected 16 chipped stone tools, three made of chert and the rest of obsidian, in order to obtain a general idea of the possible activities carried out in an open area.

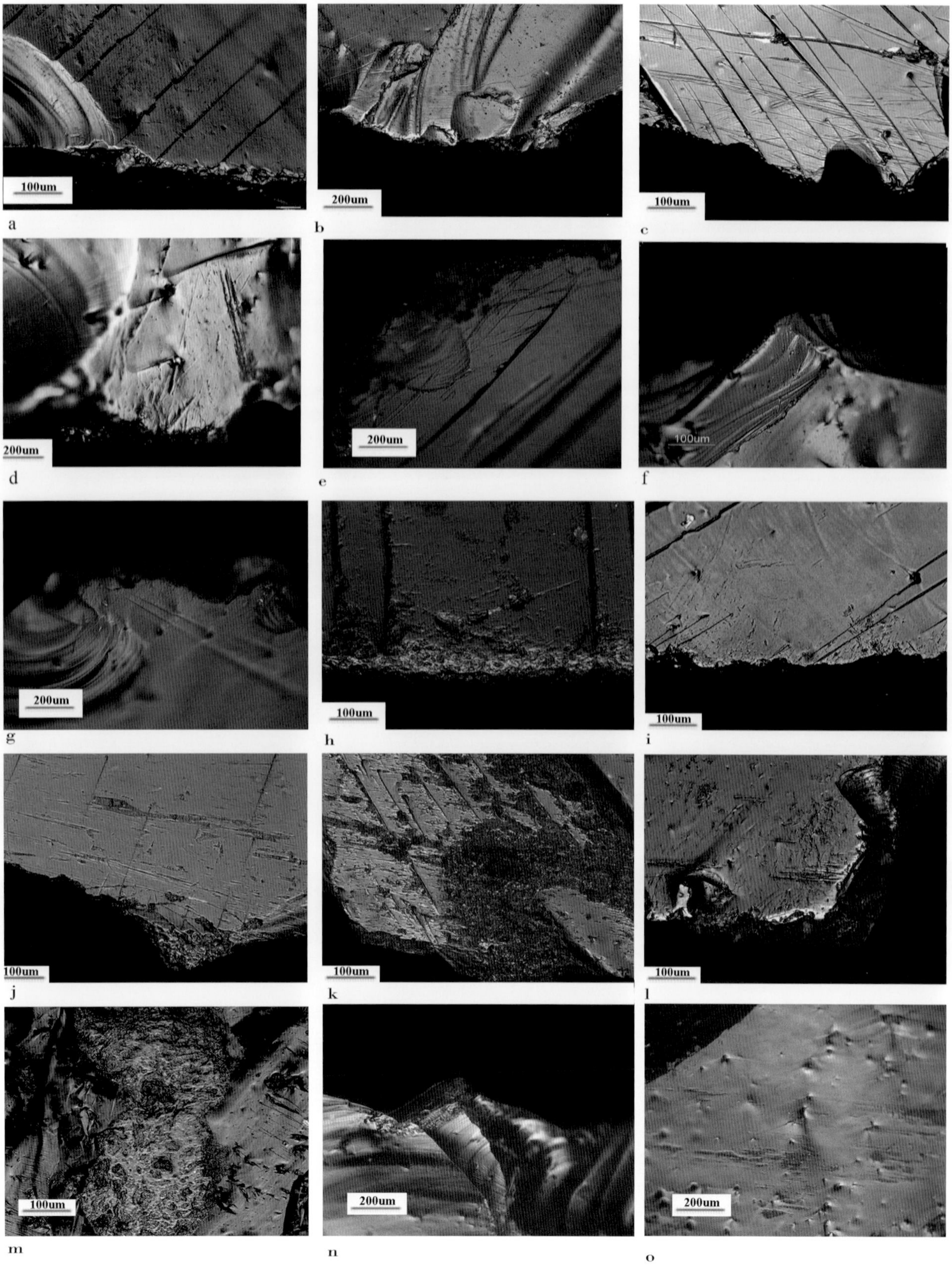

Figure 15.1. Use-wear developed on replicas of obsidian chipped stone tools: (a) butchering (contact with meat and bone); (b) cutting wool; (c) cutting fresh hide; (d) scraping semi-dry hide; (e) cutting tanned hide (vegetable tannins and linseed oil); (f) scraping antler; (g) scraping fresh hide; (h) scraping dry hide; (i) scraping dry bone; (j) carving bone; (k) carving bone; (l) incising clay; (m) carving soft stone; (n–o) technological traces produced during retouching.

Figure 15.2. Use-wear developed on replicas of obsidian chipped stone tools: (a–b) cutting reeds; (c) cutting edible herbaceous plants; (d) scraping bark; (e) cutting dry wood; (f) scraping wood; (g) cutting fresh wood; (h) carving wood; (i) cutting shrubs; (j) whittling wood; (k) harvesting Triticum dicoccum*; (l) harvesting mixed cereals; (m–n) harvesting* Triticum monococcum*; (o) cutting* Ampelodesmos mauritanicus.

Fourteen tools show use-wear on 15 active edges (table 15.1). More than half of these tools (10 out of 16) are retouched blades or flakes, and in various cases, the retouch corresponds to the reshaping of the tool to prolong its use. One tool used for cutting and scraping dry hide was retouched and reused on both edges (fig. 15.3g). One of the edges showed hafting traces that overlap the hide traces and testify to a shift in the function of the exhausted active area.

In general, this whole group of chipped stone tools shows well-developed use-wear that suggests prolonged usage. Many activities were carried out with these items (figs 15.3 and 15.4): butchering, the production of leather objects, as indicated by the cutting of dry hide, harvesting and craft activities related to the working of wood and hard animal materials (five cases). Three of these last items were utilised to scrape (fig. 15.4c) and to carve antler.

Action/Material	*Butchering*	*Hide*	*Cereals*	*Wood*	*Antler*	*Bone*	*Hard material*	*Indeterminable*	*Total*
Impact								1	1
Cutting	1	2	2			1		1	7
Scraping		1			2	1			4
Scraping/carving					1				1
Indeterminable				1			1		2
Total	1	3	2	1	3	2	1	2	15

Table 15.1. Space 623 Unit 12343 (midden) inferred data.

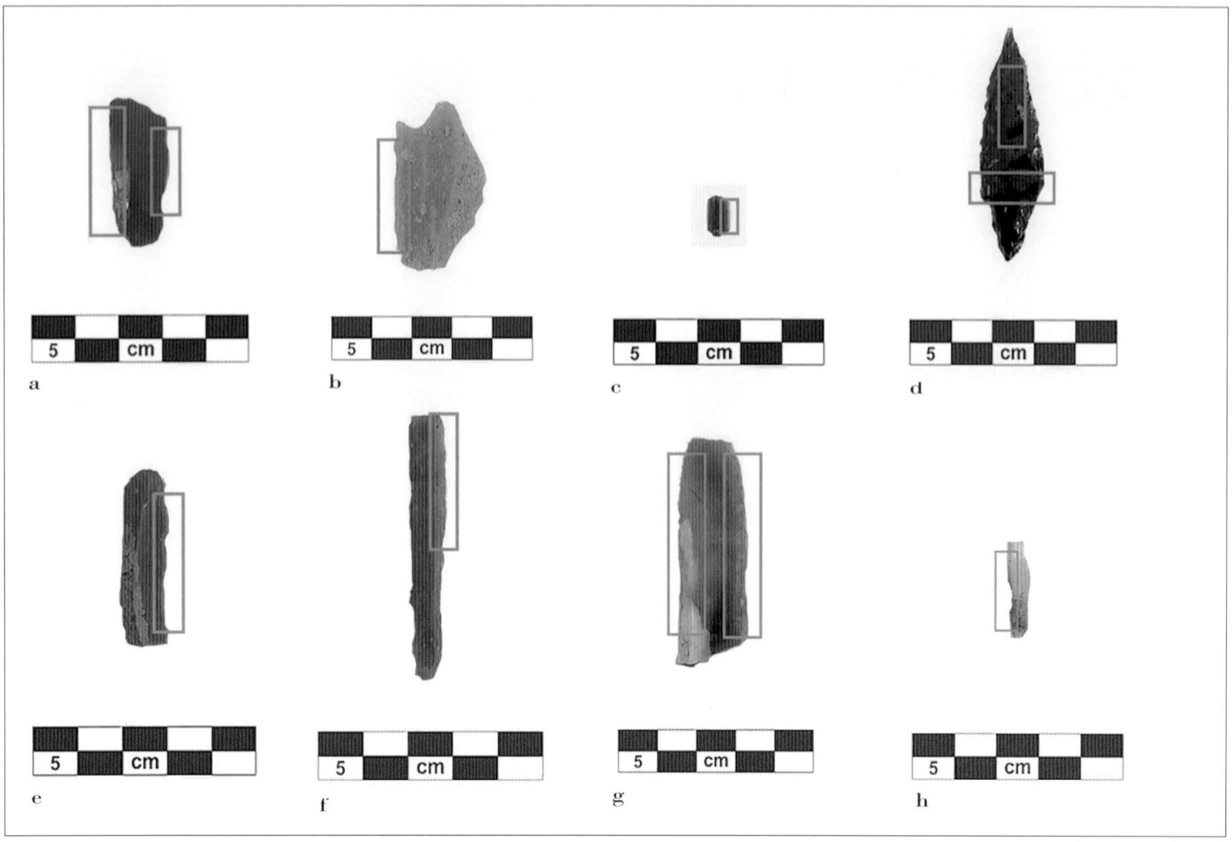

Figure 15.3. Chipped stone tools from middens and infillings with use-wear (rectangle): (a) Space 623 (23153), tool with use-wear interpreted as cutting wool (see fig. 15.4d); (b) Space 623 (23153), tool with use-wear interpreted as mixed activities on various matter; (c) Space 610 (32114), tool with use-wear interpreted as cutting meat; (d) Space 623 (23143), projectile point with use-wear interpreted as impact (see fig.15.4a); (e) Space 623 (23153), tool with use-wear interpreted as harvesting cereals; (f) Space 623 (23143), tool with remains of herbaceous plant cutting removed by re-sharpening and followed by a butchering activity; (g) Space 623 (23143), tool with use-wear interpreted as cutting semi-dry hide; (h) Space 610 (32138), burin spall with use-wear interpreted as cutting reeds.

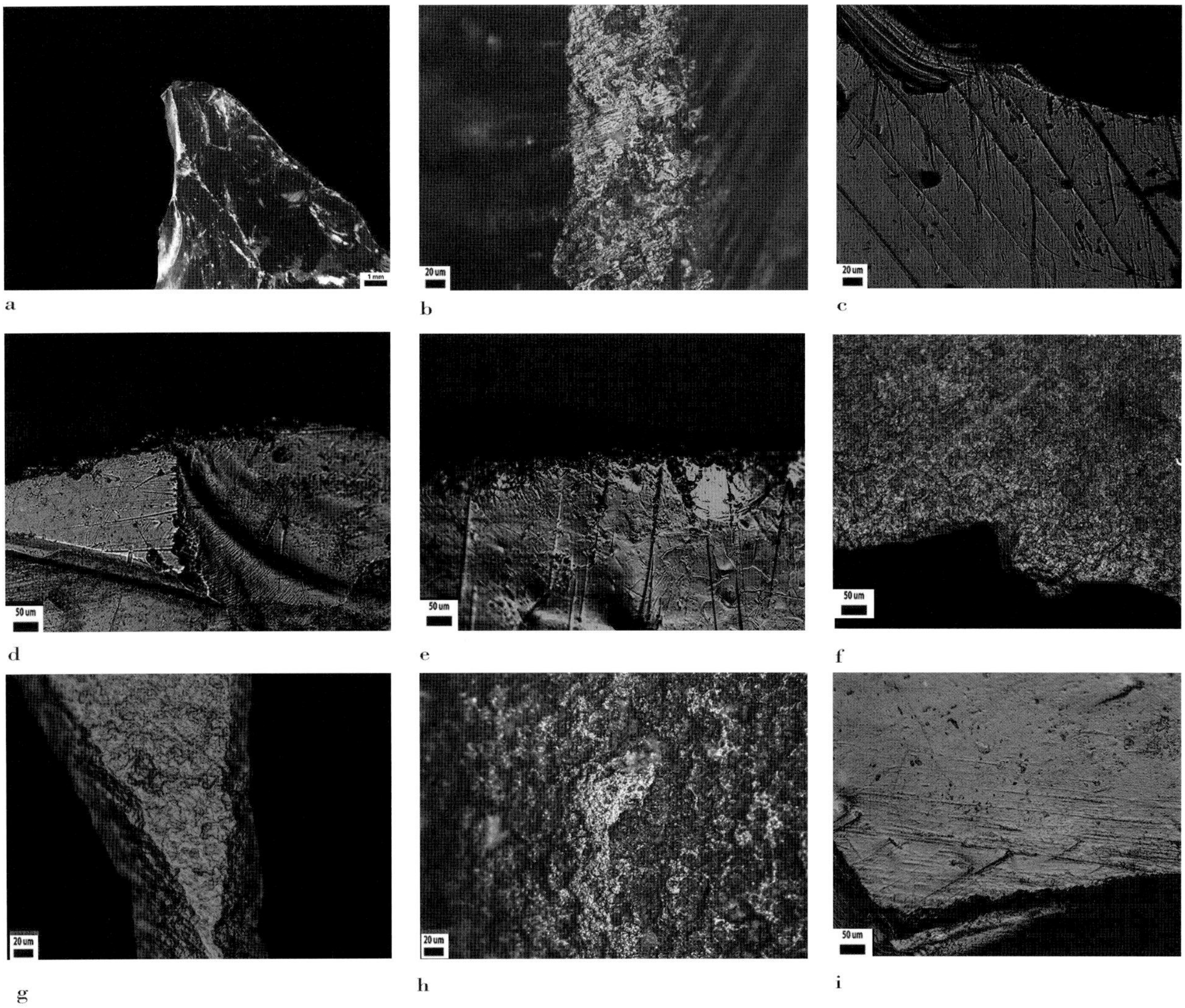

Figure 15.4. Middens and infillings, examples of use-wear observed on the chipped stone tools: (a–b) Space 623 (23143), obsidian projectile point (see fig. 15.3d) with (a) one bending-step scar interpreted as the result of an impact and (b) smooth polish on a developed abrasion interpreted as the result of the insertion in a haft; (c) Space 623 (23143), obsidian tool with edge-rounding, polish smooth, domed to flat and sleeks striations interpreted as antler processing; (d) Space 623 (23153), obsidian tool (see fig.15.3a) with slight abrasion, slight edge-rounding, smooth polish and sleeks and few intermittent striations interpreted as processing wool; (e) Space 610 (32114), obsidian tool with slight edge-rounding, developed abrasion, smooth and domed polish and intermittent striations interpreted as wood processing; (f) Space 439 (20472), chert tool with edge-rounding, smooth domed-to-flat polish interpreted as soft stone processing; (g–h) Space 439 (20472), chert tool with (g) edge-rounding, rough-to-smooth, domed-to-granular polish interpreted as hide processing and (h) patches of rough-to-smooth, granular polish on the inner surface interpreted as the result of freehand grip; (i) Space 439 (20472), obsidian tool with slight edge-rounding, developed smooth, domed-to-flat polish with striations rough-bottomed interpreted as processing wood/shrubs.

A variety of activities are represented in three distinct infilling units (23139, 23153, 23159; fig. 15.3e) of Space 623, corresponding to the filling up of a room in Building 139 (table 15.2). It is worth mentioning that in the primary room infilling (Unit 23153) a scraper made of chert (fig. 15.3b) shows on its active edge an overlapping of different actions, probably for processing different materials, that testify to its prolonged use as a multi-purpose tool. Another tool, a retouched flake made of obsidian, shows traces on both its active edges that are very similar to the traces that developed during our experiments shearing wool (figs 15.3a; 15.4d). We must be cautious, however, as more experiments are needed to try to better define the characteristics of the traces connected with this activity, since they are similar to the traces that developed during hide working or meat

Action/Material	*Butchering*	*Hide*	*Wool?*	*Cereals*	*Antler*	*Soft material*	*Overlapping of matters*	*Indeterminable*	*Total*
Impact								1	1
Cutting	1			1		1		1	4
Scraping					1			2	3
Cutting & scraping	1		2						3
Various	2	1					1		4
Indeterminable								1	1
Total	4	1	2	1	1	1	1	5	16

Table 15.2. Space 623 (23139), (23153), (23159) (infilling) inferred data.

processing. We are also aware that all the work on textiles and animal husbandry at Çatalhöyük has indicated that wool was not used at the site (this volume, Chapter 11; Volume 13, Chapters 8 and 9).

The midden of Space 631 can be tentatively correlated with the occupation phase of Building 139. Use-wear was detected on only six chipped stone tools (three made of chert and three of obsidian). Aside from tools showing the scraping or cutting of soft matter or woodworking, it is worth mentioning a retouched chert blade that represents the result of many sessions of work on different types of materials. On one lateral edge there is overlapping use-wear and retouch resulting from the processing of herbaceous plants, wood and meat; on the other retouched lateral edge, the cutting of bone is present.

Building 131

Twenty-one chipped stone tools were selected from this Level North G building directly above Building 139 (see above) for use-wear analysis following the criteria discussed in the introduction to this chapter. Fourteen items are made of obsidian and seven of chert. The sample is represented by un-retouched and retouched flakes and blades; well-defined types are rare, represented by a bifacial point, a projectile point and a scraper. Besides these chipped stone tools and a great quantity of production debitage, Building 131 had a spectacular cache in which 21 preforms of spearheads made of obsidian were recovered (Tripković 2017: 40). A survey of the whole group of preforms was carried out to verify if they were used before the final deposition or if they were deposited unused in the cache.

Apart from the big bifacial tools, which deserve a separate discussion, 15 chipped stone tools (eight made of obsidian and seven of chert) out of the 21 have use-wear. Five tools were used in two distinct areas, two lateral edges or one lateral and one proximal area; the rest of the tools show traces of use in a single area or edge. The activities carried out are balanced between food processing and craft activities (table 15.3). Butchering activities are attested to by six active edges, all used for cutting meat (two cases: figs 15.5a and 15.6a) or fresh hide and meat (four cases: figs 15.5b and 15.6b–c). These latter cases may signify the first step in butchering activity involving taking the skin off the butchered animal. However, the same combination of traces could result from the first step in the treatment of an already de-skinned fresh hide. This first step consists of the cleaning of the inner part of the hide of the remains of the fleshy tissues and fat. After this cleaning, the hide is ready to be submitted to a procedure of tanning, implying the addition to the hide of a tanning agent and various steps of scraping to soften the material. In the Building 131 assemblage, use-wear interpreted as dry-hide scraping has been detected on one scraper fragment (fig. 15.5d). The dry-hide use-wear shows a fine abrasion (fig. 15.6d) spread over the working surface that suggests the adding of a mineral additive to the hide during tanning.

It is worth mentioning that all the three formal tools in this sample – the bifacial point (fig. 15.5a), the projectile point (fig. 15.5b) and the scraper – were used for processing soft animal materials. Specifically, the bifacial point was used as a knife for cutting meat, the projectile point was used as a knife, as well, for cutting meat and fresh hide, and the scraper was used to scrape dry hide with mineral additives. The use of tools with active edges straightened and regularised by retouch for these kinds of activities suggests a dedicated toolkit for butchering and hide processing.

Another activity related to food processing is harvesting (fig. 15.6e). Two retouched blades made of chert were inserted in a haft (fig. 15.6h) and fixed with a glue still recognisable by the residues stuck on their surfaces. In one case, the glue (dark brown matter with a globular aspect; fig. 15.6g) was mixed with an oxide powder that could have been part of the glue itself, as thickener, or a colouring matter for the embellishment of the haft. The retouched blades were used to cut cereals for many hours, given the amount of use-wear on their active edges (fig. 15.6e). One of them was intensively used, then re-sharpened and intensively used again.

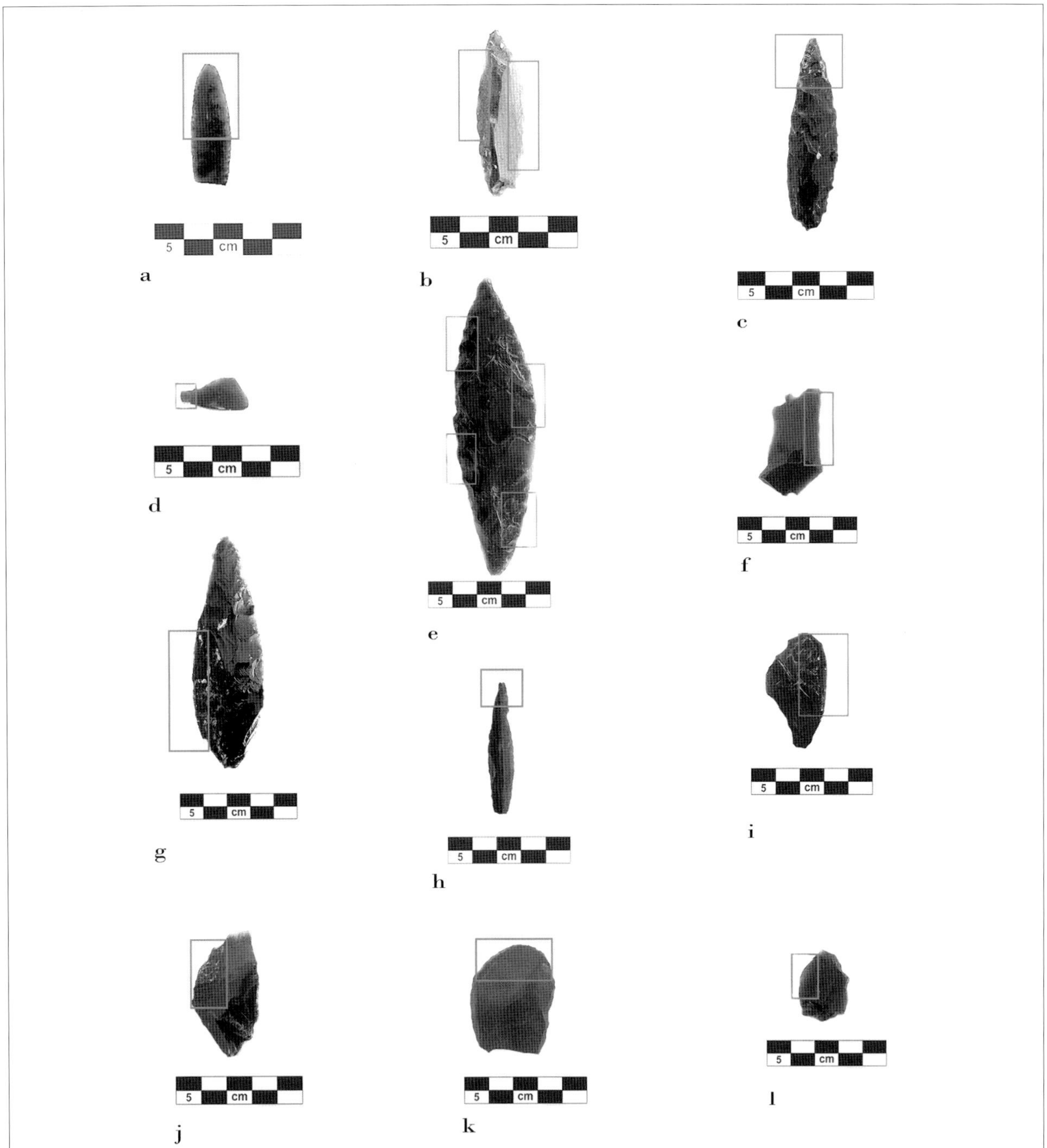

Figure 15.5. Chipped stone tools with use-wear (rectangle) from Buildings 131, 132 and 77: (a) Space 504 (23037.x2), obsidian tool use-wear interpreted as cutting meat; (b) Building 131 Space 500 (23017), chert tool with use-wear (see fig. 15.6b–c) interpreted as cutting meat and fresh hide; (c) Building 131 Space 500 (32536.x1), obsidian spearhead preform with use-wear (see fig. 15.5e) interpreted as scraping mineral matters; (d) Building 131 Space 500/506 (23110), obsidian tool with use-wear (see fig. 15.6d) interpreted as scraping semi-dry hide; (e) Building 131 Space 500 (23000.x2), obsidian spearhead preform with traces interpreted as technological blunting of the ridges and edges; (f) Building 77 Space 336 (19540), obsidian tool interpreted as cutting herbaceous plants; (g) Building 131 Space 500 (32536.x2), obsidian spearhead preform (see fig. 15.7d) with use-wear interpreted as polishing stone; (h) Building 131 Space 504 (23008), chert tool with use-wear (see fig. 15.7b) interpreted as contact with a hard mineral matter (strike-on-like activity); (i) Building 132 Space 531 (31550), obsidian tool with use-wear interpreted as cutting semi-dry hide; (j) Building 132 Space 531 (31550), obsidian tool with use-wear interpreted as scraping wood; (k) Building 77 Space 336 (16491), chert tool with use-wear (fig. 15.7g) interpreted as scraping hide; (l) Building 77 Space 336 (19540), use-wear interpreted as scraping wood.

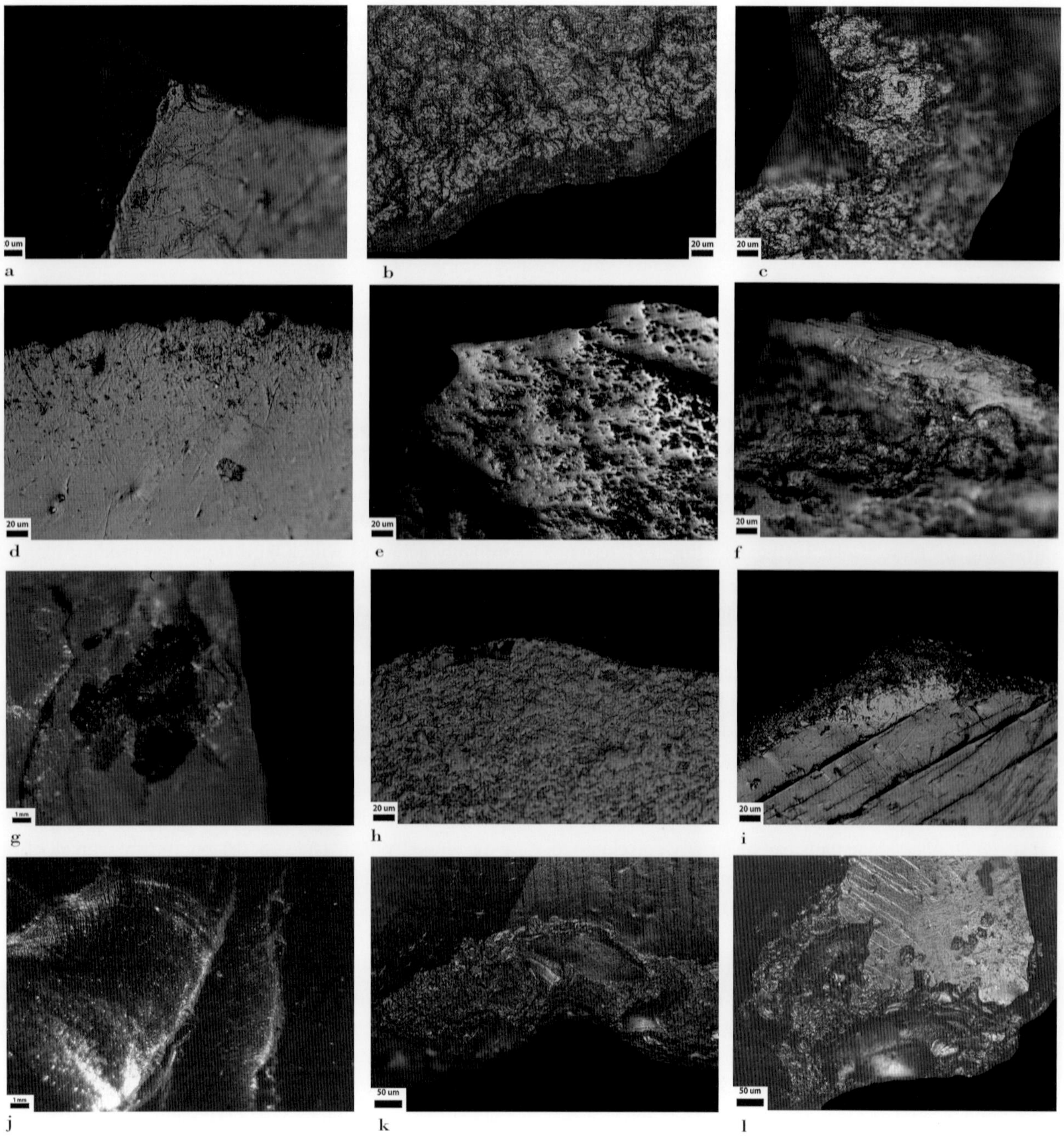

Figure 15.6. Building 131, examples of use-wear observed on the chipped stone tools: (a) Space 504 (23037), obsidian tool with slight edge-rounding, smooth polish and sleeks striations interpreted as meat processing; (b–c) Space 500 (23017), chert tool with edge-rounding, rough (b) and rough-to-smooth (c) polishes with a granular topography interpreted as meat and fresh hide processing; (d) Space 500/556 (23110), obsidian tool with slight edge-rounding, smooth polish with a flat topography, sleeks striations and a slight abrasion interpreted as dry hide and mineral additive processing; (e–h) Space 500 (23131), chert tool with (e) edge-rounding, smooth polish with a domed-to-flat topography and short sleeks striations interpreted as harvesting cereals, (f) edge-rounding, smooth, flat polish with deep and long sleeks striations interpreted as activity overlapping harvesting, (g) dark brown matter with a globular aspect mixed with an oxide powder interpreted as glue on the hafting area, (h) slight edge-rounding, smooth, flat polishes interpreted as contact with a haft; (i) Space 504 (23037), edge-rounding, strong abrasion, rough to smooth polish on abrasion, smooth, flat polish on surface, few sleeks striations interpreted as stone working; (j–k) Space 504 (23034.x20), obsidian spearhead preform with strong abrasion of its ridges interpreted as technological blunting; (l) Space 504 (23034.x10), obsidian spearhead preform with strong abrasion of its ridges interpreted as technological blunting.

Action/Material	*Butchering*	*Hide*	*Cereals*	*Wood*	*Hard animal material*	*Stone*	*Hard material*	*Indeterminable*	*Total*
Abrasion & drilling					*1*		*1*		2
Drilling						*1*			1
Abrasion						*1*			1
Scraping		1			1	*2*	1	1	6
Cutting	*6*		2	1				1	10
Indeterminable								1	1
Total	6	1	2	1	2	4	2	3	21

Table 15.3. Building 131 inferred data; in italics, the uses pertaining to the units strictly related to the building; the other uses pertain to its infilling.

Both show a final short activity that slightly overlaps with the harvesting (fig. 15.6f). It is not possible to distinguish the type of material worked in this final phase of use, but the evidence suggests a short re-purposing of these tools. In the Building 131 assemblage, craft activities are present as well. The hide treatment evidence has already been mentioned. Wood cutting was identified only on one chert flake, while the working of resistant materials such as hard animal material and stone is represented by four items used for drilling, abrading and scraping (fig. 15.6i) that testify to the production and finishing of objects.

The data that we have discussed are all assigned to Building 131. Nevertheless, only the activities inferred on seven chipped stone tools can be strictly correlated to the life of the building (in italics in table 15.3). They were found in the floor make up (23037), in posthole fill (23008) and (23017), in a plaster niche (23151) and in association with a burial (F.7963) set in the northeastern platform of the building. It is worth mentioning that all these examples suggest very focused activities: butchering and the working of stone or animal materials. The remaining chipped stone tools pertain to a foundation deposit (32536) (see below) and to the packing layers (table 15.3) of the construction phase of Building 131 (23084), (23110) and (23131). In these last cases, the use-wear data testify to a variety of activities carried out in prior phases of the life of this area. A more detailed account of the objects related to the use of the building is now provided.

Space 504 in Building 131. In Space 504, the western side room in Building 131, as mentioned above, a cache of 21 preforms for large spearheads was found (fig. 15.5e). These exceptional finds in F.8359 (23033), (23034) were probably knapped from a single obsidian core (see Chapter 14; Volume 12, Chapter 15) from Nenezi Dağ. Use-wear analysis confirmed the retouching by direct percussion technique. Moreover, we observed the quite systematic blunting of the sharp ridges and edges (fig. 15.6j) of the preforms by a punctual action of crushing, revealed by small pits along the ridges and edges and by abrasion (fig. 15.6k–l). The latter was probably achieved with a hard stone that produced large abraded areas where a very light polish developed. No use-wear was detected in the assemblage except for in one case, where the remains of the use-wear of wood cutting preceded the retouch of the preform (fig. 15.7a).

Also in the western side room of B.131 was (23008), the fill of a posthole containing baked clay, charcoal and two chipped stone tools. One of these tools, a chert point (fig. 15.5h), found below a piece of charcoal, shows very developed overlapping scars, a heavy edge-rounding and a developed polish (fig. 15.7b) on its tip typical of a strike-a-light. This evidence could be interpreted as a foundation deposit or, from a more mundane perspective, the storage of a toolkit for making fire.

Space 500 in Building 131. Unit (30536) consists of a fill below the first floor of the main room, Space 500, in Building 131. Two large spearhead preforms made of obsidian are part of this fill, which may represent a cache or a foundation deposit (Tripković, 2017: 40). The fill shows evident traces of burning. Nevertheless, the two bifacial tools do not show thermal stress, except for a small area on (30536.x2). One tool has crushing and abrasion on its ridges and edges, similar to the preform found in the cache in Space 504. But unlike the preforms in that cache, these two preforms have traces of use. (30536.x1) was briefly used for scraping a resistant material (figs 15.5c, 15.7c). Artefact (30536.x2) (fig. 15.5g) has very peculiar traces. It seems that a large flake broke away from a face of the tool and the negative surface underwent an intense and prolonged abrasion and polishing (fig. 15.7d) that produced a surface very similar to the famous Çatalhöyük mirrors, except for the absence of brightness and reflectance. It is worth mentioning that in Building 131, a mirror was found in

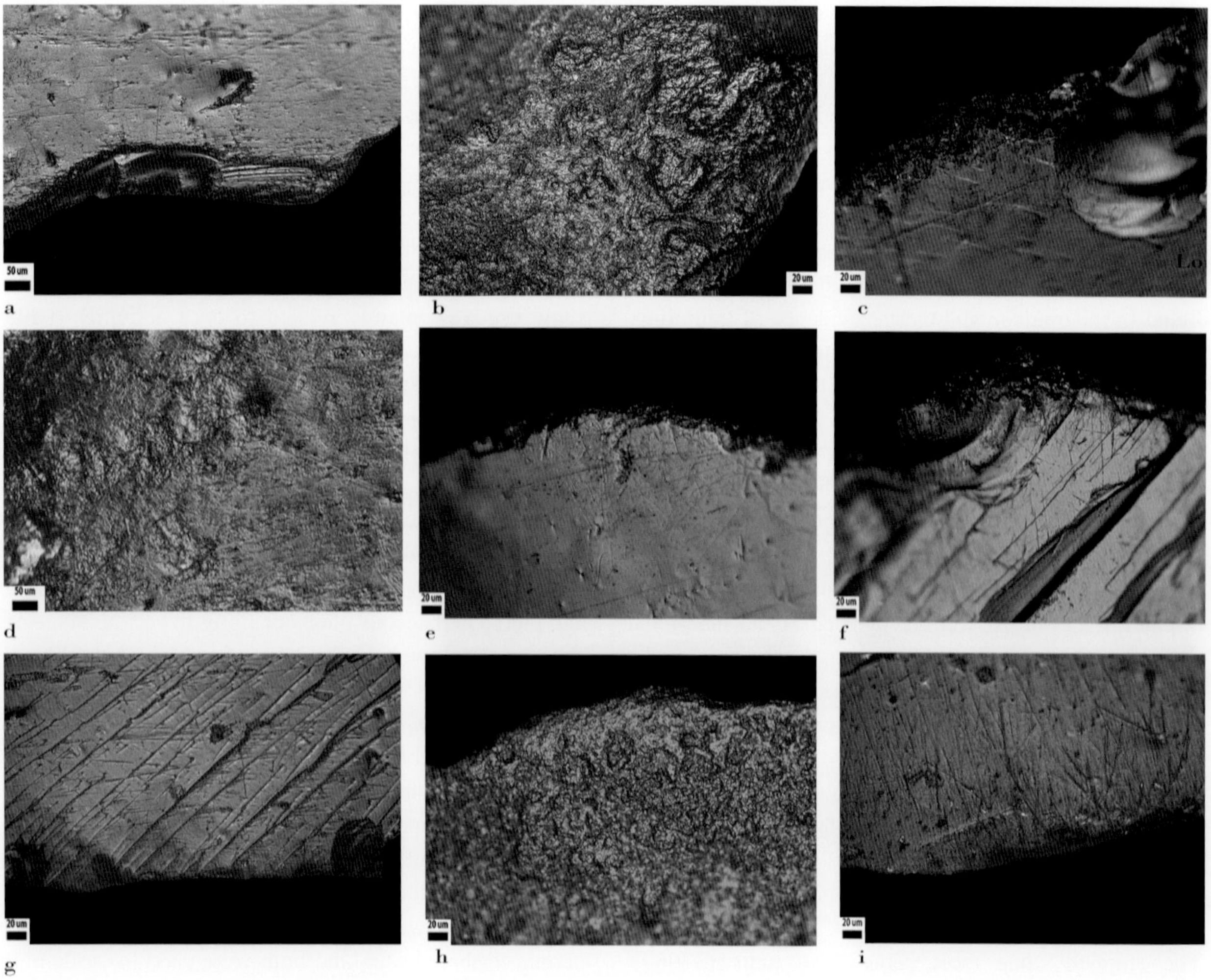

Figure 15.7. Buildings 132, 131 and 77, examples of use-wear observed on the chipped stone tools: (a) Building 131 Space 504 (23034.x17), obsidian spearhead preform with remains of pre-retouching of edge-rounding, polish smooth domed-to-flat and intermittent striations interpreted as wood processing; (b) Building 131 Space 504 (23008), chert tool (see fig. 15.5h) with developed polish smooth, domed-to-flat and short, shallow striations interpreted as the contact with a hard mineral matter (strike-on-like activity); (c) Building 131 Space 500 (32536.x1), obsidian spearhead preform (see fig. 15.5c) with developed abrasion, rough polish on abrasion and striations intermittent and rough-bottomed interpreted as mineral matters processing; (d) Building 131 Space 500 (32536.x2), obsidian spearhead preform (see fig. 15.5g) with very developed and spread polish smooth, flat with narrow striations interpreted as polishing stone; (e) Building 132 Space 532 (31550.x2), obsidian tool with slight edge-rounding, abrasion, domed polish and sleeks striations interpreted as semi-dry hide processing; (f) Building 77 Space 336 (19295), obsidian tool with edge-rounding, developed polish smooth, flat and long sleeks striations interpreted as harvesting cereals; (g) Building 77 Space 336 (16491), chert tool with strong edge-rounding, polish rough, granular interpreted as hide processing; (i) Building 77 Space 336 (22017), obsidian tool with edge-rounding, diffuse slight abrasion and sleeks striations interpreted as mineral matters processing.

a burial (F.7961) (Tripković 2017: 44). From a functional point of view, the preform with polishing traces found in (32536) could be interpreted as part of the sequence of production of one or more mirrors. It is intriguing that the preform for a spearhead, which is a highly symbolic tool certainly not pertaining to a daily-life toolkit, was recycled as a polisher, possibly for the finishing of other highly symbolic objects, the mirrors.

In the northeastern platform of Building 131, Space 500, the primary burial of a single individual, a probable male skeleton (30044), was found in F.7963. The mesial-distal portion of a retouched chert blade lay on top of the skeleton, suggesting that the tool was part of the deposition. The tool was used for gathering herbaceous plants before being retouched and re-used on both its edges for cutting meat and fresh hide.

Space 610 external area associated with Building 131. Space 610 is an external area or midden composed of three units, (32114), (32128) and (32138), that represent the progressive infilling of the foundation trench (Unit 32138) of Building 131. Each unit consists of minor dumps, small layers of ash and other deposits that testify to the repeated use of this area to dump waste. All three units are rich in organic remains, charcoal and waste from knapping activities. Many chipped stone tools found in this midden show clear evidence of trampling, attested to by scars, abrasion and striations distributed chaotically on the outline and the surface of the tools. In certain cases, the chipped stone tools are so smashed that only a small portion of the used areas remains intact. In two cases, the chipped stone tools show the remains of traces of use that precede the re-sharpening of the active edge, testifying to an occasional practice of re-use. On 12 chipped stone tools, 11 items made of obsidian and one of chert, traces of use were recognised. As no functional differences were highlighted between the three units, the results of the use-wear analysis are presented as a whole. The activities inferred (table 15.4) are butchering (meat cutting; fig. 15.3c) and the working of hard material as in Building 131. Activities related to the gathering and working of herbaceous plants (fig. 15.3h) and wood (fig. 15.4e) are present as well.

Building 132

The chipped stone tools come from units (32069), (31550), (32707), (32766), (32767) and (32776) of Space 531, which represents the first phase of occupation of this building in Level North F. The sample consists of nine items, all made of obsidian and all showing use-wear. Except for one flake (fig. 15.5j) and one fragment of blade, the remaining sample was shaped by retouches. Formal tools are well represented and consist of one scraper (fig. 15.5i), three end-scrapers, one point and one projectile point. The projectile point shows a slight crushing of its tang and the breakage of a wing that could suggest the results of an impact. The rest of the tools show use-wear mainly related to craft activities (table 15.5). Use-wear results testify to the treatment of hide (fig. 15.7e) and the working of wood (fig. 15.7f) and other soft and hard matter. It is worth mentioning that the processing of herbaceous plants for weaving purposes is suggested by the use-wear detected on one of the formal tools, the point. This tool was found in (32766), the make-up of the northwestern platform F.7722. The lateral edges and the point of this tool were used to cut, scrape and pierce herbaceous plants or fibres at various stage of dryness. All these motions inferred on a single tool have strong similarities with the motions carried out with metal knifes by modern artisans for basketry and for weaving plant fibres to produce, for example, mats. This point recovered in Building 132 is a very good example of a tool dedicated to a specific task.

Another peculiarity of Building 132 is the presence of a cluster of small objects lying in a niche (32776) in a pit (F.8330). The two end-scrapers found in Building 132 pertain to this cluster, and both were used for softening hide. Finally, a retouched blade showing use-wear from hide treatment and an end-scraper showing traces of

Action/Material	*Butchering*	*Herbaceous plants*	*Wood*	*Hard material*	*Indeterminable*	*Total*
Cutting	1	1		2	1	5
Scraping			2	1	2	5
Indeterminable		1			1	2
Total	1	2	2	3	4	12

Table 15.4. Space 610 (midden) inferred data.

Action/Material	*Hide*	*Herbaceous plants*	*Wood*	*Antler*	*Indeterminable*	*Total*
Impact					1	1
Drilling		1				1
Cutting		2				2
Scraping	2		1	1		4
Cutting/scraping	2	2				4
Total	4	5	1	1	1	12

Table 15.5. Building 132 inferred data.

antler scraping were part of a cluster of goods found in the infill (32767) of F.8331 skeleton (32770), one of the burials that characterise Building 132. Besides the two chipped stone tools, a large number of bone objects, stone beads, shell pendants and three pillared bone rings were recovered from the infill of this burial.

Building 77

Building 77 is a special structure, which differs from the other buildings of the site in the richness of its plastered and painted decoration. The abandonment phase of the building is characterised by the sealing of the living spaces with an infilling containing a huge quantity of material, especially ground stones. The sealed building was also destroyed by fire, as part of a ritual that was widespread at Çatalhöyük (Tsoraki 2018: 238–62).

Chipped stone tools are very rare in this building. Possible functional items are absent from the clusters and the infilling of the abandonment phase. The sampled chipped stone tools (25 items) come from the floors, the make-up and the graves in Space 336. Only six out of 25 chipped stone tools show traces of use. Eight other items were altered by trampling. Four items show technological traces related to their shaping. It is worth mentioning that three of them are projectiles. These projectiles were probably shaped in Space 336, where evidence of debitage, core rejuvenation and the production of projectile points was found.

The chipped stone tools with traces of use are made of obsidian (four items) and chert (two items). From floors or floor make-up, three items were recovered that were used to scrape hide (fig. 15.7h) and mineral materials (fig. 15.7i). One of them, a big chert end-scraper (fig. 15.5k) used for processing hide (fig. 15.7h), is part, with a projectile point, of a cluster (16491) where hand axes, hammer stone and bone were deposited. The other three tools were found in the infill of two burials (F.3615, F.3620). They were used for processing herbaceous plants (fig. 15.5f) and wood (fig. 15.5l). It is worth mentioning that the tool found in the infilling (U.19295) of burial F.3615 is an obsidian flake obtained from the knapping of the surface of a bifacial tool. This tool has an elongated shape and was used as a knife for harvesting (fig. 15.7g). Clear traces of the wrapping of the tool with hide were observed in its grip area.

The data coming from the use-wear analysis of Space 336 clearly show that the activities carried out in this building differ from what we observed in other buildings. Although carefully cleaned, the internal spaces of B.132 and B.131 maintain traces of the household activities carried out there. The chipped toolkit found in the living area of Building 77 has fewer traces of use, and these are related only to the working of hide or minerals. This trend is confirmed by the chipped stone tools found in Space 489, a midden close to Building 77. In this midden, five items were used, respectively, for: scraping, cutting and piercing hide (three items; fig. 15.4g–h), drilling soft stone (one item; fig. 15.4f) and cutting wood (one item; fig.15.4i).

Summarising, from the use-wear perspective, in Building 77 the usual activities carried out with chipped stone tools in a household are missing. The main activity observed is the production and shaping of chipped stone tools, including projectile points. Another relevant activity was hide processing. The lack of traces of fresh hide suggests that in Building 77 only the finishing phases of the production of leather objects were carried out.

Discussion

The results of the use-wear analysis of the chipped stone tools of Çatalhöyük reveal small pieces of a complex society. In this society, chipped stone tools had a fundamental role as a medium between humans and the world that they manipulated. Chipped stone tools are extensions of the human body (Malafouris 2013: 153–78) that allow the magnification of social interactions not only in their practical aspect of 'action on matter' but also in activities related to habit and memory.

Habituated practice may be revealed by the continuity of use of a single tool for a specific purpose. The 'weaving point' found in Building 132 and the 'butchering knives' found in Building 131 probably represent personal belongings related to skills embodied in their owners. The skill creates a special linkage between individual and tool. We can experience the same linkage between modern artisans and their toolkits that, we can suggest, was socially recognised at Çatalhöyük.

Other chipped stone tools may represent individual or household habits, such as the 'strike-a-light' carefully deposited in (23008) of Building 131 or the two end-scrapers used to process hide found in a niche (32776) in a pit (F.8330) in Building 132. From a use-wear point of view, the 'strike-a-light' and the two end-scrapers were still usable, so we can suggest that they were household belongings stored to be used as needed. Nevertheless, we also cannot discard the hypothesis that they were not stored but cached to become part of the memory of the household.

This role of social memory that connects past and present is better attested by the deposition in the burial (F.7963 (30044)) and in the foundation cache (Space 500 (32536)) of the same Building 131. The fragment of a retouched blade found directly over the bones of the right forearm of the skeleton buried in the north-eastern platform of Building 131 was used for gathering herbaceous plants and then retouched and re-used on both its edges for cutting meat and fresh hide. We do not know if this tool belonged directly to the buried individual or to the household community related to the buried. It would be suggestive to link the final use of the retouched blade to a butchering session related to rituals connected with the burial event.

The foundation ritual expressed by cache (32536) in B.131 is also important in this regard, given the deposition of two spearhead preforms. In the cache, two preforms were used not as projectiles, as their morphology suggests, but for scraping some hard material (x1) and for polishing hard material (x2). These actions could have played some role in the foundation ritual. The polishing could certainly have had some connection with the preparation of the Çatalhöyük mirrors, 'special objects' that were found in one burial in Building 131 and in two burials in Building 129, immediately above Building 131. As Hodder says (2017: 10), the mirrors of Buildings 131 and 129 are examples of 'memory' or 'history-making'. The tool deposited in the foundation cache of Building 131 could have been part of this 'memory' or 'history-making'.

The communities of Çatalhöyük lived in buildings and in their surroundings, but the open areas between the buildings were, as well, places of communal life. We can 'fill' the vivacity of these areas from the use-wear observed on the chipped stone tools coming from (12343) of Space 623, corresponding to an open area partly beneath Building 131. In these spaces, exhausted tools were abandoned that had been used for a very long time and re-sharpened several times to do a range of different tasks. We agree that in these open areas many daily activities were routinely carried out and these activities were a fundamental occasion for social exchange between households.

Finally, at Çatalhöyük not all the buildings can be defined as households. Differences in architecture and decoration, and the presence of elaborate rituals, attest that some of these spaces were dedicated to activities that went beyond the daily routine of a household and that they, probably, were a medium of social aggregation for the whole community. Building 77 is one of these buildings (Taylor et al. 2015: 127–49; Tsoraki 2018: 238–62 and references therein). The use-wear analysis of the chipped stone tools testifies to the absence of some of the range of daily activities seen in Buildings 131 and 132 and in other buildings of the South Area that we studied. This 'absence' of data is significant. It seems that some daily activities were left out of this building. This building suggests an interesting border between the inside and outside, between routine and special events. Maybe these special events included the production of special objects, as indicated by the knapping activity related to projectile points and hide finishing.

Acknowledgements
We are very grateful to Ian Hodder for inviting us to be part of the wonderful research team of the wonderful site of Çatalhöyük. We are indebted to our colleagues and friends in the 'lithic team' of Çatalhöyük for the fruitful discussions 'around stones'.

16. Data analysis and integration at Çatalhöyük

Ceren Kabukcu, Dominik Lukas and Camilla Mazzucato

Introduction

The main aim of this chapter is to offer an integrative and generalised investigation of the wide range of datasets collected by members of the Çatalhöyük team. A total of 456 units were selected for study by all participating specialists, aiming to provide a broad enough coverage of different context types, areas of excavation and occupation periods on the East Mound (for the selection process see Chapter 1). The units selected for study, and the analyses presented here, include units predominantly from the 2010–2017 excavations, as well as some units from previous years. As previous analyses, and current applications by Mazzucato (Volume 15, Chapter 2), provide a far more detailed examination of the spatial patterning in the dataset, our main venues of interpretation focus on examination of patterns in the density distributions by occupation period and context types. Further integrative analyses were carried out on specific context types: midden and fill deposits and floor deposits. A further aim of our examination was to refine and improve data quality on some of the evaluated find categories. To this end, datasets of archaeobotanical remains, representing a mix of fuel, food and food preparation waste, were evaluated under three main density categories.

Methodology

We sought to include in our analyses categories of finds reflecting a wide range of activities. For example, finds which are used and discarded on a routine basis, such as the remains of food and fuel (for example, archaeobotanical and faunal remains) were included in data synthesis. Secondly, we included finds which are used commonly across the site, are discarded routinely but tend to be more durable and more curated, such as chipped stone and ground stone. Lastly, we included find categories such as pottery, clay objects and figurines, which are normally found in lower densities, but represent unique use and discard activities. Count and weight data for each category were obtained from the specialist databases for chipped stone, faunal remains, ground stone, pottery, clay objects and figurines. Unit volume was calculated by using the recorded flotation volume and calculated sediment volume via the main excavation database, where detailed records of unit dimensions alongside the description of the excavated contexts are entered. Archaeobotanical remains were evaluated differently, as these represent the remains of routinely used fuel, plant food and food preparation waste. Data for the most commonly found residues of fuel waste, wood charcoal and charred dung, were gathered on the basis of volume of charcoal and volume of charred dung per unit (and/or flotation sample). Count data for non-wood remains, to be used in density calculations, were made separately for cereal grain, pulse, nutshell, monocot culm, wild/weedy seeds and cereal rachis/chaff remains. The calculations for these follow the quantification protocol set out by Bogaard and co-authors (2014; Volume 13, Chapter 5). The density calculations for archaeobotanical remains were made only on the soil volume recorded during flotation processing of soil samples.

Density distributions by context and temporal grouping of levels (that is, Early, Middle and Late periods) for each data category were first evaluated separately. Following this, we aimed to apply integrative analytical techniques which would enable us to evaluate wider patterns in densities across find groups and how these relate to temporal and spatial patterns observed. The analytical techniques used and the presentation of our results follow previous work closely (e.g., Cross-May 2005; Mazzucato 2013). We chose to employ a predominantly descriptive quantitative methodology (including univariate and multivariate methods) due to the nature of the datasets. In the following sections we will discuss the results of univariate and multivariate analyses, while also providing information on some of our unsuccessful attempts at combining datasets and how these relate to the structure and nature of archaeological datasets. We also sought to understand if any discrepancies exist between count and weight densities for some find categories (for example, chipped stone, ground stone, faunal remains).

Units included in the study

The units included in the present study reflect a semi-random selection of units across the excavated deposits (for further detail regarding the selection process, see Chapter 1). Firstly, a group of units studied by several specialists in common were included in the study; these consisted predominantly of either in situ and short-term activity deposits (for example, fire features, floor deposits, clusters) or fill/midden type deposits yielding rich finds in most categories, providing excellent interpretative potential. In addition, a number of other context types such as burial fills, building infill, building collapse and construction deposits were also selected for study, resulting in a list of priority units to be examined by all participating specialists, including also some units from excavations prior to 2009.

The final list of units selected represents a high number of units deriving from floor deposits and fire features, but at the same time a considerable number of deposits deriving from fills (including building infill, bin and pit fills) and middens (see fig. 16.1). In addition, a number of units derive from context types such as building collapse, burial fill, cluster and construction deposits (for example, packing layers of platforms, walls etc.). The categories included in the study, as presented here, reflect an amalgamated simplification of context descriptions, deriving predominantly from the context descriptions provided by the excavators and the assigned context categories recorded in the excavation central database. In order to facilitate the quantitative investigation and interpretation of the datasets, we re-classified some context categories under more generalised terms. For example, all contexts comprising the fills of buildings, bins, pits and other types of 'cut' features are included under the fill category. However, we differentiated the burial fill category, as a significant number of units from this context type were also included in the priority units list. Similarly, fire features encompass hearths, ovens, fire spots, and in some cases, the in situ residues of single/short-term burning events (for example, concentrated residues of fuel waste). Floor deposits encompass predominantly internal floors, but also include the floors of platforms and, in some cases, floor deposits from external (for example, courtyard/open area) contexts.

Amongst the studied units, the great majority derive from Levels South O and North G, with some from South K and North F. The remainder of the units include a few from most other occupation periods at the site (see fig. 16.2). When these units are categorised according to the occupation period, more units belong the Middle period group, while there is still an even representation of units from Early and Late periods (see fig. 16.3). By area of excavation, more of the units examined come from the South Area as compared to the North Area, and some derive from the TPC Area (see fig. 16.4).

Density distributions of find categories

Previous work on the datasets from Çatalhöyük revealed that the frequency distributions of densities of various find categories examined followed a typically right-skewed distribution (Cross-May 2005; Mazzucato 2013).

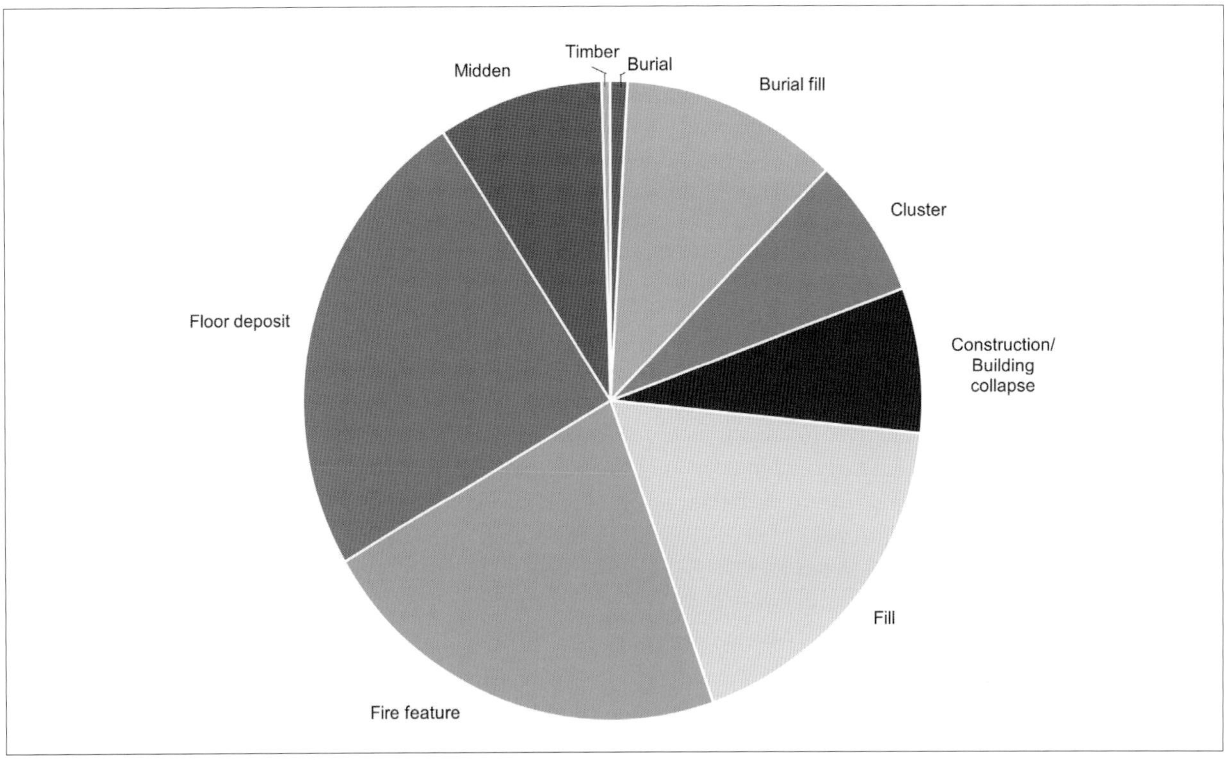

Figure 16.1. Units included in the study by context type.

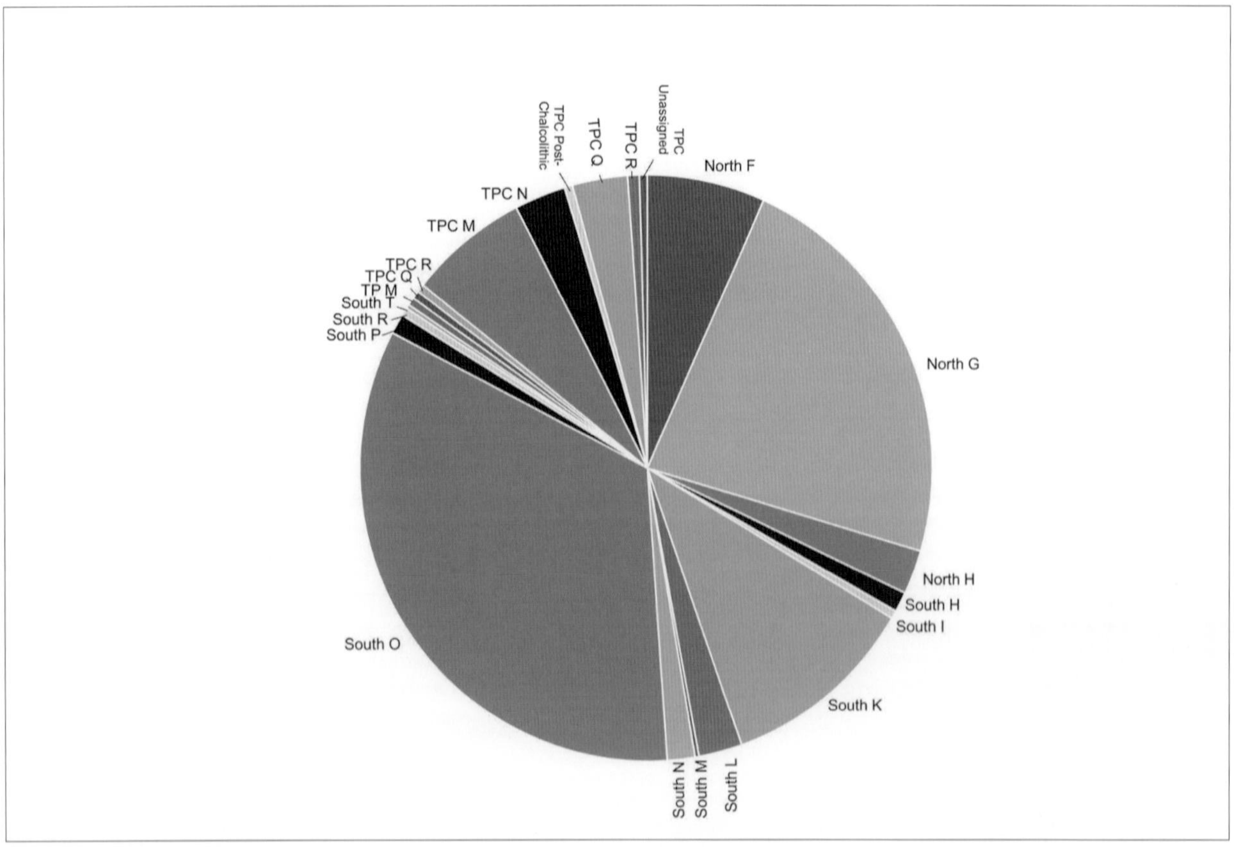

Figure 16.2. Units included in the study by level.

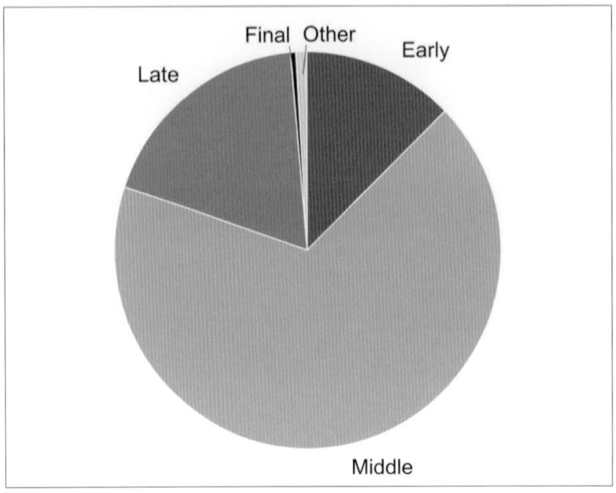

Figure 16.3. Units included in the study by period.

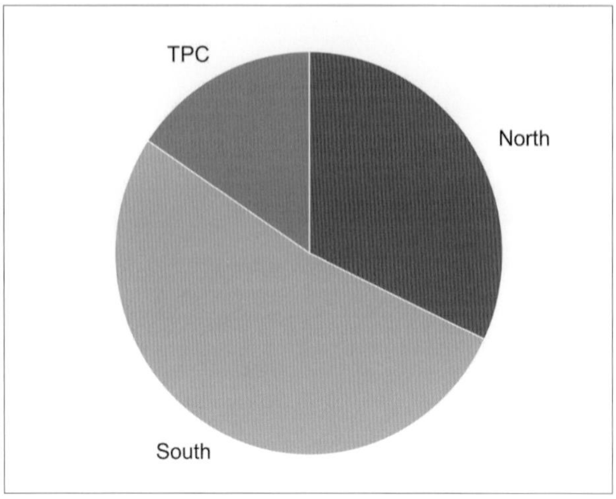

Figure 16.4. Units included in the study by area of excavation.

In other words, across several of the find categories examined, the greater majority of the units had low (and very low) density figures and much fewer units had higher density figures – often much higher than the group average. As the density distributions of faunal remains, ground stone, chipped stone and clay objects show (figs 16.5, 16.6, 16.7 and 16.8), the datasets contain very few units with unusually high concentrations of finds, and this situation holds true independently of whether density calculations are made on counts per litre or total weight of the finds per litre. Thus, the observed distribution patterns are unlikely to be the result of fragmentation during and after deposition of specific categories of finds. Additionally, all of the data categories mentioned thus far have been remains not reliant on burning and carbonisation for preservation.

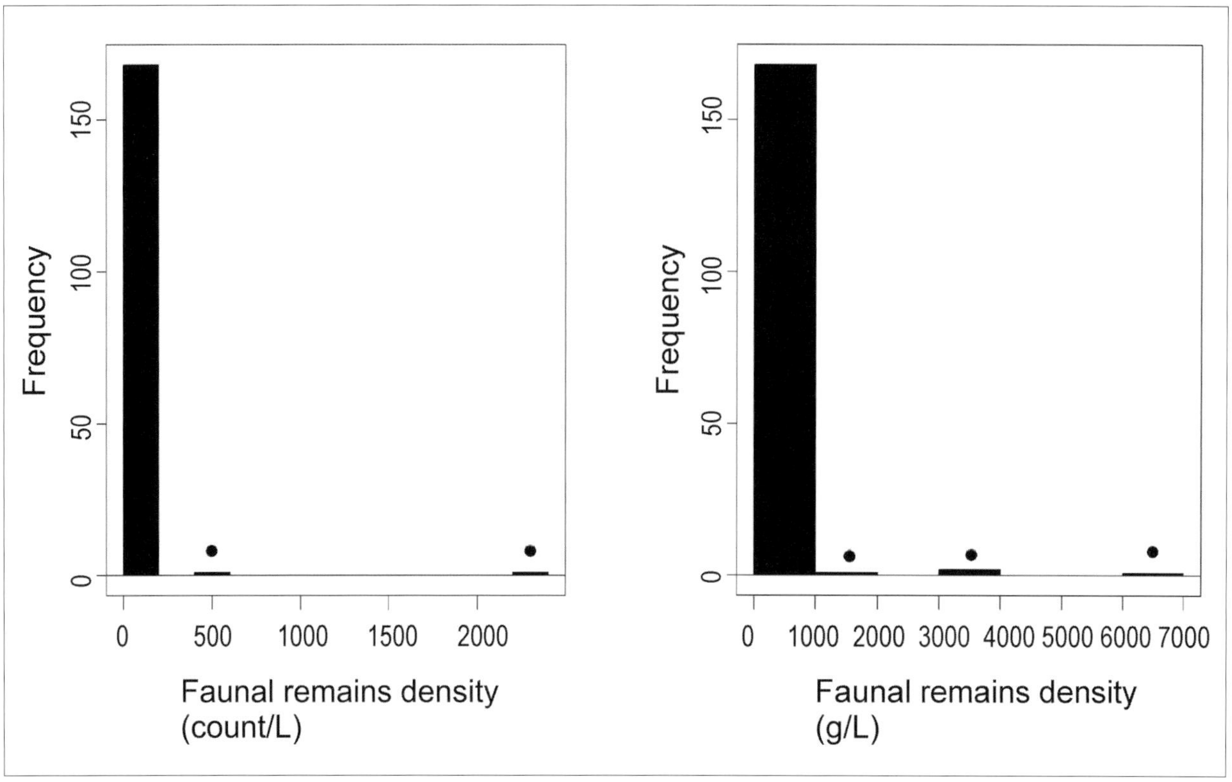

Figure 16.5. Faunal remains count density and faunal remains weight density histograms across all units included in the study. Dots represent presence.

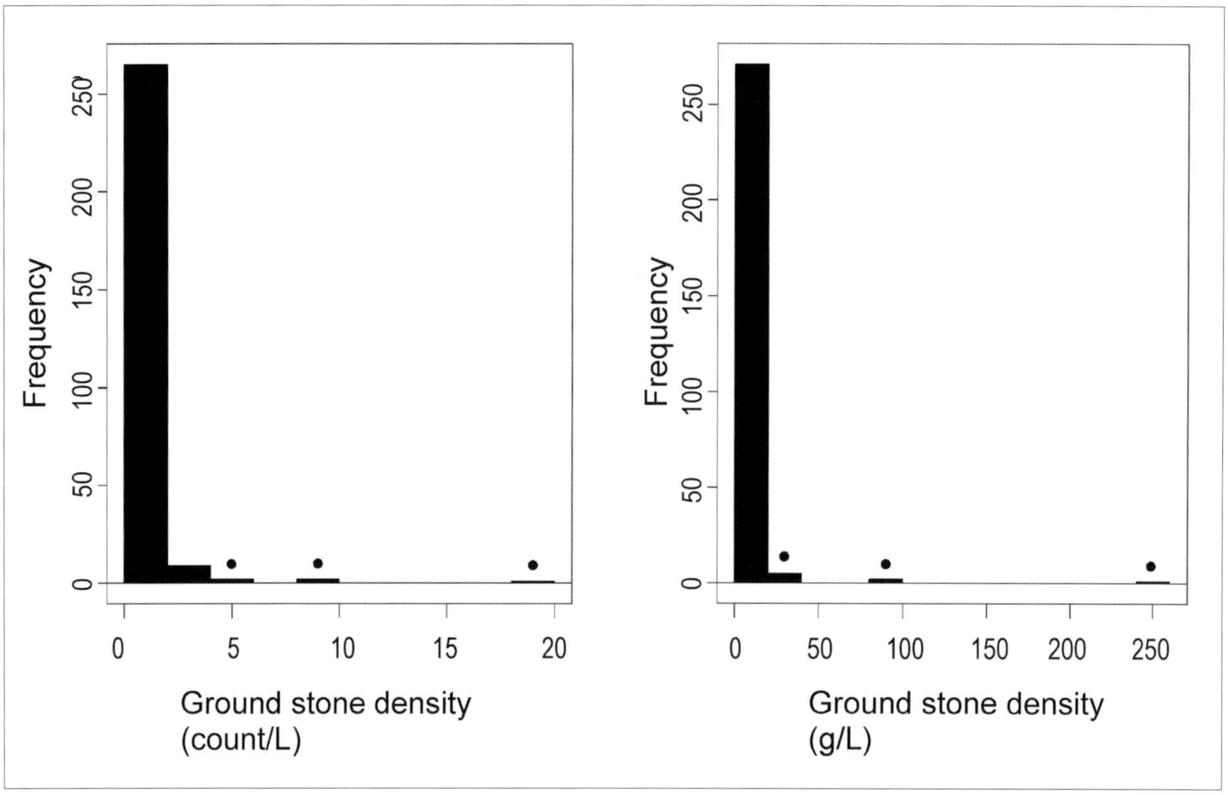

Figure 16.6. Ground stone count density and ground stone weight density histograms across all units included in the study. Dots represent presence.

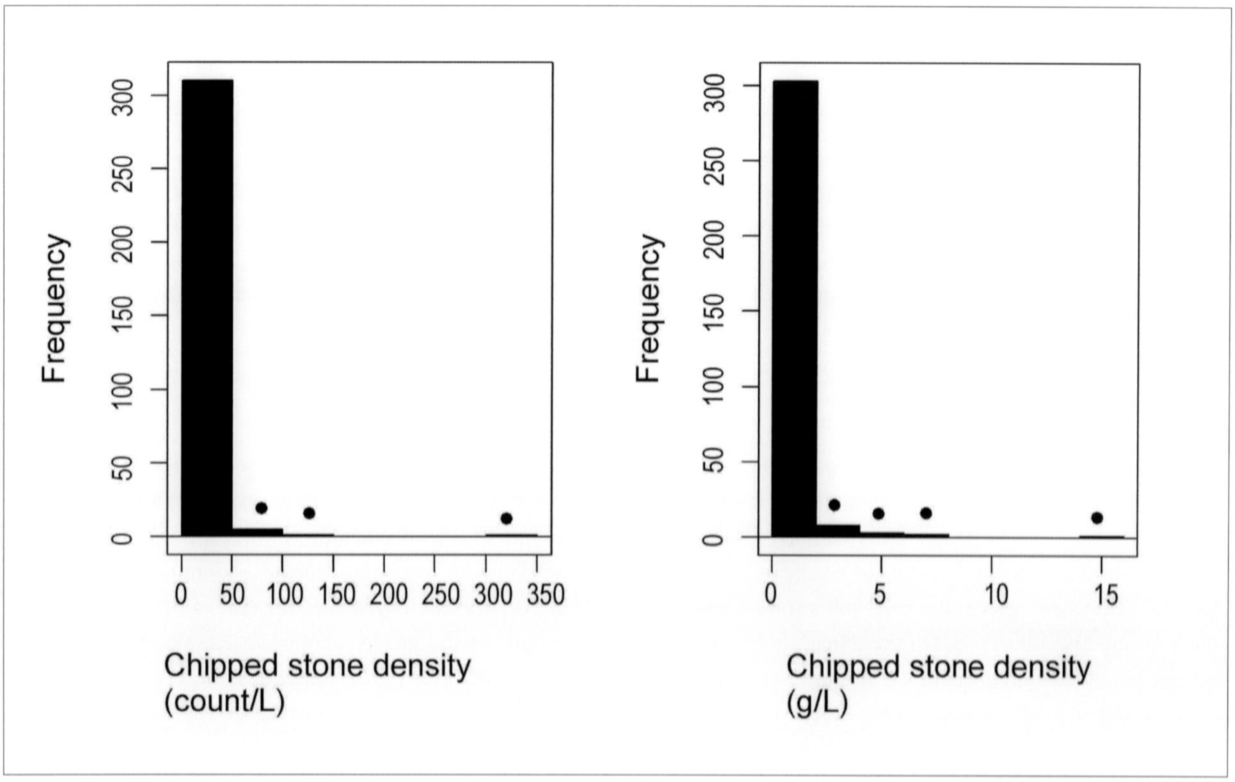

Figure 16.7. Chipped stone count density and chipped stone weight density histograms across all units included in the study. Dots represent presence.

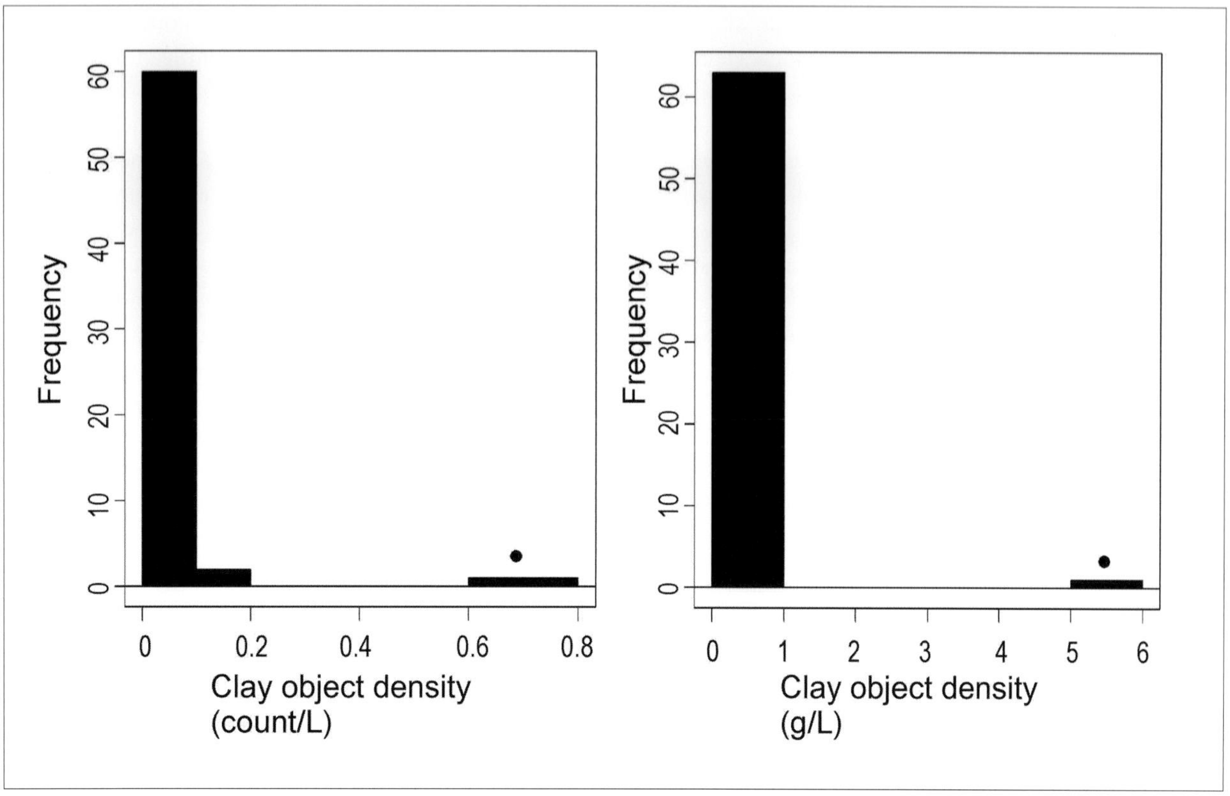

Figure 16.8. Clay objects count density and clay objects weight density histograms across all units included in the study. Dots represent presence.

Interestingly, and despite differences in taphonomic pathways, archaeobotanical remains which are preserved in a carbonised state, including wood charcoal and non-wood plant remains, also follow a similar right-skewed distribution (figs 16.9, 16.10). This type of distribution in the archaeobotanical remains is most prominent in wood charcoal, as it likely represents the waste from fuel use, and wild/weedy seeds and cereal rachis/chaff remains.

When evaluating density distributions by context type it is evident, particularly in some find categories, that activity types are one of the defining factors behind the great range of variability observed in the right-skewed distributions. For example, when examining plant remains, particularly those deriving from fuel waste, rare deposits with very high densities are found in fire features and some cluster units (fig. 16.11). In such contexts, similarly to the rest of the excavated deposits, density values are generally low, but rarely very high densities are also observed. On the other hand, non-wood archaeobotanical remains show a similar rare occurrence of high density in 'cluster' units (figs 16.12–16.14). Some of these units represent storage deposits in burnt buildings, and thus rare short-lived events, as opposed to high density accumulations over the long term.

This pattern is more evident in faunal remains, where both count and weight densities show spikes in density (rare occurrence of high-density deposition in few units) in clusters (figs 16.15, 16.16). Interestingly, ground stone densities display rare high values by count density in fire features, while by weight density very high values are observed in some floor deposits (figs 16.17, 16.18). In both cases, this is likely to be related to deposition and use within a limited time frame rather than a jump in deposition by discard due to the nature of the deposits in question. Pottery density, while notably low across the site, also shows some elevated values in clusters similar to faunal remains (fig. 16.19). On the other hand, chipped stone count densities also show spikes in fire features, somewhat similar to ground stone density values (fig. 16.20).

Interpretation of ranges of density values through time necessitates taking into account the nature of the distributions. Following previous applications, we have performed logarithmic scaling of some of the density values (that is, by taking the natural log of the raw density values). In faunal remains and ground stone (figs 16.21, 16.22), transformation of the datasets by taking the natural log (cf. Shennan 1997) yielded distributions most similar to a normal distribution.

Across all examined find categories, the untransformed density ranges are similar across occupation period (Early, Middle, Late). As mentioned, the distributions of density values are strongly influenced by the occurrence of rare spikes in density values observed only in a limited set of context types. The temporal comparisons of each find category can be seen in figures 16.23 to 16.29. One striking commonality across find groups is the limited number of 'spikes' in density during the later periods, represented in the boxplots as outliers placed outside the general spread of density values. While this could easily be interpreted as a temporal trend, thus suggesting a general decline in find densities through time, some caution is needed. As mentioned, such rare high density values are only observed in a limited set of contexts. For example, amongst the fire feature units, only 7% belong to the Late period group, while 17% make up the Middle period group. Similarly, in clusters, of the 32 units included in the study, three belong to the Late and four to the Middle period group. Amongst the floor deposits, the Late period accounts for only 5%. Thus, it is possible that the full ranges of density values across all period groups cannot be accounted for with the dataset under study. The range of contexts and the representativeness of contexts through time notwithstanding, looking at the log-transformed values of densities through time, it appears that any changes in density values through time are minimal (figs 16.30–16.32). Bartlett's test of homogeneity of variance performed on log-transformed density values for faunal remains, ground stone and chipped stone densities shows no significant difference amongst period groups in faunal weight and count densities. In ground stone and chipped stone, Bartlett's tests point to no significant differences between groups in count densities, but significant differences amongst period groups in weight density values. We also performed Levene's test of homogeneity of variance in order to investigate further any shifts in density values through time, with the tests performed on untransformed datasets. Again, any possible changes in values of faunal remains (count and weight densities) are not significant. The only significant difference between period groups according to the Levene's test is observed this time in ground stone count density values. Thus, we conclude that there is no justifiable basis to argue for any shifts through time. One of the issues faced during the analysis is the low number of units from the Early and Late periods, which may be hindering interpretations of changes in density values across occupation periods.

Data integration across find categories

In order to explore relationships between find categories, their potential patterning through time and across space, we investigated two subsets of the dataset using multivariate descriptive techniques. Following on from the earlier description of density values and their distributions, we chose midden and fill context types for one set

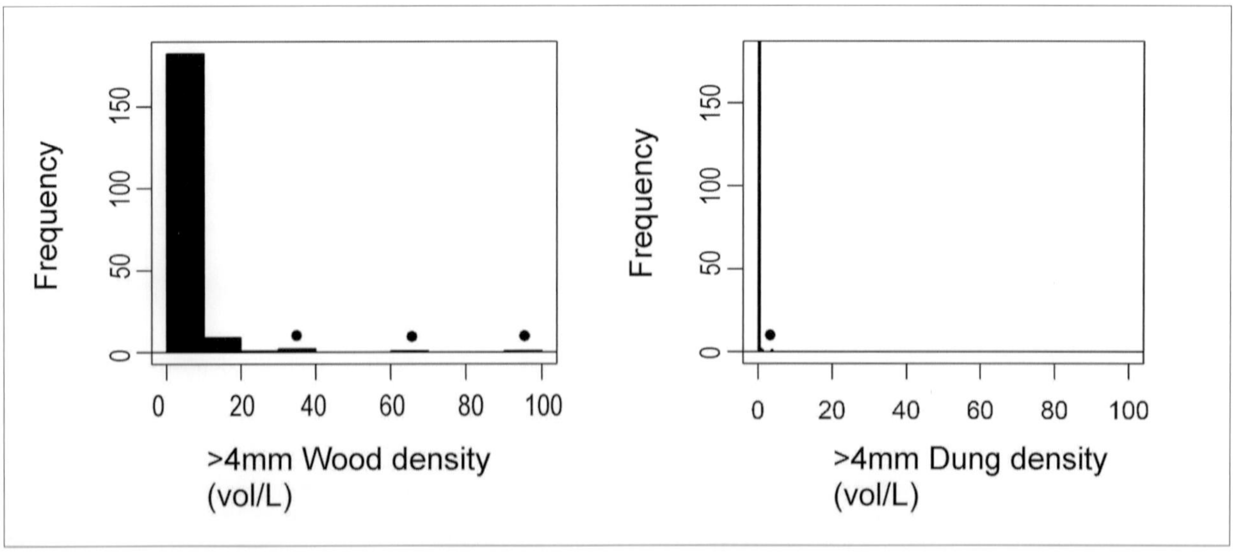

Figure 16.9. Wood charcoal density and charred dung remains density histograms across all units included in the study. Dots represent presence.

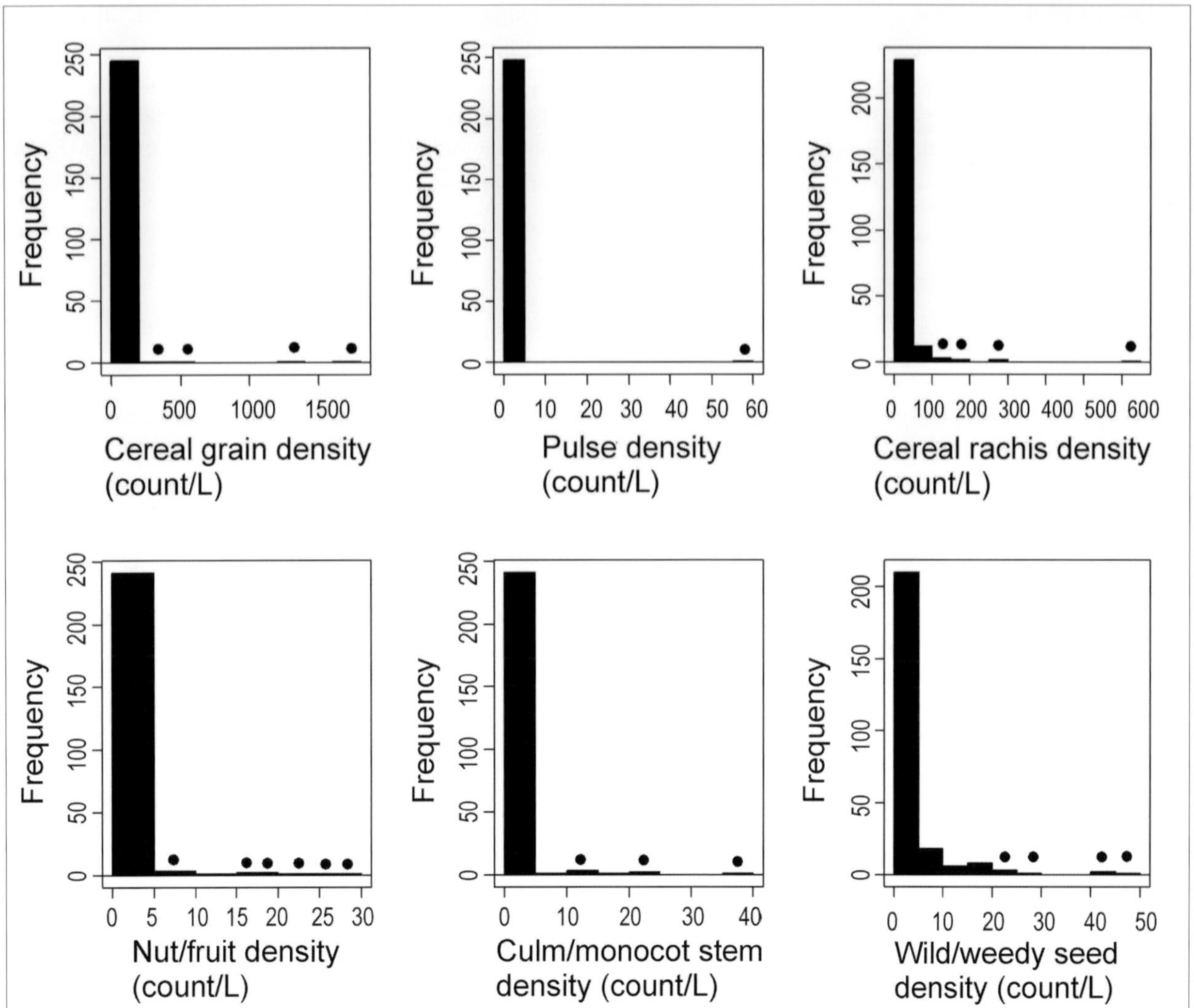

Figure 16.10. Archaeobotanical seed/non-wood remains density histograms across all units included in the study. Dots represent presence.

of analyses and floor deposits for another, using Multiple Factor Analysis (MFA). MFA, a descriptive multivariate factor analysis technique, is well suited to datasets containing multiple data groups, which sometimes contain data values in differing scales (Bécue-Bertaut, Pagès 2008; Escofier, Pagès 2008); thus, for example, count densities of chipped stone and volume densities of wood charcoal in the same units can be compared with greater ease. MFA, similarly to other factor analysis methods such as Principal Components Analysis (PCA), provides a graphical representation of relationships between variables, as well as a depiction of the relative distances/dissimilarities between individuals (that is, units) included in the analysis. Furthermore, MFA provides a summary of relationships between wider groups of variables. In this study, we grouped cereal grain and pulse density in one group (Bots large), wild/weedy seed density and cereal rachis/chaff density in one group (Bots small), chipped stone and ground stone in one group, and figurines and clay objects in one group.

One particular problem we aimed to overcome with the current dataset, as demonstrated in the description of the density distributions above, was the abundant observations of zero values and/or very low densities across several find categories in the great majority of units. This situation, alongside the numerous missing values, meant that using more conventional techniques, only a fraction of the units could be included in data integration. One motivation behind examining subsets of the dataset (see above – midden/fill and floor deposits) was to provide a reliable basis for transforming datasets to overcome abundant missing values and zeroes in the density values. Data transformation and scaling were carried out using the FactoMineR package (R version 3.5.0) prior to performing hierarchical clustering and MFA on the two subsets of the dataset.

The results of MFA on floor deposits representing the individual units in a biplot were plotted by occupation period (fig. 16.33), area of excavation (fig. 16.34) and building number (fig. 16.35). As the plots indicate, across occupation period there is very little variability in density values, as no clear clustering can be seen amongst individual units. The plot in fig. 16.34, on the other hand, shows some minor patterning by area of excavation, although this does not appear to be a particularly robust trend. However, the plot of individuals by building numbers suggests, at least in some buildings, that density values across units show some affinity with one another. For example, various units from Building 132 appear to be relatively similar to one another. The first two dimensions of the MFA on floor deposits account for 54% of the cumulative

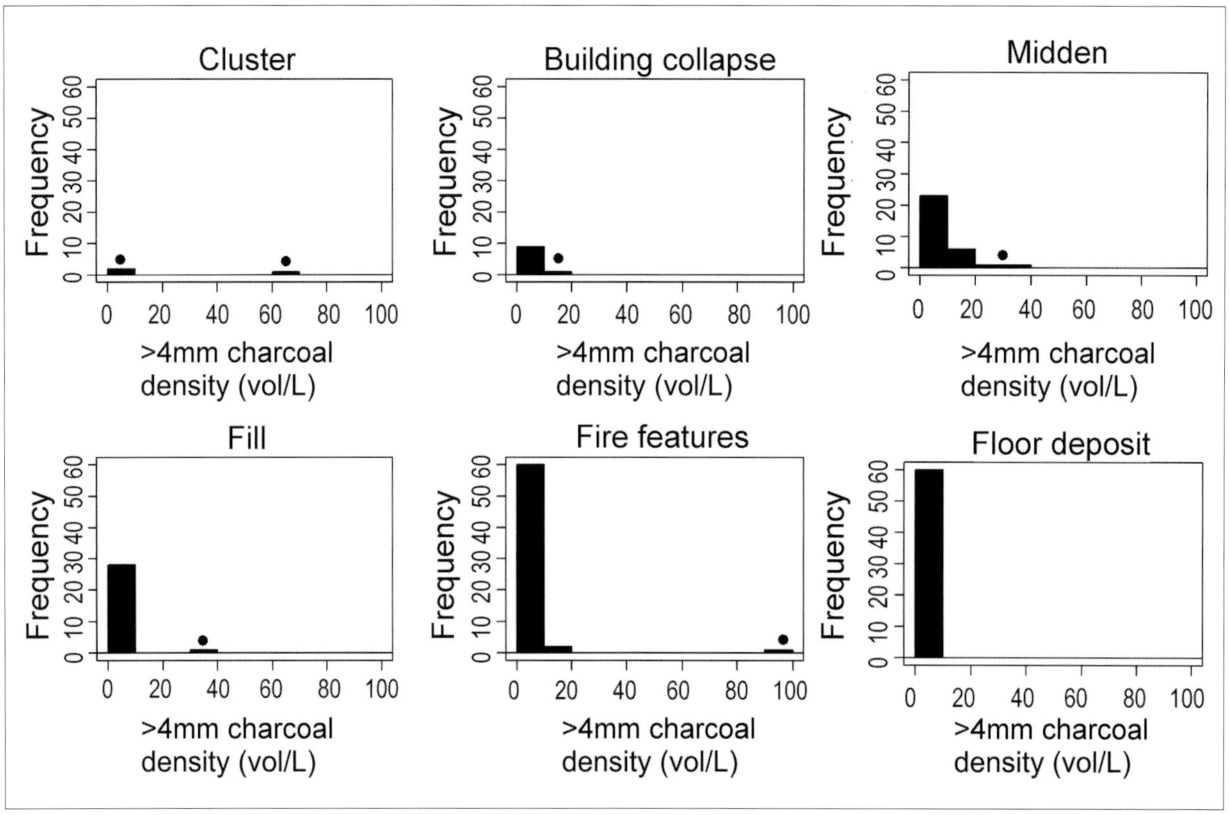

Figure 16.11. Wood charcoal density histograms for different context types included in the study. Dots represent presence.

variation in the dataset, which is driven predominantly by variation in archaeobotanical remains (including wood charcoal and the rest of the archaeobotanical remains) along Dimension 1 and figurine, clay object and pottery density in Dimension 2 (figs 16.36, 16.37).

When looking at the relationships amongst find categories in floor deposits, archaeobotanical remains are closely related, including wood charcoal density and small and large non-wood archaeobotanical remains (fig. 16.36). Figurine, clay object and pottery densities are also closely related. Faunal remains, ground stone and chipped stone contributions to any variability and/or patterning in the dataset are not captured by the current analyses.

The results of MFA on fill and midden deposits were also plotted according to occupation period (fig. 16.38) and area of excavation (fig. 16.39). Similarly to variation in floor deposits, variation in fill/midden units does not seem to be related to occupation period. Instead, there appears to be a weak clustering between different areas of excavation, where the North Area samples are found predominantly in the lower part of the biplot (fig. 16.39) and the South/TPC samples in the upper part of the biplot. The 40.95% cumulative variation explained by the first two dimensions of the MFA are driven predominantly by non-wood archaeobotanical remains, faunal remains, ground stone and clay object density (figs 16.40, 16.41). Variations in values of figurine, pottery, chipped stone and, to some extent, wood charcoal density make minor contributions to the overall patterning observed here. In addition, the density values of charcoal and the remainder of the archaeobotanical remains are not as clearly and closely related as the patterns observed in floor deposits. Instead, cereal grain density in midden units appears to be more closely associated with faunal remains. In addition, clay object and ground stone densities also appear to be more closely associated.

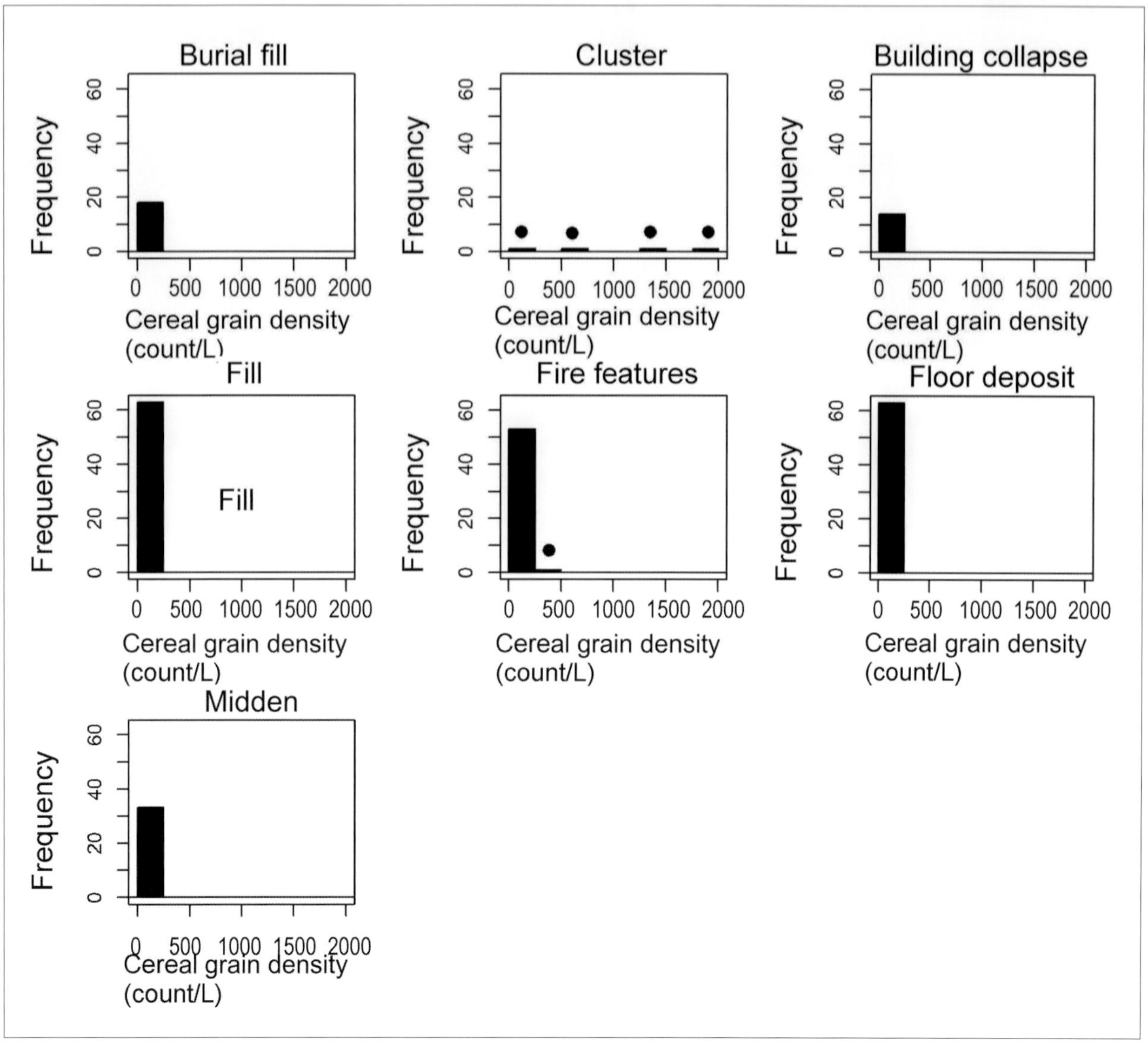

Figure 16.12. Cereal grain density histograms for different context types included in the study. Dots represent presence.

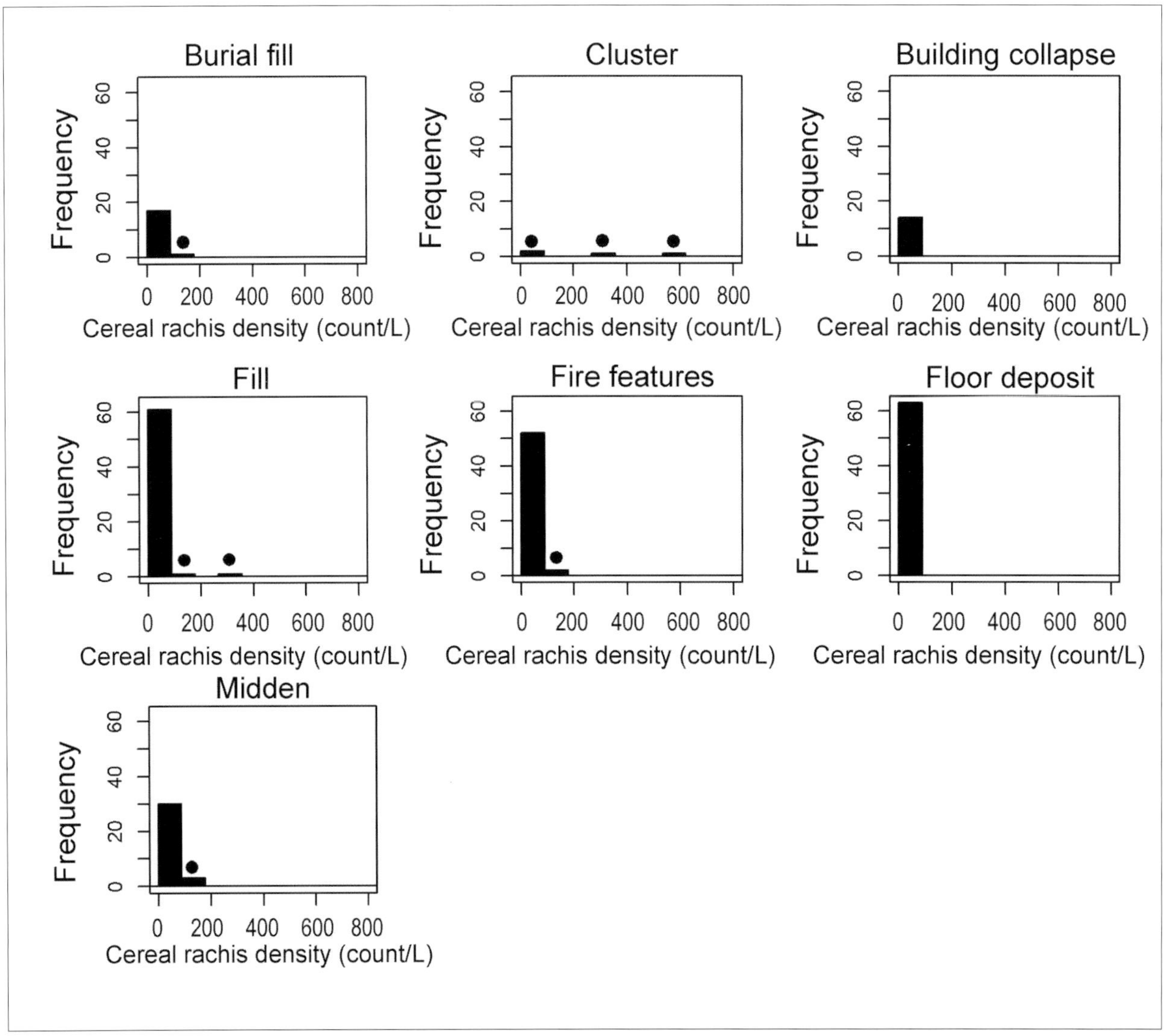

Figure 16.13. Cereal rachis/chaff remains density histograms for different context types included in the study. Dots represent presence.

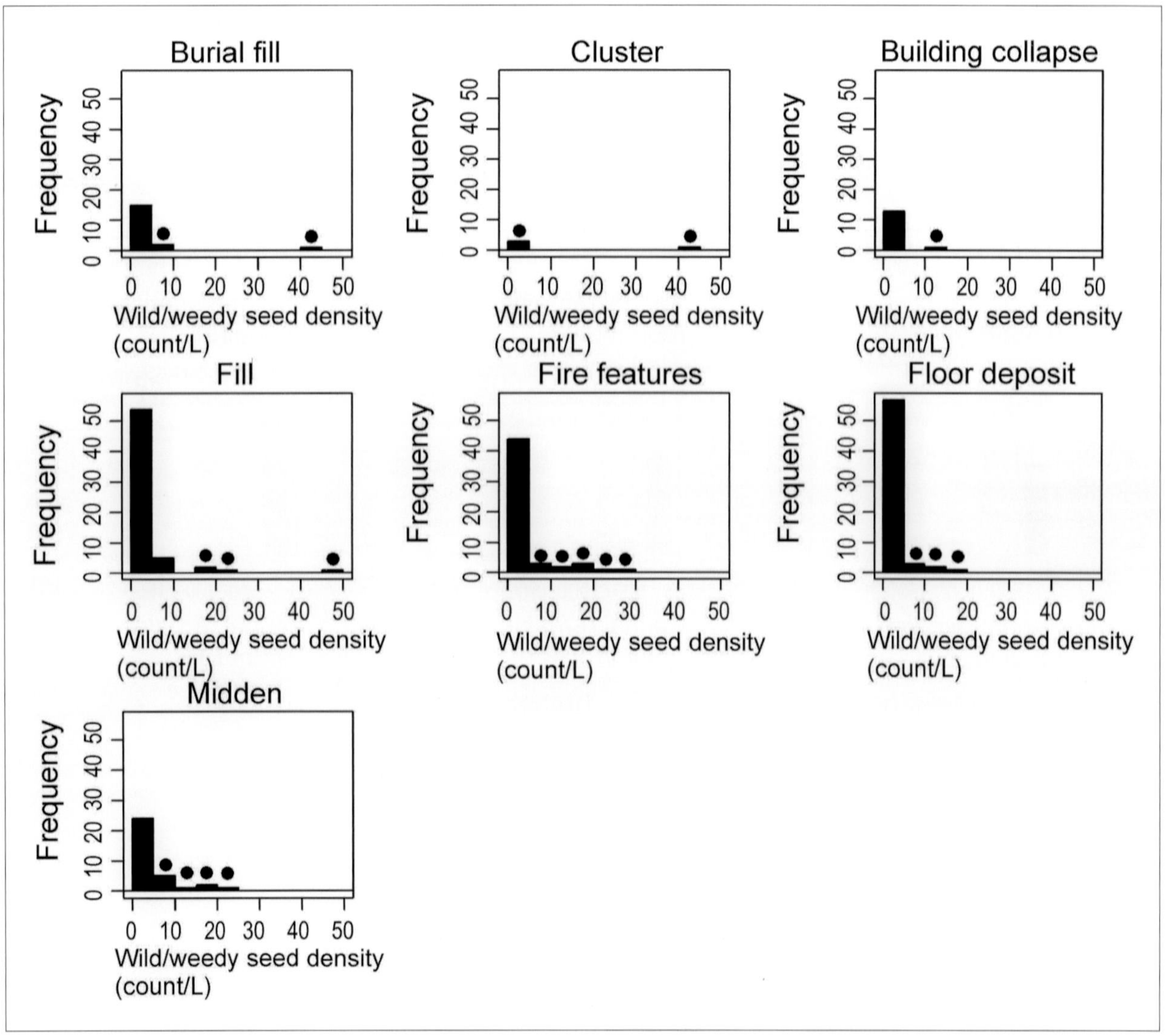

Figure 16.14. Small wild/weedy seed density histograms for different context types included in the study. Dots represent presence.

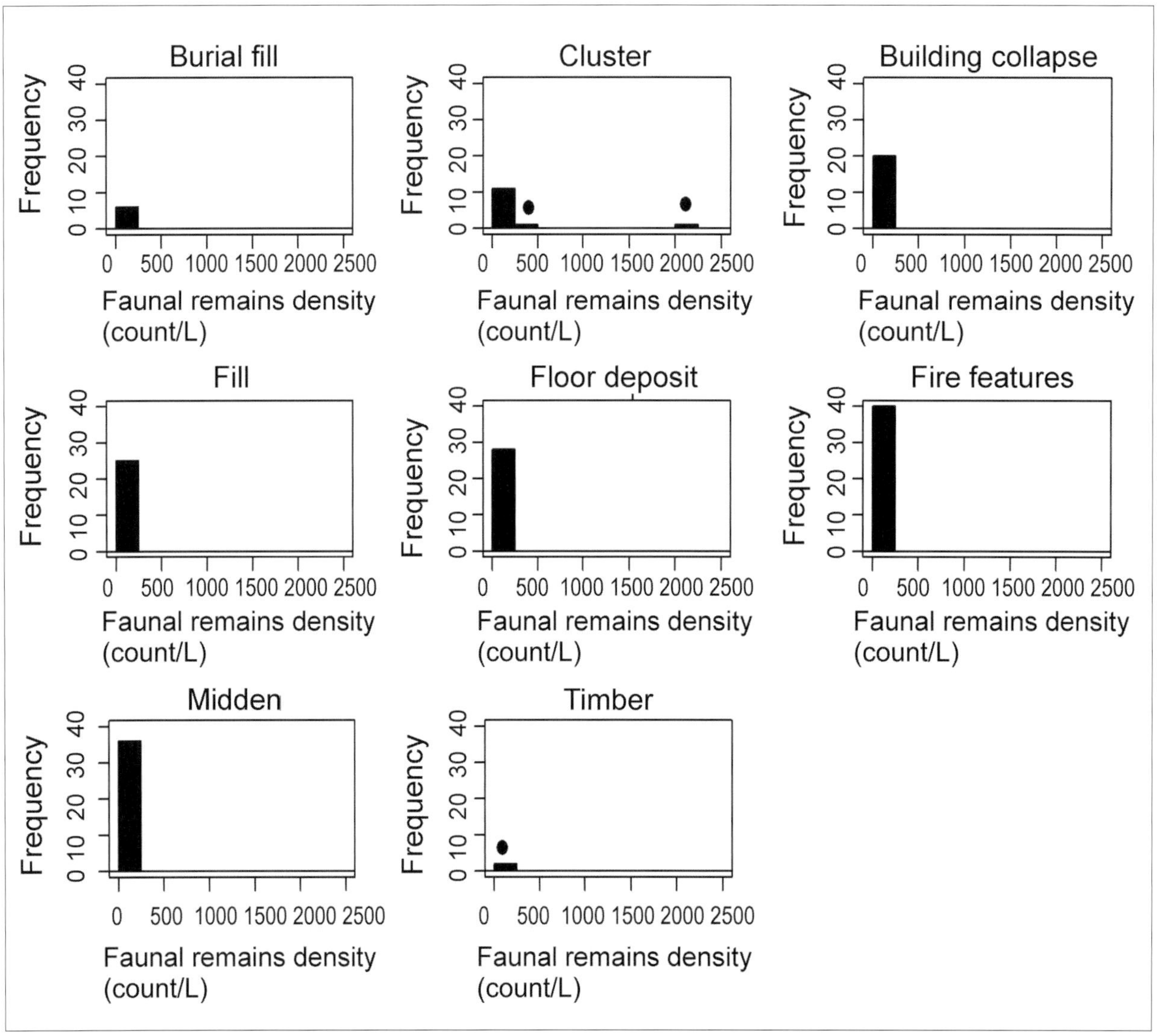

Figure 16.15. Faunal remains count density histograms for different context types included in the study. Dots represent presence.

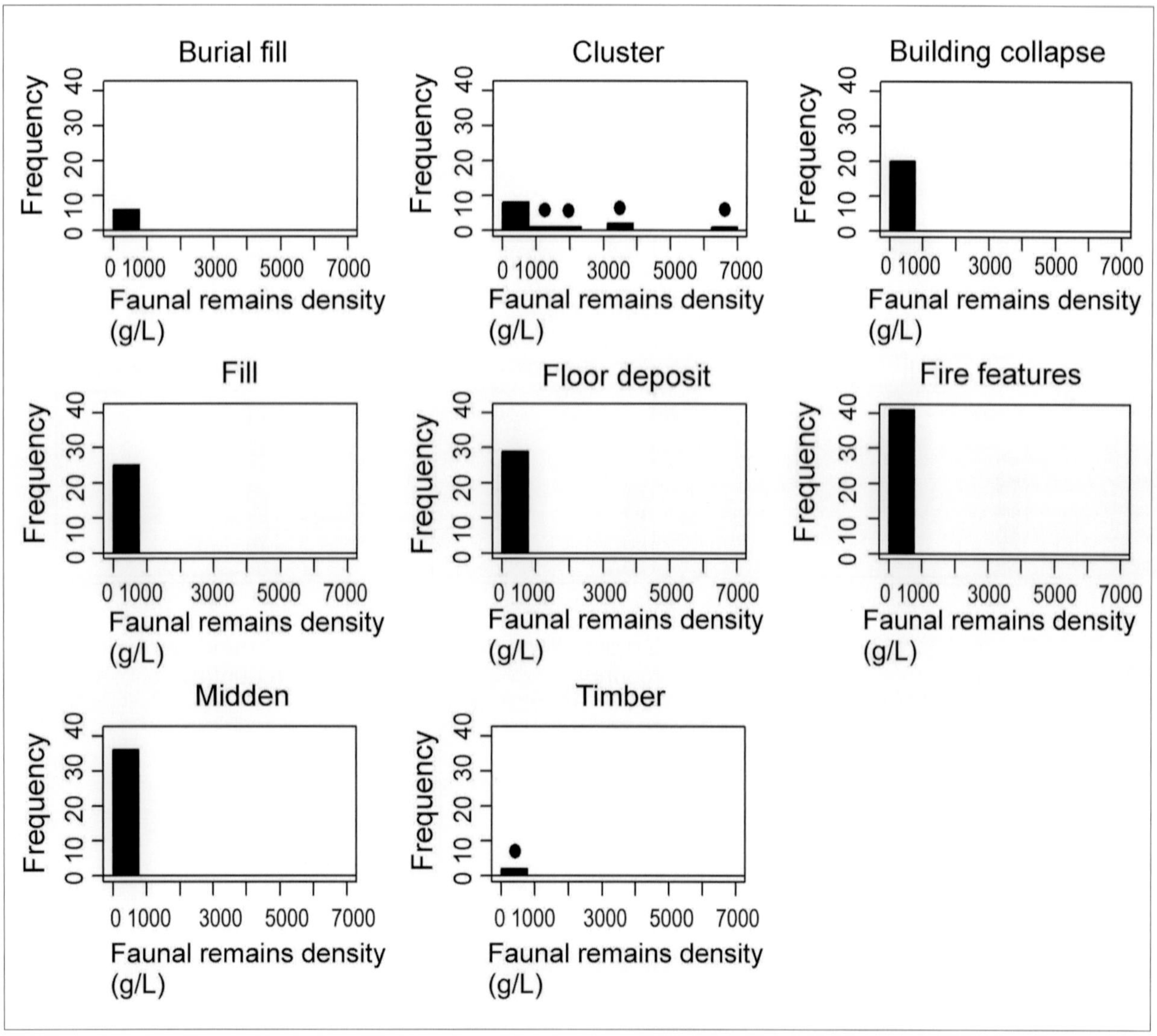

Figure 16.16. Faunal remains weight density histograms for different context types included in the study. Dots represent presence.

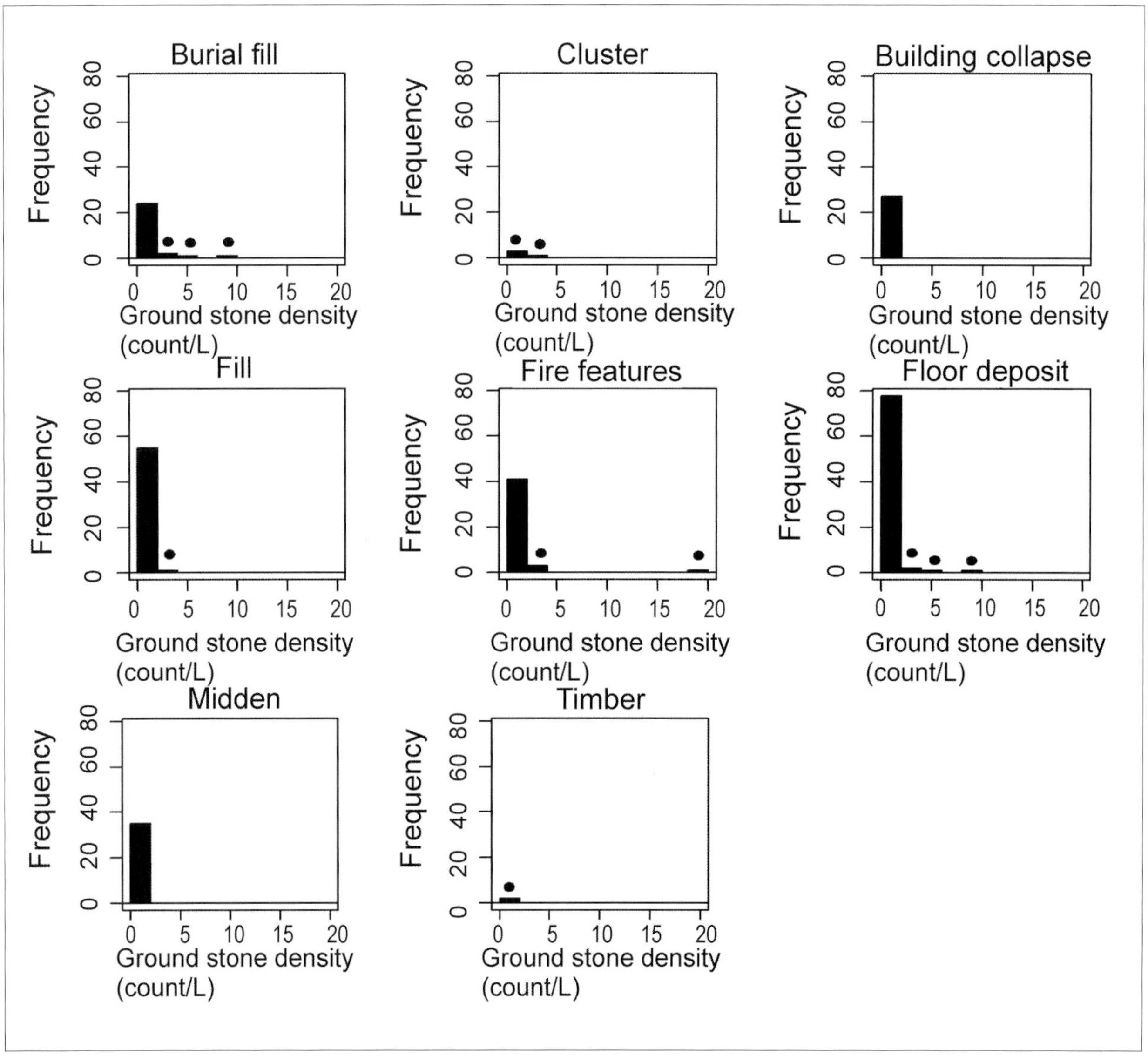

Figure 16.17. Ground stone count density histograms for different context types included in the study. Dots represent presence.

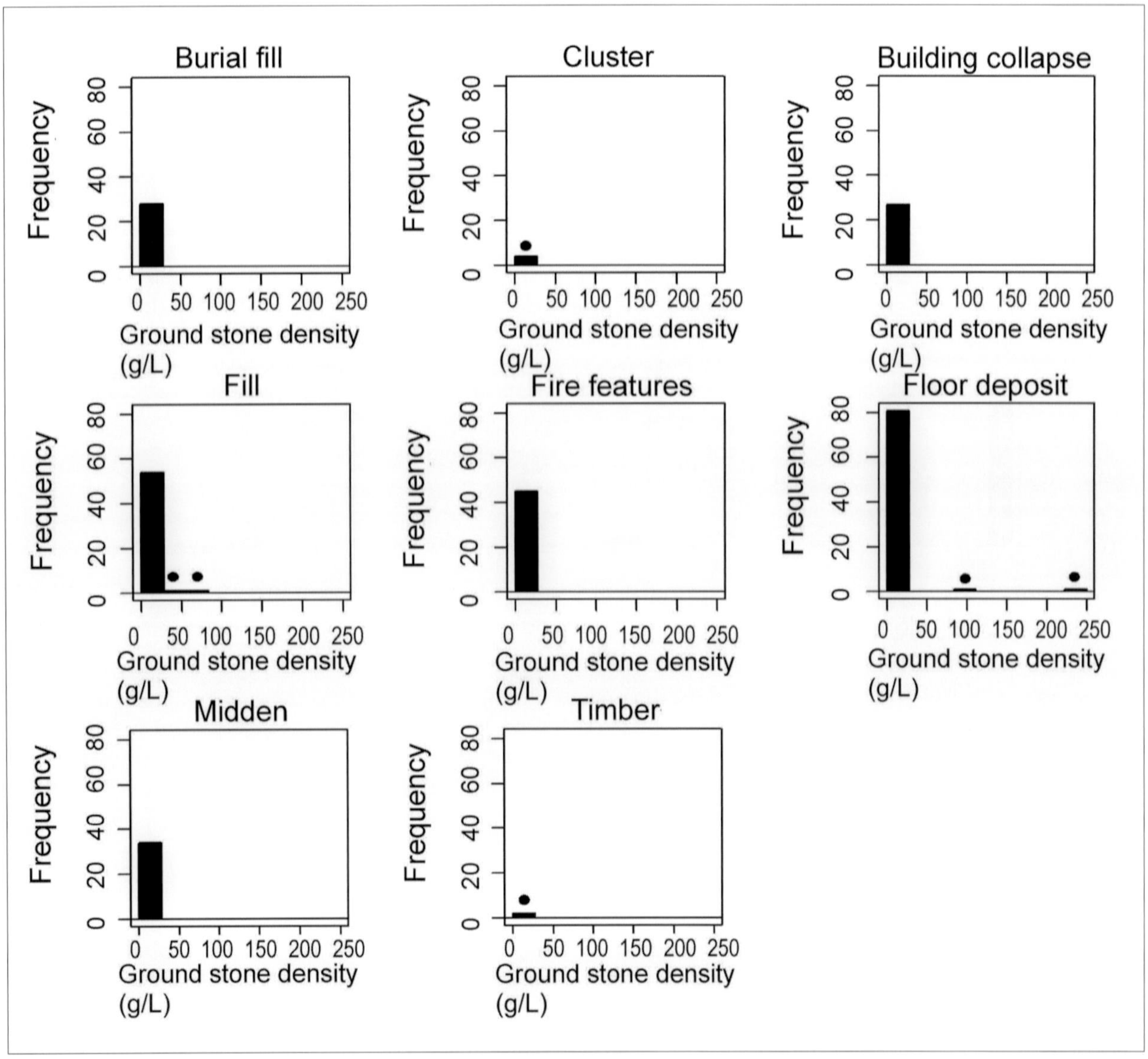

Figure 16.18. Ground stone weight density histograms for different context types included in the study. Dots represent presence.

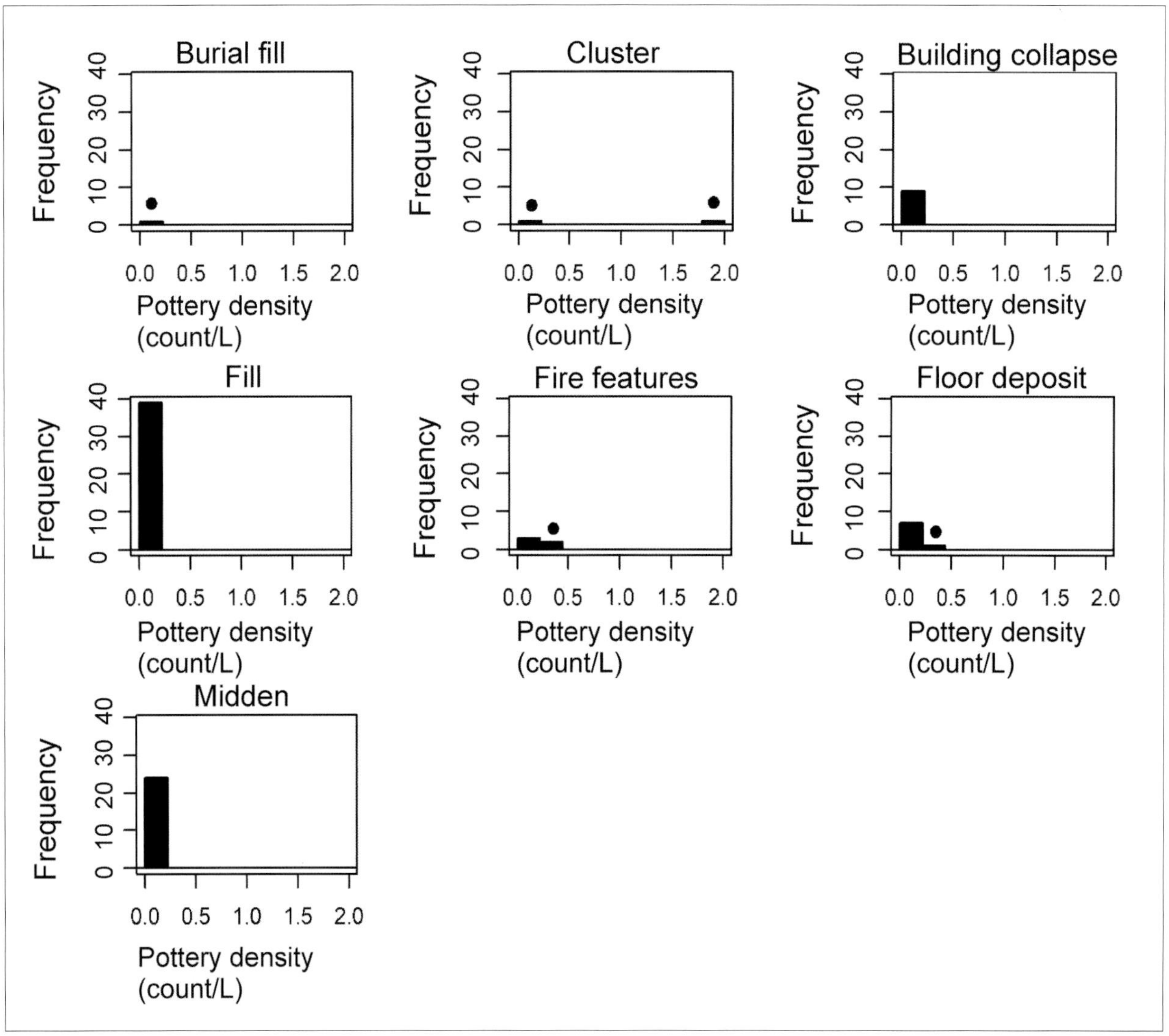

Figure 16.19. Pottery density histograms for different context types included in the study. Dots represent presence.

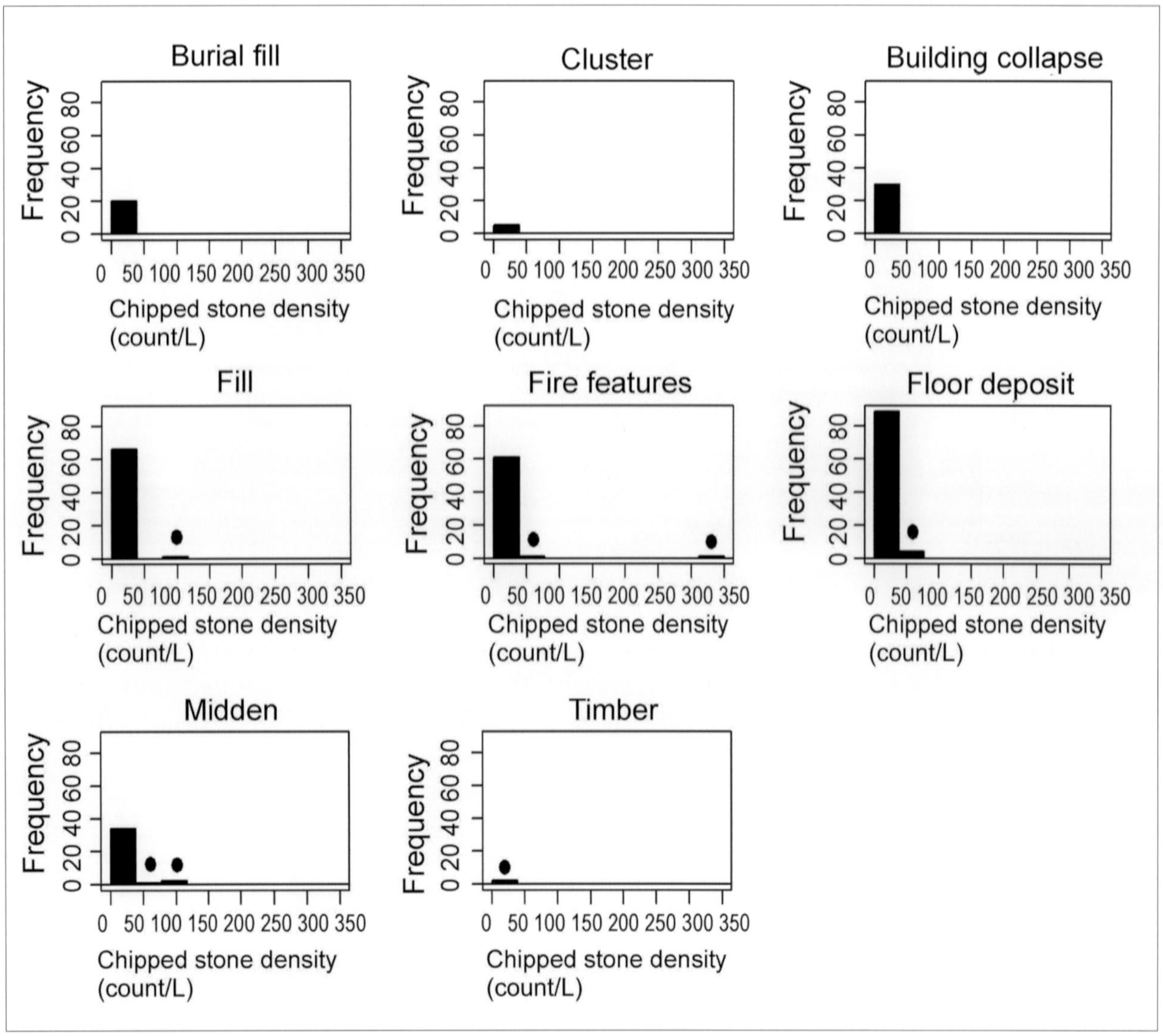

Figure 16.20. Chipped stone density histograms for different context types included in the study. Dots represent presence.

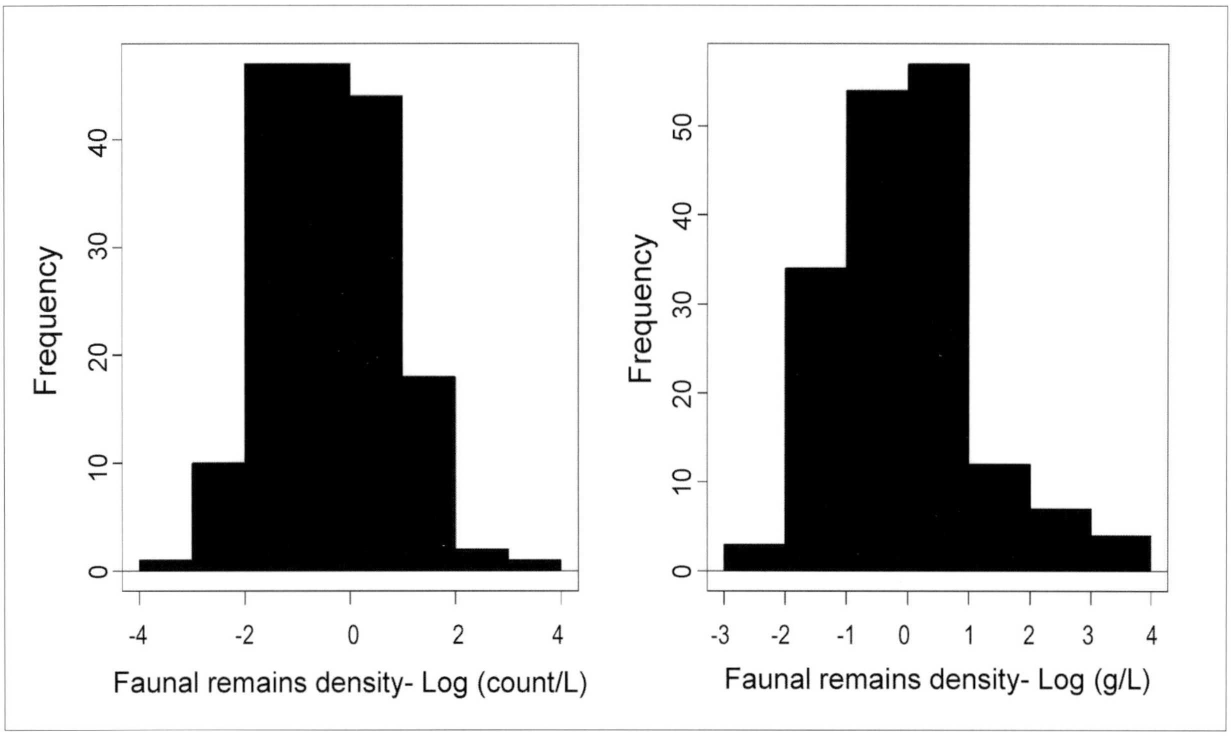

Figure 16.21. Histograms of natural log-transformed faunal remains densities across all units included in the study.

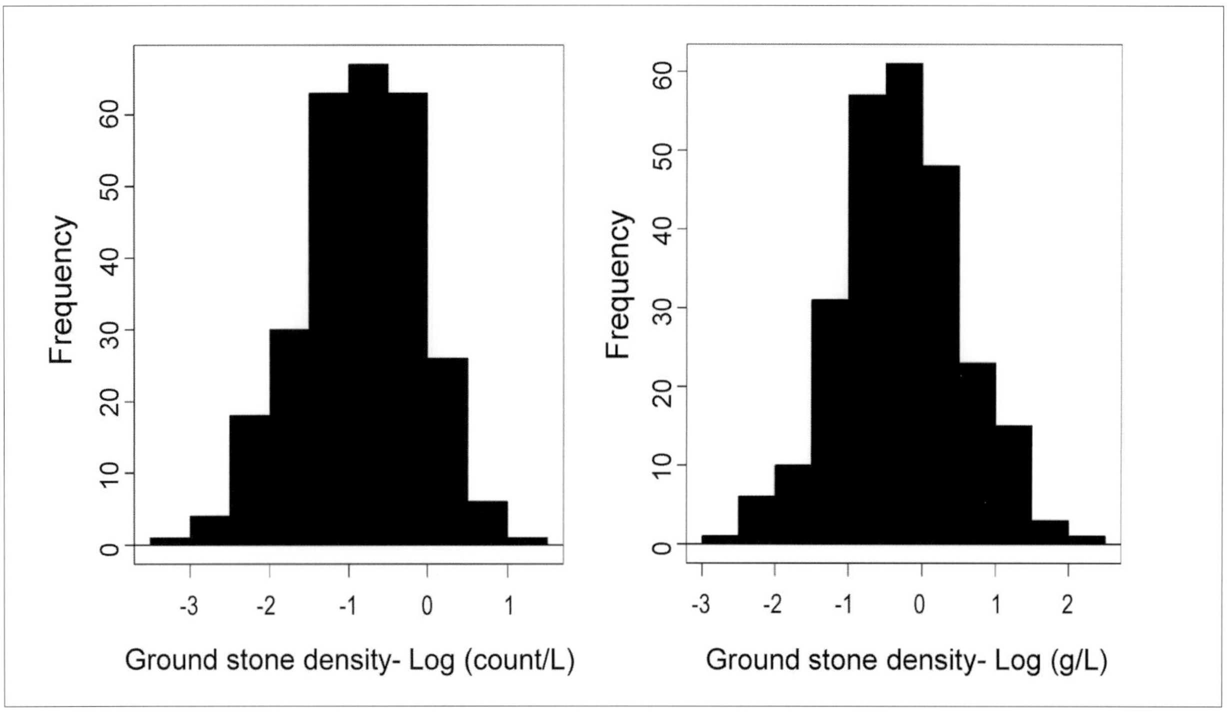

Figure 16.22. Histograms of natural log-transformed ground stone densities across all units included in the study.

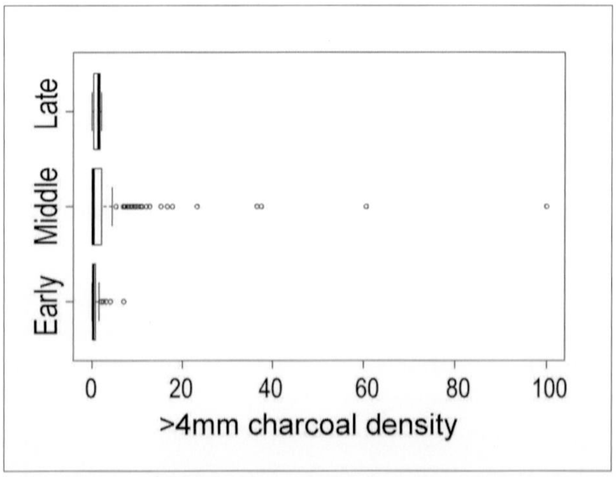

Figure 16.23. Wood charcoal density boxplots by period.

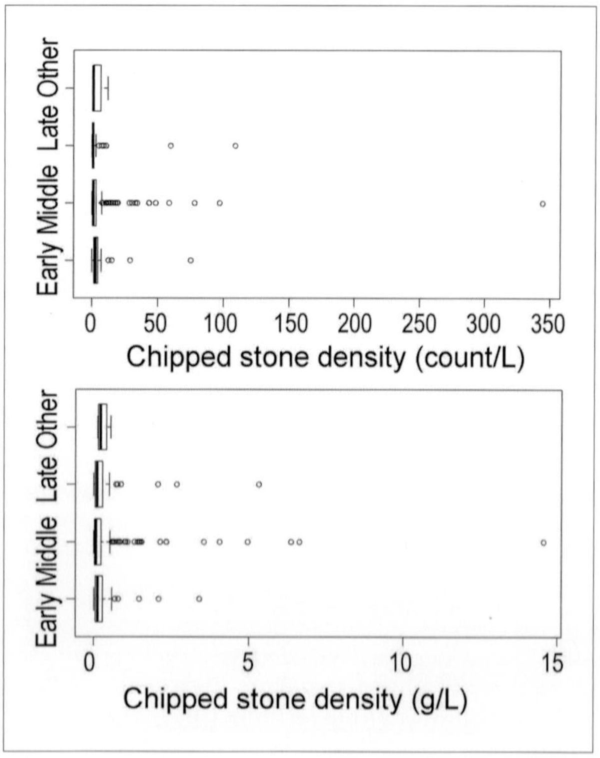

Figure 16.24. Chipped stone count and weight density boxplots by period.

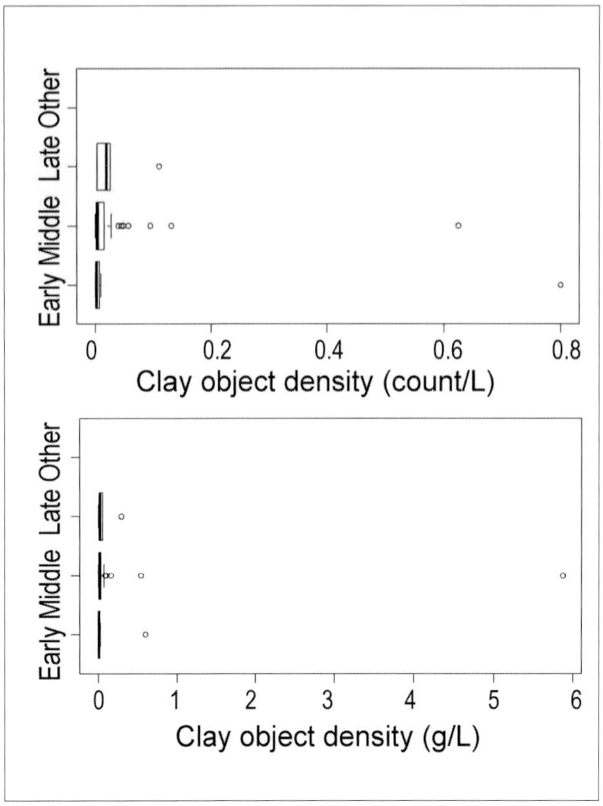

Figure 16.25. Clay object count and weight density boxplots by period.

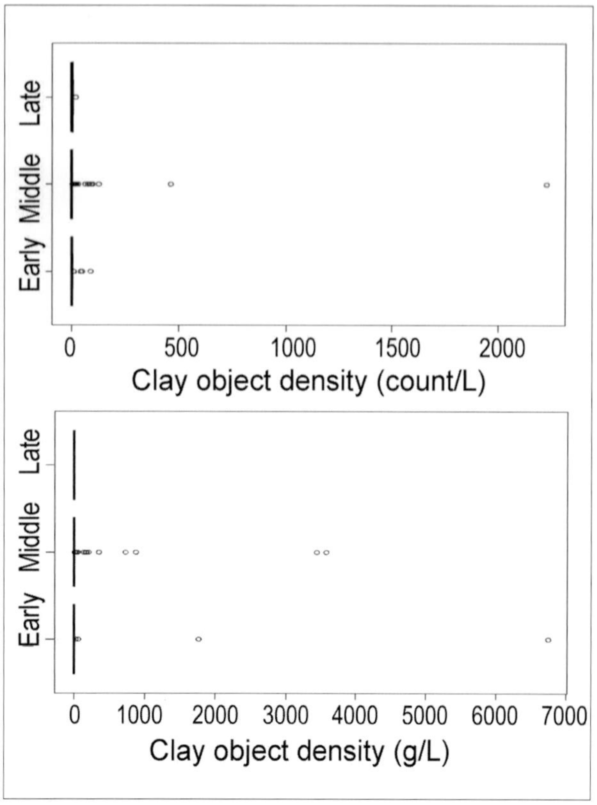

Figure 16.26. Faunal remains count and weight density boxplots by period.

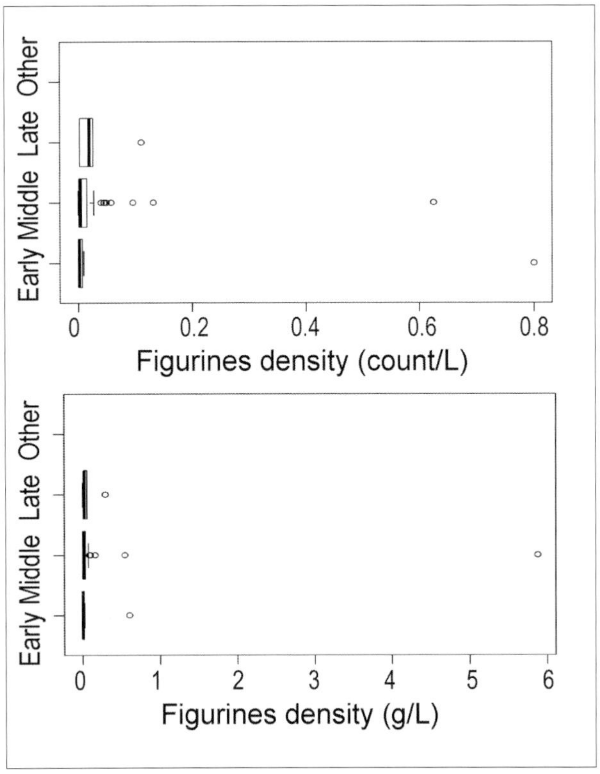

Figure 16.27. Figurines count and weight density boxplots by period.

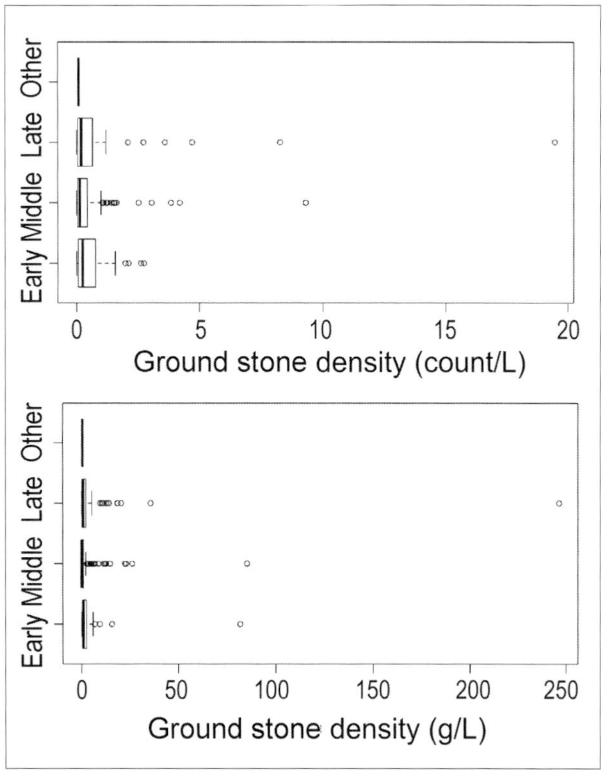

Figure 16.28. Ground stone count and weight density boxplots by period.

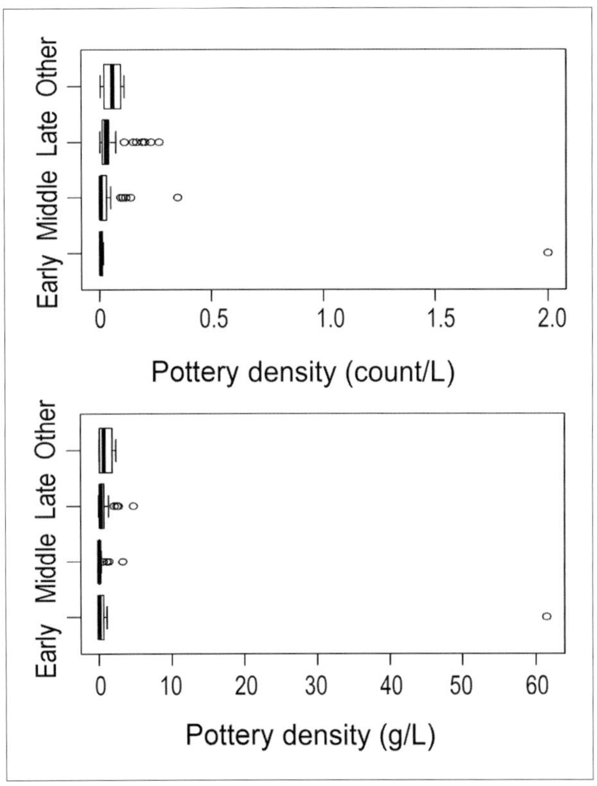

Figure 16.29. Pottery count and weight density boxplots by period.

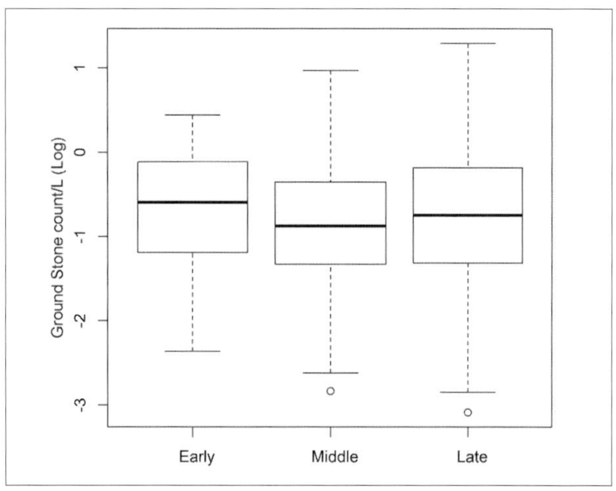

Figure 16.30. Boxplots of natural log-transformed ground stone densities by period.

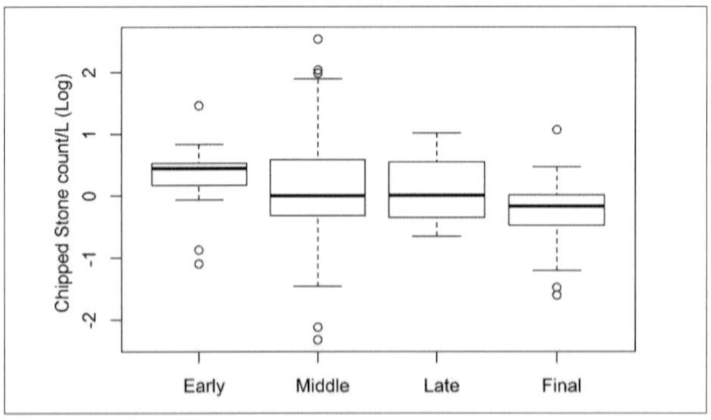

Figure 16.31. Boxplots of natural log-transformed chipped stone densities by period.

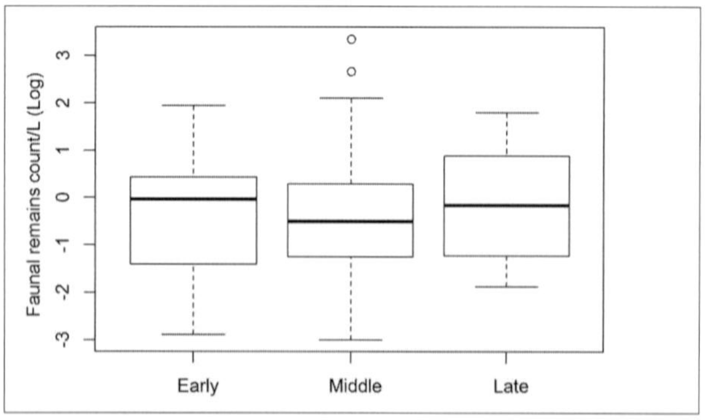

Figure 16.32. Boxplots of natural log-transformed faunal remains densities by period.

Discussion

As demonstrated in previous analyses, distributions of find categories follow a similar pattern, independent of their preservation conditions (Cross-May 2005; Mazzucato 2013; see also Vasić, this volume, Chapter 3). That is, a common right-skewed distribution of densities was observed for all examined find categories. Across context types, with regard to density, most deposits contain similar densities of materials. Some short-lived events (for example, fire features, clusters and sometimes floor deposits) rarely also contain much higher densities of material. Such a distinction was also detected in previous work (see 'activity deposits', Mazzucato 2013). We have been able to refine this further here by showing that some remains (for example, those deriving from fuel waste) can show rare high-density values in fire features. At the same time, some remains, such as animal bone and pottery, also rarely show high-density values in cluster deposits. Interestingly, both chipped stone and ground stone show a tendency for rare, high-density deposition events in fire features. Whether this situation is related to discard of production waste and/or defunct artefacts in fire features or whether the density distributions across space signify cultural practices could be investigated further.

In examinations of temporal variability of density values, we were not able to detect significant changes, through either univariate or multivariate descriptive analyses. As mentioned, the most relevant patterns were detected in ranges of densities across different context types, drawing a distinction between primary deposits/activity areas vs secondary deposits/waste discard areas. This observation appears to be in line with observations made by Vasić on density of various remains from heavy residue (Chapter 3). Following on from a detailed examination of univariate density distributions, multivariate analyses were performed using MFA on two distinct context types with interpretative potential. Multivariate analyses on floor deposits highlighted the significant contributions of all classes of archaeobotanical remains, likely representing traces of routine activities, including remnants of fuel waste, crop cleaning waste and food preparation waste. On the other hand, patterning in the dataset was also driven by variations in the densities of scarcer find categories such as figurines, clay objects and pottery, potentially signifying short-term deposits on floors. Thus it is likely that with the subset of units under study, we are able to depict variations in use and activity patterns in different buildings. Examinations of units included in the study suggested some weak relationships amongst units of the same building and a limited range of affinity amongst units in the same area of excavation (North vs South).

Multivariate analyses on midden/fill type deposits also highlighted similar patterns regarding temporal trends and spatial affiliation. Similarly, no clear temporal trends were observed in density values; and again similarly to floor deposits, there were weak affinities amongst units of the same area of excavation (North vs South/TPC) which might suggest similar patterns and densities of discard in adjacent/associated spaces across the site. In these secondary deposits, however, different relationships between find categories were observed. For example, the close association of pottery, figurine and clay objects in floor deposits is not matched by a similar association in middens. Similarly, the close correlation between wood charcoal and all classes of non-wood archaeobotanical remains is not replicated in middens/fills. Instead, some archaeobotanical remains associated with food preparation (for example, cereal grain) are more closely affiliated with faunal remains in secondary deposits. Additionally, in these deposits, discard patterns of ground stone and clay objects also appear to be more similar. It is likely that these wider patterns in find densities represent differences in the ranges of activities and discard patterns across different context types.

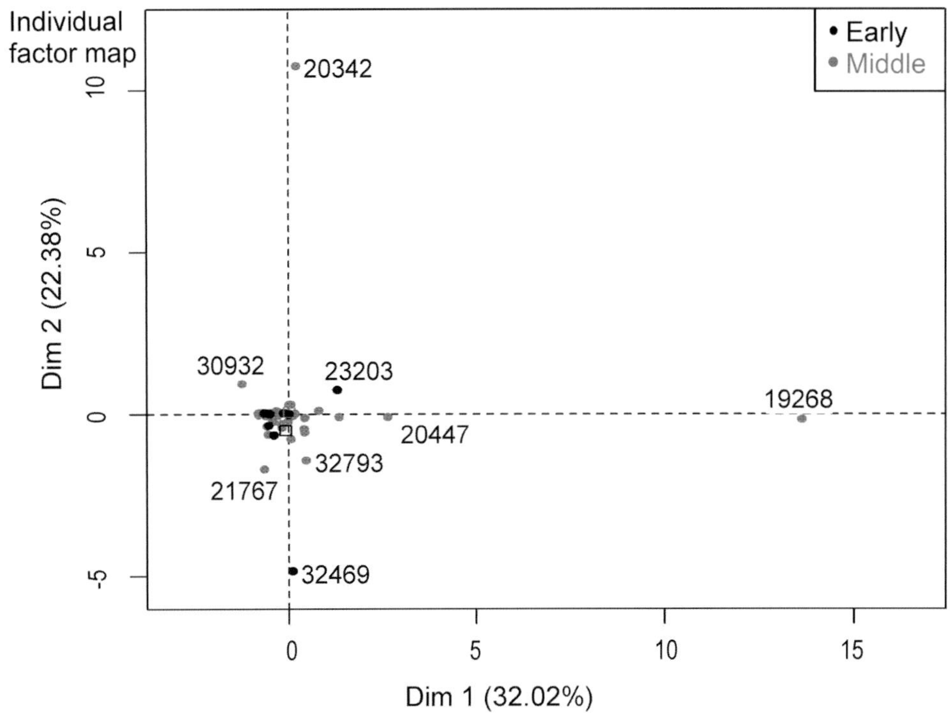

Figure 16.33. Biplot of individual units (individuals factor map), results of MFA on units from floor deposits (plotted by period); boxes represent centroids for groups of individuals (for colour version, see online supplementary material).

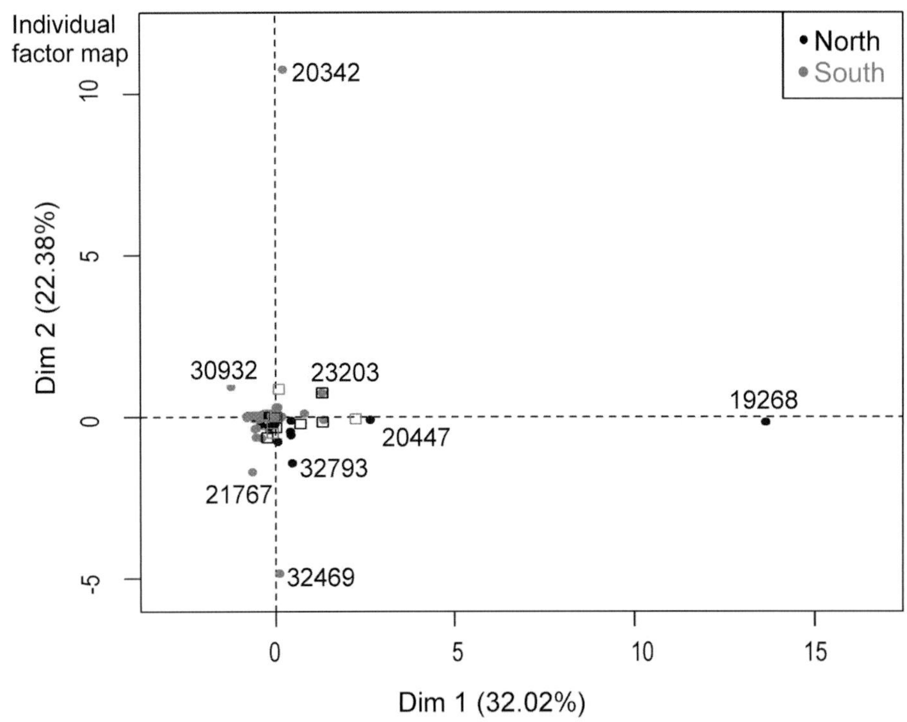

Figure 16.34. Biplot of individual units (individuals factor map), results of MFA on units from floor deposits (plotted by excavation area); boxes represent centroids for groups of individuals (for colour version, see online supplementary material).

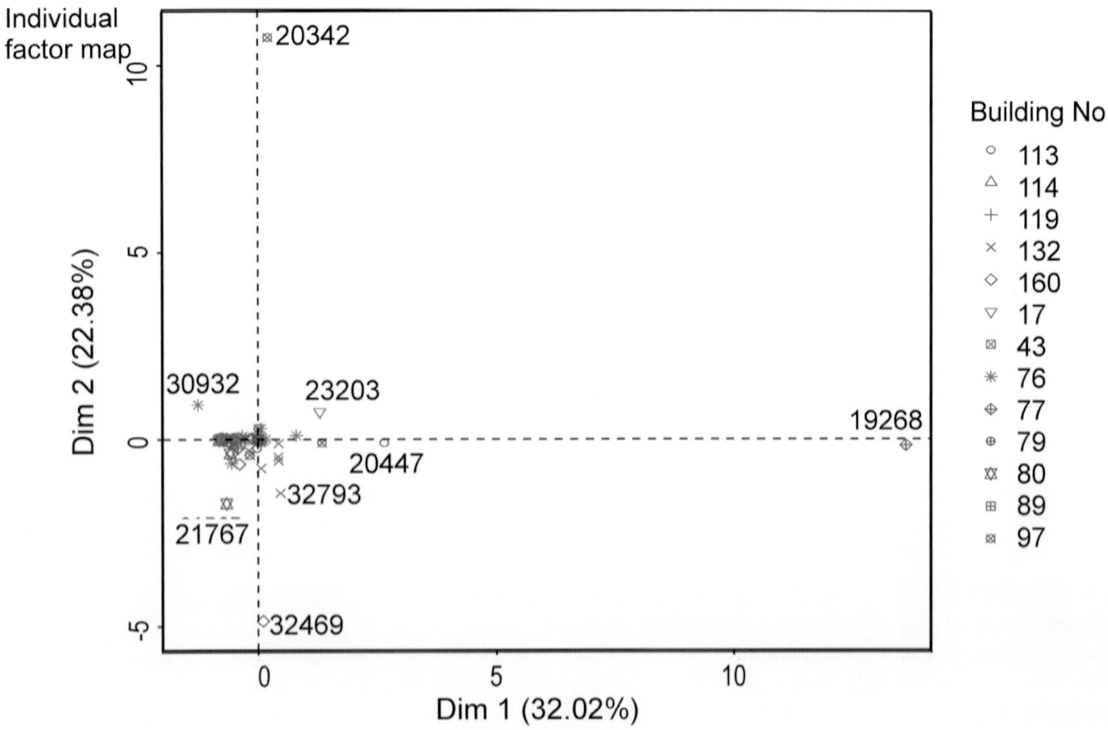

Figure 16.35. Biplot of individual units (individuals factor map), results of MFA on units from floor deposits (plotted by building) (for colour version, see online supplementary material).

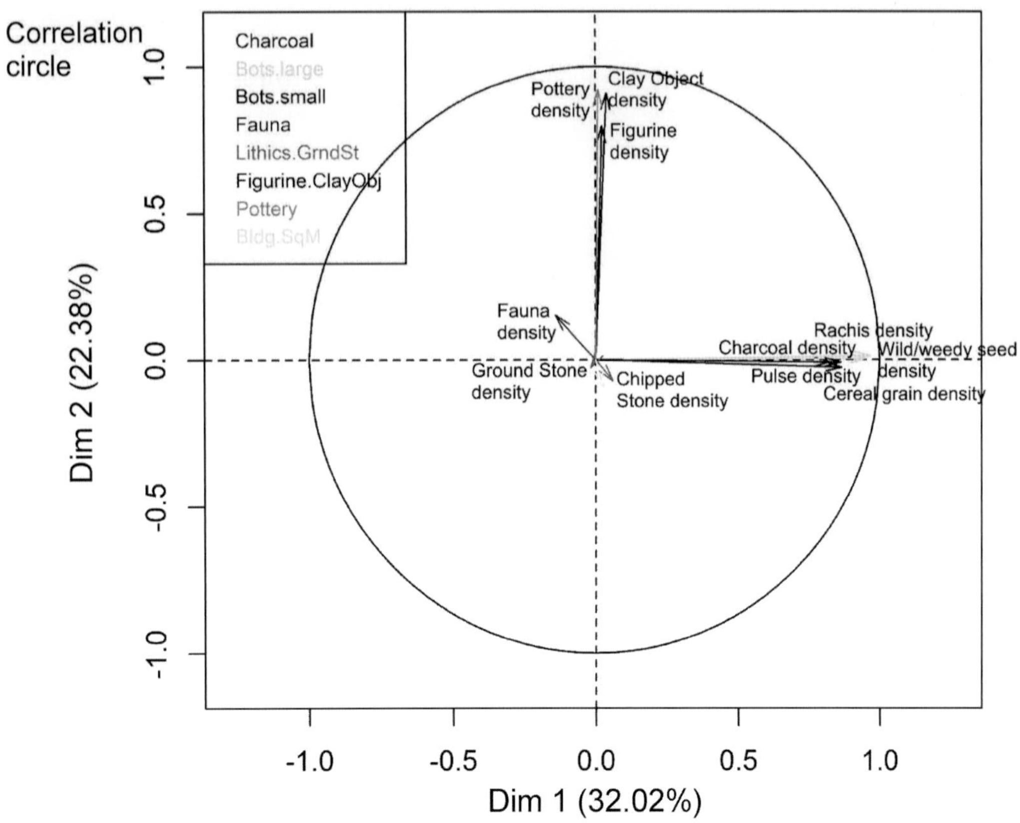

Figure 16.36. Plot of variables, results of MFA on units from floor deposits (for colour version, see online supplementary material).

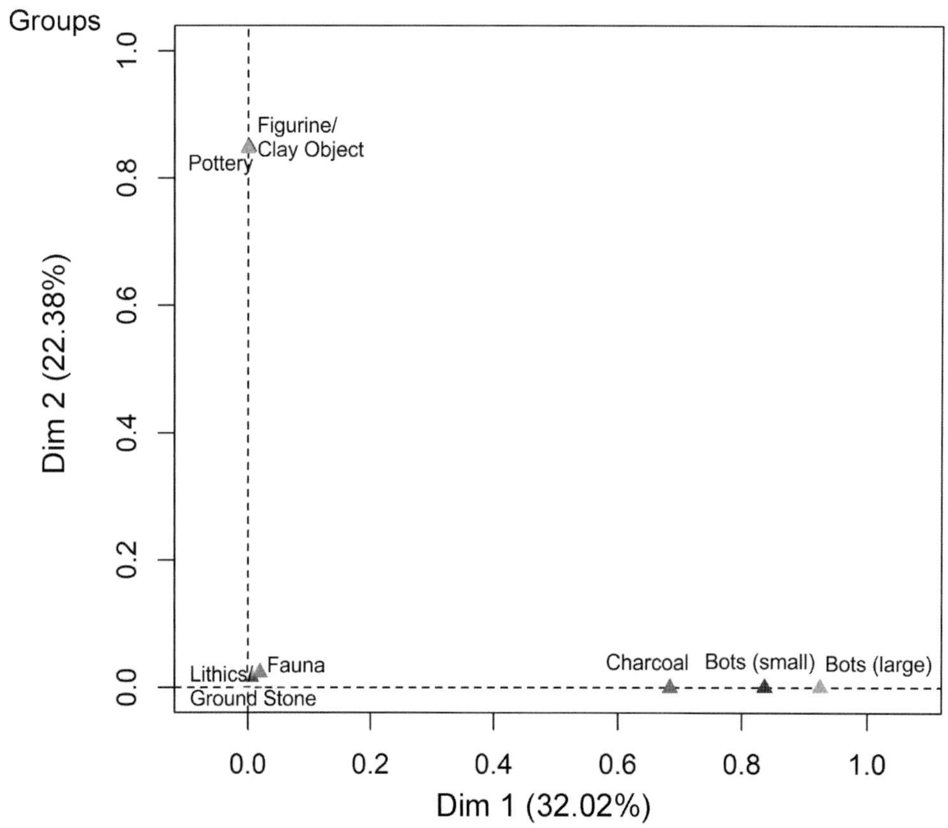

Figure 16.37. Biplot of groups of variables, results of MFA on units from floor deposits (for colour version, see online supplementary material).

These insights offer further refinement of observations of higher densities of remains from midden contexts, as reported by several authors working on various data groups such as archaeobotanical remains (Bogaard et al. 2014), faunal remains (Martin, Russell 2000) and micromorphology (Matthews 2005a; 2005b).

Conclusions

In this chapter we aimed to look at cross-cutting relationships between find categories, their deposition and accumulation through time and across various context types. Our aim was ultimately to shed light on any patterning in activities and their material residues by taking into account multiple factors relating to use, discard and preservation of various finds. We demonstrated the commonality in deposition density across time periods investigated through the study, while depicting patterns across context types and areas of occupation. Building on previous work carried out to date on the datasets from the site, in this analysis we were able to make some refinements to find categories, dividing archaeobotanical remains of multiple origin (fuel, food, crop waste) into finer groupings. Similar approaches could also be applied to chipped and ground stone datasets, which may provide further insights into the use of space and discard of artefacts, production waste etc. An interesting aspect of data integration was the detection of weak associations between units associated with the same building, suggesting consistency in discard and deposition patterns associated with specific households.

As demonstrated throughout this volume, there is clearly a change in the use and organisation of the site in the Late and Final periods (see Chapter 1). As noted by Tarkan (Chapter 4) there is an increase in pottery use overall and greater diversity of wares in the later phases of occupation, while there is a noticeable increase in the densities of edge tools in the ground stone assemblage (Tsoraki Chapter 13). In addition, various lines of evidence support the view that towards the later phases there is a general expansion in the use of the landscape, for example, for wood, clay and bead procurement alongside a range of shifts in raw material sources (this volume, Chapters 4 and 9; Kabukcu et al., Volume 13, Chapter 4) and increasing levels of mobility (e.g., Knüsel et al., Volume 13, Chapter 16). Interestingly, as discussed in this chapter, we have not detected significant temporal shifts in material densities to match any qualitative shifts in artefact types used or shifts in raw materials. Thus, it appears that despite apparent shifts in the materials themselves, the patterning of their use and discard may have remained largely similar through time.

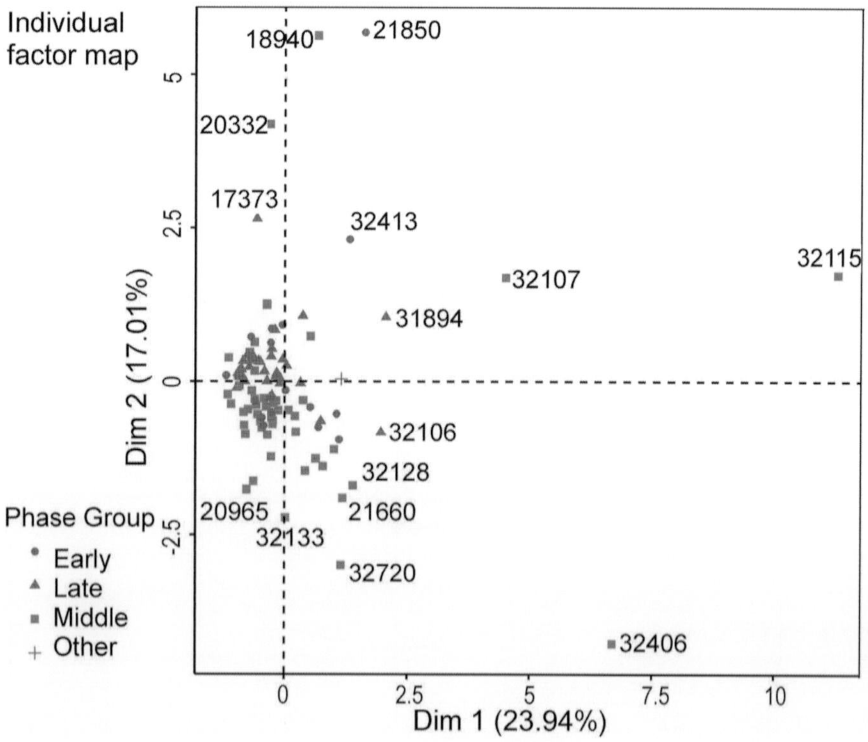

Figure 16.38. Biplot of individual units (individuals factor map), results of MFA on units from midden deposits (plotted by period) (for colour version, see online supplementary material).

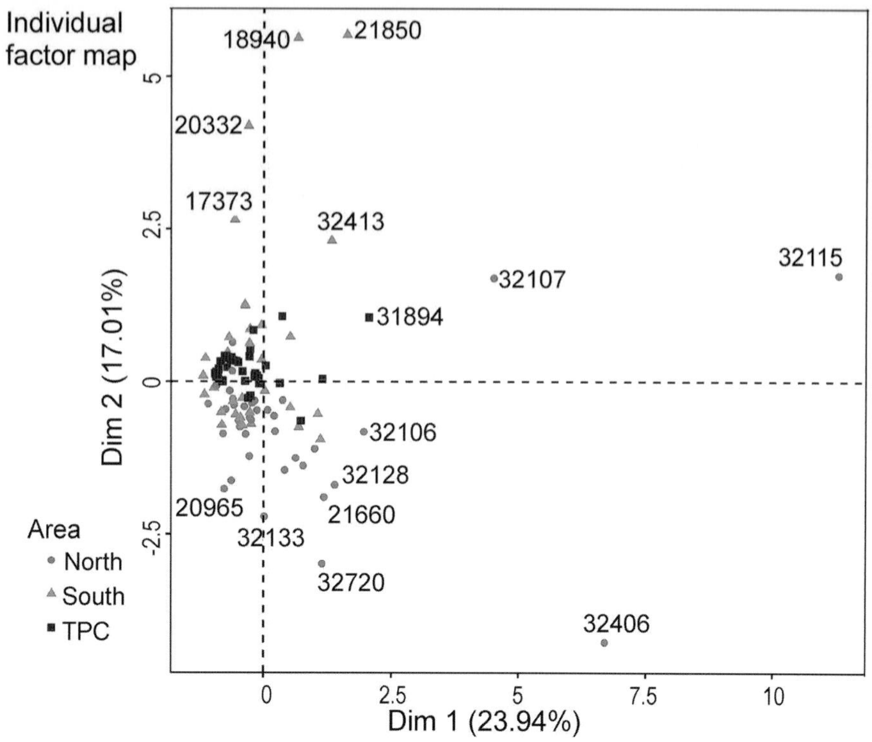

Figure 16.39. Biplot of individual units (individuals factor map), results of MFA on units from midden deposits (plotted by excavation area) (for colour version, see online supplementary material).

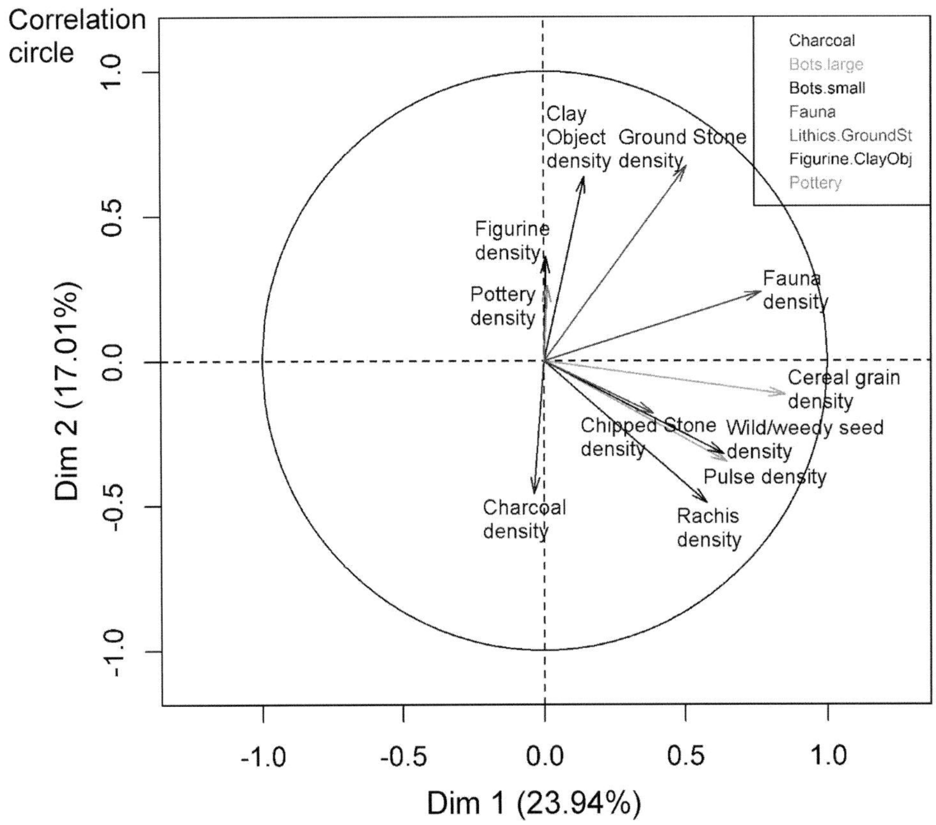

Figure 16.40. Plot of variables, results of MFA on units from midden deposits (for colour version, see online supplementary material).

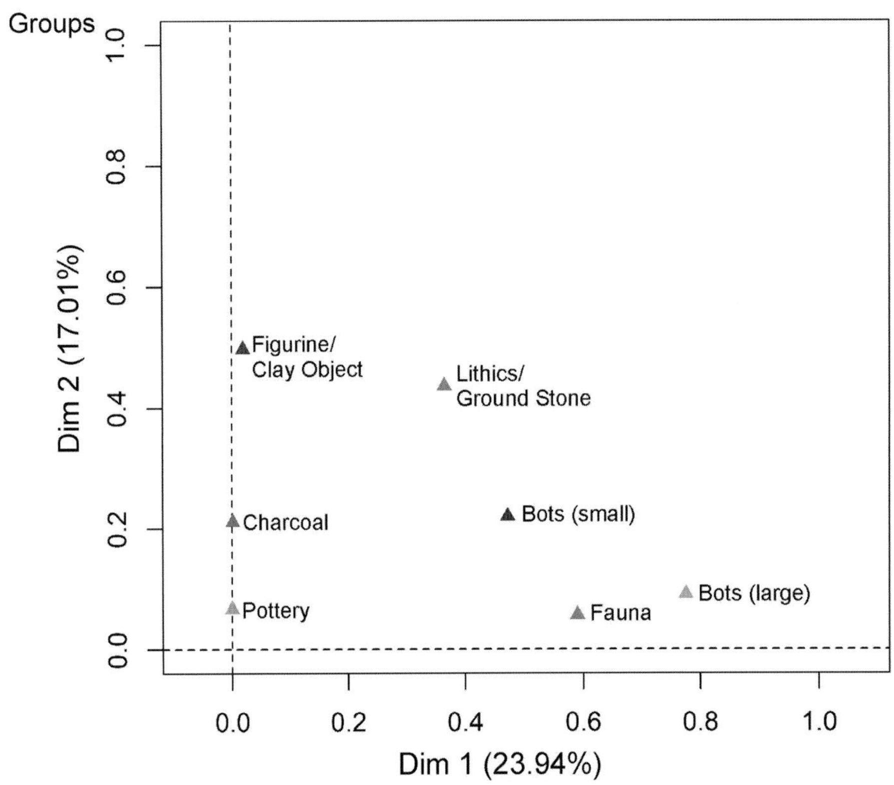

Figure 16.41. Biplot of groups of variables, results of MFA on units from midden deposits (for colour version, see online supplementary material).

Supplementary material

For supplementary material related to this chapter, please visit https://doi.org/10.18866/BIAA/e-14. It comprises colour versions of figures 16.33–16.41.

Bibliography

Adams, J.L. 2002: *Ground Stone Analysis: A Technological Approach*. Salt Lake City, University of Utah Press

Adams, R.L. 2005: 'Ethnoarchaeology in Indonesia illuminating the ancient past at Çatalhöyük?' *American Antiquity* 70.1: 181–88

Adovasio, J.M. 1977: 'The textiles and basketry impressions from Jarmo' *Paléorient* 3: 223–30

— 1983: 'Notes on the textile and basketry impressions from Jarmo' in L.S. Braidwood, R.J. Braidwood, B. Howe, C.A. Reid, P.J. Watson (eds), *Prehistoric Archaeology Along the Zagros Flanks* (University of Chicago Oriental Institute Publications 105). Chicago, University of Chicago: 425–26

Adovasio, J.M., Soffer, O., Klíma, B. 1996: 'Upper Palaeolithic fibre technology: interlaced woven finds from Pavlov I, Czech Republic, c. 26,000 years ago' *Antiquity* 70: 526–34

Akıncı, Ö. 1968: 'Seramik killeri ve jeolojisi' *Maden Tetkik ve Arama Dergisi* 71: 63–72

Akkermans, P.M.M.G. 1987: 'A late Neolithic and early Halaf village at Sabi Abyad, northern Syria' *Paléorient* 13.1: 23–40

Akkermans, P.M.M.G., Duistermaat, K. 1996: 'Of storage and nomads. The sealings from Late Neolithic Sabi Abyad, Syria [with comments and reply]' *Paléorient* 22.2: 17–44

— 2004: 'More seals and sealings from Neolithic Tell Sabi Abyad, Syria' *Levant* 36: 1–11

Albrethsen, S.E., Brinch Petersen, E. 1976: 'Excavation of a Mesolithic cemetery at Vedbæk, Denmark' *Acta Archaeologica* 47: 1–28

Alfaro, C. 2012: 'Textiles from the Pre-pottery Neolithic Site of Tell Halula (Euphrates Valley, Syria)' *Paléorient* 38.1–2: 41–54

Altınbilek-Algül, C., Astruc, L., Binder, D., Pelegrin, J. 2012: 'Pressure blade production with a lever in the Early and Late Neolithic of the Near East' in P.M. Desrosiers (ed.), *The Emergence of Pressure Blade Making: From Origin to Modern Experimentation*. New York, Springer: 157–80

Amiet, P. 1966: *Elam*. Auvers-Sur-Oise, Archée

Anderson, E., Almond, M.J., Matthews, W. 2014: 'Analysis of wall plasters and natural sediments from the Neolithic town of Çatalhöyük (Turkey) by a range of analytical techniques' *Spectrochimica Acta Part A: Molecular and Biomolecular Spectroscopy* 133: 326–34

AnimalBase Project Group. 2005–2019: AnimalBase. Early Zoological Literature Online. www.animalbase.uni-goettingen.de

Anspach, M.R. 2019: 'Paired leopards and encircled prey' in I. Hodder (ed.), *Violence and the Sacred in the Ancient Near East: Girardian Conversations at Çatalhöyük*. Cambridge, Cambridge University Press: 129–50

Appadurai, A. 1986: 'Introduction: commodities and the politics of value' in A. Appadurai (ed.), *The Social Life of Things*. Cambridge, Cambridge University Press: 3–63

Arbuckle, B.S. 2013: 'The late adoption of cattle and pig husbandry in Neolithic central Turkey' *Journal of Archaeological Science* 40: 1805–15

— 2015: 'Large game depression and the process of animal domestication in the Near East' in M.S. Kerner, R. Dann, P. Bangsgaard (eds), *Climate and Ancient Societies*. Chicago, University of Chicago Press: 215–43

Arbuckle, B.S., Makarewicz, C.A. 2009: 'The early management of cattle (*Bos taurus*) in Neolithic central Anatolia' *Antiquity* 83.321: 669–86

Arbuckle, B.S., Kansa, S.W., Kansa, E., Orton, D., Çakırlar, C., Gourichon, L., Atıcı, L., Galik, A., Marciniak, A., Mulville, J., Buitenhuis, H., Carruthers, D., de Cupere, B., Demirergi, A., Frame, S., Helmer, D., Martin, L., Peters, J., Pöllath, N., Pawłowska, K., Russell, N., Twiss, K., Würtenberger, D. 2014: 'Data sharing reveals complexity in the westward spread of domestic animals across Neolithic Turkey' *PLoS ONE* 9: e99845

Asouti, E. 2005: 'Woodland vegetation and the exploitation of fuel and timber at Neolithic Çatalhöyük: report on the wood charcoal macro-remains' in I. Hodder (ed.), *Inhabiting Çatalhöyük: Reports from the 1995–99 Seasons*. Cambridge, McDonald Institute for Archaeological Research; London, British Institute at Ankara: 213–58

— 2013: 'Woodland vegetation, firewood management and woodcrafts at Neolithic Çatalhöyük' in I. Hodder (ed.), *Humans and Landscapes of Çatalhöyük: Reports from the 2000–2008 Seasons* (Çatalhöyük Research Project Series 8). London, British Institute at Ankara; Los Angeles, Cotsen Institute of Archaeology Press: 129–62

Asouti, E., Kabukcu, C. 2012: 'Anthracology archive report' *Çatalhöyük 2012 Archive Report*. http://www.catalhoyuk.com/archive_reports/2012

Atalay, S. 1998: 'Clay ball archive report 1998' *Çatalhöyük Archive Report 1998*. http://www.catalhoyuk.com/archive_reports/1998

— 1999: 'Clays, clay balls, and other clay objects' Çatalhöyük Archive Report 1999. http://www.catalhoyuk.com/archive_reports/1999

— 2000: 'Clay balls and objects' *Çatalhöyük Archive Report 2000*. http://www.catalhoyuk.com/archive_reports/2000

— 2001: 'BACH Area clay balls, mini balls and geometric objects' *Çatalhöyük 2001 Archive Report*. http://www.catalhoyuk.com/archive_reports/2001

— 2003: *Domesticating Clay: Engaging with 'They'. The Social Life of Clay Balls from Çatalhöyük, Turkey and Public Archaeology for Indigenous Communities*. PhD thesis, University of Berkley California

— 2005: 'Domesticating clay: the role of clay balls, mini balls and geometric objects in daily life at Çatalhöyük' in I. Hodder (ed.), *Changing Materialities at Çatalhöyük: Reports from the 1995–99 Seasons* (Çatalhöyük Research Project Series 5). Cambridge, McDonald Institute for Archaeological Research; London, British Institute of Archaeology at Ankara: 139–68

— 2008: 'Clay balls research, statement' *Çatalhöyük Archive Report 2009*: 209–10. http://www.catalhoyuk.com/sites/default/files/media/pdf/Archive_Report_2008.pdf

— 2009: 'Clay balls and fire installations' *Çatalhöyük Archive Report 2009*. http://www.catalhoyuk.com/archive_reports/2009

— 2012: 'Analysis of clay balls from the BACH Area' in R. Tringham, M. Stevanović (eds), *Last House on the Hill: BACH Area Reports from Çatalhöyük, Turkey* (Çatalhöyük Research Project Volume 11). Los Angeles, Cotsen Institute of Archaeology Press: 385–90

— 2013: 'Clay balls, mini balls and geometric objects' in I. Hodder (ed.), *Substantive Technologies at Çatalhöyük: Reports from the 2000–2008 Seasons* (Çatalhöyük Research Project Volume 9). London, British Institute at Ankara; Los Angeles, Cotsen Institute of Archaeology Press: 247–52

Atalay, S., Hastorf, C.A. 2005: 'Foodways at Çatalhöyük' in I. Hodder (ed.), *Çatalhöyük Perspectives: Themes from the 1995–99 Seasons* (Çatalhöyük Research Project Series 6). Cambridge, McDonald Institute for Archaeological Research; London, British Institute at Ankara: 109–24.

— 2006: 'Food, meals, and daily activities: food habitus at Neolithic Çatalhöyük' *American Antiquity* 71.2: 283–319

Ataman, K. 1986: 'A group of projectile points from Can Hasan III' in *IV. Araştırma Sonuçları Toplantısı*. Ankara, T.C. Kültür ve Turizm Bakanlığı, Eski Eserler ve Müzeler Genel Müdürlüğü: 339–46

— 1988: *Chipped Stone Assemblage from Can Hasan III: A Study in Typology, Technology and Function*. PhD thesis, University of London, London

Attewell, P.B., Farmer, I.W. 1976: *Principles of Engineering Geology*. London, Chapman and Hall

Avis, J. 2010: 'Figurines clay composition report' *Çatalhöyük 2010 Archive Report*.www.catalhoyuk.com/archive_reports/2010

Bader, N.O. 1989: *Earliest Cultivators in Northern Mesopotamia. The Investigations of Soviet Archaeological Expedition in Iraq at Settlements Tell Magzaliya, Tell Sotto, Kül Tepe*. Moscow, Nauka

Bailey, D.W. 1994: 'Reading prehistoric figurines as individuals' *World Archaeology* 25.3: 321–31

— 2017: 'Southeast European Neolithic figurines: beyond context, interpretation and meaning' in T. Insoll (ed.), *The Oxford Handbook of Prehistoric Figurines*. Oxford, Oxford University Press: 823–50

Bains, R. 2012: *The Social Significance of Neolithic Stone Bead Technologies at Çatalhöyük*. PhD thesis, University College London, London

Bains, R., Vasić, M., Bar-Yosef Mayer, D.E., Russell, N., Wright, K.I., Doherty, C. 2013: 'A technological approach to the study of personal ornamentation and social expression at Çatalhöyük' in I. Hodder (ed.), *Substantive Technologies at Çatalhöyük: Reports from the 2000–2008 Seasons* (Çatalhöyük Research Project Series 9). London, British Institute at Ankara; Los Angeles, Cotsen Institute of Archaeology Press: 331–64

Baird, D. 2002: 'Early Holocene settlement in Central Anatolia: problems and prospects as seen from the Konya Plain' in F. Gérard, L. Thissen (eds), *The Neolithic of Central Anatolia: Internal Developments and External Relations During the 9th–6th Millennia cal BC, Proceedings of the International CANeW Round Table, Istanbul, 23–24 November 2001*. Istanbul, Ege Yayınları: 139–59

— 2005: 'The history of settlement and social landscapes in the Early Holocene in the Çatalhöyük area' in I. Hodder (ed.), *Çatalhöyük Perspectives: Themes from the 1995–99 Seasons* (Çatalhöyük Research Project Series 6). Cambridge, McDonald Institute for Archaeological Research; London, British Institute at Ankara: 55–74

— 2007a: 'The Boncuklu Project: the origins of sedentism, cultivation and herding in Central Anatolia' *Anatolian Archaeology* 13: 14–18

— 2007b: 'Pınarbaşı: from Epipalaeolithic camp site to sedentarising village in central Anatolia' in M. Özdoğan, N. Başgelen (eds.), *Neolithic in Turkey: New Excavations and New Discoveries*. Istanbul: Arkeoloji ve Sanat Yayınları: 285–311

— 2010a: 'The Boncuklu Project: investigating the beginnings of agriculture, sedentism and herding in Central Anatolia' *Anatolian Archaeology* 16: 11–13

— 2010b: 'Was Çatalhöyük a centre? The implications of a Late Aceramic Neolithic assemblage from the neighbourhood of Çatalhöyük' in D. Bolger, L.C. Maguire (eds), *Development of Pre-State Communities in the Ancient Near East*. Oxford, Oxbow Books: 207–16

— 2012: 'The Late Epipaleolithic, Neolithic and Chalcolithic of the Anatolian Plateau, 13,000–4,000 BC calibrated' in D.T. Potts (ed.), *Blackwell's Companion to Near Eastern Archaeology*. Oxford, Blackwell: 431–65

— 2016: 'Boncuklu: the spread of farming and the antecedents of Çatalhöyük' *Heritage Turkey* 6: 15–20

Baird, D., Carruthers, D., Fairbairn, A., Pearson, J. 2011: 'Ritual in the landscape: evidence from Pınarbaşı in the seventh-millennium cal BC Konya Plain' *Antiquity* 85.328: 380–94

Baird, D., Fairbairn, A., Martin, L., Middleton, C. 2012: 'The Boncuklu Project: the origins of sedentism, cultivation and herding in Central Anatolia' in M. Özdoğan, N. Başgelen (eds), *The Neolithic in Turkey* 3: *New Excavations, New Research: Central Turkey*. Istanbul, Arkeoloji ve Sanat Yayınları: 219–44

Baird, D., Asouti, E., Astruc, L., Baysal, A., Baysal, E., Carruthers, D., Fairbairn, A., Kabukcu, C., Jenkins, E., Lorentz, K., Middleton, C., Pearson, J., Pirie, A. 2013: 'Juniper smoke, skulls and wolves' tails. The Epipalaeolithic of the Anatolian plateau in its south-west Asian context: insights from Pınarbaşı' *Levant* 45.2: 175–209

Baird, D., Fairbairn, A., Martin, L. 2017: 'The animate house, the institutionalization of the household in Neolithic Central Anatolia' *World Archaeology* 49.5: 753–76

Balkan-Atlı, N. 1994: *La Néolithisation de l'Anatolie*. Istanbul: IFEA; Paris: Editions Boccard

— 2005: 'Yontmatas̨ endüstrisi' in R. Duru, G. Umurtak (eds), *Höyücek. Results of the Excavations 1989–1992*. Ankara, Türk Tarih Kurumu Yayınları: 130–37

Balkan-Atlı, N., Binder, D., Cauvin, M.-C. 1999: 'Obsidian: sources, workshops and trade in Central Anatolia' in M. Özdoğan, N. Başgelen (eds.), *Neolithic in Turkey. The Cradle of Civilization. New Discoveries*. Istanbul, Arkeoloji ve Sanat Yayınları: 133–45

Barański, M. 2016: 'GDN research on the Late Neolithic architecture' *Çatalhöyük 2016 Archive Report*. http://www.catalhoyuk.com/archive_reports/2016

Barański, M.Z., García-Suárez, A., Klimowicz, A., Love, S., Pawłowska, K. 2015: 'The architecture of Neolithic Çatalhöyük as a process: complexity in apparent simplicity' in I. Hodder, A. Marciniak (eds), *Assembling Çatalhöyük*. Leeds, Maney Publishing: 111–26

Barber, E.J.W. 1991: *Prehistoric Textiles. The Development of Cloth in the Neolithic and Bronze Ages*. Princeton, Princeton University Press

Bar-Yosef Mayer, D.E. 2005: 'The exploitation of shells as beads in the Palaeolithic and Neolithic of the Levant' *Paléorient* 31.1: 176–85

— 2013: 'Mollusc exploitation at Çatalhöyük' in I. Hodder (ed.), *Humans and Landscapes of Çatalhöyük: Reports from the 2000–2008 Seasons* (Çatalhöyük Research Project Series 8). London, British Institute at Ankara; Los Angeles, Cotsen Institute of Archaeology Press: 329–38

Bar-Yosef Mayer, D.E., Gümüş, B.A., İslamoğlu, Y. 2010: 'Fossil hunting in the Neolithic: shells from the Taurus Mountains at Çatalhöyük, Turkey' *Geoarchaeology* 25.3: 375–92

Bar-Yosef, O. 1985: *A Cave in the Desert: Nahal Hemar*. Jerusalem, The Israel Museum

— 2020: 'The Neolithic revolution in the Fertile Crescent and the origins of fibre technology 'in W. Schier, S. Pollock (eds), *The Competition of Fibres. Early Textile Production in Western Asia, South-East and Central Europe (10,000–500 BC)*. Oxford, Oxbow Books: 5–16

Bar-Yosef, O., Gopher, A. (eds) 1997: *An Early Neolithic Village in the Jordan Valley Part I: The Archaeology of Netiv Hagdud* (American School of Prehistoric Research Bulletin 43). Cambridge (MA), Peabody Museum of Archaeology and Ethnology, Harvard University

Bar-Yosef, O., Meadow, R.H. 1995: 'The origins of agriculture in the Near East' in T.D. Price, A.B. Gebauer (eds), *Last Hunters, First Farmers: New Perspectives on the Prehistoric Transition to Agriculture*. Santa Fe (NM), University of Washington Press: 39–94

Barzilai, O. 2013: 'The bidirectional blade industries of the southern Levant' in F. Borrell, J.J. Ibáñez, M. Molist (eds), *Stone Tools in Transition: From Hunter-Gatherers to Farming Societies in the Near East*. Bellaterra (Barcelona), Universitat Autònoma de Barcelona, Servei de Publicacions: 59–72

Barzilai, O., Garfinkel, Y. 2006: 'Bidirectional blade technology after the PPNB: new evidence from Sha'ar Hagolan, Israel' *Neo-lithics* 1.6: 27–31

Bayliss, A., Brock, F., Farid, S., Hodder, I., Southon, J., Taylor, R.E. 2015: 'Getting to the bottom of it all: a Bayesian approach to dating the start of Çatalhöyük' *Journal of World Prehistory* 28.1: 1–26

Bayliss, A., Barański, M.B., Czerniak, L., Marciniak, A., Goslar, T., Southon, J., Taylor, R.E. in press: 'Interpreting chronology' in A. Marciniak, L. Czerniak (eds), *Late Neolithic at Çatalhöyük in the TP Area*

Baysal, A. 1998: 'Provisional report on geological surveys in relation to groundstone study' *Çatalhöyük 1998 Archive Report*. http://www.catalhoyuk.com/archive_reports/1998

— 2009: *Social and Economic Implications of the Life Histories of Ground Stone at Neolithic Çatalhöyük*. PhD thesis, University of Liverpool

Baysal, A., Wright, K.I. 2005: 'Cooking, crafts and curation: ground-stone artefacts from Çatalhöyük' in I. Hodder (ed.), *Changing Materialities at Çatalhöyük. Reports from the 1995–99 Seasons* (Çatalhöyük Research Project 5). Cambridge, McDonald Institute for Archaeological Research; London, British Institute of Archaeology at Ankara: 307–24

Baysal, E.L. 2016a: 'Beads at the place of white earth: Late Neolithic and Early Chalcolithic Aktopraklik, north-western Turkey' *BEADS. Journal of the Society of Bead Researchers* 28: 50–59

— 2016b: 'Turkey's first evil eye? The manufacture and use of blue beads in the Neolithic' *Past* 82: 14–15

— 2017a: 'Personal ornaments in Neolithic Turkey, the current state of research and interpretation' *Journal of Archaeology and Art* 155: 1–22

— 2017b: 'Reflections of faraway places: the Chalcolithic personal ornaments of Canhasan I' *Anatolian Studies* 67: 29–49

Baysal, E.L., Erek, C.M. 2018: 'Material movement in the Near Eastern Epipalaeolithic: implications of the shell and stone beads of Direkli Cave, Turkey' *Journal of Field Archaeology* 43.8: 591–603

Becker, A.E. 1995: *Body, Self, and Society: The View from Fiji*. Philadelphia, University of Pennsylvania Press

Bécue-Bertaut, M., Pagès, J. 2008: 'Multiple factor analysis and clustering of a mixture of quantitative, categorical and frequency data' *Computational Statistics and Data Analysis* 52: 3255–68

Beek, R.J., Mason, M. 2002: *Mana Pounamu. New Zealand Jade*. Wellington (NZ), Reed

Bellot-Gurlet, L., Poupeau, G., Cauvin, M.-C., Chataigner, C., Coqueugniot, E., Keller, F. 1999: 'L'approvisionnement en obsidienne des sites néolithiques de la moyenne vallée de l'Euphrate entre 9000 et 8600 ans av. J.-C.' *Archéométrie 99 congress*, Lyon, GMPCA, 21–24 April 1999

Bender, B. 1978: 'Gatherer-hunter to farmer: a social perspective' *World Archaeology* 10.2: 204–22

Bender Jørgensen, L. 1988: 'Appendix A. An eight-thousand-year-old textile impression from Hama' in I. Thuesen (ed.), *Hama. Fouilles et recherches de la foundation Carlsberg 1931–1938* 1. *The Pre- and Protohistoric Periods*. Copenhagen, Nationalmuseet: 188

— 1992: *North European Textiles until AD 1000*. Aarhus, Aarhus University Press

— 2013: 'The textile remains from Tybrind Vig' in S.H. Andersen (ed.), *Tybrind Vig. Submerged Mesolithic Settlements in Denmark*. Højbjerg, Jutland Archaeological Society: 393–411

— 2018: 'The textiles from Mons Claudianus in a North African context' in M.S. Busana, M. Gleba, F. Meo, A.R. Tricomi (eds), *PURPUREAE VESTES VI. Textiles and Dyes in the Mediterranean Economy and Society*. Rome, Arbor Sapientiae: 451–59

Bender Jørgensen, L., Rast-Eicher, A. 2017: 'Cordage, basketry, textiles and hides' *Çatalhöyük 2017 Archive Report*. http://www.catalhoyuk.com/archive_reports/2017

Bendry, R. 2014: 'Care in the community? Interpretations of a fractured goat bone from Neolithic Jarmo, Iraq' *International Journal of Paleopathology* 7: 33–37

Bennison-Chapman, L.E. 2013: 'Geometric clay objects' in I. Hodder (ed.), *Substantive Technologies at Çatalhöyük: Reports from the 2000–2008 Seasons* (Çatalhöyük Research Project Series 9). London, British Institute at Ankara; Los Angeles, Cotsen Institute of Archaeology: 253–76

— 2014: *The Role and Function of 'Tokens' and Sealing Practices in the Neolithic of the Near East: The Question of Early Recording Systems, Symbolic Storage, Precursors to Writing, Gaming, or Monitoring Devices in the World's First Villages*. PhD thesis, University of Liverpool, Liverpool

— 2015: 'Geometric clay objects from Çatalhöyük East' *Çatalhöyük 2015 Archive Report.* http://www.catalhoyuk.com/ archive_reports/2015

— 2016: 'Geometric clay objects at Çatalhöyük East' *Çatalhöyük 2016 Archive Report.* http://www.catalhoyuk.com/ archive_reports/2016

— 2017: 'Clay balls and clay objects' *Çatalhöyük 2017 Archive Report.* http://www.catalhoyuk.com/ archive_reports/2017

Berger, R., Bendat, M., Parker, A. 1998: 'America's oldest basketry' in W.G. Mook, J. van der Plicht (eds), *Proceedings of the 16th International 14C Conference. Radiocarbon* 40.2: 615–20

Berggren, Å., Dell'Unto, N., Forte, M., Haddow, S.D., Hodder, I., Issavi, J., Lercari, N., Mazzucato, C., Mickel, A., Taylor, J.S. 2015: 'Revisiting reflexive archaeology at Çatalhöyük: integrating digital and 3D technologies at the trowel's edge' *Antiquity*, 89.344: 433–48

Bezić, A. 2007: 'Distribution of flint in Turkey from 10,000 to 6,000 cal BC. Case study – Çatalhöyük' in C. Delage (ed.), *Chert Availability and Prehistoric Exploitation in the Near East* (BAR International Series 1615). Oxford, Archaeopress: 68–86

Bialor, P.A. 1962: 'The chipped stone industry of Çatal Hüyük' *Anatolian Studies* 12: 67–110

Bıçakçı, E., Godon, M., Çakan, Y.G. 2012: 'Tepecik-Çiftlik' in M. Özdoğan, N. Başgelen, P. Kuniholm (eds), *The Neolithic in Turkey* 3: *New Excavations, New Research: Central Turkey*. Istanbul, Arkeoloji ve Sanat Yayınları: 89–134

Biehl, P.F., Franz, I., Orton, D.C., Ostaptchouck, S., Rogasch, J., Rosenstock, E. 2012: 'One community and two tells: the phenomenon of relocating tell settlements at the turn of the 7th and the 6th millennia in Central Anatolia' in R. Hofmann, F.-K. Moetz, J. Müller (eds), *Tells: Social and Environmental Space*. Bonn, Habelt: 53–65

Bille, M., Sorensen, T.F. 2007: 'An anthropology of luminosity: the agency of light' *Journal of Material Culture* 12: 263–48

Binder, D. 2002: 'Stones making sense: what obsidian could tell about the origins of the Central Anatolian Neolithic' in F. Gérard, L. Thissen (eds), *The Neolithic of Central Anatolia: Internal Developments and External Relations During the 9th–6th Millennia BC*. Istanbul: Ege Yayınları: 79–90

Binder, D., Balkan-Atlı, N. 2001: 'Obsidian exploitation and blade technology at Kömürcü-Kaletepe (Cappadocia, Turkey)' in I. Caneva, C. Lemorini, D. Zampetti, P. Biagi (eds), *Beyond Tools* (SENEPSE 9). Berlin, ex oriente: 1–16

Binford, L.R. 1973: 'Interassemblage variability: the Mousterian and the "functional" argument' in C. Renfrew (ed.), *The Explanation of Culture Change: Models in Prehistory*. London: Duckworth: 227–54

Bloch, M. 2010: 'Is there religion at Çatalhöyük...or are there just houses?' in I. Hodder (ed.), *Religion and the Emergence of Civilization: Çatalhöyük as a Case Study*. Cambridge, Cambridge University Press: 146–62

Bogaard, A., Charles, M., Twiss, K.C., Fairbairn, A., Yalman, N., Filipović, D., Demirergi, A., Ertuğ, F., Russell, N., Henecke, J. 2009: 'Private pantries and celebrated surplus: storing and sharing food at Neolithic Çatalhöyük, Central Anatolia' *Antiquity* 83.321: 649–68

Bogaard, A., Charles, M., Livarda, A., Ergun, M., Filipović, D., Jones, G. 2013: 'The archaeobotany of mid-later occupation levels at Neolithic Çatalhöyük' in I. Hodder (ed.), *Humans and Landscapes of Çatalhöyük: Reports from the 2000–2008 Seasons* (Çatalhöyük Research Project Series 8). London, British Institute at Ankara; Los Angeles, Cotsen Institute of Archaeology Press: 93–128

Bogaard, A., Ryan, P., Yalman, N., Asouti, E., Twiss, K.C., Mazzucato, C., Farid, S. 2014: 'Assessing outdoor activities and their social implications at Çatalhöyük' in I. Hodder (ed.), *Integrating Çatalhöyük: Themes from the 2000–2008 Seasons* (Çatalhöyük Research Project Series 10). London, British Institute at Ankara; Los Angeles, Cotsen Institute of Archaeology Press: 123–48

Boivin, N. 2000: 'Life rhythms and floor sequences: excavating time in rural Rajasthan and Neolithic Çatalhöyük' *World Archaeology* 31: 367–88

Bordaz, J. 1965: 'Suberde excavations 1964' *Anatolian Studies* 15: 31–33

— 1973: 'Current research in the Neolithic of south central Turkey: Süberde, Erbaba and their chronological implications' *American Journal of Archaeology* 77: 282–88

Bordaz, J., Bordaz, L. 1976: 'Erbaba excavations, 1974' *Türk Arkeoloji Dergisi* 23.2: 39–43

— 1982: 'Erbaba: the 1977 and 1978 seasons in perspective' *Türk Arkeoloji Dergisi* 26.1: 85–92

Borrell, F. 2011: 'Bi-directional Neolithic blade technology in the northern Levant during the 7th–8th millennia cal B.C.: new insights from Mamarrul Nasr 2, Syria' *Journal of Field Archaeology* 36.2: 132–50

— 2017: 'Bidirectional blade technology in the Near East during the Pre-Pottery Neolithic B' *Journal of Lithic Studies* 4.2: 129–61

Bosch, E. 2017: 'Çatalhöyük: a study of light and darkness: a photo-essay' in C. Papadopoulos, M. Holley (eds), *The Oxford Handbook of Light in Archaeology*. Oxford: Oxford University Press: DOI: 10.1093/oxfordhb/9780198788218.013.29

Boyer, P., Roberts, N., Baird, D. 2006: 'Holocene environment and settlement on the Çarşamba alluvial fan, south-central Turkey: integrating geoarchaeology and archaeological field survey' *Geoarchaeology* 21.7: 675–98

Boz, B., Hager, L.D. 2013: 'Living above the dead: intramural burial practices at Çatalhöyük' in I. Hodder (ed.), *Humans and Landscapes of Çatalhöyük: Reports from the 2000–2008 Seasons* (Çatalhöyük Research Project Series 8). London, British Institute at Ankara; Los Angeles, Cotsen Institute of Archaeology Press: 413–40

Bozyiğit, R., Güngör, Ş. 2011: 'Konya ovasının toprakları ve sorunları' *Marmara Coğrafya Dergisi* 24: 169–200

Braidwood, L.S., Braidwood, R.J., Howe, B., Reid, C.A., Watson, P.J. (eds) 1983: *Prehistoric Archaeology along the Zagros Flanks* (The University of Chicago Oriental Institute Publications 105). Chicago, The Oriental Institute of the University of Chicago

Brami M., Heyd V. 2011: 'The origins of Europe's first farmers: the role of Hacılar and western Anatolia, fifty years on' *Praehistorische Zeitschrift* 86: 165–206

Breniquet, C. 1991: 'Tell es-Sawwan: realites et problemes' *Iraq* 53: 75–90

— 2008: *Essai sur le Tissage en Mesopotamie des Premières Communautés Sédentaires au Milieu du IIIe Millénaire avant J.-C.* Paris, De Boccard

Broman, V. 1958: *Jarmo Figurines*. Masters dissertation, University of Cambridge, UK

Brown, S. 1996: 'Review of Schmandt-Besserat, D. Before Writing, Volumes I & II, 1992' *Canadian Society for Mesopotamian Studies Bulletin* 31: 35–43

Brumfiel, E.M. 1995: 'Heterarchy and the analysis of complex societies: comments' *Archeological Papers of the American Anthropological Association* 6.1: 125–31

Bubenik, G.A., Bubenik, A.B. (eds) 2012: *Horns, Pronghorns, and Antlers*. New York, Springer

Buitenhuis, H. 1997: 'Aşıklı Höyük. A "protodomestication" site' *Anthropozoologica* 25.26: 655–62

Burchell, W.J. 1822: *Travels in the Interior of Southern Africa*. London, Longman

Burnham, H.B. 1965: 'Çatal Höyük – the textiles and twined fabrics' *Anatolian Studies* 15: 169–74

Bursalı, A., Özbal, R., Baysal, E., Özbal, H., Yağcı, B. 2017: 'Neolithic blue beads in northwest Turkey: the social significance of skeumorphism' in M. Cifarelli, L. Gawlinski (eds), *What Shall I Say of Clothes? Theoretical and Methodological Approaches to the Study of Dress in Antiquity*. Boston, Archaeological Institute of America: 123–41

Busacca, G. forthcoming: *Painting Daily Life: Spatial Contexts, Temporalities and Experiences of Architectural Paintings at Çatalhöyük*. PhD thesis, Stanford University, Stanford, CA

Busacca, G., Lingle, A. 2017: 'Zoomorphic plaster heads' *Çatalhöyük 2017 Archive Report*. http://www.catalhoyuk.com/ archive_reports/2017

Butler, J. 2010: 'Performative agency' *Journal of Cultural Economy* 3.2: 147–61

Byrd, B.F. 1994: 'Public and private, domestic and corporate: the emergence of the southwest Asian village' *American Antiquity* 59.4: 639–66

Campbell, S. 2007: 'Rethinking Halaf chronologies' *Paléorient* 33.1: 103–36

Çamurcuoğlu, D. 2013: 'Çatalhöyük wall paintings: materials, technologies and artists, in I. Hodder (ed.), *Substantive Technologies at Çatalhöyük: Reports from the 2000–2008 Seasons* (Çatalhöyük Research Project Series 9). London, British Institute at Ankara; Los Angeles, Cotsen Institute of Archaeology Press: 317–30

— 2015: *The Wall Paintings of Çatalhöyük (Turkey): Materials, Technologies and Artists*. PhD thesis, University of London, London

Cardon, C. 1998: 'Neolithic textiles, matting and cordage from Charavines, lake of Paladru, France' in L. Bender Jørgensen, C. Rinaldo (eds), *Textiles in European Archaeology*. Göteborg, Göteborgs Universitet: 3–22

Carter, T. 2007: 'Of blanks and burials: hoarding obsidian at Neolithic Çatalhöyük' in L. Astruc, D. Binder, F. Briois (eds), *Systèmes Techniques et Communautés du Néolithique Précéramique au Proche-Orient* (Technical Systems and Near Eastern PPN Communities). Antibes, Éditions APDCA: 343–55

— 2009: 'Elemental characterization of Neolithic artefacts using portable X-ray fluorescence (PXRF)' *Çatalhöyük 2009 Archive Report*. http://www.catalhoyuk.com/archive_reports/2009

— 2011: 'A true gift of mother earth: the use and significance of obsidian at Çatalhöyük' *Anatolian Studies* 61: 1–19

— 2012: 'Chipped stone archive report 2012' *Çatalhöyük 2012 Archive Report*. http://www.catalhoyuk.com/archive_reports/2012

Carter, T., Cessford, C. 2005: 'Quantifying the consumption of obsidian at Neolithic Çatalhöyük, Turkey' *Journal of Field Archaeology* 30: 305–15

Carter, T., Milić, M. 2013a: 'The chipped stone' in I. Hodder (ed.), *Substantive Technologies from Çatalhöyük: Reports from the 2000–2008 Seasons* (Çatalhöyük Research Project Series 9). London, British Institute at Ankara; Los Angeles, Cotsen Institute of Archaeology: 417–78

— 2013b: 'The consumption of obsidian at Neolithic Çatalhöyük: a long-term perspective' in F. Borrell, J.J. Ibáñez, M.M. Molist (eds), *Stone Tools in Transition: From Hunter-Gatherers to Farming Societies in the Near East, 7th Conference on PPN Chipped and Ground Stone Industries of the Fertile Crescent*. Bellaterra, Universitat Autònoma de Barcelona, Servei de Publicacions: 495–508

Carter, T., Shackley, M. S. 2007: 'Sourcing obsidian from Neolithic Çatalhöyük (Turkey)' *Archaeometry* 49: 437–54

Carter, T., Unterbjerg, H.M. 2012: 'The flaked stone assemblage from the BACH Area' in R. Tringham, M. Stevanović (eds), *Last House on the Hill: BACH Area Reports from Çatalhöyük, Turkey* (Çatalhöyük Research Project Series 11). Los Angeles, Cotsen Institute of Archaeology Press: 391–416

Carter, T., Conolly, J., Spasojević, A. 2005a: 'The chipped stone' in I. Hodder (ed.), *Changing Materialities at Çatalhöyük: Reports from the 1995–99 Seasons* (Çatalhöyük Research Project Series 5). Cambridge, McDonald Institute for Archaeological Research; London, British Institute of Archaeology at Ankara: 221–83, 467–533

Carter, T., Poupeau, G., Bressey, C., Pearce, N.J.P. 2005b: 'From chemistry to consumption: towards a history of obsidian use at Çatalhöyük through a programme of inter-laboratory trace-elemental characterization' in I. Hodder (ed.), *Changing Materialities at Çatalhöyük: Reports from the 1995–99 Seasons* (Çatalhöyük Research Project Series 5). Cambridge, McDonald Institute for Archaeological Research; London, British Institute of Archaeology at Ankara: 285–305, 535–557

Carter, T., Kayacan, M., Milić, M., Waş, M., Doherty, C. 2006a: 'Chipped stone report' *Çatalhöyük 2006 Archive Report*. http://www.catalhoyuk.com/archive_reports/2006

Carter, T., Poupeau, G., Bressey, C., Pearce, N.J.P. 2006b: 'A new programme of obsidian characterization at Çatalhöyük, Turkey' *Journal of Archaeological Science* 33.7: 893–909

Carter, T., Dubernet, S., King, R., Le Bourdonnec, F.-X., Milić, M., Poupeau, G., Shakley, M.S. 2008a: 'Eastern Anatolia obsidians at Çatalhöyük and the reconfiguration of regional interaction in the Early Ceramic Neolithic' *Antiquity* 82: 900–09

Carter, T., Milić, M., Kayacan, N. 2008b: 'Çatalhöyük 2008 chipped stone report' *Çatalhöyük 2008 Archive Report*. http://www.catalhoyuk.com/archive_reports/2008

Carter, T., Haddow, S.D., Russell, N., Bogaard, A., Tsoraki, C. 2015: 'Laying the foundations: creating households at Neolithic Çatalhöyük, in I. Hodder, A. Marciniak (eds), *Assembling Çatalhöyük*. Leeds, Maney: 97–110

Cauvin, J. 2000: *The Birth of the Gods and the Origins of Agriculture*. Cambridge, Cambridge University Press

— 2002: 'The symbolic foundations of the Neolithic Revolution in the Near East' in I. Kuiht (ed.), *Life in Neolithic Farming Communities*. New York, Springer: 235–52

Cauvin, J., Watkins, T. 2007: *The Birth of the Gods and the Origins of Agriculture*. Cambridge, Cambridge University Press

Cauvin, J., Aurenche, O., Cauvin, M-C., Balkan-Atli, N. 1999: 'The Pre-pottery site of Cafer Höyük' in M. Özdoğan, N. Başgelen (eds), *Neolithic in Turkey. The Cradle of Civilization. New Discoveries*. Istanbul, Arkeoloji ve Sanat Yayınları: 87–104

Cauvin, M.-C., Gourgaud, A., Gratuze, B., Arnaud, N., Poupeau, G., Poidevin, J.-L., Chataigner, C. (eds) 1998: *L'Obsidienne au Proche et Moyen-Orient; du Volcan à l'Outil* (BAR International Series 738). Oxford, Archaeopress

Cessford, C. 2001: 'A new dating sequence for Çatalhöyük' *Antiquity* 75: 717–25

— 2003: 'Microartifactual floor patterning: the case at Çatalhöyük' *Assemblage* 7: 31–1

— 2005a: 'Absolute dating at Çatalhöyük' in I. Hodder (ed.), *Inhabiting Çatalhöyük. Reports from the 1995–99 Seasons* (Çatalhöyük Research Project Series 4). Cambridge, McDonald Institute for Archaeological Research; London, British Institute of Archaeology at Ankara: 65–99

— 2005b: 'Heavy residue analysis' in I. Hodder (ed.), *Changing Materialities at Çatalhöyük: Reports from the 1995–99 Seasons* (Çatalhöyük Research Project Series 5). Cambridge, McDonald Institute for Archaeological Research; London, British Institute of Archaeology at Ankara: 45–64

— 2007a: 'Level Pre-XII. E-A and Levels XII and XI, Spaces 181, 199 and 198' in I. Hodder (ed.), *Excavating Çatalhöyük: South, North and KOPAL Area: Reports from the 1995–99 Seasons* (Çatalhöyük Research Project Series 3). Cambridge, McDonald Institute for Archaeological Research; London, British Institute of Archaeology at Ankara: 59–101

— 2007b: 'Neolithic excavations in the North Area, East Mound, Çatalhöyük 1995–99' in I. Hodder (ed.), *Excavating Çatalhöyük: South, North and KOPAL Area: Reports from the 1995–99 Seasons* (Çatalhöyük Research Project Series 3). Cambridge, McDonald Institute for Archaeological Research; London, British Institute at Ankara: 345–552

Chapman, J. 2000: *Fragmentation in Archaeology: People, Places and Broken Objects in the Prehistory of South-eastern Europe*. London, Routledge

Chapman, J., Gaydarska, B. 2007: *Parts and Wholes: Fragmentation in Prehistoric Context*. Oxford, Oxbow Books

Choyke, A.M. 2001: 'Late Neolithic red deer canine beads and their imitation' in A.M. Choyke, L. Bartosiewicz (eds), *Crafting Bone: Skeletal Technologies Through Time and Space. Proceedings of the 2nd Meeting of the (ICAZ) Worked Bone Research Group Budapest, 31 August–5 September 1999* (BAR International Series 937). Oxford, Archaeopress: 251–66

Chyleński, M., Ehler, E., Somel, M., Yaka, R., Krzewińska, M., Dabert, M., Juras, A., Marciniak, A. 2019: 'Ancient mitochondrial genomes reveal the absence of maternal kinship in the burials of Çatalhöyük people and their genetic affinities' *Genes* 10.3: 207

Çilingiroğlu, A., Çevik, Ö., Çilingiroğlu, Ç. 2012: 'Ulucak Höyük: towards understanding the early farming communities of middle west Anatolia: the contribution of Ulucak' in M. Özdoğan, N. Başgelen, P. Kuniholm (eds), *The Neolithic in Turkey* 4*: New Excavations, New Research: Western Turkey*. Istanbul, Arkeoloji ve Sanat Yayınları: 139–75

Çilingiroğlu, Ç. 2009: 'Of stamps, loom weights and spindle whorls: contextual evidence on the function(s) of Neolithic stamps from Ulucak, Izmir, Turkey' *Journal of Mediterranean Archaeology* 22.1: 3–27

Ciszuk, M., Hammarlund, L. 2008: 'Roman looms – a study of craftsmanship and technology in the Mons Claudianus Project' in C. Alfaro, L. Karali (eds), *PURPUREAE VESTES II. Vestidos, Textiles y Tintes*. València, Universitat de València: 119–34

Clare, L. 2010: 'Pastoral clashes: conflict risk and mitigation at the Pottery Neolithic transition in the southern Levant' *Neolithics* 1.10: 13–31

Comaroff, J., Comaroff, J.L. 1991: 'How beasts lost their legs: cattle in Tswana economy and society' in J.G. Galaty, P. Bonte (eds), *Herders, Warriors and Traders: Pastoralism in Africa*. Boulder (CO), Westview Press: 33–61

— 1990: 'Goodly beasts, beastly goods: cattle and commodities in a South African context' *American Ethnologist* 17.2: 195–216

Conkey, M.W. 2010: 'Images without words: the construction of prehistoric imaginaries for definitions of "us"' *Journal of Visual Culture* 9: 272–83

Conolly, J. 1996: 'The knapped stone' in I. Hodder (ed.), *On the Surface: Çatalhöyük 1995–95*. Cambridge, McDonald Institute for Archaeological Research; London, British Institute of Archaeology at Ankara: 173–98

— 1999a: *The Çatalhöyük Flint and Obsidian Industry: Technology and Typology in Context* (BAR International Series 787). Oxford, BAR

— 1999b: 'Technical strategies and technical change at Neolithic Çatalhöyük, Turkey' *Antiquity* 73: 791–800

— 2000: 'Çatalhöyük and the archaeological "object"' in I. Hodder (ed.), *Towards Reflexive Method in Archaeology: The Example at Çatalhöyük* (Çatalhöyük Research Project Series 2). Cambridge, McDonald Institute for Archaeological Research; London, British Institute of Archaeology at Ankara: 51–57

— 2003: 'The Çatalhöyük obsidian hoards: a contextual analysis of technology' in N. Moloney, M. Shott (eds.), *Lithic Studies for the New Millennium*. London, Archetype Books: 55–78

Costello, S.K. 2000: 'Memory tools in early Mesopotamia' *Antiquity* 74.285: 475–76

— 2002: *Tools of Memory: Investigation of the Context of Information Storage in the Halaf Period*. PhD thesis, Binghamton University, New York

Cross-May, S. 2005: 'Statistical integration of contextual data' in I. Hodder (ed.), *Changing Materialities at Çatalhöyük: Reports from the 1995–99 Seasons* (Çatalhöyük Research Project Series 5). Cambridge, McDonald Institute for Archaeological Research; London, British Institute of Archaeology at Ankara: 23–44

Croutsch, C. 2005: *Techniques et Sociétés Néolithiques. Le Sciage des Roches Tenaces au Nord-Ouest des Alpes (4300–2450 av. J.-C.)* (BAR International Series 1361). Oxford, Archaeopress

Crowfoot, E. 1982: 'Textiles, matting and basketry' in K.M. Kenyon (ed.), *Excavations at Jericho* 4. London, Council for British Research in the Levant: 546–50

Crowfoot, G.M. 1931: *Methods of Hand Spinning in Egypt and the Sudan* (Bankfield Museum notes 2nd series no. 12). Halifax, County Borough of Halifax

Crumley, C.L. 1979: 'Three locational models: an epistemological assessment for anthropology and archaeology' *Advances in Archaeological Method and Theory* 2: 141–73

Czeszewska, A. 2014: 'Wall paintings at Çatalhöyük' in I. Hodder (ed.), *Integrating Çatalhöyük: Themes from the 2000–2008 Seasons* (Çatalhöyük Research Project Series 10). London, British Institute at Ankara; Los Angeles, Cotsen Institute of Archaeology Press: 185–96

d'Errico, F., Vanhaeren, M. 2002: 'Criteria for identifying red deer (*Cervus elaphus*) age and sex from their canines. Application to the study of Upper Palaeolithic and Mesolithic ornaments' *Journal of Archaeological Science* 29: 211–32

d'Errico, F., Vanhaeren, M., Barton, N., Bouzouggar, A., Mienis, H., Richter, D., Hublin, J.-J., McPherron, S.P., Lozouet, P. 2009: 'Additional evidence on the use of personal ornaments in the Middle Paleolithic of North Africa' *Proceedings of the National Academy of Sciences* 106.38: 16051–56

Daly, R. 2005: 'Worked bone' *Çatalhöyük 2005 Archive Report*. http://www.catalhoyuk.com/archive_reports/2005

Damerow, P. 1993: 'Bookkeepers invented scripture: review of Schmandt-Besserat's "Before Writing" 1992' *Rechtshistorisches Journal* 12.6: 9–35

Davis, M.K. 1982: 'The Çayönü ground stone' in L.S. Braidwood, R.J. Braidwood (eds), *Prehistoric Village Archaeology in South-Eastern Turkey: The Eighth Millennium BC Site at Çayönü: Its Chipped and Ground Stone Industries and Faunal Remains* (BAR International Series 138). Oxford, British Archaeological Reports: 73–174

Dawson, A. 1998: 'Reflections on the interactions between people and pigs' in S.M. Nelson (ed.), *Ancestors for the Pigs: Pigs in Prehistory*. Philadelphia, University of Pennsylvania Museum of Archaeology: 5–10

Dawson, P., Levy, R.M., Walls, M. 2007: 'Simulating the behaviour of light inside Arctic dwellings: implications for assessing the role of vision in task performance' *World Archaeology* 39: 17–35

De Jesus, P.S. 1985: 'Notes on the symbolism in Çatal Hüyük wall paintings' in J.L. Huot, M. Yon, T. Cavlet (eds), *De l'Indus aux Balkans: Recueil a la Memoire de Jean Deshayes*. Paris, Éditions Recherche sur les Civilisations: 127–45

De Morgan, J., Jéquier, G., de Mecquenem, R., Haussoulier, B., Graat van Roggen, D.-L. 1905: *Mémoires de la Délégation en Perse*, *7, Recherches Archéologiques, deuxième série*. Paris, Editions Ernst Le Roux

de Ridder, N.A. 1965: 'Sediments of the Konya basin, central Anatolia, Turkey' *Paleogeography, Paleoclimatology, Palaeoecology* 1: 225–54

Demir, M. 2003: 'Shells of *Mollusca* collected from the seas of Turkey' *Turkish Journal of Zoology* 27: 101–40

Demirergi, G.A. 2015: *Food Sharing during the Transition to Agriculture at Neolithic Çatalhöyük, Central Anatolia*. PhD thesis, Stony Brook University, Stony Brook

Demirergi, G.A., Twiss, K.C., Bogaard, A., Green, L., Ryan, P., Farid, S. 2014: 'Of bins, basins and banquets: storing, handling and sharing at Neolithic Çatalhöyük' in I. Hodder (ed.), *Integrating Çatalhöyük: Themes from the 2000–2008 Seasons* (Çatalhöyük Research Project Series 10). London, British Institute at Ankara; Los Angeles, Cotsen Institute of Archaeology Press: 91–108

Der, L. 2016: *The Role of Human-Animal Relations in the Social and Material Organization of Çatalhöyük, Turkey*. PhD thesis, Department of Anthropology, Stanford University

Der, L., Issavi, J. 2017: 'The urban quandary and the "mega-site" from the Çatalhöyük perspective' *Journal of World Prehistory* 30.3: 189–206

Dickson, F.P. 1981: *Australian Stone Hatchets. A Study in Design and Dynamics*. Sydney, Academic Press

Dogiama, L. 2017: *Points of Reference: Projectile Points, Hunting and Identity at the Neolithic Çatalhöyük, Turkey*. PhD thesis, McMaster University, Hamilton, Canada

— in press: 'The projectile points of Neolithic Çatalhöyük: a contextual multi-attribute analysis' in A. Baysal (ed.), *Lithic Studies from Anatolia and Beyond*. Oxford, Archaeopress

Doherty, C.J. 2008: 'Landscape coring' *Çatalhöyük 2008 Archive Report*. http://www.catalhoyuk.com/archive_reports/2008

— 2013: 'Sourcing Çatalhöyük's clays' in I. Hodder (ed.), *Substantive Technologies at Çatalhöyük: Reports from the 2000–2008 Seasons* (Çatalhöyük Research Project 9). London, British Institute at Ankara; Los Angeles, Cotsen Institute of Archaeology Press: 51–66

— 2017: *Living with Clay: Materials, Technology, Resources and Landscape at Çatalhöyük*. PhD thesis, University of Leicester, Leicester

Doherty, C.J., Tarkan, D. 2013: 'Pottery production at Çatalhöyük: a petrographic perspective' in I. Hodder (ed.), *Substantive Technologies at Çatalhöyük: Reports from the 2000–2008 Seasons* (Çatalhöyük Research Project Series 9). London, British Institute at Ankara; Los Angeles, Cotsen Institute of Archaeology Press: 183–92

Doyle, S. 2015: 'Chipped stone' *Çatalhöyük 2015 Archive Report*. http://www.catalhoyuk.com/archive_reports/2015

— 2016: 'Chipped stone from the North and South Areas' *Çatalhöyük 2016 Archive Report*. http://www.catalhoyuk.com/ archive_reports/2016

— 2017: 'Chipped Stone from the North and South Areas' *Çatalhöyük 2017 Archive Report*. http://www.catalhoyuk.com/ archive_reports/2017

Dubreuil, L., Grosman, L. 2013: 'The life history of macrolithic tools at Hilazon Tachtit Cave' in O. Bar Yosef, F. Valla (eds), *Natufian Foragers in the Levant: Terminal Pleistocene Social Changes in Western Asia*. Ann Arbor (MI), International Monographs in Prehistory: 527–43

Dubreuil, L., Savage, D., Delgado-Raack, S., Plisson, H., Stephenson, B., de la Torre, I. 2015: 'Current analytical frameworks for studies of use-wear on ground stone tools' in J. Manuel Marreiros, J.F. Gibaja Bao, N. Ferreira Bicho (eds), *Use-Wear and Residue Analysis in Archaeology*. New York, Springer: 105–58

Duistermaat, K. 1996: 'The seals and sealings' in P.M.M.G. Akkermans (ed.), *Tell Sabi Abyad, the Late Neolithic Settlement: Report on the Excavations of the University of Amsterdam (1988) and the National Museum of Antiquities Leiden (1991–1993) in Syria*. Istanbul, Istanbul Nederlands Historisch-Archaeologisch Instituut te Istanbul: 339–402

Düring, B.S. 2005: 'Building continuity in the Central Anatolian Neolithic: exploring the meaning of buildings at Aşıklı Höyük and Çatalhöyük' *Journal of Mediterranean Archaeology* 18.1: 3–29

— 2006: *Constructing Communities: Clustered Neighbourhood Settlements of the Central Anatolian Neolithic, ca. 8500–5500 cal BC*. Leiden, Nederlands instituut voor het nabije oosten

— 2007a: 'The articulation of houses at Neolithic Çatalhöyük, Turkey' in R.A. Beck (ed.), *The Durable House: House Society Models in Archaeology*. Carbondale, Center for Archaeological Investigations: 130–53

— 2007b: 'Reconsidering the Çatalhöyük community: from households to settlement systems' *Journal of Mediterranean Archaeology* 20.2: 155–82

— 2010: *The Prehistory of Asia Minor: From Complex Hunter-Gatherers to Early Urban Societies*. Cambridge, Cambridge University Press

— 2013: 'Breaking the bond: investigating the Neolithic expansion in Asia Minor in the seventh millennium BC' *Journal of World Prehistory* 26.2: 75–100

— 2014: 'The anatomy of a prehistoric community' in J. Birch (ed.), *From Prehistoric Villages to Cities: Settlement Aggregation and Community Transformation*. New York, Routledge: 23–43

Düring, B.S., Marciniak, A. 2005: 'Households and communities in the central Anatolian Neolithic' *Archaeological Dialogues* 12: 165–87

Duru, R. 1994: *Kuruçay Höyük* 1. *1978–1988 Kazılarının Sonuçları, Neolitik ve Kalkolitik Çağ Yerlesmeleri*. Ankara, Türk Tarih Kurumu

— 2007: 'Göller Bölgesi Neolitiği: Hacılar – Kuruçay Höyüğü – Höyücek – Bademağacı Höyüğü' in M. Özdoğan, N. Başgelen (eds), *Anadolu'da Uygarlığın Doğuşu ve Avrupa'ya Yayılımı. Türkiye'de Neolitik Dönem. Yeni Kazılar-Yeni Bulgular*. Istanbul, Arkeoloji Sanat Yayınları, 331–60

— 2008: *From 8000 BC to 2000 BC, Six Thousand Years of the Burdur-Antalya Region*. Antalya, Suna ve İnan Kıraç Akdeniz Medeniyetleri Araştırma Enstitüsü, Zero Yayınları

— 2012: 'The Neolithic of the Lakes Region Hacılar –Kuruçay Höyük – Höyücek – Bademağacı Höyük' in M. Özdoğan, N. Basgelen, P. Kuniholm (eds), *The Neolithic in Turkey* 4: *New Excavations, New Research: Western Turkey*. Istanbul, Arkeoloji ve Sanat Yayınları: 1–65

Duru, R., Umurtak, G. 2005: *Höyücek, 1989–1992 Yılları Arasında Yapılan Kazıların Sonuçları*. Ankara, Türk Tarih Kurumu

Earl, G., Porcelli, V., Papadopoulos, C., Beale, G., Harrison, M., Pagi, H., Keay, S. 2013: 'Formal and informal analysis of rendered space: the Basilica Portuense' in A. Bevan, M. Lake (eds), *Computational Approaches to Archaeological Spaces*. Walnut Creek (CA), Left Coast Press: 265–306

Eddisford, D. 2014: 'Building 49' in I. Hodder (ed.), *Çatalhöyük Excavations: The 2000–2008 Seasons* (Çatalhöyük Research Project Series 7). London, British Institute at Ankara; Los Angeles, Cotsen Institute of Archaeology Press: 313–56

Egami, N. 1959: *Telul eth Thalathat. The Excavations of Tell II, 1956–1957* 1. Tokyo, Yamakawa Publishing Company

Egg, M., Spindler, K. 2008: *Kleidung und Ausrüstung der Gletschermumie aus den Ötztaler Alpen*. Regensburg, Römisch-Germanisches Zentralmuseum Mainz

Ehrenreich, R.M. 1995: 'Early metalworking: a heterarchical analysis of industrial organization' *Archaeological Papers of the American Anthropological Association* 6.1: 33–39

Emery, I. 1994: *The Primary Structure of Fabrics*. Singapore, Thames and Hudson

Englund, R.K. 1993: 'The origins of script. Review of "Before Writing" by Denise Schmandt-Besserat' *Science* 260.5114: 1670–71

— 1998: 'Review of Schmandt-Besserat, "How Writing Came About"' *Written Language and Literacy* 1: 257–61

Ercan, T. 1986: 'Orta Anadolu'daki Senozoyik Volkanizması' *Maden Tetkik ve Arama Dergisi*, 107: 119–40

Erdoğu B. 2013: 'Uğurlu, a Neolithic Settlement on the Aegean Island of Gökçeada' in M. Özdoğan, N. Başgelen, P. Kuniholm (eds), *Neolithic in Turkey: New Excavations, New Research. Northwestern Turkey and Istanbul*. Istanbul, Arkeoloji ve Sanat Yayınları: 1–33

Erdoğu, B., Uysal, T., Özbek, O., Ulusoy, Ü. 2013: 'Speleothems of Çatalhöyük, Turkey' *Mediterranean Archaeology and Archaeometry* 13.1: 21–30

Escofier, B., Pagès, J. 2008: *Analyses Factorielles Simples et Multiples: Objectifs, Méthodes et Interprétation*, 4th ed. Paris, Dunod

Esin, U., Harmankaya, S. 1999: 'Aşıklı in the frame of Central Anatolian Neolithic' in M. Özdoğan, N. Başgelen (eds), *Neolithic in Turkey. The Cradle of Civilization. New Discoveries*. Istanbul, Arkeoloji ve Sanat Yayınları: 115–32

Evans, J.D., Cann, J.R., Renfrew, A.C., Cornwall, I.W., Western, A.C. 1964: 'Excavations in the Neolithic settlement of Knossos, 1957–60. Part I' *Annual of the British School at Athens* 59: 132–240

Falci, C.G., Cuisin, J., Delpuech, A., Van Gijn, A., Hofman, C.L. 2019: 'New insights into use-wear development in bodily ornaments through the study of ethnographic collections' *Journal of Archaeological Method and Theory* 26.3: 755–805

Farid, S. 2007: 'Level VII: Space 113, Space 112, Space 105, Space 109, Building 40, Space 106, Spaces 168 & 169, Buildings 8 & 20, Building 24 and relative heights of Level VII' in I. Hodder (ed.), *Excavating Çatalhöyük: South, North and KOPAL Area: Reports from the 1995–99 Seasons* (Çatalhöyük Research Project Series 3). Cambridge, McDonald Institute for Archaeological Research; London, British Institute of Archaeology at Ankara: 283–342

Farid, S., Hodder, I. 2014: 'Excavation, recording and sampling methodologies' in I. Hodder (ed.), *Çatalhöyük Excavations: The 2000–2008 Seasons* (Çatalhöyük Research Project Series 7). London, British Institute at Ankara; Los Angeles, Cotsen Institute of Archaeology Press: 35–52

Feldman, K., Norton, E. 1995: 'Niqsaq and napaaqtuq: issues in Inupiaq Eskimo life-form classification and ethnoscience' *Études/Inuit/Studies* 19.2: 77–100

Feldman, M., Fernández-Domínguez, E., Reynolds, L., Baird, D., Pearson, J., Hershkovitz, I., May, H., Goring-Morris, N., Benz, M., Gresky, J., Bianco, R.A., Fairbairn, A., Mustafaoğlu, G., Stockhammer, P.W., Posth, C., Haak, W., Jeong, C., Krause, J. 2019: 'Late Pleistocene human genome suggests a local origin for the first farmers of Central Anatolia' *Nature Communications* 10.1: 1218

Filipowicz, P. 2017: 'Semiotics in action. Neolithic imagery on the move' in I. Milevski, T.E. Levy (eds), *Framing Archaeology in the Near East. The Application of Social Theory to Fieldwork*. Sheffield, Equinox Publishing: 51–62

— 2019: *The Transformative Character of Imagery of South-Central Anatolian Communities in the Period 6500–5500 BC*. PhD thesis, Adam Mickiewicz University, Poznań

Firth, R. 1963: *Elements of Social Organization*. Boston, Beacon Press

Fletcher, A., Baird, D., Spataro, M., Fairbairn, A., 2017: 'Early ceramics in Anatolia: implications for the production and use of the earliest pottery. The evidence from Boncuklu Höyük' *Cambridge Archaeological Journal* 27.2: 351–369

Forest, J.-D. 1993: 'Çatal Höyük et son décor: pour le déchiffrement d'un code symbolique' *Anatolia Antiqua* 2: 1–42

Fowler, C. 2004: *The Archaeology of Personhood*. London, Routledge

Francis, P. 1982: 'Experiments with early techniques for making whole shells into beads' *Current Anthropology* 23.6: 713–14

— 1989: 'The manufacture of beads from shell' in C.F.I. Hayes (ed.), *Proceedings of the 1986 Shell Bead Conference*. Rochester, Rochester Museum and Science Centre: 25–35

French, D.H. 1998: *Canhasan Sites* 1. *Canhasan I: Stratigraphy and Structures*. London, British Institute of Archaeology at Ankara

— 2005: *Can Hasan Sites* 2. *Can Hasan I*: *The Pottery*. London, British Institute at Ankara

French, D.H., Hillman, G.C., Payne, S., Payne, R.J. 1972: 'Excavations at Can Hasan III 1969–1970' in E.S. Higgs (ed.), *Papers in Economic Prehistory*. Cambridge, Cambridge University Press: 181–90

Friberg, J. 1994: 'Preliterate counting and accounting in the Middle East: a constructively critical review of Schmandt-Besserat's "Before Writing"' *Orientalistische Literaturzeitung* 89.5–6: 477–502

Fukai, S., Matsutani, T. 1977: 'Excavations at Telul eth-Thalathat, 1976' *Sumer* 33.1: 48–64

Fukai, S., Horiuchi, K., Matsutani, T. 1970: *Telul eth Thalathat. The Excavations of Tell II. The Third Season (1964)* 2. Tokyo, Institute of Oriental Culture, University of Tokyo

Fuller, D., Bogaard, A. Charles, M., Filipović, D. 2014: 'Macro- and micro-botanical remains from the 2013 and 2014 seasons' *Çatalhöyük 2014 Archive Report*. http://www.catalhoyuk.com/archive_reports/2014

Gamble, C. 1998: 'Palaeolithic society and the release from proximity: a network approach to intimate relations' *World Archaeology* 29.3: 426–49

Garrard, A., Martin L., Becker, C., Ducos, P., Tchernov, E., Horwitz, K., Von den Driesch, A. 1999: 'Animal domestication in the Southern Levant' *Paléorient* 25.2: 63–80

Gero, J.M. 1989: 'Assessing social information in material objects: how well do lithics measure up?' in R. Torrence (ed.), *Time, Energy and Stone Tools*. Cambridge, Cambridge University Press: 92–105

Gerritsen, F., Özbal, R., Thissen, L. 2013: 'Barcın Höyük: the beginnings of farming in the Marmara Region' in M. Özdoğan, N. Başgelen, P. Kuniholm (eds), *Neolithic in Turkey: New Excavations, New Research. Northwestern Turkey and Istanbul*. Istanbul, Arkeoloji ve Sanat Yayınları: 93–112

Gilligan, I. 2019: *Climate, Clothing and Agriculture in Prehistory. Linking Evidence, Causes, and Effects*. Cambridge, Cambridge University Press

Gimbutas, M. 1990: 'Wall paintings of Çatal Hüyük' *The Review of Archaeology* 11: 1–5

Gleba, M., Harris, S. 2018: 'The first plant bast fibre technology: identifying splicing in archaeological textiles' *Archaeological and Anthropological Sciences* 11.5: 2329–46

Gleser, R. 2016: 'Beginnings revisited again: about the inversion of the sequence of manufacturing, the origins of early weaving and the vertical warp-weighted loom' in K. Bacvarov, R. Gleser (eds), *Southeast Europe and Anatolia in Prehistory* (Universitätsforschungen zur prähistorischen Archäologie Bd 293). Bonn, Verlag Dr. Rudolf Habelt: 79–91

Glory, A. 1958: 'Débris de corde Paléolithique à la grotte de Lascaux' *Mémoires de la Société Préhistorique Française* 5: 135–69

Goring-Morris, N., Belfer-Cohen, A. 2010: '"Great Expectations" or the inevitable collapse of the Early Neolithic in the Near East' in M.S. Bandy, J.R. Fox (eds), *Becoming Villagers: Comparing Early Village Societies*. Tucson, University of Arizona Press: 62–77

Grace, R. 1989: *Interpreting the Function of Stone Tools: The Quantification and Computerisation of Microwear Analysis* (BAR International Series 474). Oxford: British Archaeological Reports

Granger-Taylor, H. 1982: 'Weaving clothes to shape in the ncient World: the tunic and toga of the Arringatore' *Textile History* 13.1: 3–25

— 1998: 'Evidence for linen yarn preparation in Ancient Egypt – the hanks of fibre strips and the balls of prepared rove from Lahun in the Petrie Museum of Egyptian Archaeology, University College London (UC 7421, 7509 and 7510)' in S. Quirke (ed.), *Lahun Studies*. Reigate, SIA Publishing: 103–07

— 2003: 'Textile production and clothing. Technology and tools in ancient Egypt' University College London, http://www.ucl.ac.uk/museums-static/digitalegypt/textil/tools.html

Gratuze, B. 1999: 'Obsidian characterization by laser ablation ICP-MS and its application to prehistoric trade in the Mediterranean and the Near East: sources and distribution of obsidian within the Aegean and Anatolia' *Journal of Archaeological Science* 26.10: 869–81

Gratuze, B., Boucetta, S., Binder, D., Balkan-Atlı, N., Bellot-Gurlet, L., Mouralis, D. 2005: 'New investigations of the Göllü Dag obsidian lava flows system: comparison between chemical, mineralogical and fission track data' *International Association of Obsidian Studies* 33: 18–19

Grömer, K. 2006: 'Vom Spinnen und Weben, Flechten und Zwirnen. Hinweise zur neolithischen Textiltechnik an österreichischen Fundstellen' *Archäologie Österreichs* 17.2: 177–92

Grömer, K., Bender Jørgensen, L., Marić Baković, M. 2018: 'Missing link: an early wool textile from Pustopolje in Bosnia and Herzegovina' *Antiquity* 92: 351–67

Gündüz, R. 2016: 'Neolitikten Kalkolitik döneme kadar Bozkır çevresindeki yerleşmeler' in H. Bahar, H. Kuyumcu, C. Benhür, H. Gül, M. Turgut, F.N. Küçükballı (eds), *Uluslararasi Sempozyum. Geçmişten Günümüze Bozkır*. Konya, Selcuk Universitesi: 33–48

Gümüş, B.A., Bar-Yosef Mayer, D.E. 2013: 'Micro-freshwater gastropods at Çatalhöyük as environmental indicators' in I. Hodder (ed.), *Humans and Landscapes of Çatalhöyük: Reports from the 2000–2008 Seasons* (Çatalhöyük Research Project Series 8). London, British Institute at Ankara; Los Angeles, Cotsen Institute of Archaeology Press: 81–87

Güner, G. 1988: 'Anadoluda Yaşamakta Olan İlkel Çömlekçilik' *Ak Yayınları Kültür Serisi* 16.5: 81

Gutierrez, D., Sunstedt, V., Gomez, F., Chalmers, A. 2007: 'Dust and light: predictive virtual archaeology' *Journal of Cultural Heritage* 8: 209–14

Haddow, S.D., Milella, M., Tibbetts, B., Schotsmans, E., Knüsel, C.J. 2017: 'Human remains' *Çatalhöyük 2017 Archive Report*. http://www.catalhoyuk.com/archive_reports/2017

Haddow, S.D., Tsoraki, C., Vasić, M., Dori, I., Knüsel, C.J., Milella, M. 2019: 'An analysis of modified human teeth at Neolithic Çatalhöyük, Turkey' *Journal of Archaeological Science: Reports* 28: 102058

Hafner, A. 2015: *Schnidejoch und Lötschenpass. Archäologie der prähistorischen, römischen und mittelalterlichen Passübergänge in den Berner Alpen* 1. Bern, Erziehungsdirektion des Kantons Bern, Amt für Kultur

Hagan, R., Warinner, C. 2019: *Evidence for Dairying from Shotgun Proteomic Analysis of Human Dental Calculus at Çatalhöyük. Interim Report*. Jena, Max Planck Institute for the Science of Human History

Hager, L.D., Haddow. S.D. 2009: '2009 Çatalhöyük human remains' *Çatalhöyük 2009 Archive Report*. http://catalhoyuk.com/archive_reports/2009

Hald, M. 1980: *Ancient Danish Textiles from Bogs and Burials*. Copenhagen, The National Museum of Denmark

Hamilton, N. 1996: 'Figurines, clay balls, small finds and burials' in I. Hodder (ed.), *On the Surface: Çatalhöyük 1993–95* (Çatalhöyük Research Project Series 1). Cambridge, McDonald Institute for Archaeological Research; London, British Institute at Ankara: 215–63

— 2005: 'The beads' in I. Hodder (ed.), *Changing Materialities at Çatalhöyük: Reports from the 1995–99 Seasons* (Çatalhöyük Research Project Series 5). Cambridge, McDonald Institute for Archaeological Research; London, British Institute of Archaeology at Ankara: 325–32

Hamon, C., Le Gall, V. 2013: 'Millet and sauce: the uses and functions of querns among the Minyanka (Mali)' *Journal of Anthropological Archaeology* 32.1: 109–21

Hansen, S. 2014: 'Neolithic figurines in Anatolia' in M. Özdoğan, N. Başgelen, Kuniholm, P. (eds), *The Neolithic in Turkey. 10500–5200 BC: Environment, Settlement, Flora, Fauna, Dating, Symbols of Belief, with Views from North, South, East, and West*. Istanbul, Arkeoloji ve Sanat Yayınları: 265–92

Hard, R.J., Mauldin, R.P., Raymond, G.R. 1996: 'Mano size, stable carbon isotope ratios, and macrobotanical remains as multiple lines of evidence of maize dependence in the American Southwest' *Journal of Archaeological Method and Theory* 3.3: 253–318

Hastorf, C.A. 2005: 'Macrobotanical investigation: field methods and laboratory analysis procedures' in I. Hodder (ed.), *Inhabiting Çatalhöyük: Reports from the 1995–99 Seasons* (Çatalhöyük Research Project Series 4). Cambridge, McDonald Institute for Archaeological Research; London, British Institute at Ankara: 129–36

Hatziminaoğlou, Y., Boyazoğlu, J. 2004: 'The goat in ancient civilization: from the Fertile Crescent to the Aegean Sea' *Small Ruminant Research* 51: 123–29

Hauptmann, H. 1999: 'The Urfa region' in M. Özdoğan, N. Başgelen (eds), *Neolithic in Turkey. The Cradle of Civilization. New Discoveries*. Istanbul, Istanbul, Arkeoloji ve Sanat Yayınları: 65–86

Hayden, B. 1987: 'Traditional metate manufacturing in Guatemala using chipped stone tools' in B. Hayden (ed.), *Lithic Studies among the Contemporary Highland Maya*. Tucson, University of Arizona Press: 8–119

— 2012: 'Corporate groups and secret societies in the Early Neolithic: a comment on Hodder and Meskell' *Current Anthropology* 53.1: 126–27

Hayes, C.F.I. 1989: *Proceedings of the 1986 Shell Bead Conference*. Rochester, Rochester Museum and Science Centre

Hayes, E., Pardoe, C., Fullagar, R. 2018: 'Sandstone grinding/pounding tools: use-trace reference libraries and Australian archaeological applications' *Journal of Archaeological Science: Reports* 20: 97–114

Helbæk, H. 1963: 'Textiles from Çatal Höyük' *Archaeology* 16: 9–46

Helmer, D., Gourichon, L., Stordeur, D. 2004: 'À l'aube de la domestication animale. Imaginaire et symbolisme animal dans les premières sociétés néolithiques du nord du Proche-Orient' *Anthropozoologica* 39.1: 143–63

Hendy, J., Colonese, A.C., Franz, I., Fernandes, R., Fischer, R., Orton, D., Lucquin, A., Spindler, L., Anvari, J., Stroud, E., Biehl, P.F., Speller, C., Boivin, N., Mackie, M., Jersie-Christensen, R.R., Olsen, J.V., Collins, M.J., Craig, O.E., Rosenstock, E. 2018: 'Ancient proteins from ceramic vessels at Çatalhöyük West reveal the hidden cuisine of early farmers' *Nature Communications* 9.1: 4064

Henton, E. 2014: 'Oxygen stable isotope and dental microwear evidence of herding practices at Çatalhöyük' in I. Hodder (ed.), *Humans and Landscapes of Çatalhöyük: Reports from the 2000–2008 Seasons* (Çatalhöyük Research Project Series 8). London, British Institute at Ankara; Los Angeles, Cotsen Institute of Archaeology Press: 299–316

Herrero, J., Porta, J. 2000: 'The terminology and the concepts of gypsum-rich soils' *Geoderma* 96.1–2: 47–61

Hildebrandt, W.R., King, J.H. 2012: 'Distinguishing between darts and arrows in the archaeological record: implications for technological change in the American West' *American Antiquity* 77.4: 789–99

Hill, E. 2013: 'Archaeology and animal persons: toward a prehistory of human-animal relations' *Environment and Society: Advances in Research* 4.1: 117–36

Hillson, S.W., Larsen, C.S., Boz, B., Pilloud, M., Sadvari, J.W., Agarwal, S.C., Glencross, B., Beauchesne, P., Pearson, J.A., Ruff, C.B., Garofalo, E.M., Hager, L.D., Haddow, S.D. 2013: 'The human remains I: interpreting community structure, health and diet in Neolithic Çatalhöyük' in I. Hodder (ed.), *Humans and Landscapes of Çatalhöyük: Reports from the 2000–2008 Seasons* (Çatalhöyük Research Project Series 8). London, British Institute at Ankara; Los Angeles, Cotsen Institute of Archaeology Press: 339–96

Hodder, I. (ed.) 1996: *On the Surface: Çatalhöyük 1993–95* (Çatalhöyük Research Project Series 1). Cambridge, McDonald Institute for Archaeological Research; London, British Institute at Ankara.

— (ed.) 2000: *Towards Reflexive Method in Archaeology: The Example at Çatalhöyük* (Çatalhöyük Research Project Series 2). Cambridge, McDonald Institute for Archaeological Research; London, British Institute of Archaeology at Ankara

— (ed.) 2005: *Inhabiting Çatalhöyük: Reports from the 1995–99 Seasons* (Çatalhöyük Research Project Series 4). Cambridge, McDonald Institute for Archaeological Research; London, British Institute at Ankara

— (ed.) 2006a: *Changing Materialities at Çatalhöyük: Reports from the 1995–99 Seasons* (Çatalhöyük Research Project Series 5). Cambridge, McDonald Institute for Archaeological Research; London, British Institute at Ankara

— (ed.) 2006b: *Çatalhöyük Perspectives: Themes from the 1995–99 Seasons* (Çatalhöyük Research Project Series 6). Cambridge, McDonald Institute for Archaeological Research; London, British Institute at Ankara

— 2006c: *The Leopard's Tale: Revealing the Mysteries of Çatalhöyük*. London, Thames and Hudson

— (ed.) 2007: *Excavating Çatalhöyük: South, North and KOPAL Area: Reports from the 1995–99 Seasons* (Çatalhöyük Research Project Series 3). Cambridge, McDonald Institute for Archaeological Research; London, British Institute at Ankara

— (ed.) 2010a: *Religion in the Emergence of Civilization: Çatalhöyük as a Case Study*. Cambridge, Cambridge University Press

— 2010b: 'Probing religion at Çatalhöyük: an interdisciplinary experiment' in I. Hodder (ed.), *Religion in the Emergence of Civilization: Çatalhöyük as a Case Study*. Cambridge, Cambridge University Press: 1–31

— 2011a: 'Human-thing entanglement: towards an integrated archaeological perspective' *Journal of the Royal Anthropological Institute* 17: 154–77

— 2011b: 'Wheels of time: some aspects of entanglement theory and the secondary production revolution' *Journal of World Prehistory* 24.2–3: 175–87

— 2012: *Entangled. An Archaeology of the Relationships between Humans and Things*. Oxford: Wiley Blackwell

— (ed.) 2013a: *Substantive Technologies at Çatalhöyük: Reports from the 2000–2008 Seasons* (Çatalhöyük Research Project Series 9). London, British Institute at Ankara; Los Angeles, Cotsen Institute of Archaeology Press

— (ed.) 2013b: *Humans and Landscapes of Çatalhöyük: Reports from the 2000–2008 Seasons* (Çatalhöyük Research Project Series 8). London, British Institute at Ankara; Los Angeles, Cotsen Institute of Archaeology Press

— 2013c: 'Becoming entangled in things' in I. Hodder (ed.), *Substantive Technologies at Çatalhöyük: Reports from the 2000–2008 Seasons* (Çatalhöyük Research Project Series 9). London, British Institute at Ankara; Los Angeles, Cotsen Institute of Archaeology Press: 1–26

— (ed.) 2014a: *Çatalhöyük Excavations: The 2000–2008 Seasons* (Çatalhöyük Research Project Series 7). London, British Institute at Ankara; Los Angeles: Cotsen Institute of Archaeology Press

— (ed.) 2014b: *Integrating Çatalhöyük: Themes from the 2000–2008 Seasons* (Çatalhöyük Research Project Series 10). London, British Institute at Ankara; Los Angeles, Cotsen Institute of Archaeology Press

— (ed.) 2014c: *Religion at Work in a Neolithic Society: Vital Matters*. Cambridge, Cambridge University Press

— 2014d: 'Çatalhöyük: the leopard changes its spots. A summary of recent work' *Anatolian Studies* 64: 1–22

— 2014e: 'Temporal trends: the shapes and narratives of cultural change at Çatalhöyük' in I. Hodder (ed.), *Integrating Çatalhöyük: Themes from the 2000–2008 Seasons*. London, British Institute at Ankara; Los Angeles, Cotsen Institute of Archaeology Press: 169–83

— 2014f: 'Introduction: dwelling at Çatalhöyük' in I. Hodder (ed.), *Humans and Landscapes of Çatalhöyük: Reports from the 2000–2008 Seasons* (Çatalhöyük Research Project Series 8). London, British Institute at Ankara; Los Angeles, Cotsen Institute of Archaeology Press: 1–29

— 2014g: 'Mosaics and networks: the social geography of Çatalhöyük' in I. Hodder (ed.), *Integrating Çatalhöyük: Themes from the 2000–2008 Season* (Çatalhöyük Research Project Series 10). London, British Institute at Ankara; Los Angeles, Cotsen Institute of Archaeology Press: 149–68

— 2016a: 'More on history houses at Çatalhöyük: a response to Carleton et al.' *Journal of Archaeological Science* 67: 1–6

— 2016b: *Studies in Human-Thing Entanglement*. Self-published, Stanford

— 2017: 'Introduction: ending 25 years of fieldwork at Çatalhöyük' *Çatalhöyük 2017 Archive Reports*. http://www.catalhoyuk.com/archive_reports/2017

— (ed.) 2019: *Violence and the Sacred in the Ancient Near East: Girardian Conversations at Çatalhöyük*. Cambridge, Cambridge University Press

— (ed.) 2020: *Consciousness, Creativity and Self at the Dawn of Settled Life*. Cambridge, Cambridge University Press

Hodder, I., Cessford, C. 2004: 'Daily practice and social memory at Çatalhöyük' *American Antiquity* 69.1: 17–40

Hodder, I., Farid, S. 2014: 'Questions, history of work and summary of results' in I. Hodder (ed.), *Çatalhöyük Excavations: The 2000–2008 Seasons* (Çatalhöyük Research Project Series 7). London, British Institute at Ankara; Los Angeles, Cotsen Institute of Archaeology Press: 1–34

Hodder, I., Meskell, L. 2010: 'The symbolism of Çatalhöyük in its regional context' in I. Hodder (ed.), *Religion in the Emergence of Civilization: Çatalhöyük as a Case Study*. Cambridge, Cambridge University Press: 32–72

— 2011: 'A "curious and sometimes a trifle macabre artistry": some aspects of symbolism in Neolithic Turkey' *Current Anthropology* 52.2: 1–29

Hodder, I., Marciniak, A. (eds) 2015: *Assembling Çatalhöyük* (European Association of Archaeologists, Themes in Contemporary Archaeology 1). Leeds, Maney

Hodder, I., Pels, P. 2010: 'History houses: a new interpretation of architectural elaboration at Çatalhöyük' in I. Hodder (ed.), *Religion in the Emergence of Civilization: Çatalhöyük as a Case Study*. Cambridge, Cambridge University Press: 163–86

Hoffmann, M. 1964: *The Warp-Weighted Loom*. Oslo, Universitetsforlaget

Hole, F., Flannery, K. 1962: 'Excavations at Ali Kosh, Iran, 1961' *Iranica Antiqua* 2: 97–148

— 1967: 'The prehistory of southwestern Iran: a preliminary report' *Proceedings of the Prehistoric Society* N.S. 33: 147–206

Hole, F., Flannery, K., Neely, J.A. 1969: *Prehistory and Human Ecology of the Deb-Luran Plain*. Ann Arbor, University of Michigan

Horejs, B., Milić, B., Ostmann, F., Thanheiser, U., Weninger, B., Galik, A. 2015: 'The Aegean in the early 7th millennium BC: maritime networks and colonization' *Journal of World Prehistory* 28.4: 289–90

Horsfall, G.A. 1987: 'Design theory and grinding stones' in B. Hayden (ed.), *Lithic Studies among the Contemporary Highland Maya*. Tucson, University of Arizona Press: 332–77

Hoskins, J. 2006: 'Agency, biography and objects' in C. Tilley, W. Keane, S. Küchler, M. Rowlands, P. Spyer (eds), *The Handbook of Material Culture*. London, Sage Publications 74–85

House, M. 2014: 'Building 77', in I. Hodder (ed.), *Çatalhöyük Excavations: The 2000–2008 Seasons* (Çatalhöyük Research Project Series 7). London, British Institute at Ankara; Los Angeles, Cotsen Institute of Archaeology Press: 485–503

Hurcombe, L. 1992: *Use Wear Analysis and Obsidian: Theory, Experiments and Results* (Sheffield Archaeological Monographs 4). Sheffield, J.R. Collis Publications

Inizan, M.-L., Roche, H., Tixier, J. 1992: *Technology of Knapped Stone*. Paris, CNRS

Inizan, M.-L., Reduron-Ballinger, M., Roche, H., Tixier, J. 1999: *Technology and Terminology of Knapped Stone* (translated by J. Féblot-Augustins). Nanterre, CREP

Ivanova, M. 2018: '"No quern, no food?" Milling technology and the spread of farming in southeast Europe' in M. Ivanova, B. Athanassov, V. Petrova, D. Takorova, P.W. Stockhammer (eds), *Social Dimensions of Food in the Prehistoric Balkans*. Oxford, Oxbow Books: 173–89

Jenkins, E.L., Yeomans, L. 2013: 'The microfauna' in I. Hodder (ed.), *Humans and Landscapes of Çatalhöyük: Reports from the 2000–2008 Seasons* (Çatalhöyük Research Project Series 8). London, British Institute at Ankara; Los Angeles, Cotsen Institute of Archaeology Press: 259–70

Jones, A. 2000: *Rocks and Minerals*. London, Harper Collins

Kabukcu, C., Asouti, E. 2014: 'Anthracology' *Çatalhöyük 2014 Archive Report*. http://www.catalhoyuk.com/archive_reports/2014

Karul, N. 2017: *Aktopraklık. Tasarlanmış Prehistorik Bir Köy*. Istanbul, Ege Yayınları

— 2019: 'The first farmers in NW Anatolia during the 7th millennium BC' in A. Marciniak (ed.), *Concluding the Neolithic: The Near East in the Second Half of the Seventh Millennium BC*. Atlanta, Lockwood Press: 267–85

Karul, N., Avcı, M.B. 2011: 'Neolithic communities in the eastern Marmara region: Aktopraklık C' *Anatolica* 37: 1–15

Keane, W. 2010: 'Marked, absent, habitual: approaches to Neolithic religion at Çatalhöyük' in I. Hodder (ed.), *Religion in the Emergence of Civilization: Çatalhöyük as a Case Study*. Cambridge, Cambridge University Press: 187–219

Keller, J. 1974: 'Quaternary maar volcanism near Karapınar in Central Anatolia' *Bulletin Volcanologique* 38.1: 378–96

Kelterborn, P. 1991: 'Towards replicating Neolithic stone sawing' *Archéologie Expérimentale* 2. *La terre: L'os et la pierre, la maison et les champs. Actes du colloque international "Expérimentation en archéologie: bilan et perspectives", tenu à l'Archéodrome de Beaune les 6, 7, 8 et 9 avril 1988*. Paris, Editions Errance: 129–37

Kenyon, K.M. 1957: *Digging Up Jericho*. London, Ernest Benn

Kılınç, G.M., Koptekin, D., Atakuman, Ç., Sümer, A.P., Dönertaş, H.M., Yaka, R., Bilgin, C.C., Büyükkarakaya, A.M., Baird, D., Altınışık, E., Flegontov, P., Götherström, A., Togan, İ., Somel, M. 2017: 'Archaeogenomic analysis of the first steps of Neolithization in Anatolia and the Aegean' *Proceedings of the Royal Society B: Biological Sciences* 284.1867: 20172064

Kimmerer R.W. 2013: *Braiding Sweetgrass: Indigenous Wisdom, Scientific Knowledge and the Teachings of Plants*. Minneapolis: Milkweed Editions

Kirkbridge, D. 1972: 'Umm Dabaghiyah 1971: a preliminary report. An early ceramic farming settlement in marginal North Central Jazira, Iraq' *Iraq* 34: 3–15

Knapp, A.B., van Dommelen, P. 2008: 'Past practices: rethinking individuals and agents in archaeology' *Cambridge Archaeological Journal* 18: 15–34

Knüsel, C.J., Glencross, B., Milella, M. 2019: 'A Girardian framework for violent injuries at Neolithic Çatalhöyük in their Western Asian context' in I. Hodder (ed.), *Violence and the Sacred in the Neolithic: Girardian Conversations at Çatalhöyük*. Cambridge, Cambridge University Press: 60–95

Koppelson, E. 1996: *Analysis and Consolidation of Architectural Plasters from Çatalhöyük, Turkey*. Master's thesis, University of Pennsylvania

Kuijt, I. 2000: 'People and space in early agricultural villages: exploring daily lives, community size, and architecture in the Late Pre-Pottery Neolithic' *Journal of Anthropological Archaeology* 19.1: 75–102

— 2014: 'Jericho archaeological site' in C. Smith (ed.), *Encyclopaedia of Global Archaeology*. New York, Springer: 4194–98

Kurzawska, A., Bar-Yosef Mayer, D.E., Burçin, A.G. in press: 'Micro-shells from the construction material in the TP Trench of Çatalhöyük' in A. Marciniak, L. Czerniak (eds), *Late Neolithic at Çatalhöyük in the TP Area*

Kyparrisi-Apostolika, N. 2001: *Prehistoric Ornaments from Thessaly* (Publications of the Archaeological Archive 76). Athens, Fund of Archaeological Proceeds

Larsen, C.S., Hillson, S.W., Ruff, C.B., Sadvari, J.W., Garofalo, E.M. 2013: 'The human remains II: interpreting lifestyle and activity in Neolithic Çatalhöyük' in I. Hodder (ed.), *Humans and Landscapes of Çatalhöyük: Reports from the 2000–2008 Seasons* (Çatalhöyük Research Project Series 8). London, British Institute at Ankara; Los Angeles, Cotsen Institute of Archaeology Press: 397–412

Last, J. 1998: 'A design for life: interpreting the art of Çatalhöyük' *Journal of Material Culture* 3: 355–78

— 2005a: 'Pottery from the East Mound' in Hodder, I. (ed.), *Changing Materialities at Çatalhöyük: Reports from the 1995–99 Seasons* (Çatalhöyük Research Project Series 5). Cambridge, McDonald Institute for Archaeological Research; London, British Institute of Archaeology at Ankara: 101–38

— 2005b: 'Art' in I. Hodder (ed.), *Çatalhöyük Perspectives: Themes from the 1995–99 Seasons* (Çatalhöyük Research Project Series 6). Cambridge, McDonald Institute for Archaeological Research; London, British Institute of Archaeology at Ankara: 197–208

Lazaridis, I., Nadel, D., Rollefson, G., Merrett, D.C., Rohland, N., Mallick, S., Fernandes, D., Novak, M., Gamarra, B., Sirak, K., Connell, S., Stewardson, K., Harney, E., Fu, Q., Gonzalez-Fortes, G., Jones, E.R., Roodenberg, S.A., Lengyel, G., Bocquentin, F., Gasparian, B., Monge, J.M., Gregg, M., Eshed, V., Mizrahi, A.-S., Meiklejohn, C.,

Gerritsen, F., Bejenaru, L., Blüher, M., Campbell, A., Cavalleri, G., Comas, D., Froguel, P., Gilbert, E., Kerr, S.M., Kovacs, P., Krause, J., McGettigan, D., Merrigan, M., Merriwether, D.A., O'Reilly, S., Richards, M.B., Semino, O., Shamoon-Pour, M., Stefanescu, G., Stumvoll, M., Tönjes, A., Torroni, A., Wilson, J.F., Yengo, L., Hovhannisyan, N.A., Patterson, N., Pinhasi, R., Reich, D. 2016: 'Genomic insights into the origin of farming in the ancient Near East' *Nature* 536.7617: 419–24

Lazarri, I. 2005: 'The texture of things: objects, people and landscape in northwest Argentina (first millennium AD)' in L. Meskell (ed.), *Archaeologies of Materiality*. Oxford, Blackwell 126–61

Lemorini, C. 2000: *Reconnaitre des Tactiques d'Exploitation du Milieu au Paléolithique Moyen. La Contribution de l'Analyse Fonctionnelle. Etude Fonctionnelle des Industries Lithiques de Grotta Breuil (Latium, Italie) et de La Combette (Bonnieux, Vaucluse, France)*' (BAR International Series 858). Oxford, Archaeopress

Lercari, N. 2018: 'Virtually rebuilding Çatalhöyük history houses' in I. Hodder (ed.), *Religion, History and Place in the Origin of Settled Life*. Boulder, University Press of Colorado: 263–82

Leuzinger, U., Rast-Eicher, A. 2011: 'Flax processing in the Neolithic and Bronze Age pile-dwelling settlements of eastern Switzerland' *Vegetation History and Archaeobotany* 334: 535–42

Lévi-Strauss, C. 1982: *The Way of the Masks*. Seattle, University of Washington Press

Lewis-Williams, D. 2004: 'Constructing a cosmos: architecture, power and domestication at Çatalhöyük' *Journal of Social Archaeology* 4.1: 28–59

Lingle, A., Dell'Unto, N., Der, L., Doyle, S., Killackey, K., Klimowicz, A., Meskell, L., Parkes, P., Tung, B. 2015: 'Painted plaster head' *Çatalhöyük 2015 Archive Report*. http://www.catalhoyuk.com/archive_reports/2015

Lyons K.M. 2020: *Vital Decomposition: Soil Practitioners and Life Politics*. Durham, Duke University Press

Malafouris, L. 2013: *How Things Shape the Mind. A Theory of Material Engagement*. Cambridge (MA), MIT Press

Malinowski, B. 1922: *Argonauts of the Western Pacific: An Account of Native Enterprise and Adventure in the Archipelagoes of Melanesia New Guinea*. London, Routledge and Kegan Paul

Marciniak, A. 2005: *Placing Animals in the Neolithic: Social Zooarchaeology of Prehistoric Farming Communities*. London, UCL Press

— 2015a: 'A new perspective on the Central Anatolian Late Neolithic. The TPC Area excavations at Çatalhöyük' in S.R. Steadman, G. McMahon (eds), *The Archaeology of Anatolia. Recent Discoveries (2011–2014)* 1. Newcastle upon Tyne, Cambridge Scholars Publishing: 6–25

— 2015b: 'The Neolithic house as a procurement, production and consumption unit in Çatalhöyük' in K. Brink, S. Hydén, K. Jennbert, L. Larsson, D. Deborah (eds), *Neolithic Diversities. Perspectives from a Conference in Lund, Sweden*. Lund, Acta Archaeologica Lundensia: 89–97

— 2016: 'The Late Neolithic transition. The case of Çatalhöyük East' in N. Sanz (ed.), *The Origins of Food Production*. Mexico City, UNESCO Mexico Office: 78–89

— 2018: 'Bridging up Anatolia. Çatalhöyük and northwestern Anatolia in the Late Neolithic' in P. Valde-Nowak, K. Sobczyk, M. Nowak, J. Źrałka (eds), *Amici Magistro et Collegae Suo Ioanni Christopho Kozłowski Dedicant*. Kraków, Wydawnictwo UJ: 281–90

— (ed.) 2019: *Concluding the Neolithic. The Near East in the Second Half of the Seventh Millennium BC*. Atlanta, Lockwood Press

Marciniak, A., Czerniak, L. 2007: 'Social transformations in the Late Neolithic and the Early Chalcolithic periods in Central Anatolia' *Anatolian Studies* 57: 115–30

— 2012: 'Çatalhöyük unknown. The late sequence on the East Mound' in R. Matthews, J. Curtis (eds), *Proceedings of the 7th International Congress on the Archaeology of the Ancient Near East* 1: *Mega cities and Mega Sites. The Archaeology of Consumption and Disposal, Landscape, Transport and Communication*. Wiesbaden, Harrassowitz Verlag: 3–16

Marciniak, A., Barański, M.Z., Bayliss, A., Czerniak, L., Goslar, T., Southon, J., Taylor, R.E. 2015a: 'Fragmenting times: interpreting a Bayesian chronology for the Late Neolithic occupation of Çatalhöyük East, Turkey' *Antiquity* 89: 154–76

Marciniak, A., Filipowicz, P., Hordecki, J. Pettersson, P.E. 2015b: 'Excavations in the TPC Area' *Çatalhöyük 2015 Archive Report*. http://www.catalhoyuk.com/archive_reports/2015

Marciniak, A., Dembowiak, M., Filipowicz, P., Harabasz, K., Hordecki, J., Stosik, W. 2017: 'Excavations in the TPC Area' *Çatalhöyük 2017 Archive Report*. http://www.catalhoyuk.com/archive_reports/2017

Martin, L., Meskell, L. 2012: 'Animal figurines from Neolithic Çatalhöyük: figural and faunal perspectives' *Cambridge Archaeological Journal* 22.3: 401–19

Martin, L., Russell, N. 2000: 'Trashing rubbish' in I. Hodder (ed.), *Towards Reflexive Method in Archaeology: The Example at Çatalhöyük* (Çatalhöyük Research Project Series 2). Cambridge, McDonald Institute for Archaeological Research; London, British Institute of Archaeology at Ankara: 57–69

Matero, F. 2000: 'The conservation of an excavated past' in I .Hodder (ed.), *Towards Reflexive Method in Archaeology: the Example at Çatalhöyük* (Çatalhöyük Research Project Series 2). Cambridge, McDonald Institute for Archaeological Research; London, British Institute of Archaeology at Ankara: 71–88

Matsutani, T. (ed.) 1991: *Tell Kashkashok. The Excavations at Tell No. II.* Tokyo, University of Tokyo Press

Matthews, W. 1999: 'Micromorphology archive report' *Çatalhöyük 1999 Archive Report.* http://www.catalhoyuk.com/archive_reports/1999

— 2005a: 'Life-cycle and life-course of buildings' in I. Hodder (ed.), *Çatalhöyük Perspectives: Themes from the 1995–99 Seasons* (Çatalhöyük Research Project Series 6). Cambridge, McDonald Institute for Archaeological Research; London, British Institute of Archaeology at Ankara: 125–49

— 2005b: 'Micromorphological and microstratigraphic traces of uses and concepts of space' in I. Hodder (ed.), *Inhabiting Çatalhöyük: Reports from the 1995–99 Seasons* (Çatalhöyük Research Project Series 4). Cambridge, McDonald Institute for Archaeological Research; London, British Institute at Ankara: 355–98

Matthews, W., Hastorf, C., Ergenkon, B. 2000: 'Ethnoarchaeology: studies in local villages aimed at understanding aspects of the Neolithic site' in I. Hodder (ed.), *Towards Reflexive Method in Archaeology: The Example at Çatalhöyük* (Çatalhöyük Research Project Series 2). Cambridge, McDonald Institute for Archaeological Research; London, British Institute of Archaeology at Ankara: 177–88

Matthews, W., Almond, M.J., Anderson, E., Wiles, J., Williams, H., Rowe, J. 2013: 'Biographies of architectural materials and buildings: integrating high-resolution micro-analysis and geochemistry' in I. Hodder (ed.), *Substantive Technologies at Çatalhöyük: Reports from the 2000–2008 Seasons* (Çatalhöyük Research Project Series 9). London, British Institute at Ankara; Los Angeles, Cotsen Institute of Archaeology Press: 115–36

Mauldin, R. 1993: 'The relationship between ground stone and agricultural intensification in western New Mexico' *Kiva* 58.3: 317–30

Mauss, M. 1990: *The Gift.* New York, Routledge

Mazzucato, C. 2013: 'Sampling and mapping Çatalhöyük' in I. Hodder (ed.), *Humans and Landscapes of Çatalhöyük: Reports from the 2000–2008 Seasons* (Çatalhöyük Research Project Series 8). London, British Institute at Ankara; Los Angeles, Cotsen Institute of Archaeology Press: 31–64

Médard, F. 2000: *L'Artisinat Textile au Néolithique. l'Example de Delley-Portalban II (Suisse) 3272–2462 avant J.-C.* Montagnac, Éditions Moniqie Mergoil

— 2006: *Les Activités de Filage au Néolithique sur le Plateau Suisse. Analyse Technique, Économique et Sociale.* Paris, CNRS Éditions

— 2010: *Le Tissage au Néolithique sur les Sites Lavustres du Plateau Suisse. Matières Premières, Techniques et Contexte de Production.* Paris, Collection CRA Monographies

Mellaart, J. 1961: 'Early cultures of the South Anatolian Plateau' *Anatolian Studies* 11:159–84

— 1962: 'Excavations at Çatal Hüyük: first preliminary report, 1961' *Anatolian Studies* 12: 41–65

— 1963: 'Excavations at Çatal Hüyük, 1962: second preliminary report' *Anatolian Studies* 13: 43–103

— 1964: 'Excavations at Çatal Hüyük, 1963: third preliminary report' *Anatolian Studies* 14: 39–119

— 1966: 'Excavations at Catal Hüyük, 1965: fourth preliminary report' *Anatolian Studies* 16: 165–91

— 1967: *Çatal Hüyük: A Neolithic Town in Anatolia.* London: Thames and Hudson

— 1970: *Excavations at Hacılar I–II.* Edinburgh, Edinburgh University Press and British Institute at Ankara

— 1975: *The Neolithic of the Near East.* London, Thames and Hudson

Merpert, N.Y., Munchaev, R.M. 1987: 'The earliest levels at Yarim Tepe I and Yarim Tepe II in Northern Iraq' *Iraq* 49: 1–36

Meskell, L.M. 1995: 'Goddesses, Gimbutas and New Age archaeology' *Antiquity* 69.262: 74–86

— 1999: *Archaeologies of Social Life: Age, Sex, Class et cetera in Ancient Egypt.* Oxford, Blackwell

— 2007: 'Archaeologies of identities' in T. Insoll (ed.), *The Archaeology of Identities. A Reader.* London, Routledge: 23–43

— 2013: 'Dirty, pretty things: on archaeology and prehistoric materialities' in P. Miller (ed.), *Cultural Histories of the Material World.* Ann Arbor, University of Michigan Press: 92–107

Meskell, L.M., Nakamura, C. 2009: 'Figurine report 2009' *Çatalhöyük 2009 Archive Report.* http://www.catalhoyuk.com/ archive_reports/2009

Meskell, L.M., Nakamura, C., King, R., Farid, S. 2007: 'Çatalhöyük figurines' *Çatalhöyük 2007 Archive Report*. http://www.catalhoyuk.com/archive_reports/2007

— 2008: 'Figured lifeworlds and depositional practices at Çatalhöyük' *Cambridge Archaeological Journal* 18.2: 139–61

Meskell, L.M., Nakamura, C., Der, L. 2015: 'Figurine report 2015' *Çatalhöyük 2015 Archive Report*. http://www.catalhoyuk.com/archive_reports/2015

Meskell, L.M., Nakamura, C., Der, L., Tsoraki, C., Arntz, M. 2016: 'Figurine report 2016' *Çatalhöyük 2016 Archive Report*. http://www.catalhoyuk.com/archive_reports/2016

Michalowski, P. 1993: 'Review: "Tokenism: Before Writing, Volume 1: From Counting to Cuneiform" by Denise Schmandt-Besserat; "Before Writing, Volume 2: A Catalog of Near Eastern Tokens" by Denise Schmandt-Besserat' *American Anthropologist* 95.4: 996–99

Milić, M. 2014: 'PXRF characterisation of obsidian from central Anatolia, the Aegean and central Europe' *Journal of Archaeological Science* 41: 285–96

Milić, M., Brown, K., Carter, T. 2013: 'A visual characterisation of the Çatalhöyük obsidian' in I. Hodder (ed.), *Substantive Technologies at Çatalhöyük: Reports from the 2000–2008 Seasons* (Çatalhöyük Research Project 9). London, British Institute at Ankara; Los Angeles, Cotsen Institute of Archaeology Press: Chapter 21 CD Appendix 21.1

Miller, M.A. 1997: *Jewels of Shell and Stone, Clay and Bone: The Production, Function and Distribution of Aegean Stone Age Ornaments*. PhD thesis, Boston University

Mills, B. 2014: 'Relational networks and religious sodalities at Çatalhöyük' in I. Hodder (ed.), *Religion at Work in a Neolithic Society: Vital Matters*. Cambridge, Cambridge University Press: 159–86

Mithen, S. 2007: 'Did farming arise from a misapplication of social intelligence?' *Philosophical Transactions: Biological Sciences* 362.1480: 705–18

Mitrović, S., Vasić, M. 2013: 'An integrated perspective on the uses of materials at Çatalhöyük based on the analysis of heavy residues' in I. Hodder (ed.), *Substantive Technologies at Çatalhöyük: Reports from the 2000–2008 Seasons* (Çatalhöyük Research Project Series 9). London, British Institute at Ankara; Los Angeles, Cotsen Institute of Archaeology Press: 27–50

Monroe, J.S., Wicander, R. 2007: 'Fiziksel Jeoloji: Yeryuvarı'nın Araştırılması' in K. Dirik, M. Şener (eds), *TMMOB Jeoloji Mühendisleri Odası Çeviri Serisi* 1: 184

Mortensen, P. 1964: *Notes on Obsidian and Flint from Çatal Hüyük 1961–1963*. Unpublished manuscript

— 1970: *Tell Shimshara. The Hassuna Period*. Copenhagen, Kgl. Danske Videnskabers Selskab

Moundrea-Agrafioti, H.A. 1981: *La Thessalie du Sud-Est au Néolithique. Outillage Lithique et Osseux*. PhD thesis, University of Paris X

Muller, A., Clarkson, C., Baird, D., Fairbairn, A. 2018: 'Reduction intensity of backed blades: blank consumption, regularity and efficiency at the early Neolithic site of Boncuklu, Turkey' *Journal of Archaeological Science: Reports* 21: 721–32

Munn, N. 1986: *The Fame of Gawa: A Symbolic Study of Value Transformation in a Massim (Papua New Guinea) Society*. Cambridge, Cambridge University Press

Mussi, M. 2002: *Earliest Italy: An Overview of the Italian Paleolithic and Mesolithic*. New York, Kluwer Academic

Nadel, D., Danin, A., Werker, E., Schick, T., Kislev, M.E., Stewart, K. 1994: '19,000-year-old twisted fibers from Ohalo II' *Current Anthropology* 35: 451–58

Nadel, D., Weiss, E., Simchoni, O., Tsatskin, A. Danin, A., Kislev, M. 2004: 'Stone Age hut in Israel yields world's oldest evidence of bedding' *Proceedings of the National Academy of Sciences* 101: 6821–26

Nakamura, C. 2008: *The Matter of Magic: Figuring Memory and Protection in Neo-Assyrian Apotropaic Figurine Rituals (First Millennium BC)*. PhD thesis, Columbia University, New York

— 2010: 'Magical deposits at Çatalhöyük: a matter of time and place?' in I. Hodder (ed.), *Religion in the Emergence of Civilization. Çatalhöyük as a Case Study*. Cambridge: Cambridge University Press: 300–31

— 2019: 'Ritual' in A.C. Gunter (ed.), *A Companion to Ancient Near Eastern Art*. Hoboken (NJ), Wiley-Blackwell: 309–32

Nakamura, C., Meskell, L.M. 2004: 'Figurines and miniature clay objects' *Çatalhöyük 2004 Archive Report*. http://www.catalhoyuk.com/archive_reports/2014

— 2009: 'Articulate bodies: forms and figures at Çatalhöyük' *Journal of Archaeological Method and Theory* 16: 205–30

— 2013a: 'The Çatalhöyük burial assemblage' in I. Hodder (ed.), *Humans and Landscapes of Çatalhöyük: Reports from the 2000–2008 Seasons* (Çatalhöyük Research Project Series 8). London, British Institute at Ankara; London, British Institute at Ankara; Los Angeles, Cotsen Institute of Archaeology Press: 441–66

— 2013b: 'Figurine worlds at Çatalhöyük' in I. Hodder (ed.), *Substantive Technologies at Çatalhöyük: Reports from the 2000–2008 Seasons* (Çatalhöyük Research Project Series 9). London, British Institute at Ankara; Los Angeles, Cotsen Institute of Archaeology Press: 197–228

— 2017: 'Figurines and clay stamps' *Çatalhöyük Archive Report 2017*. http://www.catalhoyuk.com/archive_reports/2017

Nakamura, C., Pels, P. 2014: 'Using "magic" to think from the material: tracing distributed agency, revelation, and concealment at Çatalhöyük' in I. Hodder (ed.), *Religion at Work in a Neolithic Society: Vital Matters*. Cambridge, Cambridge University Press: 187–224

Nakamura, C., Der, L., Meskell, L. 2014: 'Çatalhöyük figurines report' *Çatalhöyük 2014 Archive Report*. http://www.catalhoyuk.com/archive_reports/2014

Nanoglou, S. 2009: 'Animal bodies and ontological discourse in the Greek Neolithic' *Journal of Archaeological Method and Theory* 16.3: 184–204

Nazaroff, A.J. 2014: *Consumption of Multiple Greenstone Resources at Neolithic Çatalhöyük: Evidence from Portable X-Ray Fluorescence Analysis*. Unpublished report.

Nazaroff, A. 2015: *Entanglement: A Study in Neolithic Resource Exploitation in the Middle East*. PhD thesis, Stanford University, Palo Alto

Nazaroff, A., Baysal, A., Çiftçi, Y. 2013: 'The importance of chert in Central Anatolia: lessons from the Neolithic assemblage at Çatalhöyük, Turkey' *Geoarchaeology* 28: 340–62

Nazaroff, A.J., Baysal, A., Çiftçi, Y., Prufer, K. 2015: 'Resilience and redundance. Resource networks and the Neolithic chert economy at Çatalhöyük, Turkey' *European Journal of Archaeology* 18.3: 402–28

Nazaroff, A., Tsoraki, C., Vasić, M. 2016: 'Aesthetic, social, and material networks: a perspective from the flint daggers at Çatalhöyük, Turkey' *Cambridge Archaeological Journal* 26.1: 65–92

Nieuwenhuyse, O.P., Akkermans, P.M.M.G., van der Plicht, J. 2010: 'Not so coarse, nor always plain. The earliest pottery of Syria' *Antiquity* 84: 71–85

Nishiaki, Y., Le Mière, M. 2005: 'The oldest pottery Neolithic of Upper Mesopotamia: new evidence from Tell Seker al-Aheimar, the Khabur, northeast Syria' *Paléorient* 31.2: 55–68

Nixon-Darcus, L., d'Andrea, A. 2017: 'Necessary for life: studies of ancient and modern grinding stones in highland Ethiopia' *African Archaeological Review* 34.2: 193–223

Oates, J. 1978: 'Religion and ritual in sixth-millennium B.C. Mesopotamia' *World Archaeology* 10.2: 117–24

O'Connor, T.P. 2013: 'Humans and animals: refuting Aquinas' *Archaeological Review from Cambridge* 28.2: 186–94

Oppenheim, A.L. 1959: 'On an operational device in Mesopotamian bureaucracy' *Journal of Near Eastern Studies* 18.22: 121–28

Orton, D., Anvari, J., Gibson, C., Last, J., Bogaard, A., Rosenstock, E., Biehl, P.F. 2018: 'A tale of two tells: dating the Çatalhöyük West Mound' *Antiquity* 92: 620–39

Ota, R., Dinçel, A. 1975: 'Volcanic rocks of Turkey' *Bulletin of the Geological Survey of Japan* 26: 19–45

Özbal, R., Gerritsen, F. 2019: 'Barcın Höyük in interregional perspective. A first assessment' in A. Marciniak (ed.), *Concluding the Neolithic: The Near East in the Second Half of the Seventh Millennium BC*. Atlanta, Lockwood Press: 287–306

Özbaşaran, M. 2011: 'The Neolithic on the plateau' in S.R. Steadman, G. McMahon (eds), *The Oxford Handbook of Ancient Anatolia*. Oxford, Oxford University Press: 99–124

— 2012: 'Aşıklı' in M. Özdoğan, N. Başgelen, P. Kuniholm (eds), *The Neolithic in Turkey* 3: *New Excavations, New Research. Central Turkey*. Istanbul, Arkeoloji ve Sanat Yayınları: 135–58

Özbek, O. 2000: 'A prehistoric stone axe production site in Turkish Thrace: Hamaylitaria' *Documenta Praehistorica* 27: 167–71

— 2011: 'Primary and secondary raw material preferences in the Neolithic societies in northwest Turkey' in V. Davis, M. Edmonds (eds), *Stone Axe Studies* 3. Oxford, Oxbow Books: 217–29

Özdoğan, A. 1999: 'Çayönü' in M. Özdoğan, N. Başgelen (eds), *Neolithic in Turkey: The Cradle of Civilisation. New Discoveries*. Istanbul, Arkeoloji ve Sanat Yayınları: 35–46

Özdoğan, E. 2015: 'Current research and new evidence for the neolithization process in western Turkey' *European Journal of Archaeology* 18.1: 33–59

Özdoğan, M. 2010: 'Westward expansion of the Neolithic way of life: sorting the Neolithic package into distinct packages' in P. Matthiae, F. Pinnock, L. Nigro, N. Marchetti (eds), *Proceedings of the 6th International Congress on the Archaeology of the Ancient Near East.* Wiesbaden, Harrassowitz Verlag: 883–98

— 2011: 'Eastern Thrace: the contact zone between Anatolia and the Balkans' in S.R. Steadman, G. McMahon (eds), *The Oxford Handbook of Ancient Anatolia*. Oxford, Oxford University Press: 657–82

Özdöl, S. 2006: *Anadolu'da Erken Neolitik Dönem Çanak Çömlek Kültürleri ve Çatalhöyük Örneği*. PhD thesis, Ege Üniversitesi Sosyal Bilimler Enstitüsü Arkeoloji Anabilim Dalı

— 2012: *The Development and Traditions of Pottery in the Neolithic of the Anatolian Plateau* (BAR International Series 2439). Oxford, Archaeopress

Özdöl-Kutlu, S., Carter, T., Czerniak, L., Marciniak, A. 2015: 'The end of the Neolithic settlement' in I. Hodder, A. Marciniak (eds), *Assembling Çatalhöyük*. Leeds, Maney: 179–95

Özkaya, V., 2009: 'Excavations at Körtik Tepe. A new Pre-Pottery Neolithic A site in southeastern Anatolia' *NEO-LITHICS* 2.9: 3–8

Öztan, A. 2012: 'Köşk Höyük: a Neolithic Settlement in Niğde-Bor Plateau' in M. Özdoğan, N. Başgelen, P. Kuniholm (eds), *The Neolithic in Turkey 3: New Excavations, New Research. Central Turkey*. Istanbul, Arkeoloji ve Sanat Yayınları: 31–70

Öztürk, B. 2011: 'Scaphopod species (*Mollusca*) of the Turkish Levantine and Aegean Seas' *Turkish Journal of Zoology* 35.2: 199–211

Pallasmaa, J. 2005: *The Eyes of the Skin: Architecture and the Senses*. Chichester, Wiley

Papadopoulos, C., Hamilakis, Y., Kyparissi-Apostolika, N. 2015: 'Light in a Neolithic dwelling: Building 1 at Koutroulou Magoula (Greece)' *Antiquity* 89: 1034–50

Papadopoulos, C., Holley, M. 2017: *The Oxford Handbook of Light in Archaeology*. Oxford, Oxford University Press

Papadopoulos, C., Sakellarakis, Y. 2010: 'Virtual windows to the past: reconstructing the "Ceramics Workshop" at Zominthos, Crete' in J. Melero, P. Cano, J. Revelles (eds), *Proceedings of Computer Applications and Quantitative Methods in Archaeology 2010*. Oxford, Archaeopress: 47–54

Pearson, J. 2013: 'Human and animal diet as evidenced by stable carbon and nitrogen isotope analysis' in I. Hodder (ed.), *Humans and Landscapes of Çatalhöyük: Reports from the 2000–2008 Seasons* (Çatalhöyük Research Project Series 8). London, British Institute at Ankara; Los Angeles, Cotsen Institute of Archaeology Press: 271–98

Pearson, J., Meskell, L.M. 2015: 'Isotopes and images: fleshing out bodies at Çatalhöyük' *Journal of Archaeological Method and Theory* 22.2: 461–82

Pearson, J.A., Buitenhuis, H., Hedges, R.E.M., Martin, L., Russell, N., Twiss, K.C., 2007: 'New light on early caprine herding strategies from isotope analysis: a case study from Neolithic Anatolia' *Journal of Archaeological Science* 34.12: 2170–79

Peccerillo, A., Taylor, J.R. 1976: 'Geochemistry of Upper Cretaceous volcanic rocks from the Pontic chain, northern Turkey' *Bulletin Volcanologique* 39.4: 557–69

Pelegrin, J. 1990: 'Prehistoric lithic technology' *Archaeological Review from Cambridge* 9.1: 116–25

— 2012: 'New experimental observations for the characterization of pressure blade production techniques" in P.M. Desrosiers (ed.), *The Emergence of Pressure Blade Making: From Origin to Modern Experimentation*. New York, Springer: 465–500

Pellant, C. 2000: *Rocks and Minerals*. London, Dorling Kindersley Limited

Pels, P. 2010: 'Temporalities of "religion" at Çatalhöyük' in I. Hodder (ed.), *Religion in the Emergence of Civilization: Çatalhöyük as a Case Study*. Cambridge, Cambridge University Press: 220–67

Peña-Chocarro, L., Zapata, L. 2014: 'Parching and dehusking hulled wheats' in A. van Gijn, J. Whittaker, P.C. Anderson (eds), *Exploring and Explaining Diversity in Agricultural Technology*. Oxford: Oxbow Books: 226–32

Pétrequin, P., Pétrequin, A.-M. 1993: *Écologie d'un Outil: la Hache de Pierre en Irian Jaya (Indonésie)* (Monographie du CRA 12). Paris, CNRS Éditions

Pilloud, M.A., Larsen, C.S. 2011: '"Official" and "practical" kin: inferring social and community structure from dental phenotype at Neolithic Çatalhöyük, Turkey' *American Journal of Physical Anthropology* 145.4: 519–30

Pitter, S. 2013: *Molecular and Stable Isotopic Analyses of the Fatty Acyl Components of the Pottery of Çatalhöyük, Turkey: Understanding the Relationships between Animal Domestication, Ceramic Technology, Environmental Variation and their Roles in the Secondary Products*. PhD thesis, Stanford University, California

Pitter, S., Yalman, N., Evershed, R. 2013: 'Absorbed lipid residues in the Çatalhöyük pottery' in I. Hodder (ed.), *Substantive Technologies at Çatalhöyük: Reports from the 2000–2008 Seasons* (Çatalhöyük Research Project Series 9). London, British Institute at Ankara; Los Angeles, Cotsen Institute of Archaeology Press: 193–200

Popenoe, R. 2012: *Feeding Desire: Fatness, Beauty and Sexuality among a Saharan People*. London, Routledge

Poppe, G.T., Goto, Y. 1991: *European Seashells (*Polyplacophora, Caudofoveata, Solenogastra, Gastropoda*),* 1. Wiesbaden, Verlag Christa Hemmen

— 1993: *European Seashells (*Scaphopoda, Bivalvia, Cephalopoda*)* 2. Hackenheim, Conchbooks

Porta, J., Herrero, J. 1990: 'Micromorphology and genesis of soils enriched with gypsum' *Developments in Soil Science* 19: 321–39

Poupeau, G., Le Bourdonnec, F.-X., Carter, T., Delerue, S., Shackley, M.S., Barrat, J.A., Dubernet, S., Moretto, P., Calligaro, T., Milić, M., Kobayashi, K. 2010: 'The use of SEM-EDS, PIXE and EDXRF for obsidian provenance studies in the Near East: a case study from Neolithic Çatalhöyük (central Anatolia)' *Journal of Archaeological Science* 37.11: 2705–20

Procopiou, H. 2014: 'Barley meal processing in the Aegean world: a look at diversity' in A. Van Gijn, J. Whittaker, P.C. Anderson (eds), *Exploring and Explaining Diversity in Agricultural Technology*. Oxford: Oxbow Books: 243–46

Prothero, D.R., Schoch, R.M. 2002: *Horns, Tusks, and Flippers: The Evolution of Hoofed Mammals*. Baltimore (MD), Johns Hopkins University Press

Pyzel, J. in press: 'Pottery in the TP Area' in A. Marciniak, L. Czerniak (eds), *Late Neolithic at Çatalhöyük in the TP Area*

Rast-Eicher, A. 2010: 'Fell- und Lederreste aus den Gräbern 2008' in C. Cooper, M. Harbeck, M. Kühn, A. Rast-Eicher, S. Schweissing, S. Ulrich-Bochsler, P. Vandorpte (eds), *Spiez-Einigen, Holleeweg 3, Naturwissenschaftliche Untersuchungen zu den bronzezeitlichen Bestattungen*. Bern, Archäologie Bern: 175–98

— 2015: 'Neolithische, bronzezeitliche und römische Geflechte und Gewebe vom Schnidejoch' in A. Hafner (ed.), *Schnidejoch und Lötschenpass. Archäologie der prähistorischen, römischen und mittelalterlichen Passübergänge in den Berner Alpen* 1. Bern, Archäologischer Dienst des Kantons Bern: 30–38

— 2016: 'Textilien' in C. Harb, N. Bleicher (eds), *Zürich-Parkhaus Opéra. Eine neolithische Feuchtbodenfundstelle* 2: *Funde* (Monographien der Kantonsarchäologie Zürich 49). Zürich, Baudirektion Kanton Zürich: 188–96

— 2018 (with contributions by T. Štolchová, H. Březinová): 'In the beginning was the fibre' in J. Sofaer (ed.), *Considering Creativity: Creativity, Knowledge and Practice in Bronze Age Europe*. Oxford, Archaeopress: 117–32

Rast-Eicher, A., Bender Jørgensen, L. 2013: 'Sheep wool in Bronze and Iron Age Europe' *Journal of Archaeological Science* 40: 1224–41

— 2018: 'News from Çatalhöyük' *Archaeological Textiles Review* 60: 100–104

Rast-Eicher, A., Dietrich, A. 2015: *Neolithische und bronzezeitliche Gewebe und Geflechte: Die Funde aus den Seeufersiedlungen im Kanton Zürich* (Monografien der Kantonsarchäologie Zürich 46). Dübendorf, Baudirektion Kanton Zürich

Rast-Eicher, A., Karg, S., Bender Jørgensen, L. forthcoming: 'The use of local fibres for textiles at Çatalhöyük (Turkey)', *Antiquity*

Reese, D.S. 1989: 'On cassid lips and helmet shells' *Bulletin of the American Schools of Oriental Research* 275: 33–39

— 2005 'The Çatalhöyük shells' in Hodder I. (ed.), *Inhabiting Çatalhöyük: Reports from the 1995–99 Seasons* (Çatalhöyük Research Project Series 4). Cambridge, McDonald Institute for Archaeological Research; London, British Institute at Ankara: 123–26

Regan, R. (with contributions by J. Taylor) 2014: 'The sequence of Buildings 75, 65, 56, 69, 44 and 10 and External Spaces 119, 129, 130, 144, 299, 314, 319, 329, 333, 339, 367, 371 and 427' in I. Hodder (ed.), *Çatalhöyük Excavations: The 2000–2008 Seasons* (Çatalhöyük Research Project Series 7). London, British Institute at Ankara; Los Angeles, Cotsen Institute of Archaeology Press: 131–90

Relke, J. 2007: 'Interpreting the bucrania of Çatalhöyük: James Mellaart, Dorothy Cameron, and beyond' *Anthrozoös* 20.4: 317–28

Renfrew, C. 2001: 'Symbol before concept: material engagement and the early development of society' in I. Hodder (ed.), *Archaeological Theory Today*. Cambridge, Blackwell: 122–40

Renfrew, C., Dixon, J.E., Cann, J.R. 1966: 'Obsidian and early culture contact in the Near East' *Proceedings of the Prehistoric Society* 32: 30–72

Rice, P.M. 1987: *Pottery Analysis*. Chicago, University of Chicago Press

Risch, R. 2002: *Recursos Naturales, Medios de Producción y Explotación Social. Un Análisis Económico de la Industria Lítica de Fuente Alamo (Almería), 2250–1400 ANE* (Iberia Archaeologica 3). Mainz, P. von Zabern

— 2008: 'Grain processing technologies and economic organisation: a case study from the south-east of the Iberian Peninsula during the Copper Age' *The Arkeotek Journal* 2.2: www.thearkeotekjournal.org

Roberts, N., Black, S., Boyer, P., Eastwood, W.J., Griffiths, H.I., Lamb, H.F., Leng, M.J., Parish, R., Reed, J.M., Twigg, D., Yiğitbaşioğlu, H. 1999: 'Chronology and stratigraphy of Late Quaternary sediments in the Konya Basin, Turkey: results from the KOPAL Project' *Quaternary Science Reviews* 18: 611–30

Robitaille, J. 2016: 'The ground stone industry of the Mursi of Maki, Ethiopia: ethnoarchaeological research on milling and crushing equipment (technique and function)' *Journal of Lithic Studies* 3.3: 429–56

Roffet-Salque, M., Marciniak, A., Valdes, P.L, Pawłowska, K., Pyzel, J., Czerniak, L., Krüger, M., Roberts, C.N., Pitter, S., Evershed, R.P. 2018: 'Synchronicity of climate and cultural proxies around 8.2 kyBP Çatalhöyük East' *Proceedings of the National Academy of Sciences* 116.9: 3345–46

Rollefson, G.O. Köhler-Rollefson, I. 1993: 'PPNC adaptations in the first half of the 6th millennium BC' *Paléorient* 19.1: 33–42

Roodenberg, J. 2011: 'Ilipinar: a Neolithic settlement in the eastern Marmara region' in S.R. Steadman, G. McMahon (eds), *The Oxford Handbook of Ancient Anatolia*. Oxford, Oxford University Press: 950–67

Roodenberg, J., von As, A., Jacobs, L., Winjen, M.H. 2003: 'Early settlement in the plain of Yenişehir (NW Anatolia). The basal occupation layers at Menteşe' *Anatolica* 29: 17–59

Rooijakkers, C.T. 2012: 'Spinning animal fibres at Late Neolithic Tell Sabi Abyad, Syria?' *Paléorient* 38.1–2: 93–109

Rosenberg, D. 2010: 'Early maceheads in the southern Levant: a "Chalcolithic" hallmark in Neolithic context' *Journal of Field Archaeology* 35.2: 204–16

— 2013: 'Not "just another brick in the wall?" The symbolism of groundstone tools in Natufian and early Neolithic Southern Levantine constructions' *Cambridge Archaeological Journal* 23.2: 185–201

Rosenberg, M., Redding, R.W. 2002: 'Hallan Çemi and early village organization in eastern Anatolia' in I. Kuijt (ed.), *Life in Neolithic Farming Communities. Social Organization, Identity, and Differentiation*. New York, Kluwer Academic Publishers: 39–62

Runnels, C.N. 1981: *A Diachronic Study and Economic Analysis of Millstones from the Argolid, Greece*. PhD thesis, University Microfilms, Michigan, Ann Arbor

Russell, N. 2003: 'Dance of the cranes: crane symbolism at Çatalhöyük and beyond' *Antiquity* 77.297: 445–55

— 2005: 'The Çatalhöyük worked bone' in I. Hodder (ed.), *Changing Materialities at Çatalhöyük: Reports from the 1995–99 Seasons* (Çatalhöyük Research Project Series 5). Cambridge, McDonald Institute for Archaeological Research; London, British Institute at Ankara: 339–68

— 2012a: *Social Zooarchaeology: Humans and Animals in Prehistory*. Cambridge, Cambridge University Press

— 2012b: 'Worked bone from the BACH Area at Çatalhöyük' in R. Tringham, M. Stevanović (eds), *Last House on the Hill: BACH Area Reports from Çatalhöyük, Turkey* (Çatalhöyük Research Project Series 11). Los Angeles, Cotsen Institute of Archaeology Press: 349–61

Russell, N., Bogaard, A. 2016: 'Subsistence actions at Çatalhöyük' in S.R. Steadman, J.C. Ross (eds), *Agency and Identity in the Ancient Near East: New Paths Forward*. London, Routledge: 63–79

Russell, N., Griffitts, J.L. 2013: 'Çatalhöyük worked bone: South and 4040 Areas' in I. Hodder (ed.) *Substantive Technologies at Çatalhöyük: Reports from the 2000–2008 Seasons* (Çatalhöyük Research Project Series 9). London, British Institute at Ankara; Los Angeles, Cotsen Institute of Archaeology Press: 277–306

Russell, N., Martin, L. 2005: 'The Çatalhöyük mammal remains' in I. Hodder (ed.), *Inhabiting Çatalhöyük: Reports from the 1995–99 Seasons* (Çatalhöyük Research Project Series 4). Cambridge, McDonald Institute for Archaeological Research; London, British Institute at Ankara: 33–98

Russell, N., Meece, S. 2005: 'Animal representations and animal remains at Çatalhöyük' in I. Hodder (ed.), *Çatalhöyük Perspectives: Themes from the 1995–99 Seasons* (Çatalhöyük Research Project Series 6). Cambridge, McDonald Institute for Archaeological Research; London, British Institute at Ankara: 209–30

Russell, N., Martin, L., Twiss, K.C. 2009: 'Building memories: commemorative deposits at Çatalhöyük' *Anthropozoologica* 44: 103–25

Russell, N., Twiss, K.C., Orton, D.C., Demirergi, A. 2013a: 'Changing animals use at Neolithic Çatalhöyük, Turkey' in E. de Cupere, V. Linseele, S. Hamilton-Dyer (eds), *Archaeology of the Near East X. Proceedings of the Tenth International Symposium on the Archaeozoology of South-Western Asia and Adjacent Areas*. Leuven, Peeters: 45–68

— 2013b: 'More on the Çatalhöyük mammal remains' in I. Hodder (ed.), *Humans and Landscapes of Çatalhöyük: Reports from the 2000–2008 Seasons* (Çatalhöyük Research Project Series 8). London, British Institute at Ankara; Los Angeles, Cotsen Institute of Archaeology Press: 213–58

Russell, N., Wright, K.I., Carter, T., Ketchum, S., Ryan, P., Yalman, N., Regan, R., Stevanović, M., Milić, M. 2014: 'Bringing down the house: house closing deposits at Çatalhöyük' in I. Hodder (ed.), *Integrating Çatalhöyük: Themes From the 2000–2008 Season* (Çatalhöyük Research Project Series 10). London, British Institute at Ankara; Los Angeles, Cotsen Institute of Archaeology Press: 109–22

Ruß-Popa, G. 2018: 'Der Gebrauch von Schaffell in der mitteleuropäischen urgeschichtlichen Bekleidung' *Annalen des Naturhistorischen Museums in Wien. Festschrift für Erich Pucher Serie A für Mineralogie und Petrographie, Geologie und Paläontologie, Anthropologie und Prähistorie* 120: 157–76

Ryan, P. 2011: 'Plants as material culture in the Near Eastern Neolithic: perspectives from the silica skeleton artifactual remains at Catalhoyuk' *Journal of Anthropological Archaeology* 30: 292–305

— 2013: 'Plant exploitation from household and landscape perspectives: the phytolith evidence' in I. Hodder (ed.), *Humans and Landscapes of Çatalhöyük: Reports from the 2000–2008 Seasons* (Çatalhöyük Research Project Series 8). London, British Institute at Ankara; Los Angeles, Cotsen Institute of Archaeology Press: 163–90

Ryder, M.L. 1965: 'Report of the textiles from Çatal Höyük' *Anatolian Studies* 15: 175–76

Sadvari, J.W., Tsoraki, C., Dogiama, L., Knüsel, C.J. 2015: 'Reading the bones, reading the stones: an integrated approach to reconstructing activity patterns at Neolithic Çatalhöyük' in I. Hodder, A. Marciniak (eds), *Assembling Çatalhöyük*. Leeds, Maney Publishing: 59–74

Safer, F.J., Gill, M.F. 1982: *Spirals from the Sea. An Anthropological Look at Shells*. New York, Clarkson N. Potter Inc.

Samuel, D.J. 2010: 'Experimental grinding and Ancient Egyptian flour production' in S. Ikram, A. Dodson (eds), *Beyond the Horizon: Studies in Egyptian Art, Archaeology and History in Honour of Barry J. Kemp*. Cairo, American University in Cairo Press: 456–77

Sato, J., Sato, T., Otomori, Y., Suzuki, H. 1969: 'University of Tokyo radiocarbon measurements II' *Radiocarbon* 11.2: 509–14

Schick, T. 1988a: 'Nahal Hemar cave. Cordage, basketry and fabrics' *'Atiqot* 18: 31–43

— 1988b: 'A Neolithic cult headdress from the Nahal Hemar cave' *The Israel Museum Journal* 7: 25–33

— 1995: 'A 10,000 year old comb from Wadi Murabba'at in the Judean Desert' *'Atiqot* 27: 199–206

— 1997: 'Miscellaneous finds: a note on the perishable finds from Netif Hagdud' in O. Bar-Yosef, A. Gopher (eds), *An Early Neolithic Village in the Jordan Valley Part I: The Archaeology of Netiv Hagdud* (American School of Prehistoric Research Bulletin 43). Cambridge (MA), Peabody Museum of Archaeology and Ethnology, Harvard University: 197–200

— 1998: *The Cave of the Warrior. A Fourth Millennium Burial in the Judean Desert*. Jerusalem, The Israel Antiquities Authority

Schiffer, M.B. 1987: *Formation Processes of the Archaeological Record*. Albuquerque, University of New Mexico Press

Schlumbaum A., Campos, P.F., Volken, S., Volken, M., Hafner, M., Schibler, J. 2010: 'Ancient DNA, a Neolithic legging from the Swiss Alps and the early history of goat' *Journal of Archaeological Science* 37: 1247–51

Schmandt-Besserat, D. 1992a: *Before Writing* 1: *From Counting to Cuneiform*. Austin, University of Texas Press

— 1992b: *Before Writing* 2: *A Catalogue of Near Eastern Tokens*. Austin, University of Texas Press

— 1996: *How Writing Came About*. Austin, University of Texas Press

Schneider, J.S. 2002: 'Milling tool design, stone textures, and function' in H. Procopiou, R. Treuil (eds), *Moudre et Broyer. L' interprétation fonctionnelle de l'outillage de mouture et de broyage dans la Préhistoire et l' Antiquité*. 1. *Méthodes. Pétrographie, chimie, tracéologie, expérimentation, ethnoarchéologie*. Paris, CTHS: 31–53

Schumann, W. 1992: *Rocks, Minerals and Gemstones*. London, HarperCollins

Scott, M.J. 1996: 'An exegesis of the curation concept' *Journal of Anthropological Research* 52.3: 259–80

Searcy, M.T. 2011: *The Life-Giving Stone: Ethnoarchaeology of Maya Metates*. Tucson, University of Arizona Press

Seiler-Baldinger, A. 1994: *Textiles: A Classification of Techniques*. Washington DC, Smithsonian Institution Press

Shamir, O., Rast-Eicher, A. 2020: 'Continuity and discontinuity in Neolithic and Chalcolithic linen textile production in the southern Levant' in W. Schier, S. Pollock (eds), *The Competition of Fibres. Early Textile Production in Western Asia, South-East and Central Europe (10,000-500 BC)*. Oxford, Philadelphia, Oxbow Books: 27–37

Shennan, S. 1997: *Quantifying Archaeology: Second Edition*. Edinburgh, Edinburgh University Press

Shillito, L.-M., Matthews, W., Almond, M.J 2013: 'Ecology, diet and discard practices: new interdisciplinary approaches to the study of middens through integrating micromorphological, phytolith and geochemical analyses' in I. Hodder (ed.), *Humans and Landscapes of Çatalhöyük: Reports from the 2000–2008 Seasons* (Çatalhöyük Research Project Series 8). London, British Institute at Ankara; Los Angeles: Cotsen Institute of Archaeology Press: 65–76

Shore, B. 1998: 'Mana and Tapu' in A. Howard, R. Borofsky (eds), *Developments in Polynesian Ethnology*. Honolulu, University of Hawaii Press: 137–73

Sidell, E.J. 1993: *A Methodology for the Identification of Archaeological Eggshell*. Philadelphia, Museum Applied Science Center of Archaeology, University of Pennsylvania

Sidéra, I. 2001: 'Domestic and funerary bone, antler and tooth objects in the Neolithic of western Europe: a comparison' in A.M. Choyke, L. Bartosiewicz (eds), *Crafting Bone: Skeletal Technologies through Time and Space* (BAR International Series 937). Oxford, Archaeopress: 221–29

Simmons, A.H. 2002: 'Villages on the edge' in I. Kuijt (ed.), *Life in Neolithic Farming Communities*. New York, Springer: 211–30

Skeates, R. 2007: 'Neolithic stamps: cultural patterns, processes and potencies' *Cambridge Archaeological Journal* 17.2: 183–98

Soffer, O., Adovasio, J.M., Hyland, D.C. 2000: 'The "Venus" figurines: textiles, basketry, gender, and status in the Upper Paleolithic' *Current Anthropology* 41.4: 511–37

Solazzo, C., Courel, B., Connan, J., van Dongen, B.E., Barden, H., Penkman, K., Taylor, S., Demarchi, B., Adam, P., Schaeffer, P., Nissenbaum, A., Bar-Yosef, O., Buckley, M. 2016: 'Identification of the earliest collagen- and plant-based coatings from Neolithic artefacts (Nahal Hemar cave, Israel)' *Nature, Scientific Reports*: 31053

Spangenberg, J.E., Ferrer, M., Tschudin, P., Volken, M., Hafner, A. 2010: 'Microstructural, chemical and isotopic evidence for the origin of late Neolithic leather recovered from an ice field in the Swiss Alps' *Journal of Archaeological Science* 37.8: 1851–65

Sperl, G. 1990: 'Zur Urgeschichte des Bleies' *Zeitschrift für Metallkunde* 81: 799–801

St. George, I. 2012: 'Çatalhöyük murals: a snapshot of conservation and experimental research' in R. Tringham, M. Stevanović (eds), *Last House on the Hill: BACH Area Reports from Çatalhöyük, Turkey* (Çatalhöyük Research Project Series 11). Los Angeles, Cotsen Institute of Archaeology Press: 473–80

Stevanović, M. 1997: 'The age of clay: the social dynamics of house destruction' *Journal of Anthropological Archaeology* 16: 334–95

— 2012a: 'Summary of the results of the excavation in the BACH Area' in R. Tringham, M. Stevanović (eds), *Last House on the Hill: BACH Area Reports from Çatalhöyük, Turkey* (Çatalhöyük Research Project Series 11). Los Angeles, Cotsen Institute of Archaeology Press: 49–80

— 2012b: 'Building the replica Neolithic House at Çatalhöyük' in R. Tringham, M. Stevanović (eds), *Last House on the Hill: BACH Area Reports from Çatalhöyük, Turkey* (Çatalhöyük Research Project Series 11). Los Angeles, Cotsen Institute of Archaeology Press: 447–72

— 2012c: 'Detailed report of the excavation of Building 3 and Spaces 87, 88, and 89 (1997–2003)' in R. Tringham, M. Stevanović (eds), *Last House on the Hill: BACH Area Reports from Çatalhöyük, Turkey* (Çatalhöyük Research Project Series 11). Los Angeles, Cotsen Institute of Archaeology Press: 81–172

Stiner, M.C., Buitenhuis, H., Duru, G., Kuhn, S.L., Mentzer, S.M., Munro, N.D., Pöllath, N., Quade, J., Tsartsidou, G., Özbaşaran, M. 2014: 'A forager-herder trade-off, from broad-spectrum hunting to sheep management at Aşıklı Höyük, Turkey' *Proceedings of the National Academy of Sciences* 111.23: 8404–09

Stone, T. 1994: 'The impact of raw-material scarcity on ground-stone manufacture and use: an example from the Phoenix Basin Hohokam' *American Antiquity* 59.4: 680–94

Stordeur, D. 1993: 'Sédentaires et nomades du PPNB final dans le désert de Palmyre (Syrie)' *Paléorient* 19.1: 187–204

— 2010: 'Domestication of plants and animals, domestication of symbols?' in D. Bolger, L.C. Maguire, E.J. Peltenburg (eds), *The Development of Pre-State Communities in the Ancient Near East: Studies in Honour of Edgar Peltenburg*. Oxford, Oxbow Books: 123–30

— 2015: *Le Village de Jerf el Ahmar (Syrie, 9500–8700 av. J.-C.): L'Architecture, Miroir d'une Société Néolithique Complexe*. Paris, CNRS

Stordeur, D., Helmer, D., Jammous, B., Khawam, R., Molist, M., Wilcox, G. 2010: 'Le PPNB de Syrie du Sud à travers les découvertes récentes à Tell Aswad', in M. al-Maqdissi, F. Braemer, J.-M. Dentzer (eds), *HAURAN V. La Syrie du Sud Néolithique à l'Antiquité Tardive. Recherches récentes. Actes du Colloque de Damas 2007* 1. Beyrouth, Institut Français du Proche-Orient: 41–67

Sykes, N. 2014: *Beastly Questions: Animal Answers to Archaeological Issues*. London, Bloomsbury

Taçon, P.S.C. 2004: 'Ochre, clay, stone and art. The symbolic importance of minerals as life-force among Aboriginal peoples of northern and central Australia' in N. Boivin, M.A. Owoc (eds), *Soils, Stones and Symbols. Cultural Perceptions of the Mineral World*. London, UCL Press: 31–42

Taniguchi, Y., Hirao, Y., Shimadzu, Y., Tsuneki, A. 2002: 'The first fake? Imitation turquoise beads recovered from a Syrian Neolithic site, Tell el-Kerkh' *Studies in Conservation* 47.3:175–83

Tarkan, D. 2015: *The Investigation of Clay Use and Resources of the Neolithic Pottery at Çatalhöyük East Mound*. MA thesis, Istanbul University

Taylor, J., Bogaard, A., Carter, T., Charles, M., Haddow, S.D., Knüsel, C.J., Mazzucato, C., Mulville, J., Tsoraki, C., Tung, B., Twiss, K.C. 2015: '"Up in flames": a visual exploration of a burnt building at Çatalhöyük in GIS' in I. Hodder, A. Marciniak (eds), *Assembling Çatalhöyük*. Leeds: Maney: 127–49

Thiessen, L. 2010: 'The Neolithic-Chalcolithic sequence in the SW-Anatolian Lakes Region' *Documenta Praehistorica* 37: 269–82

Thuesen, I. 1988: *Hama. Fouilles et Recherches de la Foundation Carlsberg 1931–1938* 1. *The Pre- and Protohistoric Periods*. Copenhagen, The National Museum of Denmark

Todd, I. 1966: 'Aşıklı Hüyük, a Protoneolithic Site in Anatolia' *Anatolian Studies* 16: 139–63

— 1976: *Çatal Hüyük in Perspective*. New York, Cummings

Todd, Z. 2014: 'Fish pluralities: human-animal relations and sites of engagement in Paulatuuq, Arctic Canada' *Études/Inuit/Studies* 38.1–2: 217–38

Toth, N., Clarke, D., Ligabue, G. 1992: 'The last stone ax makers' *Scientific American* 267.1: 66–71

Tringham, R. 2000: 'The continuous house: a view from the deep past' in R.A. Joyce, S.D. Gillespie (eds), *Beyond Kinship: Social and Material Reproduction in House Societies*. Philadelphia, University of Pennsylvania Press: 115–34

Tringham, R., Stevanović, M. (eds) 2012: *Last House on the Hill: BACH Area Reports from Çatalhöyük, Turkey* (Çatalhöyük Research Project Series 11). Los Angeles, Cotsen Institute of Archaeology Press

Trinkaus, E., Buzhilova, A.P. 2012: 'The death and burial of Sunghir I' *International Journal of Osteoarchaeology* 22: 655–66

Tripković, J. 2017: 'Building 131' *Çatalhöyük 2017 Archive Report*. http://www.catalhoyuk.com/archive_reports/2017

Tsoraki, C. 2008: *Neolithic Society in Northern Greece: The Evidence of Ground Stone Artefacts*. PhD thesis, University of Sheffield

— 2011: 'Stone-working traditions in the prehistoric Aegean: the production and consumption of edge tools at Late Neolithic Makriyalos' in V. Davis, M. Edmonds (eds), *Stone Axe Studies* 3. Oxford, Oxbow Books: 231–44

— 2012: 'Ground stone technologies at the Bronze Age settlement of Sissi. Preliminary results' in J. Driessen, I. Schoep, M. Anastasiadou, F. Carpentier, I. Crevecoeur, S. Déderix, M. Devolder, F. Gaignerot-Driessen, S. Jusseret, C. Langohr, Q. Letesson, F. Liard, A. Schmitt, C. Tsoraki, R. Veropoulidou (eds), *Excavations at Sissi* 3. *Preliminary Report on the 2011 Campaign*. Louvain-la-Neuve, Presses Universitaires de Louvain: 201–21

— 2013: 'East Mound ground stone, 2013' *Çatalhöyük 2013 Archive Report*. http://www.catalhoyuk.com/archive_reports/2013

— 2014: *Biographies of Buildings and Objects: Deconstructing the Social Significance of the Re-use of Ground Stone Artefacts as Building Material in the Neolithic*. Paper presented at the EAA 20th Annual Meeting, Istanbul, September 2014

— 2015: 'Ground stone' *Çatalhöyük 2015 Archive Report*. http://www.catalhoyuk.com/archive_reports/2015

— 2016: 'Ground stone technologies' *Çatalhöyük 2016 Archive Report*. http://www.catalhoyuk.com/archive_reports/2016

— 2017 (with contributions by M. Siebrecht): 'Ground stone technologies' *Çatalhöyük 2017 Archive Report*. http://www.catalhoyuk.com/archive_reports/2017

— 2018: 'The ritualization of daily practice: exploring the staging of ritual acts at Neolithic Çatalhöyük, Turkey' in I. Hodder (ed.), *Religion, History and Place in the Origin of Settled Life*. Louisville, University of Colorado Press: 238–62

— forthcoming: 'Stone technology under the microscope: the contribution of microwear analysis of ground stone tools to the understanding of daily activities' in A. Baysal (ed.), *Lithic Studies: Anatolia and Beyond*. Oxford, Archaeopress

Tsoraki, C., García-Granero, J.J., Madella, M. 2015a: 'Houses and Households at Neolithic Çatalhöyük: The Contribution of Microwear and Microbotanical Analyses of Ground Stone Tools to the Understanding of Household Activities'. Paper presented at the EAA 21st Annual Meeting, Glasgow, September 2015

— 2015b: 'Integrating Microwear and Microbotanical Analyses of Ground Stone Tools: Towards an Understanding of Household Activities at Neolithic Çatalhöyük, Turkey'. Paper presented at the Association of Archaeological Wear and Residue Analysts (AWRANA) Meeting, Leiden, May 2015

Tung, B. 2013: 'Building with mud: an analysis of architectural materials at Çatalhöyük' in I. Hodder (ed.) *Substantive Technologies at Çatalhöyük: Reports from the 2000–2008 Seasons* (Çatalhöyük Research Project Series 9). London, British Institute at Ankara; Los Angeles, Cotsen Institute of Archaeology Press: 67–81

Türkcan, A.U. 2005: 'Some remarks on Çatalhöyük stamp seals' in I. Hodder (ed.), *Changing Materialities at Çatalhöyük: Reports from the 1995–99 Seasons* (Çatalhöyük Research Project Series 5). Cambridge, McDonald Institute for Archaeological Research; London, British Institute of Archaeology at Ankara: 175–85

— 2013: 'Çatalhöyük stamp seals from 2000–2008' in I. Hodder (ed.), *Substantive Technologies at Çatalhöyük: Reports from the 2000–2008 Seasons* (Çatalhöyük Research Project 9). London, British Institute at Ankara; Los Angeles, Cotsen Institute of Archaeology Press: 235–47

Türkmenoğlu, A.G., Baysal, A., Toprak, V., Göncüoğlu, M.C. 2005: 'Ground-stone raw material from Çatalhöyük' in I. Hodder (ed.), *Changing Materialities at Çatalhöyük: Reports from the 1995–99 Seasons* (Çatalhöyük Research Project Series 5). Cambridge, McDonald Institute for Archaeological Research; London, British Institute of Archaeology at Ankara: 369–72

Twiss, K.C. 2008: 'Transformations in an early agricultural society: feasting in the southern Levantine Pre-Pottery Neolithic' *Journal of Anthropological Archaeology* 27.4: 418–42

Twiss, K.C., Russell, N. 2009: 'Taking the bull by the horns: ideology, masculinity, and cattle horns at Çatalhöyük' *Paléorient* 35.2: 17–29

Van Driel-Murray, C. 2000: 'Leatherwork and skin products' in P.T. Nicholson, I. Shaw (eds), *Ancient Egyptian Materials and Technology*. Cambridge, Cambridge University Press: 299–319

Van Gijn, A.L. 2010: *Flint in Focus: Lithic Biographies in the Neolithic and Bronze Age*. Leiden, Sidestone Press

— 2014: 'Science and interpretation in microwear studies' *Journal of Archaeological Science* 48: 166–69

Van Huyssteen, J.W. 2010: 'Coding the nonvisible: epistemic limitations and understanding symbolic behavior at Çatalhöyük' in I. Hodder (ed.), *Religion in the Emergence of Civilization: Çatalhöyük as a Case Study*. Cambridge, Cambridge University Press, 99–121

Van Neer, W., Gravendeel, R., Wouters, W., Russell, N. 2013: 'The exploitation of fish at Çatalhöyük' in I. Hodder (ed.), *Humans and Landscapes of Çatalhöyük: Reports from the 2000–2008 Seasons* (Çatalhöyük Research Project Series 8). London, British Institute at Ankara; Los Angeles, Cotsen Institute of Archaeology Press: 317–27

Van Pool, T.L., Leonard, R.D. 2002: 'Specialized ground stone production in the Casas Grandes region of northern Chihuahua, Mexico' *American Antiquity* 67.4: 710–30

Van Zeist, W. 2003: 'Some notes on the plant husbandry of Aşıklı Höyük' in W. Van Zeist (ed.), *Reports on Archaeobotanical Studies in the Old World*. Groningen, 115–42

Vasić, M. 2018: *Personal Adornment in the Neolithic Middle East: A Case Study of Çatalhöyük*. PhD thesis, Freie Universität, Berlin

Vasić, M., Bains, R., Russell, N. 2014: 'Dress: a preliminary study of bodily ornamentation at Çatalhöyük' in I. Hodder (ed.), *Integrating Çatalhöyük. Themes from the 2000–2008 Seasons* (Çatalhöyük Research Project Series 10). London, British Institute at Ankara; Los Angeles, Cotsen Institute of Archaeology Press: 197–220

Vedder, J.F. 2005: 'The obsidian mirrors of Çatalhöyük' in I. Hodder (ed.), *Changing Materialities at Çatalhöyük: Reports from the 1995–99 Seasons* (Çatalhöyük Research Project Series 5). Cambridge, McDonald Institute for Archaeological Research; London, British Institute at Ankara: 597–619

Verhoeven, M. 2007: 'Losing one's head in the Neolithic: on the interpretation of headless figurines' *Levant* 39.1: 175–83

— 2011: 'The birth of a concept and the origins of the Neolithic: a history of prehistoric farmers in the Near East' *Paléorient* 37.1: 75–87

Veropoulidou, R. 2011: *Shells from the Settlement of the Thermaic Gulf. Reconstructing Molluscan Exploitation during the Neolithic and the Bronze Age*. PhD thesis, Aristotle University of Thessaloniki, Thessaloniki

— 2014: 'Molluscan exploitation practices in the Neolithic and Bronze Age communities at the former Thermaic Gulf, North Aegean' in G. Touchais, R. Laffineur, F. Rougemont (eds), *PHYSIS: L'Environment Naturel et la Relation Homme-Milieu dans le Monde Égéen Protohistorique. Actes de la 14e Rencontre Égéenne Internationale, Paris, Institut National d'Histoire de l'Art (INHA), 11–14 Décembre 2012. AEGAEUM 37, Annales liégeoises et PASPiennes d'Archéologie Égéenne*. Leuven, Peeters: 415–22

— 2017: 'Shells' *Çatalhöyük 2017 Archive Report*. http://www.catalhoyuk.com/archive_reports/2017
— in press: 'Exploring the material qualities of shells: cockle artefacts from Neolithic and Bronze Age coastal sites in Northern Greece' in C. Tsoraki, S. Sherratt, R. Doonan (eds), *Material Worlds of the Aegean*. Oxford, Sheffield Studies in Aegean Archaeology
Viveiros de Castro, E. 1998: 'Cosmological deixis and Amerindian perspectivism' *Journal of the Royal Anthropological Institute* 4.3: 469–488
Vogelsang-Eastwood, G.M. 1988: 'A re-examination of the fibres from the Çatal Hüyük textiles' *Oriental Carpet and Textile Studies* 3.1: 15–19
— 1993: 'One of the oldest textiles in the world? The Çayönü textile' *Archaeological Textiles Newsletter* 16: 4–7
Voigt, M.M. 2000: 'Çatal Höyük in context: ritual at early Neolithic sites in Central and Eastern Turkey' in I. Kuijt (ed.), *Life in Neolithic Farming Communities: Social Organization, Identity, and Differentiation*. New York, Kluwer Academic/Plenum Publishers: 253–93
Völling, E. 2008: *Textiltechnik im alten Orient. Rohstoffe und Herstellung*. Würzburg, Ergon Verlag
Watts, V. 2013: 'Indigenous place-thought and agency amongst humans and non humans (First Woman and Sky Woman go on a European world tour!)' *Decolonization: Indigeneity, Education & Society* 2.1: 20–34
Weiner, A. 1992: *Inalienable Possessions: The Paradox of Keeping-While-Giving*. Berkeley, University of California Press
Wendrich, W. 1991: *Who is Afraid of Basketry? A Guide to Recording Basketry and Cordage for Archaeologists and Ethnographers*. Leiden, CNWS Centre of Non-Western Studies Leiden University
— 1999: *The World According to Basketry: An Ethno-archaeological Interpretation of Basketry Production in Egypt*. Leiden, Leiden University
— 2000: 'Basketry' in P.T. Nicholson, I. Shaw (eds), *Ancient Egyptian Materials and Technology*. Cambridge, Cambridge University Press: 254–67
— 2005: 'Çatalhöyük basketry' in I. Hodder (ed.), *Changing Materialities at Çatalhöyük: Reports from the 1995–99 Seasons* (Çatalhöyük Research Project Series 5). Cambridge, McDonald Institute for Archaeological Research; London, British Institute at Ankara: 333–38
Wendrich, W., Ryan, P. 2012: 'Phytoliths and basketry materials at Çatalhöyük (Turkey): timelines of growth, harvest and objects life histories' *Paléorient* 38.1-2: 55–63
Wiles, J. 2008: *An Analysis of Plaster Sequences from the Neolithic Site of Çatalhöyük (Turkey) by Microspectroscopic Techniques*. PhD thesis, University of Reading
Winiger, J. 1995: 'Die Bekleidung des Eismannes und die Anfänge der Weberei nördlich der Alpen' in H. Moser, W. Platzer, H. Seidler, K. Spindler (eds), *Der Mann im Eis* 2. Vienna, Springer: 119–87
Woods, C. (ed.) 2010: *Visible Language: Inventions of Writing in the Ancient Middle East and Beyond*. Chicago, Oriental Institute of the University of Chicago
WoRMS Editorial Board 2019: *World Register of Marine Species* http://www.marinespecies.org at VLIZ, https://doi.org/10.14284/170
Wright, G.A. 1969: *Obsidian Analyses and Prehistoric Near Eastern Trade: 7500 to 3500 BC*. Ann Arbor, University of Michigan Press
Wright, K.I. 1992: *Ground Stone Assemblage Variations and Subsistence Strategies in the Levant, 22,000 to 5,500 BP*. PhD thesis, Department of Anthropology, Yale University
— 1998: 'Dhuweila: ground stone' in A.V.G. Betts, S. Colledge, L. Martin, C. McCartney, K.I. Wright, V. Yagodin (eds), *The Harra and the Hamad. Excavations and Explorations in Eastern Jordan*. Sheffield, Sheffield Academic Press: 121–34
— 2008: 'Craft production and the organization of ground stone technologies' in Y.M. Rowan, J.R. Ebeling (eds), *New Approaches to Old Stones: Recent Studies of Ground Stone Artifacts*. London: Equinox Publishing: 130–43
— 2012: 'Beads and the body: ornament technologies of the BACH Area buildings at Çatalhöyük' in R. Tringham, M. Stevanović (eds), *Last House on the Hill: BACH Area Reports from Çatalhöyük, Turkey* (Çatalhöyük Research Project Series 11). Los Angeles, Cotsen Institute of Archaeology Press: 429–49
— 2013: 'The ground stone technologies of Çatalhöyük' in I. Hodder (ed.), *Substantive Technologies at Çatalhöyük: Reports from the 2000–2008 Seasons* (Çatalhöyük Research Project Series 9). London, British Institute at Ankara; Los Angeles: Cotsen Institute of Archaeology Press: 365–416
— 2014: 'Domestication and inequality? Households, corporate groups and food processing tools at Neolithic Çatalhöyük' *Journal of Anthropological Archaeology* 33: 1–33

Wright, K.I., Baysal, A. 2012: 'Ground stone tools and technologies associated with Building 3 at Çatalhöyük' in R. Tringham, M. Stevanović (eds), *Last House on the Hill: BACH Area Reports from Çatalhöyük, Turkey* (Çatalhöyük Research Project Series 11). Los Angeles, Cotsen Institute of Archaeology Press: 421–28

Yalçın, Ü., Seifert, M. 1994: 'Milet'te Arkeometrik Araştırmalar' in *X. Arkeometri Sonuçları Toplantısı* (T.C. Kültür Bakanlığı Yayınları 1736; Anıtlar ve Müzeler Genel Müdürlüğü Yayınları, sempozyum serisi 42). Ankara, T.C. Kültür Bakanlığı Milli Kütüphane Basımevi: 15–39

Yalman, N., Tarkan, D., Gültekin, H. 2013: 'The Neolithic pottery of Çatalhöyük: recent studies' in I. Hodder (ed.), *Substantive Technologies at Çatalhöyük: Reports from the 2000–2008 Seasons* (Çatalhöyük Research Project Series 9). London, British Institute at Ankara; Los Angeles, Cotsen Institute of Archaeology Press: 147–82

Yelozer, S. 2018: 'The beads from Aşıklı Höyük' in M. Özbaşaran, G. Duru, M. Stiner (eds), *The Early Settlement at Aşikli Höyük. Essays in Honor of Ufuk Esin*. Istanbul: Ege Yayınları: 383–404

Yeomans, L. 2014: 'Building 55' in I. Hodder (ed.), *Çatalhöyük Excavations: The 2000–2008 Seasons* (Çatalhöyük Research Project Series 7). London, British Institute at Ankara; Los Angeles, Cotsen Institute of Archaeology Press: 405–14

Zedeño, M.N. 2009: 'Animating by association: index objects and relational taxonomies' *Cambridge Archaeological Journal* 19.3: 407–17

Zimansky, P. 1993: 'Review: "Before Writing. Volume I: From Counting to Cuneiform" by Denise Schmandt-Besserat. "Before Writing. Volume II: A Catalogue of Near Eastern Tokens" by Denise Schmandt-Besserat' *Journal of Field Archaeology* 20.4: 513–17